Handbook of Solid-State Lighting and LEDs

SERIES IN OPTICS AND OPTOELECTRONICS

Series Editors: **E Roy Pike,** Kings College, London, UK
Robert G W Brown, University of California, Irvine, USA

Recent titles in the series

Nanophotonics and Plasmonics: An Integrated View
Dr. Ching Eng (Jason) Png and Dr. Yuriy Akimov

Handbook of Solid-State Lighting and LEDs
Zhe Chuan Feng (Ed.)

Optical Microring Resonators: Theory, Techniques, and Applications
V. Van

Optical Compressive Imaging
Adrian Stern

Singular Optics
Gregory J. Gbur

The Limits of Resolution
Geoffrey de Villiers and E. Roy Pike

Polarized Light and the Mueller Matrix Approach
José J Gil and Razvigor Ossikovski

Light—The Physics of the Photon
Ole Keller

Advanced Biophotonics: Tissue Optical Sectioning
Ruikang K Wang and Valery V Tuchin (Eds.)

Handbook of Silicon Photonics
Laurent Vivien and Lorenzo Pavesi (Eds.)

Microlenses: Properties, Fabrication and Liquid Lenses
Hongrui Jiang and Xuefeng Zeng

Laser-Based Measurements for Time and Frequency Domain Applications: A Handbook
Pasquale Maddaloni, Marco Bellini, and Paolo De Natale

Handbook of 3D Machine Vision: Optical Metrology and Imaging
Song Zhang (Ed.)

Handbook of Optical Dimensional Metrology
Kevin Harding (Ed.)

Biomimetics in Photonics
Olaf Karthaus (Ed.)

Handbook of Solid-State Lighting and LEDs

Edited by
Zhe Chuan Feng

CRC Press
Taylor & Francis Group
Boca Raton London New York

CRC Press is an imprint of the
Taylor & Francis Group, an **informa** business

CRC Press
Taylor & Francis Group
6000 Broken Sound Parkway NW, Suite 300
Boca Raton, FL 33487-2742

First issued in paperback 2019

ISBN-13: 978-1-4987-4141-5 (hbk)
ISBN-13: 978-0-367-87458-2 (pbk)

Library of Congress Cataloging-in-Publication Data

Names: Feng, Zhe Chuan, editor.
Title: Handbook of solid-state lighting and LEDs / edited by Zhe Chuan Feng.
Other titles: Series in optics and optoelectronics ; 25.
Description: Boca Raton, FL : CRC Press, Taylor & Francis Group, [2017] |
Series: Series in optics and optoelectronics ; 25
Identifiers: LCCN 2016052771| ISBN 9781498741415 (hardback ; alk. paper) |
ISBN 149874141X (hardback ; alk. paper)
Subjects: LCSH: Light emitting diodes. | LED lighting--Materials. | Solid
state electronics.
Classification: LCC TK7871.89.L53 H364 2017 | DDC 621.32--dc23
LC record available at https://lccn.loc.gov/2016052771

Visit the Taylor & Francis Web site at
http://www.taylorandfrancis.com

and the CRC Press Web site at
http://www.crcpress.com

Contents

SECTION I Overview

SECTION II GaN-Based LEDs for Lighting

SECTION III Deep Ultraviolet LEDs and Related Technologies

SECTION IV Laser Diodes

SECTION V Nano and Other Types of LEDs

SECTION VI Novel Technologies and Developments

Preface

Economic and industrial developments have advanced the world economy and culture to extreme levels, which are having a high cost in terms of overuse and starvation of energy sources leading to potential harm and destruction to the future of human existence. This threat has raised major concerns in all countries irrespective of wealth or richness of energy sources. Clearly, the promotion of new, advanced technologies aimed at saving precious energy resources is the best and only way to combat the harmful effects. Among various options, solid-state lighting (SSL), especially semiconductor-based light-emitting diodes (LEDs), should prove to be a reliable direction for the development of energy optoelectronics technologies.

More than 100 years ago, lighting technology had its major breakthrough based on the efforts of Thomas Edison and others in the invention of the electric light bulb. The recent breakthrough and development in semiconductor LEDs—in particular based on gallium nitride (GaN) and related wide gap semiconductors—have opened a new century of LED lighting, changing our daily lives, industry, and society.

The 2014 Nobel Prize in Physics was awarded jointly to Isamu Akasaki, Hiroshi Amano, and Shuji Nakamura "for the invention of efficient blue light-emitting diodes which has enabled bright and energy-saving white light sources." Energy-efficient and environmentally friendly solid-state light sources, in particular GaN-based LEDs, are revolutionizing an increasing number of applications and bring clear, widespread benefits in areas such as lighting, communications, biotechnology, imaging, and medicine. It is expected that LEDs may replace traditional light bulbs and tubes to achieve a new lighting paradigm. The III-nitride-based industry has now formed and with it there are rapid developments across the world. It is expected that III-nitride-based LEDs may replace traditional light bulbs to realize a revolution in lighting similar to the impact of Edison's invention of the electric light bulb more than 100 years ago.

Tremendous research and industry achievements and developments have been made in recent years. *Handbook of Solid-State Lighting and LEDs* reviews many of these significant results and progress, covering basic concepts and critical aspects, and serves professors, scientists, engineers, and students in the field. The handbook is organized for a wide range of audiences and covers major aspects of SSL and LED science and technology. Each chapter, written by experts in the field, reviews important topics and achievements in recent years, discusses progress made by different groups, and suggests further work needed. The handbook provides useful information on SSL–LED materials and devices; nanoscale processing; fabrication of LEDs, light diodes, photodetectors, and nanodevices; and characterization, application, and development of the various SSL and LED devices and nanoengineering.

This handbook consists of 24 review chapters, led by Prof. Hiroshi Amano, and divided into six main sections: "Overview," "GaN-Based LEDs for Lighting," "Deep UV LEDs and Related Technologies," "Laser Diodes," "Nano and Other Types of LEDs," and "Novel Technologies and Developments." The contents present the key properties of SSL and LEDs, describing current technologies and demonstrating the remaining challenges facing R&D in the twenty-first century. This book serves a diverse audience of material growers and evaluators, device design and processing engineers, potential users and newcomers, postgraduate students, engineers, and scientists in the field. If readers are interested, they are welcome to explore a recently published sister book I edited, *III-Nitride Materials, Devices and Nanostructures*.

Developments in SSL and LED materials and devices happen so quickly that invariably the handbook will miss some advances. Nonetheless, it captures the current state of the art, in particular those made from the start of this century up to the recent years. We look forward to the future when, even if the pure science remains the same, further strides will be made on the engineering and materials side.

Zhe Chuan Feng

Editor

Professor Zhe Chuan Feng earned his PhD in condensed matter physics from the University of Pittsburgh in 1987. Previously, he received his BS (1962–1968) and MS degrees (1978–1981) from the Department of Physics at Peking University. He has had positions at Emory University (1988–1992), National University of Singapore (1992–1994), Georgia Tech (1994–1995), EMCORE Corporation (1995–1997), Institute of Materials Research & Engineering, Singapore (1998–2001), Axcel Photonics (2001–2002), and Georgia Tech (2002–2003). In 2003, Professor Feng joined National Taiwan University as a professor at the Graduate Institute of Photonics & Optoelectronics and Department of Electrical Engineering, focusing on materials research and MOCVD growth of LED, III-nitrides, SiC, ZnO, and other semiconductors/oxides. He is currently a distinguished professor at the Laboratory of Optoelectronic Materials and Detection Technology, Guangxi Key Laboratory for Relativistic Astrophysics in the School of Physical Science and Technology at Guangxi University, Nanning, China.

Professor Feng has edited nine review books on compound semiconductors and microstructures, porous Si, SiC and III-nitrides, ZnO devices, and nanoengineering and has authored or coauthored more than 570 scientific papers with more than 220 indexed by the Science Citation Index and cited more than 2540 times. He has been a symposium organizer and invited speaker at different international conferences and universities and has been a reviewer for several international journals including *Physical Review Letters, Physical Review B*, and *Applied Physics Letters*. He has served as a guest editor for special journal issues and has been a visiting or guest professor at Sichuan University, Nanjing Tech University, South China Normal University, Huazhong University of Science & Technology, Nankai University, and Tianjin Normal University. He is a member of the International Organizing Committee for the Asia-Pacific Conferences on Chemical Vapor Deposition and International Conference for White LEDs and Solid-State Lighting and is on the Board of Directors for the Taiwan Association for Coating and Thin Film Technology. Professor Feng is an elected fellow of SPIE (2013).

Contributors

Hiroshi Amano
Center for Integrated Research of Future
 Electronics
Institute of Materials and Systems for
 Sustainability
Nagoya University
Nagoya, Japan

Jianwei Ben
State Key Laboratory of Luminescence and
 Applications
Changchun Institute of Optics, Fine Mechanics
 and Physics
Chinese Academy of Sciences
Changchun, People's Republic of China

Wengang Bi
Institute of Micro-Nano Photoelectron and
 Electromagnetic Technology Innovation
School of Electronics and Information
 Engineering
Hebei University of Technology
and
Key Laboratory of Electronic Materials and
 Devices of Tianjin
Tianjin, People's Republic of China

Remy Broersma
Philips Lighting
Eindhoven, The Netherlands

Changqing Chen
Wuhan National Laboratory for Optoelectronics
Huazhong University of Science and Technology
Wuhan, People's Republic of China

Hao-Tsung Chen
Department of Electrical Engineering
Institute of Photonics and Optoelectronics
National Taiwan University
Taipei, Taiwan, Republic of China

Horng-Shyang Chen
Department of Electrical Engineering
Institute of Photonics and Optoelectronics
National Taiwan University
Taipei, Taiwan, Republic of China

Jingwen Chen
Wuhan National Laboratory for Optoelectronics
Huazhong University of Science and Technology
Wuhan, People's Republic of China

Hoi Wai Choi
Department of Electrical and Electronic
 Engineering
The University of Hong Kong
Pokfulam, Hong Kong

Ching-Hsueh Chiu
Advanced Optoelectronic Technology Inc.
Hsinchu, Taiwan, Republic of China

Soo Jin Chua
Singapore-MIT Alliance for Research and
 Technology
and
Department of Electrical and Computer
 Engineering
National University of Singapore
Singapore, Singapore

Jiangnan Dai
Wuhan National Laboratory for Optoelectronics
Huazhong University of Science and Technology
Wuhan, People's Republic of China

Zhe Chuan Feng
Laboratory of Optoelectronic Materials &
    Detection Technology
Guangxi Key Laboratory for Relativistic Astrophysics
School of Physical Science & Technology
Guangxi University
Nanning, People's Republic of China

Eugene A. Fitzgerald
Singapore-MIT Alliance for Research and Technology
Singapore, Singapore

and

Department of Materials Science and Engineering
Massachusetts Institute of Technology
Cambridge, Massachusetts

Muziol Grzegorz
Institute of High Pressure Physics
Polish Academy of Sciences
Warsaw, Poland

Turski Henryk
Institute of High Pressure Physics
Polish Academy of Sciences
Warsaw, Poland

Tzu-Chien Hong
Advanced Optoelectronic Technology Inc.
Hsinchu, Taiwan, Republic of China

Ray-Hua Horng
Institute of Electronics
National Chiao Tung University
Hsinchu, Taiwan, Republic of China

Yaonan Hou
Department of Electronic and Electrical
    Engineering
University of Sheffield
Sheffield, United Kingdom

Chieh Hsieh
Department of Electrical Engineering
Institute of Photonics and Optoelectronics
National Taiwan University
Taipei, Taiwan, Republic of China

Chien-Shiang Huang
Advanced Optoelectronic Technology Inc.
Hsinchu, Taiwan, Republic of China

Chia-Yen Huang
Department of Photonics & Institute of Electro-
    Optical Engineering
National Chiao Tung University
Hsinchu, Taiwan, Republic of China

Shih-Cheng Huang
Advanced Optoelectronic Technology Inc.
Hsinchu, Taiwan, Republic of China

Jwo-Huei Jou
Department of Materials Science and Engineering
National Tsing-Hua University
Hsin-Chu, Taiwan, Republic of China

Yean-Woei Kiang
Department of Electrical Engineering
Institute of Photonics and Optoelectronics
National Taiwan University
Taipei, Taiwan, Republic of China

Hao-Chung Kuo
Department of Photonics & Institute of
    Electro-Optical Engineering
National Chiao Tung University
Hsinchu, Taiwan, Republic of China

Yang Kuo
Department of Electrical Engineering
Institute of Photonics and Optoelectronics
National Taiwan University
Taipei, Taiwan, Republic of China

Kenneth E. Lee
Singapore-MIT Alliance for Research and
    Technology
Singapore, Singapore

Dabing Li
State Key Laboratory of Luminescence and
    Applications
Changchun Institute of Optics, Fine Mechanics
    and Physics
Chinese Academy of Sciences
Changchun, People's Republic of China

Guoqiang Li
State Key Laboratory of Luminescent Materials
and Devices
and
Engineering Research Center on Solid-State
Lighting and its Informationisation of
Guangdong Province
and
Department of Electronic Materials
School of Materials Science and Engineering
South China University of Technology
Guangzhou, People's Republic of China

Kwai Hei Li
Department of Electrical and Electronic
Engineering
The University of Hong Kong
Pokfulam, Hong Kong

Chia-Feng Lin
Department of Materials Science and Engineering
National Chung Hsing University
Taichung, Taiwan, Republic of China

Chun-Han Lin
Department of Electrical Engineering
Institute of Photonics and Optoelectronics
National Taiwan University
Taipei, Taiwan, Republic of China

Tao Lin
Laboratory of Optoelectronic Materials &
Detection Technology
Guangxi Key Laboratory for Relativistic Astrophysics
School of Physical Science & Technology
Guangxi University
Nanning, People's Republic of China

Zhiting Lin
State Key Laboratory of Luminescent Materials
and Devices
and
Engineering Research Center on Solid-State
Lighting and its Informationisation of
Guangdong Province
South China University of Technology
Guangzhou, People's Republic of China

Jianping Liu
Suzhou Institute of Nano-Tech and Nano-Bionics
Key Laboratory of Nanodevices and Applications
Chinese Academy of Sciences
Suzhou, People's Republic of China

Yang Liu
School of Electronics and Information Technology
State Key Laboratory of Optoelectronic Materials
and Technologies
Sun Yat-Sen University
Guangzhou, People's Republic of China

Hao Long
School of Electronics and Information
Engineering
South-Central University for Nationalities
Wuhan, People's Republic of China

Siekacz Marcin
Institute of High Pressure Physics
Polish Academy of Sciences
and
TopGaN Ltd
Warsaw, Poland

Xuan Sang Nguyen
Singapore-MIT Alliance for Research and
Technology
Singapore, Singapore

Mart Peeters
Philips Lighting
Eindhoven, The Netherlands

Chien-Chung Peng
Advanced Optoelectronic Technology Inc.
Hsinchu, Taiwan, Republic of China

Faiz Rahman
School of Electrical Engineering and Computer
Science
Russ College of Engineering and Technology
Ohio University
Athens, Ohio

Marta Sawicka
Institute of High Pressure Physics
Polish Academy of Sciences
and
TopGaN Ltd
Warsaw, Poland

Dragan Sekulovski
Philips Lighting
Eindhoven, The Netherlands

Huafeng Shi
Department of Electrical and Electronic
    Engineering
Southern University of Science and Technology
Shenzhen, People's Republic of China

Meenu Singh
Department of Materials Science and Engineering
National Tsing-Hua University
Hsin-Chu, Taiwan, Republic of China

Czeslaw Skierbiszewski
Institute of High Pressure Physics
Polish Academy of Sciences
and
TopGaN Ltd
Warsaw, Poland

Chia-Ying Su
Department of Electrical Engineering
Institute of Photonics and Optoelectronics
National Taiwan University
Taipei, Taiwan, Republic of China

Xiao Wei Sun
Department of Electrical and Electronic
    Engineering
College of Engineering
Southern University of Science and Technology
Guangdong, People's Republic of China

Xiaojuan Sun
State Key Laboratory of Luminescence and
    Applications
Changchun Institute of Optics, Fine Mechanics
    and Physics
Chinese Academy of Sciences
Changchun, People's Republic of China

Yue Jun Sun
Philips Lighting
Eindhoven, The Netherlands

Gintautas Tamulaitis
Semiconductor Physics Department
Vilnius University
Vilnius, Lithuania

Kees Teunissen
Philips Lighting
Eindhoven, The Netherlands

Yi-Fang Tsai
Department of Materials Science and Engineering
National Tsing-Hua University
Hsin-Chu, Taiwan, Republic of China

Charng-Gan Tu
Department of Electrical Engineering
Institute of Photonics and Optoelectronics
National Taiwan University
Taipei, Taiwan, Republic of China

Po-Min Tu
Advanced Optoelectronic Technology Inc.
Hsinchu, Taiwan, Republic of China

Haiyan Wang
State Key Laboratory of Luminescent Materials
    and Devices
and
Engineering Research Center on Solid-State Lighting
    and its Informationisation of Guangdong Province
South China University of Technology
Guangzhou, People's Republic of China

Shuai Wang
Wuhan National Laboratory for Optoelectronics
Huazhong University of Science and Technology
Wuhan, People's Republic of China

Tao Wang
Department of Electronic and Electrical
    Engineering
University of Sheffield
Sheffield, United Kingdom

Rene Wegh
Philips Lighting
Eindhoven, The Netherlands

Feng Wu
Wuhan National Laboratory for Optoelectronics
Huazhong University of Science and Technology
Wuhan, People's Republic of China

Dong-Sing Wuu
Department of Materials Science and Engineering
National Chung Hsing University
Taichung, Taiwan, Republic of China

Shijie Xu
Department of Physics Shenzhen Institute of
Research and Innovation
HKU-CAS Joint Laboratory on New Materials
The University of Hong Kong
Pokfulam, Hong Kong

C.C. Yang
Department of Electrical Engineering
Institute of Photonics and Optoelectronics
National Taiwan University
Taipei, Taiwan, Republic of China

Hui Yang
Suzhou Institute of Nano-tech and Nano-bionics
Chinese Academy of Sciences
Suzhou, People's Republic of China

Yu-Feng Yao
Department of Electrical Engineering
Institute of Photonics and Optoelectronics
National Taiwan University
Taipei, Taiwan, Republic of China

Baijun Zhang
School of Electronics and Information Technology
State Key Laboratory of Optoelectronic Materials
and Technologies
Sun Yat-Sen University
Guangzhou, People's Republic of China

Jun Zhang
Wuhan National Laboratory for Optoelectronics
Huazhong University of Science and Technology
Wuhan, People's Republic of China

Li Zhang
Singapore-MIT Alliance for Research and
Technology
Singapore, Singapore

Wei Zhang
Wuhan National Laboratory for Optoelectronics
Huazhong University of Science and Technology
Wuhan, People's Republic of China

Xinhai Zhang
Department of Electrical and Electronic
Engineering
Southern University of Science and Technology
Shenzhen, People's Republic of China

Yonghui Zhang
Institute of Micro-Nano Photoelectron and
Electromagnetic Technology Innovation
School of Electronics and Information
Engineering
Hebei University of Technology
and
Key Laboratory of Electronic Materials and
Devices of Tianjin
Tianjin, People's Republic of China

Zi-Hui Zhang
Institute of Micro-Nano Photoelectron and
Electromagnetic Technology Innovation
School of Electronics and Information
Engineering
Hebei University of Technology
and
Key Laboratory of Electronic Materials and
Devices of Tianjin
Tianjin, People's Republic of China

I

Overview

1

From the Dawn of GaN-Based Light-Emitting Devices to the Present Day

Hiroshi Amano

Nagoya University

Abstract   It is believed that group III nitride semiconductors are among the most promising materials for solving global problems and realizing a sustainable society. GaN and related nitride materials have had a major impact on human lives. Their most important application is blue LEDs. Portable games machines and cellular or smart phones are very familiar items, especially to young people. Until the end of the 1990s, all the displays of portable games machines and cellular phones were monochrome. The younger generation can now enjoy full-color portable games such as Pokemon Go because of the emergence of blue LEDs. Today, the applications of blue LEDs are not limited to displays. In combination with phosphors, blue LEDs can act as a white light source and are also used in general lighting. This chapter describes the development of GaN-based blue LEDs from the early 1960s to the 1990s.

1.1 Introduction

According to the Swedish Royal Academy of Sciences, the light-emitting diode (LED) lamp holds great promise for increasing the quality of life for over 1.5 billion people around the world who lack access to an electricity grid.

On February 7, 2015, Luvsannyam Gantumur, the Minister of Science and Education of Mongolia, visited Nagoya University. He gave his hearty appreciation to researchers of LEDs for their social implementation of new artificial lighting systems. He said that a nomadic life was the traditional lifestyle of the Mongolian people. However, many young Mongolians do not want to continue a nomadic life and tend to move to urban areas such as Ulan Bator and live in permanent houses. He told me that the emergence of LED lamps has changed the attitudes of the Mongolian people and some young people are returning to a nomadic life.

On March 27, 2016, I had the opportunity to visit a Mongolian family, who lived at an hour's drive from Ulan Bator. They lived in a ger, a Mongolian tent, as shown in Figure 1.1a. In their ger, they had an LED lamp as shown in Figure 1.1b, and they used solar panels and a battery as the electricity source. It was reassuring to learn that LED lighting can not only contribute to modern life but also help maintain the traditional lifestyles of people trying to coexist with the natural environment.

LEDs are considered to be the fourth generation of artificial light sources. Before the nineteenth century, fire was the only light source, based on the chemical reaction or oxidation of flammable materials.

(a)

(b)

FIGURE 1.1   (a) Mongolian ger and person living in a ger. The panel to the left of the ger is a solar panel. The parabolic antenna to the right is for satellite TVs and smart phones. (b) LED lamp used in a ger.

However, CO_2 emission was inevitable because most flammable materials are composed of carbon and hydrogen. In 1879, the incandescent lamp was invented after a long history of development. Joseph Swan in the United Kingdom and Thomas Edison in the United States first demonstrated the use of the incandescent lamp, for which both submitted patents that were granted in 1880 [1]. The mechanism of light emission from incandescent lamps involves blackbody emission, meaning that it is based on classical quantum theory. It is not clear who invented the fluorescent light bulb, but in 1934, Arthur Compton in the United Kingdom and George Inman in the United States demonstrated the use of fluorescent light bulbs. These bulbs comprise a vacuum glass tube that is coated on the inside with several fluorescent materials. Electrons are emitted from a cathode in the tube and excite Hg vapor. The excited Hg vapor emits UV photons that excite the fluorescent materials. Then visible light is emitted from the fluorescent materials.

The mechanism of the fluorescent bulb involves energy transfer from energetic electrons to UV photons and from UV photons to visible photons. Therefore, the mechanism of light emission is actually based on quantum mechanics. Unfortunately, a fragile glass tube and Hg, which is an environmentally hazardous element, are necessary. Therefore, I believe LEDs are the ultimate light source because light emission is based on energy through band-to-band transitions in a solid, which are also a phenomenon based on quantum mechanics. In addition, no environmentally hazardous substances are necessary.

Today, portable game machines and cellular or smart phones are very familiar items, and not only young people but people of all ages can enjoy beautiful full-color displays. In the United States, portable game machines based on liquid crystal displays were first released in 1979 by the Milton Bradley Company, then cellular phones became commercially available in 1983 by Advanced Mobile Phone System, but until the end of the 1990s, the displays of all portable game machines and cellular phones were monochrome. So it should be emphasized that with the emergence of blue LEDs people can now enjoy full-color portable games and cellular/smart phones. Today, the applications of blue LEDs are not limited to displays. In combination with phosphors, blue LEDs can act as a white light source and also be used in general lighting. In 2015, the number of LED packages produced worldwide was more than 304 billion, nearly 41 times the world's population, about 54% of which were GaN-based LEDs [2]. Nitride-based devices have become a key part of human lives and are used as general lighting in the backlight units of TVs, PCs, and smart phones/cellular phones; traffic signals and street lights; huge displays and billboards in stadiums; optical storage; and broadband wireless communications.

In the following section, I will describe how blue LEDs were developed, especially in the mid-1980s and early 1990s. The contribution of blue LEDs to saving energy and the environment and their future prospects for society are also discussed.

1.2 History of GaN-Based Blue LEDs

Red, yellow, and green LEDs were commercialized in the early 1960s and 1970s using GaAsP [3] and GaP:N [4], respectively. Considering that As-based III–V compounds are used in red LEDs and P-based III–V compounds are used in green LEDs, it should be easy to predict which colors can be obtained from which element from the periodic table. At that time, many researchers believed that blue LEDs could be realized using N-based III–V compounds, that is, GaN.

To grow bulk GaN crystals from a solution, we need very high pressure and temperature, comparable to those needed for the growth of diamonds or even higher-quality crystals [5,6]. However, because it is unlikely that GaN crystals can be grown under such extreme conditions, a chemical reaction should be used to reduce the pressure and temperature required for GaN crystal growth. We also have to use a foreign substrate, that is, GaN must be grown on a different material. For GaN synthesis, we use ammonia as the nitrogen source [7] because nitrogen molecules are inert and do not react with metallic Ga. In contrast, ammonia is very reactive at around 1000°C, the temperature at which GaN can be synthesized. Therefore, a limited numbers of materials can be used as the substrate.

Sapphire is a promising substrate material because it is stable at high temperatures and does not strongly react with ammonia [8]. The most serious problem with sapphire is its large mismatch with GaN of up to 16% for the (0001) planes. Some researchers believe that for heteroepitaxial growth, the lattice mismatch should not exceed 1% [9], so a mismatch of 16% would make heteroepitaxial growth virtually impossible.

GaN powder was first synthesized in 1932 by the reaction of metallic Ga with NH_3 at high temperature [10]. However, no description of the expected applications of GaN was given in that study. In 1959, H. G. Grimmeiss and H. Koelmans from Philips Forschungslaboratorien at Aachen and Eindhoven, respectively, first reported the low-temperature luminescence properties of GaN powder grown by the same method, showing the potential of GaN as a light-emitting material [11]. This group also submitted a patent on GaN as a light-emitting material in 1960 [12].

Following these findings, in 1971, the RCA group led by J. I. Pankove developed the first GaN-based blue LEDs. These were metal-insulator-semiconductor (MIS)-type LEDs fabricated by halide vapor phase epitaxy or hydride vapor phase epitaxy or halogen vapor phase epitaxy (HVPE), which involved the chemical reaction of Ga and hydrogen chloride to form GaCl and ammonia [13]. Unfortunately, these LEDs were MIS-type, rather than pn-junction-type, and therefore their efficiency was limited, ranging only from 10^{-5} to 3×10^{-4}. Nevertheless, in the 1970s, following the success of the RCA group, many of the world's top electrical companies, such as RCA, Philips, Oki Electric, Hitachi, and Matsushita Research Institute Tokyo (MRIT), tried to commercialize MIS-type blue LEDs, but all their efforts failed [14–18]. At that time, it was believed impossible to grow p-type GaN because of self-compensation [19], which means that if we dope acceptors as an impurity, the same number of intrinsic donors such as nitrogen vacancies is generated to compensate for the doped acceptors.

In 1972, Osamura et al. deposited InGaN, a compound related to GaN, by the electron beam evaporation of metal sources in DC discharge nitrogen plasma over the full compositional range from InN to GaN [20]. They reported a large absorption edge of InN of 1.9 eV, rather than the currently accepted bandgap of 0.65 eV, which might have been caused by the unintentional formation of InON. No luminescence data were reported. It took 19 years for the first luminescence data of InGaN to be reported [21] and 30 years for the bandgap of InN to reach the present value [22].

Many researchers and companies gave up on GaN and started research on other materials, such as ZnSe and SiC. The manager of MRIT also decided to abandon the project on GaN. However, the leader of the GaN research team, Dr. Isamu Akasaki, could not abandon GaN and moved from MRIT to Nagoya University in 1981, where he became a professor. I joined his laboratory in 1982 as an undergraduate student and was highly motivated by my graduate dissertation on GaN-based blue LEDs. At that time, I was interested in microcomputer systems, now called personal computers (PCs). In 1975, Bill Gates and Paul Allen established Micro-Soft. One year later, the Apple I computer was developed by Steve Jobs and Stephen Wozniak. From these successes, the personalization of computer systems progressed rapidly and I wanted to contribute to the further development of microcomputer systems, particularly their displays. At that time, Braun tubes were used in all displays, which were large, heavy, and had high energy consumption. So, I thought that if I could make blue LEDs, I would help change the world. That is why I was so motivated and focused on my research subject of GaN.

The problem in fabricating MIS-type blue LEDs using HVPE was that the growth rate was so high that it was difficult to control the thickness of the insulating layer in the MIS-type structure. Therefore, the operating voltage could not be controlled. Professor Akasaki also noted the difficulty of growing GaN by molecular beam epitaxy (MBE) from his experience in the 1970s. Therefore, he decided to use metalorganic chemical vapor deposition (MOCVD) or metalorganic vapor phase epitaxy (MOVPE) for GaN growth.

The first report of the formation of III–V compounds by MOCVD was in 1960. A group from Union Carbide reported the formation of an $In(CH_3)_3–PH_3$ adduct [23]. The first report of the epitaxial growth of III–V compounds including group III nitrides on a substrate was in a patent filed by Monsanto Co. [24]. The North American Rockwell group first reported the experimental deposition of GaN on sapphire and SiC substrates by MOCVD in 1971 [25].

At that time, funding for research at Japanese universities was insufficient. Also, there was no commercially available MOVPE system especially designed for GaN growth. Thus, it was impossible to purchase an MOVPE system. In 1982, a Master's student 2 years older than me developed the first vertical-type MOVPE reactor [26]. At that time, the flow rate was so low that I could not grow GaN using hydrogen as the carrier gas. I tried to visualize the flow pattern by using the reaction between $TiCl_4$ and H_2O to form TiO_2 powder and found that the flow rate would be insufficient if I used hydrogen as the carrier gas [26]. In 1984, Y. Koide, a PhD student, joined Professor Akasaki's laboratory and started research on AlGaN and AlN, while I focused on growing GaN. Together, we modified the MOVPE reactor. I visited several laboratories and found that the flow rate would be insufficient if I used the old configuration of gas supply tubes in the reactor. So I merged all the gas lines into one line, which increased the flow rate from a few centimeters per second to more than 4 m/s [27]. Then, I successfully grew polycrystalline GaN on a sapphire substrate despite using hydrogen as the carrier gas. However, the surface was rather rough and the quality was very poor. I tried to grow GaN many times while varying the growth temperature, the flow rates of the source and carrier gases, the configuration of the linear tubes, the susceptor shape, and other parameters, but I could not grow high-quality GaN with a smooth surface. The large lattice mismatch (16%) was too great a problem for an inexperienced student like me to overcome. Therefore, almost 3 years passed without any success. I measured the x-ray diffraction, photoluminescence, the Hall effect, and so forth, but all the properties were poor and no promising new results were obtained.

In February 1985, I was almost nearing the completion of my Master's degree. I had decided to start a PhD program in April, and while all the other students went on a graduation trip, I conducted experiments on my own. At that time, Koide was growing Al-containing nitrides, such as AlN and AlGaN, and I was growing GaN. When I compared his Al-containing crystals with my GaN, the surface of his crystals seemed to be smoother. I concluded that AlN could be used to effectively grow GaN with a better surface morphology, so I tried to grow a thin AlN layer on a sapphire substrate immediately before the growth of GaN. At that time, I knew that the epitaxial temperature of AlN was higher than 1200°C. However, because the old oscillator did not work well, I could not get the temperature to reach 1200°C. Then I suddenly remembered a discussion in the laboratory with Dr. Sawaki, an associate professor, who had explained that the growth process of boron phosphide (BP) on Si [28], for which the lattice mismatch is as large as 16%, was similar to that of GaN and sapphire. Before Professor Akasaki started his laboratory at Nagoya University in 1981, Dr. Tatau Nishinaga studied the epitaxial growth of BP on Si by vapor phase epitaxy. Dr. Sawaki mentioned the effectiveness of using a preflow of phosphorus as a source gas immediately before BP growth on Si and pointed out that the phosphor atoms appeared to act as nucleation centers. From this, I imagined that if I supplied a small amount of AlN at a low temperature, it should provide nucleation centers. Normally, I used to look inside the reactor during growth to see whether there was an interference pattern on the substrate, whereby I could check that the source gas had been properly supplied. That time, however, I was so tired that I forgot to check the interference pattern, so when I removed the sample from the reactor and saw that it had a highly smooth surface which was perfectly transparent, I thought that I had made a mistake and forgotten to supply trimethylgallium. But thinking it over, I realized that I had not made a mistake, so I checked the surface using a Nomarski-type microscope and found that I had actually succeeded in growing atomically flat GaN. Following the advice of Professor Akasaki, I further characterized the film by determining its crystalline, optical, and electrical properties, all of which turned out to be superior to those reported in previous papers. This process is known as "low-temperature-deposited buffer layer technology" and it has been used by many researchers worldwide [27,29–45]. At that time, the majority of university professors in Japan had a negative impression of the submission of patents by university professors. However, Professor Akasaki came from a company and was aware of the importance of intellectual property rights. I was also interested in submitting a patent. So, I prepared the draft of a patent for low-temperature-deposited buffer layers. To ensure that the patent would be granted, I carefully considered similar previous results. For example, Dr. Akiyama from Oki Electric used a low-temperature-deposited GaAs buffer layer for the growth of GaAs on Si [46], so I focused on the novelty of using an AlN buffer layer, and not a GaN buffer layer. Dr. Yoshida from the Electrotechnical Laboratory used single-crystalline AlN for the growth of GaN on a sapphire substrate [47], so I focused on the novelty of low-temperature deposition, which implies that the layer is not single-crystalline but polycrystalline or amorphous. The patent was submitted in 1985 [48] and was granted in 1986. In 1989, Nichia Chemicals submitted a patent for a low-temperature-deposited buffer layer with a much wider range of compositions from GaN to close to AlN but not AlN [49]. I also submitted a paper on our breakthrough to the journal *Applied Physics Letters* [27], which accepted it for publication. I thought that the paper would attract the interest of other researchers, but I was wrong. At that time, the majority of blue LED researchers were interested in ZnSe, and GaN researchers were in the minority, although this allowed us to concentrate on research without the high stress of competition from other groups.

Anyway, I thought that our next task should be to realize p-type GaN. At that time, experienced members of the laboratory thought that nitrogen vacancies were the origin of the high donor density in nominally undoped GaN films and that by doping GaN with an acceptor impurity such as Zn, the same number of shallow donors would be automatically generated. Therefore, it appeared that p-type GaN would be

FIGURE 1.2   Second MOVPE reactor for GaN growth in Akasaki Laboratory, Nagoya University.

impossible to realize. This mechanism is called self-compensation, and even today it is a hot topic [50]. I had a different impression. I was unconvinced by the mechanism of self-compensation and I believed that p-type GaN could be realized. We were very lucky that a government organization, Japan Science and Technology (JST) Agency, was interested in our achievement of growing high-quality GaN crystals on a sapphire substrate by MOVPE using a low-temperature-deposited AlN buffer layer and decided to support us. As a result, Akasaki Laboratory was able to buy a new MOVPE reactor from ULVAC, which is shown in Figure 1.2. More importantly, JST encouraged us to start a collaboration with a private company. As a result, the rubber company Toyoda Gosei decided to collaborate with us. This partnership between the industry, the government, and academia is thought to be one of the most successful projects supported by JST.

Although we grew many Zn-doped GaN samples, they were all highly resistive or even n-type. In 1987, during my PhD program, I observed very sharp exciton emission at a cryogenic temperature from Zn-doped GaN grown on c-plane and a-plane sapphire [30]. I also measured the deformation potential of GaN. I was excited by the results and presented them at the annual fall meeting of the Japan Society for Applied Physics, held at Nagoya University in 1987. However, I was a bit disappointed to see that there were only four people in the room for my presentation: the chairman, Professor Akasaki, another guy, and myself. As I mentioned earlier, researchers were more interested in other compound semiconductors and GaN researchers were in significant minority. Moreover, in 1987, during my internship as part of my PhD program, I found that Zn-related blue emission was irreversibly enhanced during cathodoluminescence measurement [51], a process I referred to as low-energy electron beam irradiation (LEEBI) treatment. However, even after LEEBI treatment, Zn-doped GaN did not show p-type conductivity. Nevertheless, I was excited with this result, which I presented at an international conference on luminescence held at Beijing, China, in 1987 [51]. I also hoped to write a journal paper on this treatment. At that time, PhD candidates of the Department of Electric Engineering in Nagoya University were required to write at least three journal papers to obtain a PhD in addition to a doctoral thesis. However, I found that the LEEBI effect itself had been observed by researchers at Moscow University 4 years earlier [52]. Because I was unable to write a third journal paper, I abandoned the plan of obtaining a PhD within 3 years of my doctoral course, and in 1989, I became a research associate at the Akasaki Laboratory of Nagoya University without a PhD. While reading the textbook *Bonds and Bands in Semiconductors* by Dr. Phillips [53], I found one graph particularly interesting because it showed that Mg is better than Zn for the activation of acceptors in GaP, the material used in green LEDs. Although the Mg source, bis-Cp_2Mg, was very expensive, I implored Professor Akasaki to let me purchase some. He kindly gave me permission, and after waiting several months for it to arrive, I was finally able to grow many Mg-doped samples with my laboratory partner, Kito, a Master's student. Here, I would like to mention the pioneering work of Dr. Maruska in 1972 [54], who at the time was a PhD student at Stanford University. He succeeded in fabricating the world's first MIS-type violet LED using Mg-doped GaN. All our Mg-doped GaN samples were highly resistive in the as-grown state, but after LEEBI treatment, some samples showed p-type behavior when subjected to hot probe measurement. I knew that hot probes were not considered to be reliable and that no one would believe that p-type conduction had been achieved. Kito then performed Hall effect measurements on the samples, by which we finally established that we had achieved p-type GaN for the first time in the world. We also fabricated p–n junction ultraviolet/blue LEDs [55–58]. Soon after that, Dr. Nakamura's group also used LEEBI treatment [59,60]. In 1992, Dr. Nakamura claimed that p-type GaN could be obtained by simple thermal annealing [61]. Today, almost all LED companies use thermal annealing. The mechanism of p-type conduction involves the desorption of hydrogen near Mg acceptors, as first pointed out by Professor van Vechten [62], and confirmed experimentally by Dr. Nakamura [61].

For us, another important task was to realize true blue emission using a band-to-band transition, so from 1986 onward we tried to grow InGaN. However, this was also very difficult and we only succeeded in growing InGaN with an In composition of less than 1.7% [63]. In 1989, Dr. Matsuoka's group at NTT reported the successful growth of InGaN under an extremely high ammonia supply while using nitrogen as a carrier gas [64]. The only difference between our work and that of the NTT group was the carrier gas; we used hydrogen as the carrier gas. They also reported blue-violet photoluminescence at 77 K, indicating the incorporation of a high In content. However, at room temperature, deep-level-related yellow emission was observed. The mechanism of In incorporation in InGaN has been elucidated through thermodynamic analysis by Professor Koukitu et al. [65,66].

Finally, by combining high-quality-crystal growth technology using a low-temperature-deposited buffer layer, p-type growth technology by Mg doping with LEEBI or thermal annealing, and InGaN growth technology, Nichia Corporation succeeded in commercializing double-heterostructure-type InGaN blue LEDs for the first time in the world in 1993 [67]. They also fabricated single-quantum-well LEDs in 1995 [68], which are also a very important technology for enhancing the efficiency of nitride LEDs because a very narrow quantum suppresses the quantum-confined Stark effect [69], thus increasing the transition probability [70].

Let me explain how InGaN LEDs can contribute to improving the electricity situation, especially in Japan. Many people will remember the great earthquake of east Japan and the meltdown of the nuclear power plants in March 2011. For the next 3 years, none of the 48 nuclear electricity generators in Japan were in operation [71]. Although three nuclear power plants restarted operation, they were not enough to make up for the loss of 30% of Japan's generating capacity of all the nuclear power plants before 2011. The U.S. Department of Energy predicted that three quarters of lighting will have been replaced with LED lighting systems by the year 2030 in the United States, resulting in a 7% reduction in electricity use [72]. In the case of Japan, the penetration of LED lighting systems into the market is expected to be much faster. A research company in Japan has predicted that by 2020, more than 95% of the sales of general lighting systems will be LED lighting systems [73]. More importantly, we can develop and supply compact lighting systems to the younger generation, especially children in remote areas without access to electricity.

1.3 Summary

This chapter mainly gives a history of the development of GaN-based blue LEDs. I wrote this manuscript as a message to young scientists and engineers with the desire to invent something that will benefit humans. I sincerely hope that this message will encourage the next generation to tackle some of the world's problems through science and engineering.

References

1. J. W. Swan, British Patent 4933 (November 27, 1880) and T. A. Edison, US Patent US223898 A (January 27, 1880).
2. Fuji Chimera Research Institute, Inc., 2016 LED Related Market Survey (2016) p. 3 [in Japanese].
3. N. Holonyak, Jr. and S. F. Bevacqua, *Appl. Phys. Lett.*, 1 (1962) 82.
4. I. Ladany and H. Kressel, *RCA Rev.*, 33 (1972) 517.
5. F. P. Bundy, H. T. Hall, H. M. Strong, and R. H. Wentorf, *Nature*, 176 (1955) 51.
6. S. Porowski and I. Grzegory, *J. Cryst. Growth*, 178 (1997) 174.
7. W. C. Johnson, J. B. Parsons, and M. C. Crew, *J. Phys. Chem.*, 36 (1932) 7.
8. H. P. Maruska and J. J. Tietjen, *Appl. Phys. Lett.*, 15 (1969) 327.
9. F. C. Frank and J. H. van der Merwe, *Proc. R. Soc. London, Ser. A*, 198 (1949) 205.
10. W. C. Johnson, J. B. Parsons, and M. C. Crew, *J. Phys. Chem.*, 36 (1932) 2651.
11. H. G. Grimmeiss and H. Koelmans, *Z. Naturforsch. A*, 14 (1959) 264.
12. H. G. Grimmeiss, H. Koelmans, and I. B. Maak, DBP 1077330 (1960).
13. J. I. Pankove, E. A. Miller, D. Richman, and J. E. Berkeyheiser, *J. Lumin.*, 4 (1971) 63.
14. H. P. Maruska, D. A. Stevenson, and J. I. Pankove, *Appl. Phys. Lett.*, 22 (1973) 303.
15. G. Jacob and D. Bois, *Appl. Phys. Lett.*, 30 (1977) 412.
16. Y. Morimoto, *Jpn. J. Appl. Phys.*, 13 (1974) 1307.
17. A. Shintani and S. Minagawa, *J. Appl. Phys.*, 48 (1977) 1522.
18. Y. Ohki, Y. Toyoda, H. Kobayashi, and I. Akasaki, *Proc. Int. GaAs Symp.*, 63 (1981) 479.
19. G. Mandel, *Phys. Rev. A*, 134 (1964) 1073.
20. K. Osamura, K. Nakajima, and Y. Murakami, *Solid State Commun.*, 11 (1972) 617.
21. N. Yoshimoto, T. Matsuoka, T. Sasaki, and A. Katsui, *Appl. Phys. Lett.*, 59 (1991) 2251.
22. V. Yu. Davydov, A. A. Klochikhin, R. P. Seiyan, V. V. Emtsev, S. V. Ivanov, F. Bechstedt, J. Furthmüller et al., *Phys. Status Solidi B*, 229 (2002) R1.
23. R. Didchenko, J. E. Alix, and R. H. Toeniskoeter, *J. Inorg. Nucl. Chem.*, 14 (1960) 35.
24. R. A. Ruehrwein, D. Ohio, and Monsanto Co., US Patent 3312570 (Filed May 29, 1961).
25. H. M. Manasevit, F. M. Erdmann, and W. I. Simpson, *J. Electrochem. Soc.*, 118 (1971) 1864.
26. M. Hashimoto, H. Amano, N. Sawaki, and I. Akasaki, *J. Cryst. Growth*, 68 (1984) 163.
27. H. Amano, N. Sawaki, I. Akasaki, and Y. Toyoda, *Appl. Phys. Lett.*, 48 (1986) 353.
28. T. Nishinaga and T. Mizutani, *Jpn. J. Appl. Phys.*, 14 (1975) 753.
29. H. Amano, I. Akasaki, K. Hiramatsu, N. Koide, and N. Sawaki, *Thin Solid Films*, 163 (1988) 415.
30. H. Amano, K. Hiramatsu, and I. Akasaki, *Jpn. J. Appl. Phys.*, 27 (1988) L1384.
31. I. Akasaki, H. Amano, Y. Koide, K. Hiramatsu, and N. Sawaki, *J. Cryst. Growth*, 98 (1989) 209.
32. H. Amano, T. Asahi, and I. Akasaki, *Jpn. J. Appl. Phys.*, 29 (1990) L205.
33. K. Hiramatsu, H. Amano, I. Akasaki, H. Kato, N. Koide, and K. Manabe, *J. Cryst. Growth*, 107 (1991) 509.
34. K. Hiramatsu, S. Itoh, H. Amano, I. Akasaki, N. Kuwano, T. Shiraishi, and K. Oki, *J. Cryst. Growth*, 115 (1991) 628.
35. N. Kuwano, T. Shiraishi, A. Koga, K. Oki, K. Hiramatsu, H. Amano, K. Itoh, and I. Akasaki, *J. Cryst. Growth*, 115 (1991) 381.
36. S. Nakamura, *Jpn. J. Appl. Phys.*, 30 (1991) 1620.
37. H. Murakami, T. Asahi, H. Amano, K. Hiramatsu, N. Sawaki, and I. Akasaki, *J. Cryst. Growth*, 115 (1991) 648.
38. J. N. Kuznia, M. A. Khan, D. T. Olson, R. Kaplan, and J. Freitas, *J. Appl. Phys.*, 73 (1993) 4700.
39. S. T. Kim, H. Amano, I. Akasaki, and N. Koide, *Appl. Phys. Lett.*, 64 (1994) 1535.
40. T. Sasaoka and T. Matsuoka, *J. Appl. Phys.*, 77 (1995) 192.
41. Y. M. Le Vaillant, R. Bisaro, J. Oliver, O. Durand, J. Y. Duboz, S. Ruffenach-Clur, O. Briot, B. Gil, and R. L. Aulombard, *Mater. Sci. Eng. B*, 50 (1997) 32.

42. M. Iwaya, T. Takeuchi, S. Yamaguchi, C. Wetzel, H. Amano, and I. Akasaki, *Jpn. J. Appl. Phys.*, 37 (1998) L316.
43. Y. M. Le Vaillant, R. Bisaro, J. Olivier, O. Durand, J. Y. Duboz, S. Ruffenach-Clur, O. Briot, B. Gil, and R. L. Aulombard, *J. Cryst. Growth*, 189(190) (1998) 282.
44. Y. Kobayashi, T. Akasaki, and N. Kobayashi, *Jpn. J. Appl. Phys.*, 37 (1998) L1208.
45. T. Ito, K. Ohtsuka, K. Kuwahara, M. Sumiya, Y. Takano, and S. Fuke, *J. Cryst. Growth*, 205 (1999) 20.
46. M. Akiyama, Y. Kawarada, and K. Kaminishi, *Jpn. J. Appl. Phys.*, 23 (1984) L843.
47. S. Yoshida, S. Misawa, and S. Gonda, *Appl. Phys. Lett.*, 42 (1983) 427.
48. Japanese Patent Application Number S60-256806.
49. Japanese Patent Application Number H3-32259.
50. G. Miceli and A. Pasquarello, *Phys. Rev. B*, 93 (2016) 165207.
51. H. Amano, I. Akasaki, T. Kozawa, K. Hiramatsu, N. Sawaki, K. Ikeda, and Y. Ishii, *J. Lumin.*, 40–41 (1988) 121.
52. G. V. Saparin, S. K. Obyden, I. F. Chetverikova, M. V. Chukichev, and S. I. Popov, Moskovskii Universitet, Vestnik, Seriia 3 - Fizika, Astronomija, 24 (1983) 56.
53. J. C. Phillips, *Bonds and Bands in Semiconductors*, 1st edn., Bell Telephone Laboratories, Inc., Academic Press (1973).
54. H. P. Maruska, W. C. Rhines, and D. A. Stevenson, *Mater. Res. Bull.*, 7 (1972) 777.
55. H. Amano, M. Kito, K. Hiramatsu, and I. Akasaki, *Jpn. J. Appl. Phys.*, 28 (1989) L2112.
56. H. Amano, M. Kito, K. Hiramatsu, and I. Akasaki, *J. Electrochem. Soc.*, 137 (1990) 1639.
57. I. Akasaki, H. Amano, M. Kito, and K. Hiramatsu, *J. Lumin.*, 48(49) (1991) 666.
58. I. Akasaki, H. Amano, H. Murakami, M. Sassa, H. Kato, and K. Manabe, *J. Cryst. Growth*, 128 (1993) 379.
59. S. Nakamura, M. Senoh, and T. Mukai, *Jpn. J. Appl. Phys.*, 30 (1991) L1708.
60. S. Nakamura, N. Iwasa, M. Senoh, and T. Mukai, *Jpn. J. Appl. Phys.*, 31 (1992) 1258.
61. S. Nakamura, T. Mukai, M. Senoh, and N. Iwasa, *Jpn. J. Appl. Phys.*, 31 (1992) L139.
62. J. A. van Vechten, J. D. Zook, R. D. Horning, and B. Goldenberg, *Jpn. J. Appl. Phys.*, 31 (1992) 3662.
63. T. Kozawa, Master's thesis, Nagoya University, Nagoya, Japan (1987).
64. T. Matsuoka, H. Tanaka, T. Sasaki, and A. Katsui, *Inst. Phys. Conf. Ser.*, 106 (1990) 141.
65. A. Koukitu, N. Takahashi, T. Taki, and H. Seki, *Jpn. J. Appl. Phys.*, 35 (1996) L673.
66. A. Koukitu, T. Taki, N. Takahashi, and H. Seki, *J. Cryst. Growth*, 197 (1999) 99.
67. S. Nakamura, M. Senoh, and T. Mukai, *Jpn. J. Appl. Phys.*, 32 (1993) L8.
68. S. Nakamura, M. Senoh, N. Iwasa, and S. Nagahama, *Jpn. J. Appl. Phys.*, 34 (1995) L797.
69. H. Amano and I. Akasaki, *Ext. Abst. Int. Conf. Solid State Devices Mater.*, 7 (1995) 683.
70. T. Takeuchi, S. Sota, M. Katsuragawa, M. Komori, H. Takeuchi, H. Amano, and I. Akasaki, *Jpn. J. Appl. Phys.*, 36 (1997) L382.
71. http://www.enecho.meti.go.jp/category/electricity_and_gas/nuclear/001/pdf/001_02_001.pdf [in Japanese].
72. U.S. DOE Energy Savings Potential of Solid-State Lighting in General Illumination Applications, (January 2012) (http://apps1.eere.energy.gov/buildings/publications/pdfs/ssl/ssl_energy-savings-report_jan-2012.pdf).
73. Fuji Chimera Research Institute, Inc., 2016 LED Related Market Survey (2016) p. 29 [in Japanese].

2

Spectrum-Related Quality of White Light Sources

Yue Jun Sun
Philips Lighting

Dragan Sekulovski
Philips Lighting

Mart Peeters
Philips Lighting

Kees Teunissen
Philips Lighting

Remy Broersma
Philips Lighting

Rene Wegh
Philips Lighting

Abstract   Besides the merits of energy saving, form factor, long lifetime, and durability, light-emitting diodes (LEDs) also provide a much higher degree of freedom to tune the light emission spectrum compared to conventional lighting technologies. The freedom not only provides new opportunities to optimize these LED-based light sources but also introduces challenges to characterize the spectrum. Several studies have indicated that the existing international measures (and standards), which are currently widely adopted in the lighting industry for conventional lighting, are outdated, improper, or at least insufficient to specify the multidimensional light-quality properties of LED-based white light sources. The LED technology also allows for tailored solutions for different applications that require a more sophisticated description of the light-quality properties during the design and development process. Furthermore, dynamics of LED spectrum magnify the urgency of proper assessment and evaluation of light quality for LED lighting. However, this turns out to be a big challenge for the researchers and experts of color science and industrial specifiers. On the other hand, it does also bring an opportunity to reexamine the limits of existing methods and approaches and thus develop and propose more meaningful measures to describe the performance, with regard to applications in performing specific visual or even nonvisual tasks. In this chapter, we outline some key aspects of the quality of light, which are determined by the emission spectrum of white-light source. Relevant characterization methods are introduced and reviewed, and their boundaries are explored. The latest developments are also discussed.

2.1 Introduction

Visible light, or often simply called light, consists of a band of electromagnetic waves that human eyes can detect and that create a visual sensation. Vision is created by a series of complex processes that transduce these electromagnetic waves into neural signals. Intensity and colors are perceived by people as a result of the physical stimulation of light photons on the photoreceptor cells in the retina. In many applications, the preference of perceived light is caused by the physical properties of the emitted light and the reflectance of the objects and their complex interactions with individual psychological experience. In our visual system, the objective quantities are then entangled with the subjective feeling. So the visual perception is the result of processed information, deduced from individual sensations.

The influence of light on psychological mood has been studied to be systematic [1,2]. A bad lighting environment may subconsciously affect one's comfort level or lower his or her spirit. Conversely, people can be emotionally stimulated when they are immersed in an appealing lighting atmosphere.

Just recently in 1998, melanopsin was first discovered from the skin cells of a frog [3]. Later, the same team found that it mediates nonvisual photoreceptive tasks [4]. In human eyes, melanopsin is found in intrinsically photosensitive retinal ganglion cells (iPRGCs), which are particularly sensitive to bluish light. These third class photoreceptors, next to the well-known cones and rods, are connected to the suprachiasmatic nucleus (SCN), which is also known as the "body clock." Therefore, melanopsin has been considered to serve an important role in the photo entrainment of circadian rhythms [5]. Many studies on the treatment of seasonal affective disorder also unveil the impact of light on humans' mental health [6,7].

As outlined earlier, light not only stimulates vision but also is of critical importance for our psychological and physiological behaviors [8]. In modern society, the majority of people spend more time indoor and are exposed to the artificial electric lighting during most of the time. Attention should be paid to visual, emotional, and biological impacts (cf. Figure 2.1) of the electric lighting. Recently, research on these impacts has intensified under the catchphrase of human-centric lighting, which is promoted by some lighting institutions, for example, LightingEurope [9].

Compared to the rapid progress in efficiency improvements of LED sources in the last decades, the progress in specifying the multidimensional light-quality properties is, despite significant research efforts, very limited. In this chapter, we mainly focus on the visual aspects of light, particularly those aspects that are determined by the light spectrum. Several measures have been reviewed. Recent activities to propose alternatives to better specify the light quality are also outlined.

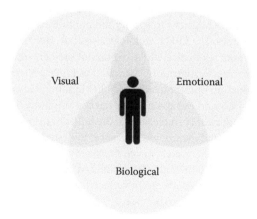

FIGURE 2.1 Impact of light on humans.

2.2 Methods of Measuring and Specifying Light Quality

To some extent, the assessment of quality of light is a subjective pathway to perceive the "real" world. However, it is often desired to be objective in describing the relevant properties in a common language. How to specify and quantify these properties and their impact is crucial for the lighting industry and designers, in order to create a specific light effect and to achieve the design targets. On the other hand, end users also require understandable and simple terms or icons to make their purchasing and application choices. Some methods to quantify light properties have been developed particularly by the International Commission on Illumination (CIE) and have been widely adopted in the lighting industry for decades. However, LED lighting, as a rapidly evolving technology, creates light and applications in a completely new way, which in turn requires a studiously look on these characterization methods and CIE e-ILV 17-550 possible alternatives.

Humans have quite complex and diverse tastes in lighting. As a result, the definition of quality of light is multidimensional and cannot be described by a single figure, that is, there is no "one-size-fits-all" solution. Conventionally, we refer to the quality of light as how good or bad the perceived light in a given scene is. The nonvisual influence is typically neglected, and only the short-term visual sense is covered, although even not completely. Table 2.1 lists the most common methods, which measure various aspects of light quality.

2.3 White Light

In most cases, light used for general lighting refers to the white light source. An equal energy spectrum (EES) has a relative spectral power distribution (SPD) equal to one for all wavelengths in the visible spectrum range (380–780 nm). Accordingly, the theoretical definition of a white object is that it diffuses

TABLE 2.1 Selected Methods for Characterizing the Quality of Light

| | | | Measure or Unit | | |
	Effect	Quantity	Name	Symbol	Standard
Visual	Brightness, lightness	Luminous flux	Lumen (=cd·sr)	lm	CIE e-ILV 17-738
		Illuminance	Lux (=lm/m²)	lx	CIE e-ILV 17-550
		Luminous intensity	Candela (=lm/sr)	cd	CIE e-ILV 17-739
		Luminance	Nit	cd/m²	CIE e-ILV 17-711
	Comfort	Glare	Unified glare rating	UGR	CIE e-ILV: 17-330
	Fidelity	General Color rendering index	CRI or R_a		CIE13.3:1995 & CIE e-ILV 17-154
	Temporal light artifacts	Flicker	Short- or long-term flicker indicator	P_{st} or P_{lt}	IEC61547-1
		Stroboscopic effect	Stroboscopic effect visibility measure	SVM	CIE TC 1-83
	Color appearance and consistency	Correlated color temperature	Kelvin	K	CIE e-ILV 17-258
Emotional	Feeling	MacAdam ellipse	Standard deviations of color matching	SDCM	CIE-1932
		Whiteness			CIE 15-2004
Biological	Photobiological safety	Retinal blue light hazard	Exposure time limit	t_{max}	IEC62471
	Circadian and neurobehavioral regulation	Cyanopic/chloropic/ erythropic/melanopic/ rhodopic illuminance	tbd	tbd	CIE JTC9

light perfectly and has a 100% spectral reflectance across the same spectrum range. CIE 1964 defines the formula of whiteness (*W*) with a 10° field of view and using a CIE Standard Illuminant D65:

$$W = Y + 800(0.3138 - x) + 1700(0.3310 - y) \tag{2.1}$$

where
  Y is the tristimulus value or relative luminance
  x and *y* are the chromaticity coordinates in the CIE 1931 color space

Apparently, if *x* = 0.3138 and *y* = 0.3310, the object has neutral color, with a whiteness that is equal to its luminance.

One can subtly, but easily, "feel" the quality of white light surrounding us. In many decades, the quality of light has been typically expressed with two characteristics, namely, correlated color temperature (CCT), combined with the distance to the blackbody locus, and color rendering index (CRI). They are well defined in international standards and are most familiar in the lighting industry and adopted by regulations across the regions. CCT and CRI together give a numerical estimate of the light appearance in comparison to a reference light source (a Planckian, or also called blackbody, radiator, or CIE illuminants representing natural daylight). As both CCT and CRI are based on colorimetry, not color appearance, they cannot reliably predict human's color perception. In particular for narrowband light sources, with "gaps" in the emission spectrum, the correspondence between colorimetric numbers and perception becomes low. Nevertheless, a quick look of these well-established methods helps us to understand where we stand now and what their intrinsic problems are and how we can cope with these issues.

2.3.1 Chromatic Aspects of White Light

All colors can roughly be divided into two parts, namely, brightness (or whiteness) and chromaticity. In color science, the chromaticity characterizes the specification of the color, regardless of its luminance. It consists of two independent parameters, namely, hue and chroma (often confused with saturation). In practice, the CIE 1931 xyY chromaticity diagram is widely used to specify colors.

CCT is the temperature of a blackbody radiator whose perceived color most closely resembles that of a given stimulus at the same brightness and under specified viewing conditions [10]. This chromatic quantity, measured in kelvin (K), is used for indicating the general appearance of a white light source. For example, a typical incandescent lamp with a CCT of 2800 K looks warm and yellowish white, while a fluorescent tube with a CCT of 6500 K appears cool and blueish. But this single number does not communicate the chromaticity when the color point is outside of the blackbody locus, or the tint of the source. This problem is addressed by Δuv, or *Duv*, the distance from the blackbody locus measured in the CIE *u,v* 1960 chromaticity diagram. The combination of CCT and *Duv*, including its sign, provides color point information intuitively. A positive *Duv* (above the blackbody locus) indicates a more greenish tint, while a negative one (below the blackbody locus) means a more purplish appearance. The CCT only gives the white tone of the light source in the chromatic space, while a given CCT value can be achieved with, in principle, countless variations in the emission spectrum. That means that light sources may show differences in color rendition, which is not disclosed by the CCT–*Duv* combination.

Table 2.2 lists the chromaticity coordinates and corresponding CCTs for most widely used CIE Standard Illuminants. These Standard Illuminants provide a basis for comparing colors recorded under different lighting sources.

To ensure a high-quality white light, manufacturers need to specify the chromaticities and tolerances of their lighting products. These categorized figures thus enable a common language across their various products and customers. For standardized CCTs, the rated values of the chromaticity coordinates are well defined in standards of IEC 60081 or ANSI C78.376 (cf. Table 2.3). It should be noted that both standards

TABLE 2.2 Chromaticity Coordinates and CCTs for CIE Standard Illuminants

Source	CIE 1964 10°		CCT (K)	Note
	x	y		
Illuminant A	0.45117	0.40594	2856	Incandescent
Illuminant D55	0.33412	0.34877	5503	Midmorning/midafternoon daylight
Illuminant D65	0.31381	0.33098	6504	Noon daylight
Illuminant F4	0.44925	0.39061	2940	Warm white fluorescent
Illuminant F10	0.35061	0.35430	5000	Philips TL850
Illuminant F11	0.38544	0.37109	4000	Philips TL840
Illuminant F12	0.44265	0.39706	3000	Philips TL830

TABLE 2.3 Standardized CCTs and Nominal Chromaticity Coordinates for Different Lamp "Colors"

IEC 60081				ANSI					Energy Star
					C78.376		C78.377		V2.0
"Color"	CCT (K)	x	y	CCT (K)	x	y	x	Y	CCT Descriptor
F6500	6400	0.313	0.337	6500	0.313	0.337	0.3123	0.3283	Daylight
F5000	5000	0.346	0.359	5000	0.346	0.359	0.3446	0.3551	Daylight
F4000	4040	0.380	0.380	4000	0.380	0.380	0.3818	0.3797	Cool white
F3500	3450	0.409	0.394	3500	0.409	0.394	0.4078	0.3930	Neutral white
F3000	2940	0.440	0.403	3000	0.440	0.403	0.4339	0.4033	Warm white
F2700	2720	0.463	0.420	2700	0.459	0.412	0.4578	0.4101	Soft white
				2500			0.4806	0.4141	Sunrise/sunset white
				2200			0.5018	0.4153	Amber white

Note: The corresponding nomenclature to describe the color term is recommended by Energy Protection Agency (EPA), as outlined in the last column.

are developed for traditional fluorescent lamps. In order to increase the overall yield of solid-state lighting products and take more applications into account, ANSI C78.377—Specifications for the Chromaticity of Solid-State Lighting Products—defines quadrangles for the tolerances of nominal CCTs, instead of MacAdam ellipses used in the conventional lighting. Figure 2.2 shows the color point of standardized 4000 K in the CIE 1931 chromaticity diagram. Note the difference of the center point and the tolerance between IEC F4000 and ANSI 4000 K.

Some applications require low-CCT light, which was addressed in the addition of the specifications for nominal CCTs of 2200 and 2500 K in ANSI C78.377 2014 revision. Although the latest Energy Star V2.0 points out that 2200 and 2500 K nominal CCTs are only applicable to filament-style lamps, it would be not surprising that these colors will become more common in "normally configured" LED bulbs in the near future, as they create a warmer and cozier atmosphere. Besides, a more dynamic scene can be achieved with a dual CCT LED configuration, in order to have the effect of sunset-like drop in terms of the color temperature when dimming down, for example, Philips Warm-Glow dimmable LED lamps from a yellowish 2700 K down to a more reddish-orange tone 2200 K. On the other hand, the lower CCT light source has, typically, less radiant energy in the short-wavelength region, especially in the spectral melanopic sensitivity range [11], which is believed to lead to a lower suppression of melatonin compared to that of the same intensity but higher CCT light. This finding enables configuring emission spectra of white-light sources that selectively enhance or reduce melatonin secretion and consequently impact our circadian rhythm and with that our sleep quality.

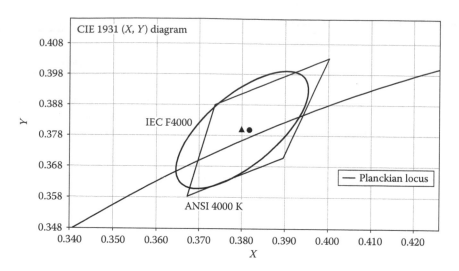

FIGURE 2.2   Standardized 4000 K in the CIE 1931 chromaticity diagram. Note the difference of the center point and the tolerance between IEC F4000 (triangle) and ANSI 4000 K (circle). 7 SDCM is used in the plot for the corresponding IEC MacAdam ellipse.

2.3.2 Color Rendition of Light

Traditionally, natural daylight is believed to offer the perfect color rendition, thanks to its continuous SPD that has a big overlap with the photopic sensitivity curve of the human visual system. In other words, we have virtually evolved and adapted to it. To assess the color rendering property of a light source, a standardized method is needed. The CIE defines the color rendering as the effect of an illuminant on the color appearance of objects by conscious or subconscious comparison with their color appearance under a reference illuminant [10].

The method of measuring and specifying color rendering properties of light sources is well described in CIE13.3-1995. The color appearance of 14 reflective test color samples (TCS) is simulated when illuminated by the test source and a CCT matched reference source. A blackbody radiator is used as the reference source if CCT of the test source is below 5000 K, otherwise a phase of daylight. With the chromatic adaptation of each sample by applying a von Kries transform, the difference in color appearance, namely, Euclidean distance ΔE_i, for each sample between the test and reference light sources is calculated in CIE 1964 ($U^*V^*W^*$) color space. The special CRI (R_i) is calculated for each reflective sample by

$$R_i = 100 - 4.6\Delta E_i \tag{2.2}$$

The most well-known term of CRI in the lighting industry actually refers to R_a, the CIE general CRI. It is the average of R_i for the first eight TCS. A high R_a is generally considered to represent high color fidelity, not to be confused with high preference, of the light source, and vice versa. It should be noted that the abbreviations R_a and CRI are normally used interchangeably, unless it is specified.

For decades, the CRI has been criticized for its limitations. The CIE $U^*V^*W^*$ color space is far from being perceptually equidistant and is considered to be outdated. The von Kries chromatic adaptation is also considered obsolete, but it is still adequate when the color point of the light source is close to the blackbody locus, which holds true when CRI computations are involved. The direction of color shifts is not taken into account.

Furthermore, all eight TCS used in the computation of R_a are pastels with relatively low color saturation. R_a alone is therefore not a good indicator for the (color rendering) quality of light in many applications where more saturated color objects are involved. As already mentioned earlier, the predictive value for the color rendition is limited when light sources are composed of narrowband spectra, for example, white LED sources composed of individual narrowband LEDs. On the other hand, by optimization of LEDs' spectra to those eight TCS, the R_a value can be optimized to meet a particular minimum threshold value, while the actual color appearance in the application could be disappointing.

Red colors are important in many applications, where R_a only represents the average color difference for the eight pastel colors. To address this issue, special CRI R_9 has been added to LED lighting specifications to supplement R_a. The Energy Star program (V2.0) in the United States indeed requests that all LED lamps should have R_9 larger than zero. It has to be noted that R_9 can be less than 100 due to an increase in saturation (more energy in the long wavelength range) or due to a decrease in saturation (less energy in the long wavelength range). This information on saturation changes is not disclosed with any of the special CRI values.

Based on all issues with CRI [12], CIE Technical Committee TC1-62, *Color Rendering of White LED Light Sources*, concluded that the CIE CRI is generally not applicable to predict the color rendering rank order of a set of light sources when white LED light sources are involved in this set [13]. The committee, however, did not recommend an alternative measure.

CIE TC1-69, *Color Rendition by White Light Sources*, was established with the goal of recommending new assessment procedures. Till now, no solution has been identified and agreed upon. Two new technical committees were then formed. TC1-90, *Color Fidelity Index*, focuses on evaluating available fidelity indices to assess the color quality of white light source with a goal of recommending a single index for industrial use. The other one, TC1-91, *New Method for Evaluating the Color Quality of White Light Sources*, is looking for other available methods while explicitly excluding color fidelity indices.

Color fidelity is only one aspect of color rendition, while it is often mistakenly interpreted as the only measure for light quality. As a single one-dimensional index value, the CRI provides no information on hue discrimination, hue-angle changes, color saturation (or vividness in a more plain language), color preference, and object attractiveness. Perceived color quality is often subjective, emotional, application, and environmental dependent, by the assessment of an individual or a group who prefers the color appearance of a given light source or the illuminated objects. In general, color preference can be gender, age, region, culture, application, and event dependent. It depends not only on the SPD of the light source but also on the reflectance property of illuminated objects. In addition to the objective fidelity index, such as CRI, a related set of index values that describes more subjective aspects of the light source is apparently needed to cope with this dilemma (see, e.g., [14]).

As a consequence, many attempts have been made to develop and standardize alternative characterization methods, such as color quality scale (CQS) [15,16], gamut area index (GAI) [17], color discrimination index [18], color rendering capacity [19], flattery index [20], color preference index [21], and color saturation index (G_a) [28]. Recently, Houser et al. [22] reviewed 22 color rendition measures and concluded that those newer measures are not essentially different from the older ones, including the CRI. The fundamental problem of the color rendition is its multidimensional aspects, which have conflicting optimization criteria. Due to the intrinsic trade-off between fidelity, discrimination, and preference, it is not realistic to have a light source simultaneously performing the highest for all aspects. In other words, no single method can fully describe these aspects with one single index value as an output. Despite significant research efforts, it is unlikely that a new fidelity measure alone will show a significant and meaningful improvement in specifying color quality over the current CRI.

Based on analyzing color rendition vectors of 1269 spectrophotometrically characterized test samples, namely, the entire Munsell palette, Zukauskas et al. [23] proposed a four-dimensional approach to assess the color quality, namely, CCT, color fidelity index, color saturation index, and hue distortion index. This four-dimensional metric is illustrated by a notation with the corresponding letter of each dimension, namely, T (CCT), F (fidelity), S (saturation), and D (distortion). For example, T30-F60R-S25-D20YP stands for a light source with a nominal CCT of 3000 K. Illuminated by this light source, 60% of the color palettes are rendered with a high fidelity and with the dominance of a red hue, 25% of the color palettes are rendered with increased saturation, and 20% of colors have distorted hues prominently for yellow and purple. With regard to the studied light sources, the hue distortion index is negatively correlated with the color fidelity index. So the hue distortion index might be an optional metric. There is no obvious correlation between the color fidelity index and the color saturation index. It indicates that no single figure of merit can combine the fidelity and the saturation. On the other hand, at least two uncorrelated metrics, say, one representing "fidelity" and the other one representing "saturation," are believed to be able to characterize the color rendering properties of the light source. The alphanumeric approach proposed by Zukauskas et al. is relatively comprehensive to better reveal the color rendition properties, as it, to some extent, quantifies the multidimensional aspects embedded in the color quality charts. However, all of their color fidelity index, color saturation index, and hue distortion index refer to the average figure of merit. Unsurprisingly, it does not tell the color preference. Moreover, it is not practical to communicate to customers of the lighting industry, especially in a manner understandable to nonexperts in the field of color science and consumers.

Researchers at the Lighting Research Center (LRC) at Rensselaer Polytechnic Institute developed a "Class A color" specification based on the study in simulated residential applications [24,25]. Fresh fruits and vegetables in the boxes illuminated under various light sources were viewed by selected subjects. The subjects were asked to rate the naturalness (fidelity) and vividness (saturation) of the display and additionally the overall acceptability of the light source. The CRI is taken as the fidelity measure, while the GAI using the same TCS as those for CRI calculation is applied for the saturation measure. The CIE 1964 $W^*U^*V^*$ color space is used for the calculation. They concluded that light sources with CRI equal or greater than 80 and GAI between 80 and 100 have good color rendering (cf. Figure 2.3). Besides these criteria, to meet the "Class A color" designation, the light source should have the minimal tint (i.e., be "white") and have a small chromaticity tolerance. Although these researchers made a good attempt to offer the lighting industry a "seal of approval" for color quality of light sources used in general lighting applications, the selected metrics to represent the color rendering properties may be not appropriate, for example, the nonuniformities of the outdated CIE 1964 $U^*V^*W^*$ color space, no chromatic adaptation is applied, and a fixed reference illuminant (an EES) is used. On the other hand, the defined values for good color rendering are to be further investigated, as the limited scenes and objectives have been studied, even in a simulated viewing environment. They may not solve the main issues from professionals like lighting designers and architects, neither the consumers facing various applications and configurations in the real world.

The latest and probably most influential campaign on promoting a new proposal for color rendition was unveiled in 2015. The Illuminating Engineering Society (IES) of North America published a technical

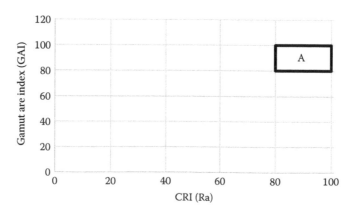

FIGURE 2.3 The area of "Class A color" in a Cartesian coordinate plot of CRI and GAI, prescribed by LRC, namely, CRI ≥ 80 and 80 ≤ GAI ≤ 100.

memorandum (IES TM-30-15), *IES Method for Evaluating Light Source Color Rendition*. Similar to other proposals mentioned earlier, it proposed a method to evaluate the color rendition of the light source by quantifying the fidelity and gamut (chroma or saturation) in a two-dimensional system. The color preference is not in the scope of that work. Unlike the LRC, this IES method takes different approaches on calculating the fidelity index and the gamut index. Instead of using 8 CIE TCS, 99 color evaluation samples (CES) are utilized to determine the color fidelity. These color samples are statistically down-selected from more than 100,000 real object colors, which colors are believed to well represent our actual world. CIE 10° standard observer color matching functions (CMFs) are applied, instead of 2° CMFs used in the CRI calculation. Another noticeable feature of this IES method is that the chromatic adaptation transform is based on CAM02UCS [26].

The details about the development of the IES metrics can be found in the publication from David et al. [27]. The IES fidelity index R_f is, similar to R_a, a numerical measure to determine the difference between the chromaticity coordinates under the test source and the reference source, but based on the new set of color samples and the most recent color space. It is an average figure of merit, not attempting to reveal the perceived fidelity. The index values for typical phosphor-converted LED light sources, obtained with R_f, are typically 3–5 unit points lower than with R_a. Although computed with 99 CES, the average fidelity index value R_f is not necessarily more representative for the actually perceived fidelity of the environment than the CIE R_a value. Also the use of the 10° CMFs for the computations is arguable, because the CAT02 and HPE matrices were defined for the 2° CMFs. The IES gamut index R_g is the measure of the *average* area spanned by the ($a_{M'}$, $b_{M'}$) coordinates of the CES in CAM02UCS. The $a_{M'}$–$b_{M'}$ colorfulness space is evenly divided into 16 bins in a radial pattern. Within each bin, the arithmetic mean of the $a_{M'}$ and $b_{M'}$ coordinates for each CES within the bin is calculated under the test source and the reference source. The area of the polygon created by these 16 points under a given illuminant is the gamut area. The ratio of the gamut areas for the test source and the reference source times 100 is the IES gamut index R_g. Same as the two-metric plot proposed by LRC, the color rendition system specified by IES TM-30-15 is characterized by IES R_f and IES R_g in a Cartesian coordinate system. There are several issues associated with the hue-angle binning method. One is that the clustering of CES into a hue-angle bin is CCT dependent, that is, different CES are combined per CCT. Another issue is that when discounting the illuminant cannot be assumed, some bins will not have a single CES to compute an average value.

Furthermore, the gamut area polygons of test source and its associated reference source created from the chromaticity coordinates of the CES with each hue-angle bin can be thus translated to a color vector graphic. It visually presents the relative saturation and hue shift in various hue regions. Again, this graphic only visualizes a kind of *average* change in each of 16 hue bins, but not the change in color appearance for individual CES. But on the other hand, visualizing these color distortions simultaneously for all 99 color samples is indeed challenging and may not be readily distinguishable at all.

Teunissen and Hoelen [28] compared the IES R_f values with the R_a values of 118 commercial LED-based light sources, included in the IES TM-30-15 calculation tool, with R_a ranging from 60 to 97 and $\Delta uv \leq 0.0054$. The correlation between R_f and R_a is revealed to be very strong, with a coefficient of 0.94. The R_f values are typically smaller than those of R_a but within the range of $R_a \pm 3$. It is yet unclear, for those light sources beyond this range, which index better predicts the correspondence with perception. This finding triggers a practical but fundamental dilemma, whether a more advanced fidelity metric, such as IES R_f, essentially adds more value to the lighting industry, if it is highly correlated to an "old-fashioned" measure. In reality, to adopt a completely new color fidelity index as a standard and in a regulation is extremely disruptive and expensive and should therefore only be considered if a demonstrated meaningful and significant improvement over the existing CRI is evident and after there is industry consensus.

IES TM-30-15 is not a required standard, and it does not provide design guidance or criteria for best practices. Meanwhile, it has been proposed to CIE for consideration as an international standard. CIE TC1-90 has modified the fidelity index (IES-R_f) to CIE-R_f, but will not recommend it as a replacement for the well-established general color rendering index (R_a).

2.3.3 Light Preference

Compared to the intricacy of color fidelity as described in the last section, how to quantify the color quality in terms of appreciation or preference is an even more complicated challenge for many lighting researchers. All efforts to supplement the color rendition by adding gamut/saturation metric (cf. Section 2.3.2) intended to provide preference-related information have not yet resulted in an international standard.

The light source with its chromaticity on the blackbody locus has been taken for granted as standard and reference white light (cf. Section 2.3.1). However, the question whether such a white light is preferred particularly for indoor lighting has not been clearly answered.

Rea and Freyssinier [29] performed a vision experiment with 20 participants, who judged the purity of hue or tint of various SPD white light via viewing the interior of a table-mount box. The inner walls of the viewing box were matte white with a reflectance of about 80% and were illuminated by the diffused light to a level of 300 lx at the center of the box floor. No other objects and colors were present in the viewing box. They concluded that the chromaticities associated with the neutral (untinted) white light sources with low CCTs (below 4100 K) lie below the Planckian locus, for example, $\Delta uv \approx -0.01$ at 2700–3500 K. In contrast, these color white points perceived to be neutral white illumination with high CCTs (above 4100 K) fall above the Planckian locus (cf. LRC curve in Figure 2.4). However, the methodology adopted and chromatic adaption conditions used in this experiment could have introduced a bias in the obtained results.

With a setup consisting of a viewing box with two white-painted chambers, Perz et al. performed a vision experiment to determine which light at a given CCT is perceived as whiter among 15 different isotemperature stimuli, after participants are fully chromatically adapted [30]. Fifteen stimuli are evenly distributed with a step of $\Delta uv = 0.003$, say, 7 stimuli with their color points positioned above the blackbody locus (BBL), 7 below, and 1 on the BBL. Subjects included in the study were both professional lighting researchers and participants without special knowledge of light or lighting (labeled as real people in Figure 2.4). It was found that all chromaticities of "whitest" points for the studied CCTs lay below the Planckian locus and no crossover was observed as presented by LRC (cf. Philips Lighting data in Figure 2.4). However, it

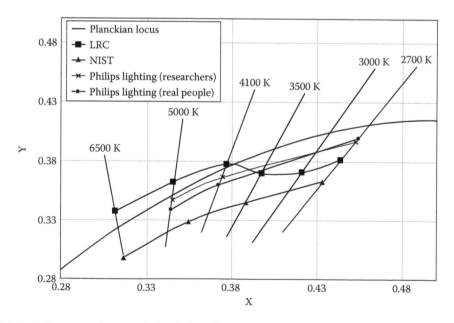

FIGURE 2.4   Color points of perceived white light in the CIE 1931 chromaticity diagram. LRC refers to whitest light when seen in isolation [29] and Philips Lighting refers to the findings of Perz et al. [30]. NIST refers to the preference of natural appearance of objects from the work done by Ohno and Fein [32].

should be noted that both psychophysical experiments conducted at LRC and Philips Lighting were designed to define the perceived pure white by subjects, while not their white preference.

To examine personal preferences of various white spectra, a 1:6 scale maquette representing an open-plan office with various colorful fittings was built up at the National Research Council Canada [31]. Thirty-two participants were first asked to evaluate 6 preset (fixed) SPDs, among which five were created by LED channels and one was the conventional fluorescent source. CCTs of five LED SPDs were evenly distributed at 2855, 3728, 4751, 5769, and 6507 K and the fluorescent source was chosen to be 3750 K. All provided approximately 500 lx on the modeled work surface. Further assessments were studied that the participants freely tuned the 5 LED channels to create their own preferred light conditions, either with or without the constraint of the illuminance. The second part of the experiment disclosed that all preferred spectra ranged from 2,850 to 14,000 K and had negative Δuv down to −0.0469, except three of 32 above the Planckian locus, for both free and constrained illuminance choices. Many points were not considered to be white according to ANSI standards. As there were no statistically significant differences of the preferred spectral properties for a large spread of illuminance, the authors thus concluded that the preference is independent of illuminance level, with a full chromatic adaption prior to judgment.

Apparently, the psychophysical experiment is more convincing as it is conducted by observing the real objects outfitted, human skin tone and the entire room, in an immersive light condition around 300 lx. This triggered Ohno and Fein [32] to study the acceptable and preferred chromaticity of white light in a real room-size setting. A total of 23 SPDs (test points) were set up at four CCTs, namely, 2700, 3500, 4500, and 6500 K, and at six different Δuv levels (−0.03, −0.02, −0.01, 0, 0.01, 0.02); 2700 K with $\Delta uv = 0.02$ was not included because it is almost a saturated yellow light and the color rendering is poor. All spectra created had a high color fidelity. The results indicated that the light with its chromaticity below the Planckian locus, around $\Delta uv \approx -0.015$, is more preferred for a natural appearance of objects in an indoor environment, in a CCT range of 2700–6500 K (cf. Figure 2.4). Similar to the findings of Dikel et al. [31], color points of most preferred SPDs were outside the chromaticity specifications of LED products according to ANSI C78.377. Note that this finding is also different from that of LRC's work, where the preferred light at higher CCT is found to be above the Planckian locus.

Anecdotally, it seems that people prefer increased color saturation, which makes perceived images or objects more vivid. A quantitative evaluation on the preferred level of saturation was conducted by Ohno et al. [33], with the same test setup as what they used for the chromaticity study [32]. However, unlike the broadband spectra applied in the study before [32], a blending of various narrowband LED emission peaks was applied for this study in order to tune the saturation. Via changing red and amber light ratio, nine different levels of saturation on the BBL were generated for each of three different CCTs: 2700, 3500, and 5000 K. Another set of SPDs at 3500 K with Δuv of −0.015 and at various saturation levels was additionally included to refer to their previous work, where off-white below the Planckian locus was found to be more preferred [32]. The results showed that subjects' preferred lights have the chroma saturation peaked at $\Delta C^*_{ab} \approx 5$ from the neutral at all CCTs for all tested objects, say, mixed fruits and vegetables, skin tone, red fruits and vegetables only, and green fruits and vegetables only. The authors thus concluded that a metric of light preference may be based on using a reference illuminant having the chroma saturation of $\Delta C^*_{ab} \approx 5$ and the deviation of the chroma saturation in either direction leads to a less preference. It is important to note that the CRI needs to be reduced in order to increase the level of saturation. Mandatory requirements for high CRI values prohibit the manufacturing and sales of user's preferred light sources with increased saturation levels.

Teunissen et al. [34] evaluated the attractiveness of object appearance for fresh food, packaged food, and skin tone in two side-by-side viewing boxes. Seven light sources at a nominal CCT of 3000 K and all on, or close to, the BBL were studied, with R_a values ranging between 70 and 100 and relative GAI (G_a, to distinguish from other GAI mentioned in paragraphs earlier, hereafter we name this term as color saturation index) values ranging between 90 and 120. Among them, one halogen lamp was used as a reference, with the nominal $R_a = 100$ and $G_a = 100$. It turned out that halogen was not the most preferred SPD in all three applications, which was in line with the early findings of Narendran and Deng [35]. Consistent with

Ohno et al.'s observation [33], it was found that there was a trend that light sources with higher color saturation index (enhanced saturation in average), with R_a values of 70 and 80, were more preferred than halogen. However, it should be also noted that two LED-created SPDs with the same R_a of 80 and the same G_a of 110 were perceived to be not equally attractive, particularly with regard to the skin tone. The study also inferred that the preference of light source was application dependent and gender specific, indicating no single or cluster of SPDs is preferred for all illuminated objects. In other words, there was no universal sweet spot identified in the two-dimensional fidelity (R_a) and GAI (G_a) system for all applications.

Both studies of Dikel et al. [31] and Ohno and Fein [32] indicated the impact of chromaticity on apparent preference of light source. The chromaticity was treated as an independent parameter from color rendition measures like color saturation. On the other hand, recently, Ohno et al. [33] also demonstrated that color preference is mainly affected by the saturation level of objects' chroma. As stated in Section 2.3.2, one of the efforts to overcome the limitations of CRI, the only color rendition index nowadays, is to add one more dimension of the color rendition, for example, CSI (G_a), which corresponds to the average change in saturation. By analyzing the preferred SPDs identified in [31,32], it was found that most of them have a gamut area scale Q_g larger than 100 [36]. Q_g is calculated as the relative gamut area formed in the CIELAB a^*, b^* coordinates of the 15 CQS color samples illuminated by the test light source. It is then normalized by the gamut area of CIE Standard Illuminant D65 and multiplied by 100. Apparently, most of those preferred lights in [31,32], which were below the Planckian locus, were also saturated. Using Gaussian-distributed RGB spectra, Wei and Houser [36] simulated a series of SPDs at the fixed CCT of 3000 and 6500 K, with Δuv ranging from +0.02 to −0.03. Q_g values of these SPDs were plotted versus their CQS fidelity measure Q_f [16]. In general, Q_g value increases as the chromaticity of spectra moves from positive Δuv to negative Δuv. Simultaneously, the number of possible SPDs first increased, while it became less when Δuv decreased further. The authors believed that the chromaticity may not be the sole attribute to the preference, and color rendition may also influence the preference.

Neodymium incandescent lamp, for example, GE *Reveal*® and Philips Lighting *Natural Daylight*®, has been successfully promoted to be more natural and superior to standard incandescent bulbs in indoor illumination. Much of the yellow portion of the visible spectrum is filtered out, which gives the lamp purplish tint. It actually has a small negative Δuv, say, −0.005, with the enhanced red saturation. Its popularity, especially in the United States, may be related to its off-white chromaticity slightly below the Planckian locus, or its enhanced red saturation, or mixed effect. Further studies to separate the impact of chromaticity and color saturation in psychophysical perception experiments may help us better understand the key driver to the preference of light.

2.3.4 Crisp White

The perceived whiteness of objects is mainly determined by brightness, contrast with surrounding, and chromaticity. In retail, many current LED light sources make white fabric look dull, while legacy ceramic discharge metal-halide lamps (CDM), for example, Philips Lighting CDM *Elite*®, offer sparkling clean white impression. Fluorescent whitening agents (FWAs) or optical brightening agents are commonly used in white paper, clothing, detergent, cosmetics, and toothpaste. They absorb light in the ultraviolet and violet part (wavelengths shorter than 430 nm) of the emission spectrum and reemit visible light in the blue-cyan region (typically 430–470 nm) by fluorescence. Figure 2.5 presents the excitation and emission spectra of several typical white objects, which contain FWAs. This process effectively shifts the color point of the reflected light toward blue tint, that is, below the BBL, thus causing a "whitening" or "crisp" effect. It may be explained by the findings that off-white below the BBL is perceived to be "whiter" in the previous section.

As normal white LEDs have the dominant blue pump greater than 450 nm, they lack energy in the part of the spectrum to excite the FWAs. The aforementioned crisp effect is thus not present and the whiteness rendering is poorer compared to natural daylight and almost all other artificial light sources, which indeed contain some (ultra) violet light in the spectrum. To circumvent this problem, a violet or deep blue pump

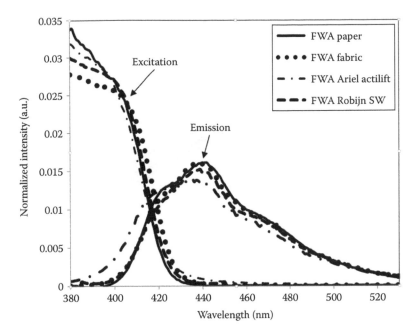

FIGURE 2.5 Excitation and emission spectra of typical white articles with FWAs. Ariel and Robijn are commercial brands of detergent manufacturers.

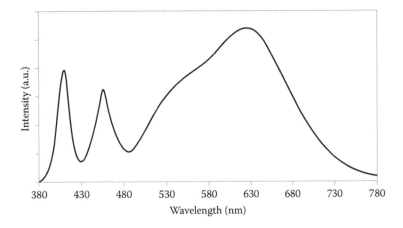

FIGURE 2.6 A typical electroluminescent spectrum of Philips Lumileds CrispWhite®.

can be added to the spectrum to activate the FWAs. An alternative solution is to tune the color point of the LED itself to be under the BBL, which will in turn also move the reflected color point.

The current Philips Lumileds CrispWhite® source combines both effects. Figure 2.6 represents a spectrum of the typical CrispWhite®. The color point of this source is slightly below BBL but within the ANSI 3000 K bin. The violet emission leads to an additional blueshift of the color point of white object by v' shift of −0.006 in the CIELUV color space. The use of violet or deep blue LEDs reduces the efficiency of the source (larger Stokes shift compared to royal blue LEDs peaked around 450 nm and lower wall-plug efficiency of 405 nm LED compared to that of 450 nm LED) and leads to an increase in the cost at the system level. By choosing the color point of the source to be below BBL, the amount of short-wavelength LEDs in the source, and in turn the efficiency loss, can be minimized.

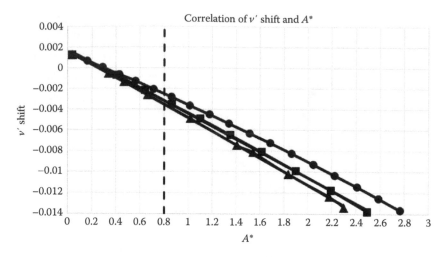

FIGURE 2.7   The plot of v' shift versus A^* of three white fabric samples.

The v' shift ($\Delta v'$) is a quantitative figure of the color shift of the reflection spectrum compared to the light source, thus a direct indicator of the crisp effect. However, to obtain it, one has to measure the color point reflected by the white material and the illuminating source. A crisp factor derived only from the light source SPD is more desired and convenient to determine the crisp effect of the light source on the object containing FWAs.

The crisp factor A^* is defined as follows [37]:

$$A^* = \frac{\int_{380}^{430} E(\lambda)(430-\lambda)d\lambda}{\int_{380}^{780} E(\lambda)d\lambda} \tag{2.3}$$

The FWA excitation spectrum is linearized and can be represented by $(430 - \lambda)$. The crisp factor simply implies a "corrected" or weighed short-wavelength spectrum ratio.

Earlier experiments demonstrated a linear correlation between A^* and $\Delta v'$ [37] (cf. Figure 2.7). Various deep blue peak wavelengths were used to generate the spectra. Three white fabric samples containing FWAs were measured. Apparently, the crisp factor A^* is a good measure to determine the v' shift, without knowing the emission property of the white object.

Empirically, a minimum color point shift $\Delta v' = -0.004$ is required to observe the crisp effect. Correspondingly, a minimum A^* of 0.8 is established. Perception studies in terms of balancing the crisp effect and other color rendition indicate that $1 < A^* < 2$ is most preferred [38].

2.3.5 Perceived Brightness of White Objects

The definition of CIE 1964 whiteness (cf. Equation 2.1) obviously discloses that the perceived whiteness is proportional to the luminance (or simply to say the brightness). As the perceived whiteness can be enhanced by the activation of FWAs as described in the previous section, it is interesting to explore the effect of increased perceived whiteness on the perception of brightness.

A brightness matching experiment using LED light sources with and without FWA activation was conducted [39]. The illuminance in the center of a simple scene was used as a correlate of brightness, and the brightness matching was done using simultaneous viewing. The light source with no FWA activation, called the reference light source, was presented at three preset light levels, and the intensity of the light source with FWA activation, called the test source, was tuned using a staircase procedure until an equal perceived brightness was achieved. Additionally, to check the measurement error of the method, the same procedure was used to match the brightness of two reference sources. Two side-by-side identical light boxes were viewed by participants from 1 m distance. The participants simultaneously observed both stimuli in viewing boxes. Each box had a size of 60° visual angle with a 10° visual angle black separator between the boxes. More details on the experimental setup can be found in [39].

The brightness matching was done at three reference illuminance levels of 150, 300, and 500 lx. For each reference light level, a range of light levels for the test light source were defined and a staircase procedure was used to traverse over the levels. Two staircases per reference light level were used: one with the reference on the right side and one with the reference on the left side. Additionally, two staircases were added with both sides showing the reference source and changing one of the sides to produce a brightness match. All staircases started from a random light level of the test source ranging from 50% to 150% of the reference light level. All staircases were presented randomly interleaved to account for possible reference bias of the participants. At each staircase step, the participants had to indicate which box was perceived as having a higher brightness using the left and the right arrow keys on a keyboard. The procedure was explained to the participants before starting the experiment. The term *brightness* was not explained such that all participants had their own interpretation of the term.

The results of the experiment are expressed as function of the illuminance level produced by the test source at the center of the scene for a given illuminance level of the reference source. Figure 2.8 shows an overview of the obtained results. The difference in illuminance between the reference source and the median illuminance of the test source for which an equal brightness was obtained was 12%, 18%, and 20% for reference illuminance levels of 150, 300, and 500 lx, respectively (cf. Figure 2.8).

The clear effect of the chromaticity change on perceived brightness, as visible in Figure 2.8, is also supported by statistical testing. All single sample t-tests for the cases with different chromaticities showed a significant difference from the mean illuminance of the reference source ($p < 0.001$ in all cases). For the additional staircases with the same test and reference light source, a single sample t-test did not show a significant effect ($p = 0.798$). The measurement error also depended on the light level and ranged from 19 lx for the lowest light level to 41 lx at the highest light level. There was also no difference between the measurement error in the regular cases and the additional case where the reference light source was used both as a reference and as a test source.

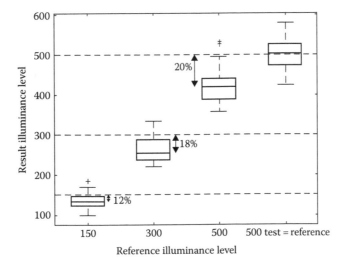

FIGURE 2.8 A boxplot of the resulting matched illuminance levels for the three reference illuminance levels, namely, 150, 300, and 500 lx.

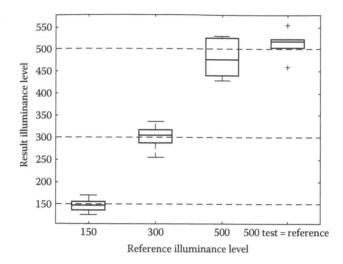

FIGURE 2.9 A boxplot of the resulting matched illuminance level for the three reference illuminance levels, with regard to the group of 3 participants not showing a significant effect.

No significant effects of age, gender, or lighting expertise were found using an ANOVA procedure. A hierarchical cluster analysis procedure identified a small group of three participants with markedly different answers than the rest of the participants. The matching values for the participants in this group were significantly higher than for the other participants and showed no significant difference from the illuminance of the reference source for all light levels. Figure 2.9 depicts the result for this group of participants. It is interesting to estimate the size of this group in the general population as the size estimated from this experiment can be biased due to the selection of the participants in the experiment.

The results show that for an average observer, the activation of FWAs increases not only the perception of whiteness of surfaces but also the overall perceived brightness of the surrounding that contains white surfaces. Consequently, an equivalent brightness impression can be obtained by using up to 20% less intensity. This result can be particularly important in applications where both the perceived brightness and the natural whiteness rendering of surfaces are important, as, for example, is the case in fashion retail. In this case, only using the illuminance in the space could be a poor indicator of the suitability of a light source. For more accurately predicting the object as well as space appearance, whiteness rendering and in particular the activation of FWAs have to be taken into account.

2.4 Conclusion

In this chapter, we review the existing and newly proposed methods to quantify the spectrum-related quality of light and the latest developments in this field. The evolution of LED lighting brings us much more dynamics and advances to tune the spectrum in order to better facilitate our visual requirements. On the other hand, it also stimulates the study of the intricacy of our visual system and corresponding measures of quality of light.

The color fidelity and color saturation are important measures for characterizing color quality. CRI, the only internationally standardized and widely accepted measure for color rendition, is on its own insufficient for characterizing the color quality of LED light sources. However, it is often mistakenly used as a measure for general light quality. An additional saturation measure, like G_a, provides, in combination with the CRI (R_a), a more meaningful description of the color rendering properties of a light source. However, the intrinsic trade-off among fidelity, color saturation, and preference makes it difficult to develop a single light source simultaneously performing the highest for all aspects. In other words, no single index can fully encapsulate these requirements into one answer. A multi-index approach proposal provides a pathway in the development of more sophisticated methods that should aim at a more complete description of the color rendering properties of white light sources. Obviously, new measures shall only be adopted after a thorough vetting process during which a superior performance is clearly and unambiguously demonstrated.

The well-accepted and standardized definition of white chromaticity seems to be not consistent with the findings of various psychophysical perception experiments. The light with color points below the Planckian locus is perceived and believed to be more natural and preferred, at least in an indoor illuminated environment. A new definition of a "white body line" may overrule the currently standardized chromaticities of white color points. On the other hand, the color saturation, especially red and green, may also be a key contributor to user preference, besides the chromaticity. A set of well-defined experiments is required to separate the effects of chromaticity and color rendition to better understand the dominant attributes of light preference.

Crisp white is a good example to show that the SPD engineering can impact the perception of end users and thus the feeling of quality of light. The study of perceived brightness versus perceived whiteness further unveils the power of the spectrum on our visual system.

References

1. C. L. B. McCloughan, P. Aspinall, and R. Webb, The impact of lighting on mood, *Lighting Research & Technology*, 31, 81–88, 1999.
2. R. Küllera, S. Ballalb, T. Laikea, B. Mikellidesc, and G. Tonellod, The impact of light and colour on psychological mood: A cross-cultural study of indoor work environments, *Ergonomics*, 49, 1496–1507, 2006.
3. I. Provencio, G. Jiang, W. J. D. Grip, W. P. Hayes, and M. D. Rollag, Melanopsin: An opsin in melanophores, brain, and eye, *Proceedings of the National Academy of Sciences of the United States of America*, 95(1), 340–345, 1998.
4. I. Provencio, I. R. Rodriguez, G. Jiang, W. P. Hayes, E. F. Moreira, and M. D. Rollag, A novel human opsin in the inner retina, *The Journal of Neuroscience*, 20(2), 600–605, 2000.
5. S. Panda, T. K. Sato, A. M. Castrucci, M. D. Rollag, W. J. DeGrip, J. B. Hogenesch, I. Provencio, and S. A. Kay, Melanopsin (Opn4) requirement for normal light-induced circadian phase shifting, *Science*, 298, 5601, 2002.
6. M. Terman, J. S. Terman, F. M. Quitkin, P. J. McGrath, J. W. Stewart, and B. Rafferty, Light therapy for seasonal affective disorder: A review of efficacy, *Neuropsychopharmacology*, 2(1), 1–22, 1989.
7. J. L. Anderson, C. A. Glod, J. Dai, Y. Cao, and S. W. Lockley, Lux vs. wavelength in light treatment of seasonal affective disorder, *Acta Psychiatrica Scandinavica*, 120(3), 203–212, 2009.
8. R. J. Lucas, S. N. Peirson, D. M. Berson, T. M. Brown, H. M. Cooper, C. A. Czeisler, P. D. G. M. G. Figueiro et al., Measuring and using light in the melanopsin age, *Trends in Neurosciences*, 37(1), 1–9, 2014.
9. www.lightingeurope.org/focus-area/human-centric-lighting.
10. CIE, International lighting vocabulary, CIE S 017/E, 2011.
11. J. Enezi, V. Revell, T. Brown, J. Wynne, L. Schlangen, and R. Lucas, A "melanopic" spectral efficiency function predicts the sensitivity of melanopsin photoreceptors to polychromatic lights, *Journal of Biological Rhythms*, 26(4), 314–323, 2011.
12. Y. Ohno, Optical metrology for LEDs and solid state lighting, in *Proceedings of the SPIE 6046, Fifth Symposium Optics in Industry*, Santiago De Queretaro, Mexico, 2006.
13. CIE Technical Report 177, 2007.
14. C. Teunissen, A framework for evaluating the multidimensional colour quality properties of white LED light sources, in *Proceedings of CIE 2016 Lighting Quality and Energy Efficiency*, Melbourne, Victoria, Australia, March 3–5, 2016.
15. Y. Ohno and W. Davis, Toward an improved color rendering metric, in *Proceedings of the SPIE 5941*, San Diego, CA, 2005.
16. W. Davis and Y. Ohno, Color quality scale, *Optical Engineering*, 49(3), 033602, 2010.

17. M. S. Rea and J. P. Freyssinier-Nova, Color rendering: A tale of two metrics, *Color Research & Application*, 33(3), 192–202, 2008.
18. W. A. Thornton, Color-discrimination index, *Journal of the Optical Society of America*, 62, 191–194, 1972.
19. H. Xu, Color-rendering capacity of light, *Color Research & Application*, 18, 267–269, 1993.
20. D. B. Judd, A flattery index for artificial illuminants, *Illuminating Engineering*, 62, 593–598, 1967.
21. W. A. Thornton, A validation of the color-preference index, *Journal of the Illuminating Engineering Society*, 4, 48–52, 1974.
22. K. W. Houser, M. C. Wei, A. David, M. R. Krames, and X. Y. Shen, Review of measures for light-source color rendition and considerations for a two-measure system for characterizing color rendition, *Optics Express*, 21, 10393–10411, 2013.
23. A. Zukauskas, R. Vaicekauskas, F. Ivanauskas, H. Vaitkevicius, P. Vitta, and M. S. Shur, Statistical approach to color quality of solid-state lamps, *IEEE Journal of Selected Topics in Quantum Electronics*, 15, 1753–1762, 2009.
24. M. S. Rea and J. P. Freyssinier, The Class A color designation for light sources, in *Proceedings of Experiencing Light 2012: International Conference on the Effects of Light on Well-Being*, Eindhoven, the Netherlands, 2012.
25. J. P. Freyssinier and M. S. Rea, Class A color designation for light sources used in general illumination, *Journal of Light & Visual Environment*, 37, 10–14, 2013.
26. C. Li, M. R. Luo, C. Li, and G. Cui, The CRI-CAM02UCS colour rendering index, *Color Research & Application*, 37, 160–167, 2012.
27. A. David, P. T. Fini, K. W. Houser, Y. Ohno, M. P. Royer, K. A. G. Smet, M. Wei, and L. Whitehead, Development of the IES method for evaluating the color rendition of light sources, *Optics Express*, 23, 15888–15906, 2015.
28. K. Teunissen and C. Hoelen, Progress in characterizing the multidimensional color quality properties of white LED light sources, in *Proceedings of the SPIE 9768, Light-Emitting Diodes: Materials, Devices, and Applications for Solid State Lighting XX*, San Francisco, CA, 976814, 2016.
29. M. Rea and J. Freyssinier, White lighting, *Color Research and Application*, 38, 82–92, 2013.
30. M. Perz, R. Baselmans, and D. Sekulovski, Perception of illumination whiteness, in Proceedings of the 4th CIE Expert Symposium on Colour and Visual Appearance, CIE x043:2016, Prague, Czech Republic, 2016.
31. E. Dikel, G. Burns, J. Veitch, S. Mancini, and G. Newshaw, Preferred chromaticity of color-tunable LED lighting, *LEUKOS*, 10, 101–115, 2014.
32. Y. Ohno and M. Fein, Vision experiment on acceptable and preferred white light chromaticity for lighting, in *Proceedings of the CIE 2014 Lighting Quality and Energy Efficiency*, Kuala Lumpur, Malaysia, 2014.
33. Y. Ohno, M. Fein, and C. Miller, Vision experiment on chroma saturation for colour quality preference, in *The 28th Session of International Commission on Illumination*, Manchester, United Kingdom, 2015.
34. C. Teunissen, F. van der Heijden, S. Poort, and E. de Beer, Characterising user preference for white LED light sources with CIE colour rendering index combined with a relative gamut area index, *Lighting Research and Technology*, 1–20, DOI: 10.1177/1477153515624484, 2016.
35. N. Narendran and L. Deng, Color rendering properties of LED light sources, in *Proceedings of the SPIE 4776, Solid State Lighting II*, Seattle, WA, 2002.
36. M. Wei and K. W. Houser, What is the cause of apparent preference for sources with chromaticity below the blackbody locus, *LEUKOS*, 12, 95–99, 2016.
37. M. Peeters, E. de Beer, D. van Kaathoven, W. Oepts, and H. Gielen, White light emitting module, PCT Patent WO2013/150470 A1, 2013.
38. M. Peeters, Study of crisp white, Philips Lighting Internal Notes.
39. D. Sekulovski, K. Teunissen, M. Peeters, Y. J. Sun, and R. Broersma, On the perceived brightness of whites, in *Proceedings of AIC 2015, Color and Image, Midterm Meeting of the International Color Association*, Tokyo, Japan, 2015.

3

Nanofabrication of III-Nitride Emitters for Solid-State Lighting

Tao Wang
University of Sheffield

Yaonan Hou
University of Sheffield

Abstract III-nitride emitters with nanostructures have generated tremendous interest in the research community due to a variety of novel properties that their planar counterparts do not have. As a result, these devices demonstrate great potentials in application in solid-state lighting with high-brightness and full color. This chapter reviews the up-to-date technologies for the nanofabrication of III-nitride light-emitting diodes. The emission properties and related mechanisms have also been discussed. At the end of this review, we display the outlook of high-efficiency III-nitride-based light-emitting diodes with nanostructures, particularly for the aim of white light illumination.

3.1 Introduction

Due to global warming and the impending energy crisis, development of energy-saving solid-state lighting could be seen as a significant contributor to addressing the need for energy consumption reduction, as 29% of the world's total electricity consumption is for general illumination. Solid-state lighting (SSL), the possibly ultimate lighting source mainly based on white light-emitting diodes (LEDs), will result in a fundamental change in the concept of illumination experiencing more than 100 years. It will massively save energy, estimated to be equivalent to $112 billion by the year 2020.[1] The SSL market has been predicted to be worth $22.2 billion by 2020, growing at a compounded annual growth rate of 7.31%.[2]

Fabrication of white LEDs is mainly based on III-nitride semiconductors, as III-nitrides have direct bandgaps across their entire composition range, covering the complete visible spectrum and a major part of the ultraviolet (UV) (from 0.7 eV for InN, through 3.43 eV for GaN, to 6.2 eV for AlN). Due to

31

a number of advantages of white LEDs over conventional light sources (potentially 10 times efficiency, 10 times lifetime, fast response, and resistance to mechanical shock), it can be expected that white LEDs will eventually replace all conventional lighting sources including currently other forms of energy-saving lighting sources that tend to contain toxic and non-environmentally friendly elements such as Cd and Hg (EU has approved a law for strict restriction of using Hg-contained lamps).[3]

In principle, there would be three main approaches for the fabrication of white LEDs needed for SSL: (1) a combination of a blue (460 nm) LED and a yellow phosphor pumped by blue light from the LED; (2) a single chip emitting ultraviolet (UV) light that is absorbed in a LED package by three phosphors (red, green, and blue) and reemitted as a broad spectrum of white light; and (3) a package of three LED chips each emitting at a different wavelength (red, green, and blue, respectively).

The "blue LED + phosphor" approach is maintaining its strong lead for the fabrication of white LEDs with several commercial successes. This type of white LEDs is denoted as a PC-LED (phosphor-converted white LEDs). However, it will soon approach its limit, which still does not meet our energy-saving requirements as shown in Figure 3.1, the U.S. road map for developing SSL.[4] The performance of current UV LEDs is far below blue LEDs and presents a major limitation to the second route.

The third approach would be ideal, and I believe that it would be the ultimate solution for the fabrication of white LEDs. The lighting source fabricated in such a manner could potentially provide not only the highest efficiency but also the best color rendering. Furthermore, the utilization of discrete wavelength LEDs would also provide a best approach to tuning color temperature in order to replicate the color temperature of the Sun. Figure 3.1 shows that such a type of white LED can achieve an efficiency beyond 250 lm/W. The current challenges would be due to the fact that longer wavelengths such as green/yellow emitters with high efficiency are missing.

There are a number of factors that can affect the final efficiency of white LEDs, which is so-called luminous efficacy (labeled as η_v) that is defined as the ratio of the luminous flux emitted by the source to the input electrical power. For simplicity, in the case of "blue LED + phosphor" (or blue LED plus any other down-conversion materials) approach, the luminous flux is mainly determined by both the optical quantum efficiency of the blue LED chips and the luminescence efficiency of the down-conversion material. The first of these is proportional to the internal quantum efficiency (η_{int}) and the extraction efficiency (η_{ext}) of the LEDs. Additionally, η_v is also affected by a factor that is related to the absorption efficiency of down-conversion material. This is the rate of

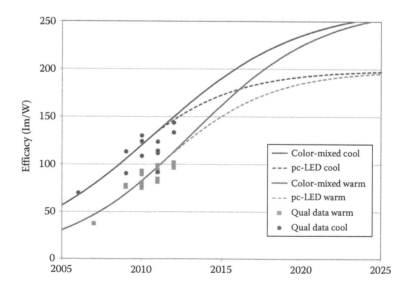

FIGURE 3.1   U.S. road map for developing solid-state lighting. (Reproduced from U.S. Department of Energy, Solid-State Lighting Research and Development: Multi-Year Program Plan, 68, 2012. With permission.)

energy transfer from the blue lighting source to the down-conversion material. There are therefore a number of challenges to overcome in order to deliver the highest luminous efficacy.

The last two decades have seen tremendous success in the field of III-nitride optoelectronics, which are mainly limited to InGaN-based emitters grown on *c-plane* GaN. The polar orientation results in strain-induced piezoelectric fields due to the large lattice mismatch between InGaN and GaN. As a result of the electric fields, emitters exhibit a reduced overlap between the electron and hole wave functions, leading to long radiative recombination times and low quantum efficiencies, the so-called quantum confined Stark effect (QCSE). This becomes a severe issue for longer wavelengths such as green/yellow emitters, as a much higher InN fraction is required. The second fundamental limitation is due to the well-known efficiency droop issue that occurs to all current InGaN-based LEDs on *c-plane* GaN, leading to a significant reduction in internal quantum efficiency (IQE) with increasing injection current.[4–8] At drive currents required for practical applications, the IQE falls by up to 50% from its peak value resulting in significant energy loss. This issue also becomes more severe as the emission wavelength shifts toward the green/yellow spectral region.[4–8]

High extraction efficiency of LEDs is essentially important. Snell's law shows the escape cone for light generated within the (In)GaN layer into free air space to have a solid angle of only ~23°. Neglecting light emitted from side walls and the backside of devices, only 4% of the generated light is expected to be extracted from the surface of standard III-nitride LEDs. The rest suffers from total internal reflection (TIR) and a risk of reabsorption unless it escapes through the sidewalls. In addition, typical III-nitride semiconductors are grown on sapphire substrates with a refractive index of ~1.7. Consequently, photons that fail to radiate into free air space are not only reflected back, but a significant portion of those photons may also be guided laterally by an air/GaN/sapphire waveguide, causing an extra TIR and then reabsorption. This generates extra heat, leading to an increase in temperature of the device itself. This elevated temperature generally results in a reduction in internal quantum efficiency, further limiting the output power of such devices and shortening the device lifetime.

An important, but often neglected, issue for the fabrication of phosphor-conversion white LEDs is due to the self-absorption of the phosphor involved in producing the white light. This puts additional limitations on the overall efficiency and also affects the color quality, that is, color rendering index (CRI). Furthermore, another factor is how to further improve the efficiency of the energy transfer from the blue LED to the down-conversion material. The intensity of the blue light generally remains much higher than the yellow emission from the down-conversion material, leading to a severe color rendering issue and the bluish tinge of most current white LEDs. This issue is becoming more and more important, as self-absorption induced poor conversion efficiency from blue to yellow and a low light quality leads to serious issues on human sleep patterns and thus mental health.[9] We could say that if the issue could not be resolved, white LEDs could not be widely applied in general illumination no matter how high the efficiency we could achieve. To some degrees, the color quality, specifically high CRI, would be more important than optical efficiency.

In order to address a number of these challenging issues, nano-engineering would offer a very promising avenue forward. The nano-engineering that I refer to here would include a wide range of techniques, including the growth and fabrication of device structures on a nanometer scale, integration of different kinds of semiconductors on a nanometer scale, etc. The last decade has seen significant effort devoted to developing a number of nano-engineering techniques, although there still exist a number of scientific and technological challenges. In theory, GaN-based nanostructures exhibit a number of potential advantages compared with conventional planar structures. As a result of the nanorod configuration, the strain in InGaN/GaN-based LEDs will be partially released and thus the strain-induced QCSE will be significantly reduced, potentially increasing the optical efficiency of LEDs. The light extraction efficiency is expected to increase as the conditions for TIR do not exist. A recent study demonstrated that the emission wavelength of InGaN/GaN structures depends on the diameter of nanorods, potentially allowing us to achieve a multiple color emitter in a single wafer through tuning the diameter of nanorods.[10]

Of course, there also exist a number of disadvantages for devices with a nanostructure. First, due to the large surface-to-volume ratios, nonradiative surface recombination will be enhanced,[11–13] and thus nanorod LEDs generally suffer from *Shockley–Read–Hall* (SRH) nonradiative recombination.[14] Therefore, surface treatment or surface passivation is required, thus leading to great challenges in device fabrication. As an extra planarization process is required prior to any further device fabrication, it also further increases difficulties in device fabrication and thus generates extra cost. Furthermore, there exists a band bending issue due to surface Fermi level pinning effect,[15,16] significantly affecting the electrical properties of nanorod LEDs.

This chapter will review recent progress on developing III-nitride-based white LEDs with different kinds of nanostructures in these areas. Section 3.2 will present a number of advanced growth and fabrication techniques for III-nitride-based nanorod or nanowire array LEDs. In Section 3.3, a number of fabrication techniques that combine III-nitride nanostructure with different kinds of down-conversion materials will be presented, where the detailed mechanisms for the nonradiative energy transfer between the III-nitride nanostructures and the down-conversion materials will be discussed. Section 3.4 will introduce the latest progress on developing phosphor-free white LEDs, possibly demonstrating a new direction for further developing white LEDs. Finally, a summary and future outlook will be provided in Section 3.5.

3.2　Growth and Fabrication of III-Nitride LEDs with Nanostructures

The research on GaN nanostructures could be dated back to as early as 1997, when the GaN nanocolumns were first grown by a radio frequency (RF) radical source molecular beam epitaxy (RF-MBE) technique.[17] So far, a large number of types of nanostructure emitters have been reported, such as single nanowire LEDs,[18–25] nanorod array LEDs,[26–33] nanowire LDs,[34–38] etc. In principle, two major approaches can be applied in the growth or fabrication of GaN-based nanowires or nanorod emitters, namely, top-down process and bottom-up process. The bottom-up process is mainly based on epitaxial growth, while the top-down process typically involves post-growth nanofabrication techniques including initial nano-patterning and subsequent etching processes.

3.2.1　Epitaxial Growth of Nanorod Array LEDs

Two major advanced epitaxial growth techniques can be used for the bottom-up growth process, namely, metal-organic chemical vapor deposition (MOCVD) and molecular beam epitaxy (MBE). Given an availability of high temperatures and high growth rates, the MOCVD technique demonstrates a major advantage in the growth of regularly arrayed nanorod LEDs or core–shell LEDs utilizing a selective overgrowth technique on nano-patterned substrates, while the MBE is commonly used for the self-organized growth of GaN-based nanowire LEDs, in particular the growth of nanowires with a high density.

MBE, in particular plasma-assisted MBE, tends to use nitrogen as a precursor for group V, where the decomposition efficiency of nitrogen molecular depends on a plasma power used instead of a growth temperature. The MOCVD technique generally utilizes NH_3 as a group V precursor, where a high temperature is required in order to decompose NH_3 efficiently and thus it is very good for the growth of planar (Al)GaN structures or selective growth of (Al)GaN. However, InGaN growth generally requires a low growth temperature; higher indium requires a lower growth temperature. In addition, for the growth of self-organized nanowires or nanorods, the growth rate along a vertical direction needs to be enhanced while the growth rate along a lateral direction needs to be suppressed. In this case, compared with standard planar GaN growth, a lower temperature would be favorable for the formation of nanowires, giving a further advantage in using the MBE to grow GaN nanowires or nanorods.

Here, I would like to first present a few typical examples of using selective overgrowth techniques by either MOCVD or MBE, in particular the growth of very regularly arrayed nanorods. In the second part of this section, I will provide an overview on the recent progress of developing the self-organized growth of nanorods by using MBE.

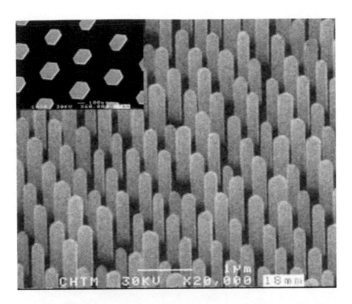

FIGURE 3.2 SEM image of GaN nanowire array; inset: plan-view SEM image. (Reprinted with permission from Hersee, S.D., Sun, X., and Wang, X., *Nano Lett.*, 6, 1808. Copyright 2006 American Chemical Society).

Hersee et al.[39] reported the growth of GaN nanowire arrays with a high uniformity by a selective overgrowth technique on a nano-patterned GaN template, as shown in Figure 3.2. Initially, they deposited a layer of SiN dielectric layer on a standard GaN layer grown on sapphire or SiC or Si and then employed a combination of an interferometric lithography technique and a dry etching process to form a hexagonal array of apertures on a 500 nm pitch, where the apertures in a circular shape have an average diameter of 221 nm. Subsequently, a selective overgrowth process was performed on the nano-patterned template by using a two-step growth approach. During the first step, both group III (i.e., TMGa) and group V (i.e., NH_3) are introduced simultaneously in order to selectively fill all available growth apertures but without the growth of any GaN on the top of the SiN mask. This is a standard GaN growth mode. Afterward, just before the nanowires emerged from the growth mask, they switched into a pulsed growth mode in order to suppress the lateral growth and enhance the vertical growth, where the TMGa and the NH_3 were supplied alternately in the sequence: TMGa with 10 sccm and flowing 20 s, and NH_3 with 500 sccm and flowing 30 s. Under such growth conditions, GaN nanowires can maintain their geometry after emerging from the growth mask. Both the timing for switching to the pulsed growth mode from the regular MOCVD growth and the duration of the steps within the pulse sequence are critical.

Based on this growth approach, they have grown nanowire LEDs,[40] which consist of a 1.5 μm n-type GaN and a 1 μm p-type GaN. After growth, spin-on glass was used to fill the space between the nanowires in order to provide electrical isolation. A further standard device fabrication process including the fabrication of n- and p-type contacts has been performed, as schematically illustrated in Figure 3.3. Finally, strong electroluminescence has been achieved at 364 nm, the GaN band-edge emission wavelength.

As mentioned earlier, it would be difficult to apply the selective overgrowth approach in MBE growth of nanorods as the selectivity at a temperature available for current MBE is not as good as that can be achieved by MOCVD. In order to address this issue, Sekiguchi et al.[41] employed Ti as a mask instead of any dielectric mask and successfully achieved GaN nanorod arrays with an accurately controlled diameter and an excellent uniformity by using rf-MBE, as shown in Figure 3.4. When the growth temperature is below 880°C, the selectivity is very low, and the GaN can be grown on the top of the mask as well. However, when the growth temperature is above 900°C, the growth rate of nanorods reduces significantly, indicating an

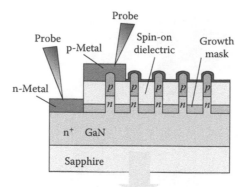

(a)   Optical detector and spectrometer

(b)

FIGURE 3.3 Schematics of nanowire array LED (a) and SEM image showing electrical probing of individual nanowire diode (b). (Reproduced with permission from Hersee, S.D., Fairchild, M., Rishinaramangalam, A.K., Ferdous, M.S., Zhang, L., Varangis, P.M., Swartzentruber, B.S., and Talin, A.A., *Electron. Lett.*, 45, 75 Copyright 2009 IEEE.)

enhancement in desorption of Ga adatoms from the surface. This also enhances the shape inhomogeneity of the nanorods and reduces the nanorod diameter. Based on the systematic study, the group further grew InGaN/GaN on the top of GaN nanorods with different diameters and achieved multiple color emission by photoluminescence (PL) measurements, covering the whole visible spectral region as demonstrated in Figure 3.5.[42] Further study indicates that the indium content sensitively depends on the diameter of the nanorods, which in return is determined by the nanomask diameter.[42]

Core–shell nanorod LEDs can also be achieved by a selective overgrowth approach on patterned templates. Generally speaking, InGaN/GaN multiple quantum wells (MQWs) as an emitting region for an LED are grown on the whole sidewalls of nanorods, and thus the emitting area is expected to significantly increase. If a nanorod is assumed cylindrical, the emitting area can be increased by a factor of 4.[43] In addition to the increased emitting area, there is another major advantage, namely, the exposed sidewalls of nanorod arrays obtained on *c-plane* sapphire are actually nonpolar facets,[44–47] and thus the QCSE that current InGaN/GaN-based LEDs on *c-plane* sapphire suffer from would be eliminated. Yeh et al. demonstrated core–shell InGaN/GaN MQWs nanorod arrays by means of using a selective overgrowth technique on a patterned template by MOCVD.[44] Figure 3.6 shows a schematic diagram of GaN nanorod

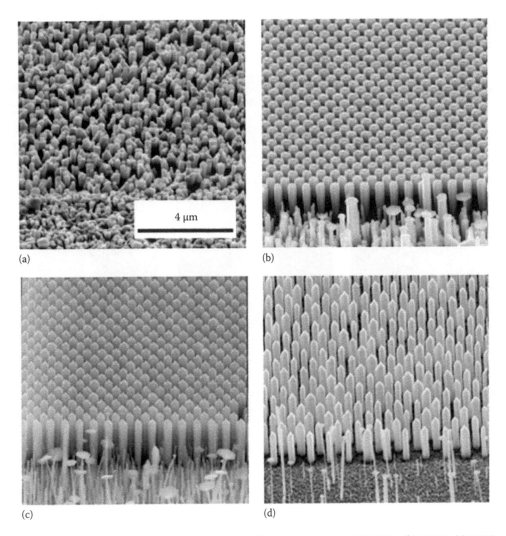

FIGURE 3.4 SEM images of GaN nanowires grown at different temperatures: (a) 880°C, (b) 900°C, (c) 915°C, and (d) 925°C. (Reproduced from Sekiguchi, H. et al., *Appl. Phys. Express*, 1, 124002, 2008. With permission from Japan Society of Applied Physics.)

arrays grown vertically, and the exposed sidewalls are {1–100} GaN, on which InGaN/GaN MQWs are grown. Experimental results have indicated that the piezoelectric fields have been eliminated. Under identical conditions, a number of InGaN/GaN MQW structures have been grown on the sidewalls of the nanorods with different diameters that are determined by the opening diameters in the mask. It has been found that the emission wavelength increases with increasing opening diameter in the mask. This means that both the indium content and the InGaN quantum well thickness can be controlled by simply tuning an opening diameter in the mask.

Almost all the impressive results on the growth of self-organized nanorod LEDs achieved so far are limited to the MBE techniques.[27,30,33,48,49] In addition to the reasons mentioned earlier, MBE growth generally takes place under a high vacuum with a very low beam flux, leading to a much lower growth rate than that of MOCVD growth, which is important for the formation of nanoseeds or nanoislands for further nanorod growth. Furthermore, MOCVD uses metal–organic precursors to supply group III elements,

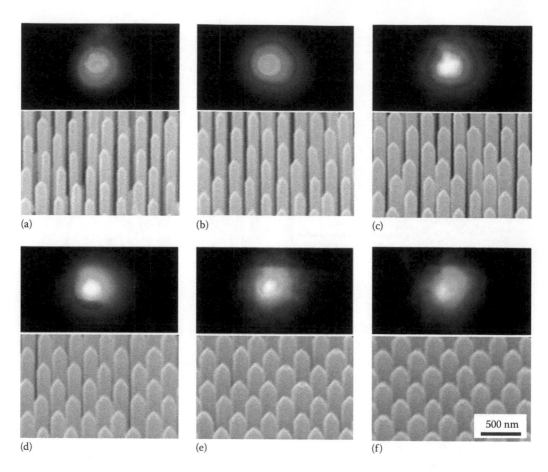

FIGURE 3.5 Bird's-eye-view SEM and PL emission images from InGaN/GaN nanowires as a function of nanowire diameter: (a) $D = 143$ nm; (b) $D = 159$ nm; (c) $D = 175$ nm; (d) $D = 196$ nm; (e) $D = 237$ nm; and (f) $D = 270$ nm. (Reproduced from Sekiguchi, H. et al., *Appl. Phys. Lett.*, 96, 231104, 2010. With permission from Japan Society of Applied Physics.)

where complicated chemical reactions with some by-products generated take place and these might affect the formation of nanorods, while MBE simply uses high-purity metals and thus all these concerns can be eliminated. Along with some *in-situ* monitoring systems such as reflection high-energy electron diffraction (RHEED) and mass spectrometer, the whole growth process, in particular the timing for the transition from a two-dimensional (2D) growth mode to a three-dimensional (3D) growth mode that is the most important parameter for the formation of nanorods, can be *in-situ* monitored, allowing us to *in-situ* tune the growth conditions. This is extremely useful for the growth of self-organized nanorods with a high density.

The growth of self-organized nanorods without any external catalyst particles can be understood using a diffusion-induced model.[50] It means that the top surface of a nanorod generally exhibits a lower chemical potential than the sidewalls of the nanorod, forming a driving force for the adatoms' diffusion to the nanorod apex from its sidewalls. A basic model for the growth of GaN nanorods on silicon has been established to describe the growth process.[51] The growth starts with a nucleation stage, during which the nucleation process will continue on the free surface. In the meantime, the nanorods that have been nucleated will continue to grow in length and diameter. The nucleation process continuously occurs until a high density of nanorod is obtained. Afterward, the nanorod grows slowly along a vertical direction with increasing diameter. The growth can be described by a semi-empirical

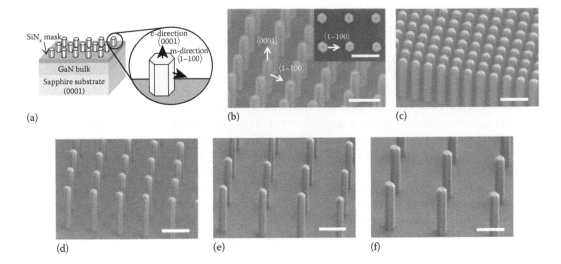

FIGURE 3.6 (a) Schematic diagram of GaN nanorod arrays; (b) SEM images of GaN nanorod arrays, showing the polar and nonpolar facets of the nanorods. Inset: top-view SEM image of hexagonal GaN nanorods; (c–f) SEM images of GaN nanorod arrays grown on 250, 500, 750 nm, and 1 μm center-to-center spacing, respectively. The scale bars are 500 nm in all figures. (Reprinted with permission from Yeh, T.W., Lin, Y.T., Stewart, L.S., Dapkus, P.D., Sarkissian, R., O'Brien, J.D., Ahn, B., and Nutt, S.R., *Nano Lett.*, 12, 3257. Copyright 2012 American Chemical Society.)

model, showing that the length of a nanorod is proportional to the reciprocal of diameter, depicted by an equation given as follows[52]:

$$l = C_1 \left(1 + C_2 / d \right)$$

where

C_1 and C_2 are constants
l and d the length and diameter, respectively

When the length of a nanorod is less than the diffusion length ($l < \delta$), the diffusion-induced growth dominates the growth process if the diffusion flux of adatoms is comparable to the deposition rate. If $l > \delta$, the adatoms cannot diffuse to the top of nanorods. As a result, the growth rate, which is due to the combination of the impinging flux on the top of nanorods and the diffused adatoms, remains constant. This model is specifically suitable for thin nanorods with a diameter of less than 80 nm, because the diffusion is less important for thick nanorods.[53] If diffusion length is much shorter than the nanorod length, a strong tapering effect will occur.[53]

The crystal quality and morphology of the self-organized nanorods are also significantly affected by growth conditions, such as beam flux, V/III ratio, temperature, and so on. Vajpeyi et al. studied the influence of temperature on the growth of nanorod array grown on Si (111).[54] They kept the V/III ratio to be 5, but changed the growth temperature from 700°C to 800°C, and found that the growth rate, diameter, and density of the nanorods decreased with increasing temperature, while the tapering effect was alleviated.

Tchernycheva et al. studied GaN nanorod growth by using RF-MBE.[20] Under a high Ga beam flux press (BFP), which means a low V/III ratio, only a 2D layer was obtained. A lateral growth rate is substantially decreased by decreasing Ga BFP, namely, by increasing V/III ratio. By optimizing the V/III ratio, very straight cylindrical nanorods without any tapering have been achieved. The nanorods also exhibit defect-free and strain homogeneity properties.

Optical properties of the self-organized nanorods grown using MBE have been intensively investigated by a number of groups.[48,49,55–57] The unique optical properties in these studies can be summarily attributed to interior strain and surface states. As a result of plastic or elastic relaxation,[49,50,56,58,59] nanorods generally exhibit a much lower strain (or even strain-free) compared with their 2D counterparts, leading to a clear blueshift in emission wavelength.[55,56] However, we also need to pay extra attention to their negative properties brought as a result of being fabricated into nanorods. For instance, surface states are formed due to absorption/desorption, defects or dislocation bending to the free surface during growth.[15,57] Proper surface treatment is essential before nanorods being fabricated into devices; otherwise, the optical properties will possibly be dominated by the surface-state effect.

A number of groups reported the growth of self-organized nanorod array-based LEDs.[30,33,60–62] In 2004, Kikuchi et al. demonstrated a prototype nanorod array LED on Si (111) substrate grown by plasma-assisted MBE.[60] The density and diameter of the nanorod arrays are 1.2×10^{10} cm^{-2} and 80–100 nm, respectively. Each nanorod contains n-GaN (750 nm) and eight pairs of InGaN (2 nm)/GaN (3 nm) MQWs, followed by a p-type layer, which was grown using different conditions in order to form a coalesced and planar surface in the final growth stage. A Ni (2 nm)/Au (3 nm) alloy has been used as the p-type ohmic contact. The current–voltage (I–V) characteristics of the LED show an injection current of 10 mA under a 5 V forward bias, which is significantly higher than that for a standard planar LED, where a typical I–V curve of blue LEDs exhibits 20 mA injection current under a forward bias of 3 V or below 3 V. The output power versus current (I–L) of the nanorod array LED shows a linear increase with increasing injection current of up to 12 mA. By varying InN content in the MQWs, the emission wavelength can be tuned from green (530 nm) to red (645 nm). Recent studies showed that defect-free nanorod array LED with a high quantum efficiency of ~30% could be obtained by the self-organized nanorod array LEDs.[30] The achievement of high-performance nanorod array LEDs further enables us to perform a detailed study of semiconductor devices. For instance, Auger recombination, which has been generally accepted as one of the major mechanisms that are responsible for the well-known efficiency droop widely occurring to conventional planar LEDs, has been studied based on a defect-free nanorod array LED using the so-called ABC model.[62] An Auger recombination coefficient of ~10^{-33} cm^6/s has been obtained,[62] concluding that the defect-assisted Auger recombination is responsible for the efficiency droop in their samples.

3.2.2 Fabrication of III-Nitride Nanorod Array LEDs

Compared with the bottom-up approach, the top-down approach has both advantages and disadvantages. For example, it would be difficult to upscale the bottom-up approach that strongly depends on detailed growth conditions and facilities used. Furthermore, there would also exist a reproducibility issue for the bottom-up approach. However, the top-down approach typically involves mask patterning processes and subsequent etching processes, both of which are fairly standard techniques widely available in the semiconductor industry. Of course, the etching-induced damages will be generated. Such damages would not be so important for the fabrication of large area emitters, but it would be extremely important for the fabrication of nanorod LEDs, as it will further significantly enhance the surface-state issue.[63] So far, almost all the nanorod LEDs do not show very good electrical properties, as indicated in Table 3.1.[30,33,60,64–68] For instance, the I–V characteristics of nanorod LEDs are far from satisfactory, typically 20 mA injection

TABLE 3.1   A Comparison of Voltages at 20 mA
of Top-Down Fabricated Nanorod Array LEDs

Year	Voltage @ 20 mA	Reference
2011	4–6 V	33
2011	>6 V	65
2010	>10 V	66
2010	> 4.5 V	30
2010	~5.8 V	67
2007	~6.5 V	68
2006	> 5 V	69
2004	> 5 V	61

current under a forward bias of >5 V in most cases, while a typical forward bias for 20 mA for a standard blue LED is around 3 V. Therefore, an extra curing process would be essential.

Given the excellent chemical stability of III-nitrides, dry etching is a typical approach for the fabrication of GaN-based nanorod LEDs. For the nano-patterning process, a number of approaches have been developed, such as electron beam lithography (EBL) that is time-consuming and expensive, nanosphere lithography (NSL)[63,69–74] that is typically limited to a scale below a millimeter, nanoimprinting lithography (NIL) that could potentially be good for mass production, and a self-organized nanomask approach that is cost-effective and up scalable for mass production.[69–74]

There are a large number of reports on using EBL for nanofabrication, mainly for the fabrication of specially designed nanostructures, such as microcavity or nanocavity. Due to its intrinsic time-consuming and high cost, it would not be a good idea to use the EBL to create a mask pattern with a large area for any further fabrication of nanorod arrays.

Both NSL and NIL approaches have been used for the fabrication of regularly arrayed nanorods with a high uniformity. The NSL approach involves the utilization of nanospheres, such as polystyrene (PS) nanospheres[75] and silica nanospheres.[73,74] The author's group has established a nanofabrication technique using the silica-based NSL approach for the fabrication of very regularly arrayed nanorods on an InGaN/GaN MQW epiwafer with emission in the green spectral region, forming a 2D array structure.[74] This leads to a significantly enhanced spontaneous emission (SE) rate of the InGaN/GaN MQWs and thus an amazing improvement in internal quantum efficiency with a factor of 88, compared with the as-grown sample. Figure 3.7 shows a detailed fabrication process of the 2D well-ordered nanodisk array structures and SEM images of the final 2D nanorod array structures with different nanorod diameters that can be controlled from 145 to 250 nm. The diameter of silica nanospheres used as masks is 274 nm. The silica nanospheres were mixed with a surfactant (sodium dodecyl sulfate) to prevent an aggregation of the nanospheres. In order to achieve a uniform nanosphere single monolayer, a SiO_2 layer with a thickness of 40 nm was initially deposited on the surface of the InGaN/GaN MQW sample by plasma-enhanced chemical vapor deposition (PECVD). The SiO_2 was then treated under O_2 plasma in a reactive ion etching (RIE) chamber to generate a hydrophilic surface, which can further improve the uniformity in the nanosphere layer. Subsequently, a nanosphere layer was deposited on the SiO_2 surface by a spin coating technique. A selective etching process was performed on the nanospheres

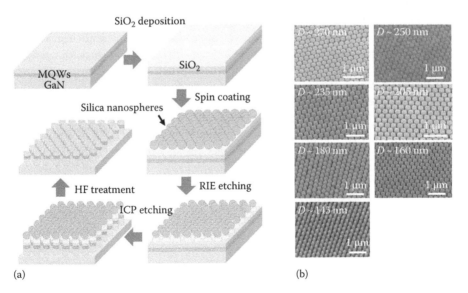

(a) (b)

FIGURE 3.7 (a) Schematics of our procedure for the fabrication of the nanodisk array structures and (b) SEM images of the nanodisk array structures with a diameter of 145, 160, 180, 205, 235, 250, and 270 nm, respectively. (Reprinted with permission from Kim, T., Liu, B., Smith, R., Athanasiou, M., Gong, Y., and Wang, T., *Appl. Phys. Lett.*, 104, 161108 Copyright 2014, American Institute of Physics.)

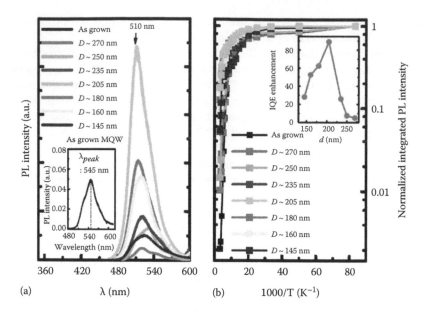

FIGURE 3.8 (a) PL spectra measured at room temperature; inset: PL spectrum of the as-grown sample measured at room temperature; (b) normalized integrated PL intensity as a function of temperature; inset: IQE enhancement factor as a function of the nanodisk diameter. The IQE enhancement is defined as the ratio of the IQE of each nanodisk array structure to that of the as-grown sample. (Reprinted with permission from Kim, T., Liu, B., Smith, R., Athanasiou, M., Gong, Y., and Wang, T., *Appl. Phys. Lett.*, 104, 161108 Copyright 2014, American Institute of Physics.)

in order to tune the diameter of the nanospheres as the final nanomasks through changing the etching time. During the etching process, the SiO_2 layer exposed between the nanospheres was etched away at the same time. Finally, a Cl_2-based inductively coupled plasma etching was used to etch the InGaN/GaN MQW sample into the 2D well-ordered nanodisk array structures with a height of 350 nm.

Figure 3.8a shows the PL spectra of all the 2D nanodisk array structures measured under identical conditions at room temperature, all exhibiting strong emission at 510 nm, in particular, a sharp jump in PL intensity for the 2D nanodisks with a diameter of 205 nm. This is due to a so-called nanocavity effect, where the emission wavelength matches the cavity mode wavelength and thus forms a resonance. Internal quantum efficiency (IQE) η can be typically estimated by means of temperature-dependent PL measurements using the ratio of an integrated intensity at room temperature to that at a low temperature, such as 12 K in the present study. Figure 3.8b shows the integrated PL intensities of all the 2D nanodisk structures as a function of temperature. For a reference, the data from the as-grown sample are also provided. Using the as-grown sample as a reference, we can plot the IQE enhancement factor (defined as the ratio of the IQE of each nanodisk sample to that of the as-grown sample) as a function of nanodisk diameter, which is given in the inset of Figure 3.8b, where *a sharp jump with an IQE enhancement factor of 88* has been achieved for the 2D nanodisk structure with a nanodisk diameter of 205 nm.

This NSL approach is fairly simple, and thus cost-effective, and the main drawback is due to its limited scale, thus restricting its wide applications in industry. NIL can also be used to fabricate GaN nanorods. For instance, Zhuang et al. developed a soft UV NIL approach along with inductively coupled plasma (ICP) dry etching for the fabrication of 2D nanorod arrays.[76] Figure 3.9 shows a schematic illustration of the detailed procedure. A 450 nm PMMA and then an 80 nm UV curable resist have been sequentially deposited on an InGaN/GaN epiwafer by means of a standard spin coating method. Subsequently, a soft UV-curing NIL was carried out to transfer a 2D nanohole mask pattern with an elliptic shape on the UV curable resist, which is followed by a dry etching process. Finally, a 2D arrayed nanorod structure has been fabricated as shown in Figure 3.9h and i. For a final device fabrication, spin-on glass (SOG) has been employed to planarize the 2D

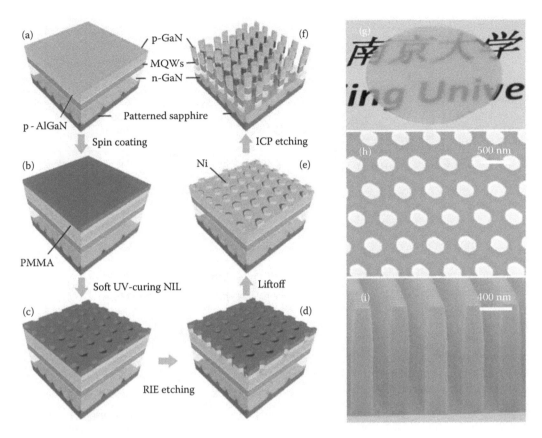

FIGURE 3.9 (a–f) Schematics of the nanorod fabrication process; (g) photograph of a 2-inch green InGaN/GaN MQW epitaxial wafer with highly ordered nanorod arrays; (h) and (i) SEM images of nanorod arrays. (Reproduced with permission from Zhuang, Z., Guo, X., Zhang, G., Liu, B., Zhang, R., Zhi, T., Tao, T., Ge, H., Ren, F., Xie, Z., and Zheng, Y., *Nanotechnology*, 24, 405303, 2013. Institute of Physics.)

arrayed nanorod structure, followed by a standard LED fabrication procedure. Figure 3.10 shows significant improvement in optical performance as a result of fabrication into a 2D nanorod LED.

Both NSL and NIL approaches demonstrate great advantages in terms of the fabrication of very regularly arrayed nanorod structures. However, there also exist a number of challenges in terms of scalability and achieving a high density of nanorod. Furthermore, for the NSL approach, it is key to form a single monolayer of closely packed silica nanoparticles, which is still a great challenge, in particular in a large scale. In this case, the self-organized metal nanomask approach demonstrates huge advantages. The self-organized metal nanomask approach has been widely employed for the fabrication of III-nitride-based nanorod array LEDs by a number of groups including the author's group.[29,65,67,77–79] Recently, the author's group developed a number of novel processes to further improve the self-organized nanomask approach and then demonstrated the first nanorod LEDs with both superior electrical performance and optical performance for practical applications.[29]

Our novel process starts with the deposition of a thin indium tin oxide (ITO) layer instead of widely used SiO_2 on an LED epiwafer, one of the crucial steps for our processing technique. The thin ITO is used to protect the top p-type layer, which will be explained later. Then, a thin film of 200 nm SiO_2 is prepared, followed by deposition of a thin Ni layer. Subsequently, the sample undergoes an annealing process at 800°C for 1 min under N_2 ambient, allowing the thin Ni layer to self-organize into nickel nanoislands as shown in Figure 3.11a. Such Ni nanoislands served as nanomasks form SiO_2 nanorod arrays using a standard reactive ion etching (RIE) process. Finally, the SiO_2 nanorod arrays as a second mask are employed to etch down to the n-type layer of the LED epiwafer as displayed in Figure 3.11b. Figure 3.11c shows a typical SEM image

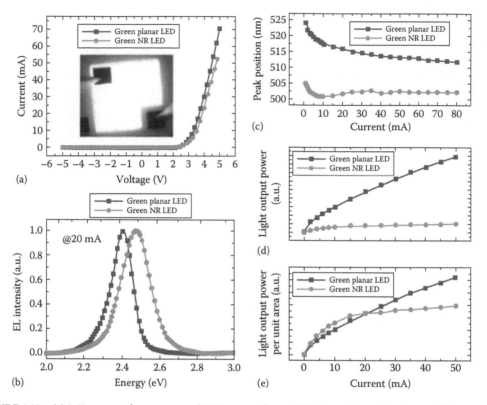

FIGURE 3.10 (a) I–V curves of green nanorod LED versus planar LED. Inset: EL emission image; (b) normalized EL spectra of green nanorod and planar LED; (c) EL wavelength of green nanorod and planar LED as a function of the injection current; (d) L–I characteristics; and (e) L–I characteristics in terms of per unit emissive area. (Reproduced with permission from Zhuang, Z., Guo, X., Zhang, G., Liu, B., Zhang, R., Zhi, T., Tao, T., Ge, H., Ren, F., Xie, Z., and Zheng, Y., *Nanotechnology*, 24, 405303, 2013. Institute of Physics.)

of our nanorod arrays, showing a cylindrical shape with straight and smooth sidewalls. The diameter and the density of the GaN nanorod arrays are estimated to be about 200 nm and 1×10^9 cm^{-2}, respectively. It is worth mentioning that inevitable damages are normally generated during the ICP dry etching process, leading to severe degradation in device performance. This is one of main reasons that the performance of InGaN-based nanorod LEDs is far from satisfactory. In our case, a treatment is carried out by boiling the nanorods in HNO$_3$ (70%) acid to remove any etching-induced damages or make the damages healed prior to a further device fabrication. In order to make ohmic contact on the p-GaN, a planarization and insulation process using SOG is carried out. Figure 3.11d shows the nanorod LED arrays covered by SOG. A subsequent etching back process is performed in order to allow for the exposure of the p-type layer of nanorods for the deposition of p-type contacts. This is a critical step in the fabrication of nanorod LEDs because the poor I–V characteristics of nanorod LEDs reported so far is mainly attributed to the etching back process. The typical gases for etching SOG or SiO$_2$ is a mixture of CHF$_3$ and either O$_2$ or SF$_6$, which can cause a severe deactivation or damage to the p-GaN layer once the p-GaN layer is exposed to either CHF$_3$ or SF$_6$. This results in a high resistance of p-type contacts, which is the case when the SOG on top of any III-nitride-based nanorods is etched away. This is a major reason why current InGaN-based nanorod LEDs (no matter what kind of techniques is used for the fabrication of III-nitride nanorods) exhibit such poor I–V characteristics.

Therefore, it is the thin ITO deposited directly at the initial stage that prevents the p-GaN from being exposed to the etching gas once the SOG on the top of the nanorods is etched away, and thus the thin ITO is effectively used as a protective layer. The advantage of utilizing ITO is that ITO can be removed easily using HCl, while HCl cannot attack any residual SOG in the gaps between the nanorods. After the SOG etching

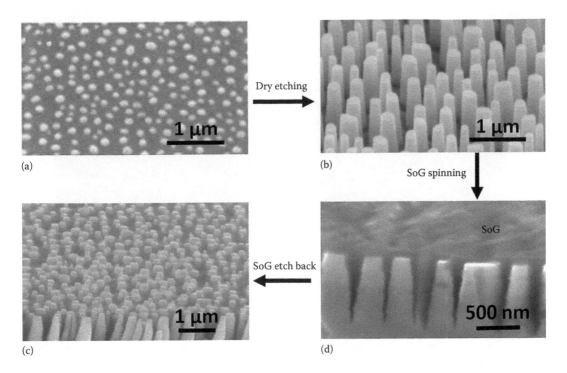

FIGURE 3.11 SEM images of (a) Ni mask; (b) SiO$_2$ mask; (c) nanorods; and (d) SoG-filled nanorods. (Reproduced with permission from Bai, J., Wang, Q., and Wang, T., *J. Appl. Phys.*, 111, 113103, 2012. Institute of Physics.)

back process and the subsequent ITO removal process, a clean and unaffected p-GaN can be obtained for further fabricating a good p-type ohmic contact. The following procedures are fairly standard, mesa etching and deposition of p and n-type contacts as usual for the fabrication of any planar InGaN-based LEDs.

Figure 3.12 shows the forward-bias I–V characteristics of the nanorod LEDs with and without the ITO protective layer, demonstrating superior electrical performance of our nanorod LEDs with initially

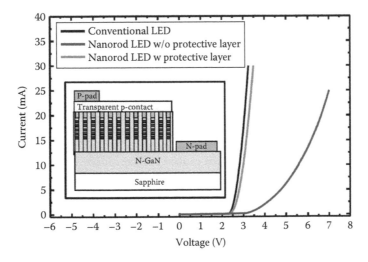

FIGURE 3.12 I–V characteristics of the nanorod array LEDs with and without ITO protective layers and conventional LED.

 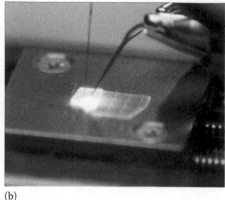

(a)   (b)

FIGURE 3.13 Images of electroluminesce of nanorod array LEDs (a) and standard planar LEDs (b), both fabricated on the same wafer.

deposited ITO. For example, a forward bias for 20 mA has been significantly decreased to 3.26 eV from the 6.6 V of the LED without the protective layer, which is very close to 3.07 V of the conventional LED. To our best knowledge, this is the best report for InGaN nanorod LEDs so far, meaning that it can be used for practical applications. At 20 mA current, the EL intensity of the nanorod LED is about 1.8 times that of the conventional one. Figure 3.13 displays typical electroluminescence (EL) photos of the nanorod LEDs and standard planar LEDs, both fabricated on the same wafer and measured under identical conditions, demonstrating significantly enhanced output power of the nanorod array LEDs.

Detailed x-ray reciprocal space mapping (RSM) measurements have quantitatively evaluated the stain relaxation in the nanorods, demonstrating that the majority of strain in InGaN/GaN MQWs has been relaxed as a result of fabrication into nanorods.[80] Further time-resolved PL measurements also confirmed that the radiative recombination process has been significantly enhanced as a result of fabrication into a nanorod structure.[81] All these results have clearly shown that a significant reduction in strain-induced QCSE has occurred to the nanorod structures, leading to significant improvement in optical performance.

3.3 Fabrication of Hybrid White LEDs

As discussed in Section 3.1, the approach of "blue LED + yellow phosphor" has a number of drawbacks, although it still remains the major technique for the fabrication of white LEDs. The nanorod array LEDs offer potential opportunities to accommodate different kinds of down-conversion materials in many ways in addition to yellow phosphor. A combination of nanorod blue LEDs with either down-conversion inorganic nanocrystals or down-conversion organic emitting polymers would be two of the most promising approaches to achieving white LEDs. Furthermore, nanorod arrayed LEDs also provide a unique configuration, potentially enabling to minimize the separation between the blue LEDs and these down-conversion materials. This would significantly enhance the energy transfer process between them, as such an energy transfer process that is basically a kind of nonradiative energy transfer sensitively depends on the separation due to a dipole–dipole coulombic interaction. This section will provide an overview on recent progress of developing III-nitrides/inorganic down-conversion materials white LEDs.

3.3.1 Fabrication of III-Nitrides/Inorganic Down-Conversion Material White LEDs

The nonradiative energy transfer based on a coulombic interaction is a fairly common phenomenon. For example, excitons in a semiconductor can couple with metal particles located in its proximity, leading to an exciton and surface plasmon interaction provided that the surface plasmon energy matches the excitonic recombination energy. In this case, the excitonic recombination rate can be significantly increased, leading to significant enhancement in optical efficiency. This has been widely used for the fabrication of novel optoelectronics, such as plasmonic laser, etc.[82–85]

Chanyawadee et al.[86] reported a hybrid structure that combines InGaN/GaN MQW-based blue LEDs with a large number of holes through the InGaN/GaN MQW and semiconductor colloidal nanocrystals (NCs) as a down-conversion material, as schematically illustrated in Figure 3.14. Two kinds of samples have been fabricated for a comparison in order to study the energy transfer from the blue LEDs to the NCs: one sample with shallow-etched holes, where the separation between the InGaN/GaN MQWs and the NCs is so large that the nonradiative energy transfer can be ignored and another sample with deep-etched holes, where the NCs are in a direct contact with the InGaN/GaN MQWs. Therefore, as long as the NCs absorption can be tuned to match the emission wavelength of the blue LEDs, the second sample is expected to show an efficient nonradiative energy transfer between the blue LED and the NCs. As a consequence, their EL measurements show that the NC emission pumped by the deep-etched LED is significantly higher than the NC emission pumped by the shallow-etched LED. Further experiment also confirmed that the down-conversion efficiency of the hybrid blue LED/NCs using the deep-etched LED is increased by 43% compared with the hybrid blue LED/NCs using the shallow-etched LED.

Jiang et al.[87] studied the nonradiative energy transfer between blue InGaN/GaN MQWs and semiconductor NCs. Instead of using nanoholes, they combined the InGaN/GaN nanorods (self-organized nanorods) and down-conversion CdSe NCs, and they achieved a significantly enhanced energy transfer between the blue nanorod LED and the CdSe NCs with an efficiency of up to 80%.

Very recently, Zhuang et al.[88] demonstrated electrically injected hybrid white LEDs using a combination of blue or violet LEDs with regularly arrayed nanoholes and various wavelength emitting NCs as down-conversion color mediums. As a result of strong coupling between the InGaN/GaN MQWs and

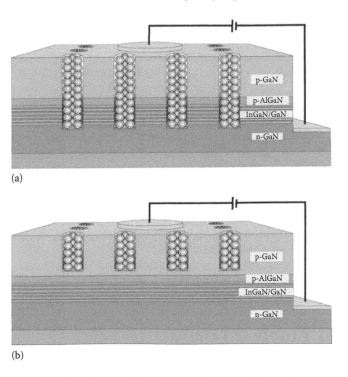

FIGURE 3.14 Schematics of (a) a hybrid device with a deep-etched LED and (b) a hybrid device with a shallow-etched LED. (From Chanyawadee, S., Lagoudakis, P.G., Harley, R.T., Charlton, M.D.B., Talapin, D.V., Huang, H.W., Lin, C.H.: *Adv. Mater.* 2009. 22. 602. Copyright Wiley-VCH Verlag GmbH & Co. KGaA. Reproduced with permission.)

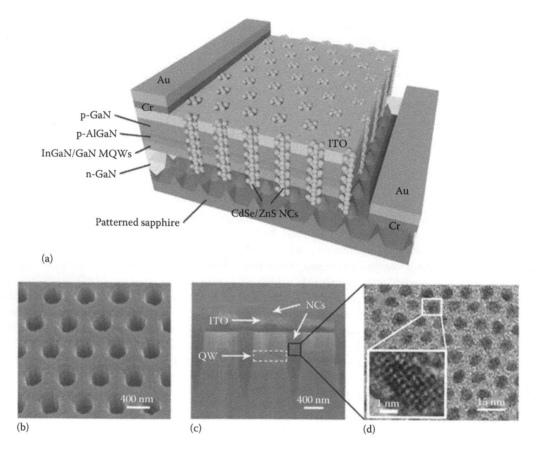

FIGURE 3.15   (a) Schematics of a hybrid LED; (b) plan-view SEM image of ordered nanohole arrays; (c) cross-sectional view SEM image of nanoholes with NCs; and (d) TEM image of CdSe/ZnS core/shell NCs. (From Zhuang, Z., Guo, X., Liu, B., Hu, F., Li, Y., Tao, T., Dai, J. et al.: *Adv. Funct. Mater.* 26. 36. 2015. Copyright Wiley-VCH Verlag GmbH & Co. KGaA. Reproduced with permission.)

the down-conversion NCs, a significantly enhanced energy transfer efficiency that is up to 80% has been achieved. Figure 3.15 shows a schematic of their hybrid white LED. The nanohole arrays are fabricated by a nano-patterning process using a UV-curing NIL technique and then a subsequent dry etching process. The nanohole arrays are designed to be 300 nm in a diameter with a hexagonal lattice of 600 nm. The CdSe/ZnS core/shell NCs are filled into the nanoholes. Figure 3.16 provides detailed characteristics of their hybrid white LEDs, such as I–V curves, E–I curves, and EL spectra. The I–V characteristics show a little higher forward bias at 20 mA implying some damages induced during the dry etching process. By optimizing the fabrication process, the hybrid white LED demonstrates very good performance in terms of both color quality and optical efficiency. For example, a hybrid violet LED/NCs white LED demonstrates a CRI of 80 with *a color temperature of* 3888 K, and a hybrid blue LED/NCs white LED exhibits a CRI of 82 with a color temperature of 3318 K.

3.3.2   Fabrication of III-Nitrides/Organic Down-Conversion Material White LEDs

Compared with existing phosphors and down-conversion NCs, a light-emitting polymer as a down-conversion material has a number of advantages. It can cover the whole visible spectrum and can have significantly higher photoluminescence efficiencies. Due to an intrinsically large Stokes shift that can

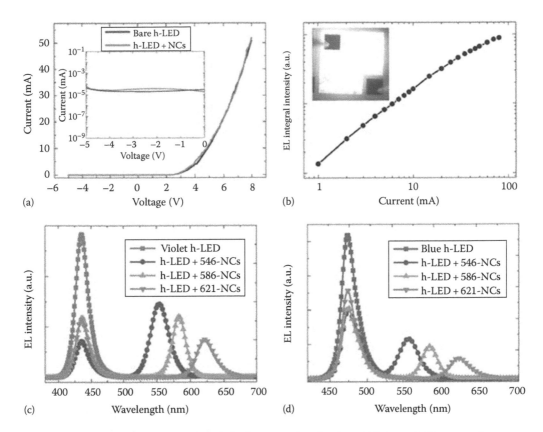

FIGURE 3.16 (a) I–V characteristics of the hybrid LEDs with and without NCs. Inset: I–V curve under a reverse bias; (b) EL-integrated intensity versus injection current. Inset: EL image of the LED; (c) EL spectra of the hybrid LED using a violet LED; (d) EL spectra of the hybrid LED using blue LED with different kinds of NCs. (From Zhuang, Z., Guo, X., Liu, B., Hu, F., Li, Y., Tao, T., Dai, J. et al.: *Adv. Funct. Mater.* 26. 36. 2015. Copyright Wiley-VCH Verlag GmbH & Co. KGaA. Reproduced with permission.)

be around 100 nm or even larger for a lighting polymer, the self-absorption issue can be eliminated. Furthermore, unlike the existing phosphors that are normally prepared in a form of grains with a typical size of tens of micrometers, lighting polymers can be dissolved in a solvent, which can be used to obtain homogeneous microstructures and also perfectly matches standard spin coating techniques widely used for the fabrication of inorganic or organic semiconductor optoelectronics. Therefore, they can simplify the process for the fabrication of white LEDs, which is particularly attractive to industry.

Of course, the major advantage for such a hybrid white LED using a lighting polymer as a down-conversion material is due to the nonradiative Förster energy transfer, namely, nonradiative resonant energy transfer (NRET),[89–99] which cannot be achieved using the existing phosphors. For such a NRET process, the distance between the InGaN-emitting layer and the proximal polymer is critical.[89–91] The nonradiative energy transfer rate labeled as Γ can be simply described as $\Gamma \sim R^{-4}$, where R is the distance between InGaN and polymer,[89–91] and the typical separation between donor and acceptor excitons needs to be less than 10 nm in order to achieve an efficient conversion efficiency.[89–92] Previous attempts to employ nonradiative RET have sacrificed the performance of InGaN blue emitters, where a lighting polymer is directly deposited on a low efficiency single quantum well structure with a thin capping layer of only several nanometers in order to reduce the separation between the polymer and InGaN layer.[89–97] However, in practical applications, an InGaN/GaN MQW structure (emitting region) with a p-type GaN capping layer of ~200 nm is required for a standard LED. Very recently, the author's group developed a nanofabrication

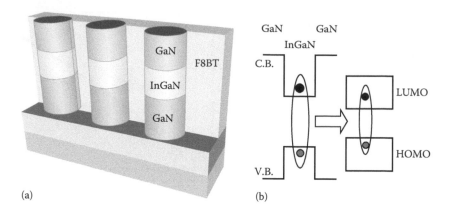

FIGURE 3.17　(a) Schematics of the hybrid structure, showing InGaN/GaN blue MQW nanorod arrays coated with the F8BT and (b) schematics of the nonradiative RET coupling between the InGaN/GaN MQWs and the F8BT. (Reprinted with permission from Smith, R., Liu, B., Bai, J., and Wang, T., *Nano Lett.*, 13, 3402. Copyright 2013 American Chemical Society.)

technique, enabling to minimize the separation of donor and acceptor excitons but without compromising InGaN/GaN LED structure.[100]

Figure 3.17a presents a schematic of the hybrid structure, showing that arrays of nanorods, on a scale of 100s of nm, are fabricated into an InGaN/GaN blue emitting MQW structure and surrounded by a down-conversion polymer. The light-emitting polymer material used is poly [(9,9-dioctylfluorenyl-2,7-diyl)-alt-co-(1,4-benzo-{2,1',3}-thiadiazole)] (F8BT), a yellow emitting polyfluorene co-polymer commercially available. In theory, such a hybrid configuration could allow the separation to approach zero, and the NRET can be massively enhanced, leading to a significantly improved efficiency of down conversion for yellow emission. Figure 3.17b illustrates a schematic of the nonradiative energy transfer between the InGaN/GaN MQWs and the F8BT light-emitting polymer: the excitons optically generated in the MQW resonantly Förster transfer their energy to the singlet excited states of the F8BT if the excitonic energy of the InGaN/GaN MQW can match the absorption energy of the F8BT, namely, the PL energy of the InGaN/GaN MQWs is coincident with the absorption maximum of the F8BT.

The InGaN/GaN nanorod array structure used was fabricated on a standard InGaN/GaN MQW sample grown on double side polished c-plane sapphire. Figure 3.18a shows the PL spectra of the InGaN/GaN MQW nanorod sample and the as-grown sample. Figure 3.18b shows the absorption and emission spectra of the F8BT film. The main absorption peak of the polymer is broad and centered at 460 nm, closely matching the emission peak of the InGaN/GaN nanorod array structure at 450 nm. A second, higher energy absorption peak at 350 nm is also present. In addition, the emission band is broad with a peak at 540 nm and exhibits a long and low energy tail giving an overall yellow emission appearance.

Figure 3.19 shows a typical cross-sectional SEM image of the hybrid structure. The samples were then transferred to an optical cryostat where they were held at a vacuum of 10^{-6} Torr (using a turbo pump) during all optical measurements in order to eliminate any photo-oxidation effect. Figure 3.20a shows the room-temperature PL spectra of the hybrid sample and the bare InGaN/GaN nanorod array structure, clearly demonstrating that there is a 64% reduction in the emission from the hybrid sample at 450 nm compared with the bare InGaN/GaN nanorod structure. The quenched emission has been transferred to the hybrid sample as a result of the nonradiative RET process from the InGaN/GaN MQWs to the F8BT, as expected.

Further investigation of the nonradiative NRET includes time-resolved PL (TRPL) measurements that were used to investigate the recombination dynamics of the InGaN/GaN excitons in these two samples. Time-correlated single photon counting (TCSPC) was used to measure the TRPL of the InGaN/GaN MQWs. A pulsed 375 nm diode laser with a pulse width of 83 ps was used as an excitation source. Figure 3.20b shows the TRPL traces of the hybrid sample and the bare nanorod structure, both measured

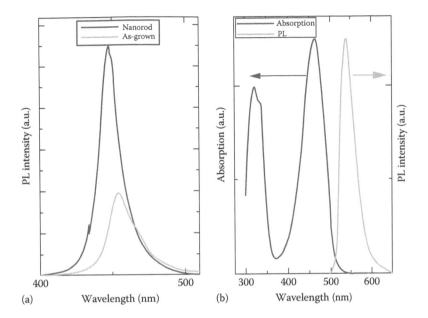

FIGURE 3.18 (a) PL spectra of the as-grown InGaN MQW epiwafer and the InGaN/GaN MQW nanorod array structure and (b) absorption and PL spectra of the F8BT. (Reprinted with permission from Smith, R., Liu, B., Bai, J., and Wang, T., *Nano Lett.*, 13, 3402. Copyright 2013 American Chemical Society.)

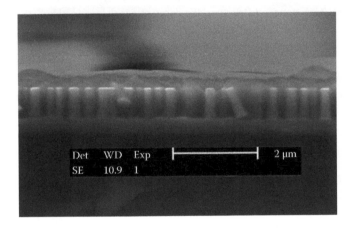

FIGURE 3.19 Cross-sectional SEM image of our hybrid sample, showing the InGaN/GaN MQW nanorod arrays coated with the F8BT. (Reprinted with permission from Smith, R., Liu, B., Bai, J., and Wang, T., *Nano Lett.*, 13, 3402. Copyright 2013 American Chemical Society.)

at room temperature under identical conditions. A standard two-exponential component model is used to study excitonic dynamics, and thus TRPL traces [$I(t)$] can be described by[101,102]

$$I(t) = a_1 \exp\left(-t/\tau_1\right) + a_2 \exp(-t/\tau_2)$$ (3.1)

where
a_1 and a_2 are constants
τ_1 and τ_2 are the decay lifetimes of the two exponential components and the decay rate is equal to $k_x = 1/\tau_x$

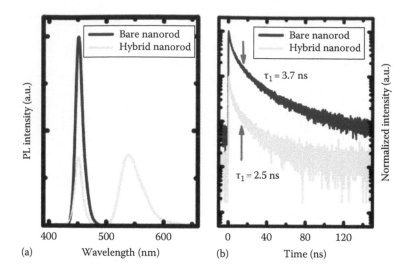

FIGURE 3.20   (a) PL spectra of the bare InGaN/GaN nanorod sample and the hybrid sample at room temperature and (b) TRPL traces of the InGaN/GaN MQWs emission from the bare InGaN/GaN nanorod sample and the hybrid sample. (Reprinted with permission from Smith, R., Liu, B., Bai, J., and Wang, T., *Nano Lett.*, 13, 3402. Copyright 2013 American Chemical Society.)

The bi-exponential fit comprises a fast decay component τ_1 describing de-localized exciton recombination and a slow decay component τ_2 describing communication between localized states.[101]

For the bare InGaN/GaN nanorod structure, the decay rate is determined by

$$k_{MQW} = k_r + k_{nr} \tag{3.2}$$

where k_{MQW}, k_r, and k_{nr} are total, radiative, and nonradiative decay rates of the MQW, respectively.

For the hybrid sample, due to an additional channel for the InGaN excitons nonradiatively and resonantly transferring their energy to the F8BT excitons, the decay rate is modified into

$$k_{hyb} = k_r + k_{nr} + k_{ET} \tag{3.3}$$

where k_{ET} is the energy transfer rate.

The decay lifetime obtained by fitting using Equation (3.1) is 2.5 ns for the hybrid sample, which is 32.4% faster than the decay lifetime τ_{MQW} of the bare InGaN/GaN nanorod structure, where τ_{MQW} is 3.7 ns. These values correspond to decay rates of $k_{hyb} = 0.4$ ns^{-1} and $k_{MQW} = 0.27$ ns^{-1}, giving an energy transfer rate of $k_{ET} = 0.13$ ns^{-1}, which can be obtained through Equations (3.2) and (3.3).

The values obtained in the preceding text allow us to estimate the efficiency η_{ET} of the nonradiative RET process given by Equation 3.4 as follows[95]:

$$\eta_{ET} = \frac{k_{ET}}{k_{ET} + k_{MQW}} \tag{3.4}$$

where k_{ET} and k_{MQW} have the same meaning stated as earlier. From Equation 3.4, an efficiency of $\eta_{ET} = 33\%$ can be obtained. Please note that this value does not take into account the fact that only part of the nanorod diameter actually makes contribution to η_{ET} due to the short range of the nonradiative RET process.

Many factors determine the energy transfer range such as dipole orientations and ordering,[103] exciton localization,[95] the wave vectors of MQW Wannier–Mott excitons and dielectric properties of any barrier materials between MQW and polymer excitons.[104] The exact determination of all of these factors in the novel nanostructure with complex geometries is very complicated, but the upper limit is generally found to be in the order of 5–10 nm.[89–92] Taking the maximum possible range for a NRET process as 10 nm, it is reasonable to consider the nanorod structures with a diameter of 220 nm in terms of two regions: in an outer region where the MQW/F8BT exciton separation is within 10 nm the NRET process can take place, while an inner region where the separation is greater than 10 nm leads to negligible possibility for the NRET to take place. If the relative volume contributions of these regions are corrected for, it is possible to determine the NRET rate from this outer region of the nanorods as the only contributor to the change in MQW lifetime. The relative volume leads to a correction factor of 5.76, giving a corrected energy transfer rate of k_{ET} =0.76 ns^{-1}. This value then gives a NRET efficiency of 73% where only the outer 10 nm of a typical nanorod contributes to the NRET process. This gives us an image how the device works: the inner parts of the nanorods do not contribute the NRET process and still remain emitting blue light with a high efficiency; and the outer parts of the nanorods, where the InGaN/GaN MQW recombination process is dominated by a high-speed NRET from the InGaN/GaN MQW to the F8BT, leading to an enhanced yellow emission. This perfectly matches the requirements for the fabrication of advanced white LEDs.

It would be interesting to compare the nonradiative decay rate in the bare InGaN/GaN nanorod structure with the nonradiative RET rate in the hybrid sample. The radiative and nonradiative decay rates, k_r and k_{nr}, which are $1/\tau_r$ and $1/\tau_{nr}$, respectively, in the bare InGaN/GaN nanorod structure, can be obtained through a combination of τ_{PL} and IQE (η_{int}) based on the following relations[102]:

$$\eta_{int} = \frac{1}{\left(1 + \tau_r/\tau_{nr}\right)} \tag{3.5}$$

$$\frac{1}{\tau_{PL}} = \frac{1}{\tau_r} + \frac{1}{\tau_{nr}} \tag{3.6}$$

IQE can be typically obtained by using the ratio of the integrated PL intensity at a low temperature (12 K) and room temperature.

As a consequence, the radiative and nonradiative rates in the bare InGaN/GaN nanorod structure are obtained as k_r = 0.01 ns^{-1} and k_{nr} = 0.26 ns^{-1}, respectively. Amazingly, the NRET rate of 0.76 ns^{-1} obtained earlier is approximately three times higher than the InGaN/GaN MQW nonradiative decay rate. This highlights the dominance of the NRET mechanism in the hybrid sample. As a result, it is highly likely that the generated excitons that otherwise would have decayed nonradiatively in the InGaN/GaN MQWs undergo a fast NRET process to the light-emitting polymer, leading to minimizing the overall nonradiative recombination losses in the device and therefore further enhancing the overall device performance.

To further confirm the nature of the NRET mechanism found in our hybrid sample, another set of samples was fabricated using green emitting (530 nm) InGaN/GaN MQW nanorod structures, where the spectral overlap of the green InGaN/GaN MQW emission and the F8BT absorption is extremely low, giving a negligible probability for any NRET process to take place. Similarly, one bare InGaN/GaN nanorod structure is used as a reference and another is fabricated into a hybrid structure using the F8BT. It is expected that we will not observe a nonradiative RET process in this particular hybrid sample fabricated using the green InGaN nanorod structure. This has also been proved by TRPL measurements, showing that the decay lifetime τ_{hyb} and τ_{MQW} from the hybrid sample and the bare InGaN/GaN nanorod structure obtained are very similar, namely, 14 and 13 ns, respectively. The tiny difference is within the error bar.

Further improving the optical performance needs to address the significantly enhanced surface-state issue that has become an even greater concern in the fabrication of hybrid LEDs due to the extremely short range of NRET coupling. Therefore, an extra passivation process is required.

Figure 3.21a provides a schematic illustration of a passivated hybrid InGaN/GaN MQW nanorod/F8BT device coated with thin silicon nitride as a passivation layer. In order to optimize the passivation process,

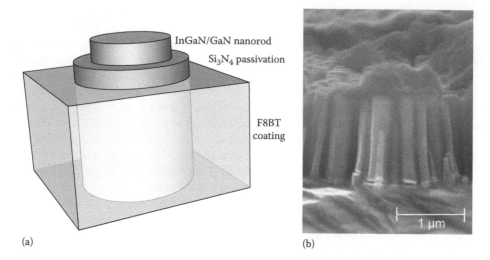

InGaN/GaN nanorod

Si$_3$N$_4$ passivation

F8BT
coating

(a)   (b)

FIGURE 3.21   (a) Schematics of the passivated hybrid InGaN/GaN MQW nanorod/F8BT structures with silicon nitride; (b) cross-sectional SEM image of cleaved GaN nanorods coated with a silicon nitride passivation layer. (Reprinted with permission from Smith, R.M., Athanasiou, M., Bai, J., Liu, B., and Wang, T., *Appl. Phys. Lett.*, 107, 121108 Copyright 2015, American Institute of Physics.)

silicon nitride layers of varying thickness were deposited on the sidewalls of the nanorods by PECVD at 300°C using silane and ammonia. The thickness of the coating was controlled by varying the deposition time, from 0 (reference sample) to 30, 60, 90, 120, 180, and 360 s, respectively. The growth rate of PECVD silicon nitride films is not expected to be linear with deposition time for short deposition times.

Figure 3.21a shows a typical cross-sectional SEM image of cleaved nanorod samples with silicon nitride films.[105] Due to the challenges associated with an accurate measurement of silicon nitride thicknesses on the 3D nanorod structures especially at low thicknesses, deposition time is used to describe silicon nitride thickness. It should be noted that this is not a linear relationship.

To investigate the passivation effect on hybrid organic/inorganic nanostructures, the samples were then cleaved into two parts, one having F8BT deposited on the top surface by spin coating from 10 mg/mL toluene solution, the other having a PMMA layer spin coated on the surface as a reference sample where there is no NRET. The polymer deposition was carried out in an air-free chamber to avoid any photo-oxidation of the F8BT. The PMMA coating on the control samples was used as an intermediate refractive index material (n = 1.5) with negligible optical absorption in the blue spectral range ruling out the possibility of NRET from the MQW to the PMMA. All samples were then transferred into an optical cryostat and held at a vacuum of 10^{-6} mBar to prevent photo-oxidation during optical characterization.

Figure 3.22a and b shows the PL spectra and TRPL traces of the unpassivated and passivated samples measured at room temperature, respectively. A clear increase in the PL intensity has been observed on the passivated sample compared to the unpassivated sample. The corresponding TRPL traces indicate that the passivated MQW sample exhibits a longer decay time than the unpassivated sample. Both of these changes indicate that the passivation process has a significant effect on the excitonic recombination processes of the InGaN/GaN MQW nanorod, improving the optical properties. The TRPL was measured at the MQW emission peak. The TRPL data were fitted with a bi-exponential function using Equation (3.1) discussed earlier.

Figure 3.22b shows the TRPL decay lifetimes of the PMMA-coated control samples as a function of silicon nitride deposition time. A general trend is clearly visible with the unpassivated sample having the fastest decay lifetime of the full set of samples. With increasing silicon nitride deposition time, the MQW PL decay lifetime increases and then remains stable for further increasing silicon nitride passivation layer thickness.

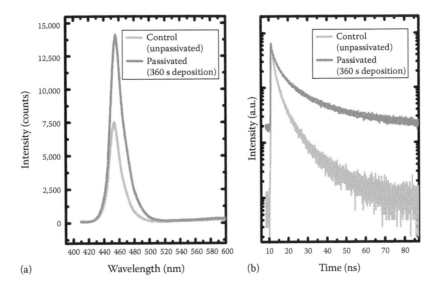

(a) Wavelength (nm) (b) Time (ns)

FIGURE 3.22 (a) PL spectrum of unpassivated and passivated (after 360 s of silicon nitride deposition) MQW nanorod arrays; (b) corresponding TRPL traces of the unpassivated and passivated samples. (Reprinted with permission from Smith, R.M., Athanasiou, M., Bai, J., Liu, B., and Wang, T., *Appl. Phys. Lett.*, 107, 121108 Copyright 2015, American Institute of Physics.)

The PL lifetime for the nanorod structures with exposed MQW sidewall surfaces (i.e., unpassivated sample), there is an additional surface recombination lifetime component labeled as $\tau_{surface}$ as follows:

$$\frac{1}{\tau_{pl}} = \frac{1}{\tau_r} + \frac{1}{\tau_{nr}} + \frac{1}{\tau_{surface}} \tag{3.7}$$

By passivating the surface, the surface recombination channel is blocked, leading to an increase in PL lifetime. Therefore, Figure 3.23 confirms that the silicon nitride deposited is effective at passivating surface recombination in the InGaN MQW nanorods.

To investigate the effect of surface passivation on the NRET process, the TRPL measurements have been performed on the corresponding hybrid (F8BT coated) and control (PMMA coated) MQW nanorod samples with varying passivation layer thicknesses. The MQW TRPL decay lifetime of the hybrid and corresponding control samples was then compared to study the NRET characteristics. As the NRET process acts as an additional nonradiative recombination process in the MQW, the decay lifetime of the hybrid MQW sample is expected to be faster than that of the control MQW decay as a result of the NRET process. The NRET rate can then be calculated as the difference between the hybrid and control decay rates[100,106]:

$$k_{NRET} = k_{hybrid} - k_{Control} \tag{3.8}$$

where $\tau_{PL} = 1/k_{PL}$ as previously demonstrated.[100,106]

Figure 3.24a presents the NRET rates of the full set of samples as a function of the thickness of silicon nitride passivation layer termed as deposition time. For the sample without any silicon nitride passivation layer between the MQW nanorods and F8BT, the NRET rate is typically 0.06 ns^{-1}. The NRET rate increases by a factor of 4 times with increasing deposition time of silicon nitride to 30 s, then decreases and finally remains close to zero with further increasing silicon nitride deposition time. As stated previously, due to the nature of a near-field coupling, NRET is highly sensitive to the separation of excitons in the MQW and the F8BT coating. Consequently, if there were no passivation effect, the presence of

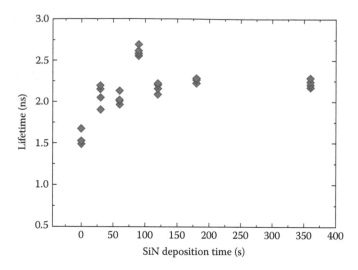

FIGURE 3.23 InGaN/GaN MQW nanorod PL lifetime versus silicon nitride passivation layer deposition time. (Reprinted with permission from Smith, R.M., Athanasiou, M., Bai, J., Liu, B., and Wang, T., *Appl. Phys. Lett.*, 107, 121108 Copyright 2015, American Institute of Physics.)

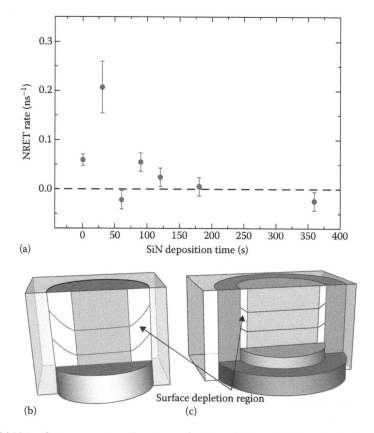

FIGURE 3.24 (a) Nonradiative energy transfer rate as a function of silicon nitride passivation interlayer deposition time; (b) schematics of the band bending and consequent surface depletion region of a nanorod; and (c) proposed reduction in surface depletion region following passivation. (Reprinted with permission from Smith, R.M., Athanasiou, M., Bai, J., Liu, B., and Wang, T., *Appl. Phys. Lett.*, 107, 121108 Copyright 2015, American Institute of Physics.)

an interlayer would purely lead to a reduction in NRET coupling rate as a result of an increase in the separation of MQW and F8BT excitons. As shown earlier in Figure 3.23, the thickness of silicon nitride with a deposition time of 360 s is >90 nm, thus precluding the possibility of NRET. For the samples with silicon nitride deposition times of greater than 30 s, the negligible NRET rate is in good agreement with the expected behavior. However, the significant increase in NRET rate for the sample with the thinnest silicon nitride passivation layer (i.e., 30 s) compared to the sample without passivation layer does not follow this trend.

To understand this increased NRET rate, the effect of surface states in III-nitrides must be considered. Using GaN nanowires, Calarco et al. demonstrated upward band bending at the surface of GaN nanowires due to Fermi level pinning.[107] This band bending leads to electron depletion at the surface of the nanowires. The InGaN MQW nanorods in this work are expected to display a similar surface electron depletion behavior at the surface for InGaN alloys with indium composition in the blue emission spectral region.[108] The presence of a surface depletion region will have a potentially significant effect on NRET in hybrid nanostructures, as the low exciton concentration close to the surface will reduce the number of excitons within the nanorod that are within the NRET coupling range of the F8BT surrounding the nanorod. This is illustrated schematically in Figure 3.24b. In order to explain the increased NRET rate for the sample with a 30 s deposition of silicon nitride compared to the unpassivated sample, it is proposed that the passivation of the surface allows the surface electron depletion to be reduced, shown schematically in Figure 3.24c. This increase in the concentration of excitons in the outer region of the nanorod reduces the effective separation of excitons between the InGaN MQWs and the F8BT at the surface, thus increasing the NRET rate. Further increase in the deposition time above 30 s results in a reduction in NRET rate compared to the unpassivated sample. From Figure 3.24a, it was seen that the passivation effect is almost unchanged for deposition times greater than 30 s, indicating that this reduction in NRET with further increasing thickness is only due to the increased separation between MQW and F8BT excitons.

3.4 Fabrication of Phosphor-Free White LEDs

It is the utmost challenge to achieve down-conversion material-free white LEDs, although a number of approaches have been proposed and attempted. So far, the optical performance of so-called phosphor-free white LEDs is still far from satisfactory. Ultimate white LEDs would require the highest color rendering index that can replicate the color temperature of the sun. Second, a high efficiency is required. Third, both color quality and optical efficiency need to remain unchanged with increasing injection current. This section will present a summary of recent progress on developing phosphor-free white LEDs using nanostructures. Hopefully, it can inspire some new ideas for further developing white LEDs.

A selective epitaxial growth would be a reasonable approach at the moment. Wang et al.[109] developed a three-step selective growth approach to obtaining nanowire LEDs with different emission wavelengths that can be achieved at each step, and eventually achieved a monolithic integration of blue, green/yellow, and orange/red InGaN nanowire LEDs on silicon. Figure 3.25 shows a schematic diagram of the three-step growth procedure. The basic idea is to take an advantage of using SiO_2 as a mask, which can be easily removed by an acid solution. A SiO_2 layer is initially deposited on a silicon substrate, and then windows are selectively opened. Subsequently, nanowire blue LEDs are grown. As the MBE growth of III-nitride LEDs takes place at a relatively low temperature compared with MOCVD, the nanowires will be grown on both the SiO_2 mask areas and the opening areas. However, the SiO_2 masks and the nanowire LEDs grown on the SiO_2 masks can be easily removed by an acid solution such as HF, leaving the nanowire blue LEDs grown on the window regions to remain unchanged. The second step is to repeat the procedure, but the nanowire green/yellow LEDs are grown instead of nanowire blue LEDs, and the final step is still to repeat the same procedure in order to grow the nanowire orange/red LEDs. Based on this method, multi-color nanowire LED arrays including red, green, and blue LEDs can be achieved on a single Si substrate.

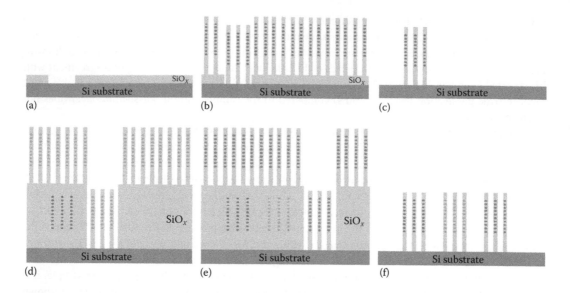

FIGURE 3.25 Schematic illustration of the three-step growth process. (a) Si substrate was patterned by a thin SiO_x layer; (b) first step: growth of blue nanowire LEDs were grown on the patterned substrate; (c) the SiO mask and the nanowires growth on the SiO_x top were selectively removed; (d) repeat the previous steps, and second step: growth of green nanowire LEDs; (d) repeat the previous step, and third step: growth of red nanowire LEDs; (e) remove all residual SiO_x. (Reproduced from Wang, R., Nguyen, H.P.T., Connie, A.T., Lee, J., Shih, I., and Mi, Z., *Opt. Express*, 22, A1768, 2014. With permission of Optical Society of America Publishing.)

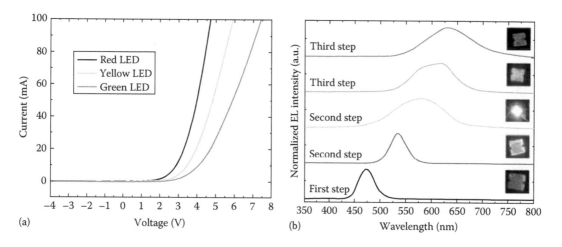

FIGURE 3.26 (a) I–V curves for red, yellow, and green nanowire LEDs; (b) EL spectra of these nanowire LEDs grown in each step. (Reproduced from Wang, R., Nguyen, H.P.T., Connie, A.T., Lee, J., Shih, I., and Mi, Z., *Opt. Express*, 22, A1768, 2014. With permission of Optical Society of America Publishing.)

Figure 3.26 shows the characteristics of the multiple-color LED array, showing their individual I–V curves and EL spectra.

Very recently, Lim et al. reported a selective overgrowth technique for the growth of multiple-color LEDs on a specially patterned template by MOCVD.[110] The original template is a standard GaN grown on *c-plane* sapphire. For patterning the template, a 30 nm Si_3N_4 layer was deposited on n-GaN templates and a standard lithography technique was employed to form a pattern as schematically

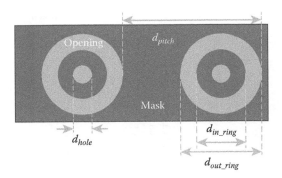

FIGURE 3.27 Top view of the designed mask pattern. (Reprinted by permission from Macmillan Publisher Ltd. *Light Sci. Appl.* Lim, S.H., Ko, Y.H., Rodriguez, C., Gong, S.H., and Cho, Y.H., 5, e16030 copyright 2016.)

shown in Figure 3.27, displaying that a concentric opening window consists of a central hole and a ring, where further overgrowth selectively takes place on the hole and ring areas. Due to the special pattern, the overgrown layer will consist of a number of facets with different orientations as indicated in Figure 3.28, which is a typical plan-view SEM image of the overgrown layer. As we know that even under identical growth conditions, InGaN/GaN MQWs on different facets will exhibit different growth rates and different indium content, leading to a wide visible spectrum and thus forming white emission. Figure 3.29 shows spatially resolved cathode luminesce (CL) spectra of the InGaN/GaN MQWs growth on the different facets, confirming that the emission wavelength depends on the orientation of the facets. The emission wavelength covers a wide spectral range from violet (at ~400 nm) to yellow (at ~560nm). Finally, based on this design, they have achieved a white LED with a color rendering index of 75.

There also exist a few other approaches to the fabrication of nanostructures.[111–113] Basically, these ideas are similar to the ones mentioned earlier, and they attempted to create a 3D structure that consists of a number of facets with different orientations. These different orientations lead to InGaN/GaN MQWs with different emission wavelengths grown on these facets, thus achieving a wide range of the visible spectrum that forms white light. A typical example would be to create a rough surface that forms multiple facets including semi-polar and nonpolar facets as shown in Figure 3.30,[114] leading to InGaN/GaN MQWs with different emission wavelengths grown on its top. Such a rough surface can be achieved by tuning growth conditions, such as growth temperature, V/III ratio, etc.

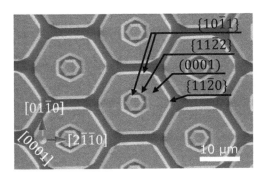

FIGURE 3.28 SEM image showing a formation of different facets after the overgrowth on the patterned substrates given in Figure 3.27. (Reprinted by permission from Macmillan Publisher Ltd. *Light Sci. Appl.* Lim, S.H., Ko, Y.H., Rodriguez, C., Gong, S.H., and Cho, Y.H., 5, e16030 copyright 2016.)

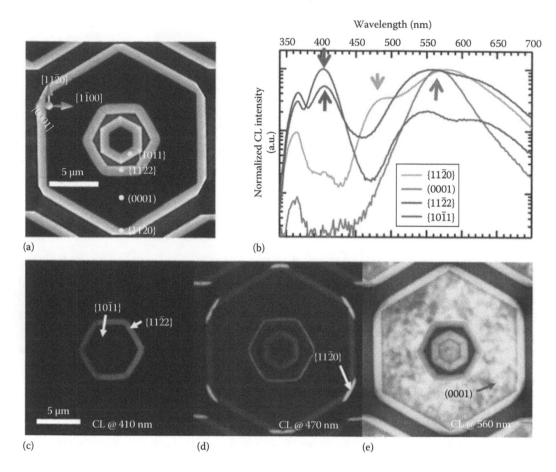

(a)   (b)   (c)   (d)   (e)

FIGURE 3.29 (a) Top-view SEM image of the overgrown sample; (b) normalized CL spectra measured on the different facets; and CL mapping images on the different facets, with emission wavelengths of 410 nm (c), 470 nm (d), and 560 nm (e). (Reprinted by permission from Macmillan Publisher Ltd. *Light Sci. Appl.* Lim, S.H., Ko, Y.H., Rodriguez, C., Gong, S.H., and Cho, Y.H., 5, e16030 copyright 2016.)

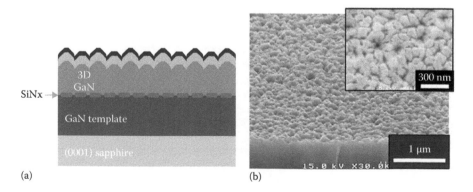

(a)   (b)

FIGURE 3.30 (a) Schematic illustration of different facets formed through roughing surface and (b) SEM image of a typical rough surface through tuning growth conditions. (Reprinted by permission from Macmillan Publisher Ltd. *Sci. Rep.* Min, D., Park, D., Jang, J., Lee, K., and Nam, O., 5, 17372, copyright 2015.)

3.5 Summary and Outlook

It is a great challenge to achieve white LEDs without involving any down-conversion materials. A number of ideas have been proposed, such as wafer bonding consisting of different emitters from different kinds of LED chips, a single wafer with different kinds of InGaN/GaN MQWs (either different InGaN well thickness or InGaN with different indium content), and a 3D structure with different facets. The wafer bonding approach would allow us to manage the output power of individual discrete wavelength LEDs through designing a proper electrical circuit, potentially providing an ideal approach to tuning color temperature in order to replicate the color temperature of the sun. However, the major issue is due to their very high costs and the currently missing green and yellow LEDs with high efficiency. The 3D structure approach potentially provides multiple facets allowing us to achieve InGaN/GaN MQWs with different wavelength, but it would be difficult to maintain a balance of the intensities from the different InGaN/GaN MQWs. A selective growth for an integration of multiple color LEDs in a controlled manner in a single wafer would be an ideal approach forward. Furthermore, real white light, for example, which is supposed to be similar to the natural white light from the sun, will have to consist of emission with continuous wavelengths covering a full visible spectral region. This would not only provide the highest color rendering index, but also be crucial for human health. In theory, InGaN alloys across their entire composition exhibit direct bandgaps covering a full range of visible spectrum. However, we understand that the optical efficiency of InGaN LEDs reduces significantly when the emission wavelength moves toward the yellow/red spectral region. Therefore, it would be worth considering including other III–V semiconductors. In this case, an integration of III-nitrides and other III–V semiconductors in a single wafer would be a new direction forward. In order to address this extremely challenging issue, new epitaxial growth technologies, in particular, selective overgrowth technologies would be required.

Furthermore, a future trend for developing white LEDs would be toward so-called smart LEDs, meaning that such LEDs will have to play some other roles in addition to general illumination, for example, visible light wireless communication. In this case, current "blue LED + yellow phosphor" white LEDs will not work efficiently due to their intrinsically long recombination lifetime of current blue LEDs and slow response time from phosphor.

In summary, considerable effort would need to be devoted to developing new technologies and generating new concepts to this extremely important area.

Acknowledgment

This work was supported by the UK Engineering and Physical Sciences Research Council (EPSRC) via Grant Nos. EP/M015181/1 and EP/L017024/1.

References

1. E. D. Jones, *Light Emitting Diodes for General Illumination*, Optoelectronics Industry Development Association (OIDA) (March 2001), http://educypedia.karadimov.info/.
2. www.marketsandmarkets.com/PressReleases/solid-state-lighting.asp.
3. http://ec.europa.eu/environment/chemicals/mercury/restriction_en.htm.
4. U.S. Department of Energy Solid-State Lighting Research and Development: Multi-Year Program Plan 68 (2012).
5. R. Vaxenburg, E. Lifshitz, and A. L. Efros, *Appl. Phys. Lett.* **102**, 031120 (2013).
6. K. J. Vampola, M. Iza, S. Keller, S. P. DenBaars, and S. Nakamura, *Appl. Phys. Lett.* **94**, 061116 (2009).
7. Y. R. Wu, R. Shivaraman, K. C. Wang, and J. S. Speck, *Appl. Phys. Lett.* **101**, 083505 (2012).
8. M. F. Schubert, S. Chhajed, J. K. Kim, E. F. Schubert, D. D. Koleske, M. F. Crawford, S. R. Lee, A. J. Fischer, G. Thaler, and M. A. Banas, *Appl. Phys. Lett.* **91**, 231114 (2007).
9. J. H. Oh, S. J. Yang, and Y. R. Do, *Light Sci. Appl.* **3**, e141 (2014).

10. Y. H. Ra, R. Wang, S. Y. Woo, M. Djavid, S. M. Sadaf, J. Lee, G. A. Botton, and Z. Mi, *Nano Lett.* **16**, 4608 (2016).

11. V. Cardin, L. I. Dion-Bertrand, P. Grégoire, H. P. T. Nguyen, M. Sakowicz, Z. Mi, C. Silva, and R. Leonelli, *Nanotechnology* **24**, 045702 (2013).

12. H. C. Wang, X. Y. Yu, Y. L. Chueh, T. Malinauskas, K. Jarasiunas, and S. W. Feng, *Opt. Express* **19**, 18893 (2011).

13. J. B. Schlager, K. A. Bertness, P. T. Blanchard, L. H. Robins, A. Roshko, and N. A. Sanford, *J. Appl. Phys.* **102**, 124309 (2008).

14. P. Parkinson, C. Dodson, H. J. Joyce, K. A. Bertness, N. A. Sanford, L. M. Herz, and M. B. Johnston, *Nano Lett.* **12**, 4600 (2012).

15. C. Pfüller, O. Brandt, F. Grosse, T. Flissikowski, C. Chèze, V. Consonni, L. Geelhaar, H. T. Grahn, and H. Riechert, *Phys. Rev. B* **82**, 045320 (2010).

16. N. Erhard, A. T. M. Golam Sarwar, F. Yang, D. W. McComb, R. C. Myers, and A. W. Holleitner, *Nano Lett.* **15**, 332 (2014).

17. M. Yoshizawa, A. Kikuchi, M. Mori, N. Fujita, and K. Kishino, *Jpn. J. Appl. Phys.* **36**, L459 (1997).

18. Y. H. Ra, R. Navamathavan, H. I. Yoo, and C. R. Lee, *Nano Lett.* **14**, 1537 (2014).

19. Y. J. Lu, H. W. Lin, H. Y. Chen, Y. C. Yang, and S. Gwo, *Appl. Phys. Lett.* **98**, 233101 (2011).

20. M. Tchernycheva, P. Lavenus, H. Zhang, A. V. Babichev, G. Jacopin, M. Shahmohammadi, F. H. Julien, R. Ciechonski, G. Vescovi, and O. Kryliouk, *Nano Lett.* **14**, 2456 (2014).

21. F. Qian, S. Gradečak, Y. Li, C. Y. Wen, and C. M. Lieber, *Nano Lett.* **5**, 2287 (2005).

22. L. Yan, S. Jahangir, S. A. Wight, B. Nikoobakht, P. Bhattacharya, and J. M. Millunchick, *Nano Lett.* **15**, 1535 (2015).

23. F. Qian, Y. Li, S. Gradečak, and D. Wang, *Nano Lett.* **4**, 1975 (2004).

24. Y. Hou, J. Bai, R. Smith, and T. Wang, *Nanotechnology* **27**, 205205 (2016).

25. S. Deshpande, J. Heo, A. Das, and P. Bhattacharya, *Nat. Commun.* **4**, 1675 (2013).

26. C. Hahn, Z. Zhang, A. Fu, C. H. Wu, Y. J. Hwang, D. J. Gargas, and P. Yang, *ACS Nano* **5**, 3970 (2011).

27. H. P. T. Nguyen, K. Cui, S. Zhang, M. Djavid, A. Korinek, G. A. Botton, and Z. Mi, *Nano Lett.* **12**, 1317 (2012).

28. S. M. Sadaf, Y. H. Ra, H. P. T. Nguyen, M. Djavid, and Z. Mi, *Nano Lett.* **15**, 6696 (2015).

29. J. Bai, Q. Wang, and T. Wang, *J. Appl. Phys.* **111**, 113103 (2012).

30. W. Guo, M. Zhang, A. Banerjee, and P. Bhattacharya, *Nano Lett.* **10**, 3355 (2010).

31. H. M. Kim, Y. H. Cho, H. Lee, S. I. Kim, S. R. Ryu, D. Y. Kim, T. W. Kang, and K. S. Chung, *Nano Lett.* **4**, 1059 (2004).

32. H. W. Lin, Y. J. Lu, H. Y. Chen, H. M. Lee, and S. Gwo, *Appl. Phys. Lett.* **97**, 073101 (2010).

33. H. P. T. Nguyen, S. Zhang, K. Cui, X. Han, S. Fathololoumi, M. Couillard, G. A. Botton, and Z. Mi, *Nano Lett.* **11**, 1919 (2011).

34. J. C. Johnson, H. J. Choi, K. P. Knutsen, R. D. Schaller, P. Yang, and R. J. Saykally, *Nat. Mater.* **1**, 106 (2002).

35. A. Das, J. Heo, M. Jankowski, W. Guo, L. Zhang, H. Deng, and P. Bhattacharya, *Phys. Rev. Lett.* **107**, 066405 (2011).

36. J. Y. Chen, T. M. Wong, C. W. Chang, C. Y. Dong, and Y. F. Chen, *Nat. Nanotechnol.* **9**, 845 (2014).

37. S. Gradečak, F. Qian, Y. Li, H. G. Park, and C. M. Lieber, *Appl. Phys. Lett.* **87**, 173111 (2005).

38. T. Frost, S. Jahangir, E. Stark, S. Deshpande, A. Hazari, C. Zhao, B. S. Ooi, and P. Bhattacharya, *Nano Lett.* **14**, 4535 (2014).

39. S. D. Hersee, X. Sun, and X. Wang, *Nano Lett.* **6**, 1808 (2006).

40. S. D. Hersee, M. Fairchild, A. K. Rishinaramangalam, M. S. Ferdous, L. Zhang, P. M. Varangis, B. S. Swartzentruber, and A. A. Talin, *Electron. Lett.* **45**, 75 (2009).

41. H. Sekiguchi, K. Kishino, and A. Kikuchi, *Appl. Phys. Express* **1**, 124002 (2008).

42. H. Sekiguchi, K. Kishino, and A. Kikuchi, *Appl. Phys. Lett.* **96**, 231104 (2010).

43. Y. Ou, D. Iida, A. Fadil, and H. Ou, *Int. J. Opt. Photonics Eng.* **1**, 001 (2016).

44. T. W. Yeh, Y. T. Lin, L. S. Stewart, P. D. Dapkus, R. Sarkissian, J. D. O'Brien, B. Ahn, and S. R. Nutt, *Nano Lett.* **12**, 3257 (2012).
45. J. R. Riley, S. Padalkar, Q. Li, P. Lu, D. D. Koleske, J. J. Wierer, G. T. Wang, and L. J. Lauhon, *Nano Lett.* **13**, 4317 (2013).
46. Y. H. Ra, R. Navamathavan, S. Kang, and C. R. Lee, *J. Mater. Chem. C* **2**, 2692 (2014).
47. H. Liao, W. M. Chang, H. S. Chen, C. Y. Chen, Y. F. Yao, H. T. Chen, C. Y. Su, S. Y. Ting, Y. W. Kiang, and C. C. Yang, *Opt. Express* **20**, 15859 (2012).
48. L. Rigutti, G. Jacopin, L. Largeau, E. Galopin, A. De Luna Bugallo, F. Julien, J. C. Harmand, F. Glas, and M. Tchernycheva, *Phys. Rev. B* **83**, 155320 (2011).
49. O. Landré, D. Camacho, C. Bougerol, Y. M. Niquet, V. Favre-Nicolin, G. Renaud, H. Renevier, and B. Daudin, *Phys. Rev. B* **81**, 153306 (2010).
50. V. Consonni, V. G. Dubrovskii, A. Trampert, L. Geelhaar, and H. Riechert, *Phys. Rev. B* **85**, 155313 (2012).
51. R. Calarco, R. J. Meijers, R. K. Debnath, T. Stoica, E. Sutter, and H. Lüth, *Nano Lett.* **7**, 2248 (2007).
52. R. K. Debnath, R. Meijers, T. Richter, T. Stoica, R. Calarco, and H. Lüth, *Appl. Phys. Lett.* **90**, 123117 (2007).
53. J. Johansson, C. P. Svensson, T. Martensson, L. Samuelson, and W. Seifert, *J. Phys. Chem. B* **109**, 13567 (2005).
54. P. Vajpeyi, A. Georgakilas, G. Tsiakatouras, K. Tsagaraki, M. Androulidaki, S. J. Chua, and S. Tripathy, *Phys. E Low Dimens. Syst. Nanostruct.* **41**, 427 (2009).
55. J. B. Schlager, K. A. Bertness, P. T. Blanchard, L. H. Robins, A. Roshko, and N. A. Sanford, *J. Appl. Phys.* **103**, 124309 (2008).
56. J. Renard, R. Songmuang, G. Tourbot, C. Bougerol, B. Daudin, and B. Gayral, *Phys. Rev. B* **80**, 121305(R) (2009).
57. O. Brandt, C. Pfüller, C. Chèze, L. Geelhaar, and H. Riechert, *Phys. Rev. B* **81**, 045302 (2010).
58. G. Tourbot, C. Bougerol, A. Grenier, M. Den Hertog, D. Sam-Giao, D. Cooper, P. Gilet, B. Gayral, and B. Daudin, *Nanotechnology* **22**, 075601 (2011).
59. X. Zhang, V. G. Dubrovskii, N. V. Sivirev, and X. Ren, *Cryst. Growth Des.* **11**, 5441 (2011).
60. A. Kikuchi, M. Kawai, M. Tada, and K. Kishino, *Jpn. J. Appl. Phys.*, **43**, L1524 (2004).
61. M. D. Brubaker, P. T. Blanchard, J. B. Schlager, A. W. Sanders, A. M. Herrero, A. Roshko, S. M. Duff et al., *J. Electron. Mater.* **42**, 868 (2013).
62. W. Guo, M. Zhang, P. Bhattacharya, and J. Heo, *Nano Lett.* **11**, 1434 (2011).
63. Q. Li, K. R. Westlake, M. H. Crawford, S. R. Lee, D. D. Koleske, J. J. Figiel, K. C. Cross, S. Fathololoumi, Z. Mi, and G. T. Wang, *Opt. Express* **19**, 25528 (2011).
64. W. Guo, A. Banerjee, P. Bhattacharya, and B. S. Ooi, *Appl. Phys. Lett.* **98**, 193102 (2011).
65. J. Zhu, J. Wang, S. Zhang, H. Wang, D. Zhao, J. Zhu, Z. Liu, D. Jiang, and H. Yang, *J. Appl. Phys.* **108**, 074302 (2010).
66. Y. Lee, C. Lee, C. Chen, T. Lu, and H. Kuo, *IEEE J. Sel. Top. Quantum Electron.* **99**, 1 (2010).
67. H. Chiu, T. C. Lu, H. W. Huang, C. F. Lai, C. C. Kao, J. T. Chu, C. C. Yu, H. C. Kuo, S. C. Wang, and C. F. Lin, *Nanotechnology* **18**, 445201 (2007).
68. A. Kikuchi, M. Tada, K. Miwa, and K. Kishino, *Proc. SPIE* **6129**, 612905-1 (2006).
69. H. Hou, S. Z. Tseng, C. H. Chan, T. J. Chen, H. T. Chien, F. L. Hsiao, H. K. Chiu, C. C. Lee, Y. L. Tsai, and C. C. Chen, *Appl. Phys. Lett.* **95**, 133105 (2009).
70. H. K. Park, J. H. Moon, S. Yoon, and Y. Rag Do, *J. Electrochem. Soc.* **158**, J143 (2011).
71. P. Dong, J. Yan, Y. Zhang, J. Wang, C. Geng, H. Zheng, X. Wei, Q. Yan, and J. Li, *Opt. Express* **22**, A320 (2014).
72. H. Zheng and K. Wu, *ECS J. Solid State Sci. Technol.* **2**, R241 (2013).
73. M. Athanasiou, T. K. Kim, B. Liu, R. Smith, and T. Wang, *Appl. Phys. Lett.* **102**, 191108 (2013).
74. T. Kim, B. Liu, R. Smith, M. Athanasiou, Y. Gong, and T. Wang, *Appl. Phys. Lett.* **104**, 161108 (2014).

75. H. W. Huang, H. C. Kuo, J. T. Chu, C. F. Lai, C. C. Kao, T. C. Lu, S. C. Wang, R. J. Tsai, C. C. Yu, and C. F. Lin, *Nanotechnology* **17**, 2998 (2006).
76. Z. Zhuang, X. Guo, G. Zhang, B. Liu, R. Zhang, T. Zhi, T. Tao et al., *Nanotechnology* **24**, 405303 (2013).
77. P. Renwick, H. Tang, J. Bai, and T. Wang, *Appl. Phys. Lett.* **100**, 182105 (2012).
78. D. J. Kong, S. Bae, C. Kang, and D. Lee, *Opt. Express* **21**, 22320 (2013).
79. H. Chiu, M. H. Lo, T. C. Lu, P. Yu, H. W. Huang, H. C. Kuo, and S. C. Wang, *J. Lightwave Technol.* **26**, 1445 (2008).
80. Q. Wang, J. Bai, Y. Gong, and T. Wang, *Phys. Status Solidi C* **9**, 620 (2012).
81. B. Liu, R. Smith, J. Bai, Y. Gong, and T. Wang, *Appl. Phys. Lett.* **103**, 101108 (2013).
82. Q. Zhang, G. Li, X. Liu, F. Qian, Y. Li, T. C. Sum, C. M. Lieber, and Q. Xiong, *Nat. Commun.* **5**, 4953 (2014).
83. Y. Wu, C. T. Kuo, C. Y. Wang, C. L. He, M. H. Lin, H. Ahn, and S. Gwo, *Nano Lett.* **11**, 4256 (2011).
84. Y. J. Lu, C. Y. Wang, J. Kim, H. Y. Chen, M. Y. Lu, Y. C. Chen, W. H. Chang et al., *Nano Lett.* **14**, 4381 (2014).
85. Y. Hou, P. Renwick, B. Liu, J. Bai, and T. Wang, *Sci. Rep.* **4**, 5014 (2014).
86. S. Chanyawadee, P. G. Lagoudakis, R. T. Harley, M. D. B. Charlton, D. V. Talapin, H. W. Huang, and C. H. Lin, *Adv. Mater.* **22**, 602 (2009).
87. B. Jiang, C. Zhang, X. Wang, M. J. Park, J. S. Kwak, J. Xu, H. Zhang, J. Zhang, F. Xue, and M. Xiao, *Adv. Funct. Mater.* **22**, 3146 (2012).
88. Z. Zhuang, X. Guo, B. Liu, F. Hu, Y. Li, T. Tao, J. Dai et al., *Adv. Funct. Mater.* **26**, 36 (2015).
89. M. Achermann, M. A. Petruska, S. Kos, D. L. Smith, D. D. Koleske, and G. C. Klimov, *Nature* **429**, 642 (2004).
90. S. Kos, M. Achermann, V. I. Klimov, and D. L. Smith, *Phys. Rev. B* **71**, 205309 (2005).
91. V. M. Agranovich, D. M. Basko, G. C. La Rocca, and F. Bassani, *J. Phys. Condens. Matter* **10**, 9369 (1998).
92. M. Achermann, M. A. Petruska, D. D. Koleske, M. H. Crawford, and V. I. Klimov, *Nano Lett.* **6**, 1396 (2006).
93. S. Blumstengel, S. Sadofev, C. Xu, J. Puls, and F. Henneberger, *Phys. Rev. Lett.* **97**, 237401 (2006).
94. G. Heliotis, G. Itskos, R. Murray, M. D. Dawson, I. M. Watson, and D. D. C. Bradley, *Adv. Mater.* **18**, 334 (2006).
95. J. Rindermann, G. Pozina, B. Monemar, L. Hultman, H. Amano, and P. Lagoudakis, *Phys. Rev. Lett.* **107**, 236805 (2011).
96. G. Itskos, C. R. Belton, G. Heliotis, I. M. Watson, M. D. Dawson, R. Murray, and D. D. C. Bradley, *Nanotechnology* **20**, 275207 (2009).
97. R. Belton, G. Itskos, G. Heliotis, P. N. Stavrinou, P. G. Lagoudakis, J. Lupton, S. Pereira et al., *J. Phys. D Appl. Phys.* **41**, 094006 (2008).
98. D. L. Andrews, *Chem. Phys.* **135**, 195 (1989).
99. L. Andrews and D. S. Bradshaw, *Eur. J. Phys.* **25**, 845 (2004).
100. R. Smith, B. Liu, J. Bai, and T. Wang, *Nano Lett.* **13**, 3402 (2013).
101. J. H. Na, R. A. Taylor, K. H. Lee, T. Wang, A. Tahraoui, P. Parbrook, M. Fox et al., *Appl. Phys. Lett.* **89**, 253120 (2006).
102. S. F. Chichibu, K. Hazu, Y. Ishikawa, M. Tashiro, H. Namita, S. Nagao, K. Fujito, and A. Uedono, *J. Appl. Phys.* **111**, 103518 (2012).
103. J. Hill, S. Heriot, O. Worsfold, T. Richardson, A. Fox, and D. Bradley, *Phys. Rev. B* **69**, 041303(R) (2004).
104. V. M. Agranovich, Y. N. Gartstein, and M. Litinskaya, *Chem. Rev.* **111**, 5179 (2011).
105. R. M. Smith, M. Athanasiou, J. Bai, B. Liu, and T. Wang, *Appl. Phys. Lett.* **107**, 121108 (2015).
106. R. M. Smith, B. Liu, J. Bai, and T. Wang, *Appl. Phys. Lett.* **105**, 171111 (2014).

107. R. Calarco, M. Marso, T. Richter, A. I. Aykanat, R. Meijers, A. V. D. Hart, T. Stoica, and H. Lüth, *Nano Lett.* **5**, 981 (2005).
108. T. D. Veal, P. H. Jefferson, L. F. J. Piper, C. F. McConville, T. B. Joyce, P. R. Chalker, L. Considine, H. Lu, and W. J. Schaff, *Appl. Phys. Lett.* **89**, 202110 (2006).
109. R. Wang, H. P. T. Nguyen, A. T. Connie, J. Lee, I. Shih, and Z. Mi, *Opt. Express* **22**, A1768 (2014).
110. S. H. Lim, Y. H. Ko, C. Rodriguez, S. H. Gong, and Y. H. Cho, *Light Sci. Appl.* **5**, e16030 (2016).
111. H. Zhang, G. Jacopin, V. Neplokh, L. Largeau, F. H. Julien, O. Kryliouk, and M. Tchernycheva, *Nanotechnology* **26**, 465203 (2015).
112. Ž. Gačević, N. Vukmirović, N. García-Lepetit, A. Torres-Pardo, M. Müller, S. Metzner, S. Albert et al., *Phys. Rev. B* **93**, 125436 (2016).
113. Y. H. Ko, J. Song, B. Leung, J. Han, and Y. H. Cho, *Sci. Rep.* **4**, 5514 (2014).
114. D. Min, D. Park, J. Jang, K. Lee, and O. Nam, *Sci. Rep.* **5**, 17372 (2015).

4

III-Nitride Deep-Ultraviolet Materials and Applications

Jianwei Ben
Chinese Academy of Sciences

Xiaojuan Sun
Chinese Academy of Sciences

Dabing Li
Chinese Academy of Sciences

Abstract   AlGaN-based materials have been recognized as one of the most promising materials for optical devices in the short-wavelength region because of the tunable direct bandgap from 3.4 to 6.2 eV by verifying the Al content from 0 to 1, as well as the thermal and chemical stability. After the two-step method was used to grow GaN-based materials, a high-quality GaN layer was obtained and several GaN-based photoelectric devices have been developed, including the GaN-based light-emitting diodes (LEDs), laser diodes (LDs), photodetectors (PDs), and so on. At present, even though GaN-based blue LEDs are available in the market, GaN-based photoelectronic devices in the deep ultraviolet (UV) region are still facing many challenges because of the relatively poor quality of AlGaN. Compared with GaN, the growth of AlGaN is more difficult due to the pre-reaction and short migration length of the Al atom. In addition, GaN buffer or substrate is not suitable for the growth of AlGaN since the strain in AlGaN will induce AlGaN to crack. Furthermore, the activation energy for Mg in AlGaN is higher than in GaN, resulting in the big problem of obtaining high-quality p-doped AlGaN, thus obstructing the application of AlGaN-based photoelectric devices. In this chapter, we will introduce the developments, challenges, and applications of AlGaN-based material and focus on the efforts that have gone into improving the quality of AlGaN.

4.1 Introduction

AlGaN-based materials own the following properties of wide tunable direct bandgap from 365 to 200 nm by tuning the Al composition from 0 to 1 as well as high electron mobility, high breakdown voltage, thermal and chemical stability, and so on. They are one of the best candidates for microelectronic, UV,

TABLE 4.1 Lattice Constant and Bandgap of AlN, GaN, and AlGaN

T = 300 K	AlN (W)	$Al_xGa_{1-x}N$	GaN (W)
a_0 (Å)	3.113		3.189
c_0 (Å)	4.982	$XC_{AlN} + (1 - X)C_{GaN}$[a]	5.185
Eg (eV)	6.2	Eg, AlN(X) + Eg, GaN(1 − X) − bX(1 − X)	3.4

[a] According to Vergard law.

and deep UV photo electronic devices, such as UV LED, UV LD, and UV PD. There are many applications of these UV and DUV photo devices. For example, the UV LED and UV LD can be used in high-density storage, anthrax agent detection, corrective eye surgery, water purification, semiconductor lithography, and so on, while DUV PDs can be used in fire detection, communication, military, and so on. Many of these applications could benefit from the compact and flexible form, high efficiency, and low heat output of AlGaN UV LED or UV LD. However, the performance of AlGaN-based photoelectronic and micro-electronic devices is still far from expectation mainly because of the quality of AlGaN materials.

Compared with the growth of GaN, obtaining high-quality AlGaN is more difficult. The mainly reasons are as follows: (1) The lack of a proper substrate: the AlGaN homo-substrate is not available at present. The GaN substrate or GaN buffer is not suitable for the growth of AlGaN materials. Since the lattice for AlGaN is shorter than for GaN, the AlGaN grown on a GaN substrate or GaN buffer will suffer from the strain, which will induce crack in the AlGaN layer. To overcome this problem, AlGaN is usually grown on AlN buffer or AlN template because AlN can provide compressive stress as it has a smaller lattice constant than AlGaN, which can reduce the crack on the AlGaN. Additionally, AlN has a broader bandgap than AlGaN, so it will not affect the emission or absorption efficiency of the active region. The characters of these three materials are shown in Table 4.1 [1–3]. Therefore, the method to obtain high-quality AlN became one of the key points in growing high-quality AlGaN. (2) More strict growth condition is needed. Because of the pre-reaction and short migration length of the Al atom, some measures should be taken to solve this problem. (3) Doping is difficult, especially for the p-type AlGaN. The activation energy for Mg increases with the increase of Al content in AlGaN, which makes it very hard to get p-type AlGaN with high carrier concentration. Up to now, even though many approaches have been employed to improve the quality of the AlGaN layer, the hole concentration is still very low in high Al content AlGaN materials, leading to considerable problem in growing high-quality p-type AlGaN.

High-quality AlGaN material determines the performance of AlGaN-based devices. Besides the doping, the defects in the semiconductor will cause carrier scattering and form an impurity energy level that will lower the efficiency and working life of devices. What is more, the impurity energy level can widen the spectrum, which is bad for devices that require high accuracy. In this chapter, we first focus on how to grow a high-quality AlN template. Then we introduce the achievement and challenges in growing and doping the AlGaN layer. Finally, we present the progress in the application of AlGaN-based materials.

4.2 Growth of AlN

As mentioned earlier, the AlN template or buffer is key to growing high-quality AlGaN, so a lot of work has been done on AlN growth. Additionally, the bandgap of AlN is 6.2 eV corresponding to 200 nm, so AlN is also an ideal material for UV devices. The growth of high-quality AlN epitaxy layer is more difficult than that of GaN. One of the reasons is that the Al–N bond is stronger than the Ga–N bond, which leads to low surface mobility of Al atoms.

Dabing Li et al. [4] adopted a high-temperature (HT) two-step growth method to grow an AlN/sapphire template and explained the mechanism of obtaining a high-quality AlN template. The two-step growth method was widely used in the growth of GaN-based materials to overcome the lattice and thermal mismatch between the epitaxy layer and the foreign substrate. The two-step growth method can be explained as follows [5]: first growing a nucleation layer on the substrate at low temperature and then growing the epitaxy layer at high temperature. Dabing Li's group investigated the influence of the initial condition on the quality of AlN. A 405 nm short-wavelength in-situ monitoring system was used to monitor the growth progress and the in-situ monitoring curve. Four different initial conditions were studied: (1) without nucleation layer or sapphire-cleaning treatment (sample A); (2) with only a nucleation layer (sample B, two-step growth method); (3) with nitridation at 950°C for 5 min and an added nucleation layer (sample C); and (4) with TMA pretreatment for 2 s and an added nucleation layer (sample D). From the in-situ monitoring curves in their article they concluded that the growth mode of sample A is 3D, because the surface of sample A was rougher with the growth, which reflected from the lower reflectance of the monitoring curve. The reflectance of sample B declined and then rose, which indicated that the surface was rough at first and then became smoother, which meant there was a transformation from 3D to 2D growth. According to the in-situ monitoring system, the growth mode of sample C was 3D with a mixture of 3D and 2D growth modes and that of sample D was 2D. The FWHM of (0002) and (10–12) for sample B are 60 and 550 arcsec, respectively, which indicated the lowest dislocation density for sample B. This results from the transition of AlN from 3D to 2D growth mode, which can bend the threading dislocations. They also presented the atomic arrangements of the four samples in their work.

Dabing Li et al. also [6] studied the effect of different nucleation layer temperatures on AlN template growth using the two-step method. The temperatures of the nucleation layer are 880°C, 950°C, and 1020°C (marked as sample A, B, and C respectively), then an HT AlN was deposited. The maximum reflectance intensity of the short-wavelength in-situ monitoring curve followed a sequence of A < B < C, which means higher temperature enhances the surface mobility of Al. It can be seen from Figure 4.1a that the growth mode of sample A was 3D, while that of samples B and C were 3D to 2D. The surface of sample B was the best among the three samples, which can be seen from the 3D AFM images in Figure 4.1b. The crystal quality of sample C is not as good as that of sample B, probably because the mobility of Al atoms is too high so that the coalescence of nano-islands takes place too early, which decreases the crystal quality of the AlN epitaxy layer.

Researchers also came up with other growth methods based on the two-step growth method, one of which is the multi-buffer method. The main idea of the multi-buffer method is to control the growth

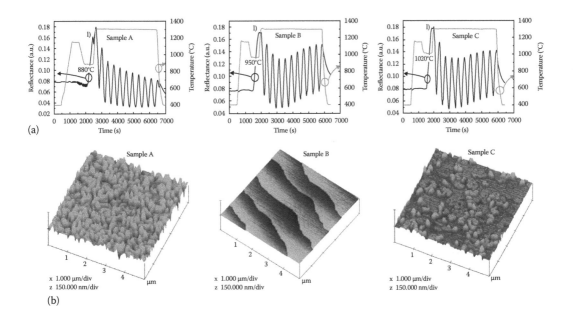

FIGURE 4.1 (a) The 405 nm wavelength in situ monitoring curve. (b) AFM images for samples A, B, and C. (Reprinted from *Mater. Lett.*, 114, Chen, Y., Song, H., Li, D. et al., Influence of the growth temperature of AlN nucleation layer on AlN template grown by high-temperature MOCVD, 26–28, Copyright 2014, with permission from Elsevier.)

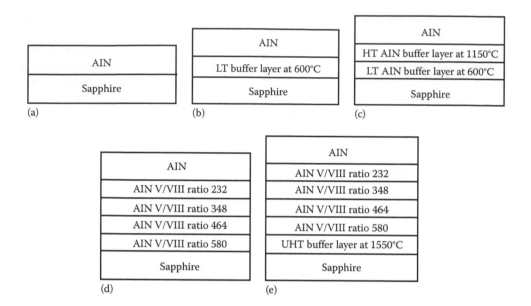

FIGURE 4.2 (a) to (e) show the five different structures of the five samples respectively. (Reprinted from *J. Cryst. Growth*, 298, Okada, N., Kato, N., Sato, S. et al., Growth of high-quality and crack free AlN layers on sapphire substrate by multi-growth mode modification, 349–353, Copyright 2007, with permission from Elsevier.)

mode gradually transitioning from 3D to 2D, during which the dislocation lines bend and the surface will become smooth. As an improvement of the two-step growth, N. Okada et al. [7] came up with the multi-buffer method and grew a series of samples combining the achievements in References 8 and 9. The structures of these samples are shown in Figure 4.2. Compared to the AlN epitaxy layer grown on sapphire directly, the insertion of a low-temperature (LT) AlN buffer layer will improve the crystal quality of the AlN epitaxy layer. The sample with LT/HT AlN buffer layers could further improve the crystal quality of the AlN epitaxy layer. The best crystal quality was achieved by the multi-buffer method. The structure of the multi-buffer sample is shown in Figure 4.2d. Peng et al. [10] grew an AlN template by the three-step growth method. The first step is to grow an AlN nucleation layer on sapphire at low temperature (600°C); the second step is to grow a middle-temperature (MT) AlN layer (1000°C); the final step is to grow an AlN epitaxy layer at high temperature (1160°C). They also grew an AlN template with MT AlN buffer layers marked as sample B and directly grew HT AlN epitaxy layers on sapphire marked as sample C for comparison. After the growth of AlN templates, a 500 nm thick $Al_xGa_{1-x}N$ epilayer was deposited under the same conditions. It was found that the higher temperature of the initial AlN buffer enhances surface migration of Al adatoms and fast relaxes biaxial stress in AlN buffer at the AlN/sapphire interface.

Although we can greatly improve the crystal quality of an epilayer on a foreign substrate by buffer layers, high-density dislocation exists in the epilayer. The lateral epitaxial overgrowth (LEO) method can reduce dislocation density efficiently. The working mechanism of LEO is to block part of the dislocation lines and bend the other part, so it can prevent many dislocation lines from reaching the surface of the epilayer. Figure 4.3 is the schematic diagram of this mechanism. The LEO method can be divided into many different ways to achieve the LEO process. Take GaN as an example: SiNx or SiO_2 can be deposited first as a mask, and then the growth windows can be exposed using the etching method. There are also many other ways to execute the GaN LEO method such as pendeoepitaxy [11], interlayer-LEO [12], air-bridged LEO [13], facet-controlled LEO (FACLEO) [14], and the use of patterned sapphire substrates (PSS) [15] and so on.

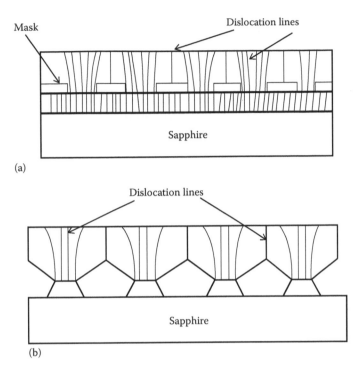

FIGURE 4.3 Behavior of dislocation lines (a) with and (b) without masks using the LEO method.

Compared to growing GaN with LEO methods, there are more limitations in growing AlN. Because Al adatoms have a much larger sticking coefficient than Ga adatoms [16], so traditional masks such as SiN$_x$ or SiO$_2$ are useless for AlN LEO growth. The LEO methods that use mask layers should be excluded. One of the LEO methods for AlN growth is to use PSS, that the LEO AlN is grown on a sapphire substrate with periodic stripes which are etched in a certain direction. Another way is by depositing AlN on sapphire first and then etching in a certain direction to get periodic stripes. The samples can be used as composite substrates for AlN or AlGaN growth.

The use of the LEO method for AlN growth can improve the quality of the surface and reduce the threading dislocation densities (TDDs). Kentaro Nagamatsu et al. [17] prepared an AlN template using a patterned AlN/sapphire composite substrate and they also deposited AlN directly on sapphire for comparison. They found that the AlN template grown by the LEO method had a better surface topography with an RMS roughness of 0.08 nm, while the RMS roughness was 0.15 nm for the AlN template grown directly on sapphire. Myunghee Kim et al. [18] grew AlN on PSS with stripes along the (10–12) direction at 1300°C. The average TDDs on the terrace and groove areas for LEO-AlN were 4.0×10^7 and 3.1×10^7 cm^{-2}.

Imura et al. [19] tried to grow AlN epilayers by the LEO method on patterned AlN/sapphire templates with (1010) and (11–20) stripe direction. In the case of the (1010) direction, the AlN layer did not coalesce, although the growth time was 120 min. On the other hand, in the case of the (11–20) direction, the AlN layer coalesced well. They also found that the increase in growth rate and thickness is helpful to decrease TDDs. The calculated dislocation density of the resulting LEO-AlN was determined to be 3.4×10^7 cm^{-2}. They also tried the multi-buffer method to grow a high-quality AlN template; then linear grooves were made along one direction such as that parallel to the (11–20) direction of the sapphire substrate. The TDDs of LEO AlN on the patterned multi-buffer AlN template were determined to be less than 10^7 cm^{-2} by plan-view TEM observation. Zeimer et al. [20] studied the effect of miscut and depth of stripes of a patterned AlN template on an AlN LEO epilayer. The substrate miscut was either 0.25° toward the a-plane or 0.25° toward the m-plane of the sapphire crystal. The dislocation densities of LEO samples are almost the same at a degree of 10^8 cm^{-2}. It has already been shown that a miscut as small as 0.25° can alter the AlN layer surface morphology significantly and determine the height and shape of the surface steps. They also deposited AlGaN on the samples and found the best material properties were observed for $Al_{0.8}Ga_{0.2}N$ grown on LEO AlN on a shallowly etched AlN/sapphire template with a miscut of 0.25° in the a-direction.

It is difficult to achieve high-quality AlN or AlGaN epilayers because of the low surface mobility of Al atoms. To increase the surface mobility of Al atoms, high growth temperature has been used as mentioned earlier. Another way to increase surface mobility is to reduce the V/III, which is applied to pulsed atomic-layer epitaxy (PALE), migration enhanced metalorganic chemical vapor deposition (MEMOCVD), and ammonia pulse-flow method. The typical flow mode of different methods is shown in Figure 4.4. The stages at which V/III is zero will appear when growing by these methods and will contribute to the maximum surface mobility, which will improve the crystal quality of the AlN and AlGaN epilayers.

An earlier study using the same principle was conducted in 1988 by Ozeki et al. [21], who grew GaAs by the atomic layer epitaxy (ALE) method. In this process, saturated monolayer by monolayer growth is achieved by alternately exposing the substrate to column III and column V gas phase reactants. During the alteration, there is a short time to clean the reaction chamber to avoid pre-reaction effectively.

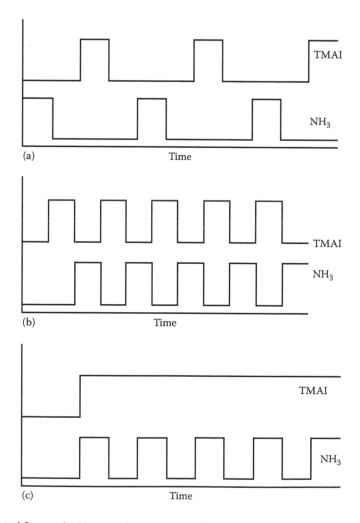

FIGURE 4.4 Typical flow mode: (a) PALE; (b) MEMOCVD; (c) NH$_3$ pulsed.

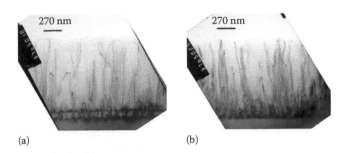

FIGURE 4.5 Behavior of dislocation lines in AlN grown by PALE. (a) g = [0002] and (b) g = [11–20]. (Reprinted with permission from Sang, L.W., Qin, Z.X., Fang, H. et al., Reduction in threading dislocation densities in AlN epilayer by introducing a pulsed atomic-layer epitaxial buffer layer, *Appl. Phys. Lett.*, 93(12), 122104 Copyright 2008, American Institute of Physics.)

Khan et al. [22] achieved high-quality AlN epitaxy by this mode in 1992. This method is also called pulsed atomic-layer epitaxy (PALE) [23].

Sang et al. [24] grew an AlN buffer layer by the PALE method and then deposited an AlN template, based on which UV LEDs with emitting wavelengths at 262, 307, and 317 nm were fabricated. They analyzed the PALE AlN buffer layer and AlN template using the SEM and XRD techniques [25]. It was found that the PALE AlN buffer is formed by a large number of hexagonal columnar hillocks covering almost the entire surface. The hexagonal columnar structure in the AlN buffer layer indicated that the vertical growth rate was much higher than the horizontal one. During the continuous growth of the AlN epilayer, the lateral growth rate was increased and the dislocation lines were redirected to annihilate in the AlN epilayer during growth. A clear sub-interface could be observed at the interface of the PALE buffer layer and the AlN epilayer, above which TDDs decreased significantly. This phenomenon indicates that the use of a PALE buffer layer can reduce the TDDs effectively. The behavior of dislocation lines is shown in Figure 4.5.

Contrary to Sang's research, Cicek et al. [26] deposited a 20 nm thick AlN buffer layer on sapphire and then grew a 600 nm thick AlN epilayer by the PALE method. They finally grew AlGaN on an AlN template to fabricate a solar-blind UV photodetector (PD) array. Researchers found that the AlN template had high surface and crystal quality. The RMS roughness was only 0.9 Å, and the FWHM of (002) was 23 and 230 arcsec for (105).

When growing material by the PALE method, the column III and column V gases alternately flow into the reaction chamber, and the cleaning process between the alternations of the two gases leads to a low growth rate. Researchers came up with the MEMOCVD method to grow III-nitride during which the cleaning process is dislodged. However, the main idea for improving the crystal quality is the same.

Khan et al. [27] grew a high-quality AlN template and AlN/AlGaN superlattice (SL) by the MEMOCVD method. The screw dislocation density was about 3×10^8 cm^{-2}, which was greatly decreased compared to the typical screw dislocation density (10^{10} cm^{-2}) by the traditional method. They finally deposited AlGaN on the sample to fabricate UV devices. Milliwatt-level power deep UV LEDs operating in the UVC region (<300 nm) have been demonstrated. LEDs of 280 nm and 295 nm with CW power exceeding 1 mW at 20 mA (with wall-plug-efficiency approaching 1%) were reported. The UV lamps produced CW power in excess of 1.5 mW at 265 nm, 11 mW at 280 nm, and pulse power over 56 mW at 280 nm. Preliminary reliability tests showed that the devices were fairly stable under low current-density operation.

Although there are many advantages of the MEMOCVD method for enhancing Al surface mobility and reducing TDDs, it is difficult to increase the growth rate and control the surface migration of Al atoms. To overcome the weakness, researchers improved the MEMOCVD method and called it the modified MEM(logogram of MEMOCVD) method. The column III and column V gases in the modified MEM

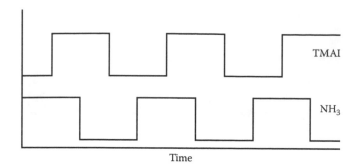

FIGURE 4.6   Flow mode of modified MEM method.

method are not separated completely. The flow mode of the modified MEM is shown in Figure 4.6. The modified MEM can increase the growth rate and enhance the control of surface migration of Al atoms.

Banal [28] has grown AlN epilayers on sapphire by MEM, modified MEM, and simultaneous source supply method. The surfaces of three samples were smooth but the crystal quality of AlN grown by modified MEM was the best. The FWHM of (0002) was 42.4 and 244.5 arcsec for (10–12). They estimated the diffusion length of Al atoms by changing the distance of the steps on sapphire and observing the density of AlN nano-islands [29]. Under certain conditions, the Al diffusion length was 44 nm for the MEM method, 42 nm for the modified MEM method, and 31 nm for the simultaneous source supply method. The (0002) FWHM of AlN grown by modified MEM was only 31.8 arcsec.

The main idea of the ammonia pulse-flow method is similar to that of PALE and MEM. The difference is that the MO source is continuous but the NH_3 source is pulsed. The ammonia pulse-flow method could reduce the pre-reaction and enhance the Al mobility on the surface of the substrate.

Hideki Hirayama et al. [1] achieved an AlN epilayer with combined ammonia pulse-flow method and simultaneous source supply method. The TDD is on the degree of 10^9 cm^{-2} and the RMS roughness is 0.16 nm. The AlGaN epilayer was deposited on an AlN template, and UV LEDs with emitting wavelengths between 231 and 261 nm were fabricated. This research group had achieved AlN templates with low TDDs during 2008–2010 [30–32]. The edge dislocation and screw dislocation density of the best AlN template are 7.5×10^8 and 3.8×10^7 cm^{-2}, respectively.

A thick epilayer will be helpful for improving the quality of crystals [33]. However, with the increase in thickness, stress is also accumulated, which will contribute to the cracks on the surface of the epilayer. Introducing an interlayer during the growth of the epilayer is thought to be an effective method to relax the stress and bend the dislocation lines because of the different lattice constant and growth conditions of the interlayer. Especially when depositing AlGaN on the GaN template, it is better to introduce an interlayer between the AlGaN epilayer and the GaN template to relax the stress [34–37].

Da-Bing Li et al. [38] compared the AlN epilayer grown on sapphire with and without an MT AlN interlayer. They found that the RMS roughness of the AlN epilayer with interlayer was only 0.84 nm and presented Al polarity; however, the RMS roughness of the AlN epilayer without interlayer was 13.54 nm with N polarity. Jianchang Yan et al. [33] also tried to apply MT AlN when growing an AlN epilayer to gain a high-quality crystal and smooth surface. The RMS roughness was less than 0.2 nm. The FWHM of (0002) and (102) was 58.9 and 383.1 arcsec, respectively, while the values were 61 and 523 arcsec for the AlN epilayer without interlayer. What's more, there was no crack on the AlN surface with MT AlN interlayer. During the growth of the interlayer, TMAl and NH3 were closed for 6 s after every 5 nm MT-AlN growth, while H_2 carrier gas was always on to etch the defects of the interlayer. When the growth temperature was increased to 1200°C, the MT AlN provided the driving force for bending the dislocation lines and also acted as elastic layers to accommodate more stress. The DUV LED fabricated on this sample has an emitting wavelength at 284.5 nm with a forward voltage of about 5.7 V.

Ren et al. [39] studied the effect of an AlN interlayer on an AlN epilayer in the MBE system. They found why an interlayer can improve the AlN epilayer. The interlayer represents 3D growth and the AlN epilayer represents 2D growth, which will bend the dislocation lines and lead to a high-quality AlN epilayer. However, the thickness should be well controlled; a low density and large number of nano-islands is good for the AlN epilayer because the reduction of the crystal boundary will decrease the edge dislocation density. A thicker interlayer will enhance the coalescence of nano-islands and decrease the amount of crystal boundary. However, too much thickness will lead to new islands appearing and stacking up on existing ones randomly, and even forming large clusters, which will reduce the crystalline quality of the main AlN layer and induce a rougher surface.

Similar to an interlayer, SL can also relax the stress and reduce TDDs to achieve a high-quality AlN template. Kim et al. [40] introduced 10 periods of AlN/Al$_x$Ga$_{1-x}$N(15/15 nm, X = 50, 60, 70, 80)SLs between LT to reduce the TDDs and gain high crystal quality. Except for the sample with x = 50%, the FWHM and RMS roughness of the other samples were all less than the sample without SL. The reason for this phenomenon is that the surface migration of Ga is larger than that of Al, which will be beneficial to bending dislocation lines. What's more, the SLs with x = 0.6 could relax stress more effectively than those with x = 0.7 and 0.8 because of a bigger difference of lattice constant between AlGaN wells and AlN barriers in the SLs. It is believed that an Al mole fraction smaller than 0.6 in the AlN/Al$_x$Ga$_{1-x}$N SLs undergoes strain relaxation via reciprocal space mapping (RSM) and would lead to the generation of new dislocations at the interface between AlN and Al$_x$Ga$_{1-x}$N. In this experiment, the combination of LT buffer and SL structures includes two aspects. The first is that it slightly relaxes the lattice mismatch between LT buffer and HT-AlN, which prevents the generation of TDDs. The second is that it functions as a stress undulation in the individual layers of the SL, which induces dislocation annihilation at the SLs' interfaces.

4.3 Growth of AlGaN

As mentioned earlier, the intrinsic long-wavelength edge can be adjusted from 200 to 365 nm by controlling the Al content of AlGaN. So to further improve UV devices, high-quality AlGaN with high Al content is needed. However, the TDDs we get nowadays are on the degree of 10^8 cm^{-2} for the best AlGaN crystals while the values are 10^9 and 10^{10} cm^{-2} for normal AlGaN crystals. The high degree of TDDs will greatly decrease the performance of AlGaN-based devices. To lower the TDDs, improving Al atoms' mobility and enhancing the blocking and bending of dislocation lines are essential measures.

Due to the success of the two-step growth method for GaN, researchers also tried to grow AlGaN on a sapphire substrate using the two-step growth method. Satoshi Kamiyama et al. [34] deposited AlGaN with Al content varying from 0 to 1 directly on sapphire using the two-step growth method when studying the effect of an AlN interlayer on the AlGaN epilayer. They found that the FWHM of (0002) would be enlarged with the increase of Al content. The RMS roughness of Al$_{0.43}$Ga$_{0.57}$N was 20 nm and got even worse with higher Al content. Some researchers impute the failure of growing AlGaN to the two-step growth mode. Unlike depositing GaN on sapphire by the two-step method, the mode will transform from 3D growth to quasi-2D growth but not to 2D mode during the growth of AlGaN by the two-step method. The quasi-2D growth mode will contribute to bad crystal quality and rough surface.

To overcome this problem, an AlN/sapphire or GaN/sapphire template is commonly used when growing the AlGaN epilayer. It means that AlN/sapphire or GaN/sapphire is used as a composite substrate for AlGaN growth. The AlN or GaN on sapphire can partly play the role of an AlN or GaN substrate to improve the quality of the epitaxy layer. However, the lattice constant of GaN is slightly larger than that of AlGaN, which will provide tensile stress that causes cracks in the epitaxy layer. Contrary to GaN, AlN will provide compressive stress, which is beneficial to the suppression of cracks. The lattice constant of AlN and AlGaN will be closer with the higher Al content in AlGaN alloy. What is more, the bandgap of GaN is 3.4 eV corresponding to the intrinsic long-wavelength edge at 365 nm, which will have a bad influence on the effectiveness of UV devices as mentioned earlier. The bandgap of AlN is 6.2 eV corresponding to the intrinsic long-wavelength edge at 200 nm, which has little effect on the effectiveness of UV devices.

In conclusion, it is better to choose AlN/sapphire as a composite substrate for III-nitride growth; there have also been studies showing that an AlN template is better for AlGaN growth [41].

However, it is easier to get high-quality GaN on sapphire, so there are many studies based on GaN samples. To relax the strain between GaN and AlGaN, many studies have applied an interlayer between GaN and AlGaN. Amano et al. [42] tried to introduce an LT AlN interlayer when growing AlGaN with different Al content on an HT-GaN template grown by the two-step method. They also compared the samples with others without an interlayer and found that there were many cracks on the surface of the AlGaN sample and that the FWHM of the XRD test was wider with the increase of Al content. However, the surface was smooth and the FWHM of the XRD test almost kept constant with various Al contents for the samples with an interlayer. This study confirms that an AlN interlayer is helpful to achieve an AlGaN epilayer with high Al content. Çörekçi et al. [37] grew a GaN template on an AlN buffer layer and inserted an AlN interlayer before the growth of AlGaN. They compared the samples with those in which AlGaN was deposited on a GaN template grown on a GaN buffer and an AlN buffer but without an interlayer. They found the dislocation density of the AlGaN sample with an interlayer was the lowest and the surface was smooth. The quality of AlGaN grown on a GaN template with an AlN buffer was better than the GaN buffer layer. This means the AlN interlayer and buffer layer both have the ability to relax strain. Studies have also shown that the $Al_yGa_{1-y}N$ interlayer can also relax the strain in an $Al_xGa_{1-x}N$ epilayer in the case of Y > X [34]. The aim of introducing an AlGaN interlayer but not AlN is to overcome the electrical separation by the LT-AlN interlayer.

Although an interlayer introduced on a GaN template can relax strain and improve crystal quality, the quality of the interlayer will also influence the quality of the AlGaN epilayer. What is more, it is difficult to achieve an AlGaN epilayer with high Al content on a GaN template in spite of the existence of an interlayer. The highest Al content of AlGaN grown on a GaN template is about 0.3 [43] by hetero facet–controlled epitaxial lateral overgrowth. So, it is essential to achieve a high-quality AlN template for a high-quality AlGaN epilayer [4]. By controlling growth temperature, the flow of Al source and Ga source, and V/III, we can achieve high-quality AlGaN with different Al contents.

Researchers have made progress in depositing a high-quality AlGaN epilayer on an LEO AlN template. Kentaro Nagamatsu et al. [17] fabricated LEDs based on AlGaN quantum well (QW) structure on a high-quality LEO AlN template. They found the dislocation density in the AlGaN active layer of the LED on LEO-AlN to be one-third less than that of the LED on planar AlN, and the output power of the UV-LED fabricated using the LEO technique was 27 times as high as that fabricated without the LEO technique. Myunghee Kim et al. [18] also grew a high-quality AlGaN alloy on an LEO AlN template and fabricated DUV LEDs with a P-I(MQWs)-N structure. The emitting wavelengths were 266 nm (Al ~ 32%) and 278 nm (Al ~ 56%), the output power was 5.3 and 8.4 mW, respectively, corresponding to 1.9% and 3.3% for external quantum efficiency (EQE). The LEDs were highly stable DUV LEDs with 70% lifetimes of over 700 h. The study showed that the quality of AlGaN was improved by using an LEO AlN template and led to high performance of DUV LEDs.

Besides growing a high-quality AlGaN alloy on an LEO AlN template, LEO AlGaN on AlN template is another method for improving the quality of an AlGaN alloy. Kueller et al. [44] studied LEO AlGaN and AlN on a patterned AlN template. They found the Al concentration corresponding to these CL wavelengths varies between around 30% at the sidewalls and around 50% at the surface due to the different surface migration of Al and Ga. During their experiment, AlGaN grown on a striped pattern oriented along the (1–100) direction reaches a flat and closed surface after 1.8 um vertical growth. However, the surface had already coalesced but was facetted and rough after 1.2 μm vertical growth on the stripes along the (11–20) direction. They also tried to grow AlN on patterned AlN/sap and then deposited AlGaN. The peak of PL spectra was 281 nm. The decrease in PL intensity with temperature was much slower for AlGaN on the AlN LEO template than on the unstructured template. Hence, the internal quantum efficiency (IQE) was successfully improved by using an AlN LEO template.

Composition modulation of AlGaN grown on LEO AlN has been mentioned in the study of Knauer et al. [45]. They tried to grow an $Al_{0.5}Ga_{0.5}N$ epilayer on (1–100) patterned AlN/sap template.

The study showed that higher Ga content at the edges is due to migration of the more mobile Ga on the c-oriented terraces and preferential incorporation near the step edges. They found that the acreage of the Ga-rich region would be smaller with the increase of growth speed but the Ga concentration would be higher. Planarization of the LEO AlN surface to step heights in the range of 5 nm avoids composition modulation.

The SL structure is considered to have functions of relaxing strain and filtering dislocations. Many researchers achieved a high-quality AlN or AlGaN epilayer with a smooth surface.

Asif Khan et al. [46] introduced AlN/AlGaN SL when growing AlGaN on a sapphire substrate and grew an Si-doping AlGaN epilayer under the same condition. They also grew an AlGaN epilayer on sapphire without SL structure but with other conditions remaining the same. The sample without SL cracked severely while the other two samples had smooth surface. The result proves that the SL can relax strain [47] and lead to a smooth surface. The edge dislocation density decreased by two orders after SL insertion, from 4.8×10^{10} to 4.5×10^{8} cm^{-2}; these data imply that the AlN/AlGaN SLs act as an effective dislocation filter during AlGaN growth. Akira Fujioka et al. [48] grew high Al content AlGaN by introducing AlN/AlGaN SL structure as a strain-relieving layer on a high-quality AlN template and fabricated DUV LEDs with emitting wavelength at 280 nm; the output power under 20 mA was 2.45 mW.

To improve the surface mobility of Al atoms, higher growth temperature and surfactant are also used in the growth procedure. The growth temperature of AlGaN is usually higher than 1200°C to enhance the surface mobility of Al atoms. A small amount of In is usually used as surfactant when growing AlGaN. The broken bonds of N may attract Al atoms, which will lead to low surface mobility. In could occupy the broken bonds of N and enhance the surface mobility of Al atoms to form a smooth surface and good crystal quality. What's more, the bond energy of InN is smaller than that of AlN and GaN so In will be replaced by Al or Ga when Al or Ga atoms are excessive, which will not affect the lattice structure. Junyong Kang et al. [49] compared the AlGaN-based MQWs structure with and without In as surfactant. They found that the AlGaN-based MQWs without In as surfactant do not coalesce completely. However, when In was further introduced as a surfactant during the growth of MQWs structure, the surface morphology and structural quality are believed to be improved due to the enhancement of the surface migration of Al, Ga, and N adatoms.

4.4 Doping of AlGaN

High-quality n-type and p-type AlGaN are the necessary elements for AlGaN photoelectronic devices, such as LEDs, LDs, and PDs. As is well known, the Nobel Prize for Physics in 2014 was awarded jointly to Isamu Akasaki, Hiroshi Amano, and Shuji Nakamura "for the invention of efficient blue light-emitting diodes which has enabled bright and energy-saving white light sources." For this invention, the obtaining of p-type GaN was key. Unfortunately, the growth of n-type and p-type AlGaN, especially the latter, is more difficult than that of GaN. The impurity energy level of AlGaN is closer to the middle of the bandgap than for GaN because of the wider bandgap, which will lead to higher activation energy and lower doping efficiency, and thus higher activation energy of the dopants, less carrier concentration, larger carrier scattering, and compensation effect. All of these effects will result in lesser electron or hole concentration and poor quality of the doped AlGaN materials. In this section, the achievement and challenges will be introduced.

4.4.1 n-type Doping

Si is usually used as the donor for n-type doping and has made much progress compared to p-type doping. The main limiting factors of n-type doping are the carrier scattering and self-compensation effect. To improve the crystal quality and suppress the formation of defects the doping efficiency of AlGaN must be improved.

Because In can be used as a surfactant to increase the surface diffusion of Al atoms during AlGaN growth and decrease the defect density of the AlGaN epilayer, In–Si co-doping is used to lower the defect density and increase doping efficiency.

During the MOCVD growth of Si-doped $Al_{0.75}Ga_{0.25}N$, it was found that as the indium flow rate increases the density of surface pits decreases and the XRD FWHM of the (002) peak is narrowed, indicating a reduction in screw dislocation density. The emission intensity of the deep-level impurity transition associated with the cation vacancy complex $(V_{III}$-complex$)^{2-}$ was found to decrease almost linearly with a decrease of the screw dislocation density [50]. Cantu et al. [51] achieved n-type doping by employing the In–Si co-doping method. The $Al_{0.65}Ga_{0.35}N$:(Si,In) layers exhibited an n-type carrier density as high as 2.5×10^{19} cm^{-3} with an electron mobility of 22 cm^2/V s, corresponding to a resistivity of 1.1×10^{-4} Ω cm. An $Al_{0.62}Ga_{0.38}N$ epilayer was also deposited for comparison with only Si doping, and a maximum carrier concentration of 1.3×10^{17} cm^{-3} and a resistivity of 6.2×10^{22} V cm were obtained in the Si doping epilayer. The result showed that In–Si co-doping can improve the doping efficiency compared to typical Si doping.

To lessen the defects in an AlGaN alloy, a high crystal quality is required. The growth method mentioned earlier to get a high-quality AlGaN alloy can also be used for the doping of AlGaN. For example, Shuchang Wang et al. [52] deposited an n-type AlGaN alloy using an HT AlN interlayer, which significantly decreased the density of defects and increased the efficiency of doping. Jeon et al. [53] achieved an n-type $Al_{0.45}Ga_{0.55}N$ epilayer on SL structure. Chenguang He et al. [54] also achieved an n-type $Al_{0.5}Ga_{0.5}N$ epilayer on SL structure; the consistency of electron under room temperature was 4.2×10^{18} cm^{-3} and the electron mobility was 48.2 cm^2/(V s).

Using the delta doping mode can also improve the quality of Si-doped AlGaN. When doping with the delta mode, the growth process is believed to be paused during the doping process, in order to stop the extension of dislocation lines along the growth direction. So the use of delta doping can improve the crystal quality and reduce the self-compensation effect, finally enhancing doping efficiency and carrier mobility [55]. A uniform-doping Al-rich n-AlGaN sample and three delta doping n-AlGaN samples with varying delta doping times were grown by MOCVD in the research of Shaoxin Zhu et al. [56]. The delta doping samples exhibited better crystal quality and electrical property with lower sheet resistance (288 Ω/$\square$ minimum), higher carrier mobility (108.7 cm^2/V·s maximum), and larger carrier density (2.17×10^{18} cm^{-3} maximum). When increasing delta doping time from 6 to 30 s, all properties were gradually improved. The result indicated the potential application of delta doping technique in a high-quality AlGaN epilayer with high Al composition.

4.4.2 p-type Doping

Compared to n-type doping, high-quality p-type AlGaN is more difficult to achieve. There could be several reasons for this [57]. First, the solubility of acceptor impurities is too small to get high-quality p-type AlGaN with high doping concentration. Second, the strong self-compensation effect arising from the presence of donor-like native defects and/or complexes involving native defects will decrease the doping efficiency. Third, the main p-type dopant nowadays is Mg. The high activation energy of Mg acceptor in $Al_xGa_{1-x}N$ rising from 160 to 500 meV for x = 0 to x = 1 will hamper the efficiency of doping. What's more, there should not be too much Mg concentration in p-type AlGaN because it will reduce the crystal quality and contribute to a strong self-compensation effect. Hyung Koun Cho et al. [58] found that the quality of p-type AlGaN could be improved with the increase of Mg flow at an earlier stage. However, the crystal quality worsens sharply when the flow beyond a certain threshold value (0.103 μmol/min under that experiment condition) and also the conductivity decrease. It has been reported that Mg acts as a surfactant, which modifies the surface mobility of the chemical species due to the reduction of surface energy in films on mismatched underlying layers. Therefore, low-concentration Mg leads to the change of Al surface mobility, increases the size of the initial AlGaN islands, and finally decreases the dislocation density.

However, excessive Mg flow might cause too much Mg to coalesce into a lattice and increase strain by the incorporation of Mg atoms with a large covalent radius. Jeon et al. [59] also found that Mg incorporation has to exceed the background (extrinsic and intrinsic) defects to achieve p-type conduction. Excessive Mg incorporation, however, causes the generation of structural defects, such as inversion domains and SFs, possibly due to the formation of Mg–N complexes, such as Mg_3N_2. These defects will enhance the scattering and the recombination of carriers, which will decrease the conductivity of the AlGaN alloy. The rapid rise of sample resistivity beyond the optimum window is likely due to (1) Mg-related compensating centers at IDBs, (2) an increased incorporation of impurities, for instance oxygen, under N-terminated growth polarity, and (3) an increased concentration of native defects due to reduced Ga diffusion on a polarity-inverted surface.

Since Amano et al. and Nakamura et al. activated Mg dopant successfully in GaN by low-energy electron beam irradiation (LEEBI) [60] and thermal annealing, respectively, devices based on GaN have seen rapid development. However, the doping of p-type AlGaN is not very good for the reasons mentioned earlier. Similar to n-type doping of AlGaN, it is necessary to get high crystal quality to achieve high efficiency in p-type doping, so many studies have tried to achieve p-type AlGaN alloy using SL [61] or interlayer or other methods to improve the crystal quality of the AlGaN epilayer.

According to the Mg dopant distribution in AlGaN, p-type doping of AlGaN includes uniform doping (including co-doping), δ-doping, modulation doping in SL, and polarization-induced doping methods. These methods can be used individually or combined when doping.

Xu Zheng-Yu et al. [62] deposited p-type $Al_{0.43}Ga_{0.57}N$ using In as assistant and found that resistivity decreased by two orders compared to that without In as assistant. Toru Kinoshita et al. [63] studied the effect of V/III on Mg doping. Photoluminescence measurements indicated that Mg doping with a high V/III ratio and a moderate Mg concentration can effectively suppress self-compensation via the formation of V_N^{3+} defects. Temperature-dependent Hall effect measurements show extremely small effective activation energies of 47–72 meV at temperatures below 500°C in $Al_{0.7}Ga_{0.3}N$ films with Mg concentrations of $2.6–3.6 \times 10^{19}$ cm^{-3}.

Similar to delta doping for n-type AlGaN, Mg delta doping in AlGaN alloy can also decrease the defect densities and increase doping efficiency. Jiang et al. [64] found that Mg delta doping could suppress TDDs on an AlGaN epilayer surface compared to uniform doping. The AFM and SEM morphologies of etched surfaces are shown in Figure 4.7. AFM and SEM images reveal that the etch pit density was significantly reduced in Mg-δ-doped p-type AlGaN compared with uniformly Mg-doped p-type AlGaN. They finally fabricated DUV LEDs using delta doping technology [55].

Delta-doping technology can be combined with other doping methods. For example, Jiang et al. [57] combined Mg delta doping with In assistant doping methods to improve the quality of p-type AlGaN. The max carrier concentration was 4.75×10^{18} cm^{-3}, which is three times that of the sample using only delta doping.

The wurtzite III-nitride is a covalent compound in which the atoms of group III and group V are separated to the two sides of the band. So there is no central symmetry that exists in a wurtzite III-nitride crystal. The crystal will present metal polarity if grown along the (0001) direction and N polarity along the (000-1) direction. The electronegativity of the atoms of group III and N is different so that III-nitride will present polarity. So the band will bend at the interface, which will make the acceptor level closer to the valence band and lower the activation energy of Mg dopant, and this will increase doping efficiency.

Waldron et al. [65] showed that Mg modulation-doped (MD, only doped in barrier layers) SL structure and Mg shifted-modulation-doped (SMD, 1/4 period shifted compared to MD) SL structure samples have superior electrical properties compared to uniformly doped (UD) SL samples. The carrier mobility in SMD and MD samples is higher than that in UD samples because of lower neutral impurity scattering in the barrier layer. The lowest activation energy was achieved in SMD samples. The schematic diagrams of the bandgap structure are shown in Figure 4.8. The self-consistently calculated free hole concentrations

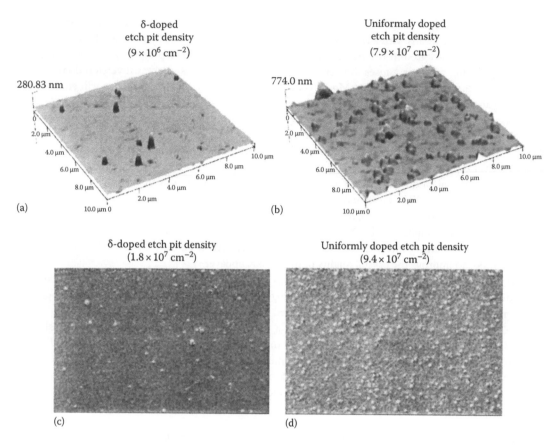

FIGURE 4.7 AFM and SEM morphologies of etched surfaces for (a and c) δ-doped p-type AlGaN and (b and d) Uniformly doped p-type AlGaN. (Reprinted with permission from Nakarmi, M. L., Kim, K. H., Li, J. et al., Enhanced p-type conduction in GaN and AlGaN by Mg-δ-doping, *Appl. Phys. Lett.*, 82(18): 3041–3043 Copyright 2003, American Institute of Physics.)

were 2.5, 3.7, and 3.7×10^{18} cm^{-3}, respectively, for MD SL, SMD SL, and UD SL, averaged over one period. The ground state is the only occupied sub-band at 90°C.

However, the band bending in these SLs is limited by the strength of polarization fields. In order to further increase the band bending and improve Mg acceptor activation efficiency, Junyong Kang et al. [66] have proposed an SL structure by introducing monoatomic layers of Mg and Si at different interfaces of Al$_{0.5}$Ga$_{0.5}$N/GaN SL, respectively, denoted as Mg-δ-co-doped and Si-δ-co-doped SL. The calculated macroscopic averaged electrostatic potential of both SLs plotted in the (0001) direction normal to the interface is shown in Figure 4.9. Due to the charge transferring from Si-doped interface to Mg-doped interface, the internal electric field in SL has been significantly intensified and results in conspicuous band bending, which will improve Mg acceptor activation efficiency. The hole concentrations and mobilities in Mg-δ-co-doped and Si-δ-codoped SLs are both larger than that of MD SL.

Although p-type doping by modulation doping in SLs could bend the energy band and lower the activation energy, the SL will introduce a barrier that contributes to low conductivity in the vertical direction.

So Debdeep Jena et al. [67] demonstrated high-efficiency p-type doping by ionizing acceptor dopants using the built-in electronic polarization in bulk uniaxial semiconductor crystals by gradually

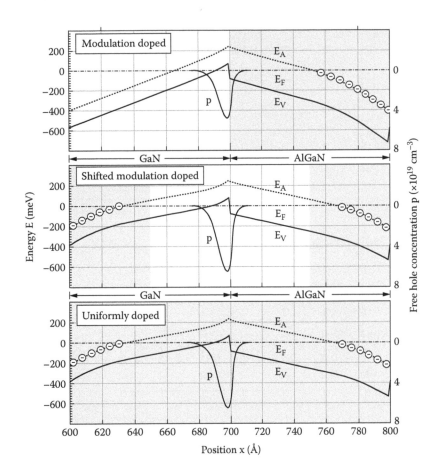

FIGURE 4.8 Self-consistent valence band diagrams of MD SL, SMD SL, and UD SL. Shaded regions are Mg doped at 10^{19} cm^{-3}. (Reprinted with permission from Waldron, E. L., Graff, J. W., Schubert, E. F. et al., Improved mobilities and resistivities in modulation-doped p-type AlGaN/GaN superlattices, *MRS Proceedings*, Cambridge University Press, Cambridge, U.K., Vol. 693, pp. I12.11.1 Copyright 2001, American Institute of Physics.)

changing the Al concentration in AlGaN alloy. A mobile 3D electron gas will be formed if the Al concentration changes from 0 to 1 along the (0001) orientation while 3D hole gas goes along the (000-1) orientation. This method is called polarization-induced doping. The schematic in Figure 4.10 shows the formation of holes using polarization-induced doping. These authors fabricated DUV LEDs based on p-type AlGaN layers using polarization-induced doping method. Compared to the LED in which the p-type AlGaN was doped by the traditional method, the LED using polarization-induced doping showed much brighter optical emission. The following are thought to be the reasons: (1) improved p-type conductivity in the vertical direction due to polarization-induced hole doping and (2) the existence of a built-in quasi-electric field imposed on minority electrons injected into the p-type layer of the graded AlGaN.

Islam et al. [68] also fabricated DUV LEDs using polarization-induced doped p-type AlGaN epitaxial layer in which the peak of the EL spectrum varies from 235 to 261 nm at both room and low temperature.

Additionally, researchers also studied other dopants. Hideo Kawanishi et al. [69] used C as dopant and fabricated UV devices in which the maximum hole concentration was 3.2×10^{18} cm^{-3} in an Al$_{0.1}$Ga$_{0.9}$N layer. Shibin Li et al. [70] used Be as a p-type dopant instead of Mg to dope an AlGaN alloy by polarization-induced doping and traditional doping methods. They also grew an Mg-doped AlGaN

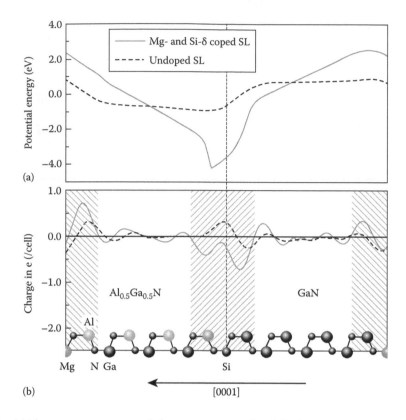

FIGURE 4.9 (a) The macroscopic averaged electrostatic potential and (b) the macroscopic averaged differential charge density of Mg- and Si-co-doped (solid lines) and undoped (dashed lines) $Al_{0.5}Ga_{0.5}N/GaN$ SLs plotted in the (0001) direction normal to the interface. (Reprinted with permission from Li, J., Yang, W., Li, S. et al., Enhancement of p-type conductivity by modifying the internal electric field in Mg-and Si-δ-codoped $Al_xGa_{1-x}N/Al_yGa_{1-y}N$ superlattices. *Appl. Phys. Lett.*, 2009, 95(15), 151113, Copyright 2009, American Institute of Physics.)

layer for comparison. The hole concentration of the graded AlGaN:Be is a function of the discrepancy of polarization owing to the Al change in the graded AlGaN. With the same thickness, the hole concentrations of the polarization-induced doping AlGaN:Be layers are 8.0×10^{18}, 2.3×10^{18}, and 2.7×10^{17} cm^{-3} at room temperature for the graded $Al_{0.7}Ga_{0.3}N$, $Al_{0.8}Ga_{0.3}N$, and $Al_{0.9}Ga_{0.1}N$ layers. However, with the same Mg flux as the graded $Al_{0.7}Ga_{0.3}N$:Be, the hole concentration in the graded $Al_{0.7}Ga_{0.3}N$:Mg is only 6.0×10^{16} cm^{-3} at room temperature, indicating that polarization-induced hole concentration in the graded $Al_{0.7}Ga_{0.3}N$:Mg is about 1% of $Al_{0.7}Ga_{0.3}N$:Be. This result demonstrates that the high activation energy of Mg in AlGaN is an obstacle to hole generation even with the inducement of polarization. However, owing to the high activation energy and compensation of Be in the AlGaN film, the conventional-doped Be impurity in the control sample (100 nm $Al_{0.7}Ga_{0.3}N$ film on AlN) cannot generate enough hole charges to make the film conductive.

Researchers also tried to replace p-type AlGaN by other materials to fabricate DUV LEDs. Dong et al. [71] tried to replace p-type AlGaN by p-type BN to fabricate DUV LEDs and compared them to LEDs with p-type AlGaN layer. The structure of an LED is shown in Figure 4.11. The light extraction efficiency and IQE of LED with p-type BN were 27.14% and 92% while for LED with p-type AlGaN they were 10% and 33%, respectively. The result shows it is possible to replace part of the AlGaN layers by other materials to achieve high-performance LED devices.

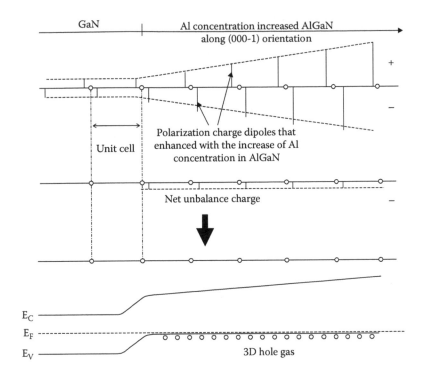

FIGURE 4.10 The schematic diagram of the formation of 3D hole gas.

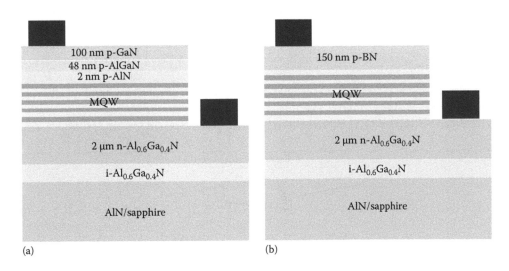

FIGURE 4.11 LED structure with (a) P-AlGaN and (b) P-BN as p-type layer. (Reprinted from *Physica E Low-Dimens. Syst. Nanostruct.*, 75, Dong, K. X., Chen, D. J., Shi, J. P. et al., Characteristics of deep ultraviolet AlGaN-based light emitting diodes with p-hBN layer, 52–55, Copyright 2016, with permission from Elsevier.)

4.5 AlGaN Low-Dimension Structure

The low-dimension structure of AlGaN includes two-dimensional (SL, MQWs, etc.), one-dimensional (nanowire, nanorod, etc.), and zero-dimensional (quantum dot, nanocone, etc.) structures.

4.5.1 Two-Dimensional Structure

The two-dimensional structure of AlGaN has been used widely in LEDs, high electron mobility transistors (HEMTs), and so on. MQWs are usually used in LEDs, LDs, and PDs. The MQW structure enhances the carrier confinement effect [72] in III-nitride. The use of MQW limits the holes and electron in the same region that increases the possibility of a recombination of carriers and work efficiency of devices.

There are many LED structures mentioned earlier that use MQW as an active region. Hideki Hirayama et al. [1,30] fabricated LEDs with emitting wavelengths between 222 and 282 nm using MQW structure. Take LEDs whose emitting wavelength is 227 nm as an example; the thickness is 1.7 nm for the well layer and 7 nm for the barrier layer. The output power is 0.15 mW under 30 mA input current intensity and the max EQE is 0.2%. Banal et al. [73] grew 10 periods of $Al_xGa_{1-x}N/AlN$ MQW structure with high Al content by the MEM method, in which the thickness of the well layer was 4.5 nm and the IQE of $Al_{0.76}Ga_{0.24}N/AlN$ MQW structure was 36%. Akira Fujioka et al. [48] also achieved DUV LEDs with 280 nm emitting wavelength using MQWs.

The polarization field exists along the (0001) orientation in III-nitride including piezoelectric polarization and spontaneous polarization. The bandgap will bend due to polarization. The bandgap bending is shown in a schematic diagram in Figure 4.12 taking the AlGaN/GaN MQW structure as an example. As shown in the figure, the bend of the bandgap will lead to the separation of electrons and holes, which will decrease the efficiency of luminescence. It is one of the main factors that have significant influence on the performance of the MQW structure. The polarization effect also makes the performance more dependent on the thickness of the well and barrier. There will be redshift with an increase in the thickness of the well layer [74].

Many methods have been used to overcome the polarization effect. For example, growing III-nitride along a-orientation [75] can avoid spontaneous polarization because there is no charge separation along a-orientation. Researchers have also proposed staggered QWs to reduce the effect of polarization. The well layer of staggered QWs contains different material concentration that will limit carriers more effectively. Min Zhang et al. [76] fabricated UV LEDs with emitting wavelengths of about 290 nm by applying staggered QWs structure. They also fabricated an LED with traditional MQWs, whose structures are shown in Figure 4.13. They found that the staggered QWs structure in shorter-wavelength AlGaN UV LEDs is proposed to reduce the influence of the electrostatic field resulting from the piezoelectric polarization, which can increase the spatial overlap of electrons and holes in the QWs and thus improve the internal quantum efficiency. The recombination rates in staggered QWs structures are much larger than the conventional structure, which means the usage of staggered QWs increases the concentration as well as the overlap of electrons and holes.

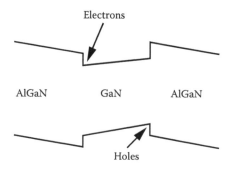

FIGURE 4.12 The bandgap of AlGaN/GaN quantum well under the influence of polarization.

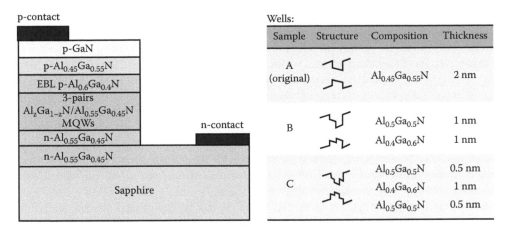

FIGURE 4.13 The LED and bandgap structures of three samples. (Reprinted from *Superlattices and Microstruct.*, 75, Zhang, M., Li, Y., Chen, S. et al., Performance improvement of AlGaN-based deep ultraviolet light-emitting diodes by using staggered quantum wells, 63–71, Copyright 2014, with permission from Elsevier.)

Researchers found that it will weaken the effect of polarization if the QW is designed to be triangular in an InGaN/GaN QW structure [77]. However, there is no related report in AlGaN QW structure as far as we know.

Although the polarization effect is hard to eliminate, the MQW structure is widely used in luminescent devices to improve the performance of photoelectronic devices. Lei Li et al. [78] grew an AlGaN QW structure with an n-type AlGaN underlayer under the QW structure and fabricated DUV LEDs. Compared to the AlGaN MQW LED without an underlayer, the turn-on voltage was decreased from 7 to 4.6 V, the reverse current under −20 V was decreased from 350 to 60 nA, and the output power and EQE were also improved. The electric field in each QW in the sample with the n-AlGaN UL was approximately 20% lower on average than that of the sample without the n-AlGaN UL. The calculated radiation recombination rates in the AlGaN MQWs in the DUV LEDs with and without the n-AlGaN UL were approximately 1.8×10^{27} and 6.9×10^{26} cm^{-3} s^{-1}. The results showed that the polarization field will be reduced and the quantum confined stark effect (QCSE) weakened by introducing n-AlGaN UL.

The other application of the two-dimensional structure is to form two-dimensional electron gas (2DEG) and a conducting channel to fabricate HEMTs. The structure of AlGaN/GaN HEMT is the heterojunction of AlGaN/GaN. The conductive band near the heterojunction is not continual and forms a deep energy well for electrons; the high-concentration 2DEG will be formed in an energy well by polarization effect and quantum effect near the interface of the heterojunction. The concentration of 2DEG is high so the ionized impurity scattering will be reduced to some extent so that high conductivity can be achieved. The band structure and charge distribution of AlGaN/AlN HEMT are shown in Figure 4.14.

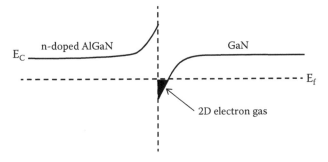

FIGURE 4.14 The bandgap and charge distribution of AlGaN/GaN HEMT.

Increasing the Al content in AlGaN is an efficient way to increase the concentration of 2DEG. However, the lattice mismatch at the interface of the heterojunction will be larger with the increase of Al concentration, so the thickness should be thinner to decrease the accumulation of stress. The defects density will be larger at the interface of the heterojunction, which will contribute to large scattering effect and low carrier mobility. To overcome the problem, researchers applied a thin AlN interlayer to release the stress while increasing Al content in the AlGaN layer and suppress the scattering effect.

Shen et al. [79] grew a series of HEMT samples: (1) conventional structure: 25 nm Al Ga N GaN; (2) novel structure with unintentionally doped (UID) cap AlGaN: UID 25 nm AlGaN/1 nm AlN/GaN; (3) novel structure with Si-doped cap AlGaN: 20 nm Si-doped AlGaN/5 nm UID AlGaN/1 nm AlN/GaN. The sample with an Si-doped AlGaN cap layer showed the best performance; the sheet charge density was 1.48×10^{13} cm^{-2} and the room temperature mobility was 1542 cm^2/(V·s). The sample with undoped AlGaN cap layer also showed high mobility 1520 cm/(V·s) and slightly higher sheet charge density of 1.22×10^{13} cm^{-2}, compared with 1200 cm/V·s and 1.1×10^{13} cm^{-2} of the conventional sample. By this, we know that the insertion of a thin AlN layer can improve the performance of HEMT. Weiguo Hu et al. [80] grew a series of novel AlGaN/AlN/GaN HEMT structures, which were at first grown on a series of vicinal sapphires with various vicinal angles of 0°, 0.25°, 0.5°, and 1°. Then they investigated the vicinal angle effects on the morphologies to determine their crystal, optical, and electrical properties. They found that the HEMT with the 0.5° off sample had the best performance.

As mentioned earlier, the III-nitride presents metal or N polarization. For N polarization GaN HEMT, there is only a GaN layer between 2DEG and source/drain electrode. However, an AlGaN barrier layer exists between 2DEG and source/drain electrode for Ga polarization. So, the application of N polarization GaN can enhance the performance of HEMT. Seshadri Kolluri et al. [81] achieved high-performance HEMT using N polarization GaN grown on sapphire. The three-terminal breakdown voltage for a leakage current of 1 mA/mm with the gate biased at pinchoff was found to be in excess of 170 V. Continuous-wave power measurements at 4 GHz at a drain bias of 50 V and a quiescent drain current density of 121 mA/mm yielded a transducer gain G_T of 9.8 dB, an output power density of 12.1 W/mm, with an associated PAE of 55%.

4.5.2 One-Dimensional and Zero-Dimensional Structures

With the development of nanotechnology, researchers tried to grow AlGaN nanowire or quantum dot (QD) because the high aspect ratio of nanostructure leads to strain relaxation of the material without the formation of dislocation lines [82]. However, there remains considerable difficulty in fabricating compositionally homogeneous Al$_x$Ga$_{1-x}$N nanowires due to phase separation and compositional modulation [83] caused by the different growth conditions and lattice mismatch between AlN and GaN. Therefore, there are many reports on the study of AlGaN thin films, but few on AlGaN nanowires [84]. The structures of nanowire-based AlGaN are mostly AlGaN/GaN core–shell structure or heterostructure. The methods can be divided into bottom-up and top-down. The bottom-up method includes a vapor–liquid–solid (VLS) growth pattern with the assistance of a catalyst and selective area growth without the assistance of a catalyst and many other methods. The top-down method is mainly achieved by photolithography or other etching technology.

Su et al. [85] grew AlGaN/GaN shell–core structure nanowires in MOCVD system. They found that samples with a high $x_{Al\text{-}gas}$ exhibited a clean background and a low density of long (4–5 μm) nanowires while samples with low $x_{Al\text{-}gas}$ were characterized by short, rodlike (1–2 μm) nanowires with nanocrystals decorating the background. The thickness and degree of tapering of the outer sheath region increased with the increase in $x_{Al\text{-}gas}$. The width of the core (GaN) region was well correlated with the physical dimension of the catalyst droplets, suggesting that the preferential incorporation into and the formation of the GaN core region are linked to catalytic growth. The presence of an Al-rich AlGaN outer sheath that is located outside the shadow projection of the nanodroplets is less likely to be attributed to catalytic growth. Separate studies of the lack of lateral epitaxial overgrowth (LEO) of AlGaN and limited surface diffusion of Al adatoms tend to support a model in which the AlGaN outer sheath is formed through conventional vapor–solid growth with negligible selectivity. Wang et al. [86] grew GaN and GaN/AlGaN core-shell structure nanowires. They found that both the optical and electric performances were improved for nanowires with the core–shell structure compared to GaN nanowires.

Kihyun Choi et al. [87,88] grew GaN QD on top of GaN/AlGaN core–shell nanowires. They first grew GaN nanowires by the selective area growth (SAG) method. Then an Al(Ga)N shell layer was deposited on the nanowires via a two-step growth method: LT-Al(Ga)N growth for 15 s at 880°C followed by Al(Ga)N growth for 75 s at 1130°C. The LT-Al(Ga)N layer growth is necessary to prevent decomposition of the GaN core at increased temperatures. Next, a thin layer of GaN was inserted to form single QDs. In order to reduce GaN deposition on the sidewall as far as possible, the growth time was limited to 8 s. Finally, the nanostructures were sequentially capped by LT- and HT-Al(Ga)N. They found that GaN QD emissions are typically observed in the energy range of 4.2–4.4 eV, which indicates that the corresponding QD height is around 1 nm. What's more, the optical properties of GaN QDs in GaN/AlGaN nanowires would improve with a smoother surface morphology of underlying shell layers.

As mentioned earlier, the interface of GaN and AlGaN will form 2DEG. Simpkins et al. [89] fabricated HEMTs using GaN/AlGaN core–shell nanowires that were grown with the assistance of Ni catalyst in the MOCVD system. They found that carrier densities increased from 2.5×10^{16} to 1.5×10^{17} cm^{-3} with the decrease of nanowire width from 200 to 50 nm. Yat Li et al. [90] inserted a 1.8 nm thick AlN interlayer between the GaN and AlGaN layers in GaN/AlGaN core–shell nanowires to improve the performance of HEMTs. The sectional view is shown in Figure 4.15.

It is difficult to grow AlGaN nanowires or QDs in the MOCVD system, so the active regions of one-dimensional AlGaN structures are always GaN, which makes the working wavelength longer than 300 nm. There is no report about DUV AlGaN nanowire LEDs fabricated via MOCVD as far as we know. Although researchers propose many theories and models on AlGaN zero-dimensional material [91–98], there is no report on growing this successfully in the MOCVD system.

Compared to the MOCVD system, the active region could be AlGaN in a one-dimensional III-nitride structure grown by the MBE system. Carnevale et al. [99] grew graded Al composition nanowires

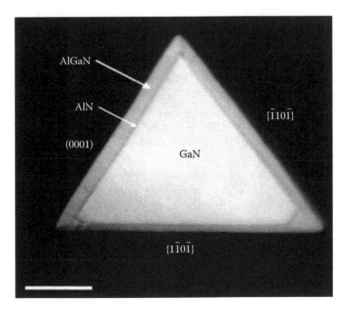

FIGURE 4.15 The sectional view of GaN/AlN/AlGaN structure nanowires. (Reprinted with permission from Li, Y., Xiang, J., Qian, F. et al., Dopant-free GaN/AlN/AlGaN radial nanowire heterostructures as high electron mobility transistors, *Nano Lett.*, 6(7), 1468–1473. Copyright 2006 American Chemical Society.)

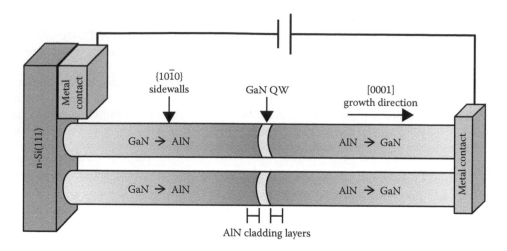

FIGURE 4.16 Polarization-induced doped nanowire LED. (Reprinted with permission from Carnevale, S. D., Kent, T. F., Phillips, P. J. et al., Polarization-induced pn diodes in wide-band-gap nanowires with ultraviolet electroluminescence, *Nano Lett.*, 12, 915–920. Copyright 2012 American Chemical Society.)

to fabricate a polarization-induced doping LED structure. The Al concentration in AlGaN alloy was changed from 0 to 1 presenting n-type polarization and was first grown with n-type doping. Then an AlGaN active layer was deposited with an AlN cap layer. Finally, AlGaN was deposited with the Al concentration changed from 1 to 0 and p-type doping was introduced as assistance. The GaN nanowires were deposited under N-rich condition. The density, average radius, and height of nanowires were adjusted by III/V and the temperature of the substrate. The substrate temperature could be increased to stop the nucleation of new nanowires when the density of nanowires is satisfactory [100]. The composition of the nanowires is controlled by changing the duty cycles of the Ga and Al shutters. Finally, the AlGaN nanowire LED was fabricated with an area of 4.25×10^{-3} cm^{-2}. The maximum power measured is just under 25 μW, while the highest EQE corresponds to 0.005% and the emission settles at a wavelength of 336.8 nm. They also fabricated polarization-induced doping LEDs with the same structure but with UID [101]. In total, 16 different polarization-induced pn junction nanowire LED samples were prepared, and all samples revealed clear diode behavior and UV light emission. EL was observed at various wavelengths from 250 nm in Al$_{0.8}$Ga$_{0.2}$N quantum disk LEDs to 360 nm in GaN quantum disk LEDs. The LED structure is shown in Figure 4.16.

Himwas et al. [102] presented structural and optical studies of Al$_x$Ga$_{1-x}$N sections and Al$_x$Ga$_{1-x}$N/AlN nanodisks (NDs) in NWs. By increasing x from 0 to 0.4 and/or reducing the ND thickness from 4 to 1 nm, the CL peak emission can be adjusted from 350 to 240 nm, with an FWHM in the range of 50–500 meV and the internal quantum efficiencies around 30%. Kent et al. [103] fabricated graded Al composition polarization-induced doping nanowire LEDs with the emitting wavelength at 250 nm. Zhao et al. [104] have demonstrated electrically injected AlGaN nanowire lasers that can operate in the UV-C band (262.1 nm). The device exhibited relatively low threshold current density (200 A/cm²), and the operation temperature was limited to 77°C.

The AlGaN zero-dimensional material can be achieved in an MBE system. Himwas et al. [105] demonstrated AlGaN/AlN QD SLs displaying room-temperature emission down to 235 nm wavelength with IQE higher than 30%. The QD SLs structure was grown under N-rich condition. The growth started two-dimensionally, with a transition into three-dimensional (3D) islands (SK growth mode) when the deposited material was beyond a certain critical thickness. The choice of N-rich conditions was motivated by the target of short-wavelength emission: the excess of nitrogen reduces the mobility of adsorbed species during growth, resulting in a high density of small QDs. The other method to achieve AlGaN QDs is to grow heterojunction in nanowires. The material in nanowire structures has already been limited to two dimensions. The zero-dimensional material can be achieved if the length is also limited. Songmuang et al. [106] achieved $GaN/Al_xGa_{1-x}N$ QDs in nanowire structures in 2011 by applying this thought.

There are many researches on GaN QDs [107–113] but few on AlGaN zero-dimensional material. So it is hard to list more reports about AlGaN zero-dimensional material.

4.6 Applications of AlGaN-Based Materials

AlGaN and AlN materials are mainly used in photoelectric devices and microelectronic devices such as LEDs, LDs, PDs, HEMTs, and so on. The wide bandgap of AlGaN and AlN allows photoelectric devices to work in the UV region and contributes to high breakdown fields, it is also ideal for the fabrication of microwave high-power devices. This section will give a brief introduction about the use of AlGaN in LEDs, LDs, PDs, and HEMTs.

The low electron inject ion efficiency (EIE) into the QW by the electron leakage caused by the low hole concentrations in the p-type AlGaN layers will lead to low EQE. In order to obtain a high EIE, an electron blocking layer (EBL) can be effectively used for the suppression of electron overflow above the QW into the p-type AlGaN layers. However, the electron–barrier height of the EBL becomes shorter in relative terms with the increase in Al content in an AlGaN alloy, which will reduce the effect of EBL. So Hideki Hirayama et al. [114] fabricated AlGaN QW LEDs with multiquantum-barrier electron-blocking layers (MQB-EBLs) on AlN/sapphire templates and investigated the effects of using the MQBs on the EQE values of the LEDs. Significant increases in the EQE values (by a factor of 2.7) were observed for a 250 nm AlGaN MQW LED by replacing the usual single-barrier EBL by an MQB-EBL. And the maximum EQE values and output powers of DUV LEDs with an MQB-EBL when measured under room temperature cw operation were 1.18% and 4.8 mW, 1.54% and 10.4 mW, for 250 and 262 nm LEDs, respectively, which are the highest achieved before 2010. High IQE of about 80% at room temperature was achieved in AlN nanowire structure LEDs in Bryan Z et al.'s work [115] in 2015. A record high IQE of 80% at a carrier density of 10^{18} cm^{-3} was achieved in UVC MQW structures emitting at 258 nm by growing them on bulk AlN substrates that contain fewer than 10^3 cm^{-2} dislocations. However, it is pretty good if the EQE of AlGaN-based LEDs could be above 10% on sapphire substrates because of the high level of TDDs [116,117].

Tan et al. [118] suppressed the parasitic peak and improved electroluminescence (EL) by both increasing the Al composition of barriers and decreasing the thickness of p-type cladding layers. Increasing Al composition could suppress electron overflow effectively by increasing the barrier height of the QWs and decreasing the barrier height for hole injection. At the same time, decreasing the thickness of the p-type cladding layer to some extent could increase the hole concentration and vertical tunneling conductivity. By increasing the Al composition from 55% to 70% and decreasing the thickness of the p-type layers from 25 to 10 nm, the light output power was increased by several orders of magnitude, for example, by a factor of 306 and 61 under 1.5 and 15 A/cm^2 DC injection, respectively.

The SiC substrate is an ideal foreign substrate for III-nitride because of its low lattice mismatch for III-nitride growth and excellent thermal conductivity. However, it is unsuitable to prepare DUV LEDs with high light extraction efficiency by using the conventional DUV structure. When transplanting the earlier-mentioned conventional DUV structure to the SiC substrate, it was found that the 6H-SiC substrate and p-GaN top layer absorb DUV emission. So Hongwei Liang et al. [119] deposited an inverted structure on an SiC substrate compared with the conventional structure, that is, the p-GaN layer was grown first and then the MQWs and n-AlGaN top layer were grown consequently. A polarization-induced backward-tunneling junction (PIBTJ) was used as the electrical connection of the n-type 6H-SiC substrate with the p-GaN hole injection layer to realize vertically conducting DUV LEDs. Simulation results indicate that the PIBTJ can carry sufficient current to be viable for UV light emitters. The DUV LED devices incorporating the PIBTJ would offer high light extraction efficiency as a result of the elimination of the GaN contact layer.

There are also many simulations on improving LED devices such as the design of sidewall-emission-enhanced DUV LEDs [120] and carrier blocking [121] or injection layers [122] and so on. As for LD devices, Martens et al. [123] have fabricated laser diode structures with p-AlGaN short-period superlattice (SPSL) cladding layers on LEO AlN/sapphire template, and with the active region emitting near 270 nm, a current density of more than 4.5 kA/cm^2 could be reached before electrical breakdown. Harumasa Yoshida et al. [124] fabricated AlGaN MQW laser diodes emitting at 354 and 360 nm, where the device exhibits a lasing characteristic with peak output power of over 80 mW, and shows a threshold current density of 8 kA/cm^2. The differential external quantum efficiency (DEQE) for the total output from both facets is estimated to be 17%. What's more, researchers also fabricated LDs with nanowire structure. Zhao et al. [125] fabricated self-organized AlGaN nanowire array structure LDs on Si substrate in 2015. The laser operates at $\sim$289 nm and exhibits a threshold of 300 A/cm^2, which is significantly smaller compared to the previously reported electrically injected AlGaN multiple QW lasers.

The AlGaN alloy with high Al concentration can also be used as solar-blind AlGaN-based photo-diodes [126] and PDs [127] as well as transistors. The deep ultraviolet (DUV) sources could be used for photoetching the light source or pumping traditional fluorescent lamp phosphors (which are conventionally excited by a mercury discharge), whereby solid-state white lamps could be realized, totally changing the world of home, commercial, and industrial lighting. In addition, DUV sources are very desirable for the development of several biological and chemical sensors and therapeutic technologies. They may also be used in communication fields combined solar-blind DUV laser with solar-blind PD.

Cicek et al. [26] grew a 600 nm thick AlN template and then a 320×256 solar-blind UV PD array was fabricated. A low-dark current density at 10 V reverse bias was less than 2×10^{-9} A/cm^2, and the ideal factor of the diode was calculated as 2.4. Under 5 V of reverse bias, at the peak detection wavelength of 275 ± 1 nm, the EQE of the PDs only varies between 77% and 89%, and the internal quantum efficiency of these optimized PDs can be estimated to be as high as 98%.

As mentioned earlier, the sheet charge density of HEMTs could reach 10^{13} cm^{-2} degrees and the room temperature mobility could be above 1500 cm^2/(V·s). In the work of Yan Tang et al. [128], the ultrahigh-speed HEMT with AlN/GaN/AlGaN epitaxial structure achieved high f_T of 454 GHz and simultaneous f_{max} of 444 GHz on a 20 nm gate HEMT with 50 nm wide gate–source and gate–drain separation. With an OFF-state breakdown voltage of 10 V, the Johnson figure of merit of this device reaches 4.5 THz-V, representing the state-of-the-art performance of GaN transistor technology.

4.7 Summary

AlGaN is one of the best candidates for photoelectronic and microelectronic devices. Up to now, a lot of effort has gone into growing high-quality AlGaN material, but there are still many challenges to obtaining high-quality AlGaN with high Al content. The TDDs in AlGaN could be lowered to 10^8 cm^{-2} degrees at best nowadays, which needs further improvement for wide usage. P-type AlGaN doping is much more difficult than n-type AlGaN doping although the hole concentration can reach 10^{18} cm^{-3} as of now. AlGaN-based devices can be applied in many fields, but due to the difficulty in growing and doping, AlGaN-based devices are not widely used in our life today.

Considering the existing problems, the research direction could be divided into three parts: to achieve high-quality AlN or GaN native substrate; to find new growth methods for improving the crystal quality of AlGaN alloy and highly efficient doping; and to reduce the polarization effect in MQW structures.

Many problems such as lattice mismatch and thermal mismatch will be solved if the native substrates are provided with suitable price and wide usage. Many methods have been proposed to improve the crystal quality of AlGaN alloy, but the defect densities are still unsatisfactory. The doping efficacy of p-type AlGaN alloy is also not ideal and still needs improving. The active region of AlGaN-based light-emitting devices is mostly MQWs structure, and the polarization effect is bad for the generation of photons, so we should find some method to weaken it. However, it is pretty difficult to come up with a completely new method to improve crystal quality, doping efficiency, and weaken polarization effect. So there is greater possibility to improve the methods that are used at present in the next few years.

References

1. Hirayama H, Yatabe T, Noguchi N et al. 231–261 nm AlGaN deep-ultraviolet light-emitting diodes fabricated on AlN multilayer buffers grown by ammonia pulse-flow method on sapphire. *Applied Physics Letters*, 2007, 91(7): 071901.
2. Ambacher O. Growth and applications of group III-nitrides. *Journal of Physics D: Applied Physics*, 1998, 31(20): 2653.
3. Jayasakthi M, Ramesh R, Prabakaran K et al. Effect of Al-mole fraction in $Al_xGa_{1-x}N$ Grown by MOCVD. *AIP Conference Proceedings*, 2014, 1591: 1458–1460.
4. Sun X, Li D, Chen Y et al. In situ observation of two-step growth of AlN on sapphire using high-temperature metal–organic chemical vapour deposition. *CrystEngComm*, 2013, 15(30): 6066–6073.
5. Han J, Ng T B, Biefeld R M et al. The effect of H_2 on morphology evolution during GaN metalorganic chemical vapor deposition. *Applied Physics Letters*, 1997, 71(21): 3114–3116.
6. Chen Y, Song H, Li D et al. Influence of the growth temperature of AlN nucleation layer on AlN template grown by high-temperature MOCVD. *Materials Letters*, 2014, 114: 26–28.
7. Okada N, Kato N, Sato S et al. Growth of high-quality and crack free AlN layers on sapphire substrate by multi-growth mode modification. *Journal of Crystal Growth*, 2007, 298: 349–353.
8. Ohba Y, Yoshida H, Sato R. Growth of high-quality AlN, GaN and AlGaN with atomically smooth surfaces on sapphire substrates. *Japanese Journal of Applied Physics*, 1997, 36(12A): L1565.
9. Fujimoto N, Kitano T, Narita G et al. Growth of high-quality AlN at high growth rate by high-temperature MOVPE. *Physica Status Solidi (c)*, 2006, 3(6): 1617–1619.
10. Peng M Z, Guo L W, Zhang J et al. Effect of growth temperature of initial AlN buffer on the structural and optical properties of Al-rich AlGaN. *Journal of Crystal Growth*, 2007, 307(2): 289–293.
11. Linthicum K, Gehrke T, Thomson D et al. Pendeoepitaxy of gallium nitride thin films. *Applied Physics Letters*, 1999, 75(2): 196–198.
12. Sakai S, Wang T, Morishima Y et al. A new method of reducing dislocation density in GaN layer grown on sapphire substrate by MOVPE. *Journal of Crystal Growth*, 2000, 221(1): 334–337.
13. Ishibashi A, Kidoguchi I, Sugahara G et al. High-quality GaN films obtained by air-bridged lateral epitaxial growth. *Journal of Crystal Growth*, 2000, 221(1): 338–344.
14. Hiramatsu K, Nishiyama K, Onishi M et al. Fabrication and characterization of low defect density GaN using facet-controlled epitaxial lateral overgrowth (FACELO). *Journal of Crystal Growth*, 2000, 221(1): 316–326.
15. Tadatomo K, Okagawa H, Ohuchi Y et al. High-output power near-ultraviolet and violet light-emitting diodes fabricated on patterned sapphire substrates using metalorganic vapor phase epitaxy, *Optical Science and Technology, SPIE's 48th Annual Meeting*, International Society for Optics and Photonics, 2004: pp. 243–249.
16. Khan M A, Shatalov M, Maruska H P et al. III–nitride UV devices. *Japanese Journal of Applied Physics*, 2005, 44(10R): 7191.
17. Nagamatsu K, Okada N, Sugimura H et al. High-efficiency AlGaN-based UV light-emitting diode on laterally overgrown AlN. *Journal of Crystal Growth*, 2008, 310(7): 2326–2329.
18. Kim M, Fujita T, Fukahori S et al. AlGaN-based deep ultraviolet light-emitting diodes fabricated on patterned sapphire substrates. *Applied Physics Express*, 2011, 4(9): 092102.
19. Imura M, Nakano K, Kitano T et al. Microstructure of epitaxial lateral overgrown AlN on trench-patterned AlN template by high-temperature metal-organic vapor phase epitaxy. *Applied Physics Letters*, 2006, 89(22): 221901.

20. Zeimer U, Kueller V, Knauer A et al. High quality AlGaN grown on LEO AlN/sapphire templates. *Journal of Crystal Growth*, 2013, 377: 32–36.
21. Ozeki M, Mochizuki K, Ohtsuka N et al. New approach to the atomic layer epitaxy of GaAs using a fast gas stream. *Applied Physics Letters*, 1988, 53(16): 1509–1511.
22. Khan M A, Kuznia J N, Skogman R A et al. Low pressure metalorganic chemical vapor deposition of AlN over sapphire substrates. *Applied Physics Letters*, 1992, 61(21): 2539–2541.
23. Khan M A, Adivarahan V, Zhang J P et al. Stripe geometry ultraviolet light emitting diodes at 305 nanometers using quaternary AlInGaN multiple quantum wells. *Japanese Journal of Applied Physics*, 2001, 40(12A): L1308.
24. Sang L W, Qin Z X, Fang H et al. AlGaN-based deep-ultraviolet light emitting diodes fabricated on AlN/sapphire template. *Chinese Physics Letters*, 2009, 26(11): 117801.
25. Sang L W, Qin Z X, Fang H et al. Reduction in threading dislocation densities in AlN epilayer by introducing a pulsed atomic-layer epitaxial buffer layer. *Applied Physics Letters*, 2008, 93(12): 122104.
26. Cicek E, McClintock R, Cho C Y et al. Al_xGa_{1-x}N-based back-illuminated solar-blind photodetectors with external quantum efficiency of 89%. *Applied Physics Letters*, 2013, 103(19): 191108.
27. Zhang J, Hu X, Lunev A et al. AlGaN deep-ultraviolet light-emitting diodes. *Japanese Journal of Applied Physics*, 2005, 44(10R): 7250.
28. Banal R G, Funato M, Kawakami Y. Initial nucleation of AlN grown directly on sapphire substrates by metal-organic vapor phase epitaxy. *Applied Physics Letters*, 2008, 92(24): 241905.
29. Banal R G, Funato M, Kawakami Y. Surface diffusion during metalorganic vapor phase epitaxy of AlN. *Physica Status Solidi (c)*, 2009, 6(2): 599–602.
30. Hirayama H, Noguchi N, Yatabe T et al. 227 nm AlGaN light-emitting diode with 0.15 mW output power realized using a thin quantum well and AlN buffer with reduced threading dislocation density. *Applied Physics Express*, 2008, 1(5): 051101.
31. Hirayama H, Fujikawa S, Noguchi N et al. 222–282 nm AlGaN and InAlGaN-based deep-UV LEDs fabricated on high-quality AlN on sapphire. *Physica Status Solidi (a)*, 2009, 206(6): 1176–1182.
32. Hirayama H, Noguchi N, Kamata N. 222 nm deep-ultraviolet AlGaN quantum well light-emitting diode with vertical emission properties. *Applied Physics Express*, 2010, 3(3): 032102.
33. Yan J, Wang J, Zhang Y et al. AlGaN-based deep-ultraviolet light-emitting diodes grown on High-quality AlN template using MOVPE. *Journal of Crystal Growth*, 2015, 414: 254–257.
34. Kamiyama S, Iwaya M, Hayashi N et al. Low-temperature-deposited AlGaN interlayer for improvement of AlGaN/GaN heterostructure. *Journal of Crystal Growth*, 2001, 223(1): 83–91.
35. Han J, Waldrip K E, Lee S R et al. Control and elimination of cracking of AlGaN using low-temperature AlGaN interlayers. *Applied Physics Letters*, 2001, 78(1): 67–69.
36. Bläsing J, Reiher A, Dadgar A et al. The origin of stress reduction by low-temperature AlN interlayers. *Applied Physics Letters*, 2002, 81(15): 2722–2724.
37. Çörekçi S, Öztürk M K, Akaoğlu B et al. Structural, morphological, and optical properties of AlGaN/GaN heterostructures with AlN buffer and interlayer. *Journal of Applied Physics*, 2007, 101(12): 123502.
38. Li D B, Aoki M, Miyake H et al. Improved surface morphology of flow-modulated MOVPE grown AlN on sapphire using thin medium-temperature AlN buffer layer. *Applied Surface Science*, 2007, 253(24): 9395–9399.
39. Ren F, Hao Z B, Zhang C et al. High quality AlN with a thin interlayer grown on a sapphire substrate by plasma-assisted molecular beam epitaxy. *Chinese Physics Letters*, 2010, 27(6): 068101.
40. Kim J, Pyeon J, Jeon M et al. Growth and characterization of high quality AlN using combined structure of low temperature buffer and superlattices for applications in the deep ultraviolet. *Japanese Journal of Applied Physics*, 2015, 54(8): 081001.
41. Kida Y, Shibata T, Miyake H et al. Metalorganic vapor phase epitaxy growth and study of stress in AlGaN using epitaxial AlN as underlying layer. *Japanese Journal of Applied Physics*, 2003, 42(6A): L572.

42. Amano H, Iwaya M, Hayashi N et al. Control of dislocations and stress in AlGaN on sapphire using a low temperature interlayer. *Physica Status Solidi B Basic Research*, 1999, 216: 683–690.

43. Yoshida H, Yamashita Y, Kuwabara M et al. A 342-nm ultraviolet AlGaN multiple-quantum-well laser diode. *Nature Photonics*, 2008, 2(9): 551–554.

44. Kueller V, Knauer A, Brunner F et al. Growth of AlGaN and AlN on patterned AlN/sapphire templates. *Journal of Crystal Growth*, 2011, 315(1): 200–203.

45. Knauer A, Zeimer U, Kueller V et al. MOVPE growth of $Al_xGa_{1-x}N$ with x ~ 0.5 on epitaxial laterally overgrown AlN/sapphire templates for UV-LEDs. *Physica Status Solidi (c)*, 2014, 11(3-4): 377–380.

46. Wang H M, Zhang J P, Chen C Q et al. AlN/AlGaN superlattices as dislocation filter for low-threading-dislocation thick AlGaN layers on sapphire. *Applied Physics Letters*, 2002, 81(4): 604–606.

47. Zhang J P, Wang H M, Gaevski M E et al. Crack-free thick AlGaN grown on sapphire using AlN/AlGaN superlattices for strain management. *Applied Physics Letters*, 2002, 80(19): 3542.

48. Fujioka A, Misaki T, Murayama T et al. Improvement in output power of 280-nm deep ultraviolet light-emitting diode by using AlGaN multi quantum wells. *Applied Physics Express*, 2010, 3(4): 041001.

49. Yang W, Li J, Lin W et al. Control of two-dimensional growth of AlN and high Al-content AlGaN-based MQWs for deep-UV LEDs. *AIP Advances*, 2013, 3(5): 052103.

50. Sedhain A, Lin J Y, Jiang H X. Si-doped high Al-content AlGaN epilayers with improved quality and conductivity using indium as a surfactant. *Applied Physics Letters*, 2008, 92(9): 092105.

51. Cantu P, Keller S, Mishra U K et al. Metalorganic chemical vapor deposition of highly conductive Al0. 65Ga0. 35N films. *Applied Physics Letters*, 2003, 82(21): 3683–3685.

52. Wang S, Zhang X, Zhu M et al. Crack-free Si-doped n-AlGaN film grown on sapphire substrate with high-temperature AlN interlayer. *Optik—International Journal for Light and Electron Optics*, 2015, 126(23): 3698–3702.

53. Jeon S R, Son S J, Park S H. Microscope investigation and electrical conductivity of Si-doped n-type Al0. 45Ga0. 55N layer grown on AlGaN/AlN superlattices/SPIE OPTO. *International Society for Optics and Photonics*, 2014: 90031S–90031S-7.

54. He C, Qin Z, Xu F et al. Growth of high quality n-Al 0.5 Ga 0.5 N thick films by MOCVD. *Materials Letters*, 2016, 176: 298–300.

55. Kim K H, Li J, Jin S X et al. III-nitride ultraviolet light-emitting diodes with delta doping. *Applied Physics Letters*, 2003, 83(3): 566–568.

56. Zhu S, Yan J, Zhang Y et al. The effect of delta-doping on Si-doped Al rich n-AlGaN on AlN template grown by MOCVD. *Physica Status Solidi (c)*, 2014, 11(3–4): 466–468.

57. Chen Y, Wu H, Han E et al. High hole concentration in p-type AlGaN by indium-surfactant-assisted Mg-delta doping. *Applied Physics Letters*, 2015, 106(16): 162102.

58. Cho H K, Lee J Y, Jeon S R et al. Influence of Mg doping on structural defects in AlGaN layers grown by metalorganic chemical vapor deposition. *Applied Physics Letters*, 2001, 79(23): 3788–3790.

59. Jeon S R, Ren Z, Cui G et al. Investigation of Mg doping in high-Al content p-type $Al_xGa_{1-x}N$ (0.3< x< 0.5). *Applied Physics Letters*, 2005, 86(8): 2107.

60. Amano H, Kito M, Hiramatsu K et al. P-type conduction in Mg-doped GaN treated with low-energy electron beam irradiation (LEEBI). *Japanese Journal of Applied Physics*, 1989, 28(12A): L2112.

61. Kim J K, Waldron E L, Li Y L et al. P-type conductivity in bulk $Al_xGa_{1-x}N$ and $Al_xGa_{1-x}N/Al_yGa_{1-y}N$ superlattices with average Al mole fraction> 20%. *Applied Physics Letters*, 2004, 84(17): 3310–3312.

62. Xu Z Y, Qin Z X, Sang L W et al. Effect of indium ambient on electrical properties of Mg-doped $Al_xGa_{1-x}N$. *Chinese Physics Letters*, 2010, 27(12): 127304.

63. Kinoshita T, Obata T, Yanagi H et al. High p-type conduction in high-Al content Mg-doped AlGaN. *Applied Physics Letters*, 2013, 102(1): 012105.

64. Nakarmi M L, Kim K H, Li J et al. Enhanced p-type conduction in GaN and AlGaN by Mg-δ-doping. *Applied Physics Letters*, 2003, 82(18): 3041–3043.

65. Waldron E L, Graff J W, Schubert E F et al. Improved mobilities and resistivities in modulation-doped p-type AlGaN/GaN superlattices, *MRS Proceedings*, Cambridge University Press, Cambridge, U.K., 2001, Vol. 693: pp. I12. 11.1.

66. Li J, Yang W, Li S et al. Enhancement of p-type conductivity by modifying the internal electric field in Mg-and Si-δ-codoped $Al_xGa_{1-x}N/Al_yGa_{1-y}N$ superlattices. *Applied Physics Letters*, 2009, 95(15): 151113.

67. Simon J, Protasenko V, Lian C et al. Polarization-induced hole doping in wide–band-gap uniaxial semiconductor heterostructures. *Science*, 2010, 327(5961): 60–64.

68. Islam S M, Protasenko V, Rouvimov S et al. Deep-UV LEDs using polarization-induced doping: Electroluminescence at cryogenic temperatures, *IEEE 73rd Annual Device Research Conference (DRC)*, 2015: pp. 67–68.

69. Kawanishi H, Tomizawa T. Carbon-doped p-type (0001) plane AlGaN (Al = 6–55%) with high hole density. *Physica Status Solidi (b)*, 2012, 249(3): 459–463.

70. Li S, Zhang T, Wu J et al. Polarization induced hole doping in graded $Al_xGa_{1-x}N$ (x = 0.7 ~ 1) layer grown by molecular beam epitaxy. *Applied Physics Letters*, 2013, 102(6): 062108.

71. Dong K X, Chen D J, Shi J P et al. Characteristics of deep ultraviolet AlGaN-based light emitting diodes with p-hBN layer. *Physica E: Low-Dimensional Systems and Nanostructures*, 2016, 75: 52–55.

72. Kishino K, Kikuchi A, Kaneko Y et al. Enhanced carrier confinement effect by the multiquantum barrier in 660 nm GaInP/AlInP visible lasers. *Applied Physics Letters*, 1991, 58(17): 1822–1824.

73. Banal R G, Funato M, Kawakami Y. Characteristics of high Al-content AlGaN/AlN quantum wells fabricated by modified migration enhanced epitaxy. *Physica Status Solidi (c)*, 2010, 7(7-8): 2111–2114.

74. Zhao D G, Jiang D S, Zhu J J et al. An experimental study about the influence of well thickness on the electroluminescence of InGaN/GaN multiple quantum wells. *Journal of Alloys and Compounds*, 2010, 489(2): 461–464.

75. Hollander J L, Kappers M J, McAleese C et al. Improvements in a-plane GaN crystal quality by a two-step growth process. *Applied Physics Letters*, 2008, 92(10): 101104.

76. Zhang M, Li Y, Chen S et al. Performance improvement of AlGaN-based deep ultraviolet light-emitting diodes by using staggered quantum wells. *Superlattices and Microstructures*, 2014, 75: 63–71.

77. Chang C Y, Li H, Lu T C. High efficiency InGaN/GaN light emitting diodes with asymmetric triangular multiple quantum wells. *Applied Physics Letters*, 2014, 104(9): 091111.

78. Li L, Tsutsumi T, Miyachi Y et al. Improved performance of AlGaN-based deep ultraviolet light-emitting diodes with n-AlGaN underlayers. *Semiconductor Science and Technology*, 2015, 30(12): 125012.

79. Shen L, Heikman S, Moran B et al. AlGaN/AlN/GaN high-power microwave HEMT. *IEEE Electron Device Letters*, 2001, 22(10): 457–459.

80. Hu W, Ma B, Li D et al. Mobility enhancement of 2DEG in MOVPE-grown AlGaN/AlN/GaN HEMT structure using vicinal (0 0 0 1) sapphire. *Superlattices and Microstructures*, 2009, 46(6): 812–816.

81. Kolluri S, Keller S, DenBaars S P et al. N-polar GaN MIS-HEMTs with a 12.1-W/mm continuous-wave output power density at 4 GHz on sapphire substrate. *IEEE Electron Device Letters*, 2011, 32(5): 635–637.

82. Sivadasan A K, Patsha A, Polaki S et al. Optical properties of monodispersed AlGaN nanowires in the single-prong growth mechanism. *Crystal Growth & Design*, 2015, 15(3): 1311–1318.

83. Chen F, Ji X, Lu Z et al. Structural and Raman properties of compositionally tunable $Al_xGa_{1-x}N$ (0.66 ≤ x≤ 1) nanowires. *Materials Science and Engineering: B*, 2014, 183: 24–28.

84. Hong L, Liu Z, Zhang X T et al. Self-catalytic growth of single-phase AlGaN alloy nanowires by chemical vapor deposition. *Applied Physics Letters*, 2006, 89(19): 193105.

85. Su J, Gherasimova M, Cui G et al. Growth of AlGaN nanowires by metalorganic chemical vapor deposition. *Applied Physics Letters*, 2005, 87(18): 183108.

86. Wang G T, Li Q, Huang J et al. III-nitride nanowires: Growth, properties, and applications, *SPIE NanoScience+ Engineering, International Society for Optics and Photonics*, 2010: pp. 77680K–77680K-9.

87. Choi K, Arita M, Kako S et al. Site-controlled growth of single GaN quantum dots in nanowires by MOCVD. *Journal of Crystal Growth*, 2013, 370: 328–331.

88. Choi K, Kako S, Holmes M J et al. Strong exciton confinement in site-controlled GaN quantum dots embedded in nanowires. *Applied Physics Letters*, 2013, 103(17): 171907.

89. Simpkins B S, Mastro M A, Eddy Jr C R et al. Space-charge-limited currents and trap characterization in coaxial AlGaN/GaN nanowires. *Journal of Applied Physics*, 2011, 110(4): 044303.

90. Li Y, Xiang J, Qian F et al. Dopant-free GaN/AlN/AlGaN radial nanowire heterostructures as high electron mobility transistors. *Nano Letters*, 2006, 6(7): 1468–1473.

91. Fonoberov V A, Balandin A A. Excitonic properties of strained wurtzite and zinc-blende GaN/Al$_x$Ga$_{1-x}$N quantum dots. *Journal of Applied Physics*, 2003, 94(11): 7178–7186.

92. Xia C X, Wei S Y, Zhao X. Built-in electric field effect on hydrogenic impurity in wurtzite GaN/AlGaN quantum dot. *Applied Surface Science*, 2007, 253(12): 5345–5348.

93. Xia C, Jiang F, Wei S et al. Hydrogenic impurity in zinc-blende GaN/AlGaN quantum dot. *Microelectronics Journal*, 2007, 38(6): 663–666.

94. Rostami A, Saghai H R, Nejad H B A. Defect-induced enhancement of absorption coefficient and electroabsorption properties in GaN/AlGaN centered defect quantum box (CDQB) nanocrystal. *Physica B: Condensed Matter*, 2008, 403(17): 2789–2796.

95. Romano G, Penazzi G, Di Carlo A. Multiscale thermal modeling of GaN/AlGaN quantum dot LEDs. *OPTO, International Society for Optics and Photonics*, 2010: pp. 75971S–75971S-8.

96. Revathi M, Yoo C K, Peter A J. Magneto-exciton in a GaN/Ga$_{1-x}$Al$_x$N quantum dot. *Physica E: Low-Dimensional Systems and Nanostructures*, 2010, 43(1): 322–326.

97. Sharkey J J, Yoo C K, Peter A J. Magnetic field induced diamagnetic susceptibility of a hydrogenic donor in a GaN/AlGaN quantum dot. *Superlattices and Microstructures*, 2010, 48(2): 248–255.

98. Kouhi M, Vahedi A, Akbarzadeh A et al. Investigation of quadratic electro-optic effects and electroabsorption process in GaN/AlGaN spherical quantum dot. *Nanoscale Research Letters*, 2014, 9(1): 1–6.

99. Carnevale S D, Kent T F, Phillips P J et al. Graded nanowire ultraviolet LEDs by polarization engineering. *SPIE NanoScience+ Engineering, International Society for Optics and Photonics*, 2012: pp. 84670L–84670L-11.

100. Carnevale S D, Yang J, Phillips P J et al. Three-dimensional GaN/AlN nanowire heterostructures by separating nucleation and growth processes. *Nano Letters*, 2011, 11(2): 866–871.

101. Carnevale S D, Kent T F, Phillips P J et al. Polarization-induced pn diodes in wide-band-gap nanowires with ultraviolet electroluminescence. *Nano Letters*, 2012, 12(2): 915–920.

102. Himwas C, den Hertog M, Dang L S et al. Alloy inhomogeneity and carrier localization in AlGaN sections and AlGaN/AlN nanodisks in nanowires with 240–350 nm emission. *Applied Physics Letters*, 2014, 105(24): 241908.

103. Kent T F, Carnevale S D, Sarwar A T M et al. Deep ultraviolet emitting polarization induced nanowire light emitting diodes with Al$_x$Ga$_{1-x}$N active regions. *Nanotechnology*, 2014, 25(45): 455201.

104. Zhao S, Liu X, Woo S Y et al. An electrically injected AlGaN nanowire laser operating in the ultraviolet-C band. *Applied Physics Letters*, 2015, 107(4): 043101.

105. Himwas C, Songmuang R, Dang L S et al. Thermal stability of the deep ultraviolet emission from AlGaN/AlN Stranski-Krastanov quantum dots. *Applied Physics Letters*, 2012, 101(24): 241914.

106. Songmuang R, Kalita D, Sinha P et al. Strong suppression of internal electric field in GaN/AlGaN multi-layer quantum dots in nanowires. *Applied Physics Letters*, 2011, 99(14): 141914.

107. Brault J, Damilano B, Kahouli A et al. Ultra-violet GaN/Al 0.5 Ga 0.5 N quantum dot based light emitting diodes. *Journal of Crystal Growth*, 2013, 363: 282–286.

108. Brown J, Wu F, Petroff P M et al. GaN quantum dot density control by rf-plasma molecular beam epitaxy. *Applied Physics Letters*, 2004, 84(5): 690–692.

109. Widmann F, Simon J, Daudin B et al. Blue-light emission from GaN self-assembled quantum dots due to giant piezoelectric effect. *Physical Review B*, 1998, 58(24): R15989.

110. Widmann F, Daudin B, Feuillet G et al. Growth kinetics and optical properties of self-organized GaN quantum dots. *Journal of Applied Physics*, 1998, 83(12): 7618–7624.

111. Yoshizawa M, Kikuchi A, Fujita N et al. Self-organization of GaN/Al$_{0.18}$Ga$_{0.82}$N multi-layer nano-columns on (0001) Al$_2$O$_3$ by RF molecular beam epitaxy for fabricating GaN quantum disks. *Journal of Crystal Growth*, 1998, 189: 138–141.

112. Adelmann C, Guerrero E M, Chabuel F et al. Growth and characterisation of self-assembled cubic GaN quantum dots. *Materials Science and Engineering: B*, 2001, 82(1): 212–214.

113. Kako S, Holmes M, Sergent S et al. Single-photon emission from cubic GaN quantum dots. *Applied Physics Letters*, 2014, 104(1): 011101.

114. Hirayama H, Tsukada Y, Maeda T et al. Marked enhancement in the efficiency of deep-ultraviolet AlGaN light-emitting diodes by using a multiquantum-barrier electron blocking layer. *Applied Physics Express*, 2010, 3(3): 031002.

115. Bryan Z, Bryan I, Xie J et al. High internal quantum efficiency in AlGaN multiple quantum wells grown on bulk AlN substrates. *Applied Physics Letters*, 2015, 106(14): 142107.

116. Shatalov M, Sun W, Jain R et al. High power AlGaN ultraviolet light emitters. *Semiconductor Science and Technology*, 2014, 29(8): 084007.

117. Hirayama H, Maeda N, Fujikawa S et al. Recent progress and future prospects of AlGaN-based high-efficiency deep-ultraviolet light-emitting diodes. *Japanese Journal of Applied Physics*, 2014, 53(10): 100209.

118. Tan S, Egawa T, Luo X D et al. Influence of barrier height and p-cladding layer on electroluminescent performance of AlGaN deep ultraviolet light-emitting diodes. *Journal of Physics D: Applied Physics*, 2016, 49(12): 125102.

119. Liang H, Tao P, Xia X et al. Vertically conducting deep-ultraviolet light-emitting diodes with inter-band tunneling junction grown on 6H-SiC substrate. *Japanese Journal of Applied Physics*, 2016, 55(3): 031202.

120. Lee J W, Kim D Y, Park J H et al. An elegant route to overcome fundamentally-limited light extraction in AlGaN deep-ultraviolet light-emitting diodes: Preferential outcoupling of strong in-plane emission. *Scientific Reports*, 2016, 6: 22537.

121. Shih Y H, Chang J Y, Sheu J K et al. Design of hole-blocking and electron-blocking layers in Al$_x$Ga$_{1-x}$N-based UV light-emitting diodes. *IEEE Transactions on Electron Devices*, 2016, 63(3): 1141–1147.

122. Xu M, Zhou Q, Zhang H et al. Improved efficiency of near-ultraviolet LEDs using a novel p-type AlGaN hole injection layer. *Superlattices and Microstructures*, 2016, 94: 25–29.

123. Martens M, Kuhn C, Ziffer E et al. Low absorption loss p-AlGaN superlattice cladding layer for current-injection deep ultraviolet laser diodes. *Applied Physics Letters*, 2016, 108(15): 151108.

124. Yoshida H, Kuwabara M, Yamashita Y et al. The current status of ultraviolet laser diodes. *Physica Status Solidi (a)*, 2011, 208(7): 1586–1589.

125. Zhao S, Woo S Y, Bugnet M et al. Three-dimensional quantum confinement of charge carriers in self-organized AlGaN nanowires: A viable route to electrically injected deep ultraviolet lasers. *Nano Letters*, 2015, 15(12): 7801–7807.

126. Biyikli N, Kimukin I, Aytur O et al. Solar-blind AlGaN-based pin photodiodes with low dark current and high detectivity. *IEEE Photonics Technology Letters*, 2004, 16(7): 1718–1720.

127. Albrecht B, Kopta S, John O et al. AlGaN ultraviolet A and ultraviolet C photodetectors with very high specific detectivity D. *Japanese Journal of Applied Physics*, 2013, 52(8S): 08JB28.

128. Tang Y, Shinohara K, Regan D et al. Ultrahigh-speed GaN high-electron-mobility transistors with of 454/444 GHz. *IEEE Electron Device Letters*, 2015, 36(6): 549–551.

II

GaN-Based LEDs for Lighting

II

GaN-Based LEDs

Applications

5

Efficiency Droop of Nitride-Based Light-Emitting Diodes

Chia-Yen Huang
National Chiao Tung University

Hao-Chung Kuo
National Chiao Tung University

Abstract Efficiency droop of nitride-based light emitting diodes (LED) was intensively studied for high-power applications. The origin of droops was attributed to many root causes, for example, carrier leakage, internal piezo-polarization field, Auger recombination, carrier delocalization, and thermal effects. The physical origins were modeled and visualized by various techniques, which will be described in detail in this chapter. Many droop alleviation strategies were proposed by an epitaxial layer structure optimization and device structure innovation, where many of them were widely applied in current commercial products.

5.1 Introduction

In the 1990s, a series of breakthroughs in crystal growth and processing techniques led to the first demonstration of candela-class blue-violet light-emitting diodes (LEDs) with a long lifetime and high reliability with Zn-doped GaN quantum wells [1]. After decades of development, the peak external quantum efficiency (EQE) was first reported to be more than 80% by Nichia Inc. [2]. The state-of-art efficacy was also announced to be ~300 lumen-per-Watt (lm/W) by Cree in 2014. In 2016, the EQE of commercial LED chips with mass production level had an average EQE 65%~70% under low injection current densities ($J \sim 10$ A/cm²). The efficacy was also more than 150 lm/W under the same injection level. In comparison, the efficacy of a compact fluorescent bulb was only 80–100 lm/W. The superior performance under low injection opened the era of LED backlight modules, resulting in an almost 100% penetration rate of LEDs for commercial LCD displays.

However, for the purpose of general illumination, the module requires to exert a much higher total lumen flux, which can be achieved with two approaches. An intuitive approach is to assemble many

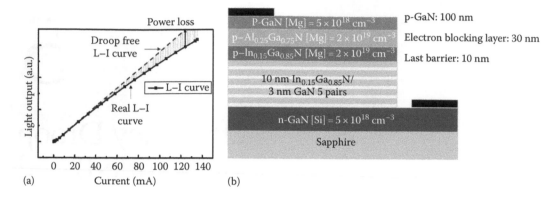

(a)   (b)

FIGURE 5.1   (a) Simulated L–I curve of the reference LED structure in (b). The Shockley–Read–Hall (SRH) lifetime is assumed to be 3 ns and Auger recombination coefficient is assumed to be $1*10^{-30}$ cm^6/s in the simulation.

LED chips in the light bulb to accumulate the total flux. In spite of the simple concept, this approach brought multiple issues for lighting purposes. The multiple point light sources impose many challenges in secondary optics such as nonuniform intensity or color distribution and multiple shadowing, but the most important issue is still the cost. The cost of materials increases proportionally with the number of chips packaged in the lighting module, causing a low profit margin and product competitiveness.

Therefore, from the point of lumen-per-dollar (lm/$) rather than lm/W, the second approach for high lumen lighting module is much preferred: driving up the current density for each single chip. If the module could provide the same amount of total lumen flux with less chips, the quality of light will be much more ideal with a lower cost. Unfortunately, in the mid-2000s, the efficiency of commercial LEDs dropped rapidly after overdriving the injection current, which is known as *efficiency droop* (or *droop*) in the community. Figure 5.1 shows a light–current (L–I) curve simulation of a conventional LED reference structure by APSYS, a commercial software developed by Crosslight Inc. According to the simulation, the power loss due to droop became much more significant under high-current injection. Therefore, overcoming droop became a research spotlight while the LEDs started to penetrate the general illumination market. From 2007, the physical mechanism of droop phenomena was studied and argued intensively. In the same period of time, many remedies to alleviate droop were also proposed.

This chapter will start with a brief review about different physical mechanisms proposed to contribute to the droop phenomena. These proposed mechanisms can be summarized into two major categories: carrier leakage (Section 5.2.1) and Auger recombination (Section 5.2.2). The party who supports carrier leakage mechanism suggested that a significant portion of carriers failed to populate in the active region and eventually vanish under high-current injection. In other words, the nonradiative recombination processes take place *outside the active region*. On the other hand, many institutes contributed to the Auger recombination process *inside the active region* as the major cause of droop (Figure 5.2). Thermal impacts due to unoptimized contact, device structure, and packaging were also attributed to the deteriorated performance under high-current injection (Section 5.2.3). Generally, thermal effects are not controversial. Studies about droop origin usually applied pulse PL excitation or pulse carrier injection to avoid self-heating.

In the final section of this chapter, we summarized various approaches to solve or mitigate droop issues. The general principle of epi-structure design is to enhance the "effective active volume" for carriers and reduce the potential carrier leakage paths (Sections 5.3.1 and 5.3.2). New device structures were also developed to overcome thermal dissipation issues under high-current injection (Section 5.3.3). Special approaches to overcome droop by growing LED on non-c-planes are also discussed (Section 5.3.3.4).

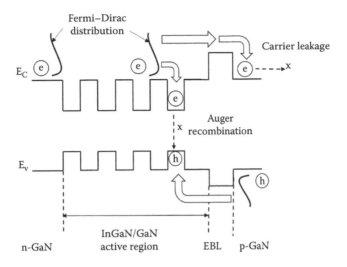

FIGURE 5.2 Schematic illustration of droop by carrier leakage and Auger recombination.

5.2 Origin of Droop Phenomenon

To quantitatively describe the internal quantum efficiency (IQE) as a function of carrier density (n), we expand the carrier recombination rate (R) as a polynomial of n:

$$R = \tau_R^{-1} + \tau_{NR}^{-1} = An + Bn^2 + Cn^3 + f(n)$$

where

τ_R and τ_{NR} are the lifetime of carriers undergoing radiative and nonradiative recombination, respectively
A, B, C are the rate coefficients
f(n) represents the high-order residuals

In most studies, f(n) is omitted for a more concise representation, giving the common *ABC model*:

$$R = An + Bn^2 + Cn^3$$

The physical interpretation of each term is as follows. Coefficient A illustrates the Shockley–Read–Hall (SRH) nonradiative recombination process due to carrier trapping. The injected carriers can be trapped by crystal imperfections such as impurity levels and threading dislocations (TD). Therefore, a high "A coefficient" value is equivalent to a short SRH lifetime (τ_{SRH}) and a poor crystal quality [3]. Bn^2 stands for the bimolecular radiative recombination of electrons and holes. Cn^3 is the dominating nonradiative recombination under high-injection current. Figure 5.3 is an illustration of common radiative and nonradiative processes in bulk materials. Therefore, we can define IQE as follows:

$$IQE(n) = \frac{\text{Radiative recombination rate}}{\text{Total recombination rate}} = \frac{Bn^2}{An + Bn^2 + Cn^3}$$

We can then derive the peak value of IQE:

$$\text{Max}(IQE) = \frac{B}{B + 2\sqrt{AC}} \quad \text{while } n = \sqrt{A/C}$$

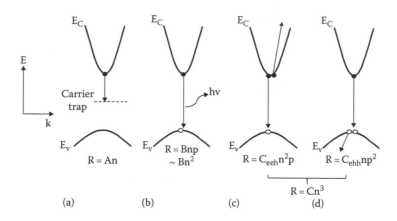

FIGURE 5.3 Schematic illustration of (a) SRH, (b) bimolecular recombination, (c) e-e-h Auger recombination, and (d) e-h-h Auger recombination.

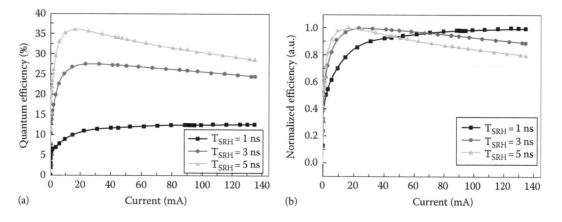

FIGURE 5.4 IQE curve with different SRH lifetime (a) before and (b) after normalization.

The values of A, B, and C coefficients were extracted or calculated in numerous studies. The typical orders of rate coefficients are reported 10^7–10^8 s^{-1} for A, 10^{-10}–10^{-11} $cm^{-3} \cdot s^{-1}$ for B, and 10^{-30}–10^{-29} $cm^{-6} \cdot s^{-1}$ for C. The reader should be aware that the carrier density for peak efficiency is also relevant to the A coefficient. Poor crystal quality also delays the IQE peak to higher carrier densities. As shown in Figure 5.4, after normalization of peak EQE to unity, it gives an illusion of droop improvement with a short τ_{SRH}. Therefore, the absolute output power of LED device is also very important for droop studies. If a droop improvement strategy does not result in actual higher output power under high-current injection, it is not valid.

In order to achieve a high efficiency under high-current injection, understanding the physical mechanisms behind the coefficient C is essential. The following sections will review theoretical infrastructures built to explain the droop phenomena.

5.2.1 Carrier Leakage

5.2.1.1 Asymmetric Carrier Injection

In nitride materials system, the physical parameters of electron and holes are highly asymmetric. The electron has a lower dopant activation energy and a higher mobility than those of holes regardless of the doping concentration [4–7] (Table 5.1).

TABLE 5.1   Activation Energy and Mobility of n-Type (Electrons) and p-Type (Holes) Carriers in Gallium Nitride

	Electrons	Holes
Activation energy, E_A (meV)	~20	~170
Carrier mobility, μ (cm²/V · s)	200~1000	5~20

According to the formula of drift current under electric field:

$$J_{drift} = en\mu E = e\mu N_0 \exp\left(-\frac{E_A}{kT}\right) * E$$

where
  n is the carrier density
  N_0 is the dopant concentration
  μ is the carrier mobility
  E_A is the activation energy
  E is the electric fields

Therefore, the electron drift current in the active region will be much higher than the hole current due to its high mobility and low activation energy, resulting in a concentrated radiative recombination profile near the p-side in multi-quantum wells (MQW). Therefore, without a proper design of electron blocking layer (EBL), the injected electron current is expected to overfly the active region and eventually vanish in the p-GaN layer or p-metal. Even with a proper EBL design, the injected electrons and holes are still expected to only populate the last one or two quantum wells (QW) next to the EBL.

The electron overflow is directly observed by Vampolar et al. by inserting an InGaN detection layer between EBL and p-GaN. The composition of the detection layer is designed to be less than those of QW in the active region, so their emission spectrum can be distinguished [8]. Without adequate Al composition in the EBL, the electroluminescence peak at shorter wavelength region was observed under high-current injection, indicating that the electrons were populated outside the active region. The nonuniform carrier distribution in the MQW is also observed by various techniques [9,10]. By analysis of the angle-resolve far-field pattern with Ag mirrors, Aurelien et al. concluded that the light emission only comes from the last one or two QWs regardless of the number of QWs in the active region. Dupuis et al. suggested that the hole transport among QWs is the limiting factor for uniform carrier distribution by analyzing dichromatic blue-green LEDs with various barrier structures.

Based on the high asymmetry between electron and hole physical parameters, Schubert et al. derived an analytical model of electron drift leakage current with fundamental diode characteristics. This model suggested that the electron drift leakage current has a cubic and fourth-order power dependence of carrier density in QWs (n_{QW}). Therefore, the Cn^3 and higher-order term in the ABC model could be well explained by the carrier leakage mechanisms [11,12].

5.2.1.2 Polarization and Hot Carrier Effects

Besides the asymmetry of physical parameters between electrons and holes, the piezoelectric nature of nitride materials also has a significant contribution in carriers leaking out of the active region. The wurtzite nitride is piezoelectric with internal polarization along the [0001] direction (c-axis) due to its low-structure symmetry. The analytical model of strain and internal polarization field as a function of alloy composition and crystallography planes can be found in Reference 13. Therefore, at the interfaces between two nitride layers with different internal polarization, sheet charges merge due to the polarization discontinuity (ΔP_z). These delta-like sheet charges result in a serrated potential profile in the band diagram. Figure 5.5 illustrates the relation between the piezoelectricity discontinuity and band bending

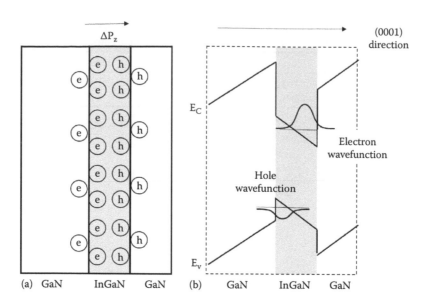

FIGURE 5.5   (a) Polarization, charge distribution, and (b) band bending of an InGaN/GaN double heterostructure.

in an InGaN/GaN double heterostructure. Carriers injected into a quantum well with strong internal electric fields suffer from a weak overlap of eigenfunctions of electrons and holes. Driving up LEDs with large ΔP_z accompanies a significant wavelength blueshift due to the Coulomb screening effect in the QW. These electroluminescence characteristics induced by polarization discontinuities is known as quantum confinement Stark effects (QCSE).

Besides influences on EL characteristics, the polarization-related electric fields also have a strong impact on carrier leakages [14,15]. According to band diagram simulations, the sheet charges in the active region strongly bend the valance band of the quantum barriers upward. Therefore, the peak energy of valance band in quantum barriers might exceed that in the EBL when the Al composition of EBL is not sufficient. Figure 5.6a shows band diagram simulations of the reference structure assuming 60% (black)

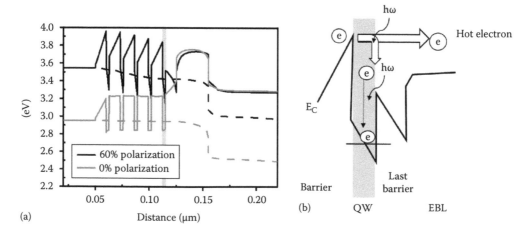

FIGURE 5.6   (a) Band diagram simulation of reference LEDs with (black) and without (gray) polarization under 100 A/cm² injection and (b) schematic illustration of phonon relaxation and hot electrons.

and 0% (red) polarizations, respectively. The electron confinement with 60% polarization is expected to be weaker due to the energy excess of the quantum barriers to the electron blocking layer.

A large internal polarization not only enhances electron overflow by failing the EBL function, but also reduces the probability of electrons to be captured by the quantum well [16,17]. Figure 5.6b illustrated the phonon relaxation process of injected electrons in the last QW. Injected electrons are eventually relaxed to lower energy states after scattering by phonons. The injected electrons sweeping through QWs without interaction with phonons are known as *hot electrons* and this carrier transport mechanism is known as *ballistic transport* [18]. In the ballistic transport model, when an electron is injected into QWs from quantum barriers, the potential energy difference in conduction band (ΔE_c) will be transformed into kinetic energy:

$$\Delta E_c = \frac{1}{2}m_e^* v_0^2,$$

where
  m_e is the effective mass of electron
  v_0 is the initial speed of electron after being injected into QWs

Although this equation is oversimplified for the real situation, it provides a quick insight of how the polarization field could influence the ballistic transport mechanism. If the dwelling time of injected carriers in the QW is less than the lifetime of phonon scattering, the injected carriers have a high probability sweeping through QW without being captured. According to band simulations, polarization discontinuity bends the valance band not only upward in the quantum barrier but also downward in the wells. The band bending results in a higher effective ΔE_c between QWs and barriers, resulting in higher initial speed and less dwelling time. Thus, the injected electrons have a higher probability to sweep through QW without being captured.

Although the ballistic transport might be suppressed by increasing well thickness, the light output power is not necessarily improved due to nonoverlapping wavefunctions. Thick InGaN QWs also suffer from strain-induced crystal defects beyond critical thickness. These limitations are more stringent for nitride-based long-wavelength devices due to its high-indium-content QWs. Sizov et al. [19] suggested that the ballistic transport is the dominating mechanism of inter-QW carrier transport for aquamarine-to-green LED and laser diodes.

In summary, the polarization discontinuities in LED epitaxy have a crucial influence on carrier leakages. The band bending deteriorated the function of EBL under forward bias. Hot carrier effects provide an additional channel for electrons to escape the active region. A significant portion of droop remedies are proposed to mitigate or circumvent polarization effects, which will be discussed in more detail in Section 5.3.

5.2.1.3 Delocalization Effects

The prosperity of nitride-based LED can be largely attributed to its defect insensitivity [20]. In phosphide-based quaternary red-IR LEDs, the threading dislocation densities (TDD) are less than 10^3 cm^{-2} for high power operation. In comparison, the InGaN-based LEDs still have good performance with TDD as high as 10^8 cm^{-2}. The defect insensitivity of InGaN LEDs is explained by the compositional inhomogeneity of InGaN QWs [21,22]. The InGaN segregation introduced many sites with local potential energy minimum in QWs. Therefore, the injected (or excited) carriers will be confined in the localized states without being trapped by point defects or line defects in the crystal. As mentioned in previous paragraph, point defect contributed to the SRH nonradiative recombination and the quantity of "A coefficient" in ABC models. The screw- and mixed-type threading dislocations are experimentally confirmed as carrier leakage paths of electronic and optoelectronic devices [23,24]. In mesoscopic dimension, the formation of pits also prevents carriers from being trapped by TDs [25]. Calculations revealed that the potential energy is high in the vicinity of the pit center, which further screens the carriers from leaking through the TD. Therefore, some groups proposed to transform all TD into V-pits to improve light output performance [26].

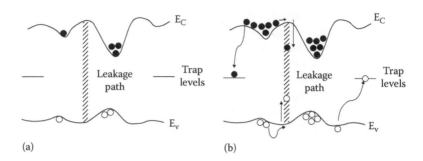

FIGURE 5.7 Schematic illustration of (a) carrier localization under low injection and (b) carrier delocalization under high injection.

However, as more carriers are injected or excited in the QWs, the localized states are eventually filled. The newly injected carriers have a higher probability to diffuse those nonradiative recombination centers (NRCs) or leakage paths. In other words, carriers are delocalized and efficiency becomes vulnerable to crystal defects under high-carrier densities. The *delocalization effect for droops* is a reverse interpretation of *defect insensitivity* of InGaN LEDs. Many theoretical models are also developed to include carrier-density-activated defect or leakage paths to explain the origin of droop [27]. Figure 5.7 shows the defect insensitivity under low injection and carrier-density-activated defects due to carrier localization and delocalization, respectively.

Delocalization of carriers in QW is experimentally observed by various techniques, such as time-resolved photoluminescence (TR-PL) [28,29], free carrier absorption (FCA) [29], and full width at half maximum (FWHM) of EL spectrum [30]. In TR-PL and FCA, enhanced carrier lateral diffusivity and carrier lifetime were observed under high pumping densities, implying a longer diffusion length of carriers. Based on the EL FWHM evolution under different forward biases, the hopping and rethermalization of carriers between shallow states and deep states were interpreted. Those experiments all correlated the delocalization with PL or EL droop phenomena decently.

In a word, the delocalization effect originates from the activation of preexisting defects, so it would be more significant when the QW crystal quality is inherently poor. Therefore, improving crystal quality is also crucial for overcoming droops.

5.2.1.4 Auger-Assisted Carrier Overflow

The Auger recombination and carrier leakage mechanism were regarded as independent or competing mechanisms in the early stage of droop studies. In 2013, Iveland et al. observed hot electrons emitting out of an operating LED, whose characteristic energy matched the Auger electrons. The community started to acknowledge that the carrier leakage can also be assisted by Auger recombination in the QW [31]. Different from the mechanism discussed in Section 5.2.1.2, the hot carriers were captured by QWs and then reemitted out of quantum well due to Auger recombination. Figure 5.8 compared the hot electrons induced by ballistic transport and Auger recombination. The detailed physical mechanism and relevant experiments of Auger-generated hot electrons will be summarized in detail in the next section.

5.2.2 Auger Recombination

5.2.2.1 PL Experiments

Auger recombination is a well-known nonradiative recombination in bulk materials that requires the participation of three carriers. The process initiates with the recombination of an electron and a hole without light emission. The energy of recombination is transferred to a third carrier, either another electron

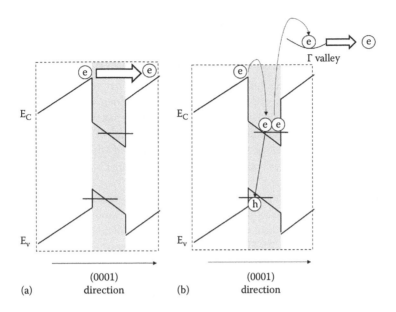

FIGURE 5.8 Physical processes of hot electrons generated from (a) ballistic transport and (b) Auger recombination.

(e-e-h) or another hole (e-h-h), exciting them to higher energy states. Therefore, the recombination rate of Auger recombination (R_{Auger}) can be expressed as

$$R_{Auger} = C_{eeh}N^2P + C_{ehh}P^2N$$

where
  N is the electron concentration
  P is the hole concentration
  C_{eeh} is the recombination rate coefficient involving two electrons and a hole
  C_{ehh} is the rate coefficient for the other case

For a single bulk material under injection or excitation, the carrier concentration of electron is assumed identical (N = P), so the R_{Auger} can be simplified in terms of carrier concentration n:

$$R_{Auger} = Cn^3,$$

where $C = C_{eeh} + C_{ehh}$. The nonradiative Auger recombination has a cubic order of the carrier concentration; therefore, it is intuitive to link the Auger recombination to the cubic term in ABC models. Auger recombination per se is not controversial, but its contribution to efficiency droop in InGaN LEDs was debated intensively by experiments and theoretical calculations.

In order to verify the existence of Auger recombination, PL experiments with increasing excitation power density were conducted by many groups [32–37]. Since the optical pumping does not involve carrier injection, the influences from the asymmetry of electrons and holes can be decoupled. Therefore, if PL efficiency droop with optical pumping density can be observed either in LED structure or in single InGaN/GaN double heterostructure (DH), the significance of Auger recombination is also unraveled.

The results of PL experiment are not consistent among different groups. Krames et al. from Philipps Lumiled observed PL efficiency decay with an onset carrier density ~5 ∗ 10^{18} cm^{-3} [32]. On the other side, Shubert et al. from Rensselaer Polytechnic Institute did not observe any PL droop phenomena with PL excitation density up to 30 kW/cm^2 [35]. The discrepancy between PL experiment results is the origin of droop controversy.

Many root causes are proposed to explain the discrepancy, such as measurement methodology, crystal quality, or emergence of stimulated emission [37]. Until 2016, the controversy remains an open question.

5.2.2.2 Estimation of Auger Coefficient of Bulk Nitride Materials

The significance of the Auger recombination process was also argued via theoretical calculations. Generally, the significance of the Auger process is not doubted in long-wavelength LEDs. However, the Auger recombination is expected to be less likely in wide bandgap materials because of their weak coupling of valance band and conduction bands and strong binding energy of electron–hole excitons. In 2008, Hader et al. estimated the Auger coefficient to be $3.5 * 10^{-34}$ cm$^{-6} \cdot$ s^{-1} by direct transition [38], which is three to four orders smaller than most experimentally extracted rate coefficients from real LED devices with ABC models (C = 10^{-30}–10^{-31} cm$^{-6} \cdot$ s^{-1}) [39–45].

In 2009 and 2010, two groups both pointed out that the direct Auger transition coefficient will be much significant if the coupling of Auger electrons to higher energy sub-bands is considered, a.k.a. interband Auger recombination [46,47]. The calculated interband Auger coefficient was reported to be ~$2 * 10^{-30}$ cm$^{-6} \cdot$ s^{-1} at bandgap energy around 2.5 eV, whose corresponding wavelength is 496 nm. The interband Auger process still could not explain the droop because the interband Auger process has almost no contribution in the blue spectral region ($\lambda = 450$–460 nm, $E_g \sim 2.75$ eV).

In 2011, Kioupakis et al. calculated the Auger coefficient incorporating the phonon scattering and alloy disorders [48]. The phonon and alloy scattering greatly loosen the criterion of momentum conservation in the Auger process. Therefore, despite the fact that more elements are incorporated in a single process for these indirect Auger recombination processes, the calculated recombination rate is actually significantly enhanced. The accumulated Auger coefficient via indirect transitions is estimated to be $2 * 10^{-31}$–$4 * 10^{-31}$ cm$^{-6} \cdot$ s^{-1} in wide visible spectral region, which is closed to experimentally extracted C-values in some studies. Therefore, the indirect Auger process is suggested to have a significant role in the droop phenomena of LEDs. Figure 5.9 is a schematic illustration of direct, interband, and indirect Auger processes involving two electrons and one hole.

5.2.2.3 Auger Recombination Rate in Devices

The Auger recombination rate is not only influenced by fundamental physical parameters, epilayer crystal quality and device structure design also have a strong interplay with the overall Auger recombination rate [49,50]. First of all, the assumption of equal concentration of electrons and holes in the quantum wells is

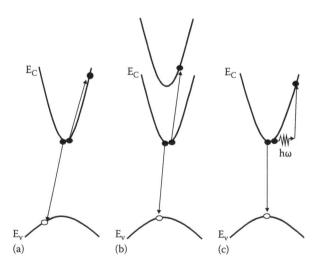

FIGURE 5.9 Schematic illustration of e-e-h Auger process via (a) direct intraband, (b) direct interband, and (c) indirect transition.

not valid in real devices. If the device structure has a poor hole injection, the hole density in QWs will be low and further reduce the Auger recombination rate. In other words, the significance of Auger recombination in efficiency droop is also influenced by carrier transport parameters. Therefore, the extracted value of coefficient C is a convoluted presentation of carrier leakage and Auger recombination.

Besides the influences from macroscopic carrier transport across the active region, the microscopic structures of a single QW also have strong influences on the Auger coefficient. In bulk InGaN materials, the Auger recombination has to be assisted by phonon for momentum conservation. Bertazzi et al. [51] suggested that direct Auger recombination can also be significant in a strained InGaN quantum well with common QW thicknesses of real devices (d = 2–2.5 nm). For example, in a 2 nm $In_{0.25}Ga_{0.75}N/GaN$ quantum well, there exists a high-energy sub-band that satisfies momentum conservation for a direct electron–electron–hole Auger recombination with a high density of state. The evaluated Auger coefficient in such a structure is $5 * 10^{-31}$ cm^{-6} · s^{-1}, which is higher than those in indirect Auger processes for bulk InGaN.

Due to the compositional inhomogeneity in the InGaN QWs, the injected carriers are prone to populate in local energy minimum sites. As a result, the Auger process will be promoted in these sites with local maximum carrier concentrations. Wu et al. [52] incorporated energy fluctuation within QWs for multidimensional current density–voltage–IQE–current (J-V-IQE) simulation. It was found that the calculated J-V-IQE behavior is more closed to experimental results after considering energy fluctuation, even assuming a relatively low Auger coefficient. Therefore, the difference between experimentally extracted C-values and theoretically estimated ones might be rooted from the total Auger recombination rate enhancement by InGaN segregation.

5.2.2.4 Visualization of Auger Electrons

Although many theoretical works have supported the significance of Auger process in InGaN alloys, there was no direct evidence of the Auger process until 2013 [31]. For the e-e-h Auger processes, electrons will be excited to high-energy sub-band, Iveland et al. characterized the energy dispersion of hot electrons emitting out of the p-GaN surface with industrial-grade LEDs. A peak with certain characteristic energy was observed under high-carrier injection, which is a signature of hot electrons populating in the side valley of GaN band structure [53]. Because the measured electron kinetic energy is much higher than those hot electrons caused by polarization fields, the contribution from ballistic transport can be excluded. Later Auger-generated hot carriers were visualized by the PL experiment of color-coded MQW stack [54,55]. In the experiment, short-wavelength QW is sandwiched by long-wavelength QWs and then pumped with long-wavelength lasers. Although the short-wavelength QW cannot be excited, the PL emission in the short-wavelength region still appeared with increasing excitation density because the Auger-generated hot carriers from long-wavelength QWs were recaptured by these short-wavelength QWs. The experiments gave a direct observation of Auger recombination in QW and successfully correlated with efficiency droop [56]. The significance of Auger recombination in InGaN LEDs is no more controversial.

5.2.3 Thermal Effects

5.2.3.1 Thermal Droops

LEDs are operated under DC operation for real applications. Therefore, high injection carrier density is inevitably accompanied with additional Joule heat. To improve the device performance for high-current density application, understanding the temperature effects on light output power (LOP) is also essential. To decouple the high-carrier density and joule heating effects, LEDs were mounted with external temperature control modules. Studies suggested that LOP performance degrades with increasing temperature under a fixed injection density except in the cryogenic region (T < 150 K). Under very low temperature, most of the holes cannot be injected into the active region, so electron leakage becomes the only dominating mechanism in LEDs [57]. For other common cases, the LOP deterioration because of high ambient temperature is known as *thermal droop* in the community [58].

According to the Fermi–Dirac distribution, carriers have a higher probability to populate in high-energy states under equilibrium. These energetic carriers have a relatively high chance to escape the local

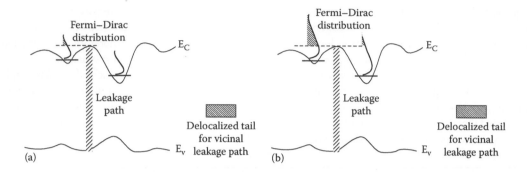

FIGURE 5.10 Schematic illustration of delocalization with energy band tail under (a) low temperature and (b) high temperature.

energy minimum and be recaptured by other nonradiative centers as illustrated in Figure 5.10. Therefore, the thermal droops can be interpreted as delocalization effects due to increased temperature [59].

It is worth noticing that the thermal effect does not favor all nonradiative processes. For example, numerical calculations suggested that high ambient temperatures actually suppress electron overflow because the hole transport enhancement is more pronounced [60,61]. Although high temperature favors the phonon-assisted Auger recombination, the experimentally extracted C values decrease under high temperature. This might be explained by the reduction of overall Auger recombination rate in devices due to carrier delocalization. Generally, reduction of radiative recombination rate is regarded as the major cause of thermal droop.

5.2.3.2 Current Crowding Effects

Metallic contacts in LED absorb light. In order to promote light extraction from the topside, transparent current-spreading layers (TCL) with narrow metallic pads, a.k.a. *fingers*, are used in common horizontal chip design. As temperatures rise, the hole injection across EBL will be enhanced due to the extension of the Boltzmann tail. On the other hand, the lateral conduction of holes is suppressed due to phonon scattering. As a result, the injected carrier will be crowded beneath the fingers, which is known as *current crowding*. Figure 5.11 is a schematic illustration of current spreading and current crowding through fingers and TCLs under different temperatures. Current crowding could be caused not only by high ambient

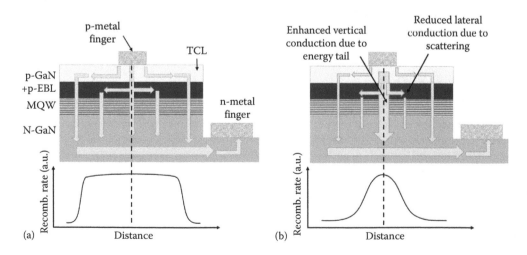

FIGURE 5.11 Current spreading and radiative recombination profile with metal finger and transparent conducting layers (TCLs) of horizontal LEDs under (a) low temperature and (b) high temperature.

temperature, but also by the inappropriate lateral current spreading path design, such as inadequate TCL thickness or over-wide spacing among fingers [62,63].

Current crowding reduced the "effective luminescent area" of a LED, and the carrier density in these areas is therefore to be enhanced. As a result, nonradiative recombination mechanisms relevant to high-carrier density will all be enhanced, such as delocalization effects and Auger processes.

5.3 Designs to Overcome Droop

5.3.1 Active Region Design

5.3.1.1 Active Volume Enhancement

Regardless of different physical mechanisms, the efficiency droop is strongly correlated to the high-carrier concentration in QWs. In order to reduce the carrier concentration under high injection current, a large "effective active volume" is strongly desired to "dilute" the carrier concentration. The most intuitive approach is to enhance the quantum well thickness. Krames et al. [64] from Luminled reported that the peak efficiency of 13 nm InGaN/GaN double heterostructure LEDs ($\lambda = 432$ nm) is delayed to J = 200 A/cm^2 while those of normal MQW LEDs is usually at J = 5–10 A/cm^2. Although the efficiency of MQW LED is superior at lower current density, the efficiency of DH LED starts to prevail under J > 100 A/cm^2.

However, DH LEDs have several critical issues when the emission is pushed to the true-blue ($\lambda = 450$–460 nm) spectral region and beyond. The high strain energy of thick InGaN could lead to additional defects; the strong polarization also causes severe QCSE. In compromise with the crystal quality, carrier distribution improvement of MQW LEDs is still much preferred. Because of the carrier injection asymmetry, radiative recombination happens only in few QWs near the p-side, making the rest of them "ineffective" for electroluminescence. Therefore, improving the uniformity of carrier distribution in MQW LEDs is a practical alternative for active volume enhancement. To balance the carrier transport between electrons and holes, several approaches are suggested to lower the energy barrier height for hole injection. For example, Morkoç et al. reduced the quantum barrier thickness down to 3 nm to couple hole wavefunctions between QWs [65]. Kuo et al. [66] proposed adding 3%–5% indium to quantum barriers to reduce the energy barrier height. Some groups increased the thicknesses of the last few QWs to enhance the effective active volume [67].

5.3.1.2 In Composition Profile Tailoring

The polarization difference between InGaN and GaN is the root cause for several droop mechanisms such as the drift leakage current and the noncapture of injected carriers. The polarization field inside quantum wells also promotes Auger processes. In many studies, the indium composition profile between quantum wells and quantum barriers was tailored to mitigate the influence of sharp InGaN/GaN interfaces. For example, Park et al. used a trapezoidal shape indium profile instead of conventional rectangular shape QWs to enhance the overlap of electron and hole wavefunctions [68]. Zhang et al. applied a step-like indium profile to increase the capture probability of injected electrons into QWs [69]. Efficiency droops were improved in both studies after tailoring the shape of QWs. Figure 5.12 illustrates various QWs in profiles mentioned in this paragraph.

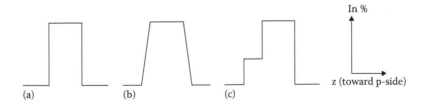

FIGURE 5.12 In compositional profile in (a) rectangular, (b) trapezoidal, and (c) staircase quantum wells.

5.3.1.3 Quaternary Quantum Barrier for Polarization Matching

The polarization difference between QW and quantum barriers can also be mitigated by introducing quaternary (Al,In,Ga)N alloys into the active region. Quaternary alloys have a higher degree of freedom in the materials parameter space of bandgap and polarization. There are many equi-polarization contours in the $Al_xIn_yGa_{1-x-y}N$ composition map with varying bandgap energy [15]. Therefore, polarization matching between InGaN quantum wells and AlInGaN barriers is possible. Shubert et al. demonstrated improved efficiency droop with polarization-matched quaternary quantum barriers. Wavelength blueshift was also observed to be smaller because of the reduced QCSE.

5.3.2 Last Barrier and EBL Design

5.3.2.1 Optimization of Bulk EBL Structure

Electron blocking layers (EBLs) are the most common measure to suppress carrier leakage in LEDs. EBLs are usually heavily p-type AlGaN to prevent electrons from drifting out of the active region. Theoretically, the Mg concentration in EBL is supposed to be as high as possible to facilitate hole injection. In practice, the Mg concentration is a compromise between hole injection and crystal quality because Mg diffusion into QWs is regarded as a luminescence killer in LEDs.

The Al composition and thickness of EBL are also a result of optimization. It is strongly relevant to the emission wavelength of LEDs. The high optimal growth temperature of EBL with high Al composition might induce quality degradation of In-rich quantum wells in long-wavelength LEDs. Therefore, the design window of EBLs in long-wavelength LEDs is quite limited because of crystal growth issues. On the contrary, the Al composition needs to be higher to suppress thermionic emission for short-wavelength LEDs with shallow QWs.

The polarization difference between AlGaN/GaN interfaces also limited the electron-blocking function of high Al composition EBLs. Strong polarization causes a highly tilted conduction band potential energy profile, which would enhance electron leakage through tunneling. Besides increasing EBL thickness, applying lattice-matched quaternary EBL is an alternative to circumvent strain-induced polarization fields [70]. $In_{0.18}Al_{0.82}N$ has a high bandgap ($\sim$5.0 eV), but its in-plane lattice constant is equal to GaN. Therefore, quaternary alloys with arbitrary linear combination between $In_{0.18}Al_{0.82}N$ and GaN are also lattice matched to GaN. The concept of lattice-matched EBL is similar to polarization-matched quaternary barriers, but with more emphasis on mechanical properties.

5.3.2.2 InGaN Last Barrier

"Last barrier" refers to the intermediate layer between EBL and the adjacent quantum well. In some reference, it is also referred to as "last quantum barrier." Several studies replaced the conventional GaN last barrier to InGaN last barrier to improve efficiency droop [71–73]. Band simulations in Figure 5.13 revealed that the InGaN last barrier bends the valance band of its vicinity toward the hole quasi-Fermi level (E_{fp}). As a result, hole equilibrium concentration in the last barrier is enhanced, facilitating hole injection into adjacent QW wells. An improved hole injection also enhances electron consumption in the active region, so the electron overflow is also reduced simultaneously.

It is worth noting that the design window of indium composition of last barrier is limited to 3%–5%. If the valance band of In last barrier penetrates E_{fp}, undesired radiative recombination from InGaN last barrier will be introduced. P-doping in the last barrier is beneficial for hole injection and also a risk of Mg diffusion into QWs.

5.3.2.3 Al Composition Profiling in EBL

There are also many studies with nonuniform Al composition in EBL for efficiency droop improvement, such as compositional grading or high-density AlGaN/GaN interfaces [74,75]. The AlGaN/GaN interface introduces sheet charges due to the abrupt change of polarization. Grading Al composition inside EBL would

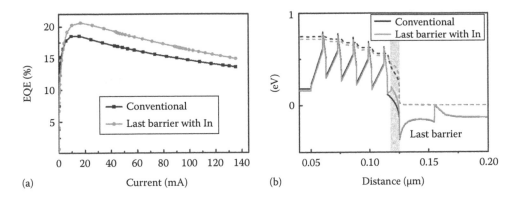

FIGURE 5.13   (a) Simulated IQE curves of reference LED with (gray) and without (black) $In_{0.03}Ga_{0.97}N$ last barrier. (b) Energy profiles around the last barrier in valance band under J = 100 A/cm².

transform sheet charge into volume charge due to the slow-varying polarization. Therefore, performance deterioration due to sheet charges could be alleviated. For convenience, we define the compositional grading into two categories: grade-up EBL and grade-down EBL. In grade-up EBLs, the Al composition monotonically increases from the last barrier side to the p-GaN side and vice versa. Kuo et al. [75] demonstrated significant efficiency droop improvement with graded-up EBL. Band diagram simulations in Figure 5.14 suggested that

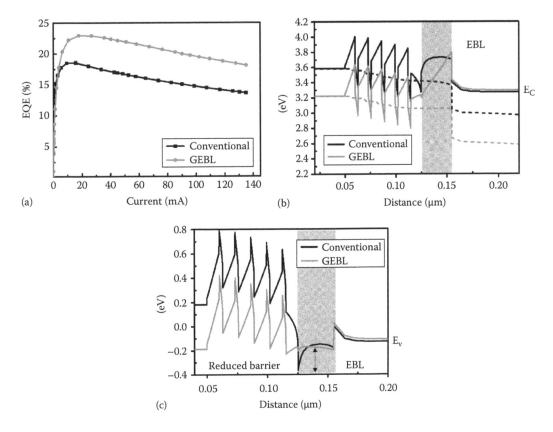

FIGURE 5.14   (a) Simulated IQE curves of reference LED with conventional EBL (black) and grade-up EBL (GEBL, gray). Energy profiles around the EBL in (b) conduction band and (c) valance band under J = 100 A/cm².

graded-up EBL not only eliminated the hole injection energy barrier but also kept the electrical confinement function of EBL. The electrical performance is also improved due to better all injection efficiency.

Another Al composition profile design is to create a high density of AlGaN/GaN interfaces with a short period in the EBLs. The design with the same thicknesses in each period is known as "superlattice EBL" (SL-EBL) [76,77]. SL-EBL benefits from "modulation doping" due to the carrier redistribution near hetero-interfaces. Modulation doping would reduce the Mg activation energy, so the hole concentration of EBL is further enhanced with the same doping concentration. The band diagram of SL-EBL is then "lifted up" to block the electrons. Some studies would use multiple SL-EBLs with different period thickness or nonperiodical AlGaN/GaN multi-stack to further suppress electron leakage through resonant tunneling [78,79]. SL-EBL design and its derivatives would be much functional when electron overflow is the primary droop mechanism in the LEDs.

5.3.3 Device Structure Innovation

5.3.3.1 Current Spreading Optimization

Uniform luminescence is highly desired to improve efficiency droop in LEDs. For conventional p-side up horizontal LEDs, the current spreading is achieved by finger pattern design with transparent electrodes, for example, indium tin oxide (ITO) or zinc oxide (ZnO), as current spreading layers [80,81]. According to multidimensional simulations, the most effective way to improve current spreading is increasing the transparent electrode thickness. However, the "transparent" electrode could still have little light absorption, and an over-thick electrode might bring few parasitic issues in processing. The optimal thickness is highly dependent on finger design and processing capabilities.

The best scenario for p-side current spreading is a blanket deposition of metal contacts [82,83]. Therefore, novel device structures with p-metal bonded to another platform such as vertical LEDs and flip-chip LEDs were developed to promote uniform hole injection. These p-side-down structures not only have a superior hole injection but also provide a direct path for fast heat dissipation via bonding metals. Therefore, the thermal droop and current crowding effects are significantly alleviated. Vertical LEDs and flip-chip LEDs are now widely used for high-power applications. The device structures of horizontal, vertical, and flip-chip LEDs are illustrated in Figure 5.15.

5.3.3.2 Homoepitaxial LED

The delocalization of carriers is responsible for the efficiency droop by activating NRCs under high-carrier injections. Some groups might intentionally control the pits depth to enhance the carrier localization. Another approach to deal with the density-activated NRCs is to eliminate them in the beginning.

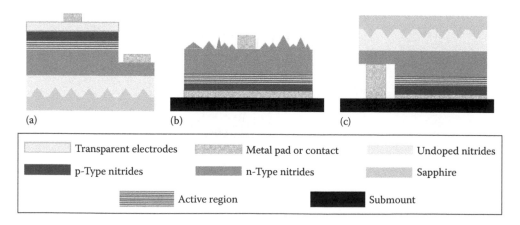

FIGURE 5.15   Schematic illustration of (a) horizontal, (b) vertical, and (c) flip-chip LED structures.

In other words, the inherent defect density has to be very low. However, this can be barely achieved for GaN/sapphire heteroepitaxy. Therefore, LED homoepitaxy on high-quality bulk GaN substrates was proposed for high-performance LEDs with low droop [84,85]. The threading dislocation density of bulk GaN substrate is reported to be as low as ~10^4 cm^{-2}, which is four orders less than common GaN heteroepitaxy on sapphire. Second, the bulk GaN substrate itself is also an excellent heat reservoir. Therefore, GaN homoepitaxial LEDs also have a superior performance under high temperature. Hurni et al. reported wall-plug efficiencies (WPE) up to 84% under temperature 85°C with excellent droop performance in GaN homoepitaxial LEDs.

5.3.3.3 Laser-Based Illumination

The carrier density in the laser diodes (LDs) is "clamped" after lasing. Because the carrier concentration remains constant, the Auger recombination rate does not increase with current density after lasing. If we neglect the parasitic thermal influence and many-body effects, the differential internal quantum efficiency LDs equals to unity after lasing. In other words, the efficiency only increases with current density so that there will be no droop [86,87]. However, the threshold current density of commercial LDs is usually in the level of kA/cm^2, which is around 100–1000 times higher for LEDs' operating condition. Therefore, the droop-free nature of LDs will be more advantageous under some extreme circumstances requiring high total light output and high directionality. For example, blue LDs with phosphor plates were proposed as a candidate for automobile headlamp and spotlights for lighthouses and ferries.

5.3.3.4 LEDs on Nonbasal Planes

The polarization effects might be dealt with by introducing quaternary alloys into LED structures, but the significant difference of optimal growth conditions among AlN/GaN/InN imposed a strong challenge for high-quality quaternary nitride epitaxy. There are also numerous studies to overcome efficiency droop by LED epi growth with crystal polarities other than c-planes.

As mentioned in Section 5.2.1.2, the polarization-induced band bending weakens the function of EBLs. Band diagram simulation suggested that a reversed polarity would turn the limitations due to polarization into advantages, which means the LED structure should be grown along the [000$\bar{1}$] direction [88,89]. Rajan et al. realized single and double QW green LEDs with [000$\bar{1}$] polarity by molecular beam epitaxy. The LED structure does not exhibit significant electron overflow even without AlGaN EBLs because the injected electrons were stopped by potential barriers built inside QWs. In practice, metal-organic chemical vapor deposition (MOCVD) growth on N-face is still difficult for scale. To date, N-polar LEDs were still in R&D stage.

Nonpolar LED planes were first demonstrated by Ponce et al. in 2000, which is grown on γ-LiAlO$_2$. The nonpolar nitride heteroepitaxy still has strong crystal quality issues [90]. After Nakamura et al. from the University of California, Santa Barbara initiated nonpolar homoepitaxy on bulk GaN substrates, nonpolar and semipolar LEDs and laser diodes were more intensively studied [91]. "Nonpolar" plans refer to the crystallography planes, which is perpendicular to the basal planes, for example, (10$\bar{1}$0) plane (*m-plane*) and (11$\bar{2}$0) plane (*a-plane*). Semipolar planes refer to all planes with an inclined angle to both basal plane and nonpolar planes. Figure 5.16 is an illustration of crystallographic planes in wurtzite nitride system, and their inclination angles to basal plane are summarized in Table 5.2. Because the polar axis has zero projection on the nonpolar planes, the nonpolar light-emitting devices are free from polarization effects. Similarly, the polarization effects should be minimal on semipolar planes that are closed to nonpolar planes, for example, (20$\bar{2}$1) and (30$\bar{3}$1) planes. Blue LEDs with low droop and decent efficiency have been demonstrated on m-plane [92,93], (20$\bar{2}$1) [94] and (30$\bar{3}$1) planes [95]. The reduced polarization also widened the design window for QW thicknesses because of the absence of QCSE. Therefore, a simple DH-LED structure with high efficiency over wide injection levels is possible. For the MQW LED structure, the polarity also played an important role in semipolar LEDs. Huang et al. [96] studied semipolar (20$\bar{2}$1) and (20$\bar{2}\bar{1}$) MQW LEDs with color-coded QWs. The interaction between p–n junction depletion field and internal polarization field also has a significant influence on hole transport between

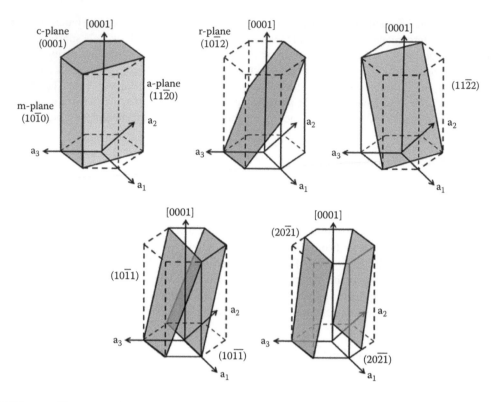

FIGURE 5.16   Illustration of nonpolar and semipolar planes in wurtzite nitride.

TABLE 5.2   Inclination Angles (θ) between Selected
Nonpolar/Semipolar Planes and the Basal Plane

Plane	θ (°)	Plane	θ (°)
$(10\bar{1}3)$	32	$(10\bar{1}2)$/r-plane	43
$(11\bar{2}2)$	58	$(10\bar{1}1)$	62
$(20\bar{2}1)$	75	$(10\bar{1}0)$/m-plane	90
$(11\bar{2}0)$/a-plane	90	$(20\bar{2}\bar{1})$	105

quantum wells. Unlike N-polar epitaxy, the MOCVD epitaxy on nonpolar and semipolar planes was more established. The major barrier for commercialization of nonpolar and semipolar LEDs was the lack of large area substrate or template with high crystal quality.

5.4 Conclusion

In conclusion, efficiency droop in nitride LED has been intensively studied since the year 2007. Various mechanisms all successfully explained the origin of droop and were supported by corresponding remedies. Although the significance of each mechanism might be still under debate, it is widely accepted that these mechanisms could happen in parallel. The proposed root causes of efficiency droop and corresponding overcoming strategies are summarized in Table 5.3.

TABLE 5.3   Summary of Root Causes of Efficiency Droop and Their Corresponding Remedies

Root Cause	Remedies
Electron/hole transport parameter asymmetry	Heavily doped p-GaN p-type electron blocking layers InGaN last barrier for hole injection
Internal polarization field	Quaternary quantum barriers and EBLs Composition tailoring in QWs and EBLs LED epitaxy on nonbasal planes
Carrier delocalization	Controlled V-pits formation High-quality LED homoepitaxy
Auger recombination	Thick QW Resonant MQW structure (thin barriers) Laser lighting
Thermal effects/current crowding	Thick current spreading layers Vertical LED Flip-chip LED

References

1. S. Nakamura, T. Mukai, and M. Senoh, Candela-class high-brightness InGaN/AlGaN double-heterostructure blue-light-emitting diodes, *Appl. Phys. Lett.* **64**, 1687 (1994).
2. Y. Narukawa, M. Ichikawa, D. Sanga, M. Sano, and T. Mukai, White light emitting diodes with super-high luminous efficacy, *J. Phys. D Appl. Phys.* **43**, 354002 (2010).
3. Q. Dai, M. F. Schubert, M. H. Kim, J. K. Kim, E. F. Schubert, D. D. Koleske, M. H. Crawford et al., Internal quantum efficiency and nonradiative recombination coefficient of GaInN/GaN multiple quantum wells with different dislocation densities, *Appl. Phys. Lett.* **94**, 111109 (2009).
4. W. Götz, N. M. Johnson, C. Chen, H. Liu, C. Kuo, and W. Imler, Activation energies of Si donors in GaN, *Appl. Phys. Lett.* **68**, 3144 (1996).
5. U. Kaufmann, P. Schlotter, H. Obloh, K. Köhler, and M. Maier, Hole conductivity and compensation in epitaxial GaN:Mg layers, *Phys. Rev. B* **62**, 10867 (2000).
6. H. M. Ng, D. Doppalapudi, T. D. Moustakas, N. G. Weimann, and L. F. Eastman, The role of dislocation scattering in n-type GaN films, *Appl. Phys. Lett.* **73**, 821 (1998).
7. C. Erin H. Kyle, S. W. Kaun, P. G. Burke, F. Wu, Y.-R. Wu, and J. S. Speck, High-electron-mobility GaN grown on free-standing GaN templates by ammonia-based molecular beam epitaxy, *J. Appl. Phys.* **115**, 193702 (2014).
8. K. J. Vampola, M. Iza, S. Keller, S. P. DenBaars, and S. Nakamura, Measurement of electron overflow in 450 nm InGaN light-emitting diode structures, *Appl. Phys. Lett.* **94**, 061116 (2009).
9. A. David, M. J. Grundmann, J. F. Kaeding, N. F. Gardner, T. G. Mihopoulos, and M. R. Krames, Carrier distribution in (0001) InGaN/GaN multiple quantum well light-emitting diodes, *Appl. Phys. Lett.* **92**, 053502 (2008).
10. J. P. Liu, J.-H. Ryou, R. D. Dupuis, J. Han, G. D. Shen, and H. B. Wang, Barrier effect on hole transport and carrier distribution in InGaN/GaN multiple quantum well visible light-emitting diodes, *Appl. Phys. Lett.* **93**, 021102 (2008).
11. Q. Dai, Q. Shan, J. Wang, S. Chhajed, J. Cho, E. Fred Schubert, M. H. Crawford, D. D. Koleske, M.-H. Kim, and Y. Park, Carrier recombination mechanisms and efficiency droop in GaInN/GaN light-emitting diodes, *Appl. Phys. Lett.* **97**, 133507 (2010).
12. G.-B. Lin, D. Meyaard, J. Cho, E. Fred Schubert, H. Shim, and C. Sone, Analytic model for the efficiency droop in semiconductors with asymmetric carrier-transport properties based on drift-induced reduction of injection efficiency, *Appl. Phys. Lett.* **100**, 161106 (2012).

13. A. E. Romanov, T. J. Baker, S. Nakamura, J. S. Speck, and ERATO/JST UCSB Group, Strain-induced polarization in wurtzite III-nitride semipolar layers, *Appl. Phys.* **100**, 023522 (2006).
14. J. H. Son and J.-L. Lee, Numerical analysis of efficiency droop induced by piezoelectric polarization in InGaN/GaN light-emitting diodes, *Appl. Phys. Lett.* **97**, 032109 (2010).
15. M.-H. Kim, M. F. Schubert, Q. Dai, J. K. Kim, E. Fred Schubert, J. Piprek, and Y. Park, Origin of efficiency droop in GaN-based light-emitting diodes, *Appl. Phys. Lett.* **91**, 183507 (2007).
16. M. F. Schubert and E. Fred Schubert, Effect of hetero interface polarization charges and well width upon capture and dwell time for electrons and holes above GaInN/GaN quantum wells, *Appl. Phys. Lett.* **96**, 131102 (2010).
17. X. Ni, X. Li, J. Lee, S. Liu, V. Avrutin, Ü. Özgür, H. Morkoç, and A. Matulionis, Hot electron effects on efficiency degradation in InGaN light emitting diodes and designs to mitigate them, *Appl. Phys.* **108**, 033112 (2010).
18. X. Ni, X. Li, J. Lee, S. Liu, V. Avrutin, A. Matulionis, Ü. Özgür, and H. Morkoç, Pivotal role of ballistic and quasi-ballistic electrons on LED efficiency, *Superlattices Microstruct.* **48**, 133 (2010).
19. D. S. Sizov, R. Bhat, A. Zakharian, K. Song, D. E. Allen, S. Coleman, and C.-E. Zah, Carrier transport in InGaN MQWs of aquamarine- and green-laser diodes, *IEEE J. Sel. Topics Quantum Electron.* **17**, 1390 (2011).
20. S. F. Chichibu, A. Uedono, T. Onuma, B. A. Haskell, A. Chakraborty, T. Koyama, P. T. Fini et al., Origin of defect-insensitive emission probability in In-containing (Al,In,Ga)N alloy semiconductors, *Nat. Mater.* **5**, 810 (2006).
21. A. Kaneta, T. Izumi, K. Okamoto, Y. Kawakami, S. Fujita, Y. Narita, T. Inoue, and M. Takashi, Spatial inhomogeneity of photoluminescence in an InGaN-based light-emitting diode structure probed by near-field optical microscopy under illumination-collection mode, *Jpn. J. Appl. Phys.* **40**, 110 (2001).
22. M. J. Galtrey, R. A. Oliver, M. J. Kappers, C. J. Humphreys, P. H. Clifton, D. Larson, D. W. Saxey, and A. Cerezo, Three-dimensional atom probe analysis of green- and blue-emitting In x Ga 1 − x N/Ga N multiple quantum well structures, *Appl. Phys. Lett.* **104**, 013524 (2008).
23. X. A. Cao, J. A. Teetsov, F. Shahedipour-Sandvik, and S. D. Arthur, Microstructural origin of leakage current in GaN/InGaN light-emitting diodes, *J. Cryst. Growth* **264**, 172 (2004).
24. J. W. P. Hsu, M. J. Manfra, R. J. Molnar, B. Heying, and J. S. Speck, Direct imaging of reverse-bias leakage through pure screw dislocations in GaN films grown by molecular beam epitaxy of GaN templates, *Appl. Phys. Lett.* **81**, 79 (2002).
25. A. Hangleiter, F. Hitzel, C. Netzel, D. Fuhrmann, U. Rossow, G. Ade, and P. Hinze, Suppression of nonradiative recombination by V-shaped pits in GaInN/GaN quantum wells produces a large increase in the light emission efficiency, *Phys. Rev. Lett.* **95**, 127402 (2005).
26. J. Kim, Y.-H. Cho, D.-S. Ko, X.-S. Li, J.-Y. Won, E. Lee, and S. Hwa, Influence of V-pits on the efficiency droop in InGaN/GaN quantum wells, *Opt. Express* **22**, 857 (2014).
27. J. Hader, J. V. Moloney, and S. W. Koch, Density-activated defect recombination as a possible explanation for the efficiency droop in GaN-based diodes, *Appl. Phys. Lett.* **96**, 221106 (2010).
28. J. Wang, L. Wang, W. Zhao, Z. Hao, and Y. Luo, Understanding efficiency droop effect in InGaN/GaN multiple-quantum-well blue light-emitting diodes with different degree of carrier localization, *Appl. Phys. Lett.* **97**, 201112 (2010).
29. R. Aleksiejūnas, K. Gelžinytė, S. Nargelas, K. Jarašiūnas, M. Vengris, E. A. Armour, D. P. Byrnes, R. A. Arif, S. M. Lee, and G. D. Papasouliotis, Diffusion-driven and excitation-dependent recombination rate in blue InGaN/GaN quantum well structures, *Appl. Phys. Lett.* **104**, 022114 (2014).
30. N. I. Bochkareva, Y. T. Rebane, and Y. G. Shreter, Efficiency droop and incomplete carrier localization in InGaN/GaN quantum well light-emitting diodes, *Appl. Phys. Lett.* **103**, 191101 (2013).
31. J. Iveland, L. Martinelli, J. Peretti, J. S. Speck, and C. Weisbuch, Direct measurement of Auger electrons emitted from a semiconductor light-emitting diode under electrical injection: Identification of the dominant mechanism for efficiency droop, *Phys. Rev. Lett.* **110**, 177406 (2013).

32. Y. C. Shen, G. O. Mueller, S. Watanabe, N. F. Gardner, A. Munkholm, and M. R. Krames, Auger recombination in InGaN measured by photoluminescence, *Appl. Phys. Lett.* **91**, 141101 (2007).

33. A. Davida and N. F. Gardner, Droop in III-nitrides: Comparison of bulk and injection contributions, *Appl. Phys. Lett.* **97**, 193508 (2010).

34. A. Laubsch, M. Sabathil, J. Baur, M. Peter, and B. Hahn, High-power and high-efficiency InGaN-based light emitters, *IEEE Trans. Electron Dev.* **57**, 79 (2010).

35. G. B. Lin, E. Fred Schubert, J. Cho, J. H. Park, and J. K. Kim, Onset of the efficiency droop in GaInN quantum well light-emitting diodes under photoluminescence and electroluminescence excitation, *ACS Photon.* **2**, 1013 (2015).

36. J. Xie, X. Ni, Q. Fan, R. Shimada, Ü. Özgür, and H. Morkoç, On the efficiency droop in InGaN multiple quantum well blue light emitting diodes and its reduction with p-doped quantum well barriers, *Appl. Phys. Lett.* **93**, 121107 (2008).

37. J. Mickevičius, J. Jurkevičius, M. S. Shur, J. Yang, R. Gaska, and G. Tamulaitis, Photoluminescence efficiency droop and stimulated recombination in GaN epilayers, *Opt. Express* **20**, 25195 (2012).

38. J. Hader, J. V. Moloney, B. Pasenow, S. W. Koch, M. Sabathil, N. Linder, and S. Lutgen, On the importance of radiative and Auger losses in GaN-based quantum wells, *Appl. Phys. Lett.* **92**, 261103 (2008).

39. M. Brendel, A. Kruse, H. Jönen, L. Hoffmann, H. Bremers, U. Rossow, and A. Hangleiter, Auger recombination in GaInN/GaN quantum well laser structures, *Appl. Phys. Lett.* **99**, 031106 (2011).

40. M. Calciati, M. Goano, F. Bertazzi, M. Vallone, X. Zhou, G. Ghione, M. Meneghini et al., Correlating electroluminescence characterization and physics-based models of InGaN/GaN LEDs: Pitfalls and open issues, *AIP Adv.* **4**, 067118 (2014).

41. W. G. Scheibenzuber, U. T. Schwarz, L. Sulmoni, J. Dorsaz, J.-F. Carlin, and N. Grandjean, Recombination coefficients of GaN-based laser diodes, *J. Appl. Phys.* **109**, 093106 (2011).

42. M. Zhang, P. Bhattacharya, J. Singh, and J. Hinckley, Direct measurement of auger recombination in $In_{0.1}Ga_{0.9}N$/GaN quantum wells and its impact on the efficiency of $In_{0.1}Ga_{0.9}N$/GaN multiple quantum well light emitting diodes, *Appl. Phys. Lett.* **95**, 201108 (2009).

43. B. Galler, P. Drechsel, R. Monnardl, P. Rode, P. Stauss, S. Froehlich, W. Bergbauer et al., Influence of indium content and temperature on Auger-like recombination in InGaN quantum wells grown on (111) silicon substrates, *Appl. Phys. Lett.* **101**, 131111 (2012).

44. D. Schiavon, M. Binder, M. Peter, B. Galler, P. Drechsel, and F. Scholz, Wavelength-dependent determination of the recombination rate coefficients in single-quantum-well GaInN/GaN light emitting diodes, *Phys. Status Solidi B* **250**, 283 (2013).

45. G. Kim, J. H. Kim, E. H. Park, D. Kang, and B. G. Park, Extraction of recombination coefficients and internal quantum efficiency of GaN-based light emitting diodes considering effective volume of active region, *Opt. Express* **22**, 1235 (2014).

46. F. Bertazzi, M. Goano, and E. Bellotti, A numerical study of Auger recombination in bulk InGaN, *Appl. Phys. Lett.* **97**, 231118 (2010).

47. K. T. Delaney, P. Rinke, and C. G. Van de Walle, Auger recombination rates in nitrides from first principles, *Appl. Phys. Lett.* **94**, 191109 (2009).

48. E. Kioupakis, P. Rinke, K. T. Delaney, and C. G. Van de Walle, Indirect Auger recombination as a cause of efficiency droop in nitride light-emitting diodes, *Appl. Phys. Lett.* **98**, 161107 (2011).

49. E. Kioupakis, Q. Yan, D. Steiauf, and C. G. Van de Walle, Temperature and carrier-density dependence of Auger and radiative recombination in nitride optoelectronic devices, *New J. Phys.* **15**, 125006 (2013).

50. J. Piprek, F. Römer, and B. Witzigmann, On the uncertainty of the Auger recombination coefficient extracted from InGaN/GaN light-emitting diode efficiency droop measurements, *Appl. Phys. Lett.* **106**, 101101 (2015).

51. F. Bertazzi, X. Zhou, M. Goano, G. Ghione, and E. Bellotti, Auger recombination in InGaN/GaN quantum wells: A full-Brillouin-zone study, *Appl. Phys. Lett.* **103**, 081106 (2013).

52. Y.-R. Wu, R. Shivaraman, K.-C. Wang, and J. S. Speck, Analyzing the physical properties of InGaN multiple quantum well light emitting diodes from nano scale structure, *Appl. Phys. Lett.* **101**, 083505 (2012).

53. M. Piccardo, J. Iveland, L. Martinelli, S. Nakamura, J. W. Choi, J. S. Speck, C. Weisbuch, and J. Peretti, Low-energy electro- and photo-emission spectroscopy of GaN materials and devices, *J. Appl. Phys.* **117**, 112814 (2015).

54. M. Binder, A. Nirsch, R. Zeisel, T. Hager, H.-J. Lugauer, M. Sabathi, D. Bougeard, J. Wagner, and B. Galler, Identification of nnp and npp Auger recombination as significant contributor to the efficiency droop in (GaIn)N quantum wells by visualization of hot carriers in photoluminescence, *Appl. Phys. Lett.* **103**, 071108 (2013).

55. A. Nirsch, M. Binder, M. Schmid, M. M. Karow, I. Pietzonka, H.-J. Lugauer, R. Zeisel, M. Sabathil, D. Bougeard, and B. Galler, Transport and capture properties of Auger-generated high-energy carriers in (AlInGa)N quantum well structures, *Appl. Phys.* **118**, 033103 (2015).

56. T. Sadi, P. Kivisaari, J. Oksanen, and J. Tulkki, On the correlation of the Auger generated hot electron emission and efficiency droop in III-N light-emitting diodes, *Appl. Phys. Lett.* **105**, 091106 (2014).

57. D.-S. Shin, D.-P. Han, J.-Y. Oh, and J.-I. Shim, Study of droop phenomena in InGaN-based blue and green light-emitting diodes by temperature-dependent electroluminescence, *Appl. Phys. Lett.* **100**, 153506 (2012).

58. D. S. Meyaard, Q. Shan, J. Cho, E. Fred Schubert, S.-H. Han, M.-H. Kim, C. Sone, S. J. Oh, and J. K. Kim, Temperature dependent efficiency droop in GaInN light-emitting diodes with different current densities, *Appl. Phys. Lett.* **100**, 081106 (2012).

59. P. Tian, J. J. D. McKendry, J. Herrnsdorf, S. Watson, R. Ferreira, I. M. Watson, E. Gu, A. E. Kelly, and M. D. Dawson, Temperature-dependent efficiency droop of blue InGaN micro-light emitting diodes, *Appl. Phys. Lett.* **105**, 171107 (2014).

60. J. Piprek, How to decide between competing efficiency droop models for GaN-based lightemitting diodes, *Appl. Phys. Lett.* **107**, 031101 (2015).

61. J. Hader, J. V. Moloney, and S. W. Koch, Temperature dependence of radiative and Auger losses in quantum wells, *IEEE J. Quantum Electron.* **44**, 185 (2008).

62. L. Wang, Z.-H. Zhang, and N. Wang, Current crowding phenomenon: Theoretical and direct correlation with the efficiency droop of light emitting diodes by a modified ABC model, *IEEE J. Quantum Electron.* **51**, 3200109 (2015).

63. V. K. Malyutenko, S. S. Bolgov, and A. D. Podoltsev, Current crowding effect on the ideality factor and efficiency droop in blue lateral InGaN/GaN light emitting diodes, *Appl. Phys. Lett.* **97**, 251110 (2010).

64. N. F. Gardner, G. O. Müller, Y. C. Shen, G. Chen, S. Watanabe, W. Götz, and M. R. Krames, Blue-emitting InGaN–GaN double-heterostructure light-emitting diodes reaching maximum quantum efficiency above 200 A/cm², *Appl. Phys. Lett.* **91**, 243506 (2007).

65. X. Ni, Q. Fan, R. Shimada, Ü. Özgür, and H. Morkoç, Reduction of efficiency droop in InGaN light emitting diodes by coupled quantum wells, *Appl. Phys. Lett.* **93**, 171113 (2008).

66. Y.-K. Kuo, T.-H. Wang, and J.-Y. Chang, Blue InGaN light-emitting diodes with multiple GaN-InGaN barriers, *IEEE J. Quantum Electron.* **48**, 946 (2012).

67. C. H. Wang, S. P. Chang, W. T. Chang, J. C. Li, Y. S. Lu, Z. Y. Li, H. C. Yang, H. C. Kuo, T. C. Lu, and S. C. Wang, Efficiency droop alleviation in InGaN/GaN light-emitting diodes by graded-thickness multiple quantum wells, *Appl. Phys. Lett.* **97**, 181101 (2010).

68. S.-H. Han, D.-Y. Lee, H.-W. Shim, G.-C. Kim, Y. S. Kim, S.-T. Kim, S.-J. Lee, C.-Y. Cho, and S.-J. Park, Improvement of efficiency droop in InGaN/GaN multiple quantum well light-emitting diodes with trapezoidal wells, *J. Phys. D Appl. Phys.* **43**, 354004 (2010).

69. F. Zhang, X. Li, S. Hafiz, S. Okur, V. Avrutin, Ü. Özgür, H. Morkoç, and A. Matulionis, The effect of stair case electron injector design on electron overflow in InGaN light emitting diodes, *Appl. Phys. Lett.* **103**, 051122 (2013).

70. S. Choi, H. J. Kim, S.-S. Kim, J. Liu, J. Kim, J.-H. Ryou, R. D. Dupuis, A. M. Fischer, and F. A. Ponce, Improvement of peak quantum efficiency and efficiency droop in III-nitride visible light-emitting diodes with an InAlN electron-blocking layer, *Appl. Phys. Lett.* **96**, 221105 (2010).

71. T. Lu, S. Li, C. Liu, K. Zhang, Y. Xu, J. Tong, L. Wu et al., Advantages of GaN based light-emitting diodes with a p-InGaN hole reservoir layer, *Appl. Phys. Lett.* **100**, 141106 (2012).

72. R.-M. Lin, S.-F. Yu, S.-J. Chang, T.-H. Chiang, S.-P. Chang, and C.-H. Chen, Inserting a p-InGaN layer before the p-AlGaN electron blocking layer suppresses efficiency droop in InGaN-based light-emitting diodes, *Appl. Phys. Lett.* **101**, 081120 (2012).

73. Z. Kyaw, Z.-H. Zhang, W. Liu, S. T. Tan, Z. G. Ju, X. L. Zhang, Y. Ji et al., Simultaneous enhancement of electron overflow reduction and hole injection promotion by tailoring the last quantum barrier in InGaN/GaN light-emitting diodes, *Appl. Phys. Lett.* **104**, 161113 (2014).

74. C. Liu, Z. Ren, X. Chen, B. Zhao, X. Wang, Y. Yin, and S. Li, Study of InGaN/GaN light emitting diodes with step-graded electron blocking layer, *IEEE Photon. Technol. Lett.* **26**, 134 (2014).

75. C. H. Wang, C. C. Ke, C. Y. Lee, S. P. Chang, W. T. Chang, J. C. Li, Z. Y. Li et al., Hole injection and efficiency droop improvement in InGaN/GaN light-emitting diodes by band-engineered electron blocking layer, *Appl. Phys. Lett.* **97**, 261103 (2010).

76. Y. Y. Zhang and Y. A. Yin, Performance enhancement of blue light-emitting diodes with a special designed AlGaN/GaN superlattice electronblocking layer, *Appl. Phys. Lett.* **99**, 221103 (2011).

77. J. H. Park, D. Y. Kim, S. Hwang, D. Meyaard, E. Fred Schubert, Y. D. Han, J. W. Choi, J. Cho, and J. K. Kim, Enhanced overall efficiency of GaInN-based light-emitting diodes with reduced efficiency droop by Al-composition-graded AlGaN/GaN superlattice electron blocking layer, *Appl. Phys. Lett.* **103**, 061104 (2013).

78. J. Piprek and Z. M. Simon Li, Origin of InGaN light-emitting diode efficiency improvements using chirped AlGaN multi-quantum barriers, *Appl. Phys. Lett.* **102**, 023510 (2013).

79. Y.-Y. Lin, R. W. Chuang, S.-J. Chang, S. Li, Z.-Y. Jiao, T.-K. Ko, S. J. Hon, and C. H. Liu, GaN-based LEDs with a chirped multi-quantum barrier structure, *IEEE Photon. Technol. Lett.* **24**, 1600 (2012).

80. C.-K. Li and Y.-R. Wu, Study on the current spreading effect and light extraction enhancement of vertical GaN/InGaN LEDs, *IEEE Trans. Electron Dev.* **59**, 400 (2012).

81. A. H. Reading, J. J. Richardson, C.-C. Pan, S. Nakamura, and S. P. DenBaars, High efficiency white LEDs with single-crystal ZnO current spreading layers deposited by aqueous solution epitaxy, *Opt. Express* **20**, A13 (2012).

82. B. Hahn, B. Galler, and K. Engl, Development of high-efficiency and high-power vertical light emitting diodes, *Jpn. J. Appl. Phys.* **53**, 100208 (2014).

83. D.-S. Han, J.-Y. Kim, S.-I. Na, S.-H. Kim, K.-D. Lee, B. Kim, and S.-J. Park, Improvement of light extraction efficiency of flip-chip light-emitting diode by texturing the bottom side surface of sapphire substrate *IEEE Photon. Technol. Lett.* **18**, 1406 (2006).

84. C. A. Hurni, A. David, M. J. Cich, R. I. Aldaz, B. Ellis, K. Huang, A. Tyagi et al., Bulk GaN flip-chip violet light-emitting diodes with optimized efficiency for high-power operation, *Appl. Phys. Lett.* **106**, 031101 (2015).

85. M. J. Cich, R. I. Aldaz, A. Chakraborty, A. David, M. J. Grundmann, A. Tyagi, M. Zhang, F. M. Steranka and M. R. Krames, Bulk GaN based violet light-emitting diodes with high efficiency at very high current density, *Appl. Phys. Lett.* **101**, 223509 (2012).

86. J. J. Wierer Jr. and J. Y. Tsao, Advantages of III-nitride laser diodes in solid-state lighting, *Phys. Status Solidi A* **212**, 980 (2015).

87. J. J. Wierer Jr., J. Y. Tsao, and D. S. Sizov, Comparison between blue lasers and light-emitting diodes for future solid-state lighting, *Laser Photon. Rev.* **7**, 963 (2013).

88. F. Akyol, D. N. Nath, S. Krishnamoorthy, P. S. Park, and S. Rajan, Suppression of electron overflow and efficiency droop in N-polar GaN green light emitting diodes, *Appl. Phys. Lett.* **100**, 111118 (2012).

89. K. Shojiki, T. Tanikawal, J.-H. Choi, S. Kuboya, T. Hanada, R. Katayama, and T. Matsuoka, Red to blue wavelength emission of N-polar (0001) InGaN light-emitting diodes grown by metalorganic vapor phase epitaxy, *Appl. Phys. Express* **8**, 061005 (2015).

90. P. Waltereit, O. Brandt, A. Trampert, H. T. Grahn, J. Menniger, M. Ramsteiner, M. Reiche, and K. H. Ploog, Nitride semiconductors free of electrostatic fields for efficient white light-emitting diodes, *Nature* **406**, 865 (2000).

91. H. Masui, S. Nakamura, S. P. DenBaars, and U. K. Mishra, Nonpolar and semipolar III-nitride light-emitting diodes: Achievements and challenges, *IEEE Trans. Electron. Dev.* **57**, 88 (2010).

92. S.-C. Ling, T.-C. Lu, S.-P. Chang, J.-R. Chen, H.-C. Kuo, and S.-C. Wang, Low efficiency droop in blue-green m-plane InGaN/GaN light emitting diodes, *Appl. Phys. Lett.* **96**, 231101 (2010).

93. X. Li, X. Ni, J. Lee, M. Wu, Ü. Özgür, H. Morkoç, T. Paskova, G. Mulholland, and K. R. Evan, Efficiency retention at high current injection levels in m-plane InGaN light emitting diodes, *Appl. Phys. Lett.* **95**, 121107 (2009).

94. C.-C. Pan, T. Gilbert, N. Pfaff, S. Tanaka, Y. Zhao, D. Feezell, J. S. Speck, S. Nakamura, and S. P. DenBaars, Reduction in thermal droop using thick single-quantum-well structure in semipolar (2021) blue light-emitting diodes, *Appl. Phys. Express* **5**, 102103 (2012).

95. D. L. Becerra, Y. Zhao, S. H. Oh, C. D. Pynn, K. Fujito, S. P. DenBaars, and S. Nakamura, High-power low-droop violet semipolar (3031) InGaN/GaN light-emitting diodes with thick active layer design, *Appl. Phys. Lett.* **105**, 171106 (2014).

96. Y. Kawaguchi, C.-Y. Huang, Y.-R. Wu, Q. Yan, C.-C. Pan, Y. Zhao, S. Tanaka et al., Influence of polarity on carrier transport in semipolar (2021) and (2021) multiple-quantum-well light-emitting diodes, *Appl. Phys. Lett.* **100**, 231110 (2012).

6

Design and Fabrication of Patterned Sapphire Substrates for GaN-Based Light-Emitting Diodes

Guoqiang Li
*South China University
of Technology*

Haiyan Wang
*South China University
of Technology*

Zhiting Lin
*South China University
of Technology*

Abstract Since the first high-brightness GaN-based blue light-emitting diode (LED) was success-fully achieved on flat *c*-plane sapphire substrate, LEDs have been widely used in diverse areas [1–3], such as displays, traffic signals, automobile headlamps, lightings, etc. The development direction of LED is high-power and high-efficiency general lighting devices based on sapphire substrates. However, further development of such devices is hampered by two main obstacles. First, the crystal-line defect density in GaN is high, resulting from the relatively large lattice mismatch between GaN and sapphire [4–6], which would impact the internal quantum efficiency (IQE) of LEDs. Second, the light extraction efficiency (LEE) is low due to severe total reflection effect between GaN and sapphire [7,8]. To alleviate these problems, significant breakthroughs have been achieved. On one hand, two main approaches are proposed to improve IQE. The first is to employ special epitaxial technologies to reduce the crystalline defect density of epitaxial wafers [9–11]. The second is to adopt advanced epitaxial structures to improve carrier radiative recombination rate and avoid car-rier leakage [12–14]. On the other hand, the approaches to improve LEE are under active research with various proposed methods, including patterned sapphire substrate (PSS) [15,16], Bragg reflec-tion layers [17], photonic crystal [18], surfaces roughing [19], flip-chip packing [20], etc. However, some of these approaches have drawbacks, increasing the hardness for them in real applications. For example, Bragg reflection layers and photonic crystal bring foreign material layers into the LED epitaxial structures, which produces extra epitaxial difficulties and thus deteriorates the crystalline quality of GaN. Among these approaches, PSS shows great potential because it is capable of improv-ing LEE and crystalline quality simultaneously. It has attracted intensive interests and become a standard procedure for GaN-based LEDs' manufacturing so far.

6.1 Working Mechanisms and Advantages of PSS

Patterned sapphire substrate (PSS) is a new substrate with dense patterns periodically distributed on the substrate surface via etching process. So far, it has been verified to be significant for improving the efficiency of LEDs owing to its superior properties.

On one hand, PSS enables one-step epitaxial growth technology to achieve the lateral growth of GaN, which is beneficial to overcome the disadvantages of the two-step epitaxial process [16]. As mentioned in the preceding text, the large lattice mismatch between sapphire and GaN of ~14% leads to high defect density of up to 10^9–10^{10} cm^{-2} in GaN films [4], which are grown on conventional flat sapphire substrates. Such massive crystalline defects serve as nonradiative recombination centers and therefore cause the decrease in carrier lifetime and radiative recombination efficiency [21,22], which eventually reduces the internal quantum efficiency (IQE) of LEDs. In this regard, employing PSS can promote the epitaxial lateral overgrowth (ELOG) mode and consequently reduce the defect density of GaN [16]. Figure 6.1 illustrates the growth mechanism of GaN on PSS [23], where the twisty black lines represent dislocations. In the initial stage of GaN growth on PSS (Figure 6.1b), the GaN grains nucleating on pattern windows grow fast in perpendicular direction, accompanied by many dislocations extending vertically. When GaN grows higher than the patterns, the slanted side facets promote the lateral growth of GaN and therefore cause some dislocations bending (Figure 6.1c). Eventually, as the thickness of as-grown GaN films increases, the dislocations propagating from the side facets of PSS would close together (Figure 6.1d). By this way, defects in the multiple quantum well (MQW) layer are effectively reduced.

On the other hand, pattern units on PSS act as light scattering elements to improve the light extraction efficiency (LEE) of LEDs. Because of the great difference in refractive index between GaN ($n_{GaN} = 2.45$) and sapphire ($n_{sapphire} = 1.78$), the critical angle for light escaping from GaN to the air is only 24.6° [24]. Hence, most of the rays are confined in LED chip, which seriously decreases the LEE of LEDs [25]. To solve this problem, PSS is adopted to alert the light path and ultimately to break the total reflection effect. As illustrated in Figure 6.2, rays are repeatedly scattered by pattern units on PSS, that is, L_1, L_2, L_3, and L_4. Once the ray enters into the escaping cones, it can be emitted into the air. As a result, the LEE of LEDs is greatly enhanced.

To sum up, PSS can not only promote the epitaxial lateral overgrowth, effectively improving the crystalline quality of GaN, but also act as the light-scattering elements to increase emitted rays, which helps to enhance the LEE of LEDs.

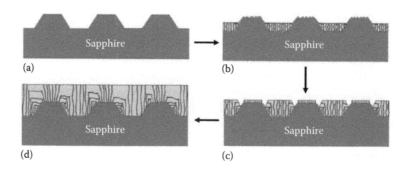

FIGURE 6.1 The growth evolution of GaN epitaxial layer and dislocations on PSS, (a) before growth, (b) during lateral growth, (c) during coalescence, and (d) after coalescence. (Reproduced with permission from Wang, M.T., Liao, K.Y., and Li, Y.L., *IEEE Photon. Technol. Lett.*, 23, 962 copyright 2011 IEEE.)

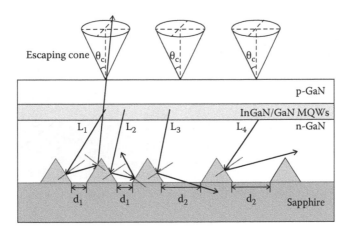

FIGURE 6.2 The light path in PSS-LED.

6.2 Development of PSS

As PSS technology developed, various patterns including groove [26,27], hexagonal frustum [28], trigonal frustum [29], hemisphere [30–32], and cone [33,34] have been proposed.

6.2.1 Groove-Shape PSS

In the early stage of development, the main limitation of PSS fabrication is the lack of advanced etching technologies. In this regard, groove-shaped patterns with simple geometrical features were first put forward. According to the shapes of pattern units, groove-shaped PSS can be mainly divided into three types, that is, straight groove [35], slanted groove [27], and V-shaped groove PSSs [27], as illustrated in Figure 6.3. Jiang et al. [27] have reported the growth mechanism of GaN grown on V-shaped groove PSSs. During the initial low-temperature growth, GaN nucleation centers are distributed uniformly on the platform and window areas of patterns, followed by the redistribution into strip-shaped areas after annealing. Subsequently, strip-shaped GaN is grown at high temperature in ELOG mode and finally coalesce with each other, leaving air gaps beneath the as-grown GaN film, as shown in Figure 6.4. Apart from the shape of pattern unit, depth is also an important factor for groove-shaped PSS. Pan et al. [26] have studied the influence of groove depth on the luminous efficiency of LEDs, as shown in Table 6.1. It is revealed that as the depth increases, the photoluminescence (PL) intensity of LEDs increases accordingly. When the depth is 0.9 μm, the PL intensity of $\langle 1-100 \rangle$ is 3.36 times larger than that of $\langle 11-20 \rangle$, indicating that the luminescence from groove-shaped PSS LED is inhomogeneous. This inhomogeneity may be caused by the difference in crystalline quality of GaN grown on pattern windows and that grown on platforms.

6.2.2 Pyramid-Shaped PSS

In order to solve the problem of luminescence inhomogeneity and further improve the performance of LED, pyramid-shaped PSS is thereby developed. This type of PSS has two-dimensionally densely distributed pattern units that enable GaN to grow homogeneously on the substrate. Meanwhile, compared to groove-shaped patterns, pyramid-shaped patterns provide more slanted planes for light

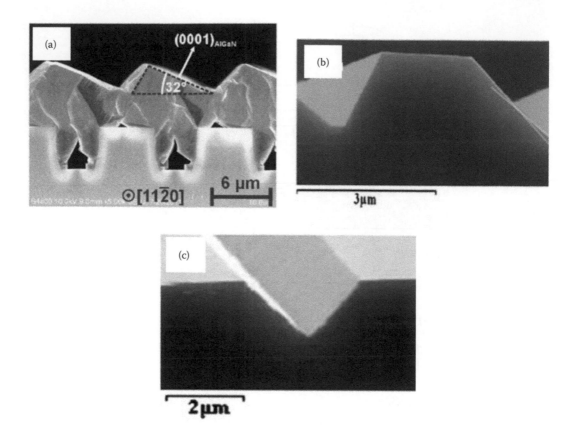

FIGURE 6.3   SEM images of the groove-shaped PSS: (a) straight, (b) slanted, (c) V-shaped groove. (a: Reprinted from *J. Cryst. Growth*, 353, Hagedorn, S., Richter, E., Zeimer, U., Prasai, D., John, W., and Weyers, M., 129, Copyright 2012, with permission from Elsevier; b and c: From Jiang, Y., Jia, H., Wang, W., Wang, L., and Chen, H., *Energy Environ. Sci.*, 4, 2625, 2011 Reproduced by permission of the Royal Society of Chemistry.)

scattering and therefore enhance the LEE of LEDs. Among the series of pyramid-shaped patterns, hexagonal frustum, triangular frustum, and triangular pyramid are easily fabricated by wet etching technology due to the hexagonal crystalline structure of sapphire. Shown in Figure 6.5 are the hexagonal frustums with 100 nm to 4.5 μm in width, 100 nm to 2 μm in height, and 10 nm to 3 μm in distance [28]. Unlike groove, one hexagonal frustum has six side facets, which can scatter incident light from all directions. If the pattern units are densely arranged with a proper density, the scattering area can be larger than that of groove-shaped PSS. In consequence, the light scattering effect is enhanced, and the LEE of LEDs is greatly improved. Figure 6.6 illustrates the typical triangular frustums fabricated by wet etching, with 1–3 μm in width, 1–3 μm in distance, and 0.2–1.5 μm in height. Depending on the etching time, triangular frustum can be transformed into triangular pyramid [34]. Figure 6.7 indicates that as the etching time increases, C-PSS and D-PSS are transformed into triangular-pyramid-shaped PSSs while A-PSS and B-PSS remain triangular-frustum-shaped PSSs. Table 6.2 shows the effect of distance on the GaN crystalline quality and the luminous efficacy of LEDs. With the distance decreasing, both the (0002) and (10–12) full width of half maximum (FWHM) of X-ray rocking curves (XRCs) decrease, which suggests that the screw and edge dislocations are reduced. The light output power (LOP) of the LED chip on D-PSS is increased by 5.6 mW under the current of 20 mA, indicating an increment of 37% compared with that of A-PSS LED.

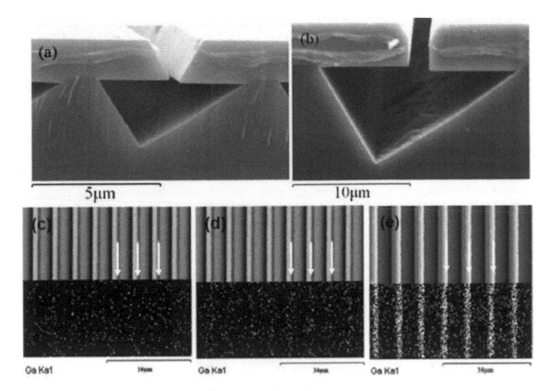

FIGURE 6.4 SEM images of the air gaps of V-shaped-groove PSSs with different pattern periods of (a) 6 and (b) 18 μm; GaN nucleation distribution of V-shaped-groove PSSs in different stages: (c) low-temperature 30 nm thick GaN nucleation; (d) annealing; (e) high-temperature 0.5 μm thick GaN layer. (From Jiang, Y., Jia, H., Wang, W., Wang, L., and Chen, H., *Energy Environ. Sci.*, 4, 2625, 2011 Reproduced by permission of the Royal Society of Chemistry.)

TABLE 6.1 Effect of Groove-Shaped PSSs on the PL Intensity and Etch Pits Density

	Room Temperature PL Intensity		Etch Pits Density ($\times 10^8$ cm^{-2})	
	$\langle 11{-}20 \rangle$	$\langle 1{-}100 \rangle$	$\langle 11{-}20 \rangle$	$\langle 1{-}100 \rangle$
Flat sapphire LED	1	1	5.56	—
0.2 μm depth	1.132	1.595	4–5	4–5
0.5 μm depth	1.4	1.61	3.3–4.2	2.1–3
0.9 μm depth	1.95	6.56	2.12–3.92	0.86–1.06

Source: Pan, C. et al., *J. Appl. Phys.* 102, 84503, 2007.

6.2.3 Hemisphere-Shaped PSS

Apart from pyramid-shaped PSS, the patterns on hemisphere-shaped PSS are also distributed two-dimensionally, with approximately hemispherical surfaces, as shown in Figure 6.8. Cho [36], Lee [37], and Jeong et al. [38] have reported that the ELOG effect can be enhanced by close-arranged hemispheres, which results in great improvement on crystalline quality. Shin et al. [39] utilized inductively coupled plasma (ICP) etching to fabricate hemispherical patterns with 2.5 μm in distance and 1.5 μm in height on sapphire substrates. The GaN films grown on these PSSs by MOCVD exhibit better crystalline quality with the dislocation density reducing to 10^7 cm^{-2}. During the initial

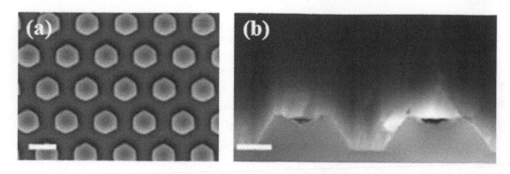

FIGURE 6.5   SEM images of hexagonal frustum: (a) top view; (b) cross-section view. (From Torma, P.T., Ali, M., Svensk, O., Suihkonen, S., Sopanen, M., Lipsanen, H., Mulot, M., Odnoblyudov, M.A., and Bougrov, V.E., *Cryst. Eng. Comm.*, 12, 3152, 2010 Reproduced by permission of the Royal Society of Chemistry.)

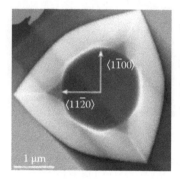

FIGURE 6.6   SEM images of triangular frustum. (Reprinted from *Solid State Electron.*, 52, Gao, H., Yan, F., Zhang, Y., Li, J., Zeng, Y., and Wang, G., 962, Copyright 2008, with permission from Elsevier.)

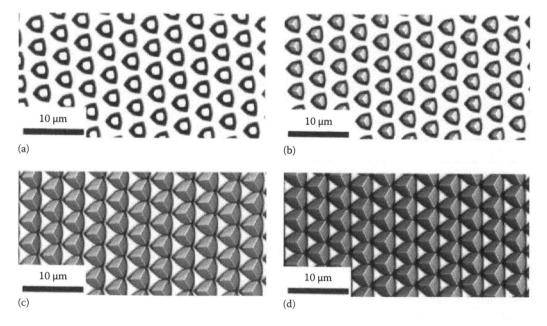

FIGURE 6.7   The evolution of pattern units with wet etching time: (a) A-PSS, (b) B-PSS, (c) C-PSS, and (d) D-PSS. (Reprinted with permission from Cheng, J., Wu, Y.S., Liao, W., and Lin, B., *Appl. Phys. Lett.*, 96, 51109, 2010, American Institute of Physics.)

TABLE 6.2 Effect of the Distance on the Crystalline Quality of GaN and the Luminous Efficacy of LEDs

	XRC FWHM (arcsec)		Etch Pits Density ($\times 10^7$ cm^{-2})	Luminous Intensity (mcd, RT, 20 mA)	LOP (mW)	IQE (%)
	(0002)	(10–12)				
A-PSS	269.3	410.3	4.32	91.2	15.2	56.5
B-PSS	264.1	353.6	1.11	121.6	17.2	60.7
C-PSS	251.5	312.6	0.87	131.2	18.5	61.6
D-PSS	243.4	301.2	0.52	140.0	20.8	66.1

Source: Cheng, J. et al., *Appl. Phys. Lett.*, 96, 51109, 2010.

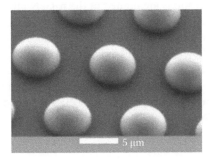

FIGURE 6.8 SEM image of hemisphere-shaped PSS. (Reprinted from *J. Cryst. Growth*, 312, Jeong, S., Kissinger, S., Kim, D., Jae Lee, S., Kim, J., Ahn, H., and Lee, C., 258, Copyright 2010, with permission from Elsevier.)

growth, GaN nucleation centers primarily grow on the platform areas between hemispheres and the top areas of hemispheres.

When GaN films grow higher than hemispherical patterns, the GaN nucleation centers on the top areas would serve as the steps to absorb Ga and N atoms, promoting the ELOG of GaN. Figure 6.9 exhibits the stacking faults formed by ELOG on the top areas of hemispheres. Such stacking faults are able to restrain threading dislocations from extending into the active region, which leads to the high quality of multiple quantum wells.

It is obvious that hemisphere-shaped PSS can greatly improve the crystalline quality of GaN and thereby enhance the IQE of LEDs. However, to improve the LEE of LEDs, PSS still needs to be further optimized.

FIGURE 6.9 TEM image of stacking faults on the top of hemisphere (pointed by white arrow). (Reprinted with permission from Bo Lee, S., Kwon, T., Lee, S., Park, J., and Choi, W., *Appl. Phys. Lett.*, 99, 211901 Copyright 2011, American Institute of Physics.)

6.2.4 Cone-Shaped PSS

Cone-shaped PSS (Figure 6.10) is developed from hemisphere-shaped PSS. The slanted facets introduced by cone-shaped patterns are beneficial to enhance the light scattering effort on incident rays and thus improve the LEE. It is reported that different parameters of cone-shaped patterns, such as height, distance, and width, make significant effects on the crystalline quality of as-grown GaN films and the luminous efficacy of LEDs. Cheng et al. used wet etching technology to fabricate cone-shaped PSSs with different heights [34]. Subsequent crystal growth reveals that the dislocation density in GaN decreases with the height decreasing, indicating that threading dislocation density is relative to the area of GaN lateral growth. When undergoing lateral growth, threading dislocations in GaN tend to bend and stop growing toward MQW layers. As the pattern height is small, the area of lateral growth correspondingly increases. As a result, more dislocations are limited in the n-GaN layer, contributing to better quality for MQWs. In addition to the height, the influence of distance on LED luminous efficiency has also been investigated. Lee et al. fabricated LED devices on cone-shaped PSSs with 3 μm in width, 1.5 μm in height, and 1–3 μm in distance [40]. Figure 6.11 exhibits the AFM images for the as-obtained PSSs, and Figure 6.12 shows the light output powers for the LED devices. The light output power for cone-shaped PSS LED increases gradually with the decrease of the distance. Compared to that on conventional flat sapphire substrate, the output power from the LED with the smallest distance is enhanced by 1.9 mW under the injection current of 20 mA.

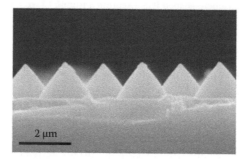

FIGURE 6.10   SEM image of the cone-shaped PSS. (Reprinted from *J. Alloy. Compd.*, 614, Xu, S.R., Li, P.X., Zhang, J.C., Jiang, T., Ma, J.J., Lin, Z.Y., and Hao, Y., 360, Copyright 2014, with permission from Elsevier.)

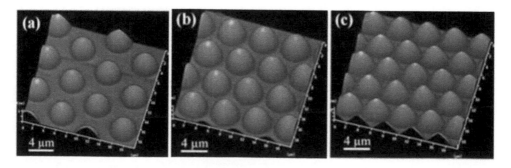

FIGURE 6.11   AFM images of cone-shaped PSSs with different distances of (a) 3 μm, (b) 2 μm, and (c) 1 μm. (Reproduced with permission from Lee, J., Oh, J.T., Kim, Y.C., and Lee, J., *IEEE Photon. Technol. Lett.*, 20, 1563 copyright 2008 IEEE.)

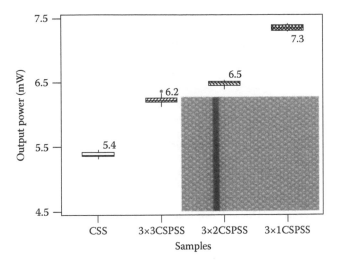

FIGURE 6.12 Light output powers from LED on conventional flat sapphire substrate and those grown on cone-shaped PSSs with different distances of 1, 2, and 3 μm. (Reproduced with permission from Lee, J., Oh, J.T., Kim, Y.C., and Lee, J., *IEEE Photon. Technol. Lett.*, 20, 1563 copyright 2008 IEEE.)

6.3 Design and Fabrication of PSS

As the critical part of PSS technology, various patterns play significant roles in the improvement of LED luminous efficacy. To explore the optimal PSS, researchers have to routinely conduct a large number of experiments to evaluate the effect of different parameters. The entire LED device fabrication process and characterization need to be run through, which is both time-consuming and financially ineffective. This lengthy procedure with many steps greatly enlarges the possibility of errors and brings inaccuracy and uncertainty of the results. To overcome these drawbacks, computer-aided simulation methodology for PSS design is usually used [42]. An optical analysis software based on light tracing is utilized to run the simulations. By this means, researchers can deploy various patterns and optimize them within a reasonable time span, which is advantageous to reduce the time for PSS design, and greatly improve research efficiency. Furthermore, subsequent crystal growth and characterizations of LED wafers have verified the effectiveness and reliability of PSS design.

6.3.1 Simulation Modeling

The simulation procedure includes four steps, that is, building LED chip model with PSS layer, defining Lambertian light source, tracing rays, and collecting data.

The first step of simulation is to build a model to present the LED chip. Figure 6.13 exhibits the LED chip model built in the first step. Diverse patterns with given parameters are subsequently built on the surface of sapphire substrate. The refractive index and thickness for each layer are shown in Table 6.3. The mole fraction of indium in InGaN/GaN MQWs, which is represented by active layer here, is set as 15%. All the parameters are set according to real LED devices to ensure the reliability of the model.

The second step is to set the top and bottom planes of the active layer as Lambertian light source. The output power and total number of rays can be set as any given values, respectively. In general, these two planes are settled with two planar Lambertian light sources with the light output power of 5000 arbitrary units (arb. units) and the total number rays as 3000. Meanwhile, it should also be noted that the LED model is usually assumed not to produce heat.

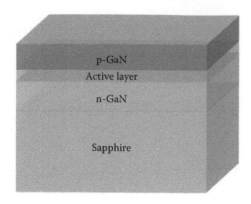

FIGURE 6.13   Schematic structure of the LED chip model.

TABLE 6.3   Refractive Index and Thickness for Each Layer of the LED Chip Model

	Sapphire	n-GaN	Active Layer	p-GaN
Refractive index	1.67	2.45	2.45	2.45
Thickness (nm)	10^5	4×10^3	50	3×10^2

Then the light emitting condition of the LED chip is simulated via the ray-tracing function integrated in the simulation software. To collect the emergent rays from LED chip, six virtual targets without any material properties are built close to their corresponding facets. After the model is complete with all geometry and parameters in position, the software will present the luminous flux from each facet according to its mathematic function.

6.3.2 Design and Optimization of Novel PSSs

By computer simulations, various PSSs such as groove-shaped PSS, hemisphere-shaped PSS, and cone-shaped PSS have been systematically optimized [42–44]. Furthermore, based on the optimization results, two novel PSSs, so-called spherical cap-shaped PSS and dome-shaped PSS, have been proposed for high-efficiency GaN-based LEDs [43,44].

6.3.2.1 Spherical Cap-Shaped PSS

Spherical cap-shaped PSS (SCPSS) is derived from hemisphere-shaped PSS (HPSS) by changing the height of its hemispherical units [43]. As shown in Figure 6.14, three parameters of spherical cap-shaped pattern have been optimized including intercepted ratio (IR), which is defined as the ratio between the spherical cap's height and the radius of the corresponding hemisphere, edge spacing, and radius of the pattern. Figure 6.15a shows total luminous fluxes of SCPSS LEDs with various radii as a function of intercepted ratio. In general, total luminous fluxes rise with the increase of intercepted height at first and then drop slowly with maximum value reaching at the intercepted ratio of 60%–85%. As for edge spacing, Figure 6.15b suggests that total luminous fluxes constantly decrease with the decrease of edge spacing. It is demonstrated that the LED grown on the optimal SCPSS with 3.4 μm in radius, 2.78 μm in height, and 1.7 μm in edge spacing achieves the maximum LEE.

To verify the accuracy of computer simulations, subsequent crystal growth and characterizations of LED wafers on these PSSs have been carried out. Two kinds of PSSs are adopted for comparison, that is, HPSS and SCPSS. The substrates are fabricated by ICP drying etching. Before etching, a uniform layer of photoresist is spread on the sapphire substrate by spin coating. The substrates are then carried out with ICP etching by the following steps, that is, exposure to ultraviolet light with photo mask, chemical

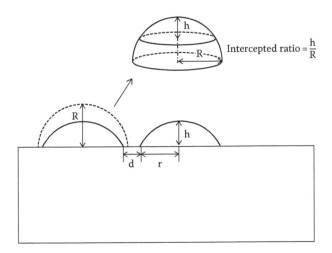

FIGURE 6.14 Schematic diagram of the definition of parameters for spherical cap-shaped PSS. (From Wang, H. et al., *Jpn. J. Appl. Phys.*, 52, 92101, 2013.)

corrosion, and heat treatment. By controlling the etching time and the flux of the plasma gas, that is, Cl_2/Ar, both HPSS and SCPSS with the specified parameters are accurately fabricated on the c-plane of sapphire substrates. To show patterns' shapes, a high-magnification LEO 1530 scanning electron microscope (SEM) is used for the cross-sectional SEM characterization. Figure 6.16a is the in-plane SEM image from SCPSS before film growth to show patterns' alignment, and Figure 6.16b is the cross-sectional TEM image from SCPSS after epitaxial growth to show its unit pattern's size of the substrate.

Subsequently, GaN-based LED epitaxial wafers with the identical structure on both DPSS and CPSS are grown by metal organic chemical vapor deposition (MOCVD) under the same experimental conditions. During the growth, trimethylindium (TMIn), trimethylgallium (TMGa), and ammonia (NH_3) are used as the source materials of In, Ga, and N, respectively. Bicyclopentadienyl magnesium (Cp_2Mg) and silane (SiH_4) are used as the p-type and n-type doping sources, respectively. Shown in Figure 6.17 is the EL spectra from LED chips grown on HPSS, SCPSS, and a commercial PSS. It reveals that the LEE of the LED grown on the optimal SCPSS is enhanced by more than 10%, compared with that on HPSS, which agrees well with the simulation results, proving the legitimacy and reliability of this new methodology.

6.3.2.2 Dome-Shaped PSS

Dome-shaped PSS (DPSS) is derived from cone-shaped PSS (CPSS), by changing the curvature of the cone-shaped pattern's generatrix, as shown in Figure 6.18a. The line (so-called L_3) represents the cone-shaped pattern's generatrix, and the curves (so-called L_1 and L_2) represent the dome-shaped patterns' generatrices. The dome-shaped pattern is described by four parameters as follows: (1) diameter of the pattern, (2) height of the pattern, (3) distance between patterns, and (4) curvature of pattern's generatrix. For intuitive description, the central angle (so-called α in Figure 6.18b) of the curve instead of curvature is used to represent the dome-shaped unit given a fixed diameter and height. In Figure 6.18a, the central angle of L_1, L_2, L_3 is in descending, that is, the curvature of L_1, L_2, L_3 is in ascending. As for the cone-shaped pattern's generatrix, that is, L_3, its central angle is 0° and the curvature becomes infinity.

Figure 6.19 represents the luminous flux from top, bottom, and side facets of LED chips grown on dome-shaped PSSs as a function of the central angle. With the central angle increment, both the top and the bottom luminous flux change in the same tendency. When the angle increases in the range of 0°–10°, they both present the slight increase of 100 and 61 a.u., respectively, and reach the maximum at 10° of the central angle. With the increase in the central angle, both the top and the bottom luminous flux decrease

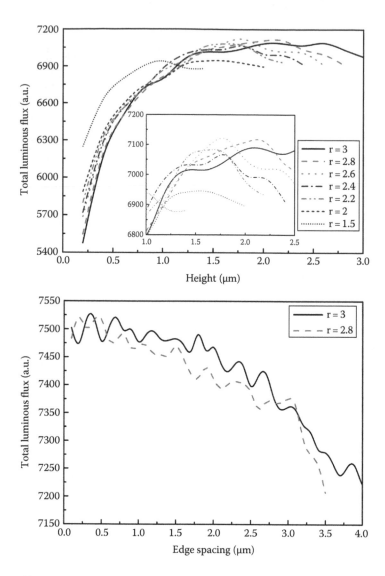

FIGURE 6.15 Total luminous fluxes of SCPSS LEDs as a function of (a) height and (b) edge spacing. (From Wang, H. et al., *Jpn. J. Appl. Phys.*, 52, 92101, 2013.)

continuously, while the top luminous flux decreases faster than the bottom. It indicates that difference in the central angle affects more on the top facet of LED chips. On the contrary, as the central angle increases, the side luminous flux shows a different tendency. It decreases by 179 a.u. quickly in the range of 0°–10°, and then increases slowly by 134 a.u. after 10°.

Figure 6.20 shows the total luminous flux from the LED chips grown on dome-shaped PSSs as a function of the central angle. As the angle increases, the total luminous flux declines in general. Detailed analysis reveals that the total flux decreases very slowly by 91.3 a.u. in the range from 0° to 40° and decreases much more quickly by 405.8 a.u. when the angle is larger than 40°. The total luminous flux between 0° and 10° in the central angles keeps at a high and constant level. In other words, if we use 10° as the central angle, both the top and the bottom luminous flux from the dome-shaped PSS-LED reach the maximum of 2191, 2474 a.u., respectively. At the same time, the total luminous flux also keeps a high value of 7448 a.u. at 10°. As we know, for the actual packaging process, LED chips are commonly packaged in surface mounted devices (SMDs).

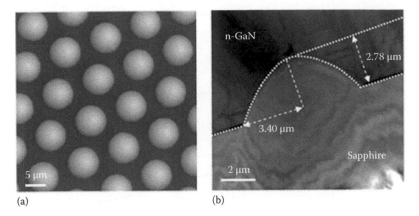

(a)   (b)

FIGURE 6.16 The as-prepared spherical cap-shaped PSS with 3.4 μm in radius, 2.78 μm in height, and 1.7 μm in edge spacing. (a) In-plane SEM image and (b) the low-magnification cross-sectional TEM image after epitaxial growth of LED structure. In practice, the patterns are distributed in hexagonal arrangement rather than the rectangular one as in the simulations.

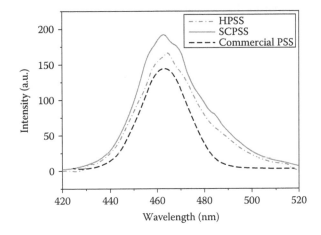

FIGURE 6.17 EL spectra from LED chips grown on HPSS, SCPSS, and a commercial PSS. The SCPSS is of the optimal dimensions with 3.4 μm in radius, 2.78 μm in height, and 1.7 μm in edge spacing, and the HPSS is with 3.45 μm in radius and 1.6 μm in edge spacing, from which the SCPSS is derived with an intercepted ratio of 0.8. (From Wang, H. et al., *Jpn. J. Appl. Phys.*, 52, 92101, 2013.)

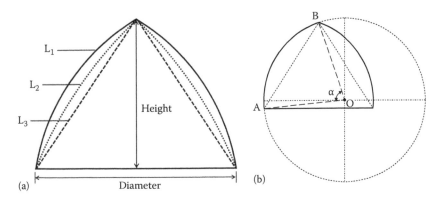

FIGURE 6.18 (a) Schematic diagram of dome-shaped pattern and (b) definition of central angle α.

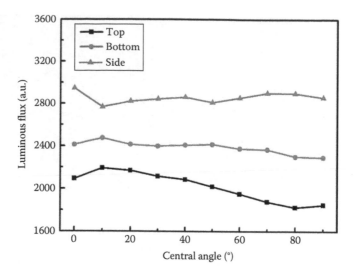

FIGURE 6.19 Luminous flux from top, bottom, and side facets of LED chips grown on dome-shaped PSSs as a function of the central angle.

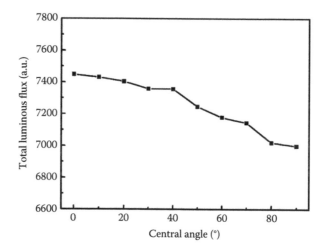

FIGURE 6.20 Total luminous flux from LED chips grown on dome-shaped PSS as a function of the central angle.

In the SMD case, side luminous flux has little effect on the LEDs' output efficiency. The LEDs' output efficiency is mainly determined by the top luminous flux (flip-chip packaging), or the bottom luminous flux (dress-chip packaging) from the LED chips. We can hence deduce that the central angle of 10° for the dome-shaped PSS is optimal to improve the luminous efficacy of LED devices with SMD packaging.

Figure 6.21 exhibits the cross-sectional morphology of DPSS. According to simulation results, the DPSS LED with 10° in generatrix's central angle presents the highest top and bottom luminous flux, and its total flux remains at a relatively high value. Furthermore, EL characterizations for LED chips shown in Figure 6.22 reveal that the luminous intensity from the optimal DPSS LED is enhanced by 19%, compared to that of CPSS LED with the same dimensions. These results straightforwardly confirm the effectiveness of optimal DPSS for improving LED luminous efficacy.

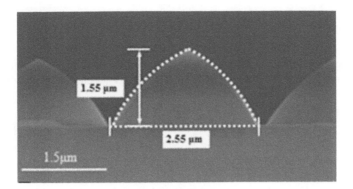

FIGURE 6.21 The cross-sectional SEM image from DPSS. (From Wang, H. et al., *RSC Adv.*, 4, 41942, 2014.)

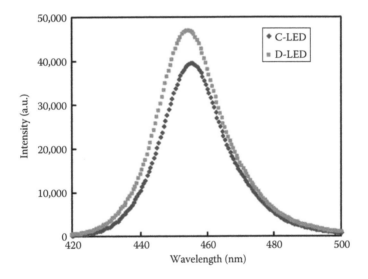

FIGURE 6.22 EL spectra of LEDs on DPSS and CPSS. (From Wang, H. et al., *RSC Adv.*, 4, 41942, 2014.)

PSS is of paramount importance for LEDs because it greatly improves the crystalline quality of GaN and the LEE of LEDs. PSS undergoes a development tendency where the pattern units become more and more complicated. The first and simplest pattern unit is groove shaped, and its one-dimensional geometrical feature makes it limited for further improvement on LEE. Afterward, various pattern units with two-dimensional geometrical feature, such as hexagonal frustum, trigonal frustum, triangular pyramid, hemisphere, and cone, are put forward. It has been verified that such complicated patterns have been very effective in improving LEDs' luminous efficiency. However, the design of novel PSSs faces a challenge that the effect of complicated geometrical features is almost impossible to be systematically studied only by experiment. Thus, a new methodology for PSS design by computer simulations is developed, which is proven to be reliable and cost-effective. By employing this method, the influence of pattern parameters can be systematically studied. Therefore, various novel patterns can be designed and optimized within a reasonable time span. The field of pattern design with such a method still needs particular attentions, since novel patterns designed by this method are potential for further improvement of the LEE of LEDs in the near future.

References

1. Pimputkar S, Speck J S, Denbaars S P, and Nakamura S 2009 *Nat. Photonics* **3** 180.
2. Xie Z L, Zhang R, Fu D Y, Liu B, Xiu X Q, Hua X M, Zhao H et al. 2011 *Chin. Phys. B* **20** 116801.
3. Liu M G, Wang Y Q, Yang Y B, Lin X Q, Xiang P, Chen W J, Han X B et al. 2015 *Chin. Phys. B* **24** 038503.
4. Liu L and Edgar J H 2002 *Mater. Sci. Eng. R Rep.* **37** 61.
5. Heying B, Wu X H, Keller S, Li Y, Kapolnek D, Keller B P, Denbaars S P, and Speck J S 1996 *Appl. Phys. Lett.* **68** 643.
6. Nakamura S 1998 *Science* **281** 956.
7. Zhmakin A I 2011 *Phys. Rep.* **498** 189.
8. Benisty H, De Neve H, and Weisbuch C 1998 *IEEE J. Quantum Electron.* **34** 1612.
9. Hiramatsu K, Nishiyama K, Onishi M, Mizutani H, Narukawa M, Motogaito A, Miyake H, Iyechika Y, and Maeda T 2000 *J. Cryst. Growth* **221** 316.
10. Marchand H, Wu X H, Ibbetson J P, Fini P T, Kozodoy P, Keller S, Speck J S, Denbaars S P, and Mishra U K 1998 *Appl. Phys. Lett.* **73** 747.
11. Hiramatsu K, Nishiyama K, Motogaito A, Miyake H, Iyechika Y, and Maeda T 1999 *Phys. Solidi State A* **176** 535.
12. Le L C, Zhao D G, Wu L L, Deng Y, Jiang D S, Zhu J J, Liu Z S et al. 2011 *Chin. Phys. B* **20** 127306.
13. Wang C H, Chang S P, Ku P H, Li J C, Lan Y P, Lin C C, Yang H C et al. 2011 *Appl. Phys. Lett.* **99** 171106.
14. Kuo Y, Chang J, Tsai M, and Yen S 2009 *Appl. Phys. Lett.* **95** 11116.
15. Huang X H, Liu J P, Fan Y M, Kong J J, Yang H, and Wang H B 2012 *Chin. Phys. B* **21** 037105.
16. Ashby C I H, Mitchell C C, Han J, Missert N A, Provencio P P, Follstaedt D M, Peake G M, and Griego L 2000 *Appl. Phys. Lett.* **77** 3233.
17. Nakada N, Nakaji M, Ishikawa H, Egawa T, Umeno M, and Jimbo T 2000 *Appl. Phys. Lett.* **76** 1804.
18. David A, Meier C, Sharma R, Diana F S, Denbaars S P, Hu E, Nakamura S, Weisbuch C, and Benisty H 2005 *Appl. Phys. Lett.* **87** 101107.
19. He A H, Zhang Y, Zhu X H, Chen X W, Fan G H, and He M 2010 *Chin. Phys. B* **19** 068101.
20. Shchekin O B, Epler J E, Trottier T A, Margalith T, Steigerwald D A, Holcomb M O, Martin P S, and Krames M R 2006 *Appl. Phys. Lett.* **89** 71109.
21. Lester S D, Ponce F A, Craford M G, and Steigerwald D A 1995 *Appl. Phys. Lett.* **66** 1249.
22. Dai Q, Schubert M F, Kim M H, Kim J K, Schubert E F, Koleske D D, Crawford M H et al. 2009 *Appl. Phys. Lett.* **94** 111109.
23. Wang M T, Liao K Y, and Li Y L 2011 *IEEE Photon. Technol. Lett.* **23** 962.
24. Belardini A, Pannone F, Leahu G, Larciprete M C, Centini M, Sibilia C, Martella C, Giordano M, Chiappe D, and Buatier De Mongeot F 2012 *Appl. Phys. Lett.* **100** 251109.
25. Fujii T, Gao Y, Sharma R, Hu E L, Denbaars S P, and Nakamura S 2004 *Appl. Phys. Lett.* **84** 855.
26. Pan C, Hsieh C, Lin C, and Chyi J 2007 *J. Appl. Phys.* **102** 84503.
27. Jiang Y, Jia H, Wang W, Wang L, and Chen H 2011 *Energy Environ. Sci.* **4** 2625.
28. Torma P T, Ali M, Svensk O, Suihkonen S, Sopanen M, Lipsanen H, Mulot M, Odnoblyudov M A, and Bougrov V E 2010 *Cryst. Eng. Commun.* **12** 3152.
29. Gao H, Yan F, Zhang Y, Li J, Zeng Y, and Wang G 2008 *Solid State Electron.* **52** 962.
30. Bo Lee S, Kwon T, Lee S, Park J, and Choi W 2011 *Appl. Phys. Lett.* **99** 211901.
31. Zhou S, Lin Z, Wang H, Qiao T, Zhong L, Lin Y, Wang W, Yang W, and Li G 2014 *J. Alloy. Compd.* **610** 498.
32. Shizhong Z A H W 2014 *Jpn. J. Appl. Phys.* **53** 25503.
33. Lee J, Lee D, Oh B, and Lee J 2010 *IEEE Trans. Electron Dev.* **57** 157.
34. Cheng J, Wu Y S, Liao W, and Lin B 2010 *Appl. Phys. Lett.* **96** 51109.
35. Hagedorn S, Richter E, Zeimer U, Prasai D, John W, and Weyers M 2012 *J. Cryst. Growth* **353** 129.
36. Cho J, Kim H, Kim H, Lee J W, Yoon S, Sone C, Park Y, and Yoon E 2005 *Phys. Status Solidi (c)* **2** 2874.
37. Lee J, Oh J, Park J, Kim J, Kim Y, Lee J, and Cho H 2006 *Phys. Status Solidi (c)* **3** 2169.

38. Jeong S, Kissinger S, Kim D, Jae Lee S, Kim J, Ahn H, and Lee C 2010 *J. Cryst. Growth* **312** 258.

39. Shin H, Kwon S K, Chang Y I, Cho M J, and Park K H 2009 *J. Cryst. Growth* **311** 4167.

40. Lee J, Oh J T, Kim Y C, and Lee J 2008 *IEEE Photon. Technol. Lett.* **20** 1563.

41. Xu S R, Li P X, Zhang J C, Jiang T, Ma J J, Lin Z Y, and Hao Y 2014 *J. Alloy. Compd.* **614** 360.

42. Lin Z, Yang H, Zhou S, Wang H, Hong X, and Li G 2012 *Cryst. Growth Des.* **12** 2836.

43. Wang H, Zhou S, Lin Z, Hong X, and Li G 2013 *Jpn. J. Appl. Phys.* **52** 92101.

44. Wang H, Zhou S, Lin Z, Qiao T, Zhong L, Wang K, Hong X, and Li G 2014 *RSC Adv.* **4** 41942.

7

Surface Plasmon–Coupled Light-Emitting Diodes

Chia-Ying Su
National Taiwan University

Chun-Han Lin
National Taiwan University

Yang Kuo
National Taiwan University

Yu-Feng Yao
National Taiwan University

Hao-Tsung Chen
National Taiwan University

Charng-Gan Tu
National Taiwan University

Chieh Hsieh
National Taiwan University

Horng-Shyang Chen
National Taiwan University

Yean-Woei Kiang
National Taiwan University

C.C. Yang
National Taiwan University

Abstract The phenomenon of surface plasmon (SP) coupling with a light emitter and its application to light-emitting didoes (LEDs) in the visible range is reviewed and the details of SP coupling in the deep-UV range with Al nanostructures are discussed. First, the fundamental principles of SP coupling with a light emitter for enhancing its emission efficiency are explained. For the application to an LED, several useful metal nanostructures for inducing SP coupling with its quantum wells (QWs) are introduced. Then, the effects of SP coupling in an LED for improving its operation functions, including enhancing internal quantum efficiency (IQE), reducing efficiency droop, and increasing modulation bandwidth, are elucidated. Next, we focus the discussion on the SP coupling in the deep-UV range with Al nanostructures. The IQE enhancement of deep-UV $Al_xGa_{1-x}N/Al_yGa_{1-y}N$ ($x < y$) QWs is shown by fabricating surface Al nanoparticles on a QW structure through thermally annealing an Al thin film for inducing SP coupling. Through temperature-dependent photoluminescence measurement, the enhancements of IQE in different emission polarizations are illustrated.

7.1 Introduction

Surface plasmon (SP) resonance represents collective electron oscillation on a metal surface and can normally be observed at the interface between a metal nanostructure and a dielectric material (including air) [1]. When the interface is smooth in a micron scale, the energy of such an electron oscillation can propagate along the interface. Such an SP is classified into the category of surface plasmon polariton (SPP). On the other hand, at the surface of a nano-scale metal structure, localized electron oscillation can occur to form a localized surface plasmon (LSP). With the propagation nature of SPP, the momentum matching condition needs to be satisfied for the energy exchange between an SPP and a photon. For a given metal and a contacting dielectric material, SPP can exist in the wavelength range longer than a resonance wavelength, at which both resonance strength and metal absorption reach the maximum levels. The density of state of SPP becomes lower as wavelength increases. The dispersion curve of an LSP (energy vs. momentum) is a horizontal line, implying that an LSP can interact with a photon of any propagation direction. No momentum matching condition needs to be satisfied in such an energy exchange process. Either SPP or LSP can produce a strong electromagnetic field distribution on the dielectric side near the interface. Such a near-field distribution covers a distance of several tens of nanometers from the interface. Such a strong field distribution can interact with a nearby light emitter for changing its emission characteristics, particularly enhancing its emission efficiency.

When a light emitter is placed near a metal structure within the aforementioned near-field distribution range, the induced SP can couple with the light emitter for increasing its radiative transition rate and hence enhancing emission efficiency [2–16]. The SP coupling mechanism can be applied to a light-emitting diode (LED) for enhancing the emission efficiency of its quantum wells (QWs). The emission efficiency enhancement of a light emitter near a metal structure can be understood from a few different viewpoints. First, it can be understood with the Purcell effect [17]. The near-field distribution of an SP mode confined around the metal structure can be regarded as an effective resonance cavity although its Q-factor is not high. When a light emitter is placed within such a resonance cavity, its emission efficiency can be enhanced by the Purcell factor. The coupling process between an SP mode and a light emitter can also be understood from the viewpoint of amplified spontaneous emission [18]. It can be regarded as a process of amplifying the near field of the coupled SP mode by the excited light emitter. In other words, the light emitter transfers energy into the SP mode for radiation. With the coupling process, an SP mode and a light emitter radiate coherently. Figure 7.1a shows schematically the top portion of a lateral LED with a metal nanostructure, which is represented by a few metal nanoparticles (NPs), on the top of its p-GaN layer. When an SP mode is induced on the metal nanostructure by the spontaneous emission of

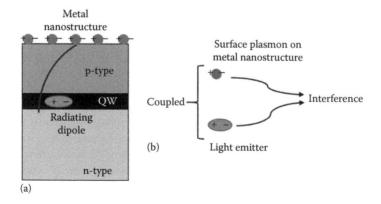

(a)   (b)

FIGURE 7.1 (a) Top portion of a lateral LED with a metal nanostructure, which is represented by a few metal NPs, on the top of its p-type layer. (b) Schematic demonstration of an SP-emitter coupling system including two interfered coherent emission sources, that is, the light emitter and the induced SP mode on the metal structure.

the QW, its near-field distribution extends into the p-GaN layer to cover the QW if the p-type layer is not thick (<100 nm), as depicted by the curve in Figure 7.1a. In this situation, carrier energy in the QW can be transferred into the SP mode for radiation through the amplification of the near field by excited QW. Therefore, in this viewpoint, the coupling creates an alternative emission channel besides the intrinsic emission through carrier recombination in the QW. However, because an SP mode loses energy through metal dissipation, the SP coupling process also creates a new energy loss channel. Hence, we can obtain an overall emission efficiency enhancement only when the new loss channel is weaker than the intrinsic energy loss in the QW through nonradiative recombination of carriers. It is noted that such a coupling process is a near-field interaction mechanism. It is not an emission-reabsorption process, which corresponds to a far-field mechanism. In this sense, a nearby light emitter can always couple with an SPP or LSP, that is, no issue of momentum matching exists, because the near-field distribution of an SP does not have a propagation direction.

The coupling of a light emitter and an SP on a metal structure does not always result in emission enhancement. Theoretically, emission enhancement can be observed only in certain spectral ranges. Outside these spectral ranges, the overall emission of the coupling system is suppressed. Such a behavior can be understood from the viewpoint of interference, as demonstrated in Figure 7.1b, in which the SP-emitter coupling system includes two coherent emission sources, that is, the light emitter and the induced SP mode on the metal structure. If the radiations from the two sources effectively interfere constructively (destructively), overall emission enhancement (suppression) is observed.

7.2 Metal Nanostructures for Inducing Surface Plasmon Coupling with a Quantum Well

Figure 7.2 schematically shows five metal structures, as denoted by A–E, for inducing SP coupling with the QW in a lateral LED. In structure A, a metal film is coated on the p-GaN layer to form a smooth metal/GaN interface [19–21]. At the interface, induced SPP can couple with the light emitters in the QW. If the metal/GaN interface is rough or periodically grooved such that the momentum-matching condition between SPP and emitted photon is satisfied, the grating-assisted SPP coupling can further enhance emission efficiency [22–26]. The simplest metal nanostructure for inducing SP coupling is a surface metal NP distribution on the top of the p-GaN layer, as structure B in Figure 7.2.

The metal NPs can be simply fabricated through a thermal annealing process of a metal thin film [27–29]. With this process, the size and position of metal NP are randomly distributed. Regularly arranged

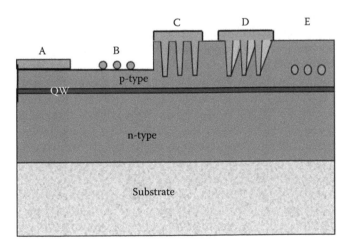

FIGURE 7.2 Schematic demonstration of five metal structures, as denoted by A–E, for inducing SP coupling with the QW in a lateral LED.

metal NPs of a uniform size can be fabricated through nano-imprint lithography, dry etching, and liftoff process [30–32]. In using metal structures A and B, to increase the SP field strength at the QW for effective coupling, the thickness of the p-type layer needs to be small, usually smaller than 100 nm. However, with such a thin p-type layer, usually the electrical property of an LED is poor unless the conductivity of p-GaN can be improved. To improve the electrical property of an LED, a thicker p-type layer needs to be used and other metal structures are desired.

With a thick p-type layer, to make the distance between the metal structure and the QW short, a metal protrusion structure like structure C shown in Figure 7.2 can be used [33–35]. To fabricate such a metal structure, a hole array is first fabricated on the p-type layer with the bottom of the hole close to the top QW. Then, the holes are filled with metal to form the structure of a metal protrusion array. LSP resonance can be induced around the protrusion tip for coupling with the QW. Such a protrusion has the advantage of making the metal structure close to the QW while maintaining a thick p-type layer and hence preserving a good LED electrical property. Besides the two-dimensional metal protrusion structure, one-dimensional metal grating is another practically useful structure for inducing effective SP coupling in an LED with a thick p-type layer [22–26]. Structure D in Figure 7.2 shows a one-dimensional metal grating structure on the p-type layer. To reach the momentum-matching condition for achieving grating-assisted SPP coupling, normally the grating period needs to be smaller than 200 nm for a green LED and even smaller for a blue or UV LED. However, it is usually expensive to fabricate such a short-period metal grating on GaN. With a larger period, LSP resonance with polarization perpendicular to the grating ridge can be induced around the tip of a metal grating ridge. When the grating ridge is deep such that its tip is close to the QWs of an LED, the induced LSP can strongly couple with the QWs for emission enhancement. Besides the LSP resonance of polarization perpendicular to the grating ridge, SPP with polarization parallel with the grating ridge induced on the stripe of a grating ridge can also couple with the QWs for emission enhancement. Embedding metal NPs in the p-type layer (see structure E in Figure 7.2) can make metal NPs close to the QW for strong coupling [36–41]. However, the fabrication of such an embedded metal-NP structure requires a regrowth process unless the metal source for forming NPs is available in the growth chamber. Also, to guarantee high-quality regrowth, the planar metal NP coverage cannot be high. In this situation, the overall SP-coupling effect can be quite weak. Meanwhile, with metal NPs completely surrounded by GaN (refractive index ~2.4), the major LSP resonance (dipole resonance) wavelength is redshifted. Although the SP coupling can rely on higher-order LSP resonance modes, the application spectral range becomes limited. In addition, the required regrowth process makes the fabrication cost of such a device higher.

In fabricating metal structures for inducing SP coupling, the metal material and morphology need to be well chosen for maximizing the SP resonance strength at the QW emission wavelength. For the application to an LED in the spectral range from shallow UV through green color, Ag is the most appropriate metal for use. In the spectral range of yellow and red colors, Au can be a good choice. For deep-UV application, Al is the better choice. The SP resonance strength at a given wavelength can be adjusted by modifying the metal morphology or its surrounding material and structure. By using the structure of surface metal NP as an example, the LSP resonance peak can be adjusted by changing the size of metal NP. A smaller NP leads to a shorter LSP resonance wavelength. However, for a given metal material, there is a short-wavelength limit of LSP resonance peak when the NP size is extremely small. In this situation, a design of the surrounding material and structure is useful for adjusting the LSP resonance behavior of a metal NP. Figure 7.3a through c show three designs.

In the structure shown in Figure 7.3a, a thin conducting layer is deposited on the p-type layer before the formation of surface metal NPs [30,42,43]. This layer must be conductive for current injection, but needs to become a dielectric in the QW emission spectral range with a refractive index lower than that of GaN (~2.4). With such a dielectric interlayer of a lower refractive index, the LSP resonance peak can be blueshifted. An oxide transparent conductor, such as Ga-doped ZnO (refractive index around 1.8 in the visible range), is a good choice for the conducting interlayer [44,45]. This conducting interlayer can

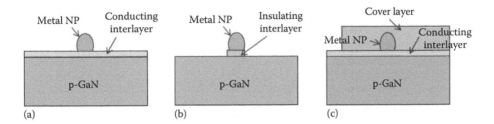

FIGURE 7.3 (a–c) Three designs of interlayer and cover-layer for adjusting the LSP resonance spectrum of a surface metal NP.

also help in current spreading. An insulating interlayer of a lower refractive index, such as SiO_2 (refractive index around 1.5), can also be used for blueshifting the LSP resonance peak [46]. However, for current injection, the interlayer region outside the coverage of metal NPs must be removed, as schematically shown in Figure 7.3b. The interlayer thickness can be changed for adjusting the LSP resonance peak. A thicker interlayer results in a larger blueshift range. However, the LSP resonance strength decreases with increasing interlayer thickness. For instance, a SiO_2 or Ga-doped ZnO interlayer of 5–10 nm in thickness can shift the LSP resonance peak of Ag NPs from the green (~520 nm) into the blue range (up to ~450 nm). When we need to redshift the LSP resonance peak, we can cover the surface metal NPs with a dielectric layer, as schematically shown in Figure 7.3c. By using SiO_2 as the cover layer, the LSP resonance peak of Ag NPs can be shifted from the green into yellow range.

7.3 Functions of Surface Plasmon Coupling in LED Operation

The SP coupling with a QW of an LED can lead to several important functions. First, the effective internal quantum efficiency (IQE) of the QW can be enhanced due to the increase of the emission efficiency of the SP-QW coupling system [19,22,30,33,42,45]. Second, the radiation pattern after the QW couples with an SP mode is changed. Therefore, with an appropriate design of device structure, the light extraction efficiency of an LED can be increased [47]. Third, during the coupling process, the carrier energy is effectively transferred into the SP mode such that the carrier density in the QW is maintained at a low level. With a low carrier density in the QW, the efficiency droop effect can be reduced [19,22,30,42,45]. Fourth, because of the stimulated emission nature in SP coupling, carrier decay in a QW is fast and hence the device response to injection modulation is enhanced such that the modulation bandwidth of an LED can be increased [48]. Finally, when the metal for inducing SP modes has an anisotropic structure, such as a one-dimensional metal grating, the LED output becomes partially polarized due to the polarized SP coupling behavior [22–24].

The efficiency droop effect of an LED at high current injection is basically caused by the high carrier density in a QW [49–52]. A typical efficiency droop behavior is schematically shown in curve I of Figure 7.4.

Although the emission efficiency of an LED at low injection current can be quite high, it droops dramatically as injection current increases. Efforts have been made for reducing the drooping slope. However, as shown in curve II of Figure 7.4, although the drooping slope can be reduced, the resultant maximum efficiency is usually decreased. With SP coupling, not only the drooping slope can be reduced, but also the maximum efficiency can be increased, as depicted in curve III of Figure 7.4. The reduction of the drooping slope is due to the decrease of carrier density in a QW. The increase of the maximum efficiency originates from the enhancement of IQE. The reduction of carrier density under SP coupling can be understood with the carrier flowchart shown in Figure 7.5.

The left portion of Figure 7.5 shows the distribution of injected carriers in a QW with carrier density N and carrier supply rate G into various carrier consumption channels before SP coupling. In this situation, the carrier consumption channels include carrier leakage with the leakage rate K_{lea}, radiative

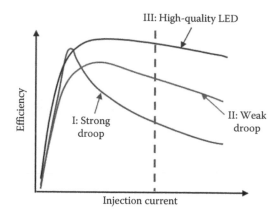

FIGURE 7.4   Schematic demonstration of three conditions of LED efficiency versus injected current, including curve I for a strong droop condition, curve II for a weak droop condition, and curve III for a condition of weak droop and high maximum efficiency (high-quality LED).

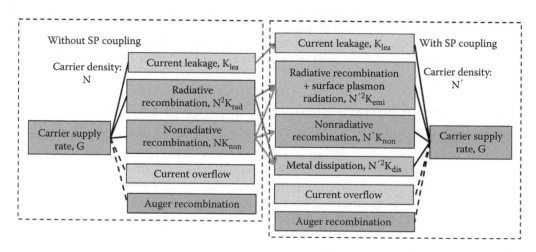

FIGURE 7.5   Carrier flowchart in the left (right) portion showing the distribution of injected carriers in a QW with carrier density N (N′) and carrier supply rate G into various carrier consumption channels without (with) SP coupling.

recombination with the rate of N^2K_{rad}, nonradiative recombination with the rate of NK_{non}, carrier overflow, and Auger recombination. Here, K_{rad} and K_{non} are the coefficients of radiative and nonradiative recombination, respectively. The right portion of Figure 7.5 shows the distribution of injected carriers in a QW with carrier density N′ and the same carrier supply rate G into various carrier consumption channels after SP coupling is applied to the device. In this situation, the energy of the carriers originally used for radiative and nonradiative recombination is redistributed into three parts, including the emission of the SP-QW coupled system with the rate of N'^2K_{emi}, the residual nonradiative recombination with the rate of $N'K_{non}$, and the metal dissipation through SP generation with the rate of N'^2K_{dis}. Here, K_{emi} is a coefficient for the emission of the SP-QW coupling system, which includes the emissions from the involved SP modes and carrier radiative recombination. The nonradiative recombination coefficient, K_{non}, is expected to be unchanged with SP coupling because it is related to the defect density in the QW. The metal dissipation rate includes a coefficient K_{dis} and is proportional to the square of carrier density since its energy comes from the transfer of carrier energy into SP resonance, which involves the annihilation of a pair of carrier. The carrier leakage rate, K_{lea}, is also unchanged after SP coupling is applied to the device. At high injection current, carrier overflow and Auger recombination also exist. However, their magnitudes can be different from those without SP coupling since the carrier density can be changed after SP coupling is applied to the device. When current injection level is not high such that carrier overflow and Auger recombination can be neglected, the steady state of carrier density without SP coupling can be expressed as

$$G = K_{lea} + NK_{non} + N^2 K_{rad}. \tag{7.1}$$

Similarly, the steady state of carrier density with SP coupling can be expressed as

$$G = K_{lea} + N'K_{non} + N'^2 \left(K_{emi} + K_{dis} \right). \tag{7.2}$$

The equality of Equations 7.1 and 7.2 leads to

$$\left(N' - N \right) K_{non} + N'^2 \left(K_{emi} + K_{dis} \right) = N^2 K_{rad}. \tag{7.3}$$

Since we can observe the enhancement of emission through SP coupling, we have

$$\left(N' - N \right) K_{non} + N'^2 \left(K_{emi} + K_{dis} \right) = N^2 K_{rad} < N'^2 K_{emi}. \tag{7.4}$$

Therefore, we can obtain the inequality as

$$\left(N' - N \right) K_{non} + N'^2 K_{dis} < 0. \tag{7.5}$$

Because $N'^2 K_{dis}$ is always positive, we can conclude that $N' < N$. In other words, with SP coupling, the carrier density in a QW is reduced. In this situation, the effects of carrier overflow and Auger recombination are weakened. Therefore, SP coupling in an LED can reduce the efficiency droop effect.

7.4 Surface Al Nanoparticles for Enhancing Deep-UV Emission

7.4.1 Current Status of AlGaN-Based Quantum Well

Because of the low crystal quality of AlGaN with high Al content, the emission efficiency of deep-UV QWs are usually not high. Although crystal growth efforts are important for improving the crystal quality and hence the IQE of the $Al_xGa_{1-x}N/Al_yGa_{1-y}N$ (x < y) QWs, other approaches are useful for further increasing the emission efficiency, including SP coupling [53–56]. For deep-UV applications, instead of Ag, Al nanostructures are preferred for enhancing the IQE of a deep-UV AlGaN-based QW. Because the split-off valence band is close to or even higher than the heavy-/light-hole valence bands [57–59], the emission polarization of such a deep-UV QW is more complicated, when compared with an InGaN/GaN QW. Electron transition between the conduction band and the heavy-/light-hole valence bands leads to transverse-electric-(TE-) polarized emission, that is, the polarization parallel with the c-plane. Electron transition between the conduction band and the split-off valence band results in transverse-magnetic-(TM-) polarized emission, that is, the polarization perpendicular to the c-plane. Because the TM-polarized emission tends to propagate along the sample-layer waveguide in the lateral direction, its light extraction efficiency is significantly lower than that of TE-polarized emission. In this situation, it is important to understand the difference of SP coupling behavior between the TE- and TM-polarized emissions.

In this section, we demonstrate the IQE enhancements of deep-UV AlGaN-based QWs by fabricating surface Al NPs on a QW structure for inducing SP coupling. By comparing the LSP resonance behaviors of Al NPs formed with different deposited Al-film thicknesses, annealing temperatures and durations, optimized fabrication conditions can be obtained for producing LSP resonance close to the QW emission wavelength in the deep-UV range. Through temperature-dependent photoluminescence (PL) measurement, the enhancements of IQE in different emission polarizations are illustrated. The strong LSP resonance at the excitation laser wavelength may lead to stronger excitation and hence higher IQE levels.

7.4.2 Sample Growth Conditions and Process Procedures

The AlGaN QW epitaxial structure is grown on c-plane sapphire substrate with metalorganic chemical vapor deposition (MOCVD). The epitaxial growth starts with a three-temperature deposition process for forming the AlN buffer layer. The low, medium, and high temperatures at 960°C, 1130°C, and 1235°C, respectively, are used successively with the V/III ratios at 1500, 300, and 75 to form the AlN layers of ~50 nm, ~150 nm, and ~2.5 μm, respectively, in thickness. Then, an AlGaN layer of ~2 μm grown at 1170°C with the V/III ratio at 1200 is deposited, followed by three periods of QW. A well layer of ~1.6 nm in thickness is grown at 1040°C with the V/III ratio at 7400. A barrier layer of ~8.4 nm in thickness is grown with the same growth conditions as those for the aforementioned AlGaN layer. On the top of the QWs, AlGaN capping layers of ~21 and ~123 nm are deposited under the same growth conditions as those of the AlGaN layer below the QWs for forming samples A and B, respectively. The pressure in the MOCVD growth chamber is fixed at 60 mbar throughout the whole growth procedure. To test the thermal annealing effect on the QW emission efficiency, sample A is thermally annealed at 600°C for 10 min with ambient nitrogen. The annealed sample is designated as sample A-anneal.

Figure 7.6 shows the transmission spectra of an AlGaN template, that is, an epitaxial structure before the growths of QWs and AlGaN capping layer, and QW samples A and B. Here, one can see that the transmittance cutoff of each curve is around 240 nm, indicating that the AlGaN layer has an Al content of around 71%. In the curves of samples A and B, the slopes in the range between 250 and 295 nm are caused by QW absorption, indicating that the QW emission wavelength at room temperature is around 295 nm.

Surface Al NPs are fabricated by first depositing a thin Al layer of a designated thickness on an epitaxial sample with an electron-gun evaporator, followed by a thermal annealing process at a designated temperature for a given duration with ambient nitrogen. Such fabrication conditions are adjusted for making LSP resonance peak as close as possible to the QW emission wavelength around 295 nm at room temperature.

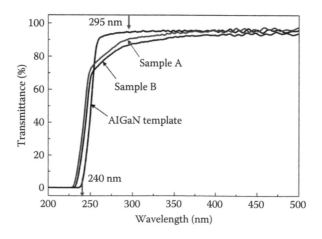

FIGURE 7.6   Transmission spectra of the AlGaN template and samples A and B.

7.4.3 Surface Plasmon Resonance Behaviors of Surface Al Nanoparticles

Before fabricating surface Al NPs on QW epitaxial structures, that is, samples A and B, we need to identify the optimized conditions for fabricating surface Al NPs on AlGaN to produce preferred LSP resonance behaviors. Three key parameters need to be optimized for fabricating desired surface Al NPs, including the thickness of the deposited Al thin film before thermal annealing, thermal annealing temperature, and annealing duration. In Figure 7.7a through c, we show the atomic force microscopy (AFM) images of the surface Al NPs with the annealing temperatures at 300°C, 400°C, and 500°C, respectively. The thickness of the deposited Al thin film and annealing duration are fixed at 5 nm and 30 min, respectively. Here, the average NP sizes in Figure 7.7a through c are ~36, ~43, and ~54 nm, respectively.

Figure 7.8 shows the transmission spectra of the surface Al NPs on AlGaN with the three annealing temperatures. The transmission spectra are normalized with respect to that of the AlGaN template, that is, a transmission spectrum shown in Figure 7.6. Therefore, the depressions in the curves of Figure 7.8 correspond to the LSP resonance features of the fabricated Al NPs. Here, one can see that among the three annealing temperatures, only the temperature of 400°C leads to a clear depression in the deep-UV range.

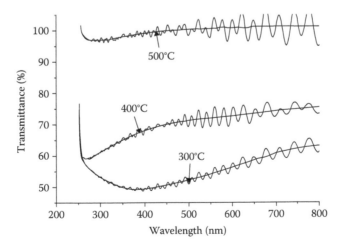

FIGURE 7.7 (a–c) AFM images of surface Al NPs when the annealing temperatures are 300°C, 400°C, and 500°C, respectively. The deposited Al-film thickness and annealing duration are fixed at 5 nm and 30 min, respectively.

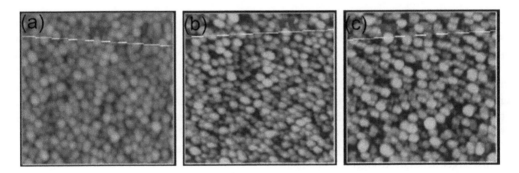

FIGURE 7.8 Transmission spectra of the surface Al NP samples with the annealing temperatures at 300°C, 400°C, and 500°C. The deposited Al-film thickness and annealing duration are fixed at 5 nm and 30 min, respectively. The smooth curves show the transmittance after the Fabry–Perot oscillations are filtered.

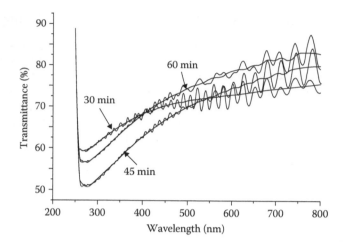

FIGURE 7.9   Transmission spectra of the surface Al NP samples with the annealing durations at 30, 45, and 60 min. The deposited Al-film thickness and annealing temperature are fixed at 5 nm and 400°C, respectively.

Figure 7.9 shows the transmission spectra of the surface Al NPs on the AlGaN template with three annealing durations at 30, 45, and 60 min. The thickness of the deposited Al thin film and annealing temperature are fixed at 5 nm and 400°C, respectively. Here, one can see that clear depressions in the deep-UV range can be observed in all the three cases of different annealing durations. The depression depth increases first and then decreases with annealing duration.

Figure 7.10 shows the transmission spectra of the surface Al NPs on the AlGaN template with three deposited Al-film thicknesses at 3, 5, and 7 nm. The annealing temperature and duration are fixed at 400°C and 30 min, respectively. Here, one can see that although a 7 nm deposited Al thin film can also lead to a clear depression in the deep-UV range, the Al NPs obtained by annealing a 5 nm Al thin film result in a larger depth or stronger LSP resonance.

We also study the effects on LSP resonance behavior of a dielectric interlayer and a cover layer in a surface Al NP structure. In Figure 7.11a through d, we schematically show four different surface Al NP structures, including structure I of surface Al NPs on AlGaN, structure II with a 8 nm

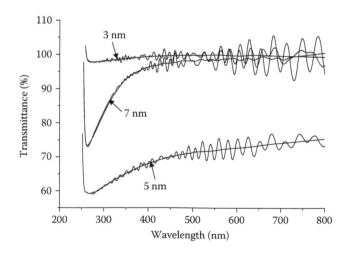

FIGURE 7.10   Transmission spectra of the surface Al NP samples with the deposited Al-film thicknesses at 3, 5, and 7 nm. The annealing temperature and duration are fixed at 400°C and 30 min, respectively.

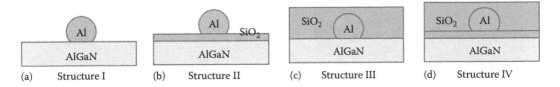

FIGURE 7.11 Four different surface Al NP structures, including (a) structure I with Al NPs on a bare AlGaN template; (b) structure II with Al NPs on an AlGaN template with an 8 nm SiO$_2$ interlayer; (c) structure III with Al NPs on an AlGaN template covered by a 30 nm SiO$_2$ layer; (d) structure IV with Al NPs on an AlGaN template with an 8 nm SiO$_2$ interlayer covered by a 30 nm SiO$_2$ layer.

SiO$_2$ interlayer between Al NPs and AlGaN, structure III of surface Al NPs on AlGaN covered by a 30 nm SiO$_2$ layer, and structure IV of surface Al NPs on a 8 nm SiO$_2$ interlayer and then covered by a 30 nm SiO$_2$ layer, respectively. The Al NPs are formed under the conditions of 5 nm in deposited Al-film thickness, 400°C in annealing temperature, and 30 min in annealing duration. The SiO$_2$ layers are grown with plasma-enhanced chemical vapor deposition.

Figure 7.12 shows the transmission spectra of the four Al NP structures shown in Figure 7.11a through d. After filtering Fabry–Perot oscillations, one can see that the spectral positions of the depression minima in structures I–IV are 295, 280, 320, and 308 nm, respectively, as indicated by the vertical dashed lines. The blueshift of LSP resonance peak from structure I to II is caused by the addition of the dielectric interlayer of SiO$_2$, which has a refractive index (~1.5) lower than that of AlGaN (~2.1). In structure III, the LSP resonance peak is redshifted, when compared with structure I, due to the increase in the surrounding refractive index (from 1 to ~1.5) through the coverage of Al NPs by SiO$_2$. The similar redshift trend can be observed when structure II is covered by a SiO$_2$ layer to become structure IV. Therefore, a SiO$_2$ interlayer or cover layer can be used to adjust the LSP resonance peak for matching the QW emission wavelength and hence maximizing the SP coupling effect.

To fabricate surface Al NPs on samples A and B for IQE measurements, we fix the deposited Al-film thickness, annealing temperature, and annealing duration at 5 nm, 400°C, and 30 min, respectively. Figure 7.13a and b shows the AFM images and scanning electron microscopy (SEM) image, respectively, of surface Al NPs on sample A. Here, the average Al NP size is ~43 nm. With surface Al NPs, this

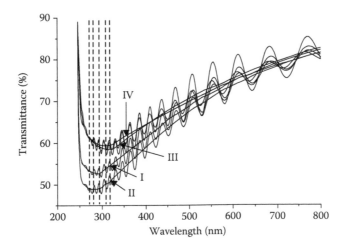

FIGURE 7.12 Transmission spectra of structures I–IV shown in Figure 7.11. The vertical dashed lines mark the spectral depression minima.

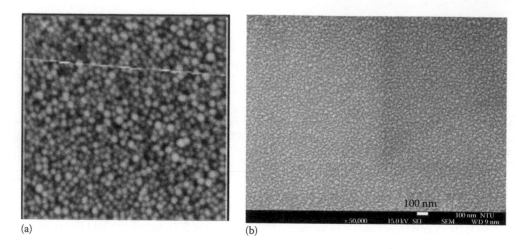

(a)   (b)

FIGURE 7.13   (a) and (b): Surface AFM and SEM images, respectively, of sample A-NP.

sample is designated as sample A-NP. The morphology of the surface Al NPs on sample B is almost the same as those in sample A-NP. With surface Al NPs on sample B, this sample is designated as sample B-NP.

Figure 7.14 shows the transmission spectra of samples A-NP and B-NP when light is incident from the sapphire side. The depression minima around 285 nm are close to the QW emission wavelength at room temperature around 295 nm, as indicated by the vertical dashed line.

7.4.4 Photoluminescence Measurement Results

In temperature-dependent PL measurement, a 266 nm Q-switched laser is used for PL excitation with the average power at ~4.5 mW. The excitation laser with its polarization in the plane of sample surface is incident from the sapphire side of a sample with an incident angle of 45° after it passes a half-wave plate. In PL measurement, edge emission is collected through a polarizer for differentiating the TE- and TM-polarized PL emissions. The TE-polarized emission originates from the electron transition between the conduction band and the heavy-/light-hole valence bands. The TM-polarized emission is generated from the electron transition between the conduction band and the split-off valence band.

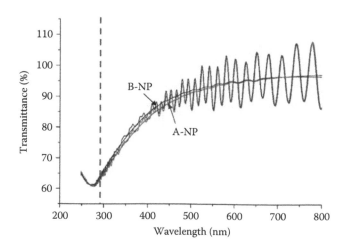

FIGURE 7.14   Transmission spectra of samples A-NP and B-NP. The vertical dashed line indicates the QW emission wavelength around 295 nm at room temperature.

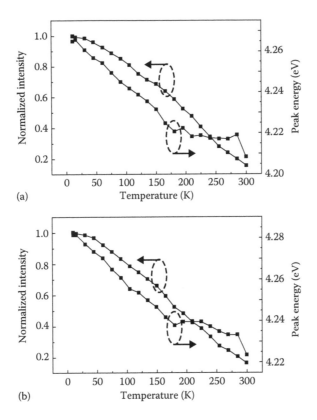

FIGURE 7.15 (a and b) Temperature-dependent integrated PL intensity (the left ordinate) and spectral peak energy (the right ordinate) of the TE- and TM-polarized emissions, respectively, of sample A.

Figure 7.15a and b shows the temperature-dependent integrated PL intensity (the left ordinate) and spectral peak energy (the right ordinate) of the TE- and TM-polarized emissions, respectively, of sample A. Here, the integrated PL intensity at 300 K, when normalized with respect to that at 10 K, is defined as the IQE of the sample. The obtained IQE values for the TE- and TM-polarized emissions are 15.5% and 15.2%, respectively. In both curves of temperature-dependent spectral peak energy, one can see a range of blueshift or smaller redshift with temperature roughly between 175°C and 275°C. Such a behavior is similar to the S-shaped variation of temperature-dependent spectral peak energy in an InGaN/GaN QW, indicating that the carrier localization behavior also exists in an AlGaN-based QW.

Data for samples B, A-anneal, A-NP, and B-NP similar to those shown in Figure 7.15a and b are also obtained. Those for sample A-NP are shown in Figure 7.16a and b.

The IQE values of all samples are listed in Table 7.1.

Here, one can see that all the IQE values are similar in the same row, indicating that the IQEs of the TE- and TM-polarized emissions are about the same even though they originate from the electron transitions of different valence bands. The IQEs of sample B are significantly higher than those of sample A. The increased IQEs in sample B are attributed to the thermal annealing effect because the thickness of the AlGaN capping layer in sample B (123 nm) is much thicker than that of sample A (21 nm). During the much longer growth time at a temperature as high as 1170°C for the capping layer, the QWs in sample B are effectively annealed for increasing their IQEs. To prove this attribution, sample A-anneal is prepared by thermally annealing sample A under the condition of 600°C in temperature for 30 min. Although the annealing temperature is much lower than the original growth temperature for sample B, we can still observe the enhancement of IQE in sample A-anneal even though the increase range is only ~2%. The IQEs of sample B can be as high as ~21%.

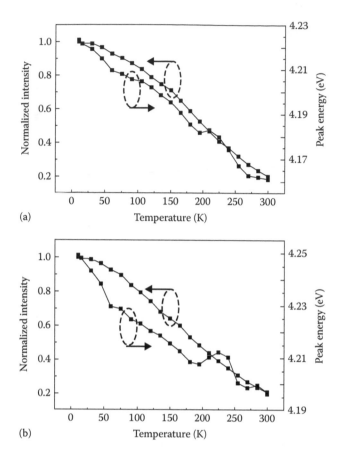

(a)

(b)

FIGURE 7.16 (a and b) Temperature-dependent integrated PL intensity (the left ordinate) and spectral peak energy (the right ordinate) of the TE- and TM-polarized emissions, respectively, of sample A-NP.

TABLE 7.1   IQEs of Different Samples under Various Polarization Conditions

	Capping Layer Thickness (nm)	IQE of TE-Polarized Emission (%)	IQE of TM-Polarized Emission (%)
Sample A	21	15.5	16.2
Sample B	123	20.1	20.9
Sample A-anneal	21	17.8	18.1
Sample A-NP	21	20.6	21.3
Sample B-NP	123	21.1	20.5

Note: The polarization of the excitation laser is along the y-axis.

As shown in row 5 of Table 7.1, with surface Al NPs for inducing SP coupling, the IQEs in sample A-NP are increased to a level of 20%–22%, which corresponds to a relative increment of ~33%. Because the distance between the surface Al NPs and the top QW is as large as 123 nm in sample B-NP, the SP coupling effect is weak such that its IQEs do not significantly change from those of sample B.

7.4.5 Discussions and Summary

The similar IQE enhancements between the TE- and TM-polarized emissions in all the samples can be attributed to the small difference in band energy level between the heavy-/light-hole and split-off valence bands. In this situation, the almost-degenerate valence band structure leads to similar IQEs. Also, simultaneous SP coupling of the radiating dipoles of different orientations may occur. In other words, the electron transitions from the conduction band to the heavy-/light-hole bands and split-off band are simultaneously coupled with hybrid SP resonance modes such that those radiating dipoles have similar emission enhancements and similar IQEs. It is noted that the LSP resonance at the excitation laser wavelength (266 nm) can be quite strong, as shown in Figure 7.14. When the distance between surface Al NPs and the QWs is not large, such as the case of sample A-NP, the LSP resonance at the excitation laser wavelength results in an enhanced excitation intensity at the QWs for producing higher measured IQEs [60]. An oxidized layer can be formed on an Al NP after it is fabricated on the sample surface. It has been reported that the oxidized layer is 2–3 nm in thickness [61]. Such a thin oxidized layer may slightly redshift the LSP resonance wavelength. However, from the results in Figure 7.14, one can see that the LSP resonance peak is still located in the deep-UV range and quite close to the QW emission wavelength. Therefore, the oxidization of an Al NP does not significantly affect the SP coupling effect.

In this section, we have demonstrated the IQE enhancement of deep-UV $Al_xGa_{1-x}N/Al_yGa_{1-y}N$ (x < y) QWs by fabricating surface Al NPs on a QW structure through thermally annealing an Al thin film for inducing SP coupling. By comparing the LSP resonance behaviors of Al NPs formed with different deposited Al-film thicknesses, annealing temperatures, and durations, optimized fabrication conditions were obtained for producing LSP resonance close to the QW emission wavelength in the deep-UV range. Through temperature-dependent PL measurement, the enhancements of IQE in different emission polarizations were illustrated. Due to the small difference in energy band level between the heavy-/light-hole and split-off valence bands, the IQEs of the TE- and TM-polarized emissions were about the same. With SP coupling, the similar IQEs between different polarizations could be attributed to the simultaneous SP couplings of the TE- and TM-polarized transitions. The strong LSP resonance at the excitation laser wavelength might lead to stronger excitation and hence higher IQE levels.

7.5 Conclusions

In this chapter, we reviewed the phenomenon of SP coupling with a light emitter and its application to LEDs in the visible range and discussed the details of SP coupling in the deep-UV range with Al nanostructures. First, we explained the fundamental principles of SP coupling with a light emitter for enhancing its emission efficiency. For the application to an LED, several useful metal nanostructures for inducing SP coupling with its QWs were introduced. Their advantages and disadvantages were also discussed. Then, the effects of SP coupling in an LED for improving its operation functions, including enhancing IQE, reducing efficiency droop, and increasing modulation bandwidth, were elucidated. Next, we focused the discussion on the SP coupling in the deep-UV range with Al nanostructures. We demonstrated the IQE enhancement of deep-UV $Al_xGa_{1-x}N/Al_yGa_{1-y}N$ (x < y) QWs by fabricating surface Al NPs on a QW structure through thermally annealing an Al thin film for inducing SP coupling. Through temperature-dependent PL measurement, the enhancements of IQE in different emission polarizations were illustrated.

References

1. S. A. Maier, *Plasmonics: Fundamentals and Applications*, Springer, New York, 2007.
2. A. Neogi, C. W. Lee, H. O. Everitt, T. Kuroda, A. Tackeuchi, and E. Yablonvitch, Enhancement of spontaneous recombination rate in a quantum well by resonant surface plasmon coupling, *Phys. Rev. B* **66**, 153305 (2002).
3. K. Okamoto, I. Niki, A. Shvartser, Y. Narukawa, T. Mukai, and A. Scherer, Surface-plasmon-enhanced light emitters based on InGaN quantum wells, *Nat. Mater.* **3**, 601–605 (2004).
4. M. L. Anderson, S. Stobbe, A. S. Sorensen, and P. Lodahl, Strongly modified plasmon–matter interaction with mesoscopic quantum emitters, *Nat. Phys.* **7**, 215 (2011).

5. C. Y. Cho, Y. Zhang, E. Cicek, B. Rahnema, Y. Bai, R. McClintock, and M. Razeghi, Surface plasmon enhanced light emission from AlGaN-based ultraviolet light-emitting diodes grown on Si (111), *Appl. Phys. Lett.* **102**, 211110 (2013).
6. C. Y. Chen, D. M. Yeh, Y. C. Lu, and C. C. Yang, Dependence of resonant coupling between surface plasmons and an InGaN quantum well on metallic structure, *Appl. Phys. Lett.* **89**(20), 203113 (2006).
7. C. Y. Chen, Y. C. Lu, D. M. Yeh, and C. C. Yang, Influence of the quantum-confined Stark effect in an InGaN/GaN quantum well on its coupling with surface plasmon for light emission enhancement, *Appl. Phys. Lett.* **90**(18), 183114 (2007).
8. Y. C. Lu, C. Y. Chen, D. M. Yeh, C. F. Huang, T. Y. Tang, J. J. Huang, and C. C. Yang, Temperature dependence of the surface plasmon coupling with an InGaN/GaN quantum well, *Appl. Phys. Lett.* **90**(19), 193103 (2007).
9. W. H. Chuang, J. Y. Wang, C. C. Yang, and Y. W. Kiang, Differentiating the contributions between localized surface plasmon and surface plasmon polariton on a one-dimensional metal grating in coupling with a light emitter, *Appl. Phys. Lett.* **92**(13), 133115 (2008).
10. W. H. Chuang, J. Y. Wang, C. C. Yang, and Y. W. Kiang, Numerical study on quantum efficiency enhancement of a light-emitting diode based on surface plasmon coupling with a quantum well, *IEEE Photon. Technol. Lett.* **20**(16), 1339–1341 (2008).
11. W. H. Chuang, J. Y. Wang, C. C. Yang, and Y. W. Kiang, Transient behaviors of surface plasmon coupling with a light emitter, *Appl. Phys. Lett.* **93**(15), 153104 (2008).
12. W. H. Chuang, J. Y. Wang, C. C. Yang, and Y. W. Kiang, Study on the decay mechanisms of surface plasmon coupling features with a light emitter through time-resolved simulations, *Opt. Express* **17**(1), 104–116 (2009).
13. C. W. Shen, J. Y. Wang, W. H. Chuang, H. L. Chen, Y. C. Lu, Y. W. Kiang, C. C. Yang, and Y. J. Yang, Effective energy coupling and preservation in a surface plasmon-light emitter coupling system on a metal nanostructure, *Nanotechnology* **20**(13), 135202 (2009).
14. G. Sun, J. B. Khurgin, and C. C. Yang, Impact of high-order surface plasmon modes of metal nanoparticles on enhancement of optical emission, *Appl. Phys. Lett.* **95**(17), 171103 (2009).
15. Y. Kuo, C. H. Lin, H. S. Chen, C. Hsieh, C. G. Tu, P. Y. Shih, C. H. Chen et al., Surface plasmon coupled light-emitting diode—Experimental and numerical studies, *Jpn. J. Appl. Phys.* **54**(2S), 02BD01 (2015).
16. C. Y. Su, C. H. Lin, P. Y. Shih, C. Hsieh, Y. F. Yao, C. G. Tu, H. T. Chen, H. S. Chen, Y. W. Kiang, and C. C. Yang, Coupling behaviors of surface plasmon polariton and localized surface plasmon with an InGaN/GaN quantum well, *Plasmonics* **11**(3), 931–939 (2016).
17. E. M. Purcell, Spontaneous emission probabilities at radio frequencies, *Phys. Rev.* **69**, 681 (1946).
18. Y. Kuo, S. Y. Ting, C. H. Liao, J. J. Huang, C. Y. Chen, C. Hsieh, Y. C. Lu et al., Surface plasmon coupling with radiating dipole for enhancing the emission efficiency of a light-emitting diode, *Opt. Express* **19**(14), 914–929 (2011).
19. C. F. Lu, C. H. Liao, C. Y. Chen, C. Hsieh, Y. W. Kiang, and C. C. Yang, Reduction in the efficiency droop effect of a light-emitting diode through surface plasmon coupling, *Appl. Phys. Lett.* **96**(26), 261104 (2010).
20. D. M. Yeh, C. F. Huang, C. Y. Chen, Y. C. Lu, and C. C. Yang, Surface plasmon coupling effect in an InGaN/GaN single-quantum-well light-emitting diode, *Appl. Phys. Lett.* **91**(17), 171103 (2007).
21. Y. C. Lu, C. Y. Cheng, K. C. Shen, and C. C. Yang, Enhanced photoluminescence excitation in surface plasmon coupling with an InGaN/GaN quantum well, *Appl. Phys. Lett.* **91**(18), 183107 (2007).
22. C. H. Lin, C. Hsieh, C. G. Tu, Y. Kuo, H. S. Chen, P. Y. Shih, C. H. Liao et al., Efficiency improvement of a vertical light-emitting diode through surface plasmon coupling and grating scattering, *Opt. Express* **22**(S3), 842–856 (2014).
23. K. C. Shen, C. Y. Chen, C. F. Huang, J. Y. Wang, Y. C. Lu, Y. W. Kiang, C. C. Yang, and Y. J. Yang, Polarization dependent coupling of surface plasmon on a one-dimensional Ag grating with an InGaN/GaN dual-quantum-well structure, *Appl. Phys. Lett.* **92**(1), 013108 (2008).

24. K. C. Shen, C. Y. Chen, H. L. Chen, C. F. Huang, Y. W. Kiang, C. C. Yang, and Y. J. Yang, Enhanced and partially polarized output of a light-emitting diode with its InGaN/GaN quantum well coupled with surface plasmons on a metal grating, *Appl. Phys. Lett.* **93**(23), 231111 (2008).

25. K. C. Shen, C. H. Liao, Z. Y. Yu, J. Y. Wang, C. H. Lin, Y. W. Kiang, and C. C. Yang, Effects of the intermediate SiO$_2$ layer on polarized output of a light-emitting diode with surface plasmon coupling, *J. Appl. Phys.* **108**(11), 113101 (2010).

26. H. L. Chen, J. Y. Wang, W. H. Chuang, Y. W. Kiang, and C. C. Yang, Characteristics of light emitter coupling with surface plasmons in air/metal/dielectric grating structures, *J. Opt. Soc. Am. B Opt. Phys.* **26**(5), 923–929 (2009).

27. D. M. Yeh, C. Y. Chen, Y. C. Lu, C. F. Huang, and C. C. Yang, Formation of various metal nanostructures with thermal annealing to control the effective coupling energy between a surface plasmon and an InGaN/GaN quantum well, *Nanotechnology* **18**(26), 265402 (2007).

28. D. M. Yeh, C. F. Huang, C. Y. Chen, Y. C. Lu, and C. C. Yang, Localized surface plasmon-induced emission enhancement of a green light-emitting diode, *Nanotechnology* **19**(34), 345201 (2008).

29. C. Y. Chen, J. Y. Wang, F. J. Tsai, Y. C. Lu, Y. W. Kiang, and C. C. Yang, Fabrication of sphere-like Au nanoparticles on substrate with laser irradiation and their polarized localized surface plasmon behaviors, *Opt. Express* **17**(16), 14186–14198 (2009).

30. C. H. Lin, C. H. Chen, Y. F. Yao, C. Y. Su, P. Y. Shih, H. S. Chen, C. Hsieh, Y. Kuo, Y. W. Kiang, and C. C. Yang, Behaviors of surface plasmon coupled light-emitting diodes induced by surface Ag nanoparticles on dielectric interlayers, *Plasmonics* **10**(5), 1029–1040 (2015).

31. C. W. Huang, H. Y. Tseng, C. Y. Chen, C. H. Liao, C. Hsieh, K. Y. Chen, H. Y. Lin et al., Fabrication of surface metal nanoparticles and their induced surface plasmon coupling with subsurface InGaN/GaN wells, *Nanotechnology* **22**(47), 475201 (2011).

32. D. Dobrovolskas, J. Mickevičius, S. Nargelas, H. S. Chen, C. G. Tu, C. H. Liao, C. Hsieh, C. Y. Su, G. Tamulaitis, and C. C. Yang, InGaN/GaN MQW photoluminescence enhancement by localized surface plasmon resonance on isolated Ag nanoparticles, *Plasmonics* **9**(5), 1183–1187 (2014).

33. H. S. Chen, C. P. Chen, Y. Kuo, W. H. Chou, C. H. Shen, Y. L. Jung, Y. W. Kiang, and C. C. Yang, Surface plasmon coupled light-emitting diode with metal protrusions into p-GaN, *Appl. Phys. Lett.* **102**(4), 041108 (2013).

34. C. Hsieh, Y. F. Yao, C. F. Chen, P. Y. Shih, C. H. Lin, C. Y. Su, H. S. Chen et al., Localized surface plasmon coupled light-emitting diodes with buried and surface Ag nanoparticles, *IEEE Photon. Technol. Lett.* **26**(17), 1699–1702 (2014).

35. Y. Kuo, Y. F. Yao, M. H. Chiu, W. Y. Chang, C. C. Yang, and Y. W. Kiang, Coupling behaviors of a radiating dipole with the surface plasmon induced on a metal protrusion, *Plasmonics* **10**(2), 241–249 (2015).

36. C. Y. Cho, J. J. Kim, S. J. Lee, S. H. Hong, K. J. Lee, S. Y. Yim, and S. J. Park, Enhanced emission efficiency of GaN-based flip-chip light-emitting diodes by surface plasmons in silver disks, *Appl. Phys. Express* **5**, 122103 (2012).

37. M. K. Kwon, J. Y. Kim, B. H. Kim, I. K. Park, C. Y. Cho, C. C. Byeon, and S. J. Park, Surface-plasmon-enhanced light-emitting diodes, *Adv. Mater.* **20**(7), 1253–1257 (2008).

38. C. Y. Cho, S. J. Lee, J. H. Song, S. H. Hong, S. M. Lee, Y. H. Cho, and S. J. Park, Enhanced optical output power of green light-emitting diodes by surface plasmon of gold nanoparticles, *Appl. Phys. Lett.* **98**(5), 051106 (2011).

39. C. Y. Cho, K. S. Kim, S. J. Lee, M. K. Kwon, H. Ko, S. T. Kim, G. Y. Jung, and S. J. Park, Surface plasmon-enhanced light-emitting diodes with silver nanoparticles and SiO$_2$ nano-disks embedded in p-GaN, *Appl. Phys. Lett.* **99**(4), 041107 (2011).

40. Y. Kuo, W. Y. Chang, H. S. Chen, Y. W. Kiang, and C. C. Yang, Surface plasmon coupling with a radiating dipole near an Ag nanoparticle embedded in GaN, *Appl. Phys. Lett.* **102**(16), 161103 (2013).

41. Y. Kuo, W. Y. Chang, H. S. Chen, Y. R. Wu, C. C. Yang, and Y. W. Kiang, Surface-plasmon-coupled emission enhancement of a quantum well with a metal nanoparticle embedded in a light-emitting diode, *J. Opt. Soc. Am. B Opt. Phys.* **30**(10), 2599–2606 (2013).

42. C. H. Lin, C. Y. Su, Y. Kuo, C. H. Chen, Y. F. Yao, P. Y. Shih, H. S. Chen, C. Hsieh, Y. W. Kiang, and C. C. Yang, Further reduction of efficiency droop effect by adding a lower-index dielectric interlayer in a surface plasmon coupled blue light-emitting diode with surface metal nanoparticles, *Appl. Phys. Lett.* **105**(10), 101106 (2014).

43. Y. Kuo, W. Y. Chang, C. H. Lin, C. C. Yang, and Y. W. Kiang, Evaluating the blue-shift behaviors of the surface plasmon coupling of an embedded light emitter with a surface Ag nanoparticle by adding a dielectric interlayer or coating, *Opt. Express* **23**(24), 30709–30720 (2015).

44. C. H. Lin, Y. F. Yao, C. Y. Su, C. Hsieh, C. G. Tu, S. Yang, S. S. Wu, H. T. Chen, Y. W. Kiang, and C. C. Yang, Thermal annealing effects on the performance of a Ga-doped ZnO transparent-conductor layer in a light-emitting diode, *IEEE Trans. Electron Devices* **62**(11), 3742–3749 (2015).

45. Y. F. Yao, C. H. Lin, C. Hsieh, C. Y. Su, E. Zhu, C. M. Weng, M. Y. Su et al., Multi-mechanism efficiency enhancement in growing Ga-doped ZnO as the transparent conductor on a light-emitting diode, *Opt. Express*, **23**(25), 32274–32288 (2015).

46. Y. C. Lu, Y. S. Chen, F. J. Tsai, J. Y. Wang, C. H. Lin, C. Y. Chen, Y. W. Kiang, and C. C. Yang, Improving emission enhancement in surface plasmon coupling with an InGaN/GaN quantum well by inserting a dielectric layer of low refractive index between metal and semiconductor, *Appl. Phys. Lett.* **94**(23), 233113 (2009).

47. Y. Kuo, H. T. Chen, W. Y. Chang, H. S. Chen, C. C. Yang, and Y. W. Kiang, Enhancements of the emission and light extraction of a radiating dipole coupled with localized surface plasmon induced on a surface metal nanoparticle in a light-emitting device, *Opt. Express* **22**(S1), 155–166 (2014).

48. C. H. Lin, C. Y. Su, E. Zhu, Y. F. Yao, C. Hsieh, C. G. Tu, H. T. Chen, Y. W. Kiang, and C. C. Yang, Modulation behaviors of surface plasmon coupled light-emitting diode, *Opt. Express* **23**(6), 8150–8161 (2015).

49. M. F. Schubert, J. Xu, J. K. Kim, E. F. Schubert, M. H. Kim, S. Yoon, S. M. Lee, C. Sone, T. Sakong, and Y. Park, Polarization-matched GaInN/AlGaInN multi-quantum-well light-emitting diodes with reduced efficiency droop, *Appl. Phys. Lett.* **93**(4), 041102 (2008).

50. G. B. Lin, D. Meyaard, J. Cho, E. F. Schubert, H. Shim, and C. Sone, Analytic model for the efficiency droop in semiconductors with asymmetric carrier-transport properties based on drift-induced reduction of injection efficiency, *Appl. Phys. Lett.* **100**(16), 161106 (2012).

51. K. T. Delaney, P. Rinke, and C. G. Van de Walle, Auger recombination rates in nitrides from first principles, *Appl. Phys. Lett.* **94**(19), 191109 (2009).

52. E. Kioupakis, Q. Yan, and C. G. Van de Walle, Interplay of polarization fields and Auger recombination in the efficiency droop of nitride light-emitting diodes, *Appl. Phys. Lett.* **101**(23), 231107 (2012).

53. Y. Kuo, C. Y. Su, C. Hsieh, W. Y. Chang, C. A. Huang, Y. W. Kiang, and C. C. Yang, Surface plasmon coupling for suppressing p-GaN absorption and TM-polarized emission in a deep-UV light-emitting diode, *Opt. Lett.* **40**(18), 4229–4232 (2015).

54. A. Taguchi, Y. Saito, K. Watanabe, S. Yijian, and S. Kawata, Tailoring plasmon resonances in the deep-ultraviolet by size-tunable fabrication of aluminum nanostructures, *Appl. Phys. Lett.* **101**, 081110 (2012).

55. N. Gao, K. Huang, J. Li, S. Li, X. Yang, and J. Kang, Surface-plasmon-enhanced deep-UV light emitting diodes based on AlGaN multi-quantum wells, *Sci. Rep.* **2**(816), 00816 (2012).

56. K. Huang, N. Gao, C. Wang, X. Chen, J. Li, S. Li, X. Yang, and J. Kang, Top- and bottom-emission-enhanced electroluminescence of deep-UV light-emitting diodes induced by localised surface plasmons, *Sci. Rep.* **4**(4380), 04380 (2014).

57. K. B. Nam, J. Li, M. L. Nakarmi, J. Y. Lin, and H. X. Jiang, Unique optical properties of AlGaNAlGaN alloys and related ultraviolet emitters, *Appl. Phys. Lett.* **84**(25), 5264–5266 (2004).

58. J. E. Northrup, C. L. Chua, Z. Yang, T. Wunderer, M. Kneissl, N. M. Johnson, and T. Kolbe, Effect of strain and barrier composition on the polarization of light emission from AlGaN/AlN quantum wells, *Appl. Phys. Lett.* **100**, 021101 (2012).

59. H. Lu, T. Yu, G. Yuan, X. Chen, Z. Chen, G. Chen, and G. Zhang, Enhancement of surface emission in deep ultraviolet AlGaN-based light emitting diodes with staggered quantum wells, *Opt. Lett.* **37**(17), 3693–3695 (2010).

60. K. Okamoto, I. Niki, A. Scherer, Y. Narukawa, T. Mukai, and Y. Kawakami, Surface plasmon enhanced spontaneous emission rate of InGaN/GaN quantum wells probed by time-resolved photoluminescence spectroscopy, *Appl. Phys. Lett.* **87**, 071102 (2005).

61. G. H. Chan, J. Zhao, G. C. Schatz, and R. P. Van Duyne, Localized surface plasmon resonance spectroscopy of triangular aluminum nanoparticles, *J. Phys. Chem. C* **112**(36), 13958–13963 (2008).

Deep Level Traps in GaN Epilayer and LED

Xuan Sang Nguyen
*Singapore-MIT Alliance
for Research and
Technology (SMART)*

Soo Jin Chua
*Singapore-MIT Alliance
for Research and
Technology (SMART)
Department of Electrical
and Computer Engineering*

Abstract   This chapter reviews the origin and characterization of electrically active deep level traps in GaN material and LED. In n-GaN, three levels $E_C - 0.25$ eV, $E_C - 0.60$ eV, and $E_C - 0.90$ eV are most commonly revealed by DLTS and $E_C - 1.25$ eV, $E_C - 2.6$ eV, and $E_C - 3.28$ eV are most commonly seen by DLOS. The levels $E_C - 0.60$ eV and $E_C - 1.25$ eV are dislocation-related levels. In GaN LED, DLOS is a more useful technique than DLTS because DLOS investigates near-mid gap levels in GaN. In fact, a combination of DLOS and lighted capacitance–voltage measurement realizes the defect density in each quantum well (QW) and quantum barrier (QB) of the GaN LED. Deep level traps within the first QW are highest in density.

8.1 Introduction

8.1.1 Origin of Traps

8.1.1.1 Point Defects and Line Defects

Study of defects or impurities in semiconductors is useful for devices because the electronic properties of a semiconductor can be significantly altered by incorporating impurities and other defects into the material. The influence of defects could be useful for device performance, such as doping and can also be deleterious, such as vacancies. Defects in semiconductors could be classified in spatial extent and in terms of their energy levels. In terms of the spatial extent of the potential and wave function created by the defects, they could be classified as point or line defects.

Point defects are created by isolated atoms, and where they interact with one another they are called complexes. Point defects include vacancies, interstitial atoms, substitutional impurities, antisite defects, and their complexes. Vacancy is a missing atom at a lattice site. Interstitial is formed by an atom in between lattice sites. It is possible to have a self-interstitial, that is, interstitials formed by atoms of the lattice. A substitutional impurity is an impurity atom C replacing a host atom A and is identified by the nomenclature C_A. Antisite defects occur only in compound semiconductors. Considering the compound semiconductor formed by two elements A and B, an antisite defect is formed when an atom B occupies a site that should have been occupied by atom A. Complexes are combinations of some point defects such as a vacancy–interstitial pair denoted by $V_A - I_A$.

In GaN semiconductor, the most common point defects are V_{Ga}, V_N, and their complexes. V_{Ga} is easily formed due to its low energy of formation [1]. V_{GA} encourages bonding with hydrogen and has been postulated to be the cause of the yellow luminescence band [1]. The substitutional or interstitial point defects in GaN are made up of residual impurities such as H, O, and C, which are incorporated from the growth environment [2]. H has the ability to passivate acceptors and is more effective in p-type GaN. O often acts as a shallow donor in GaN and influences its conductivity. C is identified to be the major residual impurity and is also postulated to be the cause of the yellow luminescence band.

Line defects are 1-D linear defects that exist when a whole row of atoms of the crystal lattice are not bonded to the other atoms in the lattice and are also known as dislocations. There are two main categories of dislocation, namely, edge dislocation and screw dislocation, as shown in Figure 8.1. An edge dislocation is formed when an incomplete plane of atoms is inserted in the lattice. A screw dislocation is formed when the extra plane of atoms forms a spiral similar to a staircase. Dislocations are often represented in terms of Burger's vectors. When the dislocations extend throughout the whole crystalline structure, they are known as threading dislocations. Line defects may also exist as a mixture of edge and screw dislocations, which are then identified by the addition of the respective Burger's vectors.

For GaN, using density functional calculations, Elsner et al. [3] claimed that the screw and edge threading dislocations are theoretically electrically inactive. He also proved that the screw dislocations exist as open-core dislocations, whereas the threading edge dislocations have filled cores. He left an open statement saying that the stress fields are, however, sufficiently large to trap impurities within the dislocation. Experiments have shown that threading dislocations and deep level defect density are closely related. The details of each deep level defect corresponding to threading dislocation will be discussed in Section 8.3.3.

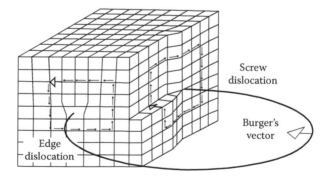

FIGURE 8.1 Illustration of edge and screw dislocation. (Taken from Passchier and Trouw, pp. 33, 2005. http://www.geology.um.maine.edu/geodynamics/AnalogWebsite/UndergradProjects2010/PatrickRyan/Pages/Background.html).

8.1.1.2 Shallow and Deep Defects

In semiconductors, shallow impurities or shallow levels typically refer to those with binding energy of <100 meV away from one of the conduction or valence band edges. The effective mass theory (EMT) has been very successful at predicting the defect energy levels. It produces results that are in excellent agreement with experiments in many materials [4].

Deep levels, in an earlier understanding, are defects that produce energy levels E where EMT is not able to quantitatively predict the defect energy levels. It was assumed that their defect levels are located in the middle of the bandgap. In recent times, the understanding is that energy levels E produced by deep centers may have a range of energies in the bandgap, which can be close to either the conduction band edge or the valence band edge. It turns out that for such defects, lattice relaxation (or distortion) effects are often important but still not the dominant effect to explain the observed defect energy levels E.

For experimental identification, deep levels are considered as defects with the binding energy greater than 150 meV from the conduction band or valence band edges. Concentrations of deep level defects can vary from 10^{12} to 10^{18} cm^{-3}. Deep levels usually have a negligible effect on the electrical conductivity when their concentrations are low. But where their concentrations are large they can either help or hinder device performance, depending on the energy level and on the type of device.

8.1.2 Effects of Deep Level Traps on GaN Device Performances

Deep level defects can act as deep traps, which have strong effects on the optical and electrical properties of the materials. Some effects that deep level defects can have on devices are the reduction of e$^-$ and h$^+$ lifetimes. This is bad for solar cells but is good for a fast switch. Deep level traps can enhance radiative recombination, which is purposely introduced in an LED to produce a specific color. However, they can degrade device performance. Therefore, deep levels need to be controlled for many device applications. A first step to controlling them is to understand the physics of deep levels.

8.1.2.1 Deep Level Traps in GaN Light-Emitting Diode

In GaN LED, point defects serve as efficient nonradioactive centers, as shown in Figure 8.2, and have been linked to the decrease in the internal quantum efficiency (IQE) [5]. The presence of dislocations and nitrogen vacancies in the active layer of LEDs is reported to cause the increase in nonradiative recombinations leading to the decrease in IQE of the GaN LED [6]. Ga vacancies are also believed to form complex defects with hydrogen, V_{Ga}-H, which is the cause of the broadband yellow emission [7], which decreases LED's efficiency [8]. In addition, trap-assisted tunneling is known to result in a higher forward subthreshold current, also contributing to reduction in LED efficiency [9].

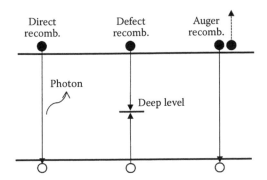

FIGURE 8.2 Point defects as nonradiative centers.

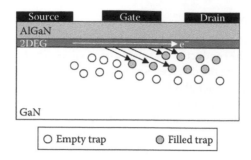

FIGURE 8.3   Current collapse during high V_{ds} operation of the device.

Experiments have been carried out to confirm that with lower dislocation and defect density, that is, lower nonradiative recombination and the efficiency of the GaN LED are increased [10].

8.1.2.2 Current Collapse due to Deep Level Traps in GaN High Electron Mobility Transistor (HEMT)

The drop in current during high-voltage and high-frequency pulsing in an HEMT device is due to the presence of traps, which limits the performance of GaN HEMT for high-power and high-frequency applications. This current drop is widely known as "current collapse."

When a large bias voltage is applied between the source and drain of GaN/AlGaN HEMT structure, as depicted in Figure 8.3, the electrons in the conducting channel are rapidly accelerated. These "hot carriers" gain enough kinetic energy from the large electric field to be injected into an adjacent region of the device structure. If this region contains a significant concentration of traps, the injected carriers can become trapped. The electric field created by these charges has the effect of increasing the resistance of the channel. The resulting reduction in drain current, referred to as "current collapse," can severely compromise the performance of a microwave HEMT [11].

In addition, bulk and surface traps can cause current drop during high-frequency pulsing. This drop in current contributes to the DC-to-RF dispersion [12].

8.1.3 Techniques of Identification of Deep Level Traps

The common methods to identify electrically active deep level traps are C–V, Hall effect measurement, deep level transient spectroscopy (DLTS), and deep level optical spectroscopy (DLOS). C–V and Hall measurement are widely used methods to characterize shallow level traps. DLTS and DLOS are used to characterize deep level traps and hyper deep level traps, respectively.

DLTS is most widely used for defect characterization because it is a sensitive and accurate method to characterize deep level defects. In addition, DLTS provides information such as density, energy level, and capture cross section of the traps, which cannot be simultaneously obtained by the other methods. Owing to its wide application, the DLTS technique is well understood, with many prior experiments producing excellent results that agree with theory.

The DLTS method uses thermal energy to excite electron emission from the trap levels to the conduction band. Due to thermal budget of the material or device, the DLTS equipment is normally limited to exciting traps up to 1.2–1.3 eV away from the conduction or valence bands. This limitation is particularly crucial for GaN, which has a wide bandgap of 3.4 eV and has a good possibility of containing traps deeper than 1.3 eV. The DLOS technique uses photon energy to further provide extra energy for carriers captured in the deeper level trap. Thus, DLOS is able to characterize hyper deep level traps located far from conduction band to valence band edges of GaN.

8.2 Parameters for Traps Characterization

8.2.1 Identifying Parameters of Deep Level Traps by Deep Level Transient Spectroscopy

Deep level transient spectroscopy (DLTS) is an experimental tool developed by D.V. Lang in 1974 [13] to study electrically active defects, also known as charge carrier traps, in semiconductors to obtain information such as energy level, capture cross section, and density of the defects. The following explanation of the physics of DLTS assumes the sample to be an n-type Schottky diode. DLTS measurements are based on the capture and emission processes of carriers from deep level traps as a function temperature. Figure 8.4 shows the electron and hole capture and emission processes in a semiconductor. For n-type semiconductors, the capture and emission of electrons from the conduction band are dominant and considered in this section. The equations given in the following follow that of Blood et al. [14].

The electron capture rate depends on the number of empty state and is given by the following equation:

$$\frac{\partial n_T}{\partial t}\bigg|_{capture} = c_n p_T \tag{8.1}$$

where

n_T is the number of electrons captured in defect states
t is the time in second
T is the temperature in Kelvin
e_n and c_n are electron emission and capture coefficients, respectively, $cm^{-3}\,s^{-1}$
p_T is the number of empty states

The electron emission rate is given by

$$\frac{\partial n_T}{\partial t}\bigg|_{emission} = e_n n_T \tag{8.2}$$

where e_n is the electron emission coefficient.

Thus, the change in the number of electrons occupied in the traps is given by

$$\frac{\partial n_T}{\partial t} = c_n p_T - e_n n_T = c_n \left[N_T - n_T\right] - e_n n_T \tag{8.3}$$

where $N_T = p_T + n_T$, is the total number of defect states.

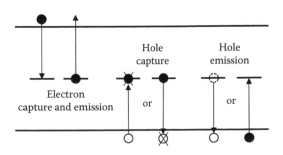

FIGURE 8.4 Carrier capture and emission processes in deep level traps.

Under equilibrium condition $\partial n_T/\partial t = 0$. The relation between emission and capture coefficients is given by

$$e_n = c_n \left(\frac{N_T - n_T}{n_T} \right) = c_n \left(\frac{N_T}{n_T} - 1 \right) \tag{8.4}$$

Moreover, in equilibrium, the relationship between n_T and N_T through the Fermi–Dirac statistics is given by

$$\frac{n_T}{N_T} = \frac{1}{1 + \exp\left(\dfrac{E_T - E_F}{kT} \right)} \tag{8.5}$$

where
  E_T is the trap energy level
  E_F is the Fermi energy level

The capture rate of the electron is given by [14] $c_n = \sigma_n \langle \nu_{th} \rangle n$ where n is the concentration of electron in conduction band; σ_n is the capture cross section, cm^2; and $\langle \nu_{th} \rangle$ is the electron thermal velocity.

The number of electrons in the conduction band is given by [14] $n = N_C \exp\left(\dfrac{E_F - E_C}{kT} \right)$ where N_C is the density of states in the conduction band.

Replace all the term in Equation 8.4:

$$e_n = c_n \langle \nu_{th} \rangle N_C \cdot \exp\left(\frac{E_T - E_F}{kT} \right) \cdot \exp\left(\frac{E_F - E_C}{kT} \right)$$

$$e_n = \sigma_n \langle \nu_{th} \rangle N_C \cdot \exp\left(\frac{E_T - E_C}{kT} \right) \tag{8.6}$$

From Reference 14, $\langle \nu_{th} \rangle = \left(\dfrac{3 k_B T}{m_n} \right)^{1/2}$ and $N_C = 2 \left(\dfrac{2\pi m_n k_B T}{h^2} \right)^{3/2}$.

For simplification, representing the term $\gamma_n = \left(\dfrac{3 k_B}{m_n} \right)^{1/2} \cdot 2 \left(\dfrac{2\pi m_n k_B}{h^2} \right)^{3/2}$

$$\rightarrow \langle \nu_{th} \rangle N_C = \gamma_n T^2$$

where k_B is Boltzmann's constant; m_n is the electron effective mass; and h is Planck's constant.

$$e_n = \sigma_n \gamma_n T^2 \cdot \exp\left(\frac{E_T - E_C}{kT} \right) \tag{8.7}$$

8.2.1.1 Identifying Deep Level Density

To perform DLTS, the Schottky device was pulsed from reverse bias to no bias condition and then returned to the previous reverse bias condition for the measurement of capacitance. Figure 8.5a represents the pulsing condition for DLTS measurement and Figure 8.5b shows the corresponding

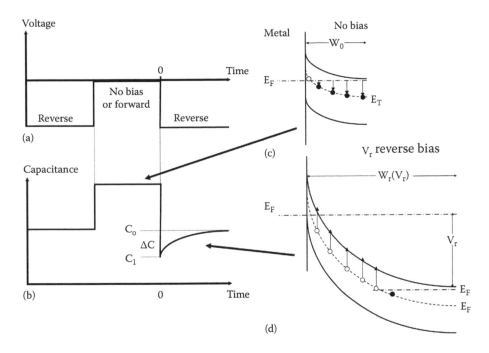

FIGURE 8.5 Energy band diagram of the Schottky diode with pulsing voltage (a). (c and d) show the band diagrams at zero bias and at reverse bias. The corresponding capacitance measured is indicated in (b).

capacitance measured. Energy band diagrams of the Schottky diode at 0 V bias and reverse bias are shown in Figure 8.5c and d, respectively. In reverse bias condition, the initial capacitance, C_o, shown in Figure 8.5c is given in the following equation: [15]

$$C_o = A \sqrt{\frac{\varepsilon\varepsilon_o e N_D}{2(V_r + V_d)}} \tag{8.8}$$

where
   N_D is the donor density of the n-type semiconductor. It is also the net charge density in the depletion region
   V_r represents the reverse bias
   V_d is the built-in diffusion voltage across the depletion region
   $\varepsilon\varepsilon_o$ is the permittivity of the semiconductor

When there is no applied bias, the depletion region is shortened to allow the capture of electrons into the deep level states as shown in Figure 8.5c. Next, the device is returned to the reverse bias condition, but the capacitance of the device does not immediately return to the C_o value due to charges in filled defect states. The net charge in the depletion region now is N_D–N_T. The capacitance can be expressed using Equation 8.9 given as follows:

$$C_1 = A \sqrt{\frac{\varepsilon\varepsilon_o e (N_D - N_T)}{2(V_r + V_d)}} \tag{8.9}$$

Next, the deep level states that have energies higher than E_F begin to emit emission electrons. During this process, the capacitance increases slowly to reach C_o value, shown in Figure 8.5d. This transient capacitance is due to the thermal emission process of the deep level traps. Concentration of deep levels is calculated as follows.

The difference in the measured capacitance is $\Delta C = C_o - C_1$ with the assumption that $N_T \ll N_D$,

$$\Delta C = A\sqrt{\frac{\varepsilon\varepsilon_o e(N_D)}{2(V_r + V_d)}} - A\sqrt{\frac{\varepsilon\varepsilon_o e(N_D - N_T)}{2(V_r + V_d)}} \cong C_o \frac{N_T}{2N_D} \qquad (8.10)$$

By manipulating Equation 8.10, the trap concentration can be expressed in the following equation:

$$N_T = \frac{2\Delta C}{C_o} N_D \qquad (8.11)$$

From this equation, the sensitivity of DLTS measurement is dependent on doping level N_D in the semiconductor and the sensitivity of capacitance measurement, ΔC.

8.2.1.2 Identifying Activation Energy and Capture Cross Section of Deep Level Traps

The method to measure activation energy and capture cross section of each deep level trap is through the measurement of the capacitance transient as shown in Figure 8.5b. With temperature variation under the same pulsing condition, the shape of capacitance transient is recorded. Figure 8.6a and b shows the typical capacitance transients at low, high, and appropriate temperatures. At high temperatures, C_1 is reverted back to C_o quickly, while at low temperatures thermal energy might be insufficient for emission and C_1 may never return to C_o.

The capacitance at any instantaneous time depends on the net charge in the depletion region, $N_D - n_T(t)$ as expressed in the following:

$$C(t) = A\sqrt{\frac{\varepsilon\varepsilon_o e(N_D - n_T(t))}{2(V_r + V_d)}} = A\sqrt{\frac{\varepsilon\varepsilon_o e N_D}{2(V_r + V_d)} - \frac{\varepsilon\varepsilon_o e n_T(t)}{2(V_r + V_d)}} = A\sqrt{\frac{\varepsilon\varepsilon_o e N_D}{2(V_r + V_d)} - \frac{\varepsilon\varepsilon_o e N_D}{2(V_r + V_d)}\frac{n_T(t)}{N_D}},$$

$$C(t) = C_o\sqrt{1 - \frac{n_T(t)}{N_D}} \approx C_o\left(1 - \frac{n_T(t)}{2N_D}\right) \qquad (8.12)$$

According to Reference 14: $n_T(t) = N_T \exp(-e_n t)$ where $n_T(t)$ is the number of traps filled with electrons; e_n is the emission coefficient of electron from the deep states.

$$C(t) = C_o\left(1 - \frac{n_T(t)}{2N_D}\right) = C_o\left[1 - \frac{N_T \exp(-e_n t)}{2N_D}\right] \qquad (8.13)$$

The times at which the capacitance is measured, t_1 and t_2 can be specified, is known as the rate window. Figure 8.6b shows the capacitance change recorded for two sets of t_1 and t_2 as a function of temperature. The DLTS signal is generated by finding the difference between the capacitance values at these rate windows and is termed as $\Delta C_{meas.}$.

$\Delta C_{measured}$ can be expressed in Equation 8.6 as

$$\Delta C_{meas.} = C(t_2) - C(t_1)$$

From Equation 8.13,

$$\Delta C_{meas.} = \frac{C_o N_T}{2N_D}\left[\exp(-e_n t_1) - \exp(-e_n t_2)\right] \qquad (8.14)$$

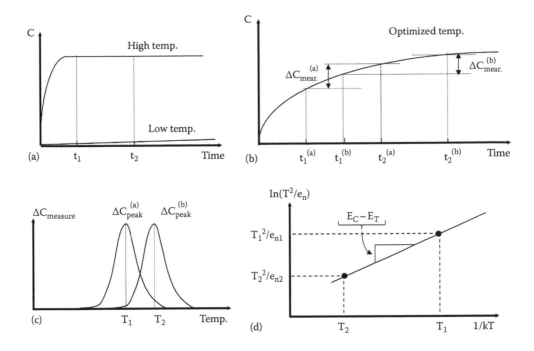

FIGURE 8.6 (a) Capacitance transient at low and high temperatures; (b) capacitance transient at an optimized temperature; (c) DLTS signal or $\Delta C_{meas.}$ versus temperature for two sets of rate windows t_1 and t_2; and (d) Arrhenius plot to identify deep level energy and capture cross section.

The $\Delta C_{meas.}$ versus temperature with each t_1 and t_2 sets are plotted as shown in Figure 8.5c. From these plots, peak temperatures with each t_1 and t_2 sets are defined. The emission rate for a given peak temperature, T_1 and T_2 in Figure 8.6c, can be obtained by taking the derivative of Equation 8.1. The detailed workings are as follows:

$$\Delta C = \frac{C_o N_T}{2 N_D} \left[\exp\left(-e_n\left(T\right)t_1\right) - \exp\left(-e_n\left(T\right)t_2\right) \right]$$

$$\frac{d\Delta C}{dT} = \frac{C_o N_T}{2 N_D} \left(-\exp\left(-e_n t_1\right)\left(\frac{d}{dT}\left(e_n\left(T\right)\right)t_1\right) + \exp\left(-e_n t_2\right)\left(\frac{d}{dT}\left(e_n\left(T\right)\right)t_2\right) \right) = 0$$

$$\exp\left(-e_n t_1\right)\left(\frac{d}{dT}\left(e_n\left(T\right)\right)t_1\right) = \exp\left(-e_n t_2\right)\left(\frac{d}{dT}\left(e_n\left(T\right)\right)t_2\right)$$

$$t_1 \exp\left(-e_n t_1\right) = t_2 \exp\left(-e_n t_2\right)$$

$$\left(-e_n t_1 + e_n t_2\right) = \ln\left(t_2 / t_1\right)$$

Therefore,

$$e_n\left(T\right) = \frac{\ln\dfrac{t_2}{t_1}}{t_2 - t_1} \tag{8.15}$$

In Equation 8.7, the emission coefficient is defined as follows:

$$e_n = \sigma_n \gamma_n T^2 \cdot \exp\left(\frac{E_T - E_C}{kT}\right)$$

Taking natural log on both sides,

$$\ln\left(\frac{T^2}{e_n}\right) = \left(E_C - E_T\right)\frac{1}{k_B T} - \ln\left(\sigma_n \gamma_n\right) \tag{8.16}$$

An Arrhenius plot $\ln(T^2/e_n)$ versus $1/kT$, where T are the temperatures T_1, T_2... at each peak, gives Figure 8.6d. From the slope of the straight line, the activation energy, E_T, of the traps can be calculated while the capture cross section, σ_n, is calculated from the y-intercept.

8.2.2 Identifying Hyper Deep Levels by Deep Level Optical Spectroscopy

Deep level optical spectroscopy (DLOS) was first introduced by Chantre et al. in 1981 to study optical capture cross section, $\sigma^o(h\nu)$, of the deep level traps in direct bandgap semiconductors [16]. Later, the DLOS technique was adopted to study the deep level trap energy and density in wide and direct bandgap semiconductors.

DLOS measurement captures and analyzes photocapacitance transient arising from the deep levels in the depletion regions of the diode. The photon flux is provided by a broadband illumination from a 1 kW Xe arc lamp or a 600 W QTH lamp, dispersed through a monochromator using appropriate mode-sorting filters. The photon energy ($h\nu$) range varies from 0.80 to 6.0 eV depending on the application.

In order to obtain photocapacitance transient, the sample is pulsed from reverse bias to forward bias or no bias V_{bias} to fill the traps with carriers as shown in Figure 8.7a. Next, it was reversed bias at $V_{meas.}$ to capture the capacitance transient, similar to the DLTS technique in Figure 8.7b. Under reverse bias, first, the sample is kept in dark to wait for the saturation of capacitance transient due to emission of carriers from traps excited by thermal energy. This is the regime where the DLTS technique works. Next, the photocapacitance transient was recorded under the illumination window shown in Figure 8.7a. The energy band diagrams of the diode at the thermal emission and optical emission of carriers are shown in Figure 8.7c and d, respectively. Next, the photocapacitance transient was recorded with increments of the photon energy in steps of 10–20 meV. More information about the DLOS technique could be found elsewhere [17,18].

With the stated pulsing condition, the optical transient as shown in Figure 8.7b mostly depends on capture and emission of electrons in n-GaN devices. The hole capture and emission processes are negligible in the depletion region. Similar to Equation 8.12 for DLTS, the capacitance at any instantaneous time in the optical emission process can be expressed as

$$C(t) = C_\infty\left(1 - \frac{n_T(t)}{2N_D}\right) \tag{8.17}$$

where
    C_∞ is the saturation capacitance (shown in Figure 8.7b)
    $n_T(t)$ is the number of traps filled with electrons
    N_D is the donor concentration

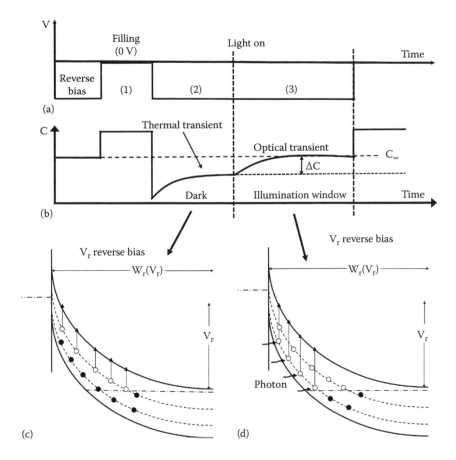

FIGURE 8.7 (a) Pulsing condition and (b) corresponding capacitance for DLOS measurement; band diagram of the diode at reverse bias condition in dark (c) and under illumination (d) to measure thermal and optical transient due to the electron emission process of deep level traps.

Differentiating both sides of Equation 8.17, with the assumption that the trap density is much smaller than the doping density, $\dfrac{n_T(t)}{2N_D} \ll 1$

$$\frac{1}{C_\infty}\frac{dC(t)}{dt}(t=0) = \frac{-1}{2N_D}\frac{dn_T(t)}{dt}(t=0) \tag{8.18}$$

Similar to Equation 8.2 in DLTS: $\dfrac{dn_T(t)}{dt}(t=0) = -e_n^0(h\upsilon)n_T$

According to Chantre et al. [16], the optical emission rate e_n^0 of electron from deep levels under excitation by flux of photons (Φ) having energy $h\upsilon$ is given as follows:

$$e_n^0 = \Phi(h\upsilon)\sigma_n^0(h\upsilon) \tag{8.19}$$

where $\sigma_n^0(h\upsilon)$ is the optical capture cross section of the hyper deep level traps with photon energy $h\upsilon$

$$\rightarrow \frac{dn_T(t)}{dt}(t=0) = -\Phi(h\upsilon)\sigma_n^0(h\upsilon)n_T \tag{8.20}$$

Rearranging the variables in Equations 8.18 and 8.20, the change in capacitance at t = 0 is directly proportional to $\Phi(h\upsilon)\sigma_n^0(h\upsilon)$, that is,

$$\sigma_n^0(h\upsilon) = \frac{2N_d}{\Phi(h\upsilon)n_T C_\infty} \frac{dC(t)}{dt}(t=0) \tag{8.21}$$

8.2.2.1 Identifying the Hyper Deep Level Density

Trap density, n_T, can be determined by the following steps.

First, the equation of trap density can be simplified as shown in Equation 8.2:

$$n_T(t=\infty) = \frac{e_p}{e_p + e_n} n_T(t=0) \tag{8.22}$$

and expressed as

$$n_T(t=\infty) = \frac{\sigma_p^0(h\upsilon)}{\sigma_p^0(h\upsilon) + \sigma_n^0(h\upsilon)}$$

Applying the change in capacitance simplifies to

$$\frac{\Delta C}{C_\infty} = \frac{N_T}{2N_D}\left(1 - \frac{\sigma_p^0(h\upsilon)}{\sigma_p^0(h\upsilon) + \sigma_n^0(h\upsilon)}\right) \tag{8.23}$$

It is typically assumed that $\sigma_p^0(h\upsilon)$ is negligible yielding the familiar equation expressed earlier in DLTS

$$N_T = \frac{2\Delta C}{C_o}(N_D - N_A) \tag{8.24}$$

8.2.2.2 Identifying Hyper Deep Level Activation Energy

a. Using steady-state photocapacitance versus photon energy: Figure 8.8a shows the relative steady-state photocapacitance ($\Delta C = C_{SS} - C_o$) versus photon energy ($h\upsilon$). The spectrum increases with photon energy ($h\upsilon$) because ΔC is proportional to the total number of charge created by emission of carriers from active deep levels. In the case of the photon energy being equal or higher than the activation energy of the traps, it is sufficient to emit carriers from the traps. Therefore, the steady-state photocapacitance (SSPC) increases abruptly for photon energy equal to the trap energy. Subsequently, the activation energies of the hyper deep level traps are revealed based as further onsets of the capacitance spectrum. Using this concept, two deep level states $E_{T3} = E_C - 1.30$ eV and $E_{T4} = E_C - 3.28$ eV were revealed in the n-GaN layer.

b. Using optical capture cross section versus photon energy: The optical capture cross section for each photon energy is calculated by using Equation 8.2. In this equation, dC(t) is measured. N_d is measured by C–V and N_t is calculated by using Equation 8.24. An example of optical capture cross section versus photon energy is shown in Figure 8.8b. According to Lucovsky [19], the dependence of optical capture cross section for the trap located at classical optical ionization energy (E_i) on photon energy ($h\upsilon$) is expressed by the following equation:

$$\sigma^0(h\upsilon) \propto \frac{1}{E_i}\left\{\left(\frac{E_i}{h\upsilon}\right)\cdot\left(1 - \frac{E_i}{h\upsilon}\right)\right\}^{3/2} \tag{8.25}$$

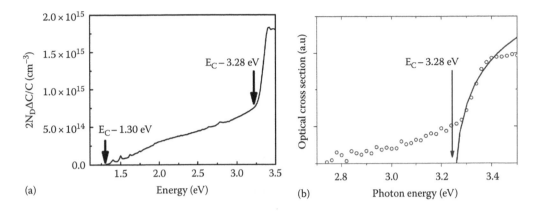

(a) Energy (eV) (b) Photon energy (eV)

FIGURE 8.8 (a) Steady-state photocapacitance for GaN grown on sapphire substrate by MBE method and (b) Lucovsky fitting for optical capture cross section to identify deep level trap at E_C – 3.28 eV.

An accurate determination of hyper deep level trap energy can be achieved by fitting the optical capture cross section. An example of using Lucovsky equation to identify deep level at E_C – 3.28 eV is shown in Figure 8.8b.

8.3 Deep Level Traps in GaN Epilayer

8.3.1 Common Deep Levels in GaN Epilayer Studied by DLTS Technique

Defect studies by DLTS and DLOS conducted on n-GaN epilayers grown on different substrates are reviewed. The common defect levels are shown in Figure 8.9. The possible nature of traps is also shown. In Figure 8.9, traps of similar energy levels are grouped into one level. The dispersion of published data on deep level energies in GaN is due to strain-induced electrical field present in the material. The density of each trap varies from sample to sample and they reflect the quality of the crystal growth.

The first trap is E_{T1} located at E_C – 0.21 to 0.26 eV. Its capture cross section falls within 8.4×10^{-17} ~ 1.4×10^{-15} cm^2. This value of capture cross section is typical of point defect. Fang et al. attributed this trap to Ga and N vacancy as it correlates with reduction in film thickness and increase in dislocation density [20]. However, Soh et al. attributed this trap to O_N [21]. In free-standing GaN, this trap is the dominant trap.

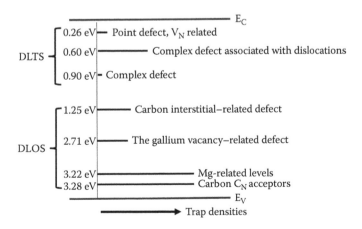

FIGURE 8.9 Common deep level defects and their energy levels in GaN materials and their possible physical origin.

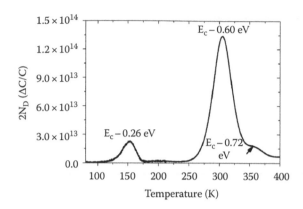

FIGURE 8.10	DLTS for GaN grown on sapphire.

The second trap is E_{T2} located at E_C – 0.51 to 0.60 eV. Its capture cross section falls within 1.8×10^{-16} to 2×10^{-12} cm^2. Tokuda et al. attributed this trap to dislocation-related defects as the trap is found to show a logarithmic dependence on filling time [22]. An earlier report by Fang et al. also made the same conclusion [23]. E_{T2} is the dominant trap in GaN grown on sapphire and SiC.

The third trap is E_{T3} located at E_C – 0.90 to 1.02 eV. Its capture cross section falls within 9.4×10^{-15} to 5.3×10^{-15} cm^2. E_{T3} is a dominant trap present in all substrates. Fang et al. attributed this trap to threading dislocations based on a logarithmic dependence of the DLTS signal with the filling pulse width [24].

The other traps observed in GaN grown on SiC and sapphire are located at E_C – 0.3 eV, E_C – 0.4 eV, E_C – 0.7 eV, E_C – 0.8 eV, and E_C – 1.3 eV. In general, these traps have lower density than the three common traps discussed earlier. Therefore, the effects of these traps on GaN device performances are less severe than the three common deep level traps. An example of DLTS for GaN on sapphire substrate is shown in Figure 8.10. Another example of DLTS for GaN grown on sapphire, GaN, and SiC substrate is given in Reference 25. In general, GaN on different substrates does not produce new defect levels but shows significant variations on their densities.

Deep level traps in p-GaN epilayer doped with magnesium and grown on sapphire substrate were reported by Götz et al. [26]. They reported three levels with activation energies for the hole emission of 0.21, 0.39, and 0.41 eV.

8.3.2 Common Deep Levels in GaN Epilayer Studied by DLOS Technique

The DLOS technique reveals four common deep levels in n-GaN, $E_{T4} = E_C$ – (1.25 – 1.45) eV, $E_{T5} = E_C$ – (2.6 – 2.8) eV, $E_{T6} = E_C$ – 3.22 eV, and $E_{T7} = E_C$ – 3.28 eV [18,27]. The level at E_C – (1.25 – 1.45) eV is attributed to C_I—a dislocation complex. The level at E_C – (2.6 – 2.8) eV contributes to yellow luminescence in GaN and is attributed to V_{Ga}-related trap. The level at E_C – 3.22 eV is commonly associated with Mg residual in the growth reactor [28,29] and E_C – 3.28 eV is carbon related [30]. An example of DLOS measurement for n-GaN Schottky diode is shown in Figure 8.11. In this sample, three levels, E_C – 1.45 eV, E_C – 2.7 eV, and $E_{T6} = E_C$ – 3.28 eV, are revealed.

DLOS measurement for p-GaN Schottky diode is more sensitive to hole traps. There are two reports of deep level centers in p-GaN layer by Götz et al. and Armstrong at al. [26,31]. Götz reported on the dominant hole trap level located at 1.8 eV in p-GaN. Armstrong reported on three hole trap levels located at E_v + 3.05 eV, E_v + 3.22 eV, and E_v + 3.26 eV and two electron trap levels located at E_c – 3.24 eV and E_c – 2.97 eV in p-GaN layer.

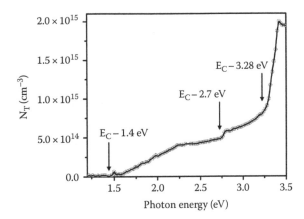

FIGURE 8.11 DLOS spectra of MBE grown n-GaN Schottky diodes. The two carbon associated levels are at $E_C - 1.4$ and $E_C - 3.28$ eV. The gallium vacancy–related level located at $E_C - 2.7$ eV.

8.3.3 Dislocation-Related Deep Levels in GaN Epilayer

The dislocation densities in GaN are in the range of 10^8 to 10^{10} cm^{-2}. It is proposed that threading dislocations may be open core that contains V_{Ga} or V_{Ga}-O acceptor-like defects [32], or may be close core that does not contain point defects [33]. In GaN LEDs, open-core threading dislocation density could be correlated to nonradiative recombination as the point defects may serve as nonradiative recombination centers [34].

In experiments, DLTS measurements pick up electrically active traps, which are normally point defects and their complexes in GaN. According to Armstrong et al. [35,36] and Arehart et al. [18,37], the level $E_C -$ 0.60 eV, revealed by DLTS, and $E_C - 1.30$ eV, revealed by DLOS, are dislocation-related defects. The level $E_C - 0.6$ eV exists in most GaN samples grown with both MOCVD and MBE on various substrates such as sapphire, GaN, and SiC. GaN grown on lateral epitaxial overgrowth (LEO) substrate and standard GaN grown on sapphire template, both by MOCVD, have threading dislocation density (TDD) of 4×10^7 cm^{-2} and 6×10^8 cm^{-2}, respectively. Comparing the TDD of these two types of samples, Arehart et al. [37] concluded that the levels $E_C - 1.30$ eV and $E_C - 0.60$ eV levels are strongly correlated to threading dislocations.

8.4 Deep Levels in GaN LED Devices

8.4.1 Deep Level Traps in GaN LED Studied by DLTS Technique

8.4.1.1 GaN LEDs Grown on Sapphire and SiC Substrates

Due to the complex structure of GaN LED devices that comprise p-GaN, n-GaN, and MQW layers, identifying and locating deep level traps of a GaN LED are not straightforward. There are few reports on deep level traps of GaN LED [38–41]. Kim et al. found a hole trap level at $E_V + 0.70$ eV in the active region of commercial MQW GaN LEDs grown on sapphire [38]. Rigutti et al. [39] found an electron trap in active region located at $E_C - (0.25 \pm 0.03)$ eV. Rossi et al. [40] reported on deep level traps of $E_C - (0.17 - 0.18)$ eV in n-GaN barrier layer located at 200 nm below the MQW. Venturi et al. [41] measured three levels $E_A \approx E_C - 0.04$ eV, $E_{A1} \approx E_C - (0.12 - 0.13)$ eV, and $E_B \approx E_C - (0.50 - 0.54)$ eV. They reported that traps A and A1 are most likely to be electron traps in the p-GaN layer while trap B is an electron trap in the n-GaN layer. The A and A1 levels have higher concentrations in LEDs grown on high dislocation density templates compared with those grown on the lower dislocation density templates. They concluded that traps A and A1 are dislocation-related energy levels.

In Kim's paper, the hole trap at $E_V + 0.70$ eV found in InGaN/GaN MQW of the LEDs grown on sapphire substrates was revealed with a reverse bias of −5 V [38]. This $E_V + 0.7$ eV hole trap is not seen in the GaN epilayer grown on sapphire. However, similar results have been reported by Fang in a GaN p-i-n photodiode structure grown by MBE on sapphire [42]. Hierro measured hole traps in p⁺-n diodes grown by MOCVD on sapphire [17] and Polyakov used optical transient current spectroscopy to measure in n-GaN samples grown by MOCVD [43]. They reported hole traps at $E_v + 0.79$ eV and $E_v + 0.85$ eV and $E_v + 0.87$ eV. Traps around these energy levels are associated with yellow luminescence band.

8.4.1.2 Deep Level Traps in GaN LED Grown on Si Substrate

Deep level traps in GaN MQW LEDs grown on Si substrate were studied by DLTS [44]. Deep level traps in the n-GaN and the p-GaN barrier layers were revealed by using different pulsing conditions. The devices were biased at −1 V and pulsed to 0 V for filling by majority carriers in the n-GaN barrier layer. The C–V characteristic of the LED at 0 and −1 V bias shows that the p-GaN layer is fully depleted while the n-GaN barrier layer is partially depleted. Therefore, with this pulsing condition electron traps are revealed in the n-GaN barrier layer only. The DLTS signals of the LED for various rate windows are shown in Figure 8.12.

The trap level detected in the n-GaN barrier layer is located at $E_C − 0.70$ eV with a capture cross section of 4×10^{-15} cm² and a density of 5.5×10^{14} cm⁻³. This trap is well documented for GaN grown on Si, measured frequently at $E_C − (0.71 − 0.72)$ eV, but with a density of 10^{13} cm⁻³, which is one order of magnitude lower than that found in an LED structure [37]. The nature of this trap is still not clear, but with a capture cross section of 4×10^{-15} cm², this level might be due to point defects. A similar level at $E_C − 0.73$ eV was found in GaN grown on SiC [25] and at $E_C − 0.67$ eV in GaN grown on sapphire [45].

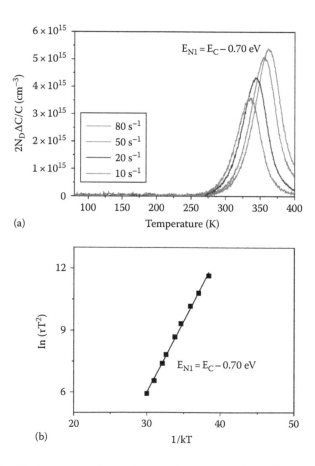

FIGURE 8.12 (a) DLTS of the GaN LED under −1.0 V reverse bias and 0 V filling voltage to reveal deep level traps in the n-GaN barrier layer and (b) Arrhenius plot of E_{N1}. (From Nguyen, X.S. et al., *Jpn. J. Appl. Phys.*, 55, 060306, 2013.)

Two common levels found in n-GaN grown on sapphire: $E_C - 0.25$ eV and $E_C - 0.60$ eV are, however, absent in the n-GaN barrier layer of the LED.

Due to a much lower doping concentration of the p-GaN layer compared with the n-GaN layer, DLTS measurement is more sensitive in picking up traps in the p-GaN layer than in the n-GaN layer as it is easier to deplete the p-layer. By selecting pulsing conditions of 2.1 and 3.5 V, the DLTS signal originates mainly from the p-GaN layer. At the higher voltage of 3.5 V, the current flowing through the junction is above 1 mA at 200 K, which is sufficient to fill all the electron traps in the p-GaN layer with carriers. Thus, with this pulsing condition, both electron and hole traps can be revealed. Two deep levels E_{P1} and E_{P2} from the p-GaN layer are found in this DLTS measurement as shown in Figure 8.13a. The negative signals show that these traps are minority carrier *electron* traps. One deep level H_{P1} is majority carrier *hole* traps.

The Arrhenius plot of the traps in the p-GaN layer is shown in Figure 8.13b. It is seen in the plot that the hole trap $H_{P1} = E_V + 0.70$ eV has a density of about 1.5×10^{14} cm^{-3} and a capture cross section of 5.3×10^{-13} cm^2. This level is commonly seen in n-GaN grown on sapphire characterized by minority carrier transient spectroscopy (MCTS) but no origin is attributed to it [25]. The second level, $E_{P1} = E_C - 0.60$ eV, is well documented in n-GaN [35]. This level is known as dislocation related, but the origin is still under debate. The dominant electron trap E_{P2} is located at $E_C - 0.79$ eV having a capture cross section of 5.9×10^{-14} cm^2 with a density of 2.3×10^{14} cm^{-3}. The value of capture cross section of this level suggests that this level originates from dislocations and their related complexes. A level similar to E_{P2} is observed in n-GaN on sapphire and is attributed to dislocation by Fang et al. [23].

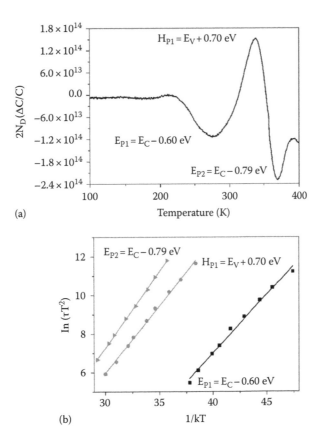

(a)

(b)

FIGURE 8.13 (a) DLTS of the GaN LED under 2.1 and 3.5 V to reveal deep level traps in the p-GaN layer and (b) Arrhenius plots of E_{P1}, E_{P2}, and H_{P1} level in the p-GaN layer. (From Nguyen, X.S. et al., *Jpn. J. Appl. Phys.*, 55, 060306, 2013.)

8.4.2 Deep Level Traps in GaN LED on Sapphire Studied by DLOS Technique

DLOS technique was used by Armstrong et al. to study hyper deep level traps in GaN LED grown on sapphire substrate [10,46,47]. A bias = −1.7 V was able to fully deplete the multi-quantum well (MQW) region [46]. Only defects located in the depletion region are sensitive in the measurement. At bias voltage of V = V_{th}, the DLOS measurement will be sensitive to MQW area while at V ≪ V_{th}, the n-GaN bulk-related deep levels are identified. By using −1.7 V bias condition, three electron traps in the MQW of the GaN LED located at E_C − 1.62 eV and E_C − 2.11 eV and E_C − 2.73 eV were revealed, as shown in Figure 8.14. The level E_C − 3.25 eV in n-GaN layer was realized by doing DLOS with bias voltage of −8 V.

In a GaN LED, the role of the first QW is more important than the other QWs because the highest recombination rate occurs at the first QW. Moreover, due to the growth and processing of GaN LEDs, the distribution of the deep level density is varied across the MQW region. Identifying defect level from quantum wells (QW) and quantum barriers (QB) is also necessary to understand the physical origin of deep level traps.

In order to determine quantitatively the spatial distribution in QWS and QBs of defect states, DLOS and lighted capacitance–voltage (LCV) [48] techniques are used. Selection of suitable bias voltages will distinguish deep levels from n-GaN bulk and MQW. Differentiation among QW- and QB-related deep levels was then achieved by comparing LED MQW-related DLOS spectra to previous studies of individual GaN and InGaN epi-layers. LCV measurements provided quantitative and nanoscale depth profiling of the deep level concentration (N_t) for individual QWs and QBs and revealed a strong depth dependence for defect incorporation in the MQW region.

In LCV measurement, the density of defects within a particular layer of the MQW region or adjacent n-GaN bulk can be determined from the difference in V (ΔV) required to achieve the same depletion depth, x_d (and therefore C) with deep levels fully occupied or empty of electrons [49]. This voltage shift is illustrated in the inset of Figure 8.15. The density of the deep level states was calculated by using the following equation: [49]

$$\Delta V = \frac{q}{\varepsilon} \int_0^{x_d} x n_t(x) dx$$

where n_t is the density of occupied defects.

The distributions of deep level density in QW and QB of the LED are shown in Figure 8.16. The increase in deep level density in each QW shows that defect incorporation happens during the QW growth.

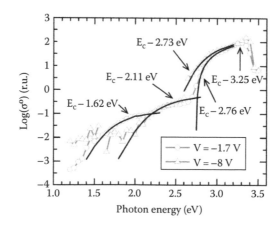

FIGURE 8.14 DLOS spectra as a function of applied bias. The levels at 1.62, 2.11, and 2.76 eV evident at V = −1.7 V are assigned to be deep levels in the MQW region. The 3.25 eV level that occurs only when V = −8 V is assigned to a deep level in the n-GaN bulk. (Reproduced from Armstrong, A. et al., *Opt. Express*, 20, 812, 2012. With permission of OSA Publishing.)

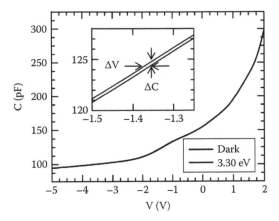

FIGURE 8.15 CV and LCV curves for the LED. (Reproduced from Armstrong, A. et al., *Opt. Express*, 20, 812, 2012. With permission of OSA Publishing.)

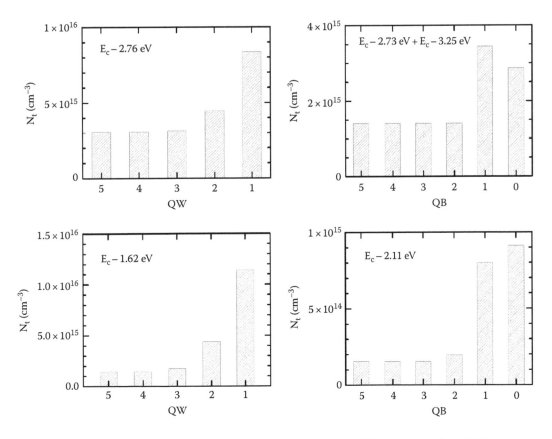

FIGURE 8.16 Distribution of the deep level density in the MQW region. QW is quantum well and QB is quantum barrier. QW 1 is grown next to the p-layer. (Reproduced from Armstrong, A. et al., *Opt. Express*, 20, 812, 2012. With permission of OSA Publishing.)

8.4.3 Deep Level Traps in Blue and Green GaN LED on Sapphire

The green GaN LEDs are well known to have lower internal quantum efficiency (IQE) than blue LED. The decrease in IQE is attributed to quantum confined Stark effect (QCSE) [50] and excess defects incorporation in the $In_xGa_{1-x}N/GaN$ MQW. Armstrong et al. [51] reported on the contribution of deep level defects to decreasing radiative efficiency of InGaN/GaN QW. On paper, deep level defects were studied by DLOS technique and IQE was obtained by photoluminescence measurement.

Deep level traps of the green LEDs with $In_{0.21}Ga_{0.81}N/GaN$ MQW and blue LEDs with $In_{0.17}Ga_{0.83}N/GaN$ MQW were measured by DLOS, as shown in Figure 8.17. The two LEDs were measured at reverse bias which the depletion region contained within the MQW region, hence DLOS picks up defect levels within the MQW region. DLOS revealed two levels E_C – 1.62 eV and E_C – 2.76 eV in the blue $In_{0.17}Ga_{0.83}N$ QWs LED (450 nm) and two levels at E_C –1.43 eV and E_C – 2.41 V in the green LED (530 nm) with the $In_{0.21}Ga_{0.79}N$ QWs. All levels are electron traps because photocapacitance is from majority carrier (electron) photoemission.

The DLOS spectra shown in Figure 8.17 do not indicate defect density. The density of each level was calculated by using lighted capacitance–voltage method. Table 8.1 lists the defect density for each level of the two LEDs. The total defect density of the $In_{0.21}Ga_{0.79}N$ QWs in green LED is one order of magnitude higher than that of the $In_{0.17}Ga_{0.83}N$ QWs of the blue LED. The reduction of about 3 times in IQE of the green compared to the blue LED is measured by photoluminescence. Therefore, deep level defect density in InGaN QWs could contribute to the reduction of IQE of GaN LED.

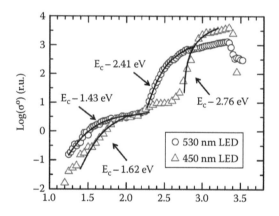

FIGURE 8.17 $In_xGa_{1-x}N$ QW DLOS spectra for the blue- and green-emitting LEDs. Note that the DLOS spectra do not indicate defect density. (Reproduced from Armstrong, A.M. et al., *Appl. Phys. Express*, 7, 032101, 2014. With permission from The Japan Society of Applied Physics.)

TABLE 8.1   Summary of $In_xGa_{1-x}N$ MQW Defect Concentration in Blue and Green GaN LED

LEDs	In Mole Fraction (x)	Activation Energy (eV)	Density (cm^{-3})
Blue LED	0.17	E_C – 1.62	4.1×10^{15}
		E_C – 2.76	5.5×10^{15}
Green LED	0.21	E_C – 1.43	1.8×10^{16}
		E_C – 2.41	5.5×10^{16}

Source: Reproduced from Armstrong, A.M. et al., *Appl. Phys. Express*, 7, 032101, 2014. With permission from The Japan Society of Applied Physics.

8.5 Conclusions

The chapter reviews the deep level trap characterization of GaN materials and LEDs by DLTS and DLOS techniques. Background of the point defect, line defect, and the characterization techniques are also introduced. The most common defect levels in GaN measured by DLTS are located at $E_C − 0.25$ eV, $E_C − 0.60$ eV, and $E_C − 0.90$ eV. The level $E_C − 0.25$ eV is point defect and to be attributed to V_N. The $E_C − 0.60$ eV level is dislocation complex–related defect. The origin of the $E_C − 0.90$ eV is still not clear. DLOS revealed three levels commonly found in n-GaN, located at $E_C − 1.25$ eV, $E_C − 2.6$ eV, and $E_C − 3.28$ eV. The $E_C − 1.25$ eV level is carbon interstitial related and $E_C − 3.28$ eV level is C_N acceptor related.

For GaN LEDs, due to their complicated structure, there are only few reports on deep level defects studied by DLTS and DLOS. Identifying deep levels in each single layer of the LED is not straightforward and depends on the LED structures. In general, DLOS reveals the highest defect density in the first QW. It is also shown that the defect density in green GaN LEDs is one order of magnitude higher than that of blue LEDs. This is one of the reasons for the lower efficiency in longer-wavelength InGaN LEDs.

Acknowledgment

This work was supported by the National Research Foundation (NRF) of Singapore through the Singapore-MIT Alliance for Research and Technology's (SMART) Low Energy Electronic Systems (LEES) IRG.

References

1. A. F. Wright, Interaction of hydrogen with gallium vacancies in wurtzite GaN, *J. Appl. Phys.* 90, 1164–1169 (2001).
2. S. J. Pearton, J. C. Zolper, R. J. Shul, and F. Ren, GaN: Processing, defects, and devices, *J. Appl. Phys.* 86, 1 (1999).
3. J. Elsner, R. Jones, P. K. Sitch, V. D. Porezag, M. Elstner, T. Frauenheim, M. I. Heggie, S. Öberg, and P. R. Briddon, Theory of threading edge and screw dislocations in GaN, *Phys. Rev. Lett.* 79, 3672–3675 (1997).
4. P. Yu and M. Cardona, *Fundamentals of Semiconductors Physics and Materials Properties*, Springer-Verlag Berlin Heidelberg, Germany, (2010).
5. M. Meneghini, L. R. Trevisanello, G. Meneghesso, and E. Zanoni, A review on the reliability of GaN-based LEDs, *IEEE Trans. Device Mater. Rel.* 8, 323–331 (2008).
6. M. Meneghini, A. Tazzoli, G. Mura, G. Meneghesso, and E. Zanoni A review on the physical mechanisms that limit the reliability of GaN-based LEDs, *IEEE Trans. Electron Devices* 57, 108–118 (2010).
7. D. Reynolds, D. Look, B. Jogai, J. Van Nostrand, R. Jones, and J. Jenny, Source of the yellow luminescence band in GaN grown by gas-source molecular beam epitaxy and the green luminescence band in single crystal ZnO, *Solid State Commun.* 106, 701–704 (1998).
8. M. A. Reshchikov and H. Morkoç, Luminescence properties of defects in GaN, *J. Appl. Phys.* 97, 061301 (2005).
9. M. Mandurrino, G. Verzellesi, M. Goano, M. E. Vallone, F. Bertazzi, G. Ghione, M. Meneghini, G. Meneghesso, and E. Zanoni, Trap-assisted tunneling in InGaN/GaN LEDs: Experiments and physics-based simulation, *IEEE International Conference on Numerical Simulation of Optoelectronic Devices*, Palma de Mallorca, Spain, pp. 13–14 (2014).
10. A. Armstrong, T. A. Henry, D. D. Koleske, M. H. Crawford, K. R. Westlake, and S. R. Lee, Dependence of radiative efficiency and deep level defect incorporation on threading dislocation density for InGaN/GaN light emitting diodes, *Appl. Phys. Lett.* 101, 162102 (2012).
11. P. B. Klein, J. A. Freitas, S. C. Binari, and A. E. Wickenden, Observation of deep traps responsible for current collapse in GaN metal–semiconductor field-effect transistors, *Appl. Phys. Lett.* 75, 4016–4018 (1999).

12. U. K. Mishra, L. Shen, T. E. Kazior, and Y. F. Wu, GaN-based RF power devices and amplifiers, *Proc. IEEE* 96, 287–305 (2008).

13. D. V. Lang, Deep-level transient spectroscopy: A new method to characterize traps in semiconductors, *J. Appl. Phys.* 45, 3023 (1974).

14. J. W. Orton and P. Blood, *The Electrical Characterization of Semiconductors: Majority Carriers and Electron States*, Academic Press (1992).

15. D. K. Schroder, *Semiconductor Material and Device Characterization*, Wiley-Interscience (2006).

16. A. Chantre, G. Vincent, and D. Bois, Deep-level optical spectroscopy in GaAs, *Phys. Rev. B* 23, 335–5359 (1981).

17. A. Hierro, D. Kwon, S. A. Ringel, M. Hansen, J. S. Speck, U. K. Mishra, and S. P. DenBaars, Optically and thermally detected deep levels in n-type Schottky and p$^+$-n GaN diodes, *Appl. Phys. Lett.* 76, 3064 (2000).

18. A. R. Arehart, A. Corrion, C. Poblenz, J. S. Speck, U. K. Mishra, and S. A. Ringel, Deep level optical and thermal spectroscopy of traps in n-GaN grown by ammonia molecular beam epitaxy, *Appl. Phys. Lett.* 93, 112101 (2008).

19. G. Lucovsky, On the photoionization of deep impurity centers in semiconductors, *Solid State Commun.* 3, 299–302 (1965).

20. Z. Q. Fang, D. C. Look, J. Jasinski, M. Benamara, Z. Liliental-Weber, and R. J. Molnar, Evolution of deep centers in GaN grown by hydride vapor phase epitaxy, *Appl. Phys. Lett.* 78, 332 (2001).

21. C. B. Soh, S. J. Chua, H. F. Lim, D. Z. Chi, S. Tripathy, and W. Liu, Assignment of deep levels causing yellow luminescence in GaN, *J. Appl. Phys.* 96, 1341 (2004).

22. Y. Tokuda, Y. Matuoka, K. Yoshida, H. Ueda, O. Ishiguro, N. Soejima, and T. Kachi, Evaluation of dislocation-related defects in GaN using deep-level transient spectroscopy, *Phys. Status Solidi C* 4, 2568–2571 (2007).

23. Z. Q. Fang, D. C. Look, X. L. Wang, J. Han, F. Khan, and I. Adesida, Plasma-etching-enhanced deep centers in n-GaN grown by metalorganic chemical-vapor deposition, *Appl. Phys. Lett.* 82, 1562–1564 (2003).

24. Z. Q. Fang, D. C. Look, D. H. Kim, and I. Adesida, Traps in AlGaN/GaN/SiC heterostructures studied by deep level transient spectroscopy, *Appl. Phys. Lett.* 87, 182115 (2005).

25. Y. Tokuda, Traps in MOCVD n-GaN studied by deep level transient spectroscopy and minority carrier transient spectroscopy, *CS MANTECH Conference*, Denver, CO (2014).

26. W. Götz, N. M. Johnson, and D. P. Bour, Deep level defects in Mg-doped, p-type GaN grown by metalorganic chemical vapor deposition, *Appl. Phys. Lett.* 68, 3470 (1996).

27. A. Armstrong, A. Chakraborty, J. S. Speck, S. P. DenBaars, U. K. Mishra, and S. A. Ringel, Quantitative observation and discrimination of AlGaN- and GaN-related deep levels in AlGaN/GaN heterostructures using capacitance deep level optical spectroscopy, *Appl. Phys. Lett.* 89, 262116 (2006).

28. A. Hierro, M. Hansen, L. Zhao, J. Speck, U. Mishra, S. DenBaars, and S. Ringel, Carrier trapping and recombination at point defects and dislocations in MOCVD n-GaN, *Phys. Status Solidi B*, 228, 937–946 (2001).

29. A. R. Arehart, *Investigation of Electrically Active Defects in GaN, AlGaN, and AlGaN/GaN High Electron Mobility Transistors*, The Ohio State University, Ohio, CO (2009).

30. A. M. Armstrong, *Investigation of Deep Level Defects in GaN: C, GaN: Mg and Pseudomorphic AlGaN/GaN Films*, The Ohio State University, Ohio, CO (2006).

31. A. Armstrong, J. Caudill, A. Corrion, C. Poblenz, U. K. Mishra, J. S. Speck, and S. A. Ringel, Characterization of majority and minority carrier deep levels in p-type GaN:Mg grown by molecular beam epitaxy using deep level optical spectroscopy, *J. Appl. Phys.* 103063722 (2008).

32. Z. Fang, D. C. Look, and L. Polenta, Dislocation-related electron capture behaviour of traps in n-type GaN, *J. Phys. Condens. Matter* 14, 13061 (2002).

33. S. Lester, F. Ponce, M. Craford, and D. Steigerwald, High dislocation densities in high efficiency GaN-based light-emitting diodes, *Appl. Phys. Lett.* 66, 1249–1251 (1995).

34. S. Evoy, H. Craighead, S. Keller, U. Mishra, and S. DenBaars, Scanning tunneling microscope-induced luminescence of GaN at threading dislocations, *J. Vac. Sci. Technol. B* 17, 29–32 (1999).

35. A. Armstrong, A. R. Arehart, B. Moran, S. P. DenBaars, U. K. Mishra, J. S. Speck, and S. A. Ringel, Impact of carbon on trap states in n-type GaN grown by metalorganic chemical vapor deposition, *Appl. Phys. Lett.* 84, 374 (2004).

36. A. Armstrong, A. Arehart, D. Green, J. S. Speck, U. K. Mishra, and S. A. Ringel, A novel method to investigate defect states in MBE grown highly resistive GaN doped with C and Si, *Phys. Status Solidi C* 2, 2413 (2005).

37. A. R. Arehart, A. Corrion,C. Poblenz, J. S. Speck, U. K. Mishra, S. P. DenBaars, and S. A. Ringel, Comparison of deep level incorporation in ammonia and rf-plasma assisted molecular beam epitaxy n-GaN films, *Phys. Status Solidi C* 5, 1750–1752 (2008).

38. J. W. Kim, G. H. Song, and J. W. Lee, Observation of minority-carrier traps in InGaN/GaN multiple-quantum-well light-emitting diodes during deep-level transient spectroscopy measurements, *Appl. Phys. Lett.* 88, 182103 (2006).

39. L. Rigutti, A. Castaldini, and A. Cavallini, Anomalous deep-level transients related to quantum well piezoelectric fields in $In_yGa_{1-y}N$/GaN heterostructure light-emitting diodes, *Phys. Rev. B* 77, 045312 (2008).

40. F. Rossi, M. Pavesi, M. Meneghini, G. Salviati, M. Manfredi, G. Meneghesso, A. Castaldini et al., Influence of short-term low current dc aging on the electrical and optical properties of InGaN blue light-emitting diodes, *J. Appl. Phys.* 99, 053104 (2006).

41. G. Venturi, A. Castaldini, A. Cavallini, M. Meneghini, E. Zanoni, D. Zhu, and C. Humphreys, Dislocation-related trap levels in nitride-based light emitting diodes, *Appl. Phys. Lett.* 104, 211102 (2014).

42. Z. Fang, D. C. Look, C. Lu, and H. Morkoç, Electron and hole traps in GaN pin photodetectors grown by reactive molecular beam epitaxy, *J. Electron. Mater.* 29, 19–23 (2000).

43. A. Polyakov, N. Smirnov, A. Usikov, A. Govorkov, and B. Pushniy, Studies of the origin of the yellow luminescence band, the nature of nonradiative recombination and the origin of persistent photoconductivity in n-GaN films, *Solid State Electron.* 42, 1959–1967 (1998).

44. X. S. Nguyen, X. L. Goh, L. Zhang, Z. Zhang, A. R. Arehart, S. A. Ringel, E. A. Fitzgerald, and S. J. Chua, Deep level traps in GaN LEDs grown by metal organic vapour phase epitaxy on an 8 inch Si(111) substrate, *Jpn. J. Appl. Phys.* 55, 060306 (2016).

45. Z. Zhang, C. A. Hurni, A. R. Arehart, J. Yang, R. C. Myers, J. S. Speck, and S. A. Ringel, Deep traps in nonpolar m-plane GaN grown by ammonia-based molecular beam epitaxy, *Appl. Phys. Lett.* 100, 052114 (2012).

46. A. Armstrong, T. A. Henry, D. D. Koleske, M. H. Crawford, and S. R. Lee: Quantitative and depth-resolved deep level defect distributions in InGaN/GaN light emitting diodes, *Opt. Express* 20, 812 (2012).

47. A. Armstrong, M. H. Crawford, and D. D. Koleske, Quantitative and depth-resolved investigation of deep-level defects in InGaN/GaN heterostructures, *J. Electron. Mater.* 40, 369–376 (2010).

48. A. Armstrong, A. R. Arehart, and S. A. Ringel, A method to determine deep level profiles in highly compensated, wide band gap semiconductors, *J. Appl. Phys.* 97, 083529 (2005).

49. S. D. Brotherton, Measurement of deep-level spatial distributions, *Solid State Electron.* 19, 341–342 (1976).

50. F. Scholz, Semipolar GaN grown on foreign substrates: A review, *Semicond. Sci. Technol.* 27, 024002 (2012).

51. A. M. Armstrong, M. H. Crawford, and D. D. Koleske, Contribution of deep-level defects to decreasing radiative efficiency of InGaN/GaN quantum wells with increasing emission wavelength, *Appl. Phys. Express* 7, 032101 (2014).

9

Photoluminescence Dynamics in InGaN/GaN Multiple Quantum Well Light-Emitting Diodes

Tao Lin
Guangxi University

Zhe Chuan Feng
Guangxi University

Abstract The recent researches referring to InGaN/GaN multiple quantum well light-emitting diodes (MQW LEDs) mainly focused on finding ways to improve the devices' performance, for the increasing demands in white light illumination and LED displays. Thus, the analysis to their photoluminescence (PL) dynamics became more and more important because these results might reveal the exact recombination mechanism inside these LED devices, which could guide us to design devices with better performance. Based upon these, this chapter reviews the studies of the PL dynamic properties of InGaN/GaN MQW LEDs, with either a typical construction or some improved structures realized in recent years, wherein exciton localization has proved to be the main reason for high-efficiency emission, and at least two different kinds of localization centers—local compositional fluctuations of indium and thickness variation of InGaN layers—existed in the MQW system and contributed to the emission; quantum-confined Stark effect (QCSE) was proven to be the main drawback in conventional

MQW LEDs. Several strategies to remove or reduce the QCSE, such as introducing charge asymmetric resonance tunneling (CART) structures or pre-strained layers, or the replacement of polar substrates with semipolar/nonpolar substrates had positive effects on the PL efficiencies. A clear picture for the photoluminescence process within InGaN/GaN MQW structure is drawn in this work by combining different kinds of PL results. This penetrative investigation was helpful to further improve the material design and growth of InGaN/GaN MQWs for the wide spectral LED applications.

9.1 Introduction

GaN-based light-emitting diodes (LEDs) have attracted much attention for solid-state lighting applications these days, owing to its much reduced energy consumption as compared to the traditional incandescent light bulbs [1–4]. In particular, InGaN active layers in LEDs are of great scientific importance to the construction of full-color displays, due to their tunable energy gaps covering the entire visible range [5–8]. Because of these revolutionary benefits, the Nobel Prize in Physics was awarded for the InGaN-based blue LEDs in 2014. Despite the high dislocation density in GaN crystals induced by the large lattice mismatch between GaN and sapphire substrates, InGaN/GaN multiple quantum well (MQW) structures demonstrated surprisingly high internal quantum efficiency (IQE) [9–13]. This was attributed to the exciton localization effect that reduces the chance of nonradiative recombination at the dislocations [14–17]. They also demonstrated other advantages, such as the lower threshold current density for laser diodes (LDs) and reduced device sensitivity to temperature change.

As a result, there have been active studies of photoluminescence dynamics and recombination mechanisms in InGaN/GaN MQW structures in recent years. Various mechanisms of exciton localization were proposed, such as compositional fluctuations of indium within InGaN wells [6,18], formation of dot-like In-rich clusters [19–21], well width fluctuations in the activated layers [22], and formation of V-shaped pits [23], aiming at the exact knowledge of its high-efficiency emission. On the other hand, the drawbacks of the present MQW LEDs were also revealed, such as the nonradiative recombination processes that induced efficiency droops under high carrier injection, and the influences of quantum-confined Stark effect (QCSE) existing in most of InGaN/GaN MQW structures grown on polar sapphire substrates that may deteriorate radiative emission. Efforts were then made to remove these disadvantages. For instance, designing LED devices with two different well widths [24,25], applying nonpolar or semipolar substrates to reduce QCSE [26–28], or inserting pre-strained layer into MQW structure to eliminate the strain-induced piezoelectric field [29,30].

Among other methods, time-resolved photoluminescence (TRPL) spectroscopy is the most important technique to investigate dynamic recombination processes in the InGaN/GaN MQW, as it provides a direct measurement of the exciton and carrier lifetimes within the semiconductor. PL lifetimes are critical to the emission efficiency and thus the overall device performance. It is therefore the purpose of this chapter to review the PL dynamic properties of InGaN/GaN MQW LEDs, with either a typical construction or an improved version, by TRPL spectroscopy in combination with steady-state (SS) PL and photoluminescence excitation (PLE) measurements, especially under a variable temperature. For example, temperature-dependent (TD)-SSPL and TD-TRPL, with a variable range of 10–300 K, were employed to extract dynamic parameters, such as activation energy, quantum efficiencies, and radiative/nonradiative decay rates, which are critical to the analysis of the recombination pathways in InGaN/GaN MQW structures.

The combined results offered a picture for the whole photoluminescence process within InGaN/GaN MQW structure: MQWs absorb incident UV photons to generate excited carriers beyond the bandgaps in both GaN barriers and InGaN wells, then parts of excited carriers formed in GaN layers may transfer to InGaN well layers. Excited carriers recombine to obtain emissions, among which the blue-to-green tunable emissions related to InGaN well are mostly expected; however, an accompanying UV emission related to GaN barriers is always observed. Theoretically, the radiative recombination rate of free excited carriers (free electron–hole pairs) in InGaN well layers is rather low because of the high density of dislocation in InGaN/GaN structure working as nonradiative trapping centers, and the QCSE that separates the different carriers in space. The strategies in MQW fabrications to remove or reduce QCSE may increase their chances to recombine, leading to enhancements of PL efficiency. Interestingly, even taking no account of QCSE, the observed radiative recombination rate of carriers in InGaN wells is far higher than we expected. So the imperfections of MQW structure, such as fluctuations of indium component inside well layers or fluctuations of well thickness, are taken into consideration. These imperfections are capable of capturing free excitons or free electron–hole pairs to form localized excitons, preventing them from reaching the nonradiative dislocations. The recombination of localized excitons dominates the emission of InGaN wells, which has much higher recombination rate and shorter lifetime than free ones. Meanwhile, the types of localization centers are various as well as their localization depths. This leads to complication of the PL dynamics. For example, the measured PL decay curves may deviate from single exponential decay. Up to now, exciton localization inducing surprisingly high PL efficiency of InGaN/GaN structure has been confirmed by many. But what are their exact origins and how they influence the PL dynamics are not completely understood and are waiting to be further clarified.

9.2 Experimental

As demonstrated in Figure 9.1, high-quality InGaN/GaN MQW LEDs were typically grown on (0001)-plane (c-facet) sapphire substrate by low-pressure (LP) metal–organic chemical vapor deposition (MOCVD) technique. Trimethylgallium (TMGa), trimethylindium (TMIn), and ammonia (NH$_3$) were used as precursors for Ga, In, and N, with carrier gas H$_2$ and N$_2$, respectively. A 30 nm thick GaN buffer layer was first grown on sapphire at low temperature (520°C), followed by a 2 µm thick GaN grown at high temperature (1020°C), then eight periods of InGaN/GaN QWs were grown at 800°C. After that, 50 nm p-type AlGaN and 150 nm GaN capping layer were grown at 1020°C in sequence to the top. A series of samples with similar structures, but mainly varied QW parameters (size, composition, and numbers), are selected to be included within this review chapter. Some improved structures, such as QWs with two different widths (dual wavelength) or inserted pre-strained layers, were also applied in certain samples involved in this chapter.

Several PL spectral techniques were used in the investigation of these LED devices. For example, a steady-state PL system with two monochrometers was used in measuring the steady-state PL/PLE spectra. A 450 W xenon lamp worked as the white light resource and a photomultiplier tube (PMT) was used as

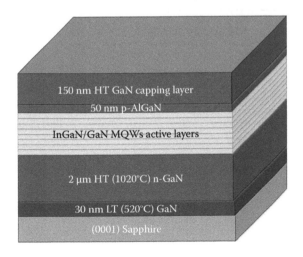

150 nm HT GaN capping layer

50 nm p-AlGaN

InGaN/GaN MQWs active layers

2 µm HT (1020°C) n-GaN

30 nm LT (520°C) GaN

(0001) Sapphire

FIGURE 9.1 Schematic of a typical InGaN/GaN MQW LED.

the photodetector. Another spectrometer equipped with a 375 nm picosecond pulse laser and a photon counter was used to measure the PL decay curves. The InGaN/GaN thin film samples were mounted in a cryostat with four UV-enhanced windows cooled by liquid helium for the purpose of measuring their PL spectra at temperature as low as ~10 K.

9.3 Overview: Typical PL Properties of InGaN/GaN MQWs

9.3.1 Steady-State PL and PLE Spectra

SS-PL and PLE spectra give us the direct view of excitation–emission properties of InGaN/GaN MQWs. In Figure 9.2, room temperature (RT) PL and PLE spectra of two InGaN-MQW samples with different indium composition (21% for sample RN9 and 24% for sample RN11) in InGaN active layers were shown in the same diagram [31]. Each PL spectrum was excited by a 325 nm (3.81 eV) light source split from a 450 W xenon lamp using a monochromater. The PL spectra for the samples RN9 and RN11 exhibit emission bands between 2.02 and 2.41 eV with multiple fine structures due to Fabry–Perot interference fringes. These emissions are identified as the yellow band (YB), which is defect related and correlated with the presence of threading dislocations and other related structural defect [32–34]. Main PL peaks related to InGaN emission locate at 2.51 eV for RN9 and 2.67 eV for RN11. The slight difference of peak position may be due to the bandgap variation of two samples. The weaker peaks at around 3.48 eV originate from band-to-band transitions of GaN layers.

PLE spectra detecting the InGaN-related PL maximum (2.51 eV for RN9 and 2.67 eV for RN11, respectively) were then examined. In each PLE spectrum, the contributions from the InGaN wells and the GaN barriers are clearly distinguishable, in which the sharp excitation peaks at 3.48 eV are ascribable to GaN band-to-band absorption and the broad bands at lower energy side are contributed by carrier generation in InGaN wells. This broadening of the PLE spectra with higher In fractions indicates that the absorption states are distributed over a wider energy range due to an increase in the degree of fluctuations in dot size and/or shape [35], or due to interface imperfection. When we moved the detected wavelength from PL maximum to the YB band, the InGaN-related broad PLE spectra were found weakened and the whole PLE spectrum was dominant by sharp 3.48 eV peaks. This indicates that YB emission mainly occurs at GaN barrier layers.

In order to analyze the Stokes shift related to InGaN emission, which is defined as the difference in energy between the effective bandgap and the emission peak energy, it is essential to have an accurate

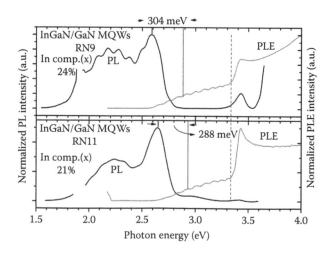

FIGURE 9.2 PL and PLE spectra of two InGaN/GaN MQWs, which demonstrate the Stark shift of the samples. (Reproduced from Lee, Z. S. et al., *Proc. SPIE*, 6669, 66690I–66690I-10, 2007. With permission by SPIE.)

description of the absorption edge that includes the effects of broadening. Martin et al. suggested fitting the PLE spectra by the sigmoidal formula [36,37],

$$\alpha = \frac{\alpha_0}{1 + e^{(E_{eff} - E)/\Delta E}} \qquad (9.1)$$

where

α_0 is a constant
E_{eff} is the effective bandgap
ΔE is the broadening parameter that indicates a distribution of absorption states
E is the excitation energy

We obtained E_{eff} values of 2.931 and 2.884 eV and ΔE values of 17.69 and 21.15 meV, for the samples with indium compositions of 21% and 24%, respectively. The Stokes shift between the PL peak energy and the "effective bandgap" ($E_{eff} - E$) was 288 and 304 meV, respectively, which indicates the large activation energy associated with the strong localization of the carriers in this MQW structure. Therefore, enlargement of this Stokes shift with increasing indium composition is ascribable to deeper localization states originating from indium fluctuations and the formation of clusters with high InN fractions.

9.3.2 Temperature-Dependent SSPL Spectra

The PL properties of InGaN/GaN MQWs change with temperature variation. Theoretically, the InGaN emission peak will monotonously redshift with increasing temperature because of the temperature-induced fundamental energy gap shrinkage of InGaN layers, which can be described by the Varshni empirical equation [38],

$$E_0(T) = E_0(0) - \alpha T^2/(\beta + T) \qquad (9.2)$$

where

$E_0(0)$ is the transition energy at 0 K
α and β are known as Varshni thermal coefficients

The real experiment results turned out that the TD-SSPL peak shift for the GaN barriers and the AlGaN capping layer was well consistent with this equation, whereas the InGaN-related PL emission did not follow the typical temperature dependence of the energy gap shrinkage. Figure 9.3 shows the evolution of

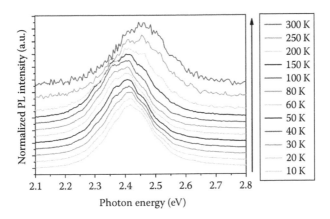

FIGURE 9.3 Temperature-dependent PL spectra of an InGaN/GaN MQW sample.

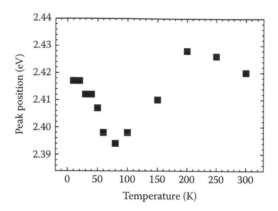

FIGURE 9.4 Peak position versus T of the InGaN/GaN MQW sample, which shows a typical "S-shaped" evolution.

InGaN-related PL spectra for the InGaN/GaN MQWs over a temperature range from 10 to 300 K. The spectra are normalized by maximum for better comparison. It can be found that they do not monotonously redshift from 10 to 300 K. Detailed indication of the peak position shifts is shown in Figure 9.4. As the temperature increases from 10 to 70 K, the emission maximum redshifts about 23 meV, much larger than the expected bandgap shrinkage of 4 meV over this temperature range [39]. In contrast, the PL peak blueshifts from 70 to 200 K. When the temperature further increases above 200 K, the peak position stops blueshifting, then redshifts again. Explanations have been given to these anomalous "S-shaped" behaviors [40]. The basic assumption was that, compared to the band-to-band transitions in GaN and AlGaN, carrier localization at deep traps originating from indium fluctuations or thickness variation of InGaN layers was the dominant pathway to emit photons in InGaN active layers. For 10–70 K, radiative recombination dominates the emission process. The carrier lifetime increases with increasing temperature, giving the carriers more opportunity to relax down into lower energy tail states before recombining. This behavior reduces the higher-energy side emission, and thus, produces a stronger redshift in the peak energy position with increasing temperature. For 70–200 K, since the dissociation rate is increased, other nonradiative processes become dominant. Thus, due to decreasing lifetime, these carriers recombine before reaching the lower energy tail states. This behavior enhances the higher-energy side emission and leads to a blueshift in the peak energy. Above 200 K, since the nonradiative recombination is the dominant process and the lifetimes are almost constant, the photogenerated carriers are less affected by the fast change of carrier lifetime so that the blueshift behavior becomes smaller and the peak position exhibits a redshift behavior again. Based on the these, "S-shaped" evolution of InGaN-related PL peak position is a sign of exciton localizations.

9.3.3 Two-Step PL Decays and Multiple Recombination Pathways

TRPL spectrum provides a good method for the direct measurement of exciton and carrier lifetimes in the InGaN/GaN QW structure. Analysis of the lifetimes is critical to understand the emission dynamics. In plenty of previous works, the obtained PL lifetimes were found diverse according to different well width and indium concentration [21,22,41–43]. Meanwhile, PL decays were found to deviate from single exponential decay: Fast decay times in the range of a few nanoseconds and slow decay times up to several hundred nanoseconds were derived from an obvious two-step PL decay process [44–47]. Three typical PL decay curves tested at 10, 120, and 300 K are shown in Figure 9.5. All of them were found to deviate from single exponential decay, and two decay stages are recognized clearly in each curve. Several explanations have been given by different researchers. Chichibu et al. [48] believed that non-Lorentzian distributions of localized states in energy space were the main courses to the decay curves deviating from

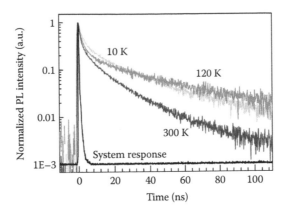

FIGURE 9.5 PL decay curves for the blue emission located at 2.68 eV tested at 10, 120, and 300 K. A 377 nm pulsed laser was used as the excitation source. All of them were found to deviate from single exponential decay, and two decay stages are recognized clearly in each curve. (Reprinted from *Mater. Lett.*, 173, Lin, T., Qiu, Z.R., Yang, J.R., Wei Ding, L., Gao, Y., and Feng, Z.C., Investigation of photoluminescence dynamics in InGaN/GaN multiple quantum wells, 170–173, Copyright 2016, with permission from Elsevier.)

single exponential decays, and a stretching parameter $\beta \sim 0.67$ was applied to fit the curves; Sun et al. [45] ascribed the fast decays to recombination at extended states and the slow decays to recombination at localized states, respectively. Anyhow, all of them indicated that multiple origins exist in InGaN/GaN QWs structures.

Here, we simply assumed that two dominant recombination pathways exist and contribute to the decay curves and used following bi-exponential function to fit the decay curves in the full range of 10–300 K:

$$\frac{I(t)}{I_0} = A_1 e^{-t/\tau_1} + A_2 e^{-t/\tau_2} \tag{9.3}$$

where
I_0 represents the PL intensity at t = 0
τ_1 and τ_2 represent the slow decay lifetime and the fast decay lifetime, respectively

The obtained prefactors A_1 and A_2 were found to be both constant and irrelated to temperature, which implied that the numbers of excitons or excited electron–hole pairs related to the slow and fast PL were independent of temperature. Based on this model, the PL decay curves were split into two exponential decays, and the corresponding decay lifetimes τ_1 and τ_2 versus temperature were summarized separately in Figure 9.6. As the temperature was raised from 10 to 120 K, the obtained slow lifetime increased. At temperatures greater than 120 K, the PL lifetime starts to decrease with decreasing temperature. It is already known that the decline of PL lifetime at high-temperature range is dominant by the increasing of nonradiative recombination rate. Accordingly, the increase of τ_1 in lower-temperature range implies that the radiative recombination rate for slow PL is not constant but decreases with increasing temperature, it is reasonable if the PL process is dominant through exciton localization, as the exciton localization rate is sensitive to temperature. Compared to slow PL lifetime, the fast lifetime is more stable with varying temperature, which exhibits only a small raise to 40 K, and monotonously decreases at 300 K. Lower-temperature value of 40 K associated with the peak lifetime than 120 K for slow PL lifetime may result from the higher sensitivity of nonradiative recombination rate to temperature.

Numerical analysis of the lifetime temperature relation is needed to visualize the role of exciton localization and nonradiative processes. Several theoretical equations have been used by Minsky et al. [49]

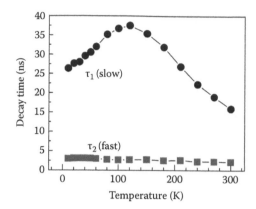

FIGURE 9.6 $\tau_1(T)$ (slow) and $\tau_2(T)$ (fast) obtained from fitting the decay time curves with biexponential decay. (Reprinted from *Mater. Lett.*, 173, Lin, T., Qiu, Z.R., Yang, J.R., Wei Ding, L., Gao, Y., and Feng, Z.C., Investigation of photoluminescence dynamics in InGaN/GaN multiple quantum wells, 170–173, Copyright 2016, with permission from Elsevier.)

and Chichibu et al. [48] to fit the lifetime data. We can extract the temperature dependence of localization lifetime and nonradiative recombination lifetime from measured TRPL results by solving the following equations:

$$1/\tau_{PL} = 1/\tau_{loc} + 1/\tau_{nr} \tag{9.4}$$

and

$$\eta_{PL} = k_{PL}/(k_{loc} + k_{nr}) = 1/(1 + \tau_{loc}/\tau_{nr}) \tag{9.5}$$

where
    k represents the rate
    τ_{nr} represents the nonradiative recombination lifetime for free excitons
    $\tau_{r,loc}$ represents the radiative recombination lifetime for localized excitons
    τ_{loc} represents the lifetime that free excitons were trapped by localized states [48], based on these assumptions that there was no free exciton emission and the nonradiative recombination lifetime for localized exciton is long enough

To completely solve these equations, the PL efficiency $\eta_{PL}(T)$ needs to be measured from TD-SSPL spectra. However, in these cases, η_{PL} corresponding to slow PL process and fast PL process is mixed in $I(T)/I_0$, making it deviate from single Arrhenius equation [50], as shown in Figure 9.7 the full squares. So a modified Arrhenius equation with two items was used to fit the $[I(T)/I_0 - T]$ spots in order to separate these two $\eta_{PL}(T)$:

$$\frac{I(T)}{I_0} = \frac{A_1}{1 + w_1 e^{-\frac{E_{act1}}{kT}}} + \frac{A_2}{1 + w_2 e^{-\frac{E_{act2}}{kT}}} \tag{9.6}$$

Note that Equation 9.6 shares the same prefactors A_1 and A_2 with Equation 9.3 as they represent the same mixed PL processes. Therefore, respective $\tau_{loc}(T)$ and $\tau_{nr}(T)$ were solved and plotted in Figure 9.8.

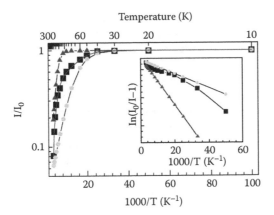

FIGURE 9.7 Temperature dependence of integrated PL intensity for the blue emission located at 2.68 eV (full squares). The intensity was normalized by the maximum at low temperature. These spots were fitted by

$$\frac{I(T)}{I_0} = \frac{A_1}{1 + w_1 e^{\frac{E_{act1}}{kT}}} + \frac{A_2}{1 + w_2 e^{\frac{E_{act2}}{kT}}}.$$ Full triangles represent $\eta_{PL}(T)$ obtained from the first item of aforementioned

equation, full circles represent $\eta_{PL}(T)$ obtained from the second item. The insert figure demonstrates $\ln(I_0/I - 1) - T^{-1}$ relations of the earlier spots. (Reprinted from *Mater. Lett.*, 173, Lin, T., Qiu, Z.R., Yang, J.R., Wei Ding, L., Gao, Y., and Feng, Z.C., Investigation of photoluminescence dynamics in InGaN/GaN multiple quantum wells, 170–173, Copyright 2016, with permission from Elsevier.)

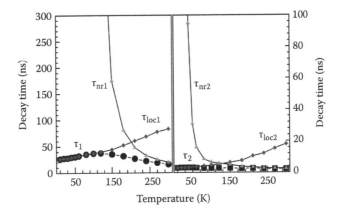

FIGURE 9.8 $\tau_{loc}(T)$ (full diamonds) and $\tau_{nr}(T)$ (full triangles) plots for slow PL process and fast PL process, respectively. They are obtained from solving $1/\tau_{PL} = 1/\tau_{loc} + 1/\tau_{nr}$ and $\eta_{PL} = k_{PL}/(k_{loc} + k_{nr}) = 1/(1 + \tau_{loc}/\tau_{nr})$. (Reprinted from *Mater. Lett.*, 173, Lin, T., Qiu, Z.R., Yang, J.R., Wei Ding, L., Gao, Y., and Feng, Z.C., Investigation of photoluminescence dynamics in InGaN/GaN multiple quantum wells, 170–173, Copyright 2016, with permission from Elsevier.)

Recombination pathway assignments can be done by detailed analysis to $\tau_{loc}(T)$ and $\tau_{nr}(T)$. It can be recognized that both τ_{loc1} and τ_{loc2} increase with temperature, which agrees with the earlier assumption that they both originate from exciton localization. Excluding the contribution of τ_{loc} to τ_{PL}, τ_{nr} for fast PL is obviously smaller than the slow one. Note that τ_{nr} represents nonradiative recombination lifetimes for free excitons, which are supposed to be identical if assuming that the circumstances for free excitons are not different. Therefore, the divergence of two τ_{nr} is ascribed to the fact that the occurrences of slow and fast PL process are spatially segregated in QW structure and can be treated as two independent processes. The obtained $\tau_{nr}(T)$ can be further fitted by $\tau_{nr}(T) = \tau_{nr\infty} \exp(-E_{act}/kT)$ to derive activation energy for nonradiative recombination, which reflects the ionization energy of surrounding nonradiative defects, or the average depth of localization states. It turns out that the E_{act} for slow and fast PL process are ~70 and ~20 meV, respectively, which also both accord the direct calculation from SSPL results fitted by Equation 9.6.

It has been reported that free excitons might be trapped in In-rich regions, obtaining radiative recombination with lifetimes in several tens of ns [40], which is also varied with indium component. In our case, the slow PL process generally accords the properties of exciton localization pathway associated with indium compositional fluctuations. Its relatively large τ_{nr} indicates that In-rich clusters are mainly distributed away from InGaN/GaN interfaces, where the density of nonradiative defect states is high. On the other hand, morphological fluctuations in the interfaces can lead to the formation of shallow radiative traps that induce fast trapping of free excitons [51], even where the indium composition is homogeneous. Their localization depths are relatively small, making the trapped excitons easily delocalized, so the behavior of excitons is more similar to free ones. Furthermore, large numbers of nonradiative defects, like the threading dislocations, exist in the InGaN/GaN interfaces [13,52]. The delocalized excitons have high chances to fall into nearby nonradiative defect states. Thus, it is explained why the τ_{nr} for the fast PL decay happens to be larger than the slow PL. Therefore, the origin of fast PL process is ascribable to morphological fluctuations in the interfaces.

9.3.4 Excitation Power and Emission Wavelength-Dependent TRPL

For a single InGaN/GaN MQW sample, its emission dynamic does not only depend on testing temperature. It also depends on several optical parameters related to PL measurements. One of these phenomena is that the emission lifetime was found to lengthen with increasing excitation power [53].

An InGaN/GaN MQW sample with the indium composition x = 0.22 was selectively excited by a pulse laser with an output wavelength at 400 nm (3.10 eV), which had a lower energy than the bandgap of GaN barriers. This selective excitation of the MQW structures allows to neglect the role of carrier diffusion from the GaN barrier into the InGaN regions, which highly simplifies the analysis of the data. As shown in Figure 9.9, the emission lifetime is observed to vary significantly as the laser intensity is varied over two orders of magnitude. The emission lifetime was found to decrease as the laser power was decreased. The observed change in emission lifetime was attributed to the saturation of recombination centers in

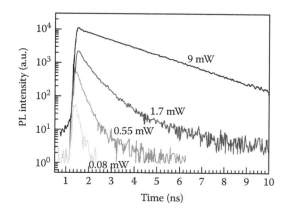

FIGURE 9.9 The intensity dependence of the PL decays for MQW measured at 480 nm (2.58 eV) for four different laser powers. The emission lifetime is found to increase with laser power. This is attributed to the saturation of recombination centers. (Reprinted with permission from Pophristic, M., Long, F.H., Tran, C., Karlicek, R.F., Feng, Z.C., and Ferguson, I.T., Time-resolved spectroscopy of InxGa₁₋ₓN/GaN multiple quantum wells at room temperature, *Appl. Phys. Lett.*, 73, 815–817 Copyright 1998, American Institute of Physics.)

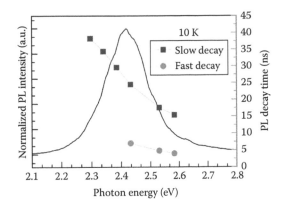

FIGURE 9.10 Photon energy dependence of PL decay times guided by PL peak. The slow decay time and fast decay time were derived from biexponential decay function. The fractions of fast decay are too small at low energy side, so only slow decay time can be obtained.

the MQW. At high excitation power, the localization states are almost fully saturated and cannot serve as efficient recombination centers. At considerably lower powers, these states are not fully populated. Excited carriers prefer to fill the localization states with smaller localization lifetime, so it significantly reduces the carrier lifetime as observed.

Besides excitation power, the InGaN-related PL lifetime was also found to depend on the detected wavelength (emission photon energy). It can be seen in Figure 9.10 that one InGaN-related PL has a main broad peak ranged from 2.25 to 2.65 eV measured at 10 K. A series of PL decay curves associated with different photon energy in this range were measured. Then each decay curve was fitted with biexponential decay to obtain two (fast and slow) lifetimes. It is worth noting that the prefactors A_1 and A_2, associated with the ratio of fast and slow decay, were found to be varying for different photon energy. The fraction of fast decay decreases with decreasing photon energy, which implies that fast decay is dominant at the high-energy side, whereas the slow decay is dominant at low-energy side.

The PL lifetime values increase with decreasing photon energy for both fast and slow decays, which is due to the energy transfer from a higher-energy localized state to a lower one. This is characteristic of the localized system, where the decay of excitons consists of both radiative recombination and the transfer process to tail states. The depth of localization can be evaluated by assuming the exponential distribution of the density of tail states and by fitting the photon energy dependence of the τ_{PL} values using the following equation [54,55]:

$$\tau_{PL} = \frac{\tau_{rad}}{1 + e^{(E - E_{me})/E_0}} \tag{9.7}$$

where τ_{rad}, E_{me}, and E_0 are the radiative lifetime, the energy similar to the mobility edge, and the depth of localization, respectively. The E_0 for slow decay was ~65 meV, which well agreed with the E_{act} value obtained in Section 9.3.3.

9.3.5 Barrier Width Modulation

It is already known that the emission wavelength of InGaN active layer is directly controlled by indium composition because of the bandgap modulation of InGaN alloy. As a matter of fact, the dimensions of MQW structures, for example, the GaN barrier widths, also have strong influences on PL properties, even with fixed indium composition. The exact mechanism for this modulation is still unclear and is considered complicated, as the possible causes include quantum confinement effect, strain fluctuation, and inter-well carrier transitions.

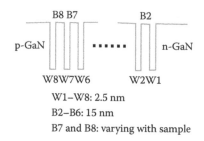

W1–W8: 2.5 nm
B2–B6: 15 nm
B7 and B8: varying with sample

FIGURE 9.11 Structures of the InGaN/GaN MQWs. The barrier width B7 and B8 are the only differences among the samples.

A series of InGaN/GaN samples were designed for investigating the barrier width modulation for their PL dynamic properties. Their designed structures are illustrated in Figure 9.11. Eight periods of InGaN/GaN quantum well layers were grown on GaN layer. The well lengths (W1–W8) and the barrier widths (B1–B6) of the samples are 2.5 and 15 nm. The barrier widths (B7 and B8) are 2.5, 4, 7.5, 15, and 30 nm, labeled as QW2585, QW2584, QW2583, QW2579, and QW2591, respectively.

Figure 9.12 gives the PL spectra of the five samples pumped by 325 He–Cd laser. The laser was focused on around 250 nm p-GaN layer upon the variational barrier layers. The emission peak, as the PL spectra shows, has a blueshift with the barrier thickness increasing across the five samples from 2.5 to 25 nm; the larger shift energy is around 40 meV, unlike the redshift in other groups' work [56]. This blueshift can be ascribed to the suppression of wave function expansion of carriers to barrier layer and better quantum confinement. If there is a strain fluctuation in the samples, the blueshift of emission peak position could also result from QCSE due to the large lattice mismatch between InGaN well layers and GaN barrier layers [57]. That is to say, strain-induced QCSE narrows down the normal bandgap of InGaN well, which leads to redshifted emission. If GaN barrier layers become thick, and the strain is partially released, which makes the QCSE weaker, blueshift of PL peak occurs. This effect can be easily examined by changing the excitation power, as excessive carriers will screen the piezoelectric field by QCSE, leading to a blueshift of PL peak.

TRPL was performed using a wavelength tunable picoseconds pulsed laser system consisting of an optical parametric amplification (OPA), mode locked Nd:YAG laser as an excitation source,

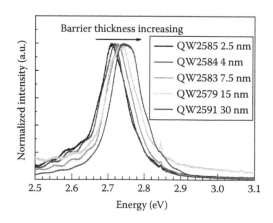

FIGURE 9.12 SSPL spectra for InGaN/GaN MQWs structure with barrier thickness of B7 and B8 range from 2.5 to 25 nm.

TABLE 9.1   Fast and Slow Decay Time Obtained from Different MQW Samples

Sample No.	B7 and B8 Barrier Thickness (nm)	Fast Decay Time (ps)	Slow Decay Time (ps)
QW2585	2.5	241.7	764.8
QW2584	4	91.6	593.9
QW2583	7.5	77.5	787.4
QW2579	15	137.2	688.6
QW2591	30	80.7	643.3

and a streak camera for detection. The wavelength, repeated frequency, and pulse width of output picoseconds laser are 325 nm, 10 Hz, and 25 ps, respectively. Then the decay curves were fitted by biexponential decay to obtain two (fast and slow) lifetimes, and the fitting results are listed in Table 9.1. The change of the decay lifetime of each sample is not evident as shown in the table. The fast lifetime and slow lifetime related to different recombination pathways seem not to be directly influenced by barrier thickness fluctuation. This phenomenon might be due to strain-induced piezoelectric field, which brings dominant control to the PL dynamics [58]. When the thickness of the barrier layers is thin enough compared to well layer, the wave function of the electrons and holes will penetrate the adjacent well layer, which leads to interwell transition happening in samples QW2585 and QW2584 with the barrier thickness of 2.5 and 4 nm, respectively. The interwell transition reduces the changes of intersubband transition, which should have meant to shorten decay lifetime. But in our case this effect is much weaker than the effect that strain-induced piezoelectric field reduces the energy of the optical transition and radiative recombination rate, which results in a redshift of PL peak and longer decay lifetime.

9.4 Photoluminescence Dynamics in Dual-Wavelength InGaN/GaN MQWs

9.4.1 CART InGaN/GaN MQW LED Structure

Recently, a type of InGaN/GaN MQW LEDs has been reported, which consists of the so-called charge asymmetric resonance tunneling (CART) structure [24,59–61]. The CART structure is to insert a wide electron emitter layer of quantum wells and a thin electron tunneling barrier between the MQW active region and the n-cladding layer to the LED. Such a structure can achieve a large electron capture rate and a large carrier confinement effect simultaneously. In fact, CART-LEDs are supposed to achieve dual wavelength emissions and better device performance. In this section, we apply the CART structure to achieve the dual light wavelength LEDs with green and yellow emission, in addition to the blue GaN emission. The optical properties of the In-rich InGaN/GaN multiple quantum well structures with CART are studied by TD-SSPL, PLE, and TRPL, to investigate the energy transfer between two different active layers.

The samples were grown on sapphire (0001) substrates by MOCVD method, similar to the one of standard InGaN/GaN MQW. The different part is that the active light emitting layers consist of six pairs of InGaN well (3.3 nm thick and grown at 846°C) and InGaN barrier (Si-doped, 8.5 nm thick, and at 945°C), plus next two periods of 7 nm GaN barrier and 2 nm InGaN wells. The schematic diagram of the InGaN/GaN CART LEDs is shown in Figure 9.13. This CART structure could decrease effectively carrier leakage from the wells and obtain the dual-wavelength emissions. Due to the CART structure, injected electrons are effectively captured by electron emitter layer, since it has six periods of InGaN wells with larger barrier width and narrower well layers in the MQW active region. The captured electrons can subsequently

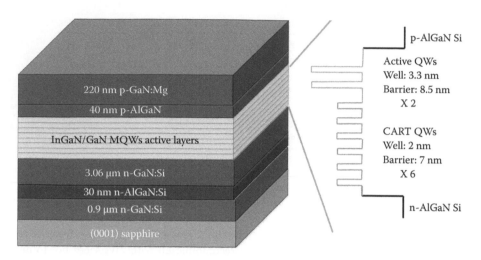

FIGURE 9.13   Schematic structure of the CART InGaN/GaN MQW LED.

tunnel through the barrier into the MQW active region (two QWs with wider well width and higher well In-composition in comparison with the six-QWs) partially so as to reach a larger electron capture rate and to obtain dual-wavelength emissions.

9.4.2 Steady-State PL Spectra

Two CART InGaN/GaN LED samples were used in the TD-SSPL measurements, which have similar designed structural parameters, but are under different well growth temperature (797°C for sample GF03 and 782°C for sample GF195, respectively).

The TD-SSPL spectra were measured at temperatures of 10–300 K, as shown in Figure 9.14. Two InGaN-related emission bands are seen with the peak energy (wavelength) at 2.36 eV (green, 525 nm) and 3.01 eV (blue, 410 nm) for sample GF03 while 2.11 eV (yellow, 590 nm) and 2.87 eV (green, 435 nm) for sample GF195, respectively. The blue emissions are found much weaker than the green and yellow bands. This may be due to the fact that the CART QWs allow the enhancement of the electron amount captured into the active layer. The profiles of the PL spectra in two representative samples do not change significantly with temperature. Fabry–Perot oscillations can be clearly observed in these PL spectra, especially in high temperature. This phenomenon implies that the interfaces of the heterojunctions are uniform. The dominant PL peak positions of these samples were evaluated by fitting the PL spectra with Gaussian curves. All emission bands show "S-shaped" behavior, that is, redshift–blueshift–redshift of the emission energy with increasing temperature. The fitting results have shown two features. First, the blueshift of the green/yellow peak position is much more obvious than that of the blue peaks for both samples with increasing temperature. This can be explained by stronger exciton localization effect in the active QWs. The reason is that lower well growth temperature results in higher indium composition and the increase of composition fluctuation in the InGaN MQW region. Second, the redshifts in CART QWs are more obvious than blueshift in these two samples. Possibly this is due to the fact that the carriers injected into the CART QWs tunnel into the active QWs mostly, which reduced the free-carrier Coulomb screening effect, predicting the stronger carrier localization effect.

Blueshifts of both green/yellow and blue peak position of samples GF03 are found stronger than the ones of GF195. This indicates that the localization effect of sample GF03 with a lower well growth temperature of 782°C is stronger than that of sample GF195 with a higher well growth temperature of 797°C. This result shows that the decrease of well growth temperature results in higher indium composition and also the increase of composition fluctuation in the InGaN MQW region, leading to the stronger carrier localization effect.

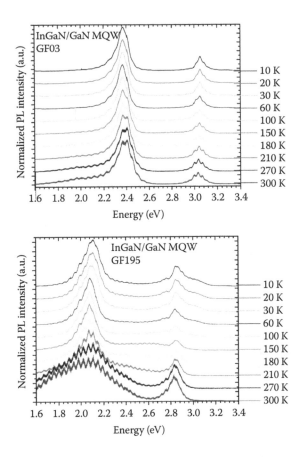

FIGURE 9.14 PL spectra of samples GF03 and GF195 at temperatures of 10–300 K. (Reprinted from *Thin Solid Films*, 529, Feng, Z.C., Zhu, L.H., Kuo, T.W., Wu, C.Y., Tsai, H.L., Liu, B.L., and Yang, J.R., Optical and structural studies of dual wavelength InGaN/GaN tunnel-injection light emitting diodes grown by metalorganic chemical vapor deposition, 269–274, Copyright 2013, with permission from Elsevier.)

9.4.3 TRPL and Energy Transfer

To further clarify the emission properties of CART InGaN/GaN MQWs, TRPL measurements were taken for two samples GF03 and GF195, respectively. The TRPL spectra for sample GF03 were measured at room temperature (RT) with detecting wavelength at λ = 515 (2.408 eV) and 550 nm (2.255 eV), respectively, while sample GF195 with detecting wavelength at λ = 565 (2.19 eV) and 590 nm (2.101 eV), respectively. TRPL measurements were done by varying the detection wavelength across the InGaN-related emission band. The photon energy dependences of τ_{PL} at RT are shown in Figure 9.15a and c for sample GF03 green part and sample GF195 yellow part, respectively. The τ_{PL} values for green/yellow parts increase with decreasing photon energy in two samples. This energy dependence was due to the energy transfer from higher-energy localized state to a lower one and is characteristic of the localized system, where the decay of excitons consists of both radiative recombination and the transfer process to tail states. The depth of localization can be evaluated by assuming the exponential distribution of the density of tail states and by fitting the photon energy dependence of the τ_{PL} values from Equation 9.7. The obtained E_0 = 47.5 meV for sample GF03, while E_0 = 64.57 meV for sample GF195. As a result, sample GF195 has a deeper localization depth and a longer lifetime. This clearly shows that the radiative lifetime and localization depth decreased with the enhancement of well growth temperature (sample GF03).

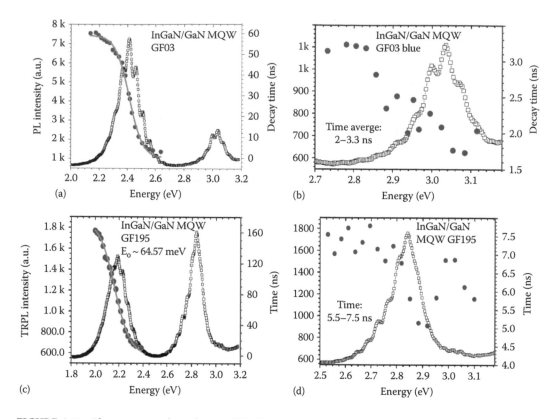

FIGURE 9.15 Photon energy dependence of PL lifetime at room temperature. (a) GF03 green part, (b) GF03 blue part, (c) GF195 yellow part, and (d) GF195 blue part. (Reproduced from Feng, Z.C. et al., *Proc. SPIE*, 7058, 70580S–70580S-12, 2008. With permission by SPIE.)

Therefore, it is reasonable to conclude that the growth with a lower well growth temperature could cause the deposition (composition) of indium better (higher).

The corresponding τ_{PL} for blue parts were also measured and shown in Figure 9.15b and d. It can be clearly seen that for both samples, blue emission had much smaller τ_{PL} and E_0 as compared to the green/yellow ones. This can be ascribed to strong delocalization in shallow wells and exciton energy transfer from shallow wells to deep wells, which results in much smaller nonradiative activation energy and recombination decay lifetimes as well.

Another three dual-wavelength samples were measured to further investigate the energy transfer effect in CART InGaN/GaN MQW LEDs. The growth temperature was similar to these three samples, but the indium content in the three samples increased gradually. Two emission peaks (low-In-content and high-In-content) were located around 406 and 496 nm for sample A, 415 and 530 nm for sample B, and 430 and 575 nm for sample C.

In order to investigate the detailed carrier transfer process and recombination dynamics in the samples, TD-TRPL curves of high-In-content MQWs for samples A, B, and C were measured at the peak photon energy ranging from 10 to 300 K, as shown in Figure 9.16. A delayed-rise part is clearly observed in the temporal profiles of samples A and B, which corresponds to an absorption process. The results indicate that carriers generated in the low-In-content well layers after photo-excitation can be effectively transferred to the high-In content part through reabsorption process [62]. As temperature rises, the delayed-rise section gradually disappears, which means that the transfer efficiency from low-In-content MQWs to high-In-content part decreases. The signal-to-noise ratio of TRPL curves for sample C is pretty

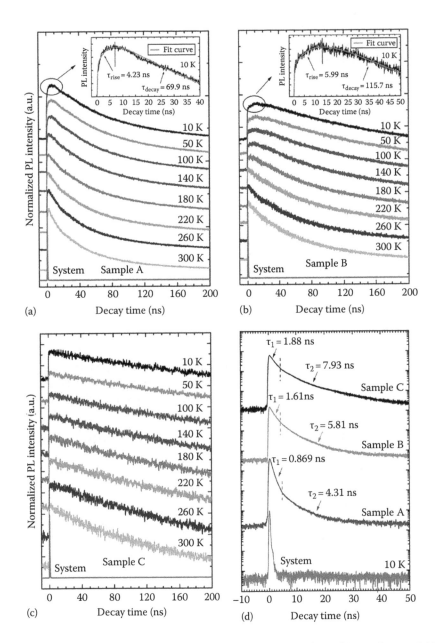

FIGURE 9.16 TRPL decay curves of high-In-content MQWs for (a) sample A, (b) sample B, and (c) sample C, measured at the PL peak photon energy over the temperature range from 10 to 300 K. The insets in (a) and (b) show the enlarged temporal profiles of samples A and B at 10 K, respectively. (d) 10 K TRPL decay curves of low-In-content MQWs for three samples measured at the peak photon energy. (Reprinted with permission from Liu, L., Wang, L., Liu, N., Yang W., Li, D., Chen, W., Feng, Z.C., Lee, Y.C., Ferguson, I., and Hu, X., Investigation of the light emission properties and carrier dynamics in dual-wavelength InGaN/GaN multiple-quantum well light emitting diodes, *J. Appl. Phys.*, 112, 083101 Copyright 2012, from American Institute of Physics.)

low as the longest delay time makes the PL intensity fall fast. Therefore, no obvious delayed-rise stage is observed for sample C. Nevertheless, the active transfer of generated carriers from low-In-content region to high-In-content region via reabsorption can induce an enhancement of luminescence intensity in long-wavelength light emission.

To further understand the interaction between two sets of different indium content MQWs, the following exponential model was used to fit the 10 K TRPL curves of high-In-content MQWs for samples A and B:

$$\frac{I(t)}{I_0} = A_1 \left[1 - e^{-\frac{t}{\tau_{rise}}} \right] + A_2 e^{-\frac{t}{\tau_{decay}}} \tag{9.8}$$

where

τ_{rise} represents the carriers transfer time from low-In-content region to high-In-content region
τ_{decay} represents the carrier decay time in high-In-content MQWs

The 10 K TRPL decay curves of low-In-content MQWs of these three samples are also measured at peak photon energy and shown in Figure 9.16d, which involve two obvious decay stages. So biexponential decay fitting (Equation 9.3) was applied to these curves and obtained two (slow and fast) decay lifetimes. The fitting results turned out that the rise time τ_{rise} in high-In-content MQWs is consistent with the slow decay lifetime in low-In-content MQWs, which confirms that there is a carrier transfer process (slow decay stage) besides the radiative recombination (fast decay stage) in the low-In-content active layer. In addition, it is clear that the fast carrier lifetime gradually increases for these three samples, indicating a reduction of transfer rate from the low-In-content MQWs to the high-In-content part. The decrease of transfer efficiency will bring the IQE of high-In-content MQWs down as the emission wavelength increases further.

9.5 InGaN/GaN MQWs Grown on Semipolar Substrates

9.5.1 Introduction on Semipolar and Nonpolar MQWs

The crystalline structure of wurtzite GaN is shown in Figure 9.17, which reveals the relations of $(1\bar{1}01)$, $(11\bar{2}2)$ and (0001) planes. The ABCD, GHCD, BLFC, and AEFC represent the nonpolar $(1\bar{1}00)$ (m-), semipolar $(1\bar{1}01)$, nonpolar $(11\bar{2}0)$ (a-), and the semipolar $(11\bar{2}2)$ planes, respectively. Figure 9.18 shows the crystal orientation dependence of polarization field in well layer; 0° and 90° represent the c-plane and the nonpolar plane, respectively. And there is a zero electric field position around 45°, besides 90°. Calculation results show that the total internal field for the semipolar sample is much weaker than the polar sample. Conventional c-plane quantum wells (QWs) suffer from strong QCSE as a result of the existence of spontaneous and strain-induced piezoelectric polarization fields parallel to (0001) c-direction [63] grown on c-plane sapphire substrate. This effect causes spatial separation of electrons and holes in QWs active regions of the optical devices, which therefore restricts the radiative recombination efficiency and induces the blueshift photon emission at a high current injection [64,65]. To overcome these inherent material problems and to push IQE to the limit, nonpolar/semipolar QWs are used to completely or partially eliminate the QCSE and to enhance the performance of LEDs.

However, the difficulty in the growth of nonpolar films lies in the planar anisotropic nature of the growing surface because of the anisotropic in-plane strain and the adatom diffusion length. Such a structural anisotropy may lead to anisotropic electrical and optical characteristics [66]. Alternatively, lateral epitaxial overgrowth (LEO) method is a promising technique in creating semipolar $(1\bar{1}01)$ facets on c-plane polar substrate [67]. Such semipolar facets own two advantages. For one thing, a reduced QCSE has been observed in $(1\bar{1}01)$, $(11\bar{2}2)$ semipolar QWs [68–71]. For another, the indium incorporation rate

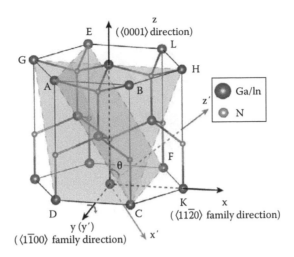

FIGURE 9.17 The relations of GaN wurtzite crystal planes, GHCD and AEFC, represent (1 $\bar{1}$01) and (11 $\bar{2}$2) semipolar planes, respectively. (Reprinted with permission from Zeng, F., Zhu, L., Liu, W., Li, X., Liu, W., Chen, B.J., Lee, Y.C., Feng, Z.C., and Liu, B., Carrier localization and phonon-assisted hopping effects in semipolar InGaN/GaN light-emitting dioses grown by selective area epitaxy, *J. Alloys Compd.*, 656, 881–886 Copyright 2016, American Institute of Physics.)

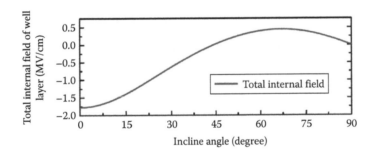

FIGURE 9.18 Crystal orientation dependence of total internal field of well layer. (Reprinted with permission from Zeng, F., Zhu, L., Liu, W., Li, X., Liu, W., Chen, B.J., Lee, Y.C., Feng, Z.C., and Liu, B., Carrier localization and pho-non-assisted hopping effects in semipolar InGaN/GaN light-emitting dioses grown by selective area epitaxy, *J. Alloys Compd.*, 656, 881–886 Copyright 2016, American Institute of Physics.)

of semipolar is higher than nonpolar GaN epilayer, which provides the possibility of enhanced radiative efficiency in green-gap region [72]. Among the semipolar facets, the advantages of (1 $\bar{1}$01) planes are their thermodynamic stability under the InGaN growth conditions and highly uniformed QWs grown on these planes. Besides, semipolar (1 $\bar{1}$01) facets forming an angle of 62° to the (0001) plane exhibit reduced polarization field and the resulting QCSE [71].

9.5.2 (1 $\bar{1}$01)-Semipolar Facet

In this section, through investigating the PL dynamic properties of the semipolar (1 $\bar{1}$01) InGaN/GaN QWs, the emission mechanism of semipolar GaN-based LEDs was explored. Semipolar (1 $\bar{1}$01) InGaN/GaN MQWs were grown on the (1 $\bar{1}$01) GaN/sapphire substrates in a 3 × 2 in. Thomasswan low-pressure MOCVD system. The epitaxial structure was grown on n-GaN templates consisting of a 2 μm thick Si-doped n-type GaN layer and a 1 μm thick undoped GaN layer on a c-plane sapphire substrate by MOCVD. After growth, ~200 nm thick SiO$_2$ mask film was then deposited on the n-GaN and patterned into stripes oriented in the

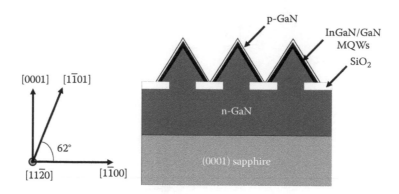

FIGURE 9.19 Schematic structure of the semipolar LED. (Reproduced from Zhu, L., Zeng, F., Liu, W., Feng, Z., Liu, B., Lu, Y., Gao, Y., and Chen, Z., Improved quantum efficiency in semipolar (1 $\bar{1}$01) InGaN/GaN quantum wells grown on GaN prepared by lateral epitaxial overgrowth, *IEEE Trans. Electron Dev.*, 60, 3753–3759 Copyright 2013 IEEE Press.)

[1 $\bar{1}$20] direction of GaN. The mask is stripe shaped with a 4 μm wide opening (window) and a 10 μm wide stripe (wing), using photolithography. After that the patterned GaN template was loaded into MOCVD reactor again for epitaxial growth. First, n-type GaN was grown selectively on the window area via the GaN seed layer to form a triangular bar. The selective area growth of GaN was then applied. After that LEO-GaN triangular stripes with (1 $\bar{1}$01) side facets and a (0001) face at the top were achieved. The bottom width was larger than the window width because of the lateral overgrowth. Then, five periods of InGaN/GaN MQWs were grown on the whole surface. Finally, on the top of the MQWs was grown an Mg-doped p-type AlGaN electron-blocking layer followed by a p-type GaN. The schematic structure of this semipolar LED sample is shown in Figure 9.19. A conventional MQW LED sample on (0001) sapphire substrate having a similar structure with the semipolar LED was also prepared as the control.

The semipolar LED sample had a PL peak at 433 nm (2.86 eV), showing a blueshift of ~30 nm with respect to the reference sample grown at the same temperature and pressure. The TRPL measurements at 10 K detecting this peak wavelength were performed to investigate the effects of internal polarization field on the carrier recombination mechanism (Figure 9.20). Because the measurement was carried out at 10 K,

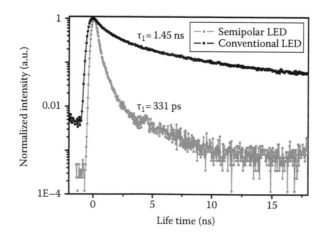

FIGURE 9.20 Low-temperature TRPL of the InGaN/GaN MQWs and (0001) InGaN/GaN MQWs. (Reproduced from Zhu, L., Zeng, F., Liu, W., Feng, Z., Liu, B., Lu, Y., Gao, Y., and Chen, Z., Improved quantum efficiency in semipolar (1 $\bar{1}$01) InGaN/GaN quantum wells grown on GaN prepared by lateral epitaxial overgrowth, *IEEE Trans. Electron Dev.*, 60, 3753–3759 Copyright 2013 IEEE.)

the influence of the nonradiative recombination process was excluded. The experimental data exhibit a non-exponential decay shape, which is a characteristic of the emission from disorder quantum structures such as the localized states [73]. Therefore, the TRPL results can be fitted with a biexponential decay function (Equation 9.7). Usually, the PL intensity is limited by the fast decay component, so we labeled fitted τ_1 results near the curves in Figure 9.20. As shown in Figure 9.20, the fitted lifetimes are $\tau_1 = 331$ ps and 1.45 ns for semi-polar and conventional LED, respectively, indicating that semipolar LED's lifetime is shorter than c-plane's at low temperature. Typically, the reduction of PL decay time could be attributed to the reduction of internal electric field that leads to the increase of electron-hole pair recombination probability [74]. This strongly implies that the highly efficient radiation is more favored in semipolar LED than that in c-plane LED.

The carrier lifetime's evolution with the temperature for the samples usually reveals the specific radiative recombination mechanism. To further analyze the origin of improved IQE of luminescence, experimental temperature-dependent carrier lifetimes were used to investigate the radiative and nonradiative behavior during the thermal processes using the relation of Equations 9.4 and 9.5. Since the overall PL intensity is limited by the fast decay component, here the fast decay time τ_1 is used to represent τ_{PL}. On the other hand, the IQE of luminescence η_{PL} was estimated by TD-SSPL spectra, assuming $\eta_{PL} = 100\%$ at T = 10 K. Thus, τ_r and τ_{nr} can be deduced from τ_{PL} and η_{PL} at different temperatures, as shown in Figure 9.21a and b.

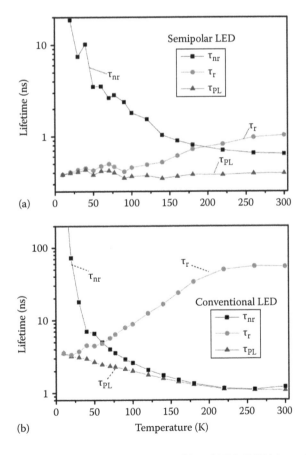

FIGURE 9.21 PL lifetimes τ of the MQWs on c-plane GaN (b) and LEO-GaN (a) as a function of temperature. Values of τ_{nr} and τ_r derived from τ_{PL} and corresponding integrated PL intensity are also plotted. (Reproduced from Zhu, L., Zeng, F., Liu, W., Feng, Z., Liu, B., Lu, Y., Gao, Y., and Chen, Z., Improved quantum efficiency in semipolar (1 $\bar{1}$01) InGaN/GaN quantum wells grown on GaN prepared by lateral epitaxial overgrowth, *IEEE Trans. Electron Dev.*, 60, 3753–3759 Copyright 2013 IEEE.)

We found that the transition from radiative to nonradiative recombination occurs at ~40 and ~100 K for the conventional and semipolar LED, respectively. Consequently, radiative recombination is dominant under the temperature of ~40 and 100 K. It is clear that nonradiative lifetimes in conventional LED drop drastically with temperature <40 K. As shown in Figure 9.21, the critical temperature, at which τ_{nr} begins to dominate τ_{PL}, for MQWs on semipolar GaN is higher than that for MQWs on c-plane GaN. This result confirms the improvement of IQE. The shorter lifetime in the semipolar QWs reflects the reduced spatial separation of electron–hole for the MQW emission in the semipolar LEO-MQWs, which could be attributed to the lower internal polarization field in semipolar QWs to c-plane QWs. The reduced spatial separation of electron–hole improved IQE for semipolar LEO-MQWs, which is consistent with the results of the temperature-dependent PL intensity. The TRPL data confirm again that the suppression of QCSE induced by such a large internal polarization field can potentially improve the luminescence properties of the InGaN QWs greatly.

9.5.3 (11$\bar{2}$2)-Semipolar Facet

Besides semipolar (1$\bar{1}$01) InGaN/GaN QW LEDs, lateral epitaxial overgrowth (LEO) method was also performed to prepare InGaN/GaN QWs on (11$\bar{2}$2) facet of GaN crystals [75]. The samples were synthesized by MOCVD method similar to the one mentioned in Section 9.5.1. The morphology of lateral epitaxial overgrowth InGaN/GaN MQWs has the shape of a prism with triangle in the cross section, as seen on SEM image in Figure 9.22. The triangles are about 5 μm in height and 7 μm in width. Based on the prism shape of MQWs, three optical detection points, A, B, and C, were defined as shown in Figure 9.23a, to perform room temperature (RT) Micro-PL measurements (with focusing size of 1–2 μm).

Figure 9.21b demonstrates the Micro-PL spectra acquired at room temperature. All the three spectra were normalized by peak value at around 3.4 eV, which relates to GaN substrate emission. These peaks for point A and point B are found slightly blueshift as compared to the one for point C. It may be ascribable to different orientations of GaN substrates. A broad emission band is found in the range of 2.2–2.9 eV for

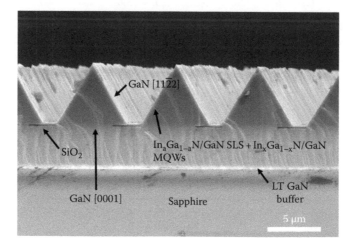

FIGURE 9.22 The morphology of epitaxial lateral overgrowth InGaN/GaN MQWs on (11$\bar{2}$2) facet of GaN, which has the shape of a prism with triangle in the cross section as seen on SEM image. (Reproduced from Huang, J.L. et al., *Proc. SPIE*, 8123, 81230C–81230C-11, 2011. With permission by SPIE.)

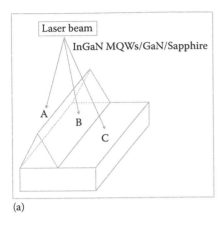

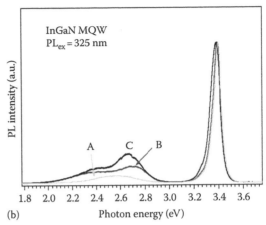

FIGURE 9.23 (a) Three optical detection points named A, B, and C for Micro-PL measurement. (b) Room temperature Micro-PL spectra. (Reproduced from Huang, J.L. et al., *Proc. SPIE*, 8123, 81230C–81230C-11, 2011. With permission by SPIE.)

point C, which may belong to the emission of InGaN epitaxial layer on (0001) facet of GaN, or surface defect states of the GaN layer. Interestingly, this emission band broadens and splits into two PL peaks at 2.4 and 2.7 eV for both point A and point C, which focus on the InGaN wells grown on (11$\bar{2}$2) facet of GaN. The PL intensities are also found to be obviously increased. This PL peak broadening and splitting were owing to spatial distribution of indium and characterized the luminescence color of the present (11$\bar{2}$2) MQWs. Theoretical references have indicated that InGaN/GaN MQWs with a crystalline tilt should have suppressed internal polar electrical fields [76], which leads to higher emission efficiency and the enhancement of InGaN-related PL intensities.

TRPL measurements were first performed at 10 K for different detection energy to examine the temporal dynamics of the luminescence. The 10 K TRPL of the sample was measured from 2.1 to 2.8 eV separated into 10 points. The photon energy dependence of τ_{PL} at 10 K is shown in Figure 9.24a. The τ_{PL} values are found to increase with decreasing photon energy in these samples, which is due to the energy transfer from higher-energy localized state to a lower one. This is characteristic of the localized system. By using Equation 9.7 to fit the temperature dependence of InGaN-related PL decay lifetimes, the depth of localization E_0 can be evaluated as ~88 meV, which is obviously larger than the ~60 meV of the counterpart sample grown on (0001) polar facet of GaN. The temperature dependence of τ_{PL} in the temperature range from 10 to 300 K is shown in Figure 9.24b. As the temperature was raised from 10 to 50 K, τ_{PL} increased. At temperatures greater than 50 K, τ_{PL} started to decrease with decreasing temperature. The decline of PL lifetime at high-temperature range is dominant by the increase in nonradiative recombination rate. So the increasing of τ_{PL} in lower-temperature range implies that the radiative recombination rate for slow PL is not constant but decreases with increasing temperature, which is reasonable if the PL process is dominated by exciton localization, as the exciton localization rate is sensitive to temperature. It is worth noting that τ_{PL} peaks at 50 K, which is much smaller than 120 K of the counterpart sample grown on (0001) GaN substrate. This indicates that nonradiative recombination processes have stronger influences over this semipolar sample, which may be ascribable to the lower quality of InGaN crystals and higher defect state density grown on semipolar substrates. Even so, the larger localization depth also indicates the better IQE of semipolar (11$\bar{2}$2) InGaN/GaN MQW sample than the conventional InGaN/GaN MQW sample as a result of partial suppression of polar electric fields.

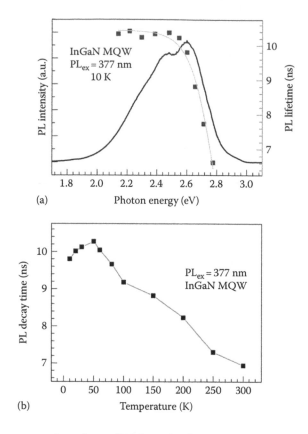

FIGURE 9.24 (a) Photon energy dependence of PL lifetime for the InGaN/GaN MQW. (b) Temperature dependence of PL lifetime. (Reproduced from Huang, J.L. et al., *Proc. SPIE*, 8123, 81230C–81230C-11, 2011. With permission by SPIE.)

9.6 InGaN/GaN MQWs with Pre-Strained Layer

9.6.1 MQW LED Structure with or without Pre-Strained Layer

It is already known that QCSE in InGaN/GaN QWs is a serious problem for c-plane LEDs, as the strong strain-related piezoelectric field induced by the large lattice mismatch between InGaN and GaN layers can strongly suppress the radiative recombination rate in the QWs and lead to a reduced IQE. Besides the replacement of c-plane polar GaN by semipolar or nonpolar GaN substrates, another strategy is to insert a pre-strained layer (or underlying layer) into the MQW structure, for the purpose of partially releasing the strain existing on the InGaN/GaN interface, as well as reducing the disadvantageous piezoelectric field. Recently, the pre-strained growth technique has attracted special interest for its significant improvement on the emission efficiency of InGaN/GaN MQWs [77,78]. As reported, the pre-strained InGaN layer deposited before the growth of subsequent InGaN/GaN QWs can create a tensile strain in the GaN barrier layer, which is beneficial for the enhancement of the incorporation of larger-sized indium atoms in the InGaN/GaN QWs; moreover, the QCSE of QWs in the pre-strained samples can be significantly reduced so that the emission efficiency will be improved.

In this section, InGaN/GaN MQW samples—with and without the introduction of pre-strained layer—were fabricated by standard MOCVD method [79]. As seen in Figure 9.25, the structure of sample without pre-strained layer (labeled as QW1505) is similar to a conventional MQW sample mentioned

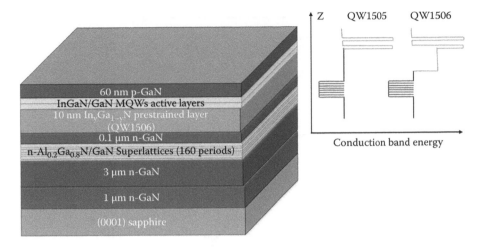

FIGURE 9.25 Schematic structures of the MQW LED samples with (QW1506) and without (QW1505) pre-strained layer.

in Section 9.2. Another sample named QW1506 has the same indium content with QW1505. The only difference between these two samples is that a thick InGaN layer with lower indium content was grown prior to the first InGaN well upon n-GaN layer in QW1506.

9.6.2 TRPL Properties and Activation Energy

The TRPL measurements were carried out using a 375 nm picosecond diode laser as the excitation source. The PL spectra and photon energy dependence of τ_{PL} for sample QW1505 and QW1506 are shown in Figure 9.26a and b at 10 K. The PL has a main peak around 2.95 and 2.93 eV for QW1505 and QW1506. It can be seen that the τ_{PL} values increase with the decrease of detection photon energy in both samples. This is the characteristic of a localized system wherein the decay of carriers consists of radiative recombination and energy transfer from higher-energy states to lower ones [80]. Then Equation 9.7 is used to fit the photon energy dependence of τ_{PL} values by assuming the exponential distribution of the density of tail states. The fitting results, including τ_{rad}, E_{me}, and E_0, are listed in Table 9.2. The results turned out that localization depth E_0 for QW1506 is larger than for QW1505, which indicates more stable and efficient localization effects. This is evidence of the decrease of piezoelectric field and the suppression of QCSE. On the other hand, the τ_{rad} for QW1506 is larger than τ_{rad} for QW1505. Considering that τ_{rad} reflects the chance of radiative recombination occurring on localization states, larger τ_{rad}, or smaller recombination possibility may be induced by the carrier transfer from InGaN well to pre-strained layer.

The temperature dependence of τ_{PL} for two samples in the temperature range from 10 to 300 K is measured to support the earlier assumptions, as shown in Figure 9.27. It clearly demonstrates that both τ_{PL} have a peak value at very low temperature point around 40 K, which implies that nonradiative recombination dominates τ_{PL} in most cases. This is a reasonable result from spatial separation of carriers mainly induced by QCSE. Another important conclusion is that, in the whole temperature range, τ_{PL} for QW1506 is obviously larger than τ_{PL} for QW1505. If we assume that τ_{rad} has little contribution to τ_{PL}, this result indicates that τ_{nr} for QW1506 is larger than τ_{nr} for QW1505. If we use $\tau_{nr}(T) = \tau_{nr\infty} \exp(-E_{act}/kT)$ to fit the high-temperature range of these two curves, the obtained E_{act} for QW1506 is also larger than that for QW1505, which accords with the results of E_0 from Equation 9.7. So here we can conclude that the main role of pre-strained layer is to reduce the piezoelectric field in InGaN active layers and to improve electron–hole wave function overlap, resulting in lower nonradiative recombination rates in the sample.

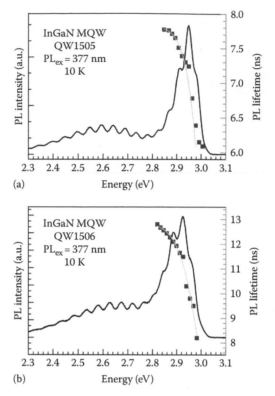

FIGURE 9.26 Photon energy dependence of PL lifetime τ_{PL} for the InGaN/GaN MQW with (b) and without (a) pre-strained layer. (Reproduced from Liu, L. et al., *Proc. SPIE*, 8484, 848412–848412-10, 2012. With permission by SPIE.)

TABLE 9.2 Radiative Lifetime, the Energy Similar to the Mobility Edge, and the Depth of Localization (τ_{rad}, E_{me}, and E_0)

Sample Name	τ_{rad} (ns)	E_{me} (eV)	E_0 (meV)
QW1505	7.83	3.02	33.4
QW1506	12.91	3.01	44.3

Source: Reproduced from Liu, L. et al., *Proc. SPIE*, 8484, 848412–848412-10, 2012. With permission by SPIE.

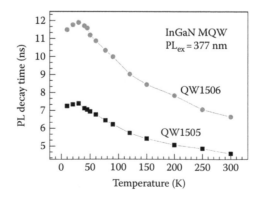

FIGURE 9.27 Temperature dependence of PL lifetime for the InGaN/GaN MQW with (QW1506) and without (QW1505) pre-strained layer. (Reproduced from Liu, L. et al., *Proc. SPIE*, 8484, 848412–848412-10, 2012. With permission by SPIE.)

9.7 Summary

This chapter offers a brief review of the PL dynamic properties of InGaN/GaN MQW LEDs, for the purpose of investigating the recombination mechanism, especially the exciton localization effect in this system. In-depth studies of the luminescent properties of a typical InGaN/GaN MQW LED have been presented via a variety of photoluminescence spectroscopic measurements. A large Stokes shift was observed from the comparison of PL and PLE measurements, which arises from the inhomogeneity in the InGaN-based materials; S-shaped (redshift–blueshift–redshift) behavior with increasing temperature of the SSPL position was shown to be related to exciton localization; two-step PL decays imply multiple recombination pathways existing in the system, and the origins of slow PL process and fast PL process were assigned to the local compositional fluctuations of indium and thickness variation of InGaN layers, respectively. The unique excitation power–dependent and temperature-dependent behaviors of PL spectra and PL lifetimes are caused by QCSE and filling effect of band-tail states within the MQW structures; GaN barrier width can be used to modulate the PL position and PL lifetime by controlling the wave function expansion of carriers to barrier layer and degree of bandgap enlargement from quantum confinement effect.

Then the PL dynamic properties of three InGaN/GaN MQW LEDs with different improved structures were discussed: (a) dual-wavelength emission induced by strong exciton localization was found in InGaN/GaN MQW LEDs with CART structures. Carrier energy transfer effects from shallow wells to deep wells and IQE improvement of lower-energy emission were confirmed by detailed TRPL measurements; (b) attempts to deposit InGaN/GaN MQWs have been made on $(1\bar{1}01)$ or $(11\bar{2}2)$-semipolar facets of GaN crystals by LEO method, with the purpose of partially avoiding the piezoelectric polarization fields parallel to [0001] c-direction. TRPL results turned out that radiative recombination lifetimes remarkably decreased with introduction of semipolar GaN templates, which is ascribable to the reduced spatial separation of electron–hole in the sample; (c) pre-strained layers were inserted into InGaN/GaN QWs for releasing the strong strain-related piezoelectric field. TRPL measurements revealed that pre-strained layers had positive effects to PL efficiencies throughout, suppressing nonradiative recombination in large temperature range.

The aforementioned penetrative investigation is helpful to better understand the luminescent mechanisms and find ways to further improve the material design and growth of InGaN/GaN MQWs for the wide spectral LED applications.

Acknowledgments

This work is supported by NSFC (61504030, 11474365 and AE0520088) and Guangxi Natural Science Foundation (2015GXNSFCA139007 and 2013GXNSFFA019001).

References

1. J. H. Choi, A. Zoulkarneev, S. I. Kim, C. W. Baik, M. H. Yang, S. S. Park et al., Nearly single-crystalline GaN light-emitting diodes on amorphous glass substrates, *Nat. Photon.*, **5**(12), 763–769 (2011).

2. J. J. Wierer, A. David, and M. M. Megens, III-nitride photonic-crystal light-emitting diodes with high extraction efficiency, *Nat. Photon.*, **3**(3), 163–169 (2009).

3. J. Kim, H. Woo, K. Joo, S. Tae, J. Park, D. Moon, S. H. Park et al., Less strained and more efficient GaN light-emitting diodes with embedded silica hollow nanospheres, *Sci. Rep.*, **3**, 3201 (2013).

4. J. W. Shon, J. Ohta, K. Ueno, A. Kobayashi, and H. Fujioka, Fabrication of full-color InGaN-based light-emitting diodes on amorphous substrates by pulsed sputtering, *Sci. Rep.*, **4**, 5325 (2014).

5. S. Nakamura, The roles of structural imperfections in InGaN-based blue light-emitting diodes and laser diodes, *Science*, **281**(5379), 956–961 (1998).

6. K. P. O'Donnell, R. W. Martin, and P. G. Middleton, Origin of luminescence from InGaN diodes, *Phys. Rev. Lett.*, **82**(1), 237–240 (1999).

7. G. Franssen, I. Gorczyca, T. Suski, A. Kamińska, J. Pereiro, E. Muñoz, E. Iliopoulos et al., Bowing of the band gap pressure coefficient in $In_xGa_{1-x}N$ alloys, *J. Appl. Phys.*, **103**(3), 033514 (2008).

8. N. Horiuchi, Light-emitting diodes: Natural white light, *Nat. Photon.*, **4**(11), 738–738 (2010).

9. A. Khan, K. Balakrishnan, and T. Katona, Ultraviolet light-emitting diodes based on group three nitrides, *Nat. Photon.*, **2**(2), 77–84 (2008).

10. S. Pimputkar, J. S. Speck, S. P. DenBaars, and S. Nakamura, Prospects for LED lighting, *Nat. Photon.*, **3**(4), 180–182 (2009).

11. Q. Dai, M. F. Schubert, M. H. Kim, J. K. Kim, E. F. Schubert, D. D. Koleske, M. H. Crawford et al., Internal quantum efficiency and nonradiative recombination coefficient of GaInN/GaN multiple quantum wells with different dislocation densities, *Appl. Phys. Lett.*, **94**(11), 111109 (2009).

12. H. Zhao, G. Liu, J. Zhang, J. D. Poplawsky, V. Dierolf, and N. Tansu, Approaches for high internal quantum efficiency green InGaN light-emitting diodes with large overlap quantum wells, *Opt. Express*, **19**(S4), A991–A1007 (2011).

13. D. P. Han, D. G. Zheng, C. H. Oh, H. Kim, J. I. Shim, D. S. Shin, and K. S. Kim, Nonradiative recombination mechanisms in InGaN/GaN-based light-emitting diodes investigated by temperature-dependent measurements, *Appl. Phys. Lett.*, **104**(15), 151108 (2014).

14. S. Nakamura, M. Senoh, S. Nagahama, N. Iwasa, T. Yamada, T. Matsushita, Y. Sugimoto, and H. Kiyoku, Room-temperature continuous-wave operation of InGaN multi-quantum-well structure laser diodes, *Appl. Phys. Lett.*, **69**(26), 4056–4058 (1996).

15. J. R. Jinschek, R. Erni, N. F. Gardner, A. Y. Kim, and C. Kisielowski, Local indium segregation and bang gap variations in high efficiency green light emitting InGaN/GaN diodes, *Solid State Commun.*, **137**(4), 230–234 (2006).

16. H. Jeong, H. J. Jeong, H. M. Oh, C.-H. Hong, E.-K. Suh, G. Lerondel, and M. S. Jeong, Carrier localization in In-rich InGaN/GaN multiple quantum wells for green light-emitting diodes, *Sci. Rep.*, **5**, 9373 (2015).

17. M. A. Sousa, T. C. Esteves, N. B. Sedrine, J. Rodrigues, M. B. Lourenço, A. Redondo-Cubero, E. Alves et al., Luminescence studies on green emitting InGaN/GaN MQWs implanted with nitrogen, *Sci. Rep.*, **5**, 9703 (2015).

18. T. J. Yang, R. Shivaraman, J. S. Speck, and Y. R. Wu, The influence of random indium alloy fluctuations in indium gallium nitride quantum wells on the device behavior, *J. Appl. Phys.*, **116**(11), 113104 (2014).

19. I.-K. Park, M.-K. Kwon, J.-O. Kim, S.-B. Seo, J.-Y. Kim, J.-H. Lim, S.-J. Park, and Y.-S. Kim, Green light-emitting diodes with self-assembled In-rich InGaN quantum dots, *Appl. Phys. Lett.*, **91**(13), 133105 (2007).

20. M. Zhang, P. Bhattacharya, and W. Guo, InGaN/GaN self-organized quantum dot green light emitting diodes with reduced efficiency droop, *Appl. Phys. Lett.*, **97**(1), 011103 (2010).

21. Y. Yang, P. Ma, X. Wei, D. Yan, Y. Wang, and Y. Zeng, Design strategies for enhancing carrier localization in InGaN-based light-emitting diodes, *J. Luminescene*, **155**, 238–243 (2014).

22. W. Liu, D. G. Zhao, D. S. Jiang, P. Chen, Z. S. Liu, J. J. Zhu, M. Shi et al., Localization effect in green light emitting InGaN/GaN multiple quantum wells with varying well thickness, *J. Alloys Compd.*, **625**, 266–270 (2015).

23. A. Hangleiter, F. Hitzel, C. Netzel, D. Fuhrmann, U. Rossow, G. Ade, and P. Hinze, Suppression of nonradiative recombination by V-shaped pits in GaInN/GaN quantum wells produces a large increase in the light emission efficiency, *Phys. Rev. Lett.*, **95**(12), 127402 (2005).

24. Z. C. Feng, L.-H. Zhu, T.-W. Kuo, C. Y. Wu, H.-L. Tsai, B.-L. Liu, and J.-R. Yang, Optical and structural studies of dual wavelength InGaN/GaN tunnel-injection light emitting diodes grown by metalorganic chemical vapor deposition, *Thin Solid Films*, **529**, 269–274 (2013).

25. L. Liu, L. Wang, N. Liu, W. Yang, D. Li, W. Chen, Z. C. Feng, Y.-C. Lee, I. Ferguson, and X. Hu, Investigation of the light emission properties and carrier dynamics in dual-wavelength InGaN/GaN multiple-quantum well light emitting diodes, *J. Appl. Phys.*, **112**(8), 083101 (2012).

26. F. Zeng, L. Zhu, W. Liu, X. Li, W. Liu, B.-J. Chen, Y.-C. Lee, Z. C. Feng, and B. I. Liu, Carrier localization and phonon-assisted hopping effects in semipolar InGaN/GaN light-emitting dioses grown by selective area epitaxy, *J. Alloys Compd.*, **656**, 881–886 (2016).

27. L. Zhu, F. Zeng, W. Liu, Z. Feng, B. Liu, Y. Lu, Y. Gao, and Z. Chen, Improved quantum efficiency in semipolar (1 $\bar{1}$01) InGaN/GaN quantum wells grown on GaN prepared by lateral epitaxial overgrowth, *IEEE Trans. Electron Dev.*, **60**(11), 3753–3759 (2013).

28. L. Wang, Z. Lu, S. Liu, and Z. C. Feng, Shallow–deep InGaN multiple-quantum-well system for dual-wavelength emission gon semipolar (11$\bar{2}$2) facet GaN, *J. Electron. Mater.*, **40**(7), 1572–1577 (2011).

29. L. Liu, L. Wang, D. Li, N. Liu, L. Li, W. Cao, W. Yang et al., Influence of indium composition in the prestrained InGaN interlayer on the strain relaxation of InGaN/GaN multiple quantum wells in laser diode structures, *J. Appl. Phys.*, **109**(7), 073106 (2011).

30. L. Liu, L. Wang, C. Lu, D. Li, N. Liu, L. Li, W. Yang et al., Enhancement of light-emission efficiency of ultraviolet InGaN/GaN multiple quantum well light emitting diode with InGaN underlying layer, *Appl. Phys. A*, **108**(4), 771–776 (2012).

31. Z. S. Lee, Z. C. Feng, A. G. Li, H. L. Tsai, J. R. Yang, Y. F. Chen, N. Li, I. T. Ferguson, and W. Lu, Photoluminescence dynamics and structural investigation of InGaN/GaN multiple quantum well light emitting diodes grown by metalorganic chemical vapor deposition, *Proc. SPIE*, **6669**, 66690I–66690I-10 (2007).

32. X. Li, P. W. Bohn, J. Kim, J. O. White, and J. J. Coleman, Spatially resolved band-edge emission from partially coalesced GaN pyramids prepared by epitaxial lateral overgrowth, *Appl. Phys. Lett.*, **76**(21), 3031–3033 (2000).

33. J. Elsner, R. Jones, M. I. Heggie, P. K. Sitch, M. Haugk, Th. Frauenheim, S. Öberg, and P. R. Briddon, Deep acceptors trapped at threading-edge dislocations in GaN, *Phys. Rev. B*, **58**(19), 12571–12574 (1998).

34. F. A. Ponce, D. P. Bour, W. Götz, and P. J. Wright, Spatial distribution of the luminescence in GaN thin films, *Appl. Phys. Lett.*, **68**(1), 57–59 (1996).

35. R. W. Martin, P. G. Middleton, K. P. O'Donnell, and W. Van der Stricht, Exciton localization and the Stokes' shift in InGaN epilayers, *Appl. Phys. Lett.*, **74**(2), 263–265 (1999).

36. M. E. White, K. P. O'Donnell, R. W. Martin, S. Pereira, C. J. Deatcher, and I. M. Watson, Photoluminescence excitation spectroscopy of InGaN epilayers, *Mater. Sci. Eng. B*, **93**(1–3), 147–149 (2002).

37. Y.-H. Cho, Y. P. Sun, H. M. Kim, T. W. Kang, E.-K. Suh, H. J. Lee, R. J. Choi, and Y. B. Hahn, High quantum efficiency of violet-blue to green light emission in InGaN quantum well structures grown by graded-indium-content profiling method, *Appl. Phys. Lett.*, **90**(1), 011912 (2007).

38. J.-H. Chen, Z.-C. Feng, H.-L. Tsai, J.-R. Yang, P. Li, C. Wetzel, T. Detchprohm, and J. Nelson, Optical and structural properties of InGaN/GaN multiple quantum well structure grown by metalorganic chemical vapor deposition, *Thin Solid Films*, **498**(1–2), 123–127 (2006).

39. W. Shan, B. D. Little, J. J. Song, Z. C. Feng, M. Schurman, and R. A. Stall, Optical transitions in $In_xGa_{1-x}N$ alloys grown by metalorganic chemical vapor deposition, *Appl. Phys. Lett.*, **69**(22), 3315–3317 (1996).

40. Y. H. Cho, G. H. Gainer, A. J. Fischer, J. J. Song, S. Keller, U. K. Mishra, and S. P. DenBaars, "S-shaped" temperature-dependent emission shift and carrier dynamics in InGaN/GaN multiple quantum wells, *Appl. Phys. Lett.*, **73**(10), 1370–1372 (1998).

41. S. Chichibu, T. Azuhata, T. Sota, and S. Nakamura, Spontaneous emission of localized excitons in InGaN single and multiquantum well structures, *Appl. Phys. Lett.*, **69**(27), 4188–4190 (1996).

42. Y. H. Cho, G. H. Gainer, J. B. Lam, J. J. Song, W. Yang, and W. Jhe, Dynamics of anomalous optical transitions in $Al_xGa_{1-x}N$ alloys, *Phys. Rev. B*, **61**(11), 7203 (2000).

43. C. T. Yu, W. C. Lai, C. H. Yen, H. C. Hsu, and S. J. Chang, Optoelectrical characteristics of green light-emitting diodes containing thick InGaN wells with digitally grown InN/GaN, *Opt. Express*, **22**(S3), A633–A641 (2014).

44. S. W. Feng, Y. C. Cheng, Y. Y. Chung, C. C. Yang, M. H. Mao, Y. S. Lin, K. J. Ma, and J. I. Chyi, Multiple-component photoluminescence decay caused by carrier transport in InGaN/GaN multiple quantum wells with indium aggregation structures, *Appl. Phys. Lett.*, **80**(23), 4375–4377 (2002).

45. G. Sun, G. Xu, Y. J. Ding, H. Zhao, G. Liu, J. Zhang, and N. Tansu, Investigation of fast and slow decays in InGaN/GaN quantum wells, *Appl. Phys. Lett.*, **99**(8), 081104 (2011).

46. Z. Li, J. Kang, B. W. Wang, H. Li, Y. H. Weng, Y.-C. Lee, Z. Liu, X. Yi, Z. C. Feng, and G. Wang, Two distinct carrier localization in green light-emitting diodes with InGaN/GaN multiple quantum wells, *J. Appl. Phys.*, **115**(8), 083112 (2014).

47. T. Lin, Z. R. Qiu, J. R. Yang, L. W. Ding, Y. H. Gao, and Z. C. Feng, Investigation of photoluminescence dynamics in InGaN/GaN multiple quantum wells, *Mater. Lett.*, **173**, 170–173 (2016).

48. S. Chichibu, T. Onuma, T. Sota, S. P. DenBaars, S. Nakamura, T. Kitamura, Y. Ishida, and H. Okumura, Influence of InN mole fraction on the recombination processes of localized excitons in strained cubic $In_xGa_{1-x}N/GaN$ multiple quantum wells, *J. Appl. Phys.*, **93**(4), 2051–2054 (2003).

49. M. S. Minsky, S. Watanabe, and N. Yamada, Radiative and nonradiative lifetimes in GaInN/GaN multiquantum wells, *J. Appl. Phys.*, **91**(8), 5176–5181 (2002).

50. R. W. Collins and W. Paul, Model for the temperature dependence of photoluminescence in a-Si:H and related materials, *Phys. Rev. B*, **25**(8), 5257 (1982).

51. A. B. Yankovich, A. V. Kvit, X. Li, F. Zhang, V. Avrutin, H. Liu, N. Izyumskaya et al., Thickness variations and absence of lateral compositional fluctuations in aberration-corrected STEM images of InGaN LED active regions at low dose, *Microsc. Microanal.*, **20**(03), 864–868 (2014).

52. S. De, A. Layek, A. Raja, A. Kadir, M. R. Gokhale, A. Bhattacharya, S. Dhar, and A. Chowdhury, Two distinct origins of highly localized luminescent centers within InGaN/GaN quantum-well light-emitting diodes, *Adv. Funct. Mater.*, **21**(20), 3828–3835 (2011).

53. M. Pophristic, F. H. Long, C. Tran, R. F. Karlicek, Z. C. Feng, and I. T. Ferguson, Time-resolved spectroscopy of $In_xGa_{1-x}N/GaN$ multiple quantum wells at room temperature, *Appl. Phys. Lett.*, **73**(6), 815–817 (1998).

54. C.-Y. Chen, D.-M. Yeh, Y.-C. Lu, and C. C. Yang, Dependence of resonant coupling between surface plasmons and an InGaN quantum well on metallic structure, *Appl. Phys. Lett.*, **89**(20), 203113 (2006).

55. S. Nagahara, M. Arita, and Y. Arakawa, No temperature dependence of spin relaxation in InGaN phase-separated quantum dots, *Appl. Phys. Lett.*, **88**(8), 083101 (2006).

56. S. K. Shee, Y. H. Kwon, J. B. Lam, G. H. Gainer, G. H. Park, S. J. Hwang, B. D. Little, and J. J. Song, MOCVD growth, stimulated emission and time-resolved PL studies of InGaN/(In)GaN MQWs: Well and barrier thickness dependence, *J. Cryst. Growth*, **221**(1–4), 373–377 (2000).

57. J. Bai, T. Wang, and S. Sakai, Study of the strain relaxation in InGaN/GaN multiple quantum well structures, *J. Appl. Phys.*, **90**(4), 1740–1744 (2001).

58. D.-J. Kim, Y.-T. Moon, K.-M. Song, and S.-J. Park, Effect of barrier thickness on the interface and optical properties of InGaN/GaN multiple quantum wells, *Jpn. J. Appl. Phys.*, **40**(5R), 3085 (2001).

59. C. H. Chen, Y. K. Su, S. J. Chang, G. C. Chi, J. K. Sheu, J. F. Chen, C. H. Liu, and Y. H. Liaw, High brightness green light emitting diodes with charge asymmetric resonance tunneling structure, *IEEE Electron Dev. Lett.*, **23**(3), 130–132 (2002).

60. Y. T. Rebane, Y. G. Shreter, B. S. Yavich, V. E. Bougrov, S. I. Stepanov, and W. N. Wang, Light emitting diode with charge asymmetric resonance tunneling, *Phys. Status Solidi (a)*, **180**(1), 121–126 (2000).

61. Z. C. Feng, T.-W. Kuo, C. Y. Wu, H.-L. Tsai, J.-R. Yang, Y. S. Huang, I. T. Ferguson, and W. Lu, Optical and structural properties of dual wavelength InGaN/GaN multiple quantum well light emitting diodes, *Proc. SPIE*, **7058**, 70580S–70580S-12 (2008).

62. Y. Sun, Y.-H. Cho, E.-K. Suh, H. J. Lee, R. J. Choi, and Y. B. Hahn, Carrier dynamics of high-efficiency green light emission in graded-indium-content InGaN/GaN quantum wells: An important role of effective carrier transfer, *Appl. Phys. Lett.*, **84**(1), 49–51 (2004).

63. T. Tetsuya, A. Hiroshi, and A. Isamu, Theoretical study of orientation dependence of piezoelectric effects in Wurtzite strained GaInN/GaN heterostructures and quantum wells, *Jpn. J. Appl. Phys.*, **39**(2R), 413 (2000).

64. P. Lefebvre, A. Morel, M. Gallart, T. Taliercio, J. Allègre, B. Gil, H. Mathieu, B. Damilano, N. Grandjean, and J. Massies, High internal electric field in a graded-width InGaN/GaN quantum well: Accurate determination by time-resolved photoluminescence spectroscopy, *Appl. Phys. Lett.*, **78**(9), 1252–1254 (2001).

65. N. Grandjean, B. Damilano, S. Dalmasso, M. Leroux, M. Laügt, and J. Massies, Built-in electric-field effects in wurtzite AlGaN/GaN quantum wells, *J. Appl. Phys.*, **86**(7), 3714–3720 (1999).

66. M. D. Craven, P. Waltereit, J. S. Speck, and S. P. DenBaars, Well-width dependence of photoluminescence emission from a-plane GaN/AlGaN multiple quantum wells, *Appl. Phys. Lett.*, **84**(4), 496–498 (2004).

67. A. Chakraborty, B. A. Haskell, S. Keller, J. S. Speck, S. P. DenBaars, S. Nakamura, and U. K. Mishra, Nonpolar InGaN/GaN emitters on reduced-defect lateral epitaxially overgrown a-plane GaN with drive-current-independent electroluminescence emission peak, *Appl. Phys. Lett.*, **85**(22), 5143–5145 (2004).

68. M. Ueda, K. Kojima, M. Funato, Y. Kawakami, Y. Narukawa, and T. Mukai, Epitaxial growth and optical properties of semipolar ($11\bar{2}2$) GaN and InGaN/GaN quantum wells on GaN bulk substrates, *Appl. Phys. Lett.*, **89**(21), 211907 (2006).

69. H. Masui, T. J. Baker, M. Iza, H. Zhong, S. Nakamura, and S. P. DenBaars, Light-polarization characteristics of electroluminescence from InGaN/GaN light-emitting diodes prepared on ($11\bar{2}2$)-plane GaN, *J. Appl. Phys.*, **100**(11), 113109 (2006).

70. C.-H. Chiu, D.-W. Lin, C.-C. Lin, Z.-Y. Li, Y.-C. Chen, S.-C. Ling, H.-C. Kuo et al., Optical properties of ($1\bar{1}01$) semi-polar InGaN/GaN multiple quantum wells grown on patterned silicon substrates, *J. Cryst. Growth*, **318**(1), 500–504 (2011).

71. H. Yu, L. K. Lee, T. Jung, and P. C. Ku, Photoluminescence study of semipolar ($1\bar{1}01$) InGaN/GaN multiple quantum wells grown by selective area epitaxy, *Appl. Phys. Lett.*, **90**(14), 141906 (2007).

72. C.-Y. Cho, S.-H. Han, S.-J. Lee, S.-C. Park, and S.-J. Park, Green light-emitting diodes on semipolar ($11\bar{2}2$) microfacets grown by selective area epitaxy, *J. Electrochem. Soc.*, **157**(1), H86–H89 (2010).

73. T. Onuma, T. Koyama, A. Chakraborty, M. McLaurin, B. A. Haskell, P. T. Fini, S. Keller et al., Radiative and nonradiative lifetimes in nonpolar m-plane $In_xGa_{1-x}N$/GaN multiple quantum wells grown on GaN templates prepared by lateral epitaxial overgrowth, *J. Vacuum Sci. Technol. B*, **25**(4), 1524–1528 (2007).

74. P. Waltereit, O. Brandt, A. Trampert, H. T. Grahn, J. Menniger, M. Ramsteiner, M. Reiche, and K. H. Ploog, Nitride semiconductors free of electrostatic fields for efficient white light-emitting diodes, *Nature*, **406**(6798), 865–868 (2000).

75. J. L. Huang, L. S. Wang, Y. S. Lai, Y. C. Lee, Z. R. Qiu, S. Liu, D. S. Wuu, and Z. C. Feng, Structural and optical properties of InGaN/GaN multiple quantum well light emitting diodes grown on (1122) facet GaN/sapphire templates by metalorganic chemical vapor deposition, *Proc. SPIE*, **8123**, 81230C–81230C-11 (2011).

76. K. Nishizuka, M. Funato, Y. Kawakami, Y. Narukawa, and T. Mukai, Efficient rainbow color luminescence from $In_xGa_{1-x}N$ single quantum wells fabricated on $(11\bar{2}2)$ microfacets, *Appl. Phys. Lett.*, **87**(23), 231901 (2005).

77. C.-F. Huang, T.-C. Liu, Y.-C. Lu, W.-Y. Shiao, Y.-S. Chen, J.-K. Wang, C.-F. Lu, and C. C. Yang, Enhanced efficiency and reduced spectral shift of green light-emitting-diode epitaxial structure with prestrained growth, *J. Appl. Phys.*, **104**(12), 123106 (2008).

78. C.-F. Huang, T.-Y. Tang, J.-J. Huang, W.-Y. Shiao, C. C. Yang, C.-W. Hsu, and L. C. Chen, Prestrained effect on the emission properties of InGaN/GaN quantum-well structures, *Appl. Phys. Lett.*, **89**(5), 051913 (2006).

79. L. Liu, W. Wang, J. L. Huang, X. Hu, P. Chen, J. J. Huang, and Z. C. Feng, Time-resolved and temperature-varied photoluminescence studies of InGaN/GaN multiple quantum well structures, *Proc. SPIE*, **8484**, 848412–848412-10 (2012).

80. T. Akasaka, H. Gotoh, H. Nakano, and T. Makimoto, Blue-purplish InGaN quantum wells with shallow depth of exciton localization, *Appl. Phys. Lett.*, **86**(19), 191902 (2005).

III

Deep Ultraviolet LEDs and Related Technologies

10

Technological Developments of UV LEDs

Ching-Hsueh Chiu
Advanced Optoelectronic Technology Inc.

Po-Min Tu
Advanced Optoelectronic Technology Inc.

Tzu-Chien Hong
Advanced Optoelectronic Technology Inc.

Chien-Chung Peng
Advanced Optoelectronic Technology Inc.

Chien-Shiang Huang
Advanced Optoelectronic Technology Inc.

Shih-Cheng Huang
Advanced Optoelectronic Technology Inc.

Abstract   The GaN-based ultraviolet light-emitting diodes (UV LEDs) have attracted great interest in several applications, such as solid-state lighting, environmental protection, and medical equipment. There have been interests in solid-state lighting by using near-UV LEDs light for the phosphor converting source. In this chapter, these novel technologies could be effective for improving UV LEDs device performance in the future.

10.1 Introduction

GaN-based ultraviolet light-emitting devices (UV LEDs) have attracted great attention in last few years due to their potential applications in photocatalytic deodorizing such as in air conditioners [1,2], and there have been interests in solid-state lighting (SSL) by using near-UV LEDs light for the phosphor converting source [3]. However, external quantum efficiency (EQE) decreases drastically below the wavelength of 400 nm [4]. However, due to the lack of localized states in the multiple quantum wells (MQWs) active regions, which is believed to have stronger carrier confinement for radiative recombination, the performance of UV LEDs is more sensitive to the threading dislocations (TDs) in the epitaxial layer than that of blue LEDs. In GaN-based materials, typical threading dislocation densities (TDDs) are as much as 10^8–10^{10} cm^{-2} [5], which are due to the mismatches of lattice constants and thermal expansion coefficient between GaN and sapphire substrate. Therefore, the means to decrease the TDD is an important subject for developing high-performance UV LEDs. Over the past years, several approaches have been proposed for improving the crystalline quality of the GaN-based epilayer, such as epitaxial lateral overgrowth (ELOG) [6], cantilever epitaxy (CE) [7], defect selective passivation [8], microscale SiN$_x$ or SiO$_x$ patterned mask [9], and the use of patterned sapphire substrate (PSS) [10]. However, these methods require additional complex photolithography and etching process. Previous theoretical and experimental studies indicated that a further reduction in defect density is possible if the lateral overgrowth approach is extended to the nanoscale template [11–13]. On the other hand, it is well known that in low indium content InGaN quantum wells (QWs), AlGaN barrier is necessary for carrier confinement. But the two materials of AlGaN and InGaN are very different in growth temperature, which affects strongly the quality of material and device performances. The enhancement of internal quantum efficiency (IQE) can be attributed to several factors: the reduction of threading dislocations, the reduction of internal polarization, a better electronic blocking layer, and also a better overlap of electron-hole wave function in the active region [14].

In the following sections, we will demonstrate major methods for fabricating high-efficient near-UV LEDs: using the selectively etched GaN (SE-GaN) template on PSS, the sputtered AlN nucleation layer on PSS, and using an Mg- and heavy Si-doping technique. Using these methods can improve the IQE and light extraction efficiency (LEE) of near-UV LEDs. On the other hand, we demonstrate the high-efficiency near-UV LEDs by InAlGaN quaternary barrier in the active region. It is not only lattice or bandgap matched in InGaN QW, but also matched in optimized AlGaN barrier for a fair investigation on the light output characteristic. Besides, the light output power of UV LEDs will be improved using side wall etching and inverted micropyramid structures.

10.2 Experimental

10.2.1 The Novel Epitaxial Template

GaN epitaxial layers are possessed of a high threading dislocation (TD) density (10^9–10^{11} cm^{-2}) when grown on lattice-mismatched substrates, such as conventional sapphire substrate (CSS) and silicon carbide substrate [15,16]. These dislocations act as nonradiative centers, resulting in low internal quantum efficiency for III–V series LEDs [17,18]. The performance of near-ultraviolet (NUV) LEDs appears to be more sensitive to the dislocation quantity than that of blue- and green-illumination LEDs [19]. Several suggestions for reduction of TDs have been reported, including epitaxial lateral overgrowth (ELOG) [6,20], cantilever epitaxy (CE) [7], defect selective passivation [8], microscale SiN$_x$ or SiO$_x$ patterned mask [9,21], and the use of patterned sapphire substrate (PSS) [22,23].

We demonstrate that microstructure, the electrical and optical properties of 410 nm InGaN-based LEDs grown on a CSS, the recess PSS, and the selectively etched (SE) GaN template were examined and compared. The SE GaN template is capable of enhancing the performance of 410 nm LEDs. The fabrication of the SE GaN template required just one lithography process, making the implementation of this method easier. The fabrication process of the SE GaN template is presented schematically in Figure 10.1a through d, and the regrowth of n-GaN is shown in Figure 10.1e.

The morphology of the recess PSS was observed by using a scanning electron microscope (SEM), as shown in Figure 10.1a. A 2 μm thick undoped GaN (u-GaN) was first grown on the recess PSS; this was followed by a defect selective etching process where the sample was submerged in a hot H$_3$PO$_4$ solution at 270°C for 10 min. Figure 10.1b shows the cross-sectional SEM photograph of the etched GaN. A 0.5 μm thick SiO$_2$ film was deposited on the etched u-GaN surface by plasma-enhanced chemical vapor deposition (PECVD). The etched pits did not entirely fill up with SiO$_2$, as shown in Figure 10.1c. The excess SiO$_2$

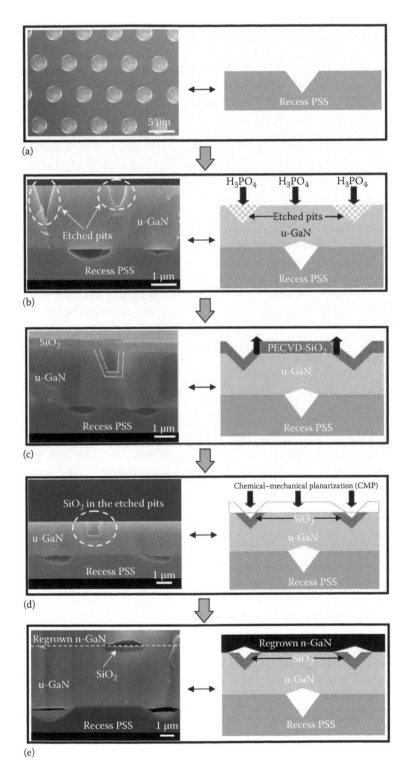

FIGURE 10.1 (a–d) Schematic flow diagram for the fabrication of a SE GaN template. (e) The regrowth of n-GaN on the SE GaN template.

which was deposited on the planar surface of u-GaN, was carefully removed by chemical–mechanical polishing (CMP). NP6502 silicon slurry was chosen, since no additional damages were exerted onto the treated u-GaN during the CMP process. The mass ratio of the slurry to pure water was 1:200, and the CMP process took nearly 3 h. Figure 10.1d shows the operational structure of the SE GaN template. Figure 10.1e shows the subsequent epitaxial growth starting with n-GaN on the SE GaN template.

The full width at half maximum (FWHM) of the rocking curve at the (002) plane was 315, 297, and 288 arcsec for n-GaN grown on a CSS, the recess PSS, and the SE GaN template. The corresponding FWHM of the (102) plane was 406, 355, and 288 arcsec. The amounts of screw type and edge type TDs were calculated from the mean FWHM of (002) and (102) planes, respectively [24], and are listed in Table 10.1. The decrease in FWHM at the (102) reflection was more prominent than in the FWHM at the (002) reflection when n-GaN was grown on either the recess PSS or the SE GaN template. In other words, both the recess PSS and the SE GaN template were quite effective in reducing the edge-type TDs. Judging from both measured and calculated results listed in Table 10.1, the SE GaN template appears to be more capable of improving the crystallinity of n-GaN than the recess PSS.

The vertical variation in the crystallinity of n-GaN grown on a CSS, the recess PSS, and the SE GaN template was revealed in cross-sectional cathodoluminescence (CL) images taken at $\lambda = 364$ nm, as shown in Figure 10.2a through c; brighter areas represent greater degrees of crystallinity [8]. In Figure 10.2a, it is clear from the numerous dark areas that the crystallinity of n-GaN/CSS was the lowest among the three samples. By concentrating TDs upon the flat areas of the recess PSS, the crystallinity of GaN grown above pyramidal holes was enhanced, as shown in Figure 10.2b. Figure 10.2c clearly shows that the CL intensity has changed across the etch pit with SiO_2 fillings since TDs were prevented from extending to the n-GaN. The diagram in Figure 10.2d reveals the distribution of TDs within the n-GaN on the SE GaN template. Then, the optical output power was enhanced by 13% and 46% in the UV LED structure on the SE GaN template, as compared with that of the 410 nm LEDs fabricated on the recess PSS and the CSS, respectively.

Next, we can demonstrate that the ultraviolet light-emitting diodes (UV LEDs) on patterned sapphire substrate (PSS) with sputtered AlN nucleation layer at 380 nm were grown by an atmospheric pressure metal–organic chemical vapor deposition (AP-MOCVD).

The structure of low temperature GaN nucleation layer followed by 18 min high temperature u-GaN layer were grown on PSS with and without sputtered AlN nucleation layer by conventional MOCVD. The SEM surface morphology of u-GaN grown on PSS without sputtered AlN layer was shown in Figure 10.3a. Obviously, the u-GaN were grown on both of the c-plane region and cone region. Therefore, the coalescence of laterally overgrown u-GaN in c-plane region and cone region occurred. However, the growth behavior of u-GaN with sputtered AlN nucleation layer was significantly different in 18 min growth time as shown in Figure 10.3b. The u-GaN grew on c-plane region rather than on the cone region. Based on these results, it was inferred that the sputtered AlN nucleation layer provided uniform nucleation layer thickness. It not only suppressed lateral growth from the cone region but also increased vertical growth from the c-plane region. On the other hand, the cross-sectional scanning electron microscope (SEM) images of whole LED structure with GaN and AlN nucleation layer are shown in Figure 10.3c and d. As we can see, the irregular air voids are observed near the interface of the u-GaN epitaxial layer and cone regions of PSS in Figure 10.3c. On the other hand, the air voids were not observed in the u-GaN/sapphire interface in Figure 10.3d.

Figure 10.4 shows bright field scanning cross-section TEM images of UV LEDs with GaN nucleation layer and sputtered AlN nucleation layer. In Figure 10.4a, it is clearly observed that the tip portion of the

TABLE 10.1 Characteristics of n-GaN Grown on CSS, PSS, and SE-GaN Templates

n-GaN Grown on	FWHM (arcsec)		EPD (cm⁻²)	Calculated Threading Dislocations (cm⁻²)	
	(002)	(102)		Screw-Type	Edge-Type
CSS	315	406	3.48×10^7	1.99×10^8	1.14×10^9
PSS	297	335	1.77×10^6	1.77×10^8	7.73×10^8
SE-GaN	288	288	1.30×10^5	1.66×10^8	5.71×10^8

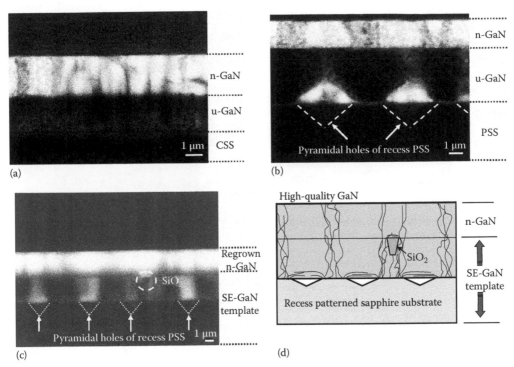

FIGURE 10.2 Cross-sectional CL images taken at λ = 364 nm for n-GaN grown on (a) CSS, (b) recess PSS, and (c) SE-GaN template. (d) Schematic diagram of the correlation between TD propagation and epilayer crystallinity for n-GaN on the SE GaN template.

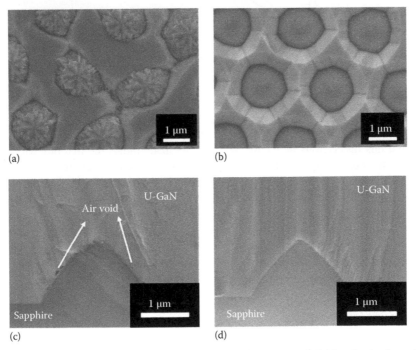

FIGURE 10.3 The top view SEM images of 1 μm GaN grew on (a) conventional GaN nucleation layer, (b) sputtered AlN nucleation layer, and crossection view SEM images of 3 μm GaN grew on (c) conventional GaN nucleation layer, and (d) sputtered AlN nucleation layer.

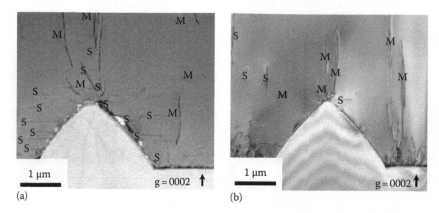

(a)   (b)

FIGURE 10.4   Bright field cross-section TEM images of the (a) UV LEDs with GaN nucleation layer. (b) UV LEDs with RPD AlN nucleation layer, g = 0002. "S" indicates screw dislocation and "M" indicates mix dislocation.

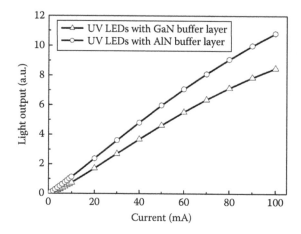

FIGURE 10.5   The output power as a function of injection current for samples with GaN and AlN buffer layer (chip size: 350 × 350 μm²).

pattern was not smooth due to the formation of 3-D GaN islands during the growth. For comparison, fewer screw and mix dislocations were observed in Figure 10.4b, indicating that film with the sputtered AlN layer has better epitaxial quality. On the other hand, the TDDs are estimated to have reduced from 6×10^7 cm^{-2} to 2.5×10^7 cm^{-2} at the interface between the u-GaN layers for conventional and AlN PSS devices, respectively. This confirmed the earlier statement that the crystalline quality of the GaN film was improved by sputtered AlN nucleation layer growth on PSS.

From the L–I curves as shown in Figure 10.5, the output power of UV LEDs with sputtered AlN nucleation layer is higher than the UV LEDs with GaN nucleation layer under all current injection conditions. In particular, the light output power intensity of UV LEDs with sputtered AlN nucleation layer is enhanced by 30% at an injection current of 20 mA.

10.2.2 Heavily Mg- and Si-Doped GaN Insertion Layer Technique

In GaN-based materials, typical threading dislocation densities (TDDs) are as much as 10^8–10^{10} cm^{-2} [5], which is due to the mismatches of lattice constants and thermal expansion coefficient between GaN and sapphire substrate. Therefore, the means to decrease the TDD is an important subject for developing high-performance UV LEDs. Over the past years, several approaches have been proposed for improving the crystalline quality of GaN-based epilayer, such as epitaxial lateral overgrowth (ELOG) [6,20], cantilever epitaxy (CE) [7], defect selective passivation [8], microscale SiN_x or SiO_x patterned mask [9,20,21], and the use of patterned sapphire substrate (PSS) [22,25,26]. Previous theoretical and experimental studies have indicated that a further reduction in defect density is possible if the lateral overgrowth approach is extended to the nanoscale template [27–30]. The enhancement of internal quantum efficiency (IQE) can be attributed to several factors: the reduction of threading dislocations, the reduction of internal polarization, a better electronic blocking layer, and also a better overlap of electron–hole wave function in the active region. Other than the first factor, most of them can improve the efficiency droop as well [31,32].

However, both of these cases require additional etching process to generate a template for the subsequent metal–organic chemical vapor deposition (MOCVD) growth of GaN epilayers. To some extent, these complicated procedures do not avoid some negative effects on the grown samples. In this study, through a single MOCVD process, we propose a heavily Mg-doped GaN insertion layer (HD-IL) technique to improve the crystalline quality of the GaN layer, which is followed by the rest of the required GaN-based LED structure.

The schematic diagram of dislocation reduction by HD-IL is shown in Figure 10.6. The figure shows an overgrowth of GaN and blocking mechanism for TDs. The improvement in crystal quality of GaN could generate high-performance 380 nm GaN-based LEDs. Details of electrical and optical properties of LED samples with and without an HD-IL will be described.

The improvements in crystalline and optical quality of GaN using the HD-IL technique are summarized in Table 10.2, where the DCXRD, EPD, and PL data were tabulated. By using an HD-IL, the DCXRD full-width at half-maximum (FWHM) is reduced by 77 and 187 arcsec for (002) and (302) planes, respectively. It has been reported that the densities of screw TDs and edge TDs are related to DCXRD FWHMs of (002) and (302) planes, respectively [33,34]. Additionally, the PL-integrated intensity ratios of near-band-edge emission to the yellow luminescence band (I_{BE}/I_{YL}) represent the optical quality of GaN epilayer. From the DCXRD and PL measurement results shown in Table 10.2, we can infer that the improvements in crystalline and optical quality of the sample with an HD-IL could be due to the efficient reduction of defect densities.

In Figure 10.7, the integrated PL intensities of both LED samples with and without an HD-IL keep to nearly constant below 100 K and decline gradually with further increase in temperature. At room temperature, the η_{int} value is about 30.72% and 21.66% for the LED samples with and without an HD-IL, respectively. The significant reduction in defect density could contribute to the present improvements in the η_{int} value.

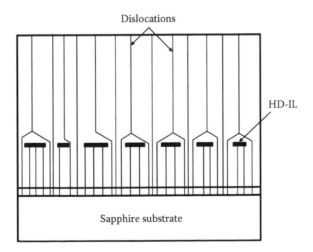

FIGURE 10.6 Schematic diagram of dislocation reduction by a heavily Mg-doped GaN insertion layer technique.

TABLE 10.2   Crystalline and Optical Performance of InGaN/AlGaN MQW LEDs Grown without and with an HD-IL

Sample	DCXRD FWHM (arcsec)		Dislocation Density (cm⁻²)		EPD (cm⁻²)	PL intensityI_{BE}/I_{YL}
	(002)	(302)	Screw	Edge		
Without HD-IL	362	504	2.6×10^8	1.7×10^9	2.5×10^8	28.3
With HD-IL	285	317	1.6×10^8	6.8×10^8	3.5×10^7	45.1

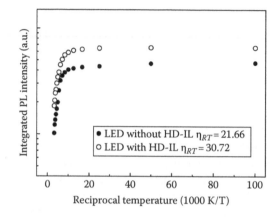

FIGURE 10.7   Temperature dependence of Arrhenius plots for InGaN/AlGaN LED samples grown with and without a heavily Mg-doped GaN insertion layer.

Figure 10.8 shows the light output power versus injection current (L–I) characteristics of these LED samples. Here, the LED chips were mounted on the epoxy-free TO-66 metal can. As shown in the inset of Figure 10.8, the electroluminescence (EL) peak positions were located at 380 nm for both the samples. It can be seen from this figure that under a 350 mA forward injection current, the output power data of the LED samples with and without the HD-IL were estimated to be 203.4 and 158.9 mW, respectively. A 28% enhancement in output power was achieved in the LED sample with the HD-IL. We attribute the enhanced output power to the reduction of dislocation density by incorporating the HD-IL structure.

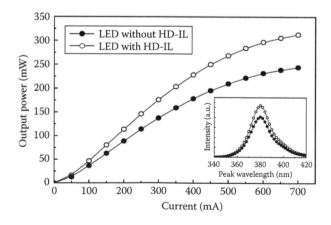

FIGURE 10.8   Output powers of 380 nm InGaN/AlGaN LED samples with and without a heavily Mg-doped GaN insertion layer. The inset shows the EL spectra of these LEDs.

On the other hand, the in situ SiN_x deposited by modulating ammonia (NH_3) and silane (SiH_4) flow can be directly grown on a nucleation layer or u-GaN without interrupting the growth process. While it offers a fast and simple method leading to TDs reduction, the SiN_x shows poor heat dissipation properties and is not desirable for inclusion in LED. Therefore, we can investigate an in situ growth mode transiting technique to achieve low TD density and high emission efficiency UV LEDs. Unlike the in situ SiN_x method given earlier, we doped an n-GaN layer heavily with Si to create a growth mode transition layer (GMTL).

Figure 10.9 shows a schematic of the UV LED structure. The inset of Figure 10.9 shows a TEM image of the GMTL region. It is clear that some TDs have bent or stopped propagating into the n-AlGaN layer due to the existence of the GMTL. The TDs, behavior was partially attributed to the relaxation of residual stresses in the u-GaN by Si incorporation [35], especially for GMTL with 10^{20} cm^{-3} Si concentrations. However, compared with Reference 49, our GMTL thickness is too thin to fully release residual stress, which induces a three-dimensional (3D) growth. The 3D growth may be another possible cause of TDs bending or stopping. Based on the previous observation, the TDs were effectively reduced by the inclusion of the GMTL.

Moreover, the TD densities of AlGaN with and without GMTL were evaluated by an etch-pit-density (EPD) measurement with KOH solution, as shown in Figure 10.10. The TD value of AlGaN grown on

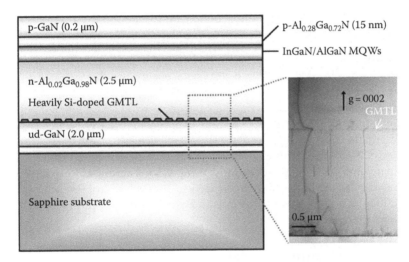

FIGURE 10.9 Schematic structure of UV LEDs. The inset shows the TEM images within GMTL.

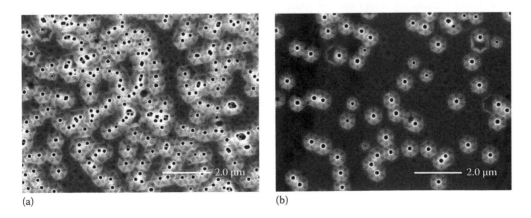

(a)

(b)

FIGURE 10.10 Typical plane-view SEM micrographs of etch pit density with KOH etched n-AlGaN surface (a) without GMTL, (b) with GMTL.

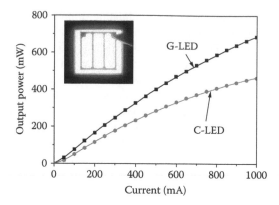

FIGURE 10.11   LED output power as functions of injection current of G-LED and C-LED. The inset shows the photograph of the G-LEDs chip alight at 350 mA.

GMTL was significantly decreased from the control sample value of 8×10^8 to 8×10^7 cm^2. Meanwhile, the full-width at FWHM of (002) and (102) x-ray diffraction peaks was measured by DCXRD. On comparison with the AlGaN without GMTL, the FWHM of AlGaN with GMTL for the (002) plane dropped from 360 to 270 arcsec. Similarly, for the (102) plane, the FWHM value dropped from 460 to 380 arcsec. From the XRD results, the AlGaN with GMTL exhibits better crystal quality than that without GMTL because the GMTL structure suppresses the TD extensions.

Figure 10.11 presents the electroluminescence light output power as a function of injection current for both C-LED and G-LED. Here, all chips were Au wire bonded and packaged using the epoxy-free metal can (TO-66). Clearly, the light output power of the G-LED is much higher than the C-LED over the injection current range of 0–1000 mA. When the vertical type LED chips were driven with a 350 mA injection current, the output power of the LEDs with and without GMTL was measured to be 286.7 and 204.2 mW, respectively. In particular, the light output power of G-LED is enhanced by a factor of approximately 40.4% at an injection current of 350 mA. The inset in Figure 10.11 shows the photograph of the vertical type G-LEDs chip lighting on.

10.2.3 Quaternary Quantum Barrier

It is difficult to fabricate near-UV LEDs with high efficiency, because the internal quantum efficiency (IQE) decreases drastically under the low indium composition [4,36,37]. Moreover, crystalline quality and light absorption of GaN are significant for short-wavelength near-UV LEDs [38,39]. It is well known that in low indium content InGaN-based quantum wells, AlGaN barrier is necessary for carrier confinement. But the two materials of AlGaN and InGaN are very different in growth temperature, which affects strongly the quality of material and device performances.

In this section, we will introduce a novel way to cut droop in 380 nm UV LEDs by using quaternary barrier in the active region. Furthermore, the efficiency droop characteristics and optical properties of high-efficiency near-UV LEDs have been measured and investigated by APSYS. The schematic of InGaN/AlGaN and InGaN/InAlGaN MQWs structures is shown in Figure 10.12. To probe the detailed properties of epitaxial layers, a 50 nm AlGaN and InAlGaN single heteroepitaxial layers were also deposited on n-AlGaN/u-GaN/sapphire substrate. The PL emission energies of these two samples are very close (about 3.594 eV), and the peak intensity of InAlGaN is slightly higher than that of AlGaN. The strong PL emission is attributed to better crystal quality.

Figure 10.13 shows the surface morphology of the two AlGaN and InAlGaN single heteroepitaxial layers with the same thickness of about 50 nm. The root mean square (RMS) roughness measured by AFM is about 0.813 and 0.595 nm, respectively. The relatively high roughness of AlGaN single heteroepitaxial layer

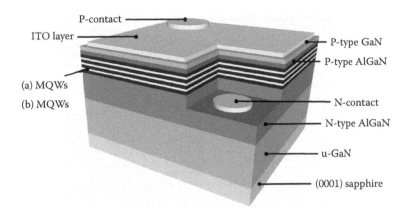

FIGURE 10.12 Schematic of (a) InGaN/AlGaN and (b) InGaN/InAlGaN MQWs structures.

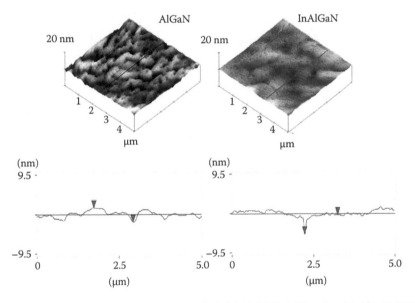

FIGURE 10.13 Surface morphology AFM over 5 × 5 µm² of AlGaN (RMS: 0.813 nm) and InAlGaN (RMS: 0.595 nm) layer with thickness of about 50 nm.

can mainly be attributed to the low deposition temperature of 830°C necessary for the adjacent InGaN well. Besides, the relatively small pits of LT InAlGaN layer can mainly be attributed to the smaller tensile strain in LT AlGaN by inserting the isoelectronic In atoms. Experiments have shown that the presence of In leads to a smooth morphology with better crystal quality and optical properties, and this result is due to the interaction between In atoms and screw dislocations [40].

Figure 10.14a shows the L–I–V characteristics for the AlGaN and InAlGaN barrier UV LEDs. The forward voltage was 3.89 and 3.98 V for InGaN/AlGaN and InGaN/InAlGaN MQWs UV LED at a forward current of 350 mA, respectively. The light output power of InGaN-based UV LED with the InAlGaN barrier is higher by 25% and 55% than the AlGaN barrier at 350 and 1000 mA, respectively. Figure 10.14b shows the normalized efficiency curves of experimental (open circles) and simulated (solid lines) as a function of forward current for the two samples. For the InGaN/AlGaN UV LEDs, when the injection current exceeds 1000 mA, the efficiency is reduced to 66% of its maximum value. In contrast, InGaN/InAlGaN UV LEDs exhibit only 13% efficiency droop when we increase the injection current to 1000 mA. Besides, the wavelength is nearly constant at about 380 nm over the entire current range.

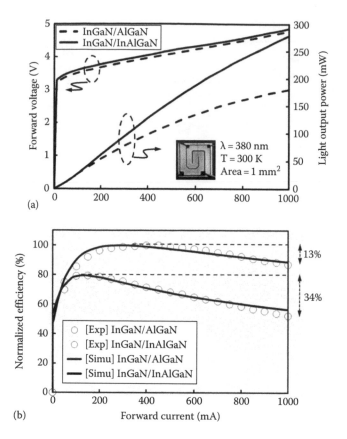

(a)

(b)

FIGURE 10.14 (a) L–I–V curves of the LEDs with AlGaN (dash) and InAlGaN (solid) barrier. (b) Normalized efficiency curves of experimental (open circles) and simulated (solid lines). Inset in figure shows the mesatype UV chip.

Figure 10.15 shows the calculated carrier distribution in these near-UV LEDs structure under a high forward current density of 100 A/cm² (1000 mA) by APSYS. When we apply the corresponding band offset ratio and the carrier mobility in InGaN/InAlGaN MQWs, the electron and hole concentration increases in the QW by about 26% and 35%, respectively, and the distribution of carrier becomes more uniform than InGaN/AlGaN case. Under high current density, the carrier distribution of both, electrons and holes, determines how efficient the photon emission process will be.

To properly assess the contributions from the IQE enhancement, we need to carry out more measurements on 365 nm UV LEDs. A general approach to evaluate the IQE of LEDs is by comparing the integrated PL intensities at low and room temperatures [41]. Figure 10.16 shows the measured IQE as a function of excitation power at 10 and 300 K for both samples. The efficiency is defined as the collected photon numbers divided by the injected photon numbers and normalized to the maximum efficiency at 10 K. Usually, the PL excitation intensity is difficult to estimate and it varies between different experiments and wafers. What we would like to do here is to convert the variable pumping intensity into the more common "carrier density." To do that, we could follow the equation below to transfer our injected power to carrier density [42]:

$$
\text{Injected carrier density} = \frac{P}{(h\nu)*\phi*d_{active}*f} * \exp\left(-\alpha_{GaN}d_{GaN}\right)* \\
\left[1-\exp\left(-\alpha_{InGaN}d_{active}\right)\right]*\left(1-R\right),
$$

(10.1)

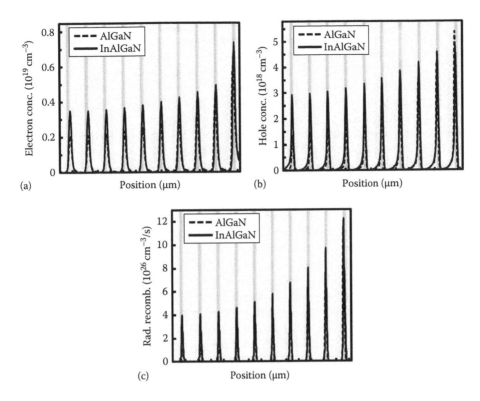

FIGURE 10.15 Distribution of (a) electron and (b) hole concentrations and (c) radiative recombination rate concentrations of the LEDs with AlGaN and InAlGaN barrier under a high forward current density of 100 A/cm².

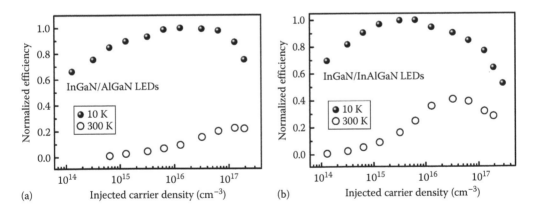

FIGURE 10.16 Relative IQE as a function of excitation power for LEDs with (a) AlGaN and (b) InAlGaN barrier layers.

The injected carrier density is determined primarily by the power of pumping laser (P), the energy of injected photon (hv), the spot size of pumping laser (ψ), the thickness of GaN and active region (d_{AlGaN}, d_{InGaN}), the repetition rate of pumping laser (f), the absorption efficiency of AlGaN and InGaN (α_{AlGaN}, α_{InGaN}), and the reflectance of pumping laser (R), as expressed by the previous equation. Experimentally, we choose $\psi = 50$ μm, $d_{GaN} = 85$ nm, $d_{active} = 150$ nm, $\alpha_{InGaN} = 10^5$ cm⁻¹, and $R = 0.17$ to calculate the injected carrier density in our samples. The power-dependent PL measurement was excited with a frequency tripled

Ti: sapphire laser at wavelength of 266 nm and the laser output power was from 0.1 to 100 mW. The IQE of 365 nm UV LEDs with AlGaN and InAlGaN barrier layers is estimated to be 22% and 41%, respectively under an excitation power of 20 mW (injected carrier density = 1×10^{17} cm^{-3}) corresponding to a current of 20 mA. The IQE of UV LED with AlInGaN barrier layers was enhanced by 86% as compared to the UV LEDs with AlGaN barrier layers. Therefore, we believe the higher IQE for the LED with InAlGaN barrier layers is due to the better crystalline quality of the multiple quantum well. We now discuss the mechanism responsible of IQE enhancement by power-dependent PL measurement.

In general, the collected PL intensity, L, is proportional to the injected carrier density, I, with a power index P, which could be expressed as [43,44]

$$L \alpha I^{P} \tag{10.2}$$

where parameter P physically reflects the various recombination processes. If P equals 1, it indicates the radiative recombination domination. On the other hand, if $P > 1$, the Shockley–Read–Hall recombination occurs, relating to the presence of nonradiative centers that provide a short path to the current. Figure 10.17 summarizes the relationship between injected carrier density and the PL intensity for both LEDs with AlGaN and InAlGaN barrier layers. At 10 K for both samples, the intensity is linearly varied with excitation power density ($P = 1$), which indicates that the radiative recombination dominates the recombination process at all injected carrier density range, and the nonradiative centers are quenched at low temperature. However, under low excitation power density at 300 K, the superlinear dependence of L on I is observed for both samples, showing that the defect-related nonradiative recombination dominates in this low carrier injection range. But as injected carriers continuously increased, the linear dependence of the PL intensity to the injected carrier density is exhibited. It must be noted here for both samples in 300 K, the value of P decreases to 1 gradually with the increasing of injected carrier density. The P values were jumping from 1.61 to 1 and 1.49 to 1, corresponding to InGaN/AlGaN and InGaN/InAlGaN LEDs samples, respectively. It means the nonradiative centers are saturated and lead to the gradual suppression of the nonradiative recombination with the injected carrier density; therefore, the radiative recombination starts to dominate the recombination process, resulting in the pronounced increasing of the internal quantum efficiency, as shown in Figure 10.17, for the region of injected carrier density less than 10^{17} cm^{-3}. In addition, since the LED with AlGaN barrier layers has higher threading dislocations than that LED with InAlGaN barrier layers, the value of P in the superlinear zone is greater for the LED with AlGaN barrier layers ($P = 1.61$) than for the LED with InAlGaN barrier layers ($P = 1.49$).

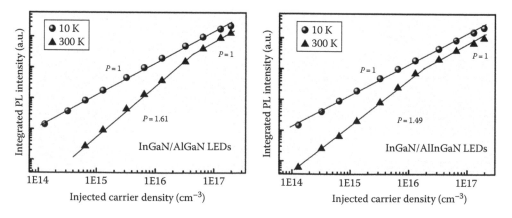

FIGURE 10.17 The relationship between the injected carrier density and the PL intensity for both LEDs with AlGaN and InAlGaN barrier layers.

In order to investigate the excitation localization effect, the temperature-dependent PL measurement was carried out. In principle, the excitation localization effect is the consequence of indium spatial fluctuations in the energy gap. Therefore, the band-tail model should be suitable for the discussion of the excitation localization effect. Based on the band-tail model, if Gaussian-like distribution of the density of states for the conduction and valence band is assumed, the temperature-dependent emission energy could be given in the following expression suggested by Eliseev et al. [45].

$$E(T) = E(0) - \alpha T^2 (T + \beta) - \sigma^2 (K_B T)^{-1} \tag{10.3}$$

The first term describes the energy gap at zero temperature; α and β are Varshni's fitting parameters. The third term comes from the localization effect, in which σ indicates the degree of localization effect, that is, the large value of σ means a strong localization effect. K_B is Boltzmann's constant. In addition, because this model is based on the assumption of nondegenerate occupation, the absolute value of σ is strongly dependent on the excitation power. Therefore, from this model, one can only obtain a relative value, that is, one can only relatively compare the values of σ in different samples under identical measurement conditions. In other words, there is no real significance if one only obtains the value of one sample without using any other as a "reference" sample. In our case, the "reference" sample is the LEDs with InGaN/AlGaN MQWs. Under identical measurement conditions, we obtain their values of σ and compare their excitation localization effects. Figure 10.18 shows the emission energy as a function of temperature for both samples; the fitting parameters based on Equation 10.3 are also given. The σ value of InGaN/AlGaN and InGaN/InAlGaN LEDs is about 5.85 and 10.02 meV, respectively. It indicates that the excitation localization effect of LEDs with InAlGaN barrier layers is much stronger than that of LEDs with AlGaN barrier layers. We believe that the indium spatial fluctuations are induced by using InAlGaN as the materials of quantum barrier layers.

In summary, we introduced and compared the performance of InGaN-based UV LEDs active region with ternary AlGaN and quaternary InAlGaN barrier layers. The crystal quality and the interface of well and barrier layers can be improved by the smooth morphology of quaternary InAlGaN layer. The EL results indicate that the light output power can be enhanced effectively when the conventional LT AlGaN barrier layers are replaced by the InAlGaN barrier layers. Besides, the InAlGaN barrier layers can enhance the IQE and the excitation localization effect for UV LEDs with such low indium content in well layers. Furthermore, simulation results show that the UV LEDs with quaternary InAlGaN barrier exhibit 62% higher radiative recombination rate and low efficiency droop of 13% at high injection current. We attribute this improvement to increasing carrier concentration and more uniform redistribution of carriers.

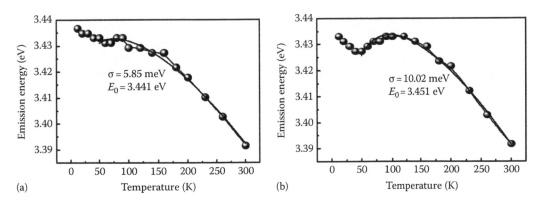

FIGURE 10.18 The emission energy as a function of temperature for both LED samples with AlGaN (a) and InAlGaN (b) quantum barrier layers.

For the quaternary InAlGaN quantum barrier layers used in 365 nm InGaN-based UV LEDs, the crystal quality and the interface of well and barrier layers can be improved. Also, the AlInGaN barrier layers can enhance the IQE and the excitation localization effect for 365 nm UV LEDs with such low indium content in well layers. The UV LED with InAlGaN quaternary barrier layers could enhance the output power by more than 60% in this study.

10.2.4 Side Well Etching Technology

In this section, we propose to enhance an output power for 380 nm UV LEDs with a hexagonal pyramid structure (HPS) on the interface of sidewall between AlGaN and AlN layers. HPSs are formed by inserting a 50 nm AlN as a sacrificial layer in n-AlGaN than using a selective wet etching process in KOH solution at 90°C for 60 min. From the SEM image, the HPS can be clearly seen on the interface of AlGaN, the facet angles and the average structure height of pyramid are 58° and 0.5 μm, respectively. According to the EL results, 12% enhancement of the light extraction efficiency can be expected in the UV LED with HPS. Furthermore, we measured the output power at 20 mA between the UV LEDs with and without HPS at 2.69 and 3.01 mW, respectively. As a result, the light extraction efficiency can be improved by this approach because of changing the routes of light reflection around the sidewall.

GaN-based UV LEDs have attracted great attention in the last few years due to their potential in applications in photocatalytic deodorizing such as in air conditioners [1], and there have been interests in solid-state lighting by using near-UV LEDs light for the phosphor converting source [2,3]. However, it is difficult to fabricate near-UV LEDs with high efficiency, because the external quantum efficiency (EQE) decreases drastically below the wavelength of 400 nm [4]. This is due to the smaller InN mole fluctuation with reduced indium composition in the near-UV quantum wells (QWs), and thus less localized energy states lead to lower efficiency of the near-UV LEDs [36,37]. Moreover, crystalline quality and light absorption of GaN are significant for short-wavelength near-UV LEDs [38,46]. Besides, the high refractive index of GaN restricts the escape angle of emitting light and results in low light extraction efficiency. To improve the light extraction efficiency of the InGaN-based LEDs, previous reports such as various surface texture [47–49], photonic crystal structure [50,51], patterned substrate [52–54], and cone shaped sidewall structure [55–58] methods have been investigated and have demonstrated significant light extraction enhancement. However, it is well known that the semiconductor will absorb light when the photon energy is larger than the bandgap. By Urbach tail law, we assume that the light absorbed inside the LED will transform to thermal energy, which will influence the total light output power of an LED device. In this study, we propose a simple selective wet etching method to form hexagonal pyramid structures in n-AlGaN with an AlN sacrificial layer, and the light extraction efficiency in the UV LEDs structure will be discussed.

The samples in this study were grown on c-plane "2" sapphire substrates by MOCVD system. The metal–organic compounds of trimethylgallium, trimethylaluminum, trimethylindium, and ammonia (NH_3) were employed as the reactant source materials for Ga, Al, In, and N, respectively. Silane and bis-cyclopentadienyl magnesium (Cp_2Mg) were used as the sources for n-type and p-type dopants, respectively. Prior to the growth, the sapphire substrates were thermal cleaned in hydrogen ambient at 1100°C. The UV LED structure with InGaN/AlGaN MQW consisted of a 30 nm thick low-temperature (500°C) GaN nucleation layer (GaN NL), a 1 μm thick u-GaN epilayer, a 1 μm Si-doped AlGaN layer, a 100 nm Si-doped AlN layer, a 2 μm Si-doped AlGaN layer, an InGaN/AlGaN MQWs active layer, a 15 nm Mg-doped AlGaN electron blocking layer (p-AlGaN), and a 0.2 μm Mg-doped GaN contact layer (p-GaN). The InGaN/AlGaN MQWs' active region consists of 10 periods of 3 nm thick $In_{0.025}Ga_{0.975}N$ well layers and 12 nm thick $Al_{0.08}Ga_{0.92}N$ barrier layers. During the growth of AlN layer, we kept the flow rates of TMAl, NH_3, and SiH_4 at 87.37 μmol/min, 0.89 mol/min, and 24.5 nmol/min. The growth rate, temperature, and pressure in AlN layer are 0.6 μm/h, 1150°C, and 15 kPa, respectively. After MOCVD growth, the sample was selectively etched through AlN layer by inductively coupled plasma (ICP) to form a mash type channel for laser cutting. Then we used ICP again to etch mesa and expose n-AlGaN layer, followed by a dry etching process, where we deposited an indium–tin–oxide (ITO) transparent contact layer to serve as the p-contact electrode. After transparent contact layer (TCL) deposit, Cr–Au contact was deposited onto the exposed n-type AlGaN layer and TCL to serve as the n-type and p-type electrode, respectively. Before chemical wet etching, the sample was separated into two areas. The left side area was protected by silicon dioxide (SiO_2) to prevent damage from the treatment of KOH base solution, and we defined the device of the area to be conventional LED (C-LED). The right surface area was exposed and AlN served as a sacrificial layer to form a hexagonal pyramid structure LED (HPS-LED). Then the sample was wet etched by potassium hydroxide (KOH) and potassium persulfate ($K_2S_2O_8$) for 60 min at 90°C. A schematic diagram of the LED without and with treatment is shown in Figure 10.19a and b, respectively.

Figure 10.20 shows a cross-sectional scanning electron microscope (SEM) image of the undercut sidewalls formed in this study. It can be seen that hexagonal pyramid structures were formed in n-AlGaN layer successfully. This fabrication process consisted of a lateral wet etching process on the AlN sacrificial layer and a crystallographic etching process on an N-face AlGaN layer. Before measurement, we used tetrafluoromethane carbon tetrafluoride (CF_4) in ICP process to remove surface SiO_2.

We want to compare the same LED device before and after KOH etch, the testing LEDs were all located near the alignment key. Figure 10.21 shows 20 mA electroluminescence (EL) spectra of the fabricated LED. It was found that EL peak positions of the LED occurred around 380 nm with the FWHM of 8.9 nm, which is before KOH etched (C-LED), and 9.5 nm, which is after KOH etched (HPS-LED).

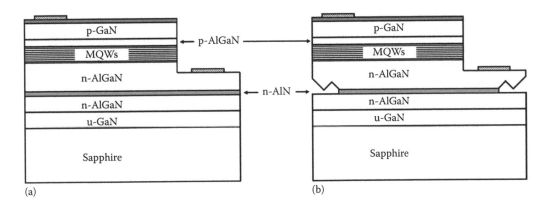

FIGURE 10.19 (a) A schematic diagram of the LED without treatment of KOH base solution, (b) a schematic diagram of the LED with treatment of KOH base solution.

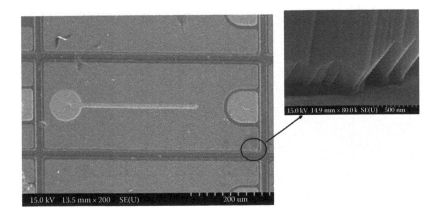

FIGURE 10.20 The scanning electron microscope (SEM) image of the HPS formed in this study.

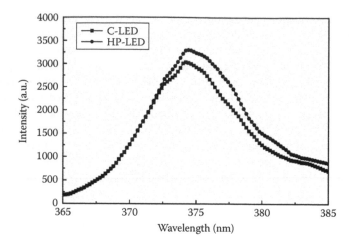

FIGURE 10.21   Electroluminescence (EL) spectra of the fabricated LEDs at a forward current of 20 mA.

Figure 10.22 shows intensity–current–voltage (L–I–V) characteristics of the LED before and after KOH treatment (C-LED and HPS-LED). Under 20 mA current injection, it was found that forward voltages (Vf) were 3.65 and 3.73 V for the C-LED and HPS-LED, respectively. It was also found that the output power measured from HPS-LED was always higher than that of C-LED under the same injection current. With 20 mA injection current, it was found that we achieved a 12% (C-LED was 2.69 mW and HPS-LED was 3.01 mW) enhancement in the output power from the HPS-LED, as compared to C-LED.

We propose a simple selective wet etching method to form hexagonal pyramid structures for 380 nm UV LED with an AlN sacrificial layer. From SEM image, the HPS and AlN sacrificial layer can be clearly seen on the interface of AlGaN layer. The measurement results indicate that the light output power can be enhanced effectively when the HPSs are fabricated in the UV LEDs. Furthermore, the electroluminescence results show that UV LEDs with HPS exhibit 12% higher output power at a forward current of 20 mA. We attribute this improvement to increased light extraction efficiency around the sidewall.

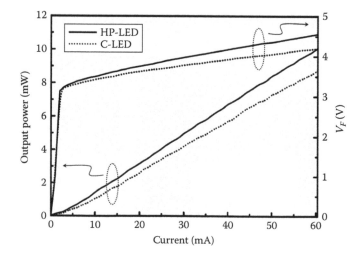

FIGURE 10.22   Intensity–current–voltage (L–I–V) characteristics of the UV LED before and after KOH treatment.

10.2.5 Inverted Micropyramid Structures

A common feature in all these different methods is having large surface variations at the GaN/air or GaN/sapphire interface. The fabrication process often involves microlithography and etching. Here, we report a fabrication process that can significantly improve both the light extraction efficiency and crystal quality without the need of photolithography substrate patterning.

The material epitaxial growth uses nominal low-pressure MOCVD. A 30 nm of low temperature GaN nucleation layer followed by a 2.5 μm GaN buffer layer was grown on a (0001) sapphire template. The GaN wafer was immersed in high-temperature molten KOH at 280°C for 12 min. The molten KOH selectively etched defects on the wafer surface and etched continuously downward opening up channels to sapphire interface. The molten KOH was led to GaN/sapphire interface through these self-assembled channel openings. The etching process then turned into a lateral direction because the defect density was high at the interface and etched away a thin layer of GaN along sapphire interface. It is known that KOH etching is typically anisotropic and preferentially etches specific crystallographic planes. A tilted view SEM image is shown in Figure 10.23a, where the inverted pyramid structure at the GaN-sapphire interface can be seen from a large opening. A large number of hexagonal pits were also formed on the surface. The etch pit density was 5×10^8 cm^{-2} from SEM image estimation. Additional GaN was grown on the etched GaN wafer to fill up both the etched openings and surface pits to provide flat top surface for the subsequent LED device growth. The LED device structure was 3.5 μm n-doped GaN, 10 pairs of Al$_{0.05}$Ga$_{0.95}$N/InGaN quantum wells 13/2.5 nm, and 100 nm of p-doped GaN cap layer. The designed emission wavelength is at 395 nm. A cross-sectional SEM image of a cleaved sample after the regrowth process is shown in Figure 10.23b.

The space and the inverted pyramid structures created at the GaN/sapphire interface are still well maintained and distributed throughout the large area. Most of the pyramid tips are still in contact with sapphire. To investigate the performance of the device structure and make fair comparisons, a reference wafer also went through exactly the same fabrication process, except for skipping the KOH etching step.

The XRD rocking curves of these two samples are shown in Figure 10.24. The line width for (102) planes was reduced from 552 to 472 arcsec. The line width for (002) planes was only reduced from 338 to 335 arcsec. The XRD line widths for (102) and (002) planes are related to edge and screw threading dislocation densities, respectively [59]. The decrease in XRD line width indicates improved material quality. The improvement is attributed to the strain relaxation of the partially relieved GaN layer by interfacial etching and the subsequent regrowth. When GaN epitaxial layer was first grown on sapphire, a compressive strain was built up in the material due to the mismatched lattice constants and thermal expansion coefficients between GaN and sapphire. The KOH interfacial etching partially relieved GaN from sapphire interface and relaxed the compressive strain. This partially relieved layer served as a buffer layer to reduce the problems of mismatched lattice constants and thermal expansion coefficients during MOCVD regrowth and led to improved crystal quality.

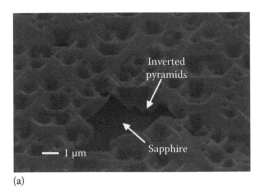

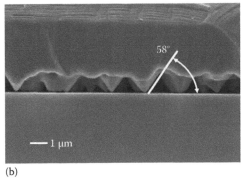

(a)   (b)

FIGURE 10.23 (a) SEM image of the etched GaN surface. (b) SEM cross-sectional image of the regrown sample.

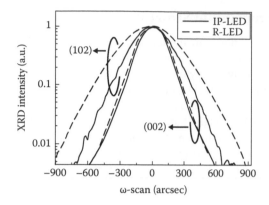

FIGURE 10.24   XRD rocking curves for IP-LED and R-LED samples.

These two samples were made into LED chips. Indium–tin–oxide was used as a current spreading layer and Ni/Au as a p-type electrode contact. Ti/Al/Ni/Au was deposited on the exposed n-GaN to serve as an n-type electrode contact. The sapphire substrates of both samples were lapped down and a 240 nm Al metal reflector coating was deposited on the sapphire back surface. The use of back reflector is common in finished LED package. Therefore, it was also included. They were finally scribed into 350 × 350 μm² LED chips.

The schematic of the LED with inverted micropyramid structures (IP-LED) is shown in Figure 10.25a. The reference LED (R-LED) has a similar structure except for a flat GaN-sapphire interface. The EL spectra of both LEDs collected in the normal to the front surface direction are shown in Figure 10.25b. The peak

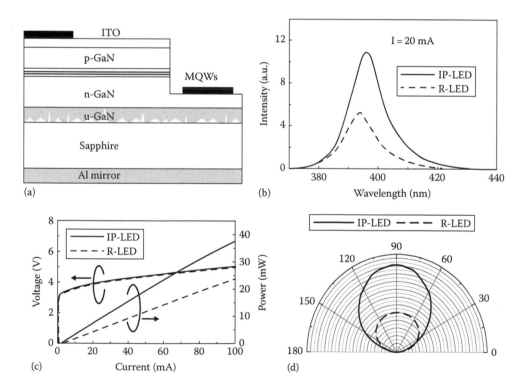

FIGURE 10.25   (a) Schematic of IP-LED structure, (b) EL spectra of IP-LED and R-LED in normal direction, (c) L–I–V curves of IP-LED and R-LED, and (d) far-field patterns of IP-LED and R-LED.

intensity of IP-LED is enhanced by 112% compared to that of R-LED. The EL spectrum of R-LED shows slight Fabry–Pérot mode ripples. The Fabry–Pérot mode spacing is about 5 nm, which is consistent with the overall GaN thickness of 6 μm and a refractive index of 2.55 at 394 nm. The EL spectrum of IP-LED on the other hand does not have the same Fabry–Pérot mode ripples. This is because the randomly distributed micropyramid structures suppress the standing wave formation between the top and bottom interfaces. The peak wavelength of IP-LED is redshifted by 2 nm compared to that of R-LED, indicating the relaxation of compressive strain on IP-LED [60].

The light–current (L–I) and voltage–current (V–I) characteristics are shown in Figure 10.25c. The forward voltages of IP LED and R-LED are 3.86 and 3.80 V, respectively, at 20 mA and increase to 5.09 and 4.93 V at 100 mA. The electric characteristic of IP-LED is still reasonably well maintained. The optical powers of IP-LED and R-LED collected by an integrating sphere are 7.31 and 3.95 mW at 20 mA and 37.5 and 23.7 mW at 100 mA, respectively. The IP-LED output power exhibits 85% and 58% enhancement at 20 and 100 mA, respectively. The decrease in output power enhancement is likely due to the lower thermal conductance from the smaller GaN sapphire contacts. The small interface contacts on the other hand are crucial for relieving the compressive strain and reducing material defects as described previously, which in turn reduces heat generation. The low thermal conductance problem is therefore not as serious as it would be. We also measured the far-field pattern at 20 mA injection current as shown in Figure 10.25d. The IP-LED emission in normal direction is much more enhanced. The emission intensity is enhanced by 120% in the normal direction and 62% in the 45° direction. The divergent angles of IP-LED and R-LED are 108° and 128°, respectively.

10.2.6 Summary

In this chapter, we demonstrate several novel technologies to produce high-efficiency UV LEDs. The high-quality UV LEDs structures were successfully fabricated onto SE-GaN templates. Besides, both the Mg- and heavy Si-doping methods have advantages for device performance and should be considered. On the other hand, the InGaN-based UV LEDs with quaternary InAlGaN barrier exhibit higher radiative recombination rate and low-efficiency droop at a high injection of current. Finally, the light output power of UV LEDs will be improved using hexagonal pyramid and inverted micropyramid structures. Based on the results obtained, using these novel technologies is suggested to be effective for improving the quality of GaN-based UV LEDs.

References

1. A. Sandhu, The future of ultraviolet LEDs, *Nat. Photon.* 1, 38 (2007).
2. Y. S. Tang, S. F. Hu, C. C. Lin, N. C. Bagkar, and R. S. Liu, Thermally stable luminescence of $KSrPO_4$: Eu_{2+} phosphor for white light UV light-emitting diodes, *Appl. Phys. Lett.* 90, 151108 (2007).
3. Y. C. Chiu, W. R. Liu, C. K. Chang, C. C. Liao, Y. T. Yeh, S. M. Jang, and T. M. Chen, Ca_2PO_4 Cl: Eu_{2+}: An intense near-ultraviolet converting blue phosphor for white light-emitting diodes, *J. Mater. Chem.* 20, 1755 (2010).
4. H. Hirayama, Quaternary InAlGaN-based high-efficiency ultraviolet light-emitting diodes, *J. Appl. Phys.* 97, 091101 (2005).
5. S. Nakamura, M. Senoh, S. Nagahama, N. Iwasa, T. Yamada, T. Matsushita, H. Kiyoku et al., InGaN/GaN/AlGaN-based laser diodes with modulationdoped strained-layer superlattices grown on an epitaxially laterally overgrown GaN substrate, *Appl. Phys. Lett.* 72, 211 (1998).
6. D. Kapolnek, S. Keller, R. Vetury, R. D. Underwood, P. Kozodoy, S. P. Den Baars, and U. K. Mishra, Anisotropic epitaxial lateral growth in GaN selective area epitaxy, *Appl. Phys. Lett.* 71, 1204 (1997).
7. D. M. Follstaedt, P. P. Provencio, N. A. Missert, C. C. Mitchell, D. D. Koleske, A. A. Allerma, and C. I. H. Ashby, Minimizing threading dislocations by redirection during cantilever epitaxial growth of GaN, *Appl. Phys. Lett.* 81, 2758 (2002).

8. M. H. Lo, P. M. Tu, C. H. Wang, Y. J. Cheng, C. W. Hung, S. C. Hsu, H. C. Kuo et al., Defect selective passivation in GaN epitaxial growth and its application to light emitting diodes, *Appl. Phys. Lett.* 95, 211103 (2009).

9. A. Sakai, H. Sunakawa, and A. Usui, Defect structure in selectively grown GaN films with low threading dislocation density, *Appl. Phys. Lett.* 71, 2259 (1997).

10. M. H. Lo, P. M. Tu, C. H. Wang, C. W. Hung, S. C. Hsu, Y. J. Cheng, H. C. Kuo et al., High efficiency light emitting diode with anisotropically etched GaN-sapphire interface, *Appl. Phys. Lett.* 95, 041109-1 (2009).

11. J.-M. Bethoux, P. Vennéguès, F. Natali, E. Feltin, O. Tottereau, G. Nataf, P. De Mierry, and F. Semond, Growth of high q quality crack-free AlGaN films on GaN templates using plastic relaxation through buried cracks, *J. Appl. Phys.* 94, 6499 (2003).

12. Y. J. Lee, J. M. Hwang, T. C. Hsu, M. H. Hsieh, M. J. Jou, B. J. Lee, T. C. Lu, H. C. Kuo, and S. C. Wang, Enhancing the output power of GaNbased LEDs grown on chemical wet etching patterned sapphire substrate, *IEEE Photon. Technol. Lett.* 18, 1152 (2006).

13. K. Kato, K. Kishino, H. Sekiguchi, and A. Kikuchi, Overgrowth of GaN on Be-doped coalesced GaN nanocolumn layer by rf-plasma-assisted molecular-beam epitaxy: Formation of high-quality GaN microcolumns, *J. Cryst. Growth* 311, 2956 (2009).

14. Q. Shan, Q. Dai, S. Chhajed, J. Cho, and E. F. Schubert, Analysis of thermal properties of GaInN light-emitting diodes and laser diodes, *J. Appl. Phys.* 108, 084504 (2010).

15. X. A. Cao, S. F. LeBoeuf, M. P. D'Evelyn, S. D. Arthur, and J. Kretchmer, Blue and near-ultraviolet light-emitting diodes on free-standing GaN substrates, *Appl. Phys. Lett.* 84, 4313 (2004).

16. M. Iwaya, T. Takeuchi, S. Yamaguchi, C. Wetzel, H. Amano, and I. Akasaki, Reduction of etch pit density in organometallic vapor phase epitaxy-grown GaN on Sapphire by insertion of a low-temperature-deposited buffer layer between high-temperature-grown GaN, *Jpn. J. Appl. Phys.* Part 2, 37, L316 (1998).

17. J. Elsner, R. Jones, P. K. Sitch, V. D. Porezag, M. Elstner, T. Frauenheim, M. I. Heggie, S. Oberg, and P. R. Briddon, Theory of threading edge and screw dislocations in GaN, *Phys. Rev. Lett.* 79, 3672 (1997).

18. P. W. Hunchinson and P. S. Dobson, Defect structure of degraded GaAlAs-GaAs double heterojunction lasers, *Philos. Mag.* 32, 745 (1975).

19. J. Han, M. H. Crawford, R. J. Shul, J. J. Figiel, L. Zhang, Y. K. Song, H. Zhou, and A. V. Nurmikko, AlGaN/GaN quantum well ultraviolet light emitting diodes, *Appl. Phys. Lett.* 73, 1688 (1998).

20. T. S. Zheleva, O.-H. Nam, M. D. Bremser, and R. F. Davis, Dislocation density reduction via lateral epitaxy in selectively grown GaN structures, *Appl. Phys. Lett.* 71, 2472 (1997).

21. D. S. Wuu, W. K. Wang, K. S. Wen, S. C. Huang, S. H. Lin, S. Y. Huang, C. F. Lin, and R. H. Horng, Defect reduction and efficiency improvement of near-ultraviolet emitters via laterally overgrown GaN on a GaN/patterned sapphire template, *Appl. Phys. Lett.* 89, 161105 (2006).

22. T. V. Cuong, H. S. Cheong, H. G. Kim, H. Y. Kim, C.-H. Hong, E. K. Suh, H. K. Cho, and B. H. Kong, Enhanced light output from aligned micropit InGaN-based light emitting diodes using wet-etch sapphire patterning, *Appl. Phys. Lett.* 90, 131107 (2007).

23. Y. J. Lee, J. M. Hwang, T. C. Hsu, M. H. Hsieh, M. J. Jou, B. J. Lee, T. C. Lu, H. C. Kuo, and S. C. Wang, Enhancing the output power of GaN-based LEDs grown on chemical wet etching patterned sapphire substrate, *IEEE Photon. Technol. Lett.* 18, 1152 (2006).

24. J. C. Zhang, D. G. Zhao, J. F. Wang, Y. T. Wang, J. Chen, J. P. Liu, and H. Yang, The influence of AlN buffer layer thickness on the properties of GaN epilayer, *J. Cryst. Growth* 268, 24 (2004).

25. Y. J. Lee, J. M. Hwang, T. C. Hsu, M. H. Hsieh, M. J. Jou, B. J. Lee, T. C. Lu, H. C. Kuo, and S. C. Wang, Enhancing the output power of GaN-based LEDs grown on chemical wet etching patterned Sapphire substrate, *IEEE Photon Technol. Lett.* 18, 1152 (2006).

26. Z. H. Feng, Y. D. Qi, Z. D. Lu, and K. M. Lau, GaN-based blue light-emitting diodes grown and fabricated on patterned sapphire substrates by metalorganic vapor-phase epitaxy, *J. Cryst. Growth*, 272, 327 (2004).

27. Y.-K. Ee, J. M. Biser, W. Cao, H. M. Chan, R. P. Vinci, and N. Tansu, Metalorganic vapor phase epitaxy of III-nitride light-emitting diodes on nano-patterned AGOG Sapphire substrate by abbreviated growth mode, *IEEE J. Sel. Top. Quantum Electron.* 15, 1066 (2009).

28. Y.-K. Ee, X.-H. Li, J. Biser, W. Cao, H. M. Chan, R. P. Vinci, and N. Tansu, GaN-based blue light-emitting diodes grown and fabricated on patterned sapphire substrates by metalorganic vapor-phase epitaxy, *J. Cryst. Growth* 312, 1311 (2010).

29. Y. Li, S. You, M. Zhu, Z. Liang, T. Wenting Hou, Y. Detchprohm, N. Taniguchi, S. T. Tamura, and C. Wetzel, Defect-reduced green GaInN/GaN light-emitting diode on nanopatterned sapphire, *Appl. Phys. Lett.* 98, 151102 (2011).

30. T. Jung, L. K. Lee, and P.-C. Ku, Novel epitaxial nanostructures for the improvement of InGaN LEDs efficiency, *IEEE J. Sel. Top. Quantum Electron.* 15, 1073 (2009).

31. R. M. Farrell, D. F. Feezell, M. C. Schmidt, D. A. Haeger, K. M. Kelchner, K. Iso, H. Yamada et al., Continuous-wave operation of AlGaN-cladding-free nonpolar m-plane InGaN/GaN laser diodes, *Jpn. J. Appl. Phys., Part 2* 46, L761 (2007).

32. R. A. Arif, Y. K. Ee, and N. Tansu, Polarization engineering via staggered InGaN quantum wells for radiative efficiency enhancement of light emitting diodes, *Appl. Phys. Lett.* 91, 091110 (2007).

33. H. Heinke, V. Kirchner, S. Einfeldt, and D. Hommel, X-ray diffraction analysis of the defect structure in epitaxial GaN, *Appl. Phys. Lett.* 77, 2145 (2000).

34. B. Heying, X. H. Wu, S. Keller, Y. Li, D. Kapolnek, B. P. Keller, S. P. DenBaars, and J. S. Speck, Role of threading dislocation structure on the x-ray diffraction peak widths in epitaxial GaN films, *Appl. Phys. Lett.* 68, 643 (1996).

35. I. H. Lee, I. H. Choi, C. R. Lee, and S. K. Noh, Evolution of stress relaxation and yellow luminescence in GaN/sapphire by Si incorporation, *Appl. Phys. Lett.* 71, 1359 (1997).

36. I. H. Ho and G. B. Stringfellow, Solid phase immiscibility in GaInN, *Appl. Phys. Lett.* 69, 2701 (1996).

37. T. Mukai and S. Nakamura, Ultraviolet InGaN and GaN single-quantum-well-structure light-emitting diodes grown on epitaxially laterally overgrown GaN substrates, *Jpn. J. Appl. Phys.* 38, 5735 (1999).

38. R. H. Horng, W. K. Wang, S. C. Huang, S. Y. Huang, S. H. Lin, C. F. Lin, and D. S. Wuu, Growth and characterization of 380-nm InGaN/AlGaN LEDs grown on patterned sapphire substrates, *J. Cryst. Growth* 298, 219 (2007).

39. D. Morita, M. Yamamoto, K. Akaishi, K. Matoba, K. Yasutomo, Y. Kasai, M. Sano, S. I. Nagahama, and T. Mukai, Watt-class high-output-power 365 nm ultraviolet light-emitting diodes, *Jpn. J. Appl. Phys.* 43, 5945 (2004).

40. S. Yamaguchi, M. Kariya, T. Kashima, S. Nitta, M. Kosaki, Y. Yukawa, H. Amano, and I. Akasaki, Control of strain in GaN using an In doping-induced hardening effect, *Phys. Rev. B.* 64, 035318 (2001).

41. S. Watanabe, N. Yamada, M. Nagashima, Y. Ueki, C. Sasaki, Y. Tamada, T. Taguchi, and H. Kudo, Internal quantum efficiency of highly-efficient $In_xGa_{1-x}N$-based near-ultraviolet light-emitting diodes, *Appl. Phys. Lett.* 83, 4906 (2003).

42. Y. J. Lee, C. H. Chiu, C. C. Ke, P. C. Lin, T. C. Lu, H. C. Kuo, and S. C. Wang, Study of the excitation power dependent internal quantum efficiency in InGaN/GaN LEDs grown on patterned sapphire substrate, *IEEE J. Sel. Top. Quantum Electron.* 15, 1137 (2009).

43. I. Mártil, E. Redondo, and A. Ojeda, Influence of defects on the electrical and optical characteristics of blue light-emitting diodes based on III–V nitrides, *J. Appl. Phys.* 81, 2442 (1997).

44. X. A. Cao, E. B. Stokes, P. M. Sandvik, S. F. LeBoeuf, J. Kretchmer, and D. Walker, Diffusion and tunneling currents in GaN/InGaN multiple quantum well light-emitting diodes, *IEEE Electron. Dev. Lett.* 23 (2002) 535.

45. P. G. Eliseev, The red σ2/kT spectral shift in partially disordered semiconductors, *J. Appl. Phys.* 93, 5404 (2003)

46. D. Morita, M. Yamamoto, K. Akaishi, K. Matoba, K. Yasutomo, Y. Kasai, M. Sano, S. I. Nagahama, and T. Mukai, Fabrication of nano-patterned sapphire substrates and their application to the improvement of the performance of GaN-based LEDs, *Jpn. J. Appl. Phys.* 43, 5945 (2004).

47. T. Fujii, Y. Gao, R. Sharma, E. L. Hu, S. P. DenBaars, and S. Nakamura, Increase in the extraction efficiency of GaN-based light-emitting diodes via surface roughening, *Appl. Phys. Lett.* 84, 855 (2004).

48. C. E. Lee, Y. C. Lee, H. C. Kuo, T. C. Lu, and S. C. Wang, High-brightness InGaN–GaN flip-chip light-emitting diodes with triple-light scattering layers, *IEEE Photon. Technol. Lett.* 20, 659 (2008).

49. S. J. Chang, C. F. Shen, W. S. Chen, C. T. Kuo, T. K. Ko, S. C. Shei, and J. K. Sheu, Nitride-based light emitting diodes with indium tin oxide electrode patterned by imprint lithography, *Appl. Phys. Lett.* 91, 013504 (2007).

50. A. David, T. Fujii, E. Matioli, R. Sharma, S. Nakamura, S. P. DenBaars, and C. Weisbuch, Photonic crystal laser lift-off GaN light-emitting diodes, *Appl. Phys. Lett.* 88, 133514 (2006).

51. T. A. Truong, L. M. Campos, E. Matioli, I. Meinel, C. J. Hawker, C. Weisbuch, and P. M. Petroff, Light extraction from GaN-based light emitting diode structures with a noninvasive two-dimensional photonic crystal, *Appl. Phys. Lett.* 94, 023101 (2009).

52. Y. J. Lee, H. C. Kuo, T. C. Lu, B. J. Su, and S. C. Wang, Fabrication and characterization of GaN-based LEDs grown on chemical wet-etched patterned sapphire substrates, *J. Electrochem. Soc.* 153, G1106 (2006).

53. A. Bell, R. Liu, F. A. Ponce, H. Amano, I. Akasaki, and D. Cherns, Light emission and microstructure of Mg-doped AlGaN grown on patterned sapphire, *Appl. Phys. Lett.* 82, 349 (2003).

54. J. Lee, S. Ahn, S. Kim, D. U. Kim, H. Jeon, S. J. Lee, and J. H. Baek, GaN light-emitting diode with monolithically integrated photonic crystals and angled sidewall deflectors for efficient surface emission, *Appl. Phys. Lett.* 94, 101105 (2009).

55. C. F. Lin, C. C. Yang, J. F. Chien, C. M. Lin, K. T. Chen, and S. K. Yen, Fabrication of the InGaN-based light-emitting diodes through a photoelectrochemical process, *IEEE Photon. Technol. Lett.* 21, 1142 (2009).

56. D. S. Kuo, S. J. Chang, T. K. Ko, C. F. Shen, S. J. Hon, and S. C. Hung, Nitride-based LEDs with phosphoric acid etched undercut sidewalls, *Photon. Technol. Lett.* 21, 510 (2009).

57. C. F. Lin, C. M. Lin, C. C. Yang, W. K. Wang, Y. C. Huang, J. A. Chen, and R. H. Horng, InGaN-based light-emitting diodes with a cone-shaped sidewall structure fabricated through a crystallographic wet etching process, *Electrochem. Solid-State Lett.* 12, 233 (2009).

58. H. P. Shiao, C. Y. Wang, M. L. Wu, and C. H. Chiu, Enhancing the brightness of GaN light-emitting diodes by manipulating the illumination direction in the photoelectrochemical process, *IEEE Photon. Technol. Lett.* 22, 1653 (2010).

59. H. Heinke, V. Kirchner, S. Einfeldt, and D. Hommel, X-ray diffraction analysis of the defect structure in epitaxial GaN, *Appl. Phys. Lett.* 77, 2145 (2000).

60. P. P. Paskov, R. Schifano, T. Malinauskas, T. Paskova, J. P. Bergman, B. Monemar, S. Figge et al., Photoluminescence of a-plane GaN: Comparison between MOCVD and HVPE grown layers, *Phys. Status Solidi C* 3, 1499 (2006).

11

Influence of Carrier Localization on Efficiency Droop and Stimulated Emission in AlGaN Quantum Wells

Gintautas
Tamulaitis
Vilnius University

Abstract Carrier localization is a phenomenon substantially affecting the carrier dynamics in AlGaN epitaxial layers and quantum well structures. In this chapter, the origins of the carrier localization in AlGaN are discussed, the key experimental techniques for the study of this phenomenon and its influence on the efficiency of the radiative and nonradiative recombination of nonequilibrium carriers are presented, and the features important for practical applications of AlGaN are studied. The current status in the study of the peak photoluminescence efficiency and the efficiency droop at elevated excitation intensities in AlGaN and AlGaN-based heterostructures is presented. The importance of stimulated emission on carrier dynamics at high densities of nonequilibrium carriers is revealed. From the application point of view, the review is focused on the influence of carrier localization on the luminescence efficiency droop, which is of especial importance for the development of efficient deep UV LEDs, and on the peculiarities of the stimulated emission in high-aluminum-content AlGaN, which is prospective for the development of deep UV laser diodes still encountering substantial problems.

11.1 Carrier Localization in AlGaN Epitaxial Layers and Quantum Wells

11.1.1 Carrier Localization due to Composition Fluctuations

The carriers in ternary compound $Al_xGa_{1-x}N$ might be localized due to bandgap fluctuations caused by fluctuations in aluminum content x. It is generally accepted that the lattice constants of $Al_xGa_{1-x}N$, as well as of many others semiconductor compounds, follow the Vegard's law and linearly increase from their values in GaN to those in AlN as x in $Al_xGa_{1-x}N$ increases from 0 to 1. Meanwhile, the corresponding increase in the bandgap is nonlinear and is usually described as

$$E_g(x) = (1-x)E_g(\text{GaN}) + xE_g(\text{AlN}) - bx(1-x), \tag{11.1}$$

where

$E_g(\text{GaN})$ and $E_g(\text{AlN})$ are the bandgaps of GaN and AlN (at room temperature, they are equal to 3.42 and 6.13 eV, respectively)

b is the bowing parameter

The b values reported for AlGaN are scattered in a wide range from -0.8 to $+2.6$ eV [1]. Currently, values close to $b = 1$ are used in most of AlGaN studies (e.g., $b = 0.94$ eV is suggested in the recent study of bandgap bowing of AlGaN [2].)

The alloy AlGaN has intrinsic composition fluctuations due to random spatial distribution of Ga and Al cations in AlGaN lattice. Broadening of the PL band at low temperatures is the most straightforward signature of the fluctuations. The increase in full width at half maximum (FWHM) of the PL band is proportional to the standard deviation of the bandgap, which can be expressed as

$$\sigma_{BG} = \frac{\partial E_g}{\partial x} F\big(x, m_e, m_h, \varepsilon(x), a\big). \tag{11.2}$$

Here, the first factor reflects the change in the bandgap when the compound composition x is changed and can be easily calculated using Equation 11.1. The second factor depends on x and material parameters: electron and hole masses, m_e and m_h, composition-dependent static dielectric constant $\varepsilon(x)$, and lattice constant a. Two approaches are suggested to calculate the second factor $F(x, m_e, m_h, \varepsilon(x), a)$ [3,4]. It is estimated [5,6] that the change in FWHM calculated according to the expression suggested in Reference 3 is larger than that calculated according to Reference 4, however, there is no clear evidence on which approach provides a better accuracy for the width of the PL band in AlGaN. For undoped $Al_xGa_{1-x}N$, the first approach describes the composition dependence of the low-temperature PL band FWHM in the range studied ($x = 0.1$–0.5) quite well, while doping of the compound by silicon results in additional broadening [7] (see Figure 11.1). The PL bandwidths corresponding to a homogeneous broadening and a random cation distribution have been observed in $Al_xGa_{1-x}N$ with $0.6 < x < 0.7$ using the scanning near-field photoluminescence spectroscopy [8].

Significant fluctuations in local composition might occur due to decomposition of the ternary compound. This phenomenon is important in InGaN exhibiting a large miscibility gap (see, e.g., the seminal work [9] or a more recent report [10]), which strongly hinders the deposition of InGaN epilayers and MQWs with In content above ~20%. Meanwhile, the calculations show that AlGaN system is completely miscible [11], and $Al_xGa_{1-x}N$ epilayers are grown with Al content spanning from $x = 0$ to 1. Nevertheless, the spatial distribution of Al and Ga cations in AlGaN depends on growth conditions and, as discussed later, might significantly deviate from the content fluctuations due to statistical disorder.

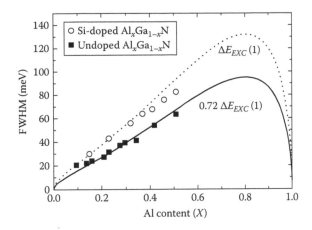

FIGURE 11.1 Full width at half maximum of PL band at the temperature of 12 K as a function of Al content measured in (dots) and calculated for (lines) undoped and Si-doped $Al_xGa_{1-x}N$ epilayers. The indicated multiplication factor is applied to match the units in calculations and experimental data. (After James, R., Leitch, A.W.R., Omnès, F., and Leroux, M., The effect of exciton localization on the optical and electrical properties of undoped and Si-doped $Al_xGa_{1-x}N$, *Semicond. Sci. Technol.*, 21, 744–750, 2006. Reproduced with permission from IOP Publishing.)

11.1.2 Carrier Localization due to Well Width Fluctuations in Quantum Wells

The active regions of light-emitting diodes (LEDs) and laser diodes (LDs) are fabricated as multiple quantum wells (MQWs). In comparison with the epilayers, growing the heterostructures encounters additional problems, and the MQWs exhibit many undesirable features deteriorating the emission efficiency. However, these disadvantages are considerably outweighed by the increase in the effective density of injected carriers. On the other hand, the problems occurring due to the introduction of the heterostructures are being solved, at least partly. Several problems in AlGaN heterostructures have their origin in the mismatch of the lattice constants in the adjacent well and barrier layers containing different percentages of aluminum. A thin layer initially grows by adopting the lattice constant of the previous layer. Consequently, such a layer becomes strongly strained. Since AlGaN is piezoelectric material, the strain results in a built-in electric field, which is directed perpendicularly to the layers. This field occurs in addition to the field due to the spontaneous polarization caused by different surface charges in the materials forming the well and barrier. The built-in electric field causes the quantum confined Stark effect, that is, the redshift in light absorption and emission due to distortion of the potential profile in the quantum well. In the presence of an electric field, the rectangular quantum well transforms into a triangular shape, and the distance between the lowest energy states for electrons and holes becomes smaller than that in the rectangular quantum well. More importantly, the transformation of the well into the triangular shape separates electrons and holes in the direction perpendicular to the QW, the spatial overlap of the electron and hole wave functions decreases, and, as a result, the rate of radiative recombination also decreases. This decrease might be substantial, especially in wide QWs. To diminish this effect and to increase the emission quantum efficiency, narrow QWs, usually 2–3 nm thick, are fabricated in AlGaN-based MQWs designed for light emitters. Thus, the quantum confinement in such QWs is quite strong, and the effective bandgap, that is, the distance between the lowest energy states for electron and hole, strongly depends on the well width. The confinement-induced shift of the lowest energy state from the bottom of the quantum well is described in the effective mass approximation, which works well even at the well widths as small as 2 nm, which is well within the range of the spatial expansion of the wave functions of electrons and holes:

$$\Delta E = \frac{\hbar^2 \pi^2}{2L_w^2 m^2}. \tag{11.3}$$

Here, L_w is the well width and m is the effective mass of electron or hole. According to Equation 11.3, the energy shift due to variation of the well width by one monolayer equals, for example, 7 and 47 meV for the well widths of 5 and 1.65 nm, respectively. Thus, the potential fluctuations due to the well width variation are substantial in AlGaN even in comparison with the thermal energy at room temperature (25 meV). As demonstrated in AlGaN quantum wells with different QW widths ranging from 1.65 to 5 nm [12], the increase in the carrier lifetime in narrower QWs due to stronger localization might be more pronounced than its decrease expected due to the increasing radiative recombination rate caused by a better overlap of the electron and hole wave functions.

11.2 Signatures of Carrier Localization in AlGaN

11.2.1 Studies of Composition Fluctuations by Structural Analysis

The direct studies of composition fluctuations in III-N compounds have been started in the early days of the "blue revolution" by the extensive investigations of the inhomogeneous distribution of indium in InGaN by transmission electron microscopy (TEM). The interpretation of most of these results is based on the assumption that the local strain fields are caused by the distribution of In in the alloy. The observations of indium clusters were reported in numerous publications. Later on, it was demonstrated that the electron beam used in TEM experiments might easily impose substantial damage to the sample under study. Thus, the TEM experiments should be carried out with due care and the danger of misinterpretation of the TEM results is quite high [13].

The TEM technique is also used to characterize the AlGaN epilayers and structures. TEM images showed Al segregation in low-Al-content AlGaN [14], the combined application of TEM with electron beam–induced current (EBIC) technique enabled revealing phase separation in Al-rich AlGaN [15]. TEM is an especially useful tool for the characterization of heavily doped AlGaN epilayers [16] or AlGaN-based nanostructures [17]. Nevertheless, the TEM studies of AlGaN do not lead to any generalized conclusions about the Al distribution in this compound and its heterostructures, though the technique is a useful tool for case-by-case characterization.

The scanning electron microscopy (SEM), as well as the atomic force microscopy (AFM), is often exploited for the characterization of AlGaN surface. The SEM images can often be matched with the images of emission characteristics obtained by cathodoluminescence spectroscopy in the same experimental facility, while the AFM technique can be productively used together with the photoluminescence spectroscopy. These experiments provide some information on composition distribution in AlGaN, as partially discussed in the following.

11.2.2 Inhomogeneous Distribution of Luminescence Parameters

The direct measurement of the distribution of photoluminescence (PL) parameters on the sample surface by scanning the surface with tightly focused excitation light beam is limited by light self-diffraction. For a light beam with wavelength λ, the smallest spot diameter in the focus d depends on the numerical aperture of the focusing optics $NA = n\sin\theta$, where θ is the half angle between the marginal converging rays of the light cone and n is the refraction index of the medium between the lens and the focal plane:

$$d_R = 0.61\,\lambda/NA. \tag{11.4}$$

The technique exploiting PL spectroscopy by using a tightly focused light beam is called microphotoluminescence (μ-PL) spectroscopy and usually employs an appropriate optical microscope. According to Equation 11.4, the spatial resolution in μ-PL measurements is limited by the wavelength of the PL light, since the wavelength of the excitation light is shorter. In the visible range, the spatial resolution in μ-PL experiments is usually above 1 μm.

The resolution can be improved by exploiting the scanning confocal microscopy [18–20]. Moreover, application of confocal microscopy enables scanning the distribution of the PL characteristics in three dimensions. In confocal microscope, the excitation light beam is focused by an objective to a small, nearly diffraction-limited spot on the sample surface. The luminescence signal is collected usually using the same objective. The image of the luminescence collected from the focal point is focused on a screen containing a pinhole confocally aligned with the excited spot of the sample under study. Thus, only the light emitted from the excited spot can pass the pinhole and reach the detection system, while the light from outside the spot is focused off the pinhole and is blocked by the screen. Moreover, the light emitted closer or further on the optical axis than the focal plane is focused in front of the pinhole or behind it. Thus, only a small fraction of this light can pass the pinhole and reach the detector, while the detected signal is dominated by the light from the focal plane. Shifting the excitation spot on the sample surface (usually, by shifting the sample in respect to the excitation spot) and moving the focal plane in the direction perpendicular to the surface enable a 3D scanning.

The best resolution in the sample surface plane xy can be approximated as

$$d_{xy} \approx 0.4 \, \lambda/NA. \tag{11.5}$$

Thus, under appropriate conditions, the xy spatial resolution in confocal microscopy exceeds that in μ-PL by a factor of ~1.4 [20]. The spatial resolution along the z-axis (perpendicular to the surface) is lower and can be expressed as

$$d_z \approx 1.4 \, \lambda n/NA^2. \tag{11.6}$$

The spatial resolution can be considerably improved by using near-field spectroscopy based on the light property to penetrate an aperture of the size considerably smaller than the light wavelength. In this case, the light intensity decays as the sixth power of the distance from the aperture and, thus, penetrates actually only to a distance comparable to the aperture diameter. Consequently, the size of the spot excited through the aperture or of the spot, wherefrom the light is collected though the aperture, is limited not by the wavelength, as in the far-field microscopy, but rather by the aperture diameter. In the practical scanning near-field optical microscopy (SNOM), the aperture is formed at the apex of a tip. Tapered metal-coated fibres with typical aperture diameter of 10–100 nm or hollow silicon tips with a channel tapering to a diameter of approximately 100 nm at the apex are used in SNOM microscopy. The SNOM microscope is equipped with a system to monitor the tip, which is similar to that used in the atomic force microscope (AFM). Thus, the SNOM technique provides the capability to simultaneously scan two spatially matched mapping images providing information on the distribution of PL characteristics and the surface morphology. The SNOM system can operate in different excitation and illumination modes. In the illumination (I) mode, the sample is excited through the tip aperture in the near field, while the luminescence is detected in the far field. In the collection (C) mode, the sample is excited in the far field, but the luminescence is collected in the near field via the tip. The illumination–collection (I–C) mode, when the sample is excited and the luminescence is collected through the tip aperture, is also being exploited. The difference in the experimental conditions caused by using the different excitation and signal collection modes was exploited to study the inhomogeneous diffusivity of the carriers in InGaN QWs [21]. The areas with stronger carrier diffusivity exhibited a higher PL intensity in I mode than that in I-C mode. In both modes, the carriers are excited in near field and diffuse outside the excited spot. Emission of all the carriers is detected in I mode with collection in the far field, while only the emission by the carriers left within the excitation spot is probed through the tip in I-C mode. The interpretation is supported by the PL spectra recorded in the SNOM experiment.

Recently, apertureless SNOM techniques have become increasingly popular. In this case, the tip is a metal needle in close proximity of the sample surface. The tip area of the sample is photoexcited in the far field, but the light induces electron oscillations in the metal tip, and the light field directly below the tip is enhanced by many orders of magnitude. The luminescence light is collected in a far-field regime from the entire excited spot, but the signal is dominated by the luminescence from the small area below the tip, where the excitation is considerably larger. The apertureless SNOM provides a better spatial resolution, but the interpretation of the results is more complicated than that by using the tips with apertures.

Additional experimental information can be extracted using SNOM systems containing two probes with capability of controlling each of the tips and changing the distance between the tips. Such a dual SNOM system was used to reconstruct the potential profile in InGaN single QWs [22]. The SNOM spectroscopy has been successfully exploited for the study of spatial inhomogeneities in AlGaN emission, which will be discussed in more detail in Section 11.3.2.

In cathodoluminescence (CL), the emission is excited by a focused beam of high-energy electrons. The ability to focus the electron beam is limited rather by the instrumental capabilities than by physical limits, however, the spatial resolution in CL experiments is often determined by spreading of secondary excitations induced by the fast electrons in the bulk of the material. The excited area inside the crystal depends on the electron energy and usually has a drop-like shape. The CL is often excited using the electron beam produced by a scanning electron microscope (SEM) or a scanning transmission electron microscope (STEM), while the emission is detected and analyzed using the conventional spectroscopic equipment as in PL spectroscopy experiments. Thus, the CL images can be coupled with the SEM or STEM images reflecting the structural properties. The CL images served as the first direct evidence of carrier localization of InGaN epilayers and QWs [23].

11.2.3 Temperature Dependence of Luminescence Characteristics

An indirect evidence of the carrier localization can be obtained by studying the temperature dependence of the PL band parameters: peak position and bandwidth. In semiconductors without carrier localization, the peak position of the band-to-band emission follows the bandgap shrinkage and smoothly shifts to the long-wavelength side as the temperature is increased. The redshift is weak at low temperatures (typically below 20–30 K) and becomes more pronounced (typically equals ~0.4 meV/K) at elevated temperatures. This dependence is caused by the lattice thermal expansion and, as will be discussed in more detail later, is described by the empirical Varshni formula. The dependence becomes "abnormal" in semiconductors with carrier localization. In this case, the PL peak position dependence on temperature exhibits an S-shaped form (see Figure 11.2), while the bandwidth versus temperature dependence often has a W shape. This behavior has been revealed in ternary and quaternary III-nitrides (InGaN [24,25], AlGaN [26], and AlInGaN [27–29]) already in the early years of their intensive study and are usually considered as a signature of exciton hopping [30–32].

At low temperatures, the photoexcited carriers have low probability for hopping out of the localized sites they occupy after the initial relaxation and their radiative recombination occurs predominantly from these states. Consequently, a considerable part of the localized states with smaller energies remains unpopulated though there are nonequilibrium carriers in the higher localized states. The increase in temperature facilitates hopping of the carriers down the localized states and shifts the carrier population to the low-energy side. As a result, the PL band experiences a redshift. At certain temperature, all the lowest localized states are occupied. The further increase in temperature increases the probability for the carriers to hop up, and the carrier population moves to the high-energy side causing a blueshift for the PL band position. At elevated temperatures, the conventional redshift of the PL band due to thermal expansion of the lattice starts to dominate.

The S-shaped temperature dependence of the PL band position proceeds in parallel to the temperature dependence of the bandgap. The temperature dependence of the bandgap, or more accurately, the exciton energy in the crystal can be extracted from the absorption spectra (see. e.g., Figure 11.3).

The temperature dependence of the PL band position depends on the photoexcitation power density (carrier density), as illustrated for AlGaN and AlInGaN epitaxial layers in Figure 11.4. Two features are

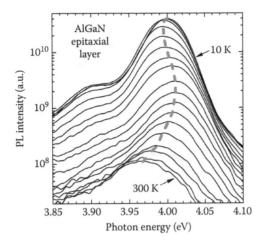

FIGURE 11.2 Photoluminescence spectra of $Al_{0.26}Ga_{0.74}N$ epitaxial layer as a function of temperature. Dashed line is a guide for the eye to indicate the band peak position. (Reprinted with permission from Kazlauskas, K., Žukauskas, A., Tamulaitis, G., Mickevičius, J., Shur, M.S., Fared, R.S.Q., Zhang, J.P., and Gaska, R., Exciton hopping and nonradiative decay in AlGaN epilayers, *Appl. Phys. Lett.*, 87, 172102. Copyright 2005, American Institute of Physics.)

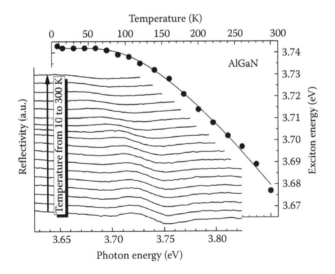

FIGURE 11.3 Reflectivity spectra of AlGaN at different temperatures (lines) and temperature dependence of exciton energy extracted from these spectra (points). (From Kazlauskas, K., Tamulaitis, G., Žukauskas, A., Khan, M.A., Yang, J.W., Zhang, J., Simin, G., Shur, M.S., and Gaska, R.: Localization and hopping of excitons in quaternary AlInGaN. *Phys. Stat. Sol. (c)*. 2002. 0. 512. Copyright Wiley-VCH GmbH & Co. KGaA. Reproduced with permission.)

worth noting in Figure 11.4: (1) the PL band peak position dependence on temperature might not have the explicit S-shape but still might be well below the exciton energy due to localization; (2) the S-shape is smoothened by increasing carrier density (which is feasible due to occupation of the lowest localized states at high carrier density).

An example of the temperature dependence of the full width at half maximum (FWHM) is presented in Figure 11.5. In the temperature range corresponding to the redistribution of carriers down to the lowest localized states, the FWHM does not change significantly. In some samples it decreases slightly, while it

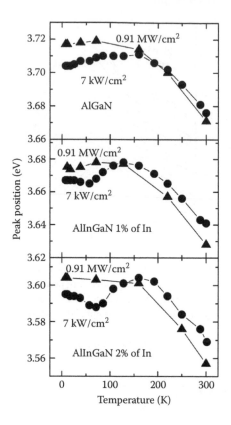

FIGURE 11.4 Temperature dependence of photoluminescence band peak position under excitation power densities of 7 kW/cm² (dots) and 910 kW/cm² (triangles) for AlInGaN samples with different content of indium (indicated). (From Kazlauskas, K., Tamulaitis, G., Žukauskas, A., Khan, M.A., Yang, J.W., Zhang, J., Simin, G., Shur, M.S., and Gaska, R.: Localization and hopping of excitons in quaternary AlInGaN. *Phys. Stat. Sol. (c).* 2002. 0. 512. Copyright Wiley-VCH GmbH & Co. KGaA. Reproduced with permission.)

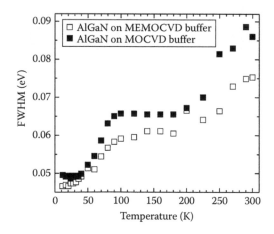

FIGURE 11.5 Temperature dependences of photoluminescence band width in $Al_{0.26}Ga_{0.74}N$ epitaxial layers grown on buffers deposited using MOCVD (solid points) and MEMOCVD (open points) techniques. (Reprinted with permission from Kazlauskas, K., Žukauskas, A., Tamulaitis, G., Mickevičius, J., Shur, M.S., Fared, R.S.Q., Zhang, J.P., and Gaska, R., Exciton hopping and nonradiative decay in AlGaN epilayers, *Appl. Phys. Lett.*, 87, 172102. Copyright 2005, American Institute of Physics.)

slowly increases with temperature in other samples. This slow temperature dependence of the bandwidth is consistent with the redshift of the entire carrier population. The band considerably broadens at elevated temperatures, when higher and higher localized states become populated and the carrier distribution spreads on the energy scale. The further plateau in the temperature dependence of FWHM corresponds to the situation, when a significant part of the carriers becomes thermally delocalized. The free carriers can move further in the crystal than the localized ones can and have a greater probability to reach nonradiative recombination centers and recombine nonradiatively there. Thus, the contribution of the free carriers to the radiative recombination and, consequently, to the emission band is small within the temperature range corresponding to the plateau in the W-shaped dependence. The plateau transforms into an increase in FWHM at elevated temperatures, when a predominant fraction of the carriers is delocalized, and the PL band is formed mainly by the recombination of free carriers.

11.3 Description of Carrier Localization

11.3.1 Exciton Hopping via Localized States

The carrier localization parameter σ can be estimated experimentally using the temperature dependence of PL band peak position. The PL band position in semiconductors without carrier localization should follow the bandgap, which shrinks due to increasing distance between the atoms in the lattice as the temperature increases. The bandgap of semiconductors is quite accurately calculated at zero temperature, but the calculations at elevated temperatures become more complicated and less accurate. The bandgap as a function of temperature is usually described by empirical formulas.

A Bose–Einstein-like expression

$$E(T) = E(0) - \frac{\lambda}{\exp(\theta/T) - 1},\tag{11.7}$$

with $E(0)$ as the bandgap at zero temperature and λ and θ as fitting parameters, often fits well the experimental results. Since the bandgap is nearly constant at temperatures below ~20 K, $E(0)$ can be obtained experimentally with sufficient accuracy. The adjustable parameters λ and θ for GaN and AlN are reviewed by Nam et al. [35].

More often, the empirical Varshni formula is exploited to describe the temperature dependence of the bandgap and, as a result, of the PL peak position [36]:

$$E_g(T) = E_g(0) - \frac{\alpha T^2}{\beta + T}.\tag{11.8}$$

Here, $E_g(0)$ is, as in Equation 11.7, the effective bandgap at $T = 0$, while the Varshni coefficients α and β are free fitting parameters. In line with the Eliseev model, the dependence can be modified to take into account carrier localization described by the ratio of the dispersion parameter σ to the thermal energy $k_B T$ [24]:

$$E_{peak}(T) = E_g(0) - \frac{\alpha T^2}{\beta + T} - \frac{\sigma^2}{k_B T}.\tag{11.9}$$

Equation 11.9 can be exploited to estimate the fluctuation dispersion parameter σ. However, it should be taken into account that (11.9) does not describe the initial redshift of the PL band position at increasing temperature and could be used to fit the bandgap temperature dependence only for thermalized carriers, that is, at the temperatures above the temperature corresponding to the dip in the S-shaped dependence.

Since the temperature corresponding to the dip has a specific physical meaning, sometimes it is useful to use the description of the bandgap dependence on the temperature suggested in [37]:

$$
E_{peak}(T) = \begin{cases} E_g(0) - \dfrac{\alpha T^2}{\beta + T} - \dfrac{\sigma^2}{k_B T}, & (T > T_0), \\[3mm] E_g(0) - \dfrac{\alpha T^2}{\beta + T} - \dfrac{\sigma^2}{k_B T_0}, & (T \le T_0). \end{cases}
\tag{11.10}
$$

Here, the parameter T_0 approximately corresponds to the temperature of the dip in the S-shaped dependence.

11.3.2 Double-Scaled Potential Fluctuations

The PL band parameters can be simulated numerically using the Monte Carlo technique. The comparison of the temperature dependence of simulated and experimental results is shown to be informative for characterization of the system of localized carriers in III-nitrides in general and in AlGaN in particular. To perform the simulations, an assumption on whether the electrons and holes move in the crystal independently or in exciton-like pairs should be accepted. As discussed earlier, the minima in the potential profile for electrons and holes form at the same place in real space. Thus, electrons and holes have conditions to find localized states at the same location. Coulomb attraction stabilizes the pair by lowering its energy and by influencing carrier hopping from one localized state to another. Though tunneling might be important at high density of the localized states and/or at low temperatures, it is reasonable to assume that the carriers move in AlGaN by hopping between the neighboring localized states. However, if one of the carriers, for example, the electron, of the pair acquires enough thermal energy and hops to the neighboring localized state, the other carrier (hole) still experiences the Coulomb attraction, thus the probability for the hole to hop to the site, where the electron is already located, is higher than to hop to any other neighboring localized site.

Under assumption that carriers move in the crystal by hopping in exciton-like pairs, the probability of exciton transition from the state i to the state j can be described by the Miller–Abrahams expression for phonon-assisted exciton hopping between the initial and final states i and j with the energies E_i and E_j, respectively:

$$
\nu_{i \rightarrow j} = \nu_0 \exp\left(-\frac{2r_{ij}}{\alpha} - \frac{E_j - E_i + |E_j - E_i|}{2k_B T} \right).
\tag{11.11}
$$

Here,

    r_{ij} is the distance between the states
    ν_0 is the attempt-to-escape frequency
    α is a parameter reflecting the decay length of the exciton wave function

Exciton hopping can be simulated over a randomly generated set of localized states with certain density N and localization energies dispersed in accordance with a Gaussian distribution around the central position coinciding with the exciton energy and having the dispersion parameter σ. For each exciton, the hopping processes compete with exciton recombination, the probability of which can be described by the exciton lifetime as τ_0^{-1}. The energy of the localized state, where the exciton ends its hopping due to radiative recombination, can be recorded, and scoring these energies for many excitons can be used to construct the emission spectrum $S_0(h\nu)$. The variation of the spectrum with temperature is caused by the thermal enhancement of exciton hopping within the energy distribution and is basically a function of the spatial and temporal parameters described by the products $N\alpha^3$ in 3D or $N\alpha^2$ 2D case and $\nu_0\tau_0$, respectively. The temperature dependences of the emission band parameters obtained by simulation and experimentally can be compared to extract the information on carrier localization potential profile. Both the S-shaped dependence of the band peak position and W-shaped dependence of the band width on temperature can be exploited. The latter dependence turned out to be more sensitive in fitting. The comparison of experimental and simulated results for AlInGaN [38] is illustrated in Figure 11.6.

A straightforward application of the Monte Carlo simulations in the low-temperature range up to the temperatures corresponding to exciton thermalization (the second FWHM increase after the plateau) results in a similar shape of the bandwidth dependence on temperature. However, the simulation curve is shifted parallel down in the FWHM versus temperature plot indicating an additional band broadening. In line with the experimental results, the additional broadening is attributed to a double-scale potential profile. According to this model, there are regions in the crystal, where the average potential (the average bandgap) is considerably lower than in the rest of the crystal. The excitons accumulated in these regions experience potential fluctuations on smaller spatial and energy scales. The situation is schematically depicted in Figure 11.7.

The potential fluctuations of the bottom of the low-potential region are characterized by dispersion parameter σ, while the average potential in different low potential regions fluctuates with dispersion parameter Γ. The inhomogeneous PL band broadening due to the double-scaled potential profile can be described by convoluting the spectra, $S_0(\nu')$, obtained directly from the Monte Carlo simulation with a Gaussian curve, $G(\Gamma, \nu)$, which describes the distribution of the average potential in different low-potential regions:

$$S(\nu) = \int S_0(\nu')G(\Gamma, \nu - \nu')d\nu'. \tag{11.12}$$

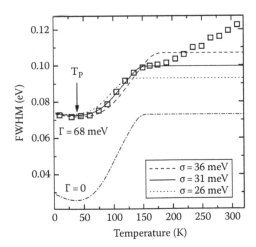

FIGURE 11.6 Experimental (points) and calculated (lines) temperature dependences of full width at half maximum (FWHM) in $Al_{0.1}In_{0.01}Ga_{0.89}N$ epitaxial layer. Dot–dashed line corresponds to a straightforward Monte Carlo simulation for the potential roughness $\sigma = 16$ meV; other solid lines represent the results for double-scaled potential profile with an additional broadening ($\Gamma = 42$ meV) and different parameter σ. (From Kazlauskas, K., Tamulaitis, G., Žukauskas, A., Khan, M.A., Yang, J.W., Zhang, J., Simin, G., Shur, M.S., and Gaska, R.: Localization and hopping of excitons in quaternary AlInGaN. *Phys. Stat. Sol. (c)*. 2002. 0. 512. Copyright Wiley-VCH GmbH & Co. KGaA. Reproduced with permission.)

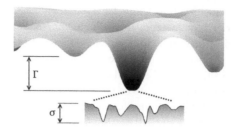

FIGURE 11.7 Schematic picture of double-scaled potential profile in AlGaN. Dispersion parameters of the small scale potential fluctuations σ and of the distribution of the average energy in low-potential regions are indicated.

As seen in Figure 11.6, an appropriate selection of the dispersion parameter Γ results in a fairly good fit of experimental and simulated temperature dependences of FWHM. The dispersion σ defines the FWHM difference between its minimal value and the plateau in the W-shaped dependence. According to the simulation results, σ is related with the temperature T_p corresponding to the minimum in the dependence as

$$\sigma = 2k_B T_p \tag{11.13}$$

The total dispersion parameter Γ can be expressed as

$$\Gamma = \sqrt{\Gamma_P^2 / \ln 4 - \sigma^2}, \tag{11.14}$$

where Γ_P is the experimental line width in the plateau region.

11.3.3 Modified Model of Double-Scaled Potential Fluctuations

The carrier density decay is extensively studied and is usually investigated by time-resolved photoluminescence (TRPL) spectroscopy. Nonexponential PL kinetics are usually observed. The spectral dependence of the PL decay rate has been extensively studied in InGaN [39–42] and AlGaN [43–48] epilayers and QWs. It is revealed that the effective decay time decreases as the detection energy is shifted from low to high energy across the PL band [40,42–44,46–49], as it could be expected due to the energy-loss hopping [31,50]. In QWs, the recombination dynamics is affected also by the built-in electric field [40,49,51]. The PL decay time dependence on spectral position is also sensitive to temperature [42,44] and excitation intensity [29,38,39]. The recent study of AlGaN with different conditions for carrier localization in the temperature range from 10 to 300 K and in a wide range of the density of carriers initially photoexcited by a short (30 ps) laser pulse revealed new features in emission behavior, which cannot be explained either by random potential fluctuations or within the framework of the double-scaled potential profile model exploited before [52].

Typical PL kinetics are demonstrated in Figure 11.8. Three decay ranges can be distinguished in the decay. The initial fast decay is better pronounced at high initial carrier densities and could be attributed to free carrier recombination. The decay in the intermediate range proceeds at a rate being constant at a fixed temperature without showing dependence on the initial carrier density, but becoming faster at elevated temperatures. This component is attributed to recombination of localized carriers. The effective decay times for the fast and intermediate decay components become equal at room temperature. This is an indication that a strong coupling of the systems of localized and free carriers is established at room temperature. The third, slow decay component is persistent at all temperatures and excitation intensities. Its origin will be discussed in Section 11.5.2.

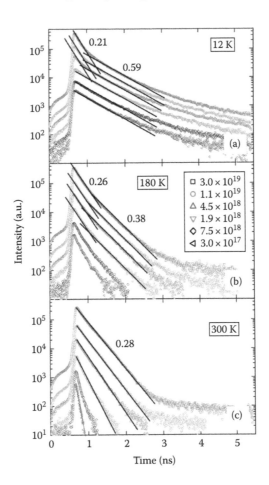

FIGURE 11.8 Time evolution of PL intensity in the vicinity of the band peak position (at 4.9 eV) in $Al_{0.6}Ga_{0.4}N$ epitaxial layer at different initial carrier densities (indicated) and temperatures of 12 K, (a) 180 K (b), and 300 K (c). Straight lines depict the decay slopes corresponding to fast and intermediate decay components. (Reprinted with permission from Saxena, T., Nargelas, S., Mickevičius, J., Kravcov, O., Tamulaitis, G., Shur, M., Shatalov, M., Yang, J., and Gaska, R., Spectral dependence of carrier lifetime in high aluminum content AlGaN epitaxial layers, *J. Appl. Phys.*, 118, 085705. Copyright 2015, American Institute of Physics.)

The dependence of the decay times for the first two decay components is presented at different temperatures in Figure 11.9. Each point in the figure corresponds to the decay time estimated in a narrow spectral range centered at the position of the point on the energy scale. Note that the decay constant of the intermediate decay component shows a considerable dependence on the spectral position, that is, on the energy of the carriers, the recombination of which results in the emission of a photon of the corresponding energy. As pointed out earlier, the decrease of the decay time at the high energy side of the PL band is often observed in InGaN and AlGaN epitaxial layers and QWs and is attributed to weaker localization of the carriers responsible for the high-energy part of the PL band. The decrease in the decay time on the low-energy slope of the PL band has been previously observed in a few publications [40,43,46,49], but no explanation of the feature has been attempted till recently.

The decrease of the PL decay time at the low-energy side of the spectrum was interpreted by introducing a modified double-scale potential profile model [52]. The main difference from the previous model is the assumption that the edges of the regions, where the average potential is lower, are not abrupt, as supposed earlier, but have slopes at the length scale comparable with the distance, which the carrier can

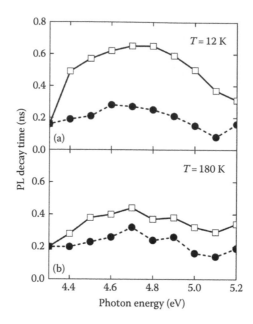

FIGURE 11.9 Dependence of effective times of fast (dashed line) and intermediate (solid line) PL decay components on photon energy at temperatures of 12 K (a) and 180 K (b). (Reprinted with permission from Saxena, T., Nargelas, S., Mickevičius, J., Kravcov, O., Tamulaitis, G., Shur, M., Shatalov, M., Yang, J., and Gaska, R., Spectral dependence of carrier lifetime in high aluminum content AlGaN epitaxial layers, *J. Appl. Phys.*, 118, 085705. Copyright 2015, American Institute of Physics.)

move during its lifetime. The sketch of the fluctuations on the slope of a region with lower average potential is presented in Figure 11.10. A measured PL spectrum is attached in the figure at the corresponding position on the energy scale. Note that the edge of a single low-potential region is depicted in Figure 11.10. Meanwhile, the measured spectrum is formed by emission from many such regions with slightly different average bottom energies. Thus, the measured spectrum is additionally broadened.

The emission intensity at certain spectral position for the intermediate decay component depends on the population of the localized exciton states with the corresponding energy. Radiative and nonradiative recombination and relaxation to lower localized states diminish the population, while it is increased by the carrier relaxation to the states from above. It is reasonable to assume that the rates of the radiative and nonradiative recombination do not significantly depend on the location of the localized exciton on the slope of the low-potential region. Meanwhile, the balance between the numbers of the excitons relaxing to the state from above and those relaxing down from the state is sensitive to the position of the state on the slope.

As explained by the modified model, the effective exciton lifetime is the longest in the middle of the slope depicted in Figure 11.10 due to accumulation of excitons in this region. This accumulation is caused mainly by two effects. First of all, it is feasible that the characteristic length of the slope to the low-potential region in real space is larger than the effective distance, which the carriers can travel during their lifetime. This distance is approximately the same as the exciton diffusion length, which is shown to be in the submicrometer range in AlGaN [53]. Therefore, most of the excitons recombine while hopping down the slope before reaching the end of the slope. On the other hand, the steepness of the slope of the wall of the low-potential region becomes smaller in its lower part, closer to the bottom of the potential well. As a result, the distance between the localized states in real space increases in the lower part of the slope, and, according to the Miller–Abrahams expression 11.11, the probability of hopping between two neighboring states decreases (exponentially with the distance between the states). Thus, the hopping slows down

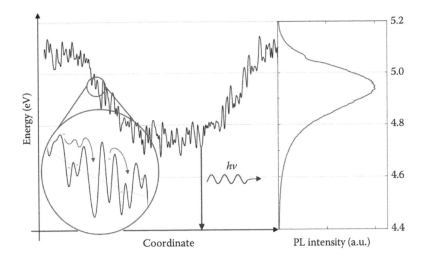

FIGURE 11.10　Sketch of potential fluctuations on the slope of a region with lower average potential (left) and PL spectrum (right). (Reprinted with permission from Saxena, T., Nargelas, S., Mickevičius, J., Kravcov, O., Tamulaitis, G., Shur, M., Shatalov, M., Yang, J., and Gaska, R., Spectral dependence of carrier lifetime in high aluminum content AlGaN epitaxial layers, *J. Appl. Phys.*, 118, 085705. Copyright 2015, American Institute of Physics.)

closer to the bottom of the low-potential region, and the excitons accumulate higher on the slope. Under assumption that the exciton recombination does not have significant dependence of their position on the slope, the accumulation results in longer PL decay time on the central part of the slope, that is, in the vicinity of the PL band peak.

The hopping of excitons down the slope also affects the decrease of the PL decay time on the high-energy part of the band even at low temperatures when contribution of free carriers might be negligibly small. Hopping down the slope of the excitons on the upper part of the slope, which corresponds to the high-energy slope of the PL band, decreases the density of such excitons. Consequently, the effective PL decay time in this energy range is smaller than that down the slope.

The existence of the slope to the low-potential region is also confirmed by the observation that the fast component is present along the entire PL band spectrum. In the conventional model, it should be expected that the fast component is significant for free or weakly localized carriers, which are responsible for the high-energy part of the PL band. However, the experiments show that the fast component is significant throughout the entire spectrum. This feature is easy to interpret by simultaneous generation of carriers on the entire slope of the low-potential region. Consequently, a part of them contribute to the fast component before getting localized to the deeper localized states.

At elevated temperatures, the thermal coupling of all the localized and free states becomes stronger, and the exciton density becomes governed by the fastest recombination channel. It is feasible that the nonradiative recombination of free carriers is the fastest recombination channel. Thus, the two recombination components merge into one, and the PL decay rate does not depend on the spectral position.

The potential profile in AlGaN epilayers and MQWs depends on the substrate and/or template used for the deposition of the active layer, the technological conditions of the deposition, and the aluminum content, and can be quite different in different samples. Anyway, the double-scale potential profile seems to be quite feasible in many cases. The two scales might have quite different influences on carrier dynamics. The small-scale potential fluctuations result in localization of the nonequilibrium carriers, while the large-scale fluctuations facilitate accumulation of carriers in the macroscopic regions, where the average potential is lower than that in the background areas.

11.4 Stimulated Recombination in AlGaN

In spite of the considerable progress in efficiency of deep UV LEDs, the development of deep UV laser diodes (LDs) is still a challenge. Thus, the investigation of light amplification due to stimulated recombination of nonequilibrium carriers is currently one of the hot topics in the study of AlGaN.

The shortest wavelength reported so far in UV LDs is 336 nm [54]. The room temperature stimulated emission under optical pumping has been observed at the peak wavelengths of down to ~241 nm in the AlGaN-based structures grown on sapphire [55–64] and SiC [65–69]; however, the AlN substrates turned out to be considerably more favorable for stimulated emission [70–77] than sapphire and SiC. Though stimulated emission at record-short wavelengths was demonstrated in AlGaN deposited on bulk AlN a decade ago [70], the progress in the development of LD structures homoepitaxially grown on AlN has been hindered by lack of such AlN substrates, which would be large enough at reasonable price and commercially available. The recent breakthrough in LDs based on homoepitaxially grown on bulk AlN is promising and is discussed in detail in Chapter 12. Developing novel buffer layers to enable the deposition of low-dislocation-density AlGaN on sapphire is another approach in searching for the means to fabricate AlGaN-based LDs emitting in deep UV. For example, application of the hetero facet-controlled epitaxial lateral overgrowth (hetero-FACELO) enabled fabrication of an LD lasing at a wavelength of 342.3 nm at room temperature in pulsed current mode [78]. However, no substantial breakthrough in this approach is evident up to now.

There are only a few techniques for measuring the optical gain in semiconductors in general and in III-nitrides and their structures in particular. The most straightforward technique is pump-and-probe spectroscopy [79,80]. The sample is excited by one laser pulse and the changes to the optical transmittance are probed by the second laser beam. If a white-light (or broad spectrum) pulse is used for probing, the spectrum of the changes in transmittance can be obtained. If the optical gain due to stimulated transitions is stronger than the losses due to the light absorption and scattering, the spectrum of the net gain can be obtained in the pump–probe experiments. However, thin samples (thinner than approximately 1 μm) have to be used in the pump-and-probe experiments to ensure sufficient transparency. On the other hand, if the gain region is too thin, the light propagating through this region is not amplified strongly enough for the gain study. Moreover, the AlGaN-based active regions, for example, consisting of MQWs, can have the thickness appropriate for the pump-and-probe experiments, but are deposited on thick buffer layers and substrates, and the background light absorption, though minute per unit of length but strong due to a large thickness of the buffer and substrate, might unacceptably deteriorate the experimental results. For materials contained in cavities, the Hakki–Paoli technique [81,82] can be used. This technique is based on the analysis of the longitudinal Fabry–Perot modes.

The most popular method for measuring the optical gain is the variable stripe length technique [83,84]. The excitation light is focused into a narrow stripe, which is perpendicular to the sample edge. The light propagating along the stripe is amplified due to stimulated recombination and is detected from the edge of the sample. Using the intensity dependence of this light on the length of the stripe, which is changed by blocking the rear part of the stripe at different distances from the edge, the gain coefficient can be extracted in absolute values. Application of this method encounters two major problems: applicability of assumption on one-dimensional light propagation and gain saturation. To extract the absolute value of the gain coefficient and its spectrum, a one-dimensional light propagation model is used. This assumption might be too crude for epilayers without waveguiding layers and for short stripe lengths. Furthermore, the extraction of the gain coefficient is based on a small-signal gain analysis. However, the light propagating along the stipe might be amplified strongly enough to reduce the density of photoexcited carriers, and saturation of the optical gain occurs. The gain saturation occurs when the product of the optical gain coefficient and the stripe length reaches the critical value [85–87], which typically equals ~5 in semiconductor materials and was estimated to be between 4 and 10 in InGaN laser structure, depending on the peak value of the modal gain and the number of quantum wells in the structure [85].

The narrow emission line due to stimulated recombination in AlGaN is observed on both sides of the spontaneous emission band, as illustrated in Figure 11.11. The stimulated emission on the high-energy side

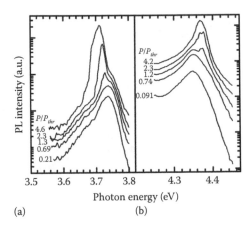

FIGURE 11.11 Edge photoluminescence spectra at 20 K temperature under several excitation power densities in $A_xGa_{1-x}N$ MQWs with Al content in the well of $x = 8\%$ (a) and $x = 35\%$ (b). The indicated values are normalized to the thresholds of stimulated emission $P_{thr} = 171$ kW/cm² and $P_{thr} = 1$ MW/cm² in (a) and (b), correspondingly. (Reprinted with permission from Mickevičius, J., Jurkevičius, J., Kazlauskas, K., Žukauskas, A., Tamulaitis, G., Shur, M.S., Shatalov, M., Yang, J., and Gaska, R., Stimulated emission in AlGaN/AlGaN quantum wells with different Al content, *Appl. Phys. Lett.*, 100, 081902. Copyright 2012, American Institute of Physics.)

of the spontaneous emission band is usually observed in bulk semiconductors. This position is typical of the AlGaN epilayers and structures with low Al content. In high-Al-content AlGaN, the stimulated emission band is observed on the low-energy side of the spontaneous emission band. This position is usually observed in the semiconductor materials with strong carrier localization, where the spontaneous emission band is observed in the spectral region corresponding to emission from localized states, while the stimulated emission occurs in the vicinity of the mobility edge.

The temperature dependence of the edge emission spectra evidences that the carrier dynamics in AlGaN is strongly affected by carrier localization. Figure 11.12 depicts the temperature dependences of the peak positions of the spontaneous and stimulated emission bands for three samples with different Al content [58]. Since the threshold for stimulated emission depends on temperature, the points in Figure 11.12 were obtained at different excitation power densities (indicated) to keep the excitation approximately at the same level above the threshold.

The solid lines in Figure 11.12 are calculated using the Varshni equation [see expression (11.8)] and show the bandgap shrinkage caused by increasing temperature without carrier localization taken into account. The experimental points deviate from the calculated dependence as a result of carrier localization. This deviation becomes more pronounced at higher Al contents. The peak position of the stimulated emission band observed in low-Al-content MQWs also follows the Varshni dependence, but with an approximately constant redshift of ~50 meV. These features are consistent with the behavior of stimulated emission due to free carrier recombination. At elevated Al contents (35% in the data presented in Figure 11.12) the stimulated emission peak follows the Varshni dependence but is below the spontaneous emission peak at high temperature and shifts to the high-energy side of it at temperatures below ~160 K. This is an indication that the nonequilibrium carriers in AlGaN MQWs with considerable Al content are predominantly localized at low temperatures. The stimulated emission occurs approximately at the mobility edge due to a high density of localized states at this energy. At low temperatures, the majority of carriers occupy states below the mobility edge and their spontaneous recombination results in the emission band below the mobility edge. As the temperature increases, the spontaneous emission band blueshifts due to occupation of deep localized states, and an increasing part of the carriers becomes free and provides emission at energies above the mobility edge. Thus, the stimulated emission band at temperatures close to room temperature occurs on the low-energy side of the spontaneous emission band.

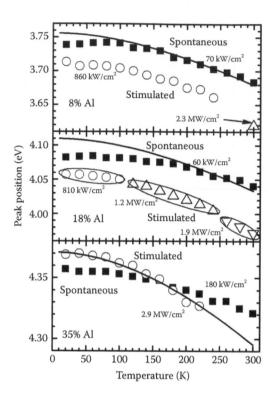

FIGURE 11.12 Temperature dependences of the peak positions of the spontaneous (filled squares) and stimulated emission (open circles and triangles) bands in AlGaN MQWs with different Al content in the wells (indicated in each panel): 8%, 18%, and 35%. (Reprinted with permission from Mickevičius, J., Jurkevičius, J., Kazlauskas, K., Žukauskas, A., Tamulaitis, G., Shur, M.S., Shatalov, M., Yang, J., and Gaska, R., Stimulated emission in AlGaN/AlGaN quantum wells with different Al content, *Appl. Phys. Lett.*, 100, 081902. Copyright 2012, American Institute of Physics.)

As demonstrated in a set of $Al_{0.35}Ga_{0.65}N$ MQWs with quantum wells of different width resulting in different strength of carrier localization [59], the temperature corresponding to the shift of the stimulated emission band position from low-energy to high-energy side in respect to the spontaneous emission band increases with increasing strength of carrier localization. In the samples with strong carrier localization, the stimulated emission occurs on the high-energy side of the spontaneous emission up to room temperature. It is also observed that the increase of the stimulated emission threshold with temperature becomes faster with increasing strength of carrier localization [59]. In AlGaN with strong carrier localization, it is impossible to reach the threshold and observe stimulated emission at room temperature. Thus, the conditions of carrier localization can be intentionally designed, for example, by a proper choice of the well width, to achieve a lower threshold and higher efficiency of the stimulated emission in AlGaN.

11.5 Emission Efficiency and Efficiency Droop

11.5.1 Internal Quantum Efficiency

In spite of the recent progress, the internal quantum efficiency (IQE) is still an important factor limiting the efficiency of AlGaN-based LEDs. IQE is defined by the ratio of the number of emitted photons to the number of electron–hole pairs injected into or photoexcited in the active region under study. The room temperature IQE is usually estimated by measuring the ratio of PL intensities at room and low temperatures. This estimation is based on the assumption that the nonradiative recombination at low temperatures is negligible, and consequently, the IQE equals 100% [88–90]. However, the assumption is valid only at low carrier densities, since the emission efficiency strongly depends on carrier density even at low temperatures [91–94]. Experimentally extracted carrier lifetimes at low and room temperatures, instead of PL intensities, can be used to calculate the IQE; however, this approach is also based on the assumption that the low-temperature IQE equals 100%, while the reliable extraction of the carrier lifetime is more problematic than the measurement of the PL intensity.

Indirectly, IQE is often estimated by fitting the experimental results with the results of the calculations based on the rate equation taking into account linear, bimolecular, and Auger recombination. This description of carrier dynamics is usually referred to as the ABC model [95–99]. The ABC model is a useful simplification, but its application is strongly limited. First of all, the coefficients A, B, and C at high carrier densities become density dependent. For example, at the high densities of nonequilibrium carriers, the term of the linear recombination might become superlinear due to saturation of nonradiative recombination centers, the rate of bimolecular recombination might decrease due to degeneration of the electron–hole system or carrier heating, while the carrier recombination rate might increase much faster than the carrier density squared above the threshold for stimulated recombination. Mathematically, the ABC model can be used as a power series expansion anyway; however, the coefficients A, B, C lose their physical meaning.

In semiconductors with strong carrier localization including AlGaN, where the carrier localization is strong even at room temperature, the application of the ABC model is also strongly limited by the definition of the carrier density. The free carriers become only a fraction of the nonequilibrium carriers. The rest of the carriers are localized, and the rates of the radiative and nonradiative recombination of the localized carriers are different from those for free carriers and should be described by different rate equations. Moreover, the systems of free and localized carriers are linked. Thus, a set of linked rate equations for different kinds of nonequilibrium carriers, free and localized, should be written to properly describe the dynamics of the entire system. At least two coupled rate equations have to be solved to follow the carrier dynamics in such a system containing both free and localized carriers. Moreover, the carriers are localized at different depths in respect to the band edge or, to be more accurate, to the mobility edge separating localized and free states. The description of the system might be improved by decomposing the entire spectrum of the localized states and describing the system by a set of many equations, each describing the rate of carrier transitions into and from certain localized state. However, this approach is not exploited, since such a set of equations would contain many coupling parameters, which are not known. Instead, the exciton hopping via localized states might be described using Monte Carlo simulation, as discussed in Section 11.3.

Another possibility to estimate the IQE is to measure the external quantum efficiency (EQE) and calculate the light extraction efficiency (LEE) by using proprietary or commercial ray tracing software. Assuming that the injection efficiency is 100% or taking it into account when the efficiency is known, the IQE can be calculated from the ratio of EQE and LEE.

Anyway, the estimation of the IQE by using the temperature dependence of PL intensity is the most straightforward and simple technique. However, the technique should be applied with due care. Figure 11.13 depicts the temperature dependences of PL intensity measured at different excitation intensities in AlGaN epilayer and two MQW structures with QW width of 2.5 and 5 nm. The PL efficiency units are the same for all the curves presented in Figure 11.13.

It is generally accepted that the increase in PL efficiency with decreasing temperature is caused by the decreasing rate of nonradiative recombination, and the constant value at low temperatures is considered as an indication of the negligible influence of the nonradiative recombination resulting in IQE = 100%. However, Figure 11.13 evidences that the PL efficiency might be constant at the temperatures below 100 K at considerably different values. For high excitation power densities, the low-temperature value of PL efficiency might be by an order of magnitude lower than the maximum value observed at low excitation intensities. As the excitation intensity decreases, the low-temperature efficiency approaches a

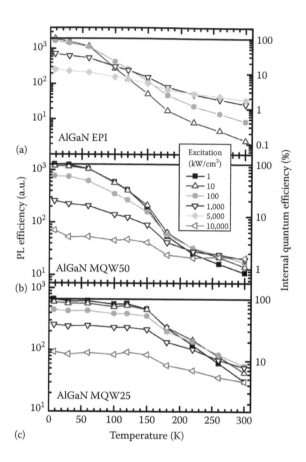

FIGURE 11.13 Temperature dependence of photoluminescence efficiency at different excitation power densities (indicated) in $Al_{0.33}Ga_{0.67}N$ epilayer (a) and MQWs with well width of 5.0 nm (b) and 2.5 nm (c). Internal quantum efficiency calculated under assumption that the highest efficiency observed at low temperatures corresponds to IQE = 100% is indicated on the right-hand side of the plots. (Reprinted with permission from Mickevičius, J., Jurkevičius, J., Kazlauskas, K., Žukauskas, A., Tamulaitis, G., Shur, M.S., Shatalov, M., Yang, J., and Gaska, R., Stimulated emission in AlGaN/AlGaN quantum wells with different Al content, *Appl. Phys. Lett.*, 100, 081902. Copyright 2012, American Institute of Physics.)

constant value. This behavior can be explained by carrier localization, which is quite feasible in AlGaN. The IQE is close to 100% at low excitation intensities when all the photoexcited carriers are localized and their ability to reach nonradiative recombination centers is strongly limited. Increasing excitation intensity results in population of the localized states; thus, an increasing part of the nonequilibrium carriers are free, and their probability of nonradiative recombination increases. Consequently, the PL efficiencies rather than intensities have to be compared to estimate the IQE at elevated temperatures and the PL efficiency at low temperature should be measured at sufficiently low excitation intensities to reliably assume a 100% IQE.

On the other hand, the PL efficiency at room temperature also depends on the excitation intensity. This should be taken into account when comparing the IQEs determined in different samples. In general, to estimate the IQE corresponding to the conditions for carrier dynamics in real LED, the PL efficiency has to be estimated at the temperature and carrier density corresponding to those in the LED for the injection current density at the operating point. This PL efficiency divided by the PL efficiency at low temperature measured at sufficiently low excitation intensity provides the IQE value characterizing the LED performance.

11.5.2 Dynamics of Localized Carriers

The carrier dynamics in AlGaN is experimentally studied usually either under current injection or in quasi-steady-state photoexcitation conditions, or under photoexcitation by short pulses with duration considerably shorter than the carrier lifetime. The current injection experiments are most realistic in view of application of the material in real light-emitting or detection devices, but are less flexible in variability of excitation conditions in respect to photoexcitation and require formation of the entire LED or LD structure consisting of n-type, active, and p-type regions, and Ohmic contacts. Photoluminescence spectroscopy is a contactless investigation tool. The quasi-steady-state photoexcitation, that is, the photoexitation with laser pulses with pulse duration exceeding the carrier lifetime, creates nonequilibrium carriers in the conditions, which are quite similar to those under current injection. The short-pulse excitation enables application of time-resolved PL spectroscopy techniques providing information on the kinetics of the nonequilibrium carrier decay. In both the quasi-steady-state and time-resolved modes, the relationship between the PL characteristics and carrier density depends on the mechanisms and rates of nonequilibrium carrier recombination via radiative and nonradiative recombination channels.

The direct probe for nonequilibrium carrier density can be experimentally obtained by the light-induced transient grating (LITG) technique [100]. In the LITG experiments, two coherent short (picosecond or femtosecond) laser beams of strongly absorbed light (with photon energy above the bandgap) are focused at a certain incidence angle onto the same spot on the sample surface and create an interference pattern (see Figure 11.14).

As a result, a spatially modulated nonequilibrium carrier distribution $N(x) = N_0 + \Delta N \cos(2\pi x/\Lambda)$ with the period Λ occurs. According to the Drude–Lorentz model, the change in the carrier density induces changes in refractive index, which can be expressed as [100]

$$\Delta n(x) = \Delta N(x) n_{eh} = -\frac{e^2}{2n_0 \omega^2 \varepsilon_0} \Delta N(x) \left[\frac{1}{m_e} + \frac{1}{m_h} \right], \qquad (11.15)$$

where
    ΔN is the density of the photoexcited carriers
    ω is the angular frequency of the light creating the grating
    n_0 is the material refractive index before excitation
    ε_0 is the static dielectric permittivity
    n_{eh} is the change in the refractive index due to one electron–hole pair per unit volume

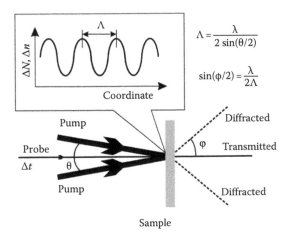

FIGURE 11.14 Configuration of the light beams in light-induced transient grating (LITG) experiments.

Since the electron effective mass in most inorganic semiconductors including AlGaN is considerably smaller than the hole effective mass, the refractive index modulation is dominated by the inhomogeneous distribution of electrons. The decay of the photoinduced carrier grating is monitored by the diffraction of a probe beam in the transparency region of the semiconductor. The diffraction efficiency, that is, the ratio between the intensities of the diffracted and transmitted parts of the probe beam, is proportional to the square of the nonequilibrium carrier density integrated over the excited depth [100]:

$$\eta \propto \Delta n(t)^2 \propto \left[\int \Delta N(z,t)dz\right]^2. \tag{11.16}$$

Thus, the diffraction efficiency is explicitly linked with the carrier density.

The grating decays due to carrier recombination and their ambipolar diffusion, that is, the diffusion of the carrier system consisting of the faster diffusing electrons dragging behind the heavier holes via Coulomb attraction. The grating efficiency decays as

$$\eta = \eta_0 \exp\left(-\frac{2t}{\tau_G}\right). \tag{11.17}$$

The grating decay time constant τ_G can be expressed as

$$\frac{1}{\tau_G} = \frac{1}{\tau_R} + \frac{4\pi^2 D_a}{\Lambda^2}. \tag{11.18}$$

Here, τ_R and $4\pi^2 D_a/\Lambda^2$ are the carrier lifetime and the grating decay time depending on the ambipolar diffusion coefficient D and the grating period Λ. The grating period can be varied by changing the angle between the laser beams creating the grating. Thus, a plot of the grating decay rate on the inverse square of the grating period (see Figure 11.15) enables simultaneous determination of the carrier lifetime and their ambipolar diffusion coefficient. The carrier diffusion length can also be calculated as $L_D = \sqrt{D\tau_R}$.

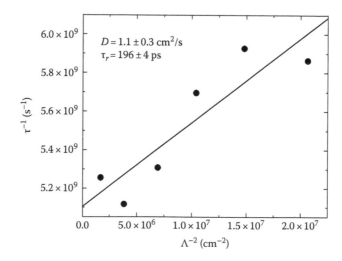

FIGURE 11.15 Decay rate of the light-induced transient grating in AlGaN as a function of the grating period. Diffusion coefficient and carrier lifetime are indicated.

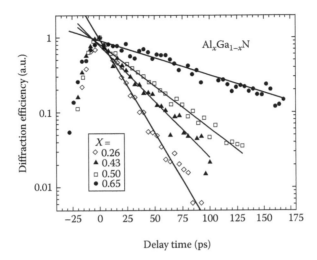

FIGURE 11.16 Decay kinetics of the light-induced transient gratings at a large grating period of 10.2 μm in AlGaN samples with different aluminum content (points) as indicated and the best linear fits of the grating decays (lines). The curves are normalized to their peak values.

If the LITG period is large enough, the carrier recombination dominates over diffusion in the grating decay, the second term in Equation 11.18 can be neglected, and the carrier lifetime can be directly extracted from the LITG decay time. The LITG decay in AlGaN samples with different carrier decay times is illustrated in Figure 11.16.

The decay of the carrier density obtained from the LITG decay after a short pulse excitation of two n-$Al_{0.6}Ga_{0.4}N$ epilayers with different threading dislocation densities (2–5×10^8 cm^{-2} and 2–5×10^9 cm^{-2}) is presented in Figure 11.17. The PL decay kinetics are also presented in Figure 11.17. The PL intensity in the high dislocation density epilayer decays exponentially approximately by three orders of magnitude and has a slow decay component with the time constant substantially exceeding the time range covered in these experiments. In the low dislocation density sample, a fast decay component occurs at high excitation pulse energy, in addition to the exponential and slow decay components. The three decay components can be interpreted by recombination of free and localized carriers and the carriers captured at the states well below the bandgap, respectively [101].

The most interesting feature in the comparison is that the carrier density decays initially at the same rate as the PL intensity, but reaches a slow decay stage at considerably higher carrier density than in the PL decay. This is quite direct evidence that the luminescence at the initial stage of the decay after short-pulse excitation is directly proportional to the carrier density. Linear recombination is expected for exciton-like pairs of localized electrons and holes. However, the luminescence intensity decays at the same rate as the carrier density at the initial relaxation stage after strong short-pulse excitation, when the decay is nonexponential and cannot be described only by exciton recombination.

This behavior can be described by using a set of three coupled rate equations for the densities of free electrons and holes, n and p, and exciton-like pairs of localized electrons and holes n_{ex}. As discussed earlier, the description of the localized electrons and holes as exciton-like pairs is quite reasonable. There are some experimental results, though indirect, indicating that the carriers are localized in such pairs [93,102].

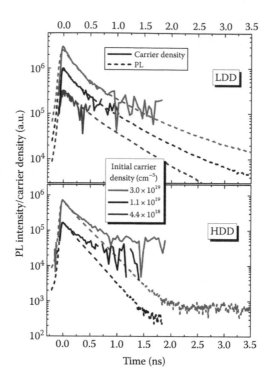

FIGURE 11.17 Decay of the carrier density obtained from LITG experiments (solid lines) and PL intensity kinetics (dashed lines) after short pulse excitation at similar initial carrier densities (indicated). (Reproduced from Saxena, T., Shur, M., Nargelas, S., Podlipskas, Ž., Aleksiejūnas, R., Tamulaitis, G., Shatalov, M., Yang, J., and Gaska, R., Dynamics of nonequilibrium carrier decay in AlGaN epitaxial layers with high aluminum content, *Opt. Express*, 23, 19646–19655, 2015. With permission from The Optical Society.)

After short pulse excitation generating certain initial density of free electron hole pairs, the dynamics of the coupled system is basically governed by the following set of equations:

$$\frac{dn}{dt} = -\frac{n}{\tau_{ne}} - Bnp - \gamma_{down}np + \gamma_{up}n_{ex} \tag{11.19}$$

$$\frac{dp}{dt} = -\frac{p}{\tau_{np}} - Bnp - \gamma_{down}np + \gamma_{up}n_{ex} \tag{11.20}$$

$$\frac{dn_{ex}}{dt} = \gamma_{down}np - \frac{n_{ex}}{\tau_{rad}} - \frac{n_{ex}}{\tau_{nrad}} - \gamma_{up}n_{ex} \tag{11.21}$$

This set is strongly simplified, since only one equation for excitons is included, though the excitons are localized in the random potential profile and have different localization energies and, consequently, rate coefficients. Instead, only one rate equation with certain effective values for capture and escape coefficients as well as radiative and nonradiative lifetimes γ_{down}, γ_{up}, τ_{rad}, τ_{nrad}, respectively, is used in Equations 11.19 through 11.21. The first terms in the first two equations describe the nonradiative decay of electrons and holes with the lifetimes τ_{ne} and τ_{np}, respectively, which differ due to the different thermal velocities and the rates of capture to the nonradiative recombination centers. The second term describes the bimolecular recombination with the rate coefficient B. The last two terms describe the exchange between the free carriers and excitons with the corresponding rate coefficients. Equation 11.21 also contains terms describing exciton radiative and nonradiative recombination with the characteristic time constants τ_{rad}, τ_{nrad}, respectively. The initial conditions based on the assumption that the nonequilibrium carriers are generated by a short pulse predominantly in free states can be written as $p = p_{pg}$, $n = n_{pg} + n_0$, and $n_{ex} = 0$. Here, $n_{pg} = p_{pg}$ are the densities of photogenerated nonequilibrium electrons and holes, and n_0 is the equilibrium electron density (if the sample under study is n-type doped).

The luminescence contains contributions of both free electron–hole pairs and excitons. Thus, the spectrally integrated PL intensity can be expressed as

$$I_L \propto \frac{n_{ex}}{\tau_{rad}} + Bnp = \frac{n_{ex}}{\tau_{rad}} + B n_0 p + Bnp. \tag{11.22}$$

The term reflecting the bimolecular recombination is split into two parts. The second part is proportional to the square of the density of photogenerated carrier, while the first part exhibits a linear dependence.

The emission bands caused by the radiative recombination of free carriers and excitons spectrally overlap due to large bandgap fluctuations and, correspondingly, large bandwidths. Thus, the two contributions cannot be separated in the PL spectrum of AlGaN.

The contributions of the three terms to the total carrier density decay rate and to the PL intensity are illustrated in Figure 11.18. The emission rates described by the three terms in Equation 11.22 are presented in Figure 11.18a. The decay kinetics of n, p, and n_{ex} were calculated by solving the set of Equations 11.19 through 11.21 with the following parameters: $B = 10^{-10}$ cm^3 s^{-1}, $\tau_{ne} = \tau_{np} = 0.36$ ns , $\gamma_{down} = 0.5 \times 10^{-9}$ cm^3 s^{-1},

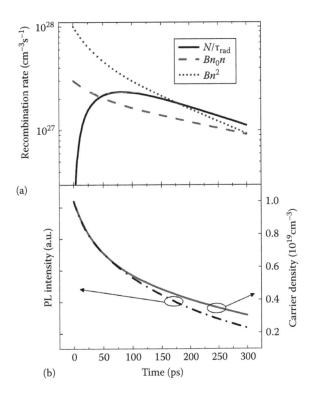

FIGURE 11.18 Recombination rates, described by first (solid line), second (dashed), and third (dotted) terms in Equation 11.22 (a), and calculated decay of PL intensity (dashed line) and free electron density (solid line) (b). (Reproduced from Saxena, T., Shur, M., Nargelas, S., Podlipskas, Ž., Aleksiejūnas, R., Tamulaitis, G., Shatalov, M., Yang, J., and Gaska, R., Dynamics of nonequilibrium carrier decay in AlGaN epitaxial layers with high aluminum content, *Opt. Express*, 23, 19646–19655, 2015. With permission from The Optical Society.)

$\gamma_{up} = 0.07$ ns, $\tau_{rad} = 0.7$ ns, $\tau_{nrad} = 5$ ns, $n_0 = 3 \times 10^{18}$ cm^{-3} (obtained using Hall measurements), and $n_{pg} = p_{pg} = 10^{19}$ cm^{-3} estimated according to excitation conditions. The quantitative fitting of the calculation results with the experimental data is not reliable due to the involvement of many rate coefficients, the values of which are not well known. Nevertheless, the calculations demonstrate that there is a reasonable set of the parameters ensuring the coincidence of the initial decay rates of PL intensity and carrier density, which is observed experimentally. The key feature in the PL decay dynamics is the partial compensation of the decrease in the intensity of PL due to bimolecular recombination of free carriers by the increasing contribution of the emission due to localized carriers, which is enhanced by increasing the density of the localized carriers due to their relaxation from the pool of free carriers.

Another interesting feature in the comparison of the kinetics of PL intensity and carrier density presented in Figure 11.17 is the substantial difference in the decay rates at the late stages of the decays. The PL intensity decays nearly exponentially by two–three orders of magnitude, while the LITG signal shows that the carrier density, after the fast initial decay, decreases extremely slow on the time scale of the lifetimes of the majority of non-equilibrium carriers. This effect is more pronounced in AlGaN with higher dislocation densities (and, probably, higher densities of point defects), where the fraction of the carriers in the slow recombination channel can be of the order of 10%. According to [101], these carriers are captured at the states well below the bandgap. They exhibit a weak PL component with the spectrum redshifted in respect to the main PL band. Carrier capture to these centers might have considerable influence on the internal quantum efficiency of AlGaN.

A valuable insight into the influence of carrier localization on PL efficiency can be obtained by PL spectroscopy at low temperatures. It is demonstrated that the excitation power density dependence of the PL band peak position has a nonmonotonous shape, as illustrated in Figure 11.19 [103]. The peak redshifts at low excitations reaches the minimum and starts blueshifting afterward. The origin of the redshift with increasing carrier density can be caused by two processes: the bandgap renormalization and the carrier redistribution within localized states. The bandgap renormalization due to many-body interaction can be excluded, since the effect is strongly reduced by carrier localization [104,105], which is important in AlGaN at low temperatures and excitations. Moreover, according to the experimental conditions, the red-shift starts already at quite low carrier density of $\sim 10^{17}$ cm^{-3}, that is, is well below the density sufficient for significant bandgap renormalization. Consequently, the carrier redistribution within localized states is a more probable origin of the band redshift than the bandgap renormalization.

The mechanism of the redshift as the carrier density increases is schematically depicted in Figure 11.20. The increasing occupation of the localized states increases the distance the carriers can move during

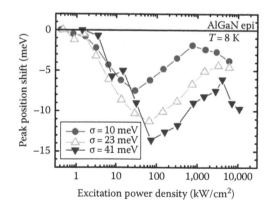

FIGURE 11.19 Excitation power density dependence of PL peak position shift in AlGaN epilayers with different localization parameters (indicated) at temperature of 8 K. Arrows indicate the onset of efficiency droop. (Reproduced from Mickevičius, J., Jurkevičius, J., Kadys, A., Tamulaitis, G., Shur, M., Shatalov, M., Yan, J., and Gaska, R., Low-temperature redistribution of non-thermalized carriers and its effect on efficiency droop in AlGaN epilayers, *J. Phys. D Appl. Phys.*, 48, 275105, 2015. With permission from IOP Publishing.)

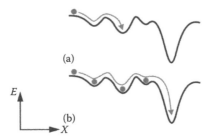

FIGURE 11.20 Carrier localization at low (a) and high (b) carrier density. Population of localized states enhances the probability of reaching the lowest localized states. (Reproduced from Mickevičius, J., Jurkevičius, J., Kadys, A., Tamulaitis, G., Shur, M., Shatalov, M., Yan, J., and Gaska, R., Low-temperature redistribution of non-thermalized carriers and its effect on efficiency droop in AlGaN epilayers, *J. Phys. D Appl. Phys.*, 48, 275105, 2015. With permission from IOP Publishing.)

their lifetime and facilitates the occupation of the lowest localized states. This redistribution causes the band redshift. A similar redshift has been previously observed and correspondingly interpreted in InAs/AlAs quantum dots [106] and semipolar InGaN/GaN quantum wells [107]. The further increase in carrier density after all the lowest localized states are occupied results in an increasing population of higher localized states and, correspondingly, to a blueshift of the PL band. The redshift observed again at the highest excitation power densities might be attributed to bandgap renormalization.

The significant redistribution of carriers induced by their increasing density at low temperatures corroborates with the decrease in IQE discussed in Section 11.5.1. The occupation-enhanced carrier redistribution increases the carrier's ability to reach nonradiative recombination centers.

11.5.3 Efficiency Droop

The efficiency of III-nitride LEDs showed a tremendous growth during the last two decades of intense research and development and has still more promising prospects for further improvement. However, in many applications, especially in application of III-nitride based LEDs for general lighting, not only efficiency but also the total light flux emitted per chip is of major importance. Increasing the driving current is the most straightforward way to increase the flux. However, it turned out that the increase in driving current above a certain threshold results in decreasing efficiency, and the light output as a function of driving power is strongly sublinear. This effect is referred to as the efficiency droop. In spite of considerable efforts of many research groups, there is still no general consensus on the efficiency droop origin and the means to diminish the undesirable effect even in InGaN-based LEDs. The industry partially solved the problem by the fabrication of multichip light sources; many LED chips are fabricated in a closely packaged matrix, and each single chip is driven at a small current corresponding to high emission efficiency. Nevertheless, a better understanding of droop origin could be useful for further development of LEDs based on InGaN and AlGaN.

The droop might be caused by decreased efficiency of the carrier injection into the active region, which in all high-brightness III-nitride LEDs consist of MQWs and of the internal quantum efficiency. Carrier leakage from the active region is quite feasible effect deteriorating the emission efficiency, as demonstrated in numerous publications [108–111]. However, the decrease in light emission efficiency is also observed under photoexcitation (this effect is also often referred to as the efficiency droop). Thus, the decrease in IQE is also important.

The most popular interpretation of the internal droop effect is Auger recombination. As a result of this three particle interaction, electron and hole recombine nonradiatively, and their energy and quasi-momentum is transferred to the third interacting carrier (electron or hole). Depending on the type of the process, the probability of the Auger recombination depends on the density of electrons and holes as n^2p or np^2. At high

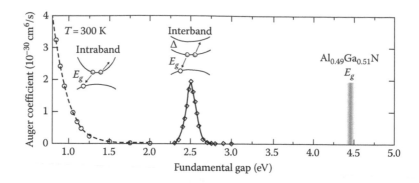

FIGURE 11.21 Dependence of Auger coefficient on bandgap provided in [116] to the spectral region corresponding to the bandgap of AlGaN-based LEDs. (Reprinted with permission from Delaney, T., Rinke, P., and Van der Walle, C.G., Auger recombination rates in nitrides from first principles, *Appl. Phys. Lett.*, 94, 191109. Copyright 2009, American Institute of Physics.)

carrier densities $p = n$, thus the rate of Auger recombination is proportional to n^3. Initially, the explanation of the efficiency droop by Auger processes came into the scene by introducing third term Cn^3 describing the Auger processes with certain rate coefficient C into the rate equation containing the terms of linear and bimolecular recombination. The third recombination term with stronger dependence on n than the first two decreases the carrier density at high n without contributing to emission and causes the emission efficiency droop. Study of carrier dynamics in InGaN provided some evidences in favor of this interpretation [112], while the recent detection of Auger electrons by electron emission spectroscopy from InGaN-based LED added a substantial support [113]. However, the value of the coefficient C is still disputed. The rate of Auger recombination is important and evidenced in numerous studies of narrow bandgap semiconductors; however, C strongly depends on the bandgap [114]. Interaction with phonons facilitates the momentum conservation and significantly enhances the rate of Auger processes. Involvement of defect states in the Auger processes is quite feasible in InGaN [115]. The Auger coefficient for InGaN was calculated from first principles [116]. The calculations revealed significant enhancement of Auger recombination rate in the spectral region peaked at 2.5 eV due to resonant interband transitions. There are no calculations of C for AlGaN. A simple extension of the graph depicting the dependence of the Auger coefficient on the bandgap provided in [116] to the spectral region corresponding to the bandgap of AlGaN-based LEDs (see Figure 11.21) shows that the expected C value should be quite small and that resonant enhancement could not be expected, since the interband gap in AlGaN (2.65 eV) is nearly as small as in GaN (2.5 eV).

Most of the studies considering importance of Auger recombination in AlGaN provide only indirect evidences. Thus, further investigation of this problem is still on the agenda. The problem is fundamental: if the droop is caused by Auger recombination, any increase in carrier density will lead to the droop. The Auger processes might be suppressed in nanocrystals [117], carrier localization might also have an influence on Auger recombination. It is quite feasible that several effects contribute to the efficiency droop in AlGaN, and their importance depends on peculiarities in material properties and structure design. Phase space filling [118], incomplete carrier localization [119,120], nonradiative recombination activated by carrier delocalization [121–123], and stimulated carrier recombination [124,125] might be considered as possible mechanisms for the efficiency droop in AlGaN.

11.5.4 Efficiency Droop due to Carrier Delocalization

The potential fluctuations experienced by carriers in AlGaN have been directly observed using spatially-resolved photoluminescence spectroscopy techniques. Different features have been detected in different epilayers and MQW structures.

Spatially nonuniform emission intensity and PL band peak position distribution due to compositional inhomogeneity were observed in CL images of $Al_xGa_{1-x}N$ with $0.2 < x < 0.5$ grown by plasma-assisted molecular-beam epitaxy [126]. An inhomogeneous distribution of PL parameters and correlation between the PL intensity and band peak position were observed using SNOM technique in low-Al-content AlGaN QWs and were attributed to the inhomogeneous screening of the built-in electric field [127]. Meanwhile, quite homogeneous distribution of PL parameters (e.g., the peak wavelength standard deviations of only ~2 meV) have been observed in $Al_xGa_{1-x}N$ epitaxial layers with $0.6 < x < 0.7$ using the SNOM spectroscopy [8]. These small variations in PL parameters can be attributed, as discussed earlier, to a random distribution of Al and Ga ions in the ternary compound. Another study using SNOM spectroscopy of AlGaN epilayers revealed dual localization patterns [128], which are consistent with the double-scaled potential profile model suggested by analyzing the temperature dependence of PL characteristics (see Section 11.3.2). The mapping images of the localization depth parameter σ_L describing the fine fluctuations of the bandgap values are presented in Figure 11.22.

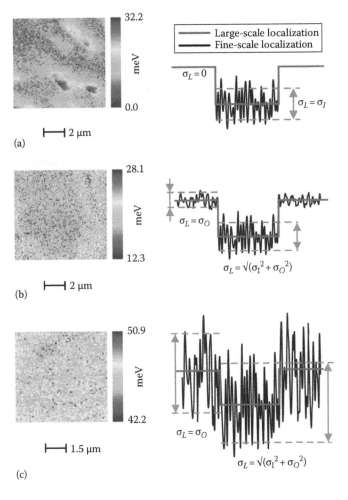

FIGURE 11.22 Maps of localization parameter σ_L and sketches of the corresponding potential profiles on large (red-line) and small (black line) scales for AlGaN epilayers with different Al content: (a) 30%, (b) 42%, and (c) 50%. (Reprinted with permission from Pinos, A., Liuolia, V., Marcinkevicius, S., Yang, J., Gaska, R., and Shur, M.S., Localization potentials in AlGaN epitaxial films studied by scanning near-field optical spectroscopy, *J. Appl. Phys.*, 109, 113516. Copyright 2011, American Institute of Physics.)

Potential fluctuations on a small spatial scale of <100 nm were estimated from the PL bandwidth. The fluctuation depths increasing with Al content up to 51 meV were detected. These small-scale compositional fluctuations were interpreted by stress variations, dislocations, and existence of Al-rich grains. In addition, larger-scale potential variations of 25–40 meV in depth were observed in the mapping images and attributed to Ga-rich regions close to grain boundaries or atomic layer steps.

A double-scale potential profile has also been revealed in Al-rich AlGaN/AlN QWs by combining results obtained using custom-built confocal microscopy photoluminescence apparatus with a reflective system (spatial resolution of 1.8 µm) and CL measurements with spatial resolution of ~100 nm [129].

As discussed earlier, the large-scale fluctuations in potential profile serve for accumulation of the nonequilibrium carriers in the regions with a lower average potential. Meanwhile, the carrier localization occurs due to the small-scale potential fluctuations. The enhanced delocalization from the localized states with increasing total density of the nonequilibrium carriers (due to stronger injection current or higher photogeneration rate) has a dual effect on the carrier dynamics and IQE [102,103]. The delocalized carriers are able to move further distances in the crystal during their lifetimes and, thus, have a higher probability to reach the nonradiative recombination centers and recombine there. This effect decreases the IQE. On the other hand, the radiative recombination rate of free carriers is proportional to the carrier density squared and is more efficient in competition with linear nonradiative recombination than the radiative recombination of exciton-like localized electron–hole pairs with linear dependence of the recombination rate on their density. Competition of these two effects of an opposite sign determines the dependence of IQE on excitation intensity in AlGaN. As illustrated in Figure 11.23, the slope of the increase in IQE as the excitation power density increases below the droop onset might be very different in various AlGaN epilayers [130].

The dependence of the rate of the efficiency increase with the carrier generation rate G can be described by the expression $IQE \propto G$ in the power of alpha. As discussed earlier, in semiconductors without carrier localization (e.g., in GaN), the dynamics of carrier density n under steady-state conditions can be described using a simplified rate equation (ABC model):

$$\frac{dn}{dt} = G - An - Bn^2 - Cn^3 = 0. \tag{11.23}$$

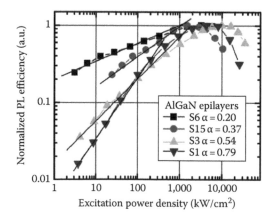

FIGURE 11.23 Room temperature normalized photoluminescence efficiency dependences on excitation power density in AlGaN epilayers. The efficiency growth rates are indicated. (With kind permission from Springer Science+Business: *J. Electron. Mater.*, Nonradiative recombination, carrier localization, and emission efficiency of AlGaN epilayers with different Al content media, 44, 2015, 4706, Mickevičius, J., Podlipskas, Ž., Aleksiejūnas, R., Kadys, A., Jurkevičius, J., Tamulaitis, G., Shur, M.S., Shatalov, M., Yang, J., and Gaska, R.)

Here, A, B, and C are the rate coefficients for linear, bimolecular, and Auger recombination. If the nonradiative term is dominant in the rate equation, and the Auger recombination is neglected (these assumptions are definitely valid at low carrier densities), $n \propto G$, $I_{LUM} \propto G^2$, and $IQE \propto G$. If the radiative recombination becomes dominant, $n \propto \sqrt{G}$, $I_{LUM} \propto G$, and IQE reaches 100%. In AlGaN, the increase in IQE at low G is significantly smaller (α varies between 0.17 and 0.8). The difference might be explained by the influence of localized states affecting the carrier dynamics via carrier trapping–detrapping processes. The localized carriers, as discussed earlier, can also recombine radiatively. It might be expected that the efficiency increase rate and, thus, the parameter α should be determined by the localization conditions, which can be described by the dispersion parameter σ, and by the density of nonradiative recombination centers, which is reflected in the carrier lifetime τ_R. Figure 11.24 presents the correlation between the carrier lifetime τ_R at room temperature and the localization parameter σ in many AlGaN epilayers. Each point in the plot corresponds to a different sample.

For σ values above the thermal energy at room temperature (26 meV), the lowest limit of the carrier lifetime τ_R increases with localization parameter σ, because the carrier localization prevents the carriers from reaching the nonradiative recombination centers. The upper limit of the points in Figure 11.24 is caused by the density of nonradiative recombination centers. The slight downward slope of the upper limit for samples with higher σ can be interpreted by a correlation between the density of nonradiative recombination centers and the depth of potential fluctuations, which is usually observed in high-Al-content AlGaN epilayers [131,132].

Both the localization parameter σ and the density of nonradiative recombination centers determine the IQE. The correlation of the peak IQE value, that is, its value at the point of droop onset, the efficiency increase rate α, and the carrier lifetime τ_R is presented in Figure 11.25. AlGaN epilayers with carrier lifetimes in a wide range were selected to better reveal the effects under study.

The peak IQE in the AlGaN epilayers exhibits a strong dependence on carrier lifetimes. Meanwhile, no significant correlation between IQE and the efficiency increase rate α is detected for the epilayers with similar τ_R. This is an indication that the localization strength is important for the rate of the increase in IQE with increasing excitation at low carrier densities, but has a small effect on the peak value of the IQE. Instead, the carrier lifetimes at high carrier densities and the peak IQE depend predominantly on the balance between the bimolecular radiative recombination and the nonradiative recombination, the rate of which depends predominantly on the density of nonradiative recombination centers. These studies show

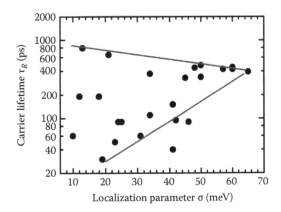

FIGURE 11.24 Correlation between carrier lifetime τ_R at room temperature and localization parameter σ in many AlGaN epilayers (each point corresponds to a different sample). Solid lines are just guides for the eye. (With kind permission from Springer Science+Business: *J. Electron. Mater.*, Nonradiative recombination, carrier localization, and emission efficiency of AlGaN epilayers with different Al content media, 44, 2015, 4706, Mickevičius, J., Podlipskas, Ž., Aleksiejūnas, R., Kadys, A., Jurkevičius, J., Tamulaitis, G., Shur, M.S., Shatalov, M., Yang, J., and Gaska, R.)

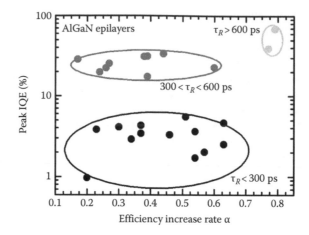

FIGURE 11.25 Correlation between peak IQE, efficiency increase rate α, and carrier recombination time τ_R in different AlGaN epilayers. (With kind permission from Springer Science+Business: *J. Electron. Mater.*, Nonradiative recombination, carrier localization, and emission efficiency of AlGaN epilayers with different Al content media, 44, 2015, 4706, Mickevičius, J., Podlipskas, Ž., Aleksiejūnas, R., Kadys, A., Jurkevičius, J., Tamulaitis, G., Shur, M.S., Shatalov, M., Yang, J., and Gaska, R.)

that the nonequilibrium carriers in AlGaN at room temperature are mobile enough to reach the nonradiative recombination centers even by hopping via localized states. As a result, the strong carrier localization alone is insufficient to ensure a high IQE. The peak IQE value is determined by the rate of the bimolecular recombination of free carriers, which is favorable for a higher IQE, and the rate of enhanced nonradiative recombination of the free carriers, which diminishes the IQE.

11.5.5 Efficiency Droop due to Stimulated Emission

The stimulated recombination stabilizes the carrier density as the excitation intensity increases above the threshold of stimulated emission. Thus, the efficiency of spontaneous luminescence measured as the ratio of the spontaneous emission intensity to the excitation power density decreases. This phenomenon might be responsible for the decrease in PL efficiency observed in the conventional configuration used in luminescence spectroscopy as well as in the LED structures under electrical carrier injection. In the conventional front-face configuration in photoluminescence spectroscopy, the focused laser beam excites a spot of a diameter of approximately several hundreds of micrometers. If the photon energy is sufficient for band-to-band excitation, the effective depth of the excited volume is approximately 0.1 μm. Thus, the light propagating perpendicularly to the surface and detected in this configuration experiences only negligible amplification even at reasonably high gain coefficient (e.g., 1% at 10^3 cm^{-1} gain). Meanwhile, the light propagating parallel to the sample surface might be significantly amplified due to stimulated recombination, since the distance of propagation through the excited material in this direction is higher by a factor of ~10^4. Consequently, the influence of stimulated recombination on the carrier dynamics might be substantial even under conditions when the luminescence collected in front-face configuration contains a negligible contribution of stimulated luminescence due to a small fraction of the strongly amplified light propagating parallel to the surface and scattered in the detection direction by inhomogeneities of the surface or in the bulk of the sample under study.

The possible influence of stimulated carrier recombination on the decrease of the detected PL efficiency, which also can be referred to as the efficiency droop, was tested in the simplest III-nitride system: GaN epilayers [124]. The PL efficiency dependence on the excitation power density in GaN epitaxial layers with different carrier lifetimes is presented in Figure 11.26. The efficiency was calculated

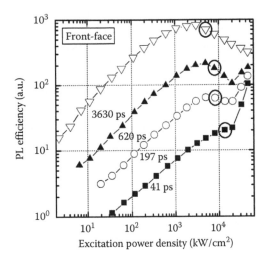

FIGURE 11.26 Photoluminescence efficiency dependence on excitation power density in GaN epitaxial layers with different carrier lifetimes (indicated) measured in front-face configuration. Encircled points correspond to the threshold of stimulated emission, which was measured in edge emission configuration. (Reproduced from Mickevičius, J., Jurkevičius, J., Shur, M.S., Yang, J., Gaska, R., and Tamulaitis, G., Photoluminescence efficiency droop and stimulated recombination in GaN epilayers, *Opt. Express*, 20, 25195, 2012. With permission from The Optical Society.)

according to the PL light detected in front-face configuration under quasi-steady-state photoexcitation. As expected, the efficiency at a fixed excitation power density strongly depends on the carrier lifetime. However, the efficiency as a function of excitation intensity experiences a decrease after a certain threshold in all the epilayers. This droop onset occurs at lower excitation power densities for the epilayers with longer carrier lifetimes.

The excitation power density dependence of stimulated luminescence was measured in the edge configuration. The light propagating in parallel to the sample surface along a narrow excited stripe perpendicular to the sample edge was collected in this configuration. Its spectrum above the stimulated recombination threshold was dominated by a narrow band of stimulated emission. The encircled points in Figure 11.26 correspond to the stimulated emission threshold. Due to the uncertainties in spot size determination and different edge qualities at the selected stripe position, the accuracy of the comparison of the excitation intensities in the two configurations is not high, but sufficient to conclude that the droop onset is close to the stimulated emission threshold.

The coincidence of the thresholds for droop and stimulated emission in the samples of substantially different PL efficiency shows that the observed droop effect might be caused by the contribution of stimulated recombination. It is worth noting that the stimulated optical transitions became important in the carrier recombination earlier than their signature as a separate narrow emission peak appears in the PL spectra recorded in front-face configuration.

Since the rate of stimulated recombination above the threshold is considerably faster than that of the spontaneous recombination, the stimulated emission might be the mechanism limiting the increase of carrier density at elevated excitation intensities and result in the efficiency droop observed in the front-face configuration. Obviously, the total luminescence, both spontaneous and stimulated, emitted in all directions does not experience the efficiency droop above the stimulated emission threshold.

One more interesting feature worth noting in Figure 11.26 is the S-shape of the PL efficiency dependence on the excitation power density. After some decrease above the droop onset, the PL efficiency starts increasing at further increase of excitation intensity. This feature is easy to explain by increasing the contribution of stimulated luminescence into the total detected PL intensity. Meanwhile, such an increase cannot be explained under the popular assumption that the droop is caused by Auger processes, since the nonradiative Auger recombination becomes increasingly pronounced at increasing carrier densities, and no increase in IQE is expected after the Auger processes become dominant in the carrier recombination. The increase in the efficiency of the detected PL at further increase of excitation intensity above the droop onset is an indication that the carrier density necessary for the Auger recombination to cause a significant decrease in PL efficiency is above the carrier density corresponding to the stimulated emission threshold in GaN.

The influence of stimulated emission on the droop in room-temperature efficiency of the emission detected perpendicularly to the photoexcited surface of the sample should be weaker in AlGaN than that in GaN due to carrier localization and a larger rate of their nonradiative recombination. As discussed in Section 11.4, the strong carrier localization results in a faster increase of the stimulated emission threshold with increasing temperature so that the room-temperature threshold in AlGaN with strong carrier localization might become larger than the threshold for photomodification of the material [133]. Thus, the stimulated emission in such samples is not observed or is observed only during the first few pulses of intense photoexcitation, before the sample is optically damaged. This effect becomes more pronounced with increasing aluminum content. However, it is demonstrated that carrier accumulation in submicrometer-size clusters in AlGaN QWs is favorable for achieving the stimulated emission even at short wavelengths down to 230 nm [67]. Thus, in AlGaN active regions with certain potential profile, the stimulated emission might also play a considerable role in stabilizing the carrier density and diminishing the efficiency of spontaneous luminescence.

11.5.6 Efficiency Droop due to Carrier Heating

The internal quantum efficiency is deteriorated by nonradiative recombination, which is enhanced by increasing temperature. Thus, carrier heating, which is expected at high densities of nonequilibrium carriers, might also decrease the IQE. Moreover, the rate of the radiative bimolecular recombination decreases with increasing temperature T as $T^{-3/2}$ or T^{-1} in epilayers and MQWs, respectively.

The carrier heating might occur as a direct result of the excess energy of the nonequilibrium carriers or due to hot phonons [134]. The photoexcited carriers might acquire a significant excess energy due to the difference between the energy of excitation photon and the bandgap of the semiconductor under photoexcitation, as well as due to the difference between the effective bandgap of the barrier and the quantum well in QW structures both under current injection and photoexcitation.

The carrier heating can be detected by the PL band shape. The high-energy side of the PL band is determined mainly by the carrier distribution function, which, far enough from the band peak, decreases exponentially with the corresponding photon energy. Thus, the high-energy PL band slope is determined by the carrier temperature T_e. Electrons and holes at the densities high enough for the heating to be feasible are well thermalized and can be characterized by a single temperature. The effect is illustrated in Figure 11.27, where the PL spectra of a $Al_{0.17}Ga_{0.83}N$ epilayer measured in quasi-steady-state conditions under different excitation power densities are presented [135]. The nearly exponential slopes are fitted in Figure 11.27 by straight dashed lines, and the temperatures corresponding to the fit slopes are indicated.

The carrier temperatures estimated as in Figure 11.27 are presented as a function of excitation power density of $Al_xGa_{1-x}N$ epilayers with different Al content in Figure 11.28.

In low-Al-content epilayers (see the dependence for $Al_{0.17}Ga_{0.83}N$ in Figure 11.28), the carrier temperature is close to lattice temperature (300 K in this experiment) at low excitation intensities and increases by nearly 200 K at elevated excitation. In $Al_xGa_{1-x}N$ with higher Al content, the carrier temperature increases at high excitation power densities, but does not approach the lattice temperature as the excitation is diminished. Moreover, the dependence becomes nonmonotonous and has a dip. This effect is more pronounced in the epilayers containing higher Al content and can be interpreted by inhomogeneous broadening, which is caused mainly by the potential fluctuations due to inhomogeneous Al distribution.

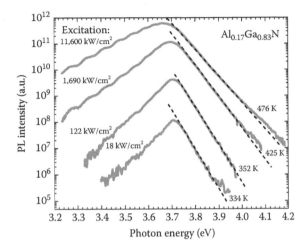

FIGURE 11.27 Photoluminescence spectra of $Al_{0.17}Ga_{0.83}N$ epilayer at different excitation power densities (indicated). Carrier temperatures corresponding to the slopes fitted by the straight dashed lines are also indicated. (From Tamulaitis, G., Mickevičius, J., Dobrovolskas, D., Kuokštis, E., Shur, M.S., Shatalov, M., Yang, J., and Gaska, R.: Carrier dynamics and efficiency droop in AlGaN epilayers with different Al content. *Phys. Stat. Solidi C.* 2012. 9. 1677–1679. Copyright Wiley-VCH Verlag GmbH & Co. KGaA with permission.)

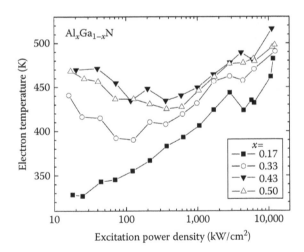

FIGURE 11.28 Excitation power density dependence of the carrier temperature estimated according to the high-energy slope of photoluminescence bands in $Al_xGa_{1-x}N$ epilayers with different aluminum content x. (From Tamulaitis, G., Mickevičius, J., Dobrovolskas, D., Kuokštis, E., Shur, M.S., Shatalov, M., Yang, J., and Gaska, R.: Carrier dynamics and efficiency droop in AlGaN epilayers with different Al content. *Phys. Stat. Solidi C.* 2012. 9, 1677–1679. Copyright Wiley-VCH Verlag GmbH & Co. KGaA with permission.)

The initial decrease in the high-energy slope of the PL band as the excitation power density is increased might be interpreted by the redistribution of the carriers downward the lower localized states when the carrier density increases. The high-energy slope of the PL band usually reflects the carrier temperature. In high-Al-content samples, this is the case only at high excitation intensities, since the slope is substantially influenced by inhomogeneous broadening at low carrier densities. As shown by the time-resolved PL investigations, the carrier heating in AlGaN is observed only for free carriers. Nevertheless, the carrier heating might be a significant factor enhancing the droop effect in AlGaN.

References

1. S. R. Lee, A. F. Wright, M. H. Crawford, G. A. Petersen, J. Han, and R. M. Biefeld, The band-gap bowing of $Al_xGa_{1-x}N$ alloys, *Appl. Phys. Lett.* 74, 3344–3346 (1999).
2. C. Coughlan, S. Schulz, M. A. Caro, and E. P. O'Reilly, Band gap bowing and optical polarization switching in $Al_{1-x}Ga_xN$ alloys, *Phys. Status Solidi B* 252, 879 (2015).
3. E. F. Schubert, E. O. Gobel, Y. Horikoshi, K. Ploog, and H. J. Queisser, Alloy broadening in photoluminescence spectra of $Al_xGa_{1-x}As$, *Rev. B* 30, 813 (1984).
4. R. Pässler, Dispersion-related assessments of temperature dependences for the fundamental band gap of hexagonal GaN, *J. Appl. Phys.* 90, 3956 (2001).
5. B. K. Meyer, G. Steude, A. Gölder, A. Hoffmann, H. Amano, and I. Akasaki, Photoluminescence investigations of AlGaN on GaN epitaxial films, *Phys. Status Solidi B* 216, 187 (1999).
6. K. K. B. Coli, J. Li, J. Y. Lin, and H. X. Jiang, Linewidths of excitonic luminescence transitions in AlGaN alloys, *Appl. Phys. Lett.* 78, 1829 (2001).
7. R. James, A. W. R. Leitch, F. Omnès, and M. Leroux, The effect of exciton localization on the optical and electrical properties of undoped and Si-doped $Al_xGa_{1-x}N$, *Semicond. Sci. Technol.* 21, 744–750 (2006).
8. S. Marcinkevičius, R. Jain, M. Shatalov, J. Yang, M. Shur, and R. Gaska, High spectral uniformity of AlGaN with a high Al content evidenced by scanning near-field photoluminescence spectroscopy, *Appl. Phys. Lett.* 105, 241108 (2014).
9. I. Ho and G. B. Stringfellow, *Appl. Phys. Lett.* 69, 2701 (1996).
10. J. Adhikari and D. A. Kofke, Molecular simulation study of miscibility in $In_xGa_{1-x}N$ ternary alloys, *J. Appl. Phys.* 95, 4500 (2004).
11. H. Ho and G. B. Springfellow, Incomplete solubility in nitride alloys. In: III-V Nitrides, *MRS Proc.* 449, 871–880 (1997).
12. J. Mickevičius, G. Tamulaitis, E. Kuokštis, K. Liu, M. S. Shur, J. P. Zhang, and R. Gaska, Well-width-dependent carrier lifetime in AlGaN/AlGaN quantum wells, *Appl. Phys. Lett.* 90, 131907 (2007).
13. T. M. Smeeton, M. J. Kappers, J. S. Barnard, M. E. Vickers, and C. J. Humphreys, Electron-beam-induced strain within InGaN quantum wells: False indium "cluster" detection in the transmission electron microscope, *Appl. Phys. Lett.* 83, 5419 (2003).
14. L. Chang, S. K. Lai, F. R. Chen, and J. J. Kai, Observations of Segregation of Al in AlGaN Alloys, *Phys. Status Solidi A* 188, 811 (2001).
15. A. Cremdes, M. Albrecht, J. Krinke, R. Mimitrov, M. Stutzmann, and H. P. Strunk, Effects of phase separation and decomposition on the minority carrier diffusion length in $Al_xGa_{1-x}N$ films, *J. Appl. Phys.* 87, 2357 (2000).
16. K. Forghani, L. Schade, U. T. Schwarz, F. Lipski, O. Klein, U. Kaiser, and F. Scholz, Strain and defects in Si-doped (Al)GaN epitaxial layers, *J. Appl. Phys.* 112, 093102 (2012).
17. C. Himwas, M. den Hertog, L. S. Dang, E. Monroy, and R. Songmuang, Alloy inhomogeneity and carrier localization in AlGaN sections and AlGaN/AlN nanodisks in nanowires with 240–350 nm emission, *Appl. Phys. Lett.* 105, 241908 (2014).
18. T. R. Corle and G. S. Kino, Introduction, a book chapter in *Confocal Scanning Optical Microscopy and Related Imaging Systems*, Academic Press, San Diego, CA, Vol. 1, pp. 1–66 (1996).
19. J. B. Pawley (ed.), *Handbook of Biological Confocal Microscopy*, Plenum, New York, p. 227 (1995).
20. D. B. Murphy, *Fundamentals of Light Microscopy and Electronic Imaging*, Wiley-Liss, New York, 388pp (2001).
21. A. Kaneta, M. Funato, and Y. Kawakami, Nanoscopic recombination processes in InGaN/GaN quantum wells emitting violet, blue, and green spectra, *Phys. Rev. B* 78, 125317 (2008).
22. A. Kaneta, T. Hashimoto, K. Nishimura, M. Funato, and Y. Kawakami, Visualization of the local carrier dynamics in an InGaN quantum well using dual-probe scanning near-field optical microscopy, *Appl. Phys. Express* 3, 102102 (2010).

23. S. Chichibu, K. Wada, and S. Nakamura, Spatially resolved cathodoluminescence spectra of InGaN quantum wells, *Appl. Phys. Lett.* 71, 2346 (1997).

24. P. G. Eliseev, P. Perlin, J. Lee, and M. Osinski, Blue temperature-induced shift and band-tail emission in InGaN-based light sources, *Appl. Phys. Lett.* 71, 569 (1997).

25. Y.-H. Cho, G. H. Gainer, A. J. Fisher, J. J. Song, S. Keller, U. K. Mishra, and S. P. DenBaars, "S-shaped" temperature-dependent emission shift and carrier dynamics in InGaN/GaN multiple quantum wells, *Appl. Phys. Lett.* 73, 1370 (1998).

26. J. Li, K. B. Nam, J. Y. Lin, and H. X. Jiang, Optical and electrical properties of Al-rich AlGaN alloys, *Appl. Phys. Lett.* 79, 3245 (2001).

27. H. Hirayama, A. Kinoshita, T. Yamabi, Y. Enomoto, A. Hirata, T. Araki, Y. Nanishi, and Y. Aoyagi, Marked enhancement of 320-360 nm ultraviolet emission in quaternary $In_xAl_yGa_{1-x-y}N$ with In-segregation effect, *Appl. Phys. Lett.* 80, 207 (2002).

28. T. Wang, Y. H. Liu, Y. B. Lee, J. P. Ao, J. Bai, and S. Sakai, 1 mW AlInGaN-based ultraviolet light-emitting diode with an emission wavelength of 348 nm grown on sapphire substrate, *Appl. Phys. Lett.* 81, 2508 (2002).

29. K. Kazlauskas, G. Tamulaitis, A. Žukauskas, M. A. Khan, J. W. Yang, J. Zhang, E. Kuokstis, G. Simin, M. S. Shur, and R. Gaska, Double-scaled potential profile in a group-III nitride alloy revealed by Monte Carlo simulation of exciton hopping, *Appl. Phys. Lett.* 82, 3722 (2003).

30. D. Monroe, Hopping in exponential band tails, *Phys. Rev. Lett.* 54, 146 (1985).

31. S. D. Baranovskii, R. Eichmann, and P. Thomas, Temperature-dependent exciton luminescence in quantum wells by computer simulation, *Phys. Rev. B* 58, 13081 (1998).

32. S. A. Tarasenko, A. A. Kiselev, E. L. Ivchenko, A. Dinger, M. Baldauf, C. Klingshirn, H. Kalt, S. D. Baranovskii, R. Eichmann, and P. Thomas, Energy relaxation of localized excitons at finite temperature, *Semicond. Sci. Technol.* 16, 486 (2001).

33. K. Kazlauskas, A. Žukauskas, G. Tamulaitis, J. Mickevičius, M. S. Shur, R. S. Q. Fared, J. P. Zhang, and R. Gaska, Exciton hopping and nonradiative decay in AlGaN epilayers, *Appl. Phys. Lett.* 87, 172102 (2005).

34. K. Kazlauskas, G. Tamulaitis, A. Žukauskas, M. A. Khan, J. W. Yang, J. Zhang, G. Simin, M. S. Shur, R. Gaska, Localization and hopping of excitons in quaternary AlInGaN, *Phys. Stat. Solid (C)* 0, 512 (2002).

35. K. B. Nam, J. Li, J. Y. Lin, and H. X. Jiang, Optical properties of AlN and GaN in elevated temperatures, *Appl. Phys. Lett.* 85, 3489–3491 (2004).

36. Y. P. Varshni, Band-to-band radiative recombination in groups IV, VI, and III-V semiconductors, *Phys. Status Solidi B* 19, 459–514 (1967).

37. C.-Z. Zhao, B. Liu, D.-Y. Fu, H. Chen, M. Li, X.-Q. Xiu, Z.-L. Xie, S.-L. Gu, and Y.-D. Zheng, Investigation of localization effect in GaN-rich InGaN alloys and modified band-tail model, *Bull. Mater. Sci.* 36, 619 (2013).

38. K. Kazlauskas, G. Tamulaitis, P. Pobedinskas, A. Žukauskas, M. Springis, C. F. Huang, Y. C. Cheng, and C. C. Yang, Exciton hopping in InxGa1-xN multiple quantum wells, *Phys. Rev. B* 71, 085306 (2005).

39. T. J. Badcock, P. Dawson, M. J. Davies, M. J. Kappers, F. C.-P. Massabuau, F. Oehler, R. A. Oliver, and C. J. Humphreys, Low temperature carrier redistribution dynamics in InGaN/GaN quantum wells, *J. Appl. Phys.* 115, 113505 (2014).

40. A. Satake, Y. Masumoto, T. Miyajima, T. Asatsuma, F. Nakamura, and M. Ikeda, Localized exciton and its stimulated emission in surface mode from single-layer $In_xGa_{1-x}N$, *Phys. Rev. B* 57, R2041 (1998).

41. A. Morel, P. Lefebvre, S. Kalliakos, T. Taliercio, T. Bretagnon, and B. Gil, Donor-acceptor-like behavior of electron-hole pair recombinations in low-dimensional (Ga,In)N/GaN systems, *Phys. Rev. B* 68, 045331 (2003).

42. Y. Narukawa, W. Kawakami, S. Fujita, S. Fujita, and S. Nakamura, Recombination dynamics of localized excitons in In0.20Ga0.80N-In0.05Ga0.95N multiple quantum wells, *Phys. Rev. B* 55, R1938 (1997).

43. Y. Narukawa, Y. Kawakami, S. Fujita and S. Nakamura, Dimensionality of excitons in laser-diode structures composed of $In_xGa_{1-x}N$ multiple quantum wells, *Phys. Rev. B* 59, 10283 (1999).

44. H. S. Kim, R. A. Mair, J. Li, J. Y. Lin, and H. X. Jiang, Time-resolved photoluminescence studies of AlxGa1-xN alloys, *Appl. Phys. Lett.* 76, 1252 (2000).

45. Y.-H. Cho, G. H. Gainer, J. B. Lam, J. J. Song, and W. Yang, Temperature dependence of transmission and emission spectra in MOCVD-grown AlGaN ternary alloys, *Phys. Status Solidi A* 188, 815 (2001).

46. T. Onuma, S. F. Chichibu, A. Uedono, T. Sota, P. Cantu, T. M. Katona, J. F. Keading et al., Radiative and nonradiative processes in strain-free $Al_xGa_{1-x}N$ films studied by time-resolved photoluminescence and positron annihilation techniques, *J. Appl. Phys.* 95, 2495 (2004).

47. H. Murotani, Y. Yamada, T. Taguchi, A. Ishibashi, Y. Kawaguchi, and T. Yokogawa, Temperature dependence of localized exciton transitions in AlGaN ternary alloy epitaxial layers, *J. Appl. Phys.* 104, 053514 (2008).

48. H. Murotani, R. Kittaka, S. Kurai, Y. Yamada, H. Miyake, and K. Hiramatsu, Recombination dynamics of localized excitons in $Al_xGa_{1-x}N$ (0.37< x< 0.81) ternary alloys, *Phys. Status Solidi C* 8, 2133 (2011).

49. D. M. Graham, A. Soltani-Vala, P. Dawson, M. J. Godfrey, T. M. Smeeton, J. S. Barnard, M. J. Kappers, C. J. Humphreys, and E. J. Thrush, Optical and microstructural studies of InGaN/GaN single-quantum-well structures, *J. Appl. Phys.* 97, 103508 (2005).

50. C. Gourdon and P. Lavallard, Exciton transfer between localized states in $CdS_{1-x}Se_x$ alloys, *Phys. Status Solidi B* 153, 641 (1989).

51. N. Grandjean, B. Damilano, S. Dalmasso, M. Leroux, M. Laugt, and J. Massies, Built-in electric-field effects in wurtzite AlGaN/GaN quantum wells, *J. Appl. Phys.* 86, 3714 (1999).

52. T. Saxena, S. Nargelas, J. Mickevičius, O. Kravcov, G. Tamulaitis, M. Shur, M. Shatalov, J. Yang, and R. Gaska, Spectral dependence of carrier lifetime in high aluminum content AlGaN epitaxial layers, *J. Appl. Phys.* 118, 085705 (2015).

53. J. Mickevičius, R. Aleksiejūnas, M. S. Shur, G. Tamulaitis, R. S. Qhalid Fareed, J. P. Zhang, R. Gaska, and M. A. Khan, Lifetime of nonequilibrium carriers in high-Al-content AlGaN epilayers, *Phys. Status Solidi A* 202, 126 (2005).

54. H. Yoshida, Y. Yamashita, M. Kuwabara, and H. Kan, Demonstration of an ultraviolet 336 nm AlGaN multiple-quantum-well laser diode, *Appl. Phys. Lett.* 93, 241106 (2008).

55. Q. Wang, Y. P. Gong, J. F. Zhang, J. Bai, F. Ranalli, and T. Wang, Stimulated emission at 340 nm from AlGaN multiple quantum well grown using high temperature AlN buffer technologies on sapphire, *Appl. Phys. Lett.* 95, 161904 (2009).

56. V. N. Jmerik, A. M. Mizerov, A. A. Sitnikova, P. S. Kop'ev, S. V. Ivanov, E. V. Lutsenko, N. P. Tarasuk, N. V. Rzheutskii, and G. P. Yablonskii, Low-threshold 303 nm lasing in AlGaN-based multiple-quantum well structures with an asymmetric waveguide grown by plasma-assisted molecular beam epitaxy on c-sapphire, *Appl. Phys. Lett.* 96, 141112 (2010).

57. V. N. Jmerik, A. N. Mizerov, T. V. Shubina, A. A. Toropov, K. G. Belyaev, A. A. Sitnikova, M. A. Yagovkina et al., Optically pumped lasing at 300.4 nm in AlGaN MQW structures grown by plasma-assisted molecular beam epitaxy on c-Al2O3, *Phys. Status Solidi A* 207, 1313 (2010).

58. J. Mickevičius, J. Jurkevičius, K. Kazlauskas, A. Žukauskas, G. Tamulaitis, M. S. Shur, M. Shatalov, J. Yang, and R. Gaska, Stimulated emission in AlGaN/AlGaN quantum wells with different Al content, *Appl. Phys. Lett.* 100, 081902 (2012).

59. J. Mickevičius, J. Jurkevičius, K. Kazlauskas, A. Žukauskas, G. Tamulaitis, M. S. Shur, M. Shatalov, J. Yang, and R. Gaska, Stimulated emission due to localized and delocalized carriers in $Al_{0.35}Ga_{0.65}N/Al_{0.49}Ga_{0.51}N$ quantum wells, *Appl. Phys. Lett.* 101, 041912 (2012).

60. V. Lutsenko, N. V. Rzheutskii, V. N. Pavlovskii, G. P. Yablonskii, D. V. Nechaev, A. A. Sitnikova, V. V. Ratnikov, V. Kuznetsova, V. N. Zhmerik, and S. V. Ivanov, Spontaneous and stimulated emission in the mid-ultraviolet range of quantum-well heterostructures based on AlGaN compounds grown by molecular beam epitaxy on c-sapphire substrates, *Phys. Solid State* 55, 2173 (2013).

61. T. Oto, R. G. Banal, M. Funato, and Y. Kawakami, Optical gain characteristics in Al-rich AlGaN/AlN quantum wells, *Appl. Phys. Lett.* 104, 181102 (2014).

62. X.-H. Li, T. Detchprohm, T.-T. Kao, M. M. Satter, S.-C. Shen, P. D. Yoder, R. D. Dupuis et al., Low-threshold stimulated emission at 249 nm and 256 nm from AlGaN-based multiple-quantum-well lasers grown on sapphire substrates, *Appl. Phys. Lett.* 105, 141106 (2014).

63. X.-H. Li, T.-T. Kao, M. M. Satter, Y. O. Wei, S. Wang, H. Xie, S.-C. Shen et al., Demonstration of transverse-magnetic deep-ultraviolet stimulated emission from AlGaN multiple-quantum-well lasers grown on a sapphire substrate, *Appl. Phys. Lett.* 106, 041115 (2015).

64. Y. Tian, J. Yan, Y. Zhang, X. Chen, Y. Guo, P. Cong, L. Sun et al., Stimulated emission at 288 nm from silicon-doped AlGaN-based multiple-quantum-well laser, *Opt. Express* 23, 11334 (2015).

65. T. Takano, Y. Narita, A. Horiuchi, and H. Kawanishi, Room-temperature deep-ultraviolet lasing at 241.5 nm of AlGaN multiple-quantum-well laser, *Appl. Phys. Lett.* 84, 3567 (2004).

66. H. Kawanishi, M. Senuma, and T. Nukui, Anisotropic polarization characteristics of lasing and spontaneous surface and edge emissions from deep-ultraviolet ($\lambda \approx 240nm$) AlGaN multiple-quantum-well lasers, *Appl. Phys. Lett.* 89, 041126 (2006).

67. F. Pecora, W. Zhang, A. Y. Nikiforov, L. Zhou, D. J. Smith, J. Yin, R. Paiella, L. Dal Negro, and T. D. Moustakas, Sub-250 nm room-temperature optical gain from AlGaN/AlN multiple quantum wells with strong band-structure potential fluctuations, *Appl. Phys. Lett.* 100, 061111 (2012).

68. F. Pecora, W. Zhang, J. Yin, R. Paiella, L. Dal Negro, and T. D. Moustakas, Polarization properties of deep-ultraviolet optical gain in Al-rich AlGaN structures, *Appl. Phys. Express* 5, 032103 (2012).

69. E. F. Pecora, W. Zhang, A. Y. Nikiforov, J. Yin, R. Paiella, L. Dal Negro, and T. D. Moustakas, Sub-250 nm light emission and optical gain in AlGaN materials, *J. Appl. Phys.* 113, 013106 (2013).

70. R. Gaska, C. Chen, J. Yang, E. Kuokstis, A. Khan, G. Tamulaitis, I. Yilmaz, M. S. Shur, J. C. Rojo, and L. J. Schowalter, Deep-ultraviolet emission of AlGaN/AlN quantum wells on bulk AlN, *Appl. Phys. Lett.* 81, 4658 (2002).

71. M. Kneissl, Z. Yang, M. Teepe, C. Knollenberg, O. Schmidt, P. Kiesel, and N. M. Johnson, Ultraviolet semiconductor laser diodes on bulk AlN, *J. Appl. Phys.* 101, 123103 (2007).

72. T. Wunderer, C. L. Chua, Z. Yang, J. E. Northrup, N. M. Johnson, G. A. Garrett, H. Shen, and M. Wraback, Pseudomorphically grown ultraviolet C photopumped lasers on bulk AlN substrates, *Appl. Phys. Express* 4, 092101 (2011).

73. Z. Lochner, T.-T. Kao, Y.-S. Liu, X.-H. Li, M. M. Satter, S.-C. Shen, P. D. Yoder et al., Deep-ultraviolet lasing at 243 nm from photo-pumped AlGaN/AlN heterostructure on AlN substrate, *Appl. Phys. Lett.* 102, 101110 (2013).

74. J. Xie, S. Mita, Z. Bryan, W. Guo, L. Hussey, B. Moody, R. Schlesser et al., Lasing and longitudinal cavity modes in photo-pumped deep ultraviolet AlGaN heterostructures, *Appl. Phys. Lett.* 102, 171102 (2013).

75. T.-T. Kao, Y.-S. Liu, M. M. Satter, X.-H. Li, Z. Lochner, P. D. Yoder, T. Detchprohm et al., Sub-250 nm low-threshold deep-ultraviolet AlGaN-based heterostructure laser employing HfO$_2$/SiO$_2$ dielectric mirrors, *Appl. Phys. Lett.* 103, 211103 (2013).

76. Z. Lochner, X.-H. Li, T.-T. Kao, M. M. Satter, H. J. Kim, S.-C. Shen, P. D. Yoder et al., Stimulated emission at 257 nm from optically-pumped AlGaN/AlN heterostructure on AlN substrate, *Phys. Status Solidi A* 210, 1768 (2013).

77. W. Guo, Z. Bryan, J. Xie, R. Kirste, S. Mita, I. Bryan, L. Hussey et al., Stimulated emission and optical gain in AlGaN heterostructures grown on bulk AlN substrates, *J. Appl. Phys.* 115, 103108 (2014).

78. H. Yoshida, Y. Yamashita, M. Kuwabara, and H. Kan, A 342-nm ultraviolet AlGaN multiplequantum-well laser diode, *Nat Photon.* 2, 551 (2008).

79. Y. Kawakami, Y. Narukawa, K. Omae, and S. Fujita, Dynamics of optical gain in In$_x$Ga$_{1-x}$N multi-quantum-well-based laser diodes, *Appl. Phys. Lett.* 77, 2151 (2000).

80. K. Omae, Y. Kawakami, Y. Narukawa, Y. Watanabe, T. Mukai, and S. G. Fujita, Nondegenerated pump and probe spectroscopy in InGaN-based semiconductors, *Phys. Status Solidi A* 190, 93–98 (2002).

81. W. Hakki and T. L. Paoli, CW degradation at 300°K of GaAs double-heterostructure junction lasers. II. Electronic gain, *J. Appl. Phys.* 44, 4113 (1973).

82. U. T. Schwarz, E. Sturm, W. Wegscheider, V. Kümmler, A. Lell, and V. Härle, Gain spectra and current-induced change of refractice index in (In/Al)GaN diode lasers, *Phys. Status Solidi A* 200, 143 (2003).

83. K. L. Shaklee, R. E. Nahory, and R. F. Leheny, Optical gain in semiconductors, *J. Lumin.* 7, 284 (1973).

84. M. Röwe, P. Michler, J. Gutowski, V. Kümmler, A. Lell, and V. Härle, Influence of the carrier density on the optical gain and refractive index change in InGaN laser structures, *Phys. Status Solidi A* 200, 135 (2003).

85. M. Vehse, P. Michler, O. Lange, O. Röwe, J. Gutowski, S. Bader, H.-J. Lugauer et al., Optical gain and saturation in nitride-based laser structures, *Appl. Phys. Lett.* 79, 1763 (2001).

86. K. Kyhm, R. A. Taylor, J. F. Ryan, T. Someya, and Y. Arakawa, Analysis of gain saturation in In0.02Ga0.98N/In0.16Ga0.84N multiple quantum wells, *Appl. Phys. Lett.* 79, 3434 (2001).

87. M. Vehse, J. Meinertz, O. Lange, P. Michler, J. Gutowski, S. Bader, A. Lell, and V. Härle, Analysis of gain saturation behavior in GaN based quantum well lasers, *Phys. Status Solidi C* 0, 43 (2002).

88. S. F. Chichibu, M. Sugiyama, T. Onuma, T. Kitamura, H. Nakanishi, T. Kuroda, A. Tackeuchi, T. Sota, Y. Ishida, and H. Okumura, Localized exciton dynamics in strained cubic In0.1Ga0.9N/GaN multiple quantum wells, *Appl. Phys. Lett.* 79, 4319 (2001).

89. M. S. Minsky, S. Watanabe, and N. Yamada, Radiative and nonradiative lifetimes in GaInN/GaN multiquantum wells, *J. Appl. Phys.* 91, 5176 (2002).

90. H. Wang, Z. Ji, S. Qu, G. Wang, Y. Jiang, B. Liu, X. Xu, and H. Mino, Influence of excitation power and temperature on photoluminescence in InGaN/GaN multiple quantum wells, *Opt. Express* 20, 3932 (2012).

91. S. Watanabe, N. Yamada, M. Nagashima, Y. Ueki, C. Sasaki, Y. Yamada, T. Taguchi, K. Tadatomo, H. Okagawa, and H. Kudo, Internal quantum efficiency of highly-efficient In$_x$Ga$_{1-x}$N-based near-ultraviolet light-emitting diodes, *Appl. Phys. Lett.* 83, 4906 (2003).

92. A. Hangleiter, D. Fuhrmann, M. Grewe, F. Hitzel, G. Klewer, S. Lahmann, C. Netzel, N. Riedel, and U. Rossow, Towards understanding the emission efficiency of nitride quantum wells, *Phys. Status Solidi A* 201, 2808 (2004).

93. M. Shatalov, J. Yang, W. Sun, R. Kennedy, R. Gaska, K. Liu, M. Shur, and G. Tamulaitis, Efficiency of light emission in high aluminum content AlGaN quantum wells, *J. Appl. Phys.* 105, 073103 (2009).

94. T. Kohno, Y. Sudo, M. Yamauchi, K. Mitsui, H. Kudo, H. Okagawa, and Y. Yamada, Internal quantum efficiency and nonradiative recombination rate in InGaN-based near-ultraviolet light-emitting diodes, *Jpn. J. Appl. Phys.* 51, 072102 (2012).

95. H.-Y. Ryu, H.-S. Kim, and J.-I. Shim, Rate equation analysis of efficiency droop in InGaN light-emitting diodes, *Appl. Phys. Lett.* 95, 081114 (2009).

96. J. Wang, L. Wang, L. Wang, Z. Hao, Y. Luo, A. Dempewolf, M. Muller, F. Bertram, and J. Christen, An improved carrier rate model to evaluate internal quantum efficiency and analyze efficiency droop origin of InGaN based light-emitting diodes, *J. Appl. Phys.* 112, 023107 (2012).

97. H. Y. Ryu, K. H. Ha, J. H. Chae, K. S. Kim, J. K. Son, O. H. Nam, Y. J. Park, and J. I. Shim, Evaluation of radiative efficiency in InGaN blue-violet laser-diode structures using electroluminescence characteristics, *Appl. Phys. Lett.* 89, 171106 (2006).

98. Q. Dai, M. F. Schubert, M. H. Kim, J. K. Kim, E. F. Schubert, D. D. Koleske, M. H. Crawford et al., Internal quantum efficiency and nonradiative recombination coefficient of GaInN/GaN multiple quantum wells with different dislocation densities, *Appl. Phys. Lett.* 94, 111109 (2009).

99. H. Yoshida, M. Kuwabara, Y. Yamashita, K. Uchiyama, and H. Kan, Radiative and nonradiative recombination in an ultraviolet GaN/AlGaN multiple-quantum-well laser diode, *Appl. Phys. Lett.* 96, 211122 (2010).

100. A. Miller, Transient grating studies of carrier diffusion and mobility in semiconductors, a book chapter in *Nonlinear Optics in Semiconductors II*, Vol. 59 (Semiconductors and Semimetals), Academic Press, Elsevier, New York, Chapter 5, pp. 287–312 (1998).

101. T. Saxena, M. Shur, S. Nargelas, Ž. Podlipskas, R. Aleksiejūnas, G. Tamulaitis, M. Shatalov, J. Yang, and R. Gaska, Dynamics of nonequilibrium carrier decay in AlGaN epitaxial layers with high aluminum content, *Opt. Express* 23, 19646–19655 (2015).

102. G. Tamulaitis, J. Mickevičius, J. Jurkevičius, M. S. Shur, M. Shatalov, J. Yang, and R. Gaska, Photoluminescenceefficiency in AlGaN quantum wells, *Phys. B* 453, 40–42 (2014).

103. J. Mickevičius, J. Jurkevičius, A. Kadys, G. Tamulaitis, M. Shur, M. Shatalov, J. Yan, and R. Gaska, Low-temperature redistribution of non-thermalized carriers and its effect on efficiency droop in AlGaN epilayers, *J. Phys. D Appl. Phys.* 48, 275105 (2015).

104. F. A. Majumder, S. Shevel, V. G. Lyssenko, H. E. Swoboda, and C. Klingshirn, Luminescence of and gain spectroscopy of disordered $CdS_{1-x}Se_x$ under high excitation, *Z. Phys. B* 66, 409 (1987).

105. D. Hirano, T. Tayagaki, and Y. Kanemitsu, Disorder-induced rapid localization of electron-hole plasmas in highly excited InxGa1− xN mixed crystals, *Phys. Rev. B* 77, 073201 (2008).

106. Z. Ma and K. Pierz, Carrier thermalization and activation within self-assembled InAs/AlAs quantum dot states, *Surf. Sci.* 511, 57 (2002).

107. S. Marcinkevičius, K. Gelžinytė, Y. Zhao, S. Nakamura, S. P. DenBaars, and J. S. Speck, Carrier redistribution between different potential sites in semipolar (20-21) InGaN quantum wells studied by near-field photoluminescence, *Appl. Phys. Lett.* 105, 111108 (2014).

108. U. Ozgur, H. Liu, X. Li, X. Ni, and H. Morkoc, GaN-based light-emitting diodes: Efficiency at high injection levels, *IEEE Proc.* 98, 1180 (2010).

109. B.-J. Ahn, T.-S. Kim, Y. Dong, M.-T. Hong, J.-H. Song, J.-H. Song, H.-K. Yuh, S.-C. Choi, D.-K. Bae, and Y. Moon, Experimental determination of current spill-over and its effect on the efficiency droop in InGaN/GaN blue-light-emitting-diodes, *Appl. Phys. Lett.* 100, 031905 (2012).

110. G. Verzellesi, D. Saguatti, M. Meneghini, F. Bertazzi, M. Goano, G. Meneghesso, and E. Zanoni, Efficiency droop in InGaN/GaN blue light-emitting diodes: Physical mechanisms and remedies, *J. Appl. Phys.* 114, 071101 (2013).

111. E. Jaehee Cho, F. Schubert, and J. K. Kim, Efficiency droop in light-emitting diodes: Challenges and countermeasures, *Laser Photon. Rev.* 7, 408 (2013).

112. Y. C. Shen, G. O. Mueller, S. Watanabe, N. F. Gardner, A. Munkholm, and M. R. Krames, Auger recombination in InGaN measured by photoluminescence, *Appl. Phys. Lett.* 91, 141101 (2007).

113. J. Iveland, L. Martinelli, J. Peretti, J. S. Speck, and C. Weisbuch, Direct measurement of auger electrons emitted from a semiconductor light-emitting diode under electrical injection: Identification of the dominant mechanism for efficiency droop, *Phys. Rev. Lett.* 110, 177406 (2013).

114. A. Haug, Auger recombination in direct-gap semiconductors: Bend-structure effects, *J. Phys. C Solid State Phys.* 16, 4159 (1983).

115. S. Nargėlas, R. Aleksiejūnas, M. Vengris, T. Malinauskas, K. Jarašiūnas, and E. Dimakis, Dynamics of free carrier absorption in InN layers, *Appl. Phys. Lett.* 95, 162103 (2009).

116. T. Delaney, P. Rinke, and C. G. Van der Walle, Auger recombination rates in nitrides from first principles, *Appl. Phys. Lett.* 94, 191109 (2009).

117. G. E. Cragg and A. L. Efros, Suppression of auger processes in confined structures, *Nano Lett.* 10, 313 (2010).

118. G. Bourdon, I. Robert, I. Sagnes, and J. J. Abram, Spontaneous emission in highly excited semiconductors: Saturation of the radiative recombination rate, *J. Appl. Phys.* 92, 6595 (2002).

119. N. I. Bochkareva, Y. T. Rebane, and Y. G. Shreter, Efficiency droop and incomplete carrier localization in InGaN/GaN quantum well light-emitting diodes, *Appl. Phys. Lett.* 103, 191101 (2013).

120. N. I. Bochkareva, Y. T. Rebane, and Y. G. Shreter, Efficiency droop in GaN LEDs at high current densities: Tunneling leakage currents and incomplete lateral carrier localization in InGaN/GaN quantum wells, *Semiconductors* 48, 1079 (2014).

121. J. Wang, L. Wang, W. Zhao, Z. Hao, and Y. Luo, Understanding efficiency droop effect in InGaN/GaN multiple-quantum-well blue light-emitting diodes with different degree of carrier localization, *Appl. Phys. Lett.* 97, 201112 (2010).

122. S. Hammersley, D. Watson-Parris, P. Dawson, M. J. Godfrey, T. J. Badcock, M. J. Kappers, C. McAleese, R. A. Oliver, and C. J. Humphreys, The consequences of high injected carrier densities on carrier localization and efficiency droop in InGaN/GaN quantum well structures, *J. Appl. Phys.* 111, 083512 (2012).

123. J. Mickevičius, G. Tamulaitis, M. Shur, M. Shatalov, J. Yang, and R. Gaska, Correlation between carrier localization and efficiency droop in AlGaN epilayers, *Appl. Phys. Lett.* 103, 011906 (2013).

124. J. Mickevičius, J. Jurkevičius, M. S. Shur, J. Yang, R. Gaska, and G. Tamulaitis, Photoluminescence efficiency droop and stimulated recombination in GaN epilayers, *Opt. Express* 20, 25195 (2012).

125. J. Mickevičius, J. Jurkevičius, G. Tamulaitis, M. S. Shur, M. Shatalov, J. Yang, and R. Gaska, Influence of carrier localization on high-carrier-density effects in AlGaN quantum wells, *Opt. Express* 22, A491 (2014).

126. J. Collins, A. V. Sampath, G. A. Garrett, W. L. Sarney, H. Shen, M. Wraback, A. Y. Nikiforov, G. S. Cargill III, and V. Dierolf, Enhanced room-temperature luminescence efficiency through carrier localization in $Al_xGa_{1-x}N$ alloys, *Appl. Phys. Lett.* 86, 031916 (2005).

127. H. Murotani, T. Saito, N. Kato, Y. Yamada, T. Taguchi, A. Ishibashi, Y. Kawaguchi, and T. Yokogawa, Localization-induced inhomogeneous screening of internal electric fields in AlGaN-based quantum wells, *Appl. Phys. Lett.* 91, 231910 (2007).

128. A. Pinos, V. Liuolia, S. Marcinkevicius, J. Yang, R. Gaska, and M. S. Shur, Localization potentials in AlGaN epitaxial films studied by scanning near-field optical spectroscopy, *J. Appl. Phys.* 109, 113516 (2011).

129. Y. Iwata, T. Oto, D. Gachet, R. G. Banal, M. Funato, and Y. Kawakami, Co-existence of a few and sub micron inhomogeneities in Al-rich AlGaN/AlN quantum wells, *J. Appl. Phys.* 117, 115702 (2015).

130. J. Mickevičius, Ž. Podlipskas, R. Aleksiejūnas, A. Kadys, J. Jurkevičius, G. Tamulaitis, M. S. Shur, M. Shatalov, J. Yang, and R. Gaska, Nonradiative recombination, carrier localization, and emission efficiency of AlGaN epilayers with different Al content, *J. Electron. Mater.* 44, 4706 (2015).

131. Q. Sun, Y. Huang, H. Wang, J. Chen, R. Q. Jin, S. M. Zhang, H. Yang, D. S. Jiang, U. Jahn, and K. H. Ploog, Lateral phase separation in AlGaN grown on GaN with a high-temperature AlN interlayer, *Appl. Phys. Lett.* 87, 121914 (2005).

132. M. Gao, S. T. Bradley, Y. Cao, D. Jena, Y. Lin, S. A. Ringel, J. Hwang, W. J. Schaff, and L. J. Brillson, Compositional modulation and optical emission in AlGaN epitaxial films, *J. Appl. Phys.* 100, 103512 (2006).

133. T. Saxena, G. Tamulaitis, M. Shatalov, J. Yang, R. Gaska, and M. S. Shur, Low threshold for optical damage in AlGaN epilayers and heterostructures, *J. Appl. Phys.* 114, 203103 (2013).

134. G. Tamulaitis, A. Žukauskas, J. W. Yang, M. A. Khan, M. S. Shur, and R. Gaska, Heating of photogenerated electrons and holes in highly excited GaN epilayers, *Appl. Phys. Lett.* 75, 2277–2279 (1999).

135. G. Tamulaitis, J. Mickevičius, D. Dobrovolskas, E. Kuokštis, M. S. Shur, M. Shatalov, J. Yang, and R. Gaska, Carrier dynamics and efficiency droop in AlGaN epilayers with different Al content, *Phys. Stat. Solidi C* 9, 1677–1679 (2012).

12

Solar-Blind AlGaN Devices

Jiangnan Dai
Huazhong University of Science and Technology

Jingwen Chen
Huazhong University of Science and Technology

Jun Zhang
Huazhong University of Science and Technology

Wei Zhang
Huazhong University of Science and Technology

Shuai Wang
Huazhong University of Science and Technology

Feng Wu
Huazhong University of Science and Technology

Changqing Chen
Huazhong University of Science and Technology

Abstract There has been a growing interest in the development of solar-blind ultraviolet photodetectors for a variety of applications, including detection of missiles, precision guidance, biochemical analysis, lithography aligners, secure space-to-space communications, and ozone monitor. $Al_xGa_{1-x}N$ has emerged as the most promising material system for such devices. $Al_xGa_{1-x}N$-based solar-blind photodetectors with different structures had been successfully fabricated, and their working principle has been demonstrated. However, the material quality and the device process of AlGaN devices still need to be further improved to enhance performance, including the development of ohmic metal contacts to both n-type and p-type $Al_xGa_{1-x}N$ films with high Al component.

12.1 Introduction

Following the infrared and laser detecting technology, ultraviolet (UV) detecting technology is one of the dual-use photoelectric detection technologies with great application potential in various areas such as detection of missiles, precision guidance, biochemical analysis, lithography aligners, secure space-to-space communications, and ozone monitoring [1–3]. The UV photodetectors (PDs) have been fabricated for commercial purposes such as photodetection system of ultrafast imaging, measuring speckle motion to monitor ultrasonic vibrations, remote optical diagnostics of nonstationary aerosol media in a wide range of particle sizes, atmospheric clock transfer based on femtosecond frequency combs, measurement of the polarization state of a weak signal field by homodyne detection, and so on [3]. And, in many of these applications, it is especially important to detect UV light without response to infrared or visible light, especially from the sun, in order to minimize the chances of false detection or high background. According to the standard from the International Commission on Illumination (CIE), the deep UV (DUV) wavelengths from 220 to 280 nm are in the solar-blind region. This value of wavelength has been established from the fact that there are few photons from the Sun with a wavelength shorter than 285 nm that reach the surface of the Earth due to atmospheric absorption [4]. Therefore, the solar-blind UV PDs can be defined as PDs that are blind to photons with wavelengths longer than ~290 nm [5].

To absorb photons with energy in the solar-blind region, semiconductor with bandgap larger than ~4.43 eV is required for solar-blind UV PDs. The AlGaN ternary alloys have a wide direct bandgap that can be easily tuned into solar-blind wavelength range by varying the Al content of the alloy, making it a promising candidate for the solar-blind PDs. This fundamental property essentially makes detectors fabricated from $Al_xGa_{1-x}N$ materials with x > 0.45 intrinsically solar blind, whereas alternative materials such as Si, GaAs, GaP, and SiC are not intrinsically blind to solar portions of the spectrum. In addition, because of their stronger internal chemical bonds, AlGaN crystals are physically and chemically more robust than most others. The first AlGaN-based UV PDs were made by Walker et al. [6] in 1996 and applied in aerospace, automotive, petroleum, and military industries. Additionally, with a smaller lattice mismatch between AlN and GaN, these materials are used to develop heterostructures to improve the device performance. Furthermore, the wider bandgap also results in a lower dark current because of the lower concentrations of thermally generated carriers. In spite of these obviously advantageous physical properties, there remained a number of technical issues that needed to be investigated. Therefore, solar-blind UV PDs with different structures are fabricated to achieve an efficient solar-blind detector from these materials.

In this chapter, we review the recent progress in solar-blind UV PDs based on III-nitride material and focus on various advances that have been discussed in related articles. For more knowledge of their advantages, different types of $Al_xGa_{1-x}N$-based solar-blind UV PDs are presented in this section. The organization of this chapter is as follows. First, AlGaN-based photoconductor is presented. Then, AlGaN-based metal–semiconductor–metal (MSM) PD and Schottky barrier photodiode are discussed. This is followed by AlGaN-based p-i-n photodiode and, last, AlGaN-based avalanche PD is discussed. Finally, we conclude this chapter with some future research directions in this field.

12.2 Various Types of Solar-Blind AlGaN Photodetectors

12.2.1 AlGaN-Based Photoconductor

A photoconductor, also termed *photoconductive detector*, consists of a pair of ohmic electrodes processed on AlGaN surface. Different from the MSM AlGaN-based detector, the electrodes are ohmic. The wavelength of detectable light entirely depends on the Al content of the AlGaN layer. A large detection area can be realized utilizing an interdigital arrangement of multiple pairs of electrodes. The structural schematic diagram is shown in Figure 12.1.

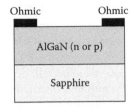

FIGURE 12.1 The schematic diagram of AlGaN-based solar-blind photoconductor.

The current in the device is driven by the electric field between a pair of electrodes, which is adjusted by an externally applied bias, U_b. In the dark, the dark current Id is given by Id $= U_b/R_d$, where R_d is the resistance given by the dark resistivity of the material. When the biased photoconductor is illuminated and the photons have enough energy to create free carriers, the resistivity decreases to a value $R_i < r_d$. The photocurrent is then $I_p = U_b \cdot (R_i^{-1} - r_d^{-1})$. Since the mobilities of electron and hole in AlGaN materials are different ($\mu_e > \mu_h$), the gain factor of the AlGaN-based photoconductor can be written as $g = \tau_h/t_e$, where τ_h represents the hole recombination lifetime and t_e the electron transit time. Thus, in photoconductors, a photoconductive gain mechanism occurs for certain values of τ_h and $t_e = d^2/\mu_e U_b$, since the latter depends on the electrode spacing d, the electron mobility μ_e, and the bias voltage U_b.

When scaling a photoconductor for an application, the limitations to the photocurrent due to electric field effects, such as space charge limited current, avalanche, and dielectric breakdown, must be taken into consideration. Furthermore, a compromise between a high gain factor and a fast response has to be found, since the device becomes slow when τ_h or τ_e is long. The main drawback of a photoconductor is the necessity of a nonzero bias operation because the dark current levels give rise to additional noise and thus decrease the detectivity of the device. Furthermore, the formation of ohmic contacts on AlGaN layer is very difficult, when the Al content of the AlGaN materials exceeds 50%. Therefore, the AlGaN-based solar-blind photoconductors are seldom studied [7–12].

12.2.2 AlGaN-Based MSM Photodetector

The metal–semiconductor–metal (MSM) photodiodes consist of two interdigitated Schottky contacts deposited on highly resistive material, as shown in Figure 12.2. Electron–hole pairs are generated when photons are absorbed near the depletion regions, which are formed at the Schottky junctions. Then the carriers are separated by the biased electrical field. These MSM photodiodes are simple to fabricate in the case of AlGaN materials, because there is no need to achieve highly doped semiconductors and to achieve ohmic contacts. The MSM photodiodes exhibit low dark current, high speed, large bandwidth, and high sensitivity. The performance of AlGaN MSM photodiodes depends on the crystal quality, applied bias, and geometry, for example, the spacing between the interdigitated fingers and the thickness of the active epilayer.

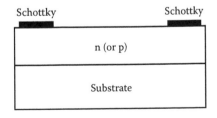

FIGURE 12.2 General structure of the AlGaN-based Schottky metal–semiconductor–metal photodiodes.

In 2006, J. Li et al. reported the first AlN-based MSM PDs with a peak responsivity at 200 nm and a sharp cutoff wavelength at around 207 nm. In the same year, Giovanni Mazzeo et al. realized a linear array of 300 pixels for solar-blind UV imaging based on AlGaN MSM PDs. Each pixel had an active area of 30×26 µm². The detector showed a peak responsivity of 12 mA/W and a dark current smaller than 1 fA at the bias of 4 V [13]. In 2012, MutluGökkavas et al. demonstrated the monolithically integrated quadruple back-illuminated UV MSM PDs with four different spectral responsivity bands [14]. In 2013, Andrea Knigge et al. compared the performance of $Al_{0.4}Ga_{0.6}N$ MSM PDs grown on conventional planar AlN templates and epitaxial laterally overgrown (ELO) AlN templates. They found that PDs on ELO templates with contact fingers oriented perpendicular to the etched stripe pattern exhibit photoconductive gain leading to external quantum efficiencies of up to 77 at 30 V applied bias, surpassing that of the planar grown PDs by a factor of 100 [15]. In recent years, some groups have utilized localized surface plasmon to further improve the performance of AlGaN MSM PDs [16,17]. Especially, the Al nanoparticles embedded on the device surface have been proven to greatly improve the responsivity. The spectral response of AlGaN-based MSM detectors with and without Al nanoparticles is shown in Figure 12.3.

Normally, the MSM photodiodes are operated under front illumination, that is, the incident light reaches the device from the epilayer side. Back-illuminated solar-blind MSM detectors have been reported in an effort to move the technology toward focal plane arrays in which the epilayers (front) side of the device will be connected to the readout circuits. In the case of back-illuminated design, the devices have the advantage of avoiding the blockage of incident photons by the interdigitated metal contacts and enhancing the quantum efficiency [18].

The AlGaN MSM photodiodes hold a great promise for the realization of commercial solar-blind detectors. For focal plane array applications, it remains to be determined whether these devices can be suitably integrated into readout circuitry and systems.

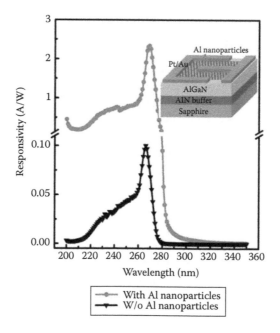

FIGURE 12.3 Spectral response of AlGaN-based MSM detectors with and without Al nanoparticles under 20 V applied bias. (Reprinted with permission from Zhang, W., Xu, J., Ye, W., Li, Y., Qi, Z., Dai, J., Wu, Z. et al., High-performance AlGaN metal–semiconductor–metal solar-blind ultraviolet photodetectors by localized surface plasmon enhancement, *Appl. Phys. Lett.*, 106, 21112 Copyright 2015, American Institute of Physics Publishing.)

12.2.3 AlGaN-Based Schottky Barrier Photodiode

Schottky photodiodes consist of a semitransparent Schottky contact and an ohmic contact. They present a flat responsivity for excitation above the bandgap and a UV/visible rejection of 10^3. Usually, the device photocurrent increases linearly with the incident optical power and applied bias. The detector speed is usually determined by exciting with short (ns) laser pulses and by varying the external load resistor in a proper low-capacitance, high-frequency electronics setup. Extrapolating results for zero external resistance, an estimation of the device internal delay response is obtained. It was found that the time response of these Schottky photodiodes was RC limited.

AlGaN-based Schottky photodiodes have been fabricated for a range of Al compositions. In 2007, H. Jiang reported the high-performance AlGaN solar-blind Schottky photodiodes, as shown in Figure 12.4. The devices exhibited dark current densities as low as 1.6×10^{-11} A/cm^2 at −5 V bias. The zero-bias peak responsivity was 41 mA/W at 256 nm, corresponding to an external quantum efficiency of 20%. The responsivity increases with the reverse bias and reaches 54 mA/W (external QE = 26%) at −10 V [19].

R. Dahal et al. have demonstrated deep UV Schottky barrier PDs by exploiting the epitaxial growth of high-quality AlN epilayer on n-type SiC substrate [20]. These PDs exhibit very high reverse breakdown voltage (>200 V) and extremely low dark current (less than 10 fA at a reverse bias of −50 V). The peak responsivity could reach 0.125 A/W at 200 nm under −25 V bias. Recently, the inverted Schottky structure in submicron-thin AlGaN layers was designed especially for optimized backside sensitivity illuminated two-dimensional arrays with a very small pixel-to-pixel pitch [21]. When the device is illuminated from the backside, detrimental absorption and recombination in the doped layer occur so that the conventional Schottky structure is not suitable for EUV detection. Cutoff wavelength was at 280 nm and intrinsic rejection ratio was three orders of magnitude of the visible radiation. The Schottky barrier photodiodes have received much less attention than other types of PDs, mainly because of the rapid interest and progress in MSM detectors. But these AlGaN-based Schottky barrier photodiodes have been commercially used in flame detection and other applications.

12.2.4 AlGaN-Based p-i-n Photodiode

In recent years, AlGaN-based p-i-n solar-blind photodiodes have attracted considerable attention for the following intrinsic advantages: (1) a relative low dark current due to large potential barrier; (2) a high operation speed; (3) a direct control of the quantum efficiency and response speed through the control

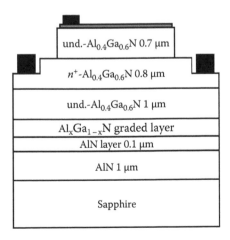

FIGURE 12.4 Schematic structure of the $Al_{0.4}Ga_{0.6}N$ Schottky photodiodes on AlN/sapphire template. (Reprinted with permission from Jiang, H. and Egawa, T., High quality AlGaN solar-blind Schottky photodiodes fabricated on AlN/sapphire template, *Appl. Phys. Lett.*, 90, 121121 Copyright 2007, American Institute of Physics Publishing.)

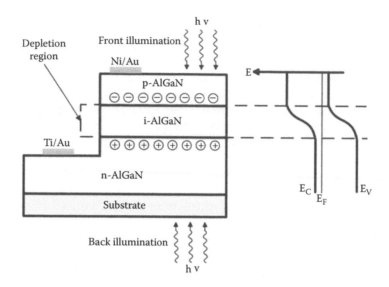

FIGURE 12.5 Schematics of device and energy band diagram of AlGaN p-i-n photodiodes, showing the incident light under front and back illumination.

thickness of the intrinsic layer; (4) a high impedance suitable for FPA readout circuitry; (5) the device can operate under low bias [22]. The typical structure of AlGaN-based p-i-n solar-blind photodiodes and the corresponding energy band diagram can be illustrated in Figure 12.5. The cutoff wavelength of the photodiodes can be well manipulated by adjusting the Al molar fraction of the ternary alloys.

For p-i-n photodiodes operation with photovoltaic mode, under no bias, the devices exhibit the lowest dark current corresponding to the reverse saturation current. When they are operated under a reverse bias with photoconductive mode, the depletion region becomes wide, and the dark current can be controlled at a very low level. What is more, the response time is much shorter. The operation mode of the photodiodes depends on the different application scenarios.

A wide range of $Al_xGa_{1-x}N$ (0.4 < x < 1) p-i-n solar-blind photodiodes have been demonstrated with the incident light illuminated from the p-type contact side. Because the ionization energy of Mg dopants in AlGaN with high Al content is relatively high [23], it results in a low hole concentration in AlGaN, which will strongly influence the performance of the photodiodes. To solve the problem, a p-type GaN layer is usually grown on the p-AlGaN to collect the photoexcited holes. It is disappointing that the GaN layer will absorb the incident solar-blind UV light with wavelength below 280 nm, indicating serious vertical diffusion losses. The incident light that can reach the depletion region will decrease greatly. In 1988, the back-illuminated photodiodes were first developed in AlGaN material system to enhance the quantum efficiency by avoiding the absorption of photons in the p-typeGaN top layer [24], thanks to the transparency property of sapphire substrate in solar-blind region. Meanwhile, an AlN layer with larger bandgap energy than incident phonons instead of GaN was grown between the n-type AlGaN layer and the sapphire to act as the buffer layer. To ensure that the photons near the desired peak responsivity wavelength will be able to reach the depletion region with minimal absorption, it is preferred to use a p-i-n heterostructure device, in which the bottom n-type AlGaN layer has a higher Al content, larger bandgap than the i- and p-type AlGaN layer. The typical Al molar fraction in the bottom n-type AlGaN layer is ~0.6, while it is ~0.4 in the i- and p-type AlGaN layers. Furthermore, to minimize the compressive stress between p-type AlGaN and p-type GaN, the Al content is sometimes graded at the interfaces.

Typically, the spectral response of AlGaN-based p-i-n solar-blind photodiodes shows the bandpass behavior under back illumination mode [25]. A record external quantum efficiency of 80% under zero bias at 275 nm and a high responsivity of 176 mA/W have been obtained, increasing to 89% and 192 mA/W under 5 V of reverse bias [26]. The UV/visible rejection ratio is measured to be more than six orders of magnitude. The inspiring results can be attributed to the high quality of an Si-In co-doped n-type $Al_{0.5}Ga_{0.5}N$ layer with resistivity, mobility, and carrier concentrations of 1.55×10^{-2} Ω cm, 74 cm^2/(V s), and 4.41×10^{18} cm^{-3}, respectively. Specific detectivities in the range are at the order of magnitude of 10^{14} cm $Hz^{1/2}$/W for AlGaN-based p-i-n solar-blind photodiodes [27,28].

12.2.5 AlGaN-Based Avalanche Photodetector

An avalanche photodetector (APD) is a highly sensitive semiconductor electronic device that exploits the photoelectric effect to convert light to electricity. APDs can be thought of as PDs that provide built-in first stage of gain through avalanche multiplication [29]. By applying a high reverse bias voltage, APDs internally multiply the primary photocurrent before entering following circuitry. In order to make carrier multiplication take place, the photogenerated carriers must traverse along a high field region. In this region, photogenerated electrons and holes gain enough energy to ionize bound electrons in VB upon colliding with them. This multiplication is known as impact ionization. The newly created carriers in the presence of high electric field result in more ionization, which is called avalanche effect. The multiplication factor (current gain) M for all carriers generated in the photodiode is defined as

$$M = \frac{I_M}{I_P}$$

where I_M and I_P are the average values of the total multiplied output current and the primary photocurrent, respectively.

It is well known that the UV background radiation is extremely low, which is beneficial to the solar-blind APD. Owing to this advantage, solar-blind APDs are in high demand for numerical applications such as flame monitoring, missile warning and guidance, and biological diagnostics. Currently, photomultiplier tube dominates the solar-blind PDs due to high sensitivity through internal gain, typically 10^6. However, these detectors are costly, bulky, fragile glass tubes that require large biases, typically 1000 V, to operate effectively. For applications such as early missile threat detection and interception, secure nonline-of-sight communication, portable chemical and biological threat detection, and UV environmental monitoring, the size and voltage requirements of PMTs can be serious drawbacks, and it is highly desirable to have a smaller semiconductor-based PD capable of similar sensitivity.

Among the solar-blind PDs with various structures, avalanche photodiode has the highest gain [30]. The first AlGaN-based solar-blind APD was demonstrated simultaneously and separately by Turgut Tut et al. [31] and McClintock et al. [32] in 2005. In Tut's work, they obtained a reproducible gain of 25 at 72 V, as shown in Figure 12.6a. The devices exhibited a maximum quantum efficiency of 55% and a peak responsivity of 0.11 A/W at a wavelength of 254 nm, as shown in Figure 12.6b and c.

In McClintock's work, the device was designed for back illumination and uses double-side polished sapphire as the substrate. The key of this device is the adopting of high-quality AlN template layer grown by atomic layer epitaxy. Figure 12.6 shows the photocurrent and corresponding gain as a function of reverse bias voltage. The dashed line is the fit by the following formula:

$$M = \exp\left\{ W \times \alpha_0 \exp\left\{ \frac{-C \cdot W}{V} \right\} \right\}$$

The fitted results show a good agreement with the experimental data. At 60 V reverse bias, the gain reaches a maximum at 700. Later on, Turgut Tut reported high-performance solar-blind PDs with reproducible

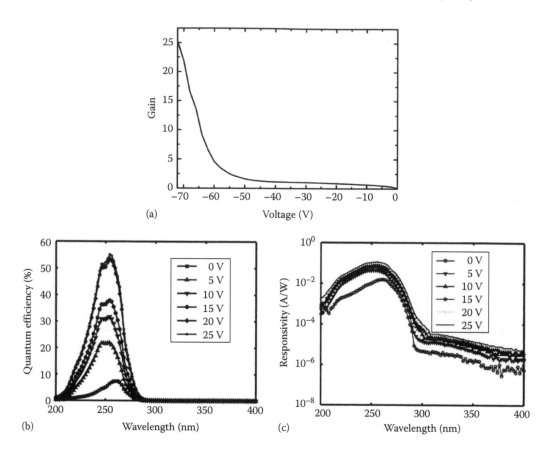

FIGURE 12.6 (a) Avalanche gain as a function of reverse bias, (b) quantum efficiency measured at different reverse biases, and (c) responsivity of the device measured at different reverse biases. (Reprinted with permission from Tut, T., Butun, S., Butun, B., Gokkavas, M., Yu, H., and Ozbay, E., Solar-blind $Al_xGa_{1-x}N$-based avalanche photodiodes, *Appl. Phys. Lett.*, 87, 223502 Copyright 2005, American Institute of Physics Publishing.)

avalanche gain as high as 1560 at reverse bias of 68 V under UV illumination in 2007, by reducing the device size to decreasing the leakage current [33]. To further increase the multiplication gain, many efforts have been made to improve the material quality and doping efficiency. In 2010, Sun et al. from Sun Yat-sen University reported a high-performance AlGaN-based solar blind PD through improving the crystal quality of the AlGaN layer [34]. The fabricated solar-blind APDs exhibited a maximum responsivity of 79.8 mA/W at 270 nm and zero bias, corresponding to an external quantum efficiency of 37%. Multiplication gain as high as more than 2500 had been obtained under 62 V reverse bias voltage, which was the highest value reported for solar-blind AlGaN-based APDs at that time. Later, the same group reported a new world record of the gain as high as 4000 at reverse bias of 177 V, as shown in Figure 12.7. They used Schottky barrier contact instead of ohmic contact for the p-electrode to increase the breakdown voltage [35].

Another group from Nanjing University also did a lot of work on AlGaN-based solar-blind APDs and obtained a very high impacted result. By adopting separated absorption and multiplication structure, a maximum multiplication gain up to 3000 at the reverse bias of 91 V was achieved [36], as shown in Figure 12.8. Recently, by applying high-quality AlN template and introducing a photoelectrochemical treatment process after mesa etching to reduce damage induced by etching, the leakage current of the fabricated devices was reduced obviously and a high gain of 1.2×10^4 at the reverse bias of 84 V was achieved, which is presently the highest value for the AlGaN-based solar-blind APDs [37].

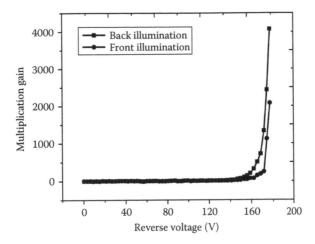

FIGURE 12.7 M–V curves of the AlGaN solar-blind APD under front and back illumination conditions. (Reproduced from Huang, Z. et al., *Appl. Phys. Express*, 6, 2013. With permission from the Japan Society of Applied Physics.)

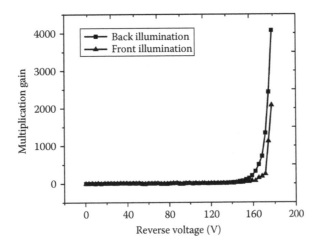

FIGURE 12.8 Reverse I–V curves in dark and under UV illumination. The right-hand axis indicates the gain. (Reprinted with permission from Huang, Y., Chen, D.J., Lu, H., Dong, K.X., Zhang, R., Zheng, Y.D., Li, L., and Li, Z.H., Back-illuminated separate absorption and multiplication AlGaN solar-blind avalanche photodiodes, *Appl. Phys. Lett.*, 101, 253516 Copyright 2012, American Institute of Physics Publishing.)

12.3 Summary

This overview on AlGaN-based solar-blind photodetectors covers the general structure and mechanism of photoconductor, MSM, Schottky, p-i-n, and avalanche PDs. For the different types of PDs, the ohmic contact of high Al content AlGaN and the challenges in the epitaxy of AlGaN layer structures as well as related processing issues and a variety of performance data were reported. Through increasing the Al content of AlGaN active layer, the response wavelength of these kinds of PDs can be adjusted to solar-blind area, accompanied by difficulties, such as ohmic contact of AlGaN, epitaxial of high Al content AlGaN, and absorption of incident photos in p-GaN layer. Because there is no need to achieve highly doped semiconductors and to achieve ohmic contacts, AlGaN-based MSM photodiodes hold a great promise for the realization of commercial solar-blind detectors and focal plane array applications. For solar-blind numerical applications, APDs exhibit great advantages due to the highest gain.

In conclusion, a variety of $Al_xGa_{1-x}N$-based solar-blind PDs have been realized. Although there is still room for improvement concerning material quality and process technology, AlGaN-based solar-blind PDs are promised to be commercially available.

References

1. A. Rogalski and M. Razeghi, Semiconductor ultraviolet detectors, *J Appl Phys* 79, 7433–7473 (1996).
2. E. Monroy, F. Omnès, and F. Calle, Wide-bandgap semiconductor ultraviolet photodetectors, *Semiconduct Sci Amp Technol* 18, R33 (2003).
3. Z. Alaie, S. Mohammad Nejad, and M. H. Yousefi, Recent advances in ultraviolet photodetectors, *Mater Sci Semiconduct Proc* 29, 16–55 (2015).
4. R. McClintock, A. Yasan, K. Mayes, D. Shiell, S. R. Darvish, P. Kung, and M. Razeghi, High quantum efficiency AlGaN solar-blind p-i-n photodiodes, *Appl Phys Lett* 84, 1248 (2004).
5. G. Parish, S. Keller, P. Kozodoy, J. P. Ibbetson, H. Marchand, P. T. Fini, S. B. Fleischer, S. P. DenBaars, U. K. Mishra, and E. J. Tarsa, High-performance (Al,Ga)N-based solar-blind ultraviolet p-i-n detectors on laterally epitaxially overgrown GaN, *Appl Phys Lett* 75, 247–249 (1999).
6. D. Walker, X. Zhang, P. Kung, A. Saxler, S. Javadpour, J. Xu, and M. Razeghi, AlGaN ultraviolet photoconductors grown on sapphire, *Appl Phys Lett* 68, 2100 (1996).
7. C. Johnson, J. Y. Lin, H. X. Jiang, M. Asif Khan, and C. J. Sun, Metastability and persistent photoconductivity in Mg-doped p-type GaN, *Appl Phys Lett* 68, 1808 (1996).
8. K. S. Stevens, M. Kinniburgh, and R. Beresford, Photoconductive ultraviolet sensor using Mg-doped GaN on Si(111), *Appl Phys Lett* 66, 3518 (1995).
9. D. Walker, X. Zhang, A. Saxler, P. Kung, J. Xu, and M. Razeghi, $Al_xGa_{1-x}N$ ($0 \leq x \leq 1$) ultraviolet photodetectors grown on sapphire by metal-organic chemical-vapor deposition, *Appl Phys Lett* 70, 949 (1997).
10. M. Asif Khan, J. N. Kuznia, D. T. Olson, J. M. Van Hove, M. Blasingame, and L. F. Reitz, High-responsivity photoconductive ultraviolet sensors based on insulating single-crystal GaN epilayers, *Appl Phys Lett* 60, 2917 (1992).
11. E. Monroy, F. Calle, J. A. Garrido, P. Youinou, and E. Muoz, Si-doped $Al_xGa_{1-x}N$ photoconductive detectors, *Semiconduct Sci Technol* 14, 685–689 (1999).
12. J. A. Garrido, E. Monroy, I. Izpura, and E. Muoz, Photoconductive gain modelling of GaN photodetectors, *Semiconduct Sci Amp Technol* 13, 563–568 (1998).
13. J. Li, Z. Y. Fan, R. Dahal, M. L. Nakarmi, J. Y. Lin, and H. X. Jiang, 200 nm deep ultraviolet photodetectors based on AlN, *Appl Phys Lett* 89, 213510 (2006).
14. M. Gokkavas, S. Butun, P. Caban, W. Strupinski, and E. Ozbay, Integrated AlGaN quadruple-band ultraviolet photodetectors, *Semiconduct Sci Technol* 27, 065004 (2012).
15. A. Knigge, M. Brendel, F. Brunner, S. Einfeldt, A. Knauer, V. Kueller, U. Zeimer, and M. Weyers, AlGaN metal–semiconductor–metal photodetectors on planar and epitaxial laterally overgrown AlN/Sapphire templates for the ultraviolet C spectral region, *Jpn J Appl Phys* 52, 3J–8J (2013).
16. W. Zhang, J. Xu, W. Ye, Y. Li, Z. Qi, J. Dai, Z. Wu et al., High-performance AlGaN metal–semiconductor–metal solar-blind ultraviolet photodetectors by localized surface plasmon enhancement, *Appl Phys Lett* 106, 21112 (2015).
17. M. Brendel, M. Helbling, A. Knauer, S. Einfeldt, A. Knigge, and M. Weyers, Top- and bottom-illumination of solar-blind AlGaN metal-semiconductor-metal photodetectors, *Phys Status Solidi (a)* 212, 1021–1028 (2015).
18. G. Bao, D. Li, X. Sun, M. Jiang, Z. Li, H. Song, H. Jiang, Y. Chen, G. Miao, and Z. Zhang, Enhanced spectral response of an AlGaN-based solar-blind ultraviolet photodetector with Al nanoparticles, *Opt Express* 22, 24286 (2014).

19. H. Jiang and T. Egawa, High quality AlGaN solar-blind Schottky photodiodes fabricated on AlN/sapphire template, *Appl Phys Lett* 90, 121121 (2007).

20. R. Dahal, T. M. Al Tahtamouni, Z. Y. Fan, J. Y. Lin, and H. X. Jiang, Hybrid AlN–SiC deep ultraviolet Schottky barrier photodetectors, *Appl Phys Lett* 90, 263505 (2007).

21. P. E. Malinowski, J.-Y. Duboz, P. De Moor, K. Minoglou, J. John, S. M. Horcajo, F. Semond et al., Extreme ultraviolet detection using AlGaN-on-Si inverted Schottky photodiodes, *Appl Phys Lett* 98, 141104 (2011).

22. M. Razeghi, Short-wavelength solar-blind detectors—Status, prospects, and markets, *Proc IEEE* 90, 1006–1014 (2002).

23. M. Katsuragawa, S. Sota, M. Komori, C. Anbe, T. Takeuchi, H. Sakai, H. Amano, and I. Akasaki, Thermal ionization energy of Si and Mg in AlGaN, *J Cryst Growth* 189–190, 528–531 (1998).

24. W. Yang, T. Nohova, S. Krishnankutty, R. Torreano, S. McPherson, and H. Marsh, Back-illuminated GaN/AlGaN heterojunction photodiodes with high quantum efficiency and low noise, *Appl Phys Lett* 73, 1086 (1998).

25. R. McClintock, P. Sandvik, K. Mi, F. Shahedipour, A. Yasan, C. Jelen, P. Kung, and M. Razeghi, $Al_xGa_{1-x}N$ materials and device technology for solar blind ultraviolet photodetector applications, in *Proceedings of the Society of Photo-Optical Instrumentation Engineers (SPIE)*, Symposium on Integrated Optics, G. J. Brown and M. Razeghi, eds., International Society for Optics and Photonics, pp. 219–229 (2001).

26. E. Cicek, R. McClintock, C. Y. Cho, B. Rahnema, and M. Razeghi, $Al_xGa_{1-x}N$-based back-illuminated solar-blind photodetectors with external quantum efficiency of 89%, *Appl Phys Lett* 103, 191108 (2013).

27. H. Jiang and T. Egawa, Low-dark-current high-performance AlGaN solar-blind p-i-n photodiodes, *Jpn J Appl Phys* 47, 1541–1543 (2008).

28. N. Biyikli, I. Kimukin, O. Aytur, and E. Ozbay, Solar-blind AlGaN-based p-i-n photodiodes with low dark current and high detectivity, *IEEE Photonic Technol* 16, 1718–1720 (2004).

29. F. Capasso, Chapter 1: Physics of avalanche photodiodes, in *Semiconductors and Semimetals*, W. T. Tsang, ed. Elsevier, the Netherland, pp. 1–172 (1985).

30. L. Sang, M. Liao, and M. Sumiya, A comprehensive review of semiconductor ultraviolet photodetectors: From thin film to one-dimensional nanostructures, *Sensors-Basel* 13, 10482–10518 (2013).

31. T. Tut, S. Butun, B. Butun, M. Gokkavas, H. Yu, and E. Ozbay, Solar-blind $Al_xGa_{1-x}N$-based avalanche photodiodes, *Appl Phys Lett* 87, 223502 (2005).

32. R. McClintock, A. Yasan, K. Minder, P. Kung, and M. Razeghi, Avalanche multiplication in AlGaN based solar-blind photodetectors, *Appl Phys Lett* 87, 241123 (2005).

33. T. Tut, M. Gokkavas, A. Inal, and E. Ozbay, $Al_xGa_{1-x}N$-based avalanche photodiodes with high reproducible avalanche gain, *Appl Phys Lett* 90, 163506 (2007).

34. S. Lu, J. Chen, J. Li, and H. Jiang, AlGaN solar-blind avalanche photodiodes with high multiplication gain, *Appl Phys Lett* 97, 191103 (2010).

35. Z. Huang, J. Li, W. Zhang, and H. Jiang, AlGaN solar-blind avalanche photodiodes with enhanced multiplication gain using back-illuminated structure, *Appl Phys Express* 6, 4101 (2013).

36. Y. Huang, D. J. Chen, H. Lu, K. X. Dong, R. Zhang, Y. D. Zheng, L. Li, and Z. H. Li, Back-illuminated separate absorption and multiplication AlGaN solar-blind avalanche photodiodes, *Appl Phys Lett* 101, 253516 (2012).

37. Z. G. Shao, D. J. Chen, H. Lu, R. Zhang, D. P. Cao, W. J. Luo, Y. D. Zheng, L. Li, and Z. H. Li, High-gain AlGaN solar-blind avalanche photodiodes, *IEEE Electronic Dev* 35, 372–374 (2014).

IV

Laser Diodes

13

Laser Diode–Driven White Light Sources

Faiz Rahman
Ohio University

Abstract In recent years, solid-state lighting, based on light-emitting diodes (LEDs), has emerged as one of the most prominent developments in lighting technology. Their applications have proliferated over the course of the past decade and LED "bulbs" can now be found in most homes and offices. This transformation from traditional incandescent and fluorescent luminaires to LED-based lighting has been greatly celebrated. However, it has become evident that LED-based luminaires also have a serious weakness that limits their electricity-to-light conversion efficiency. This limitation arises from the "droop effect" in LEDs that reduces their conversion efficiency at high drive currents. This is a very serious shortcoming because it makes it impossible to construct compact, low-cost, high-brightness LED-based light sources. The way forward is then to replace LEDs with laser diodes (LDs) for pumping wavelength down-conversion phosphors, as LDs do not show the droop effect. This chapter discusses the advantages of this approach for solid-state lighting. The history of electric lighting and LEDs is briefly discussed, followed by a discussion on possible strategies for implementing solid-state lighting with LDs. Various components of LD-based luminaires such as pump LDs, optics, electronics, and phosphors are described. The design and simulation of typical LD-pumped phosphor-converted light sources are described, followed by a look at possible future developments.

13.1 Introduction

It has been nearly 50 years since the first semiconductor diodes with the ability to emit visible light were first demonstrated. During this half century, light-emitting diodes (LEDs) have evolved from tiny devices emitting only a glimmer of light to veritable light engines capable of putting out several watts of visible light. A remarkable development over the past two decades has been that of phosphor-converted light sources pumped by blue-emitting gallium nitride (GaN) LEDs. These "white" LEDs have ushered in the era of solid-state lighting. Over the past few years, solid-state lighting has become well established with

LED-based retrofit bulbs now being widely used in both homes and offices. The initial reluctance in their acceptance now seems to have largely disappeared with their widespread availability at prices that are competitive with that of fluorescent light bulbs. The many advantages cited for LED-based luminaires include their energy efficiency, lack of mercury and other hazardous materials, availability in different shades of white, and their very long life (Zukaukas et al., 2002). Commercially, these well-known advantages compete against the somewhat higher cost of these luminaires. In spite of great advances in solid-state lighting technology in recent years, there is still a long way to go in terms of further enhancing LED energy conversion efficiency and bringing down their cost of production. A very visible trend in these developments has been the steadily increasing brightness of LED-based emitters. At first, this was achieved simply by increasing the area of LED chips. Larger area chips can be driven at higher current densities and thus can emit proportionately more light. Later, heat dissipation from single-die devices became so challenging that manufactures started producing multiple-die single phosphor encapsulation modules. All high-wattage, high-brightness LED modules currently available on the market are of this type. While ongoing technical advances will keep improving such multichip modules, the electrical-to-optical energy conversion efficiency will likely reach a plateau soon. Fundamental physical effects, mentioned later, will cap LED efficiency figures such that it will become gradually more difficult to continue to raise LED brightness levels. The slowing down in LED efficiency improvement has already begun to be noticed and will likely become more pronounced in the coming years. A way out of this predicament is to turn to non-LED solid-state light sources for phosphor pumping. Laser diodes (LDs) can provide that alternative. In contrast to LEDs, LDs have not been extensively used for lighting applications although they offer high output power levels and small form factors. This has been largely due to their higher cost. Other factors that have made LDs take a back seat when it comes to illumination applications include their beam characteristics (intense, highly directional beams are less suitable for direct lighting compared to the spread-out, divergent beam from LEDs) and high coherence that makes laser light show undesirable speckle effects. Also, LDs are not available in all colors that may be needed for lighting applications. Thus, whereas red and blue LDs have been available for quite some time, green LDs are a recent development and are still somewhat more expensive and less widely available than other common LDs. As we will see in this chapter, these limitations are not holding back laser-based illumination any longer and soon we will see LDs making significant inroads into the illumination market.

13.2 Development of LED-Based Solid-State Lighting

Artificial, electrically driven lighting accounts for a very significant portion of humankind's total energy usage on this planet. Since the advent of electrical lighting in the late nineteenth century, illuminating our homes, workplaces, and outdoor spaces has been consuming an increasingly larger proportion of electric power. During this time, other electrical devices and appliances have gained in efficiency and benefited from the solid-state electronics revolution. However, until quite recently, mainstream electric lighting remained firmly in the vacuum tube era with devices based on gas-filled incandescent and fluorescent bulbs illuminating our homes and workplaces. It is only over the last two decades that solid-state light-generating devices in the form of light-emitting diodes (LEDs) have become available as an alternative for lighting up living spaces. Their invention and evolution has been driven largely by considerations of energy efficiency as electricity is transformed to light.

That lighting represents an inefficient use of electric power dawned very early on and efforts were directed toward developing more efficient light sources. Tungsten filament–based incandescent lamps initially had efficiencies as low as 2% but were greatly improved in later years through the incorporation of such technologies as inert gas filling and double-coiled filaments, which resulted in severalfold increase in their efficiency to figures of around 8% today. The development of tungsten–halogen lamps, incorporating the halogen regeneration technology, in the 1950s, was a significant step toward further improving the nearly century-old incandescent lamp technology. Filling bulbs up with small amounts of halogens allows the lamp filament to operate at much higher temperatures without the risk of shortening its operating life. Higher filament temperatures increase the efficiency of electrical-to-light energy conversion and tungsten–halogen lamps boast efficiencies of as high as 32%. There have been no further advances in the intervening years, and it is very likely that incandescent lamps have reached the pinnacle of practically achievable efficiency. Moving to a radically different technology is then the only way to further increase the efficiency of lighting devices.

At about the same time that tungsten–halogen lamps were making their first commercial appearance, another lighting technology—this one based on electrical discharge in gases at low pressure—was also being commercialized. The fluorescent lamp, developed at General Electric, was a very different kind of light source—based as it was on light emission from excited gas molecules, energized by an electric discharge. In this lamp, mercury atoms are excited by collisions with energetic argon atoms that, in turn, are fed energy from an electric discharge. The excited mercury atoms emit ultraviolet radiation as they shed their excess energy and relax to the ground state. The ultraviolet radiation is converted to visible light as it strikes a fluorescent coating on the inside wall of the lamp. Due to the high efficiencies of the atomic radiation emission process and the fluorescent wavelength conversion in lamp phosphors, the fluorescent lamp is a relatively efficient converter of electrical energy to light. Modern mercury-vapor-filled fluorescent lamps have typical efficiencies in the range of 60%–65%. Just like its predecessor, the tungsten filament incandescent lamp, fluorescent lamps have also reached the zenith of perfection and significant further improvements in either their lifetime or energy efficiency are hardly likely. In order to achieve higher-efficiency figures for electric lighting, a radically different technology is required. Fortunately, such a technology has been in existence for a long time but was not considered suitable for space illumination applications until two decades ago.

During the 1950s and 1960s, semiconductor-based pn-junction diodes capable of light emission in the infrared and visible regions of the electromagnetic spectrum were developed. Gradually, these LEDs transformed from mere scientific curiosities to practical light sources. After that, for the next several decades, there was not much development in LED technology. During the early 1990s, LEDs were still tiny solid-state light emitters, mostly used for indicator applications in electrical and electronic systems. At that time, they were still much the same in physical form as in the 1960s—the decade of their invention. This began to change in the later years of 1990s as higher-powered devices were developed with bigger chips. The main problem with such devices was the necessity to remove heat as efficiently as possible so that the diodes did not overheat and destroy themselves. Once suitable device packaging technologies were developed, this became a reality. A concomitant development was the realization that blue LEDs can be used to generate white light by the same technique that enables fluorescent lamps to emit visible light, that is, phosphor-based wavelength down-conversion. Initial trials were quite successful and thus the first lighting-quality white LEDs were born (Schubert et al., 2006).

Almost every commercial white LED on the market today employs a GaN/InGaN blue LED that pumps a coating of one or more phosphors, placed either directly on top of the LED chip (proximity pumping) or some distance away from it (remote pumping). Most such white LEDs are based around a core of one or more blue LED dies emitting somewhere in the range of 460–480 nm. The phosphor, excited by the short-wavelength light, then converts the radiation to longer wavelengths, with a broad spectral distribution peaking in the yellow region of the spectrum (Chen et al., 2010). The resulting yellow light, together with the unconverted blue light, gives the appearance of white light to human eyes. Light from LED-driven white light sources is often inferior to that from thermal sources such as incandescent lamps because of its uneven spectral profile. Gaps in the spectral coverage of the visible region have been a well-known characteristic of light from white LEDs. This leads to unnatural color reproduction and even perceptible shifts in the color tone as the phosphor heats up during device operation. LED phosphors have improved greatly over the past two decades and thus the modern white LED produces light of much better quality than its predecessors. However, the most critical concern with white LEDs is due to the relative inefficiency of pump LEDs when driven at high current levels. The decrease of electrical-to-optical energy conversion efficiency at high LED drive currents is a well-known but poorly understood effect. This so-called droop effect has been widely debated in recent literature and, while considerable progress has been made, an unequivocal explanation of its origin is yet to emerge. Among the likely candidates causing droop are effects such as Auger recombination (where an energetic electron shares its energy with another electron rather than undergoing radiative recombination), over-the-barrier leakage in LED quantum wells, and various polarization-induced effects. There is overall consensus, however, that droop will likely result in a fundamental limit to the operating efficiency of III-nitride LED-based light sources. Unless ways are found to overcome this phenomenon, the brightness of individual LED chips will hit a brick wall. This has set the stage for investigating other possible ways of building solid-state white light sources.

13.3 Need for LD-Based Solid-State Lighting

Several parameters are of importance when considering LED-based lighting systems. These include the brightness of the luminaire, the quality of emitted light, the projected lifetime of the device, and the energy efficiency of the system. These parameters have been much explored in recent years in the context of phosphor-converted LEDs. There is increasing realization, however, that with respect to these parameters, phosphor-based LEDs have some serious limitations. The brightness of a single-chip LED is limited by the heat dissipation capability of the chip substrate material (sapphire, for present-day commercial LEDs) so that for single-die LEDs, the power dissipation for most present-day LEDs is a maximum of about 5 W. Multichip single-encapsulation LEDs with chip-on-board (COB) construction can go up to several tens of watts. Still, it is not possible to find single LED packages with electrical power ratings of 50 W and above. The quality of light emitted by LEDs also shows considerable device-to-device variation, which is controlled by "binning" LEDs at the point of manufacture, that is, by grouping them in different intensity and peak wavelength classes. Then there is the shift in color coordinates to contend with, as an LED warms up and its color point changes. While LEDs rarely suffer from catastrophic instantaneous failure, they do lose their brightness over time due to deleterious effects going on in their phosphor coatings. These effects are only accentuated at high temperatures so that devices with poor thermal management show a more pronounced reduction in emission intensity. Finally, the mentioned "droop" effect severely limits the maximum attainable wall-plug efficiency of gallium nitride (GaN)-based LEDs so that both the maximum attainable brightness and efficiency are capped by the diminution of internal quantum efficiency at high drive currents (Kim et al., 2007). This is clearly seen in a comparison of the external efficiency of typical phosphor pump LED and LD, as a function of drive current (see Figure 13.1). The external efficiency of an LED, operated at a constant temperature, falls steadily with increasing drive current, whereas that of a comparable LD, operating beyond threshold current of 30 mA, rises quickly to a large value and then stays nearly constant until the maximum allowable current rating is reached. The decrease in LD efficiency seen in this figure after reaching a peak value of 80 mA is due to heating effects, not droop. With better thermal management, this decrease can be minimized and thus efficiency maintained at even large drive currents.

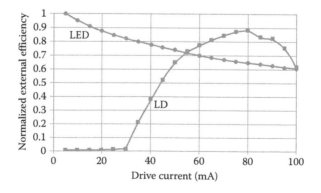

FIGURE 13.1 External efficiency of an LED and an LD plotted against drive current.

Due to these and other reasons, attention is now turning to semiconductor LDs as an alternative device for pumping phosphors for white light generation. LDs do not suffer from droop and can provide efficient, remote pumping of high-efficiency near-UV phosphors (Denault et al., 2013).

LD-based phosphor-converted white light sources offer several advantages over their LED-based counterparts. For starters, absence of droop means that while their electricity-to-light conversion efficiency drops at high drive currents, this is a much smaller effect because it arises only from internal device heating. This can be effectively reduced through clever heat dissipation schemes and thus very high conversion efficiencies are potentially possible. Thus, LD pumping has no intrinsic limitation when it comes to the ultimate brightness of the light source. These devices are also superior for pumping down-conversion phosphors, as LD-based luminaires feature remote pumping of phosphors deposited on extended glass or plastic surfaces. This arrangement reduces or eliminates phosphor heating, removing a key cause of chromaticity shift—both short term (during device startup) and long term (over the lifetime of the device). Furthermore, absence of appreciable phosphor heating also makes for extremely long phosphor lifetimes. This greatly reduces the gradual diminution in light intensity observed with LED-pumped luminaires. The spectral characteristics of light from LD-pumped phosphor sources are also significantly superior to that from LED-pumped sources. The reason for this lies in the wider spectral coverage that is possible with the former. Pumping at a shorter wavelength, such as 405 nm, allows effective generation of spectrally rich light with broadband spectral coverage through the use of trichromatic phosphors. Additionally, LED-pumped luminaires, in contrast, are pumped at around 460 nm and their light thus does not contain wavelengths in the 400–450 nm region. With 405 nm LD pumping, much fuller spectral coverage is possible with violet light being supplied by the pump diode itself. If LD pumping in the near UV is used (typically, 380–400 nm), then broadband near-UV-excited trichromatic phosphors can be used to again generate full visible spectral coverage. Both UV-emitting and near-UV-emitting LDs can also pump the growing class of single-phosphor multichromatic phosphors for improved spectral coverage. These facts show that LD-pumped phosphor-converted white light sources can be significantly superior to their LED-based counterparts (Ryu and Kim, 2010).

13.4 Strategies for LD-Based Lighting

There are two principal ways in which LD-based white lighting systems can be implemented. These are very similar to approaches used with LEDs to build sources of illumination quality white light.

A conceptually simpler but operationally more complicated strategy is to simply combine light from separate red-, green-, and blue-emitting lasers to generate light that appears white to our eyes. This color-mixing approach has been used with color LEDs as well. This route, whether implemented with LDs or LEDs, offers the great benefit of the ability to tune the color or shade of resulting light by altering the relative proportions of the red, green, and blue lights. Thus, this is the preferred approach where partial or full control over illumination color is needed. The downside of this approach is the need for properly mixing the different lights to generate fully mixed output light. This is quite hard to do with high efficiency, that is, without losing a big fraction of light in the mixing process. The simplest light-mixing elements consist of translucent polymer or glass plates that randomize the direction of light rays that travel through them and in the process mix different light components. However, such devices necessarily also absorb a good fraction of light propagating through them and thus are only used in applications where low efficiencies can be tolerated. Other, more efficient mixing devices consist of mirror baffles and multifaceted mixing chambers that take more volume but lose less of the lights that are being mixed.

Neumann et al. (2011) have shown that red, yellow, green, and blue light from four different lasers can be mixed together using chromatic thin-film beam combiners. After further mixing in a volumetric spatial diffuser, the resulting light appeared white in color. The diffuser also served to reduce the speckle in laser light. Very significantly, for illumination applications, they found that in spite of the resulting light's spectrum being composed of four narrow spectral lines, it was able to render the color of everyday objects quite faithfully. This is an important observation for laser-based illumination systems,

FIGURE 13.2 Type PL 520 green-emitting LD from Osram.

especially those that employ only lasers (no phosphor-based down-conversion) or make use of narrow-band phosphors. It turns out that due to the nature of the spectral responsivity function of human eyes even white-appearing light with "spiky" spectrum is able to render actual physical colors quite effectively. Of course, the use of broadband phosphors leads to spectrum-filling white light with much better color rendering characteristics, but "spiky" white light also performs reasonably well, somewhat contrary to expectations. True RGB LD-based white light sources have now become available with the development of green-emitting LDs from Osram and other companies. Typical PL 520 type green LDs from Osram, for example (see Figure 13.2), emit 30 or 50 mW of output optical power in the 510–530 nm range and can be combined with red and blue LDs emitting about 100 mW of optical power to make compact white light sources.

The other approach for generating white light from LDs is more recent but has started picking up in popularity. This is through down-conversion of LD radiation, used as a pump source, to one or more longer wavelengths through the use of single or multicomponent phosphors (Xu et al., 2008). This is similar to the approach used with phosphor-converted white LEDs. The only difference here is that an LD instead of an LED is used as the pump source. Most of the rest of this chapter describes this technique in more detail. Just like the case of LEDs, LD-pumped phosphor down-conversion is a much simpler scheme for generating white light. In contrast to LED-based systems, this approach is exclusively based on a remote phosphor arrangement, which is described in more detail in later sections.

It should be mentioned here that there are other strategies too for generating white light from monochromatic LD light. Because of system cost and complexity, these approaches are much less prevalent and are only met with in research environments. One of these approaches is through spectral stretching of optical energy through the use of special photonic crystal optical fibers. This so-called super-continuum white light generation is a valuable method for generating "flat," that is, equi-energy visible radiation. Systems available from NKT Photonics, Denmark, for example, provide continuum light in the 400–2400 nm region (see Figure 13.3).

FIGURE 13.3 SuperK Extreme supercontinuum white light source from NKT photonics. (Courtesy of NKT Photonics A/S, Birkerød, Denmark.)

FIGURE 13.4 Energetiq LDLS xenon plasma white light source from Energetiq Technology, Inc. (Courtesy of Energetiq Technology Inc., Woburn, MA.)

The other approach makes use of plasma generation by focusing high-power LD light to a tight spot (~100 μm diameter) inside a rarefied gas (Xe) atmosphere. This produces a small bright broadband source of light that can be coupled to both free space and optical fibers. Figure 13.4 shows a commercial laser plasma-based white light source from Energetiq Technology Inc.

13.5 LDs for LD-Based Lighting

LDs at a number of different wavelengths can be used for pumping down-conversion phosphors. Traditional, blue LED down-conversion phosphors (designed for excitation in the 345–370 nm region) can be pumped by blue-emitting LDs. These and near-UV excitable phosphors can often also be pumped with violet-emitting LEDs. These latter devices, emitting at 405 nm wavelength, are often well suited to this task because they are the most mature and cheapest of all "blue" LDs. This is because they were extensively developed during the 1990s for CD and DVD applications. Both low- and high-power LDs at this wavelength are easily available from multiple manufacturers (see Figure 13.5).

A further advantage of these LDs is that pumping at this wavelength also provides the violet light component in white light, which is missing in light from white LEDs. Sub-400 nm LDs (such as devices

FIGURE 13.5 Typical violet-emitting LDs in TO-style metal can packages.

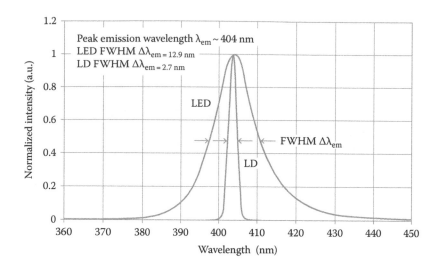

FIGURE 13.6 Spectra of an LED and an LD emitting at the same peak wavelength, showing the difference in spectral emission width.

emitting at 390 nm) can also be used for exciting near-UV phosphors. In this case, all the three primary color components have to be generated from individual phosphor components in a trichromatic phosphor blend. This allows precise spectral shaping for high color rendering index (CRI) light sources. The 405 nm violet LD pumping approach is especially attractive because of its wide availability at low cost and the availability of phosphors that can be excited in the 400–420 nm region (DenBaars et al., 2013). As seen in Figure 13.6, both LEDs and LDs are available with peak wavelengths around 405 nm. The differences between these two pump sources are the different spectral full width at half maximum (FWHM), angular beam profile, wavefront structure, and coherence. LD light is significantly narrower in wavelength spread (~2–3 nm) compared to LED light (~12–30 nm). It is also highly coherent, which can lead to interesting effects in certain situations. A number of companies such as Sony and Osram produce LDs suitable for solid-state phosphor-pumping applications. Most LDs, useful for phosphor pumping, come packaged in TO-style metal cans for good heat dissipation. Such leaded packages can be mounted directly on a printed circuit board (PCB). Suitable heat sinking must be provided to remove the heat generated during LD operation. The metal can packages have a transparent glass window integrated at the top through which the laser light escapes. Other than that, there is, usually, no beam-shaping optics provided. Optical elements needed to appropriately shape the beam for efficient phosphor pumping (see the next section) must be provided external to the LD package. Devices emitting in the UV region may have a fused silica (highly pure glass) window instead of an ordinary float glass window. Usually, a photodiode is packaged with the LD chip inside the same package. This photodiode is used to monitor the emission intensity of the LD so that closed-loop feedback control can be used to keep the emission intensity stable against changes due to temperature and so on. Thus, most LD packages have three leads that come out of the metal can. Two of them connect to the photodiode and LD anodes, whereas the third one is a common cathode lead. Some LDs also come mounted on a Peltier cooler, but this form of device is not used for phosphor pumping.

A number of parameters are important for selecting LDs suitable for exciting phosphors. These include the emission wavelength, threshold current, operating current, operating voltage, and power conversion efficiency. All of these parameters are carefully measured by LD manufacturers and listed on datasheets. Violet-light-emitting LDs are the most common for pumping phosphors because of their wide availability and low price. These devices nominally emit at 405 nm (±1%) with a narrow spectral width of around

2 nm FWHM. Their beam quality is often not very good, but this is not important for phosphor-pumping applications. As long as their operating current is within the range bounded by the threshold current and the maximum allowable current, these devices perform very well.

Unlike LEDs, LDs are more susceptible to gradual degradation through mechanisms related to dislocation multiplication in the active region. Threading dislocations that inevitably exist in LDs tend to multiply at high current densities and operating temperatures. These dislocations are nonradiative recombination sites that appear as so-called dark line defects (DLDs). With prolonged device operation, especially, at high current levels, DLDs tend to proliferate and the LDs can gradually lose their brightness. This is a serious degradation mechanism for many types of LDs, which can place an upper limit on the longevity of LD-pumped phosphor-converted white light sources. Short-wavelength visible light-emitting LDs have been much improved over the past several years but more work needs to be done to further enhance their long-term reliability. Growing LD device layers on native (free-standing) GaN wafers, which are now commercially available, can go a long way toward reducing the density of threading dislocations and thus DLDs.

13.6 Optics for LD-Based Lighting

There are two aspects of optics that are relevant for LD-pumped solid-state lighting systems. One relates to shaping the light beam that comes off the pump LD so that the phosphor can be properly illuminated. The other aspect concerns the shaping of the down-converted light that comes from the phosphor plate or screen. Note that the first of these is not of concern in LED-based proximity-pumped phosphor-converted light sources. However, this is of primary importance in all laser-pumped luminaires. Both lenses and mirrors can be used in LD-pumped phosphor-converted luminaires. Lenses are usually used for shaping the LD beam, while a suitable mirror, with or without an accompanying lens, can be used for final output light beam shaping.

Light coming out of semiconductor LDs is almost always divergent and with non-Gaussian intensity profile. The reason for this is the shape of the resonant cavity in typical side-emitting LDs. The long, narrow cavity with rectangular cross section causes the light output to diffract asymmetrically. The wavefronts diverge more along the rather small thickness of the multiquantum well (MQW) structure and less along the wider lateral rib that defines the width of the lasing cavity. Light exiting such a rectangular aperture assumes a divergent fan shape with elliptical cross section. This is far from the structure of the nice round TEM_{00} beams that come out of most gas lasers. The two elliptical axes of an LD light beam are called slow and fast. Rounding and collimating such a laser beam are very challenging. Nevertheless, effective techniques have been developed to shape this type of light beam. The first task is to circularize the cross section of light from side-emitting LDs. A common way to achieve this is to use cylindrical lenses. These lenses have cylindrical shapes at right angle to the optical axis and tend to shape light fields such that light is spread out into a line at right angle to the cylinder's axis. Two separate plano-convex cylindrical lenses with different focal lengths can be used to circularize the output of an LD by shaping light along both the slow and fast axis. Cemented and mounted cylindrical doublets are available for this purpose. Once the beam has attained a circular cross section, any curvature of the wavefront can be removed through the use of a collimating lens assembly. An appropriate plano-convex or plano-concave lens can achieve this. A well-collimated laser light beam has maximally parallel rays that imply plane parallel wavefronts. Such a light beam is in a sense "well behaved" and can be further shaped to suit a given application. For solid-state lighting application, the pump LD light needs to be expanded so as to fill the entire area of the remote phosphor plate. This reduces the aerial optical power density incident on the phosphor. Without this approach, the laser power density will be too high to be effectively used for phosphor-based wavelength downconversion. High incident light power causes phosphor to heat up, thus greatly shortening its life. It also causes both short-term and long-term drifts in the chromaticity coordinates of down-converted light. However, the most serious problem is that excessive optical power causes phosphor to saturate, which both reduces its down-conversion efficiency and increases pump light pass-through (bleed). In order to avoid such deleterious effects, laser pump light has to be expanded and projected on to the remote phosphor plate or screen. With properly circularized and collimated light beam, expansion is easily done by inserting a single, short focal length, plano-concave or bi-concave lens in the beam path. This lens is chosen to be of small diameter—just a little bigger than the cross section of the pump beam. The beam expansion lens is placed immediately after the collimating lens. Diverging wavefronts coming out of this lens are made to fall on the remote phosphor plate. The beam expansion lens is placed a suitable distance in front of the phosphor plate such that the expanded beam just fills the area of the plate. This scheme is further discussed in the subsequent section on luminaire design.

Down-converted light from the remote phosphor plate has to be collected and brought out of the lamp assembly by a parabolic mirror that surrounds the phosphor plate. This element can be made out of alu-minized plastic formed into the desired shape. This is also further described in the following luminaire design section.

13.7 Electronics for LD-Based Lighting

Diodes that emit light, both LEDs and LDs, are usually operated in a constant current mode such that they are driven by a source that appears as a current source with high internal Norton resistance. An ordinary voltage source with a series resistance can make a tolerable current source but current-mode transistor-controlled power supplies are much preferable. Such supplies are now also available as integrated circuits that require only a few external components to make excellent LED and LD drivers. An example is MAX16834 chip from Maxim Integrated Inc.

Pulse width modulation (PWM) is the most widely used scheme for controlling the intensity of light emission from all kinds of LEDs and LDs. By varying the duty cycle of a pulse waveform, used to drive an LD, between 0% and 100%, it is possible to control the light output from the LD. Typical PWM frequencies are around a few kilo Hertz, and these waveforms can be generated by a variety of devices such as microcontrollers, complex programmable logic devices (CPLDs), field programmable gate arrays (FPGAs), and special LED-driving integrated circuits. Normally, the PWM output is filtered with a simple one-pole RC filter before it is used to drive LEDs, either directly or with a power transistor. A better alternative has now become possible with the availability of LTC 2645 from Linear Technology Inc. This IC is a PWM-to-DC level converting chip that Linear Technology calls a PWM to VOUT DAC. This device takes in up to four channels of PWM waveforms and produces corresponding DC voltage levels. Unlike an RC filter, this is accomplished with no latency and very high accuracy. Linear Technology also makes available a reference board for this device, which plugs directly onto Arduino UNO form factor boards, like a shield, to make PWM-based multichannel lighting systems very simple to implement (see Figure 13.7). Single-channel PWM circuits can control the brightness of phosphor-converted LD-pumped white light sources, whereas multiple-channel PWM circuits can individually control the red, green, and blue channels of a color-mixing-type white light source, pumped by three (or more) LDs. The latter technique allows the spectral profile of resulting mixed light to be tuned as desired. The combination of multiple LDs (either emitting band edge light or down-converted light) with individual PWM controls enables the generation of full-spectrum white light with desired chromaticity point and correlated color temperature (CCT).

Several semiconductor companies also make special LD drivers that can be used to drive LDs in both continuous wave (CW) and pulsed mode. An example of this kind of device is the EL6934 from Intersil Corporation. This IC is targeted at CD and DVD drives but can also control LDs for phosphor pumping, with currents up to 450 mA. iC-Haus of Germany also makes a range of CW and pulsed LD drivers that are suitable for operating blue and violet GaN LDs for illumination applications.

In addition to the LD driver, a complete LD-based illumination system also needs a power supply to convert mains power to low-voltage DC power that can be used by the LD and the LD driver. This is accomplished by a switch-mode power converter. These systems use step-down converter ICs with a few

(a)

(b)

FIGURE 13.7 An Arduino UNO board (a) and an LTC2645 reference board from Linear Technologies Inc. (b).

external converters and can convert any voltage in the 80–300 V AC range to a low DC voltage. Such "universal" power supplies are small, light, and fairly inexpensive.

While it is easy to develop and implement the electronics needed for LD-based white light systems, their long-term reliability is usually inferior to that of other components in the system. Thus, a light source can fail because an electronic component, instead of a light path component, has degraded, gradually or catastrophically. This kind of reliability issue is also seen in LED-based white luminaires.

13.8 Phosphors for LD-Based Lighting

Since the advent of LED-based solid-state lighting, around a decade and a half ago, inorganic phosphors of various compositions pumped by blue LEDs have remained the primary scheme for generating quasi-white light. Almost every commercial white LED on the market today employs a GaN/InGaN blue LED that pumps a coating of one or more phosphors, placed either directly on top of the LED (proximity pumping) or some distance away from it (remote pumping). Most LED phosphors can be used for LD-based lighting as well, as long as they have high absorption at the LD emission wavelength. Another important consideration for suitable LD-pumped phosphors is high saturation intensity so that phosphor particles do not stop down-conversion of photons too soon, as the incident optical flux increases. This is an especially important concern because phosphor pumping by LDs results in much higher aerial power densities than are achievable with even very high brightness blue LEDs.

LD-pumped white light sources can be produced with either a single broadband yellow-emitting phosphor or a mixture of red, green, and blue phosphors—a so-called trichromatic phosphor. This latter approach is preferred for its better spectrum-filling light quality. A trichromatic white-emitting phosphor capable of pumping with LDs emitting at 405 nm was developed by mixing three rare-earth-doped phosphors: (1) potassium europium tungstate phosphor ($KEu(WO_4)_2$, type: CPK63/N-U1)—emission color: red, (2) europium and manganese-doped barium magnesium aluminate phosphor ($BaMg_2Al_{16}O_{27}$:Eu,Mn type: KEMK63M/F-U1)—emission color: green, and (3) europium-doped strontium magnesium silicate phosphor ((Sr,Mg)$_2$SiO$_4$:Eu, type: HEBK63/N-D1)—emission color: blue. The blue phosphor—a double alkaline–earth silicate—had roughly equal amounts of Mg and Sr ions at cation sites. All of these phosphors had europium ion concentrations in the range of 0.5%–1%. These phosphors were chosen because of their ease of availability (all three are reasonably priced phosphors available from stock but none is a standard LED phosphor) and their general ability for being excited at short wavelengths around 400 nm. Most LED phosphors, on the other hand, are designed to be excited from blue GaN LED emissions in the range of 450–470 nm and thus will not be suitable for pumping with easily available inexpensive GaN LDs emitting at 405 nm. As the 405 nm wavelength used for pumping phosphors is not a standard wavelength for LED-based white lighting applications, it remains a task for phosphor manufacturers to develop phosphors that are better suited to pump wavelengths in this region. All three phosphors described were obtained as powders from Phosphor Technology Limited (PTL), England, United Kingdom.

Phosphor samples were produced for luminescence and chromaticity measurements by mixing phosphor powders with a potassium silicate solution (KASIL 2132) and coating the slurry on small BK7 glass substrates. The phosphor coatings had thicknesses in the range of 100–200 μm. Once dry, the samples were placed inside an integrating sphere with conduits for laser radiation entry and light exit to an Ocean Optics spectrometer model USB2000 through a long pass filter (type: OG-435) and optical fiber cable. Due to this arrangement, the light from the phosphor was measured in both reflection and transmission modes simultaneously. The OG-435 filter, used to stop unconverted laser light from reaching and overwhelming the detector, had a sharp cutoff at 435 nm. It allowed all light at wavelengths longer than 435 nm to pass through while blocking radiation at shorter wavelengths. The blue phosphor emission peaked at 460 nm and thus only a negligible amount of blue light was cut off by this filter and nothing was blocked from the emissions of green and red phosphors (other than the flat reduction at all wavelengths due to the neutral density of the filter in its pass band). Thus, the use of the long pass filter did not significantly affect the perception of the color generated from the phosphors. In real LD-driven lamps, no long pass filter is used, but the light produced in such lamps has only a very small amount of violet photons from the laser. This is because almost all of the light is absorbed by a thicker phosphor, containing internal light-scattering particles. This is also aided by the fact that in a real lamp the pump laser is usually projected onto the phosphor after passing through a beam expansion lens so as to reduce the per unit area power density on the phosphor.

The trichromatic white-emission phosphor was made by mixing appropriate amounts of the three color phosphors in order to produce balanced white light with chromaticity point near the center of the *CIE* chromaticity diagram. The phosphor mixture was balanced to obtain a chromaticity point as close to the equal power point ($x = 0.3$, $y = 0.3$) as possible. It has been found that in order to obtain a good white point, phosphors have to be mixed in inverse proportions to the amount of radiometric energy they contribute to the white light's spectrum. Thus, significantly more red phosphor, of the type mentioned earlier, is needed in the phosphor mix due to its relatively low conversion efficiency. The best trichromatic white phosphor contained 57.5%, 25.4%, and 17.1% of red, green, and blue phosphors, respectively, by weight fraction. Even for narrowband phosphors it is possible to produce very good white lighting by proper mixing of individual phosphors (Lei et al., 2007).

Photoluminescence excitation (PLE) spectra have been measured using light generated by combined xenon and tungsten lamps, dispersed with a monochromator, equipped with different holographic gratings, and the resulting luminescence was detected at specific wavelengths using a photomultiplier. The measured PLE spectra were corrected for the spectral response of the setup and were normalized to the pump energy of the excitation source. The primary source of illumination for measuring PL spectra (emission spectra) and color spectral response, in this case, was a violet-emitting InGaN/GaN LD module operating at 404 nm. The module operated from a 5 V output power supply and generated a maximum of 96 mW of continuous wave (CW) power at room temperature. In order to keep the output power constant and to avoid overheating of the module, it was operated with an automatic 50 s cutoff switch. An Optotune laser speckle reducer model LSR-3000-17S-VIS with a clear aperture of 5 mm diameter and light diffusion

FIGURE 13.8 Speckle pattern on a two-dimensional screen produced by the emission from a 405 nm violet LD. Top shows as-emitted light, whereas bottom shows light with reduced speckle after it was scrambled by a dynamic laser beam speckle reducer.

angle of 17° was used for converting coherent laser light into incoherent light to assess the role coherence plays in luminescence from phosphors. The laser speckle reducer is a dynamic light diffuser using an electrically driven electro-active polymer membrane to diffuse laser light without significantly absorbing it. With its operation, reduced optical coherence with no change in peak wavelength and spectral width can be obtained. The light does become uncollimated after passing through this device but it can be collected and refocused with a lens system. Figure 13.8 shows the effect of laser speckle reducer, with the upper panel showing the distribution of local intensity pattern with the speckle reducer switched off and the lower panel showing the distribution with the speckle reducer switched on. The reduction in speckle is easily seen. Reduction of speckle causes the far-field pattern of laser light to become more uniform, which avoids saturation of phosphor particles. Saturated (and under-illuminated) phosphor particles produce reduced down-converted light, which adversely affects system performance. Observations have shown that for reasonably thick phosphor coatings, speckle does not reduce wavelength down-conversion efficiency.

The red and blue phosphors, described earlier, are typical of a large class of near-UV pumped phosphors containing Eu as the active luminescent ion whereas the green phosphor belongs to a class of Eu/Mn co-doped phosphors where Eu^{2+} ions act as sensitizers, transferring energy nonradiatively to Mn^{2+} activator ions that produce the main green luminescence. These phosphors show efficient luminescence when excited at selected wavelengths: between 250 and 450 nm (blue phosphor), at 465 nm (green phosphor) and at 550 nm (red phosphor) at 300 K. The typical excitation spectra of these phosphors appear in Figure 13.9 (dashed lines). The excitation spectrum of the red phosphor (see Figure 13.9a) monitored at 615 nm can be divided into two parts. One is a charge transfer band from 230 to 320 nm centered at 275 nm. This band is attributed to the $O^{2-}W^{6+}$ charge transfer within the WO_4^{2-} groups. When monitored by the red emission of Eu^{3+} (615 nm, $^5D_0-^7F_2$), the PLE spectrum of the red phosphor exhibits peaks at 323, 364, 385, 395, 416, 466, 480, and 530 nm, corresponding to the intra-4f transitions of Eu^{3+} ion from the ground level $^7F_0$ to the $^5H_3$, $^5D_4$, $^5L_7$, $^5L_6$, $^5D_3$, and $^5D_2$ excited terms ground and Stark levels, respectively. The red phosphor, emitting primarily in the 610–630 nm spectral regions, can be effectively excited due to the intra-configurational f–f transitions of Eu^{3+} ion by photons with energies from near UV. It is seen that the excitation wavelengths at 395 and 465 nm are the most efficient ones. However, the red phosphor is also excited by wavelengths in the 355–420 nm spectral range, but with lower efficiency, including at 404 nm—the chosen LD output wavelength. The representative PL spectrum when excited at 404 nm is shown in Figure 13.9a. The intra-4f-shell transitions in the Eu^{3+} ion transitions are split into components depending on the host matrix crystal field. These emission lines of the Eu^{3+} ion cover the orange and red spectral region, with less intense $^5D_0-^7F_1$ magnetic–dipole transition emission centered at 595 nm and the dominant $^5D_0-^7F_2$ electric–dipole transition lines at 614 and 618 nm, respectively. The observed $^5D_0-^7F_2$ to $^5D_0-^7F_1$ transition intensity ratio confirms that the Eu^{3+} ions occupy predominantly sites without inversion symmetry in the $KEu(WO_4)_2$ phosphor. However, the intensity of the electric–dipole transition of the Eu^{3+} ions is almost 10 times higher than that of the magnetic–dipole transition, implying that the Eu^{3+} ions predominantly occupy low symmetry sites.

Figure 13.9b shows the excitation spectrum of the green phosphor excited with photons of wavelength between 250 and 470 nm and monitored at 515 nm. The PLE spectrum consists of a broadband tail ascribed to the charge transfer of Mn–O unit centered at ~240 nm (not shown here), a broadband spanning the region from 250 to 430 nm with maximum at 320 nm and a pronounced shoulder on the low-energy side of the band at ~380 nm. A low-intensity broad band peaking at 450 nm is also seen. The dominant broad

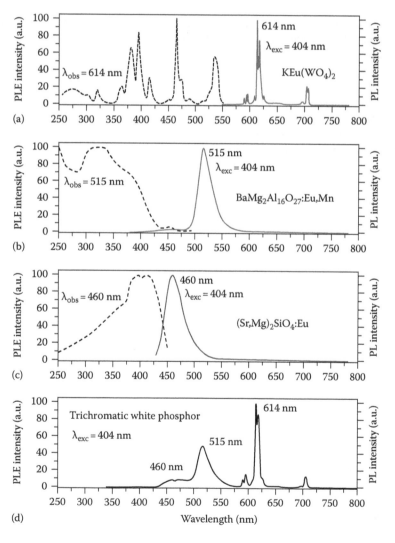

FIGURE 13.9 Excitation and photoluminescence (PL) spectra of (a) red, (b) green, (c) blue, and (d) trichromatic white phosphors.

PLE band when monitored at 515 nm (emission from the Mn^{2+} ion due to the $^4T_1$–$^6A_1$ transition) is composed of spectral features resulting from the direct excitation of Eu^{2+} and Mn^{2+} ions and energy transfers between them. The broadband that peaks at 320 nm is due to the $4f^7$–$4f^65d^1$ transitions of Eu^{2+} ions that replaced Ba^{2+} ions in the host crystal. The blue emission band at 450 nm is attributed to the $4f^65d^1$–$4f^7$ transition of Eu^{2+} ions. The broadband high-energy wing peaking at ~385 nm is due to the $^6A_1$–$^4T_2$ transitions of Mn^{2+} ions that replaced Al^{3+} ions in the host crystal. When an Mn^{2+} ion replaces an Al^{3+} ion, a negative charge is developed in the lattice because of the nonequivalent replacement of ions involved. To maintain electrical neutrality of the phosphor, in the absence of extrinsic impurities, it is expected that positive charge compensation is induced by structural defects and grain boundaries. The observed divalent manganese emission critically depends on the size of the crystallographic cation site that Mn^{2+} is likely to occupy (here Mn_{Al}) and the coordination number of Mn^{2+} ion in the $BaMg_2Al_{16}O_{27}$:Eu,Mn host matrix. In the present case, the Mn^{2+} ion occupies a tetrahedrally coordinated site. It is known that the mechanism of energy transfer in phosphors basically requires spectral overlap between the sensitizer (donor) emission and the activator (acceptor) excitation. In the case of $BaMg_2Al_{16}O_{27}$:Eu,Mn, the Eu^{2+} ion is a sensitizer and the Mn^{2+} ion is an activator. It has been shown that there exists efficient nonradiative energy transfer between the Eu^{2+} and Mn^{2+} ions in $BaMg_2Al_{16}O_{27}$ and other barium aluminate hosts. The presence of Eu^{2+} ion emission band in the PL spectrum at 450 nm, when excited at 404 nm (see Figure 13.9b), and the large intensity ratio of the green band of Mn^{2+} to the blue band of Eu^{2+} indicates that the aforementioned sensitization process is very effective. In the green phosphor, Eu^{2+} ions act as sensitizers and transfer their energy to Mn^{2+} ions that act as activators (luminescent ions). Often energy migrates over a chain of Eu^{2+} ions through radiation-less transfer processes until it is transferred to a Mn^{2+} ion, which then undergoes radiative de-excitation. This energy transfer process is not 100% efficient and thus not all of the absorbed energy results in down-converted luminescence generation.

Figure 13.9c shows the excitation spectrum for the blue phosphor while monitoring at 460 nm. It can be seen that the excitation spectrum is composed of a few broad bands in the region spanning from 250 to 450 nm, with a well-defined maximum at 410 nm and a shoulder at ~350 nm on the high-energy wing. There are two major crystallographic sites in monoclinic Sr_2SiO_4, namely, 10-coordinated Sr(I) and 9-coordinated Sr(II). Thus, it is expected that Eu^{2+} ions in monoclinic Sr_2SiO_4:Eu^{2+} will replace primarily both the Sr(I) and Sr(II) sites due to the similar radius of Sr^{2+} (0.131–0.136 nm) and Eu^{2+} (0.130–0.135 nm) ions when 9- or 10-coordinated. This will happen instead of their being incorporated into the $Mg_2(Si_2O_4)$ sublattice due to the big difference in the radius of Mg^{2+} (0.065 nm) and Eu^{2+} ions. The PLE spectrum can be assigned to the absorptions from the Eu^{2+} ions occupying Sr(I) (minority) and Sr(II) (majority) sites. The short-wavelength band shoulder at ~350 nm corresponds to the Eu^{2+} ions occupying Sr(I) sites, whereas the band at 410 nm corresponds to the Eu^{2+} ions occupying Sr(II) sites. The observed luminescence spectrum when excited at 404 nm shows emission predominantly from Sr(I) site with emission from Sr(II) site strongly suppressed. There exists effective energy migration between the Eu^{2+} ions occupying Sr(I) and the Eu^{2+} ions occupying Sr(II) sites. Relying on the arguments of the intensity ratio between 350 nm (Sr(I)) and 410 nm (Sr(II)) peaks and the absence of emission from Eu^{2+} ions at Sr(II) site, one can assume that the emission from Eu^{2+} at Sr(I) sites dominates. Furthermore, it can be speculated that Eu ions are effectively excited when occupying both Sr(I) and Sr(II) sites. However, observed emission at 460 nm originates from Eu^{2+} ions occupying Sr(I) sites with strong energy transfer from Eu^{2+} occupying Sr(II) site to Eu^{2+} occupying Sr(I) site and weak back energy transfer between these two centers. It is worth noting that typical sharp line emissions corresponding to the transitions of Eu^{3+} ions are not observed in the PL spectrum of the blue phosphor (see Figure 13.9c), indicating that Eu is primarily present in (2+) ionization state in the Sr_2SiO_4 host.

On room-temperature excitation with a laser source emitting at 404 nm, the three phosphors produced emission spectra that have also been depicted in Figure 13.9. The wavelength peaks for the red, green, and blue phosphors were located at 616, 515, and 460 nm, respectively. The FWHM of the primary emission peaks were 18, 38, and 62 nm for the red, green, and blue phosphors, respectively. In the case of the trichromatic white phosphor, the measured correlated color temperature (CCT) was 5037 K (see Figure 13.9d).

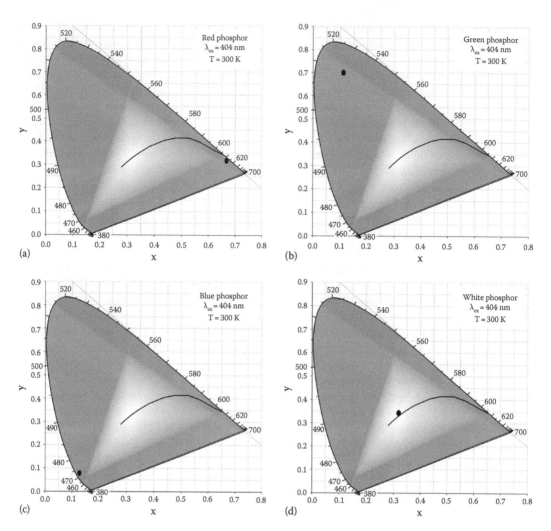

FIGURE 13.10 Chromaticity diagrams showing the location of chromaticity point corresponding to light emission from (a) red, (b) green, (c) blue, and (d) trichromatic white phosphors. The curved black line in each diagram is the Planckian locus of chromaticity points of an ideal black body.

There was no measurable increase in phosphor temperature just from exposure to laser pump light. The position of the chromaticity coordinate for each phosphor under the same measurement conditions as that used for taking their spectra (λ_{ex} = 404 nm, T = 300 K) is shown in Figure 13.10. The RGB phosphors generate their characteristic colors from respective atomic transitions in europium (Eu^{3+} (red phosphor), Eu^{2+} (blue phosphor)) ions, and manganese (Mn^{2+} (green phosphor)) ions. As discussed previously, in the case of the green phosphor, the Eu^{2+} ions act as sensitizers–transferring energy gained from the absorbed pump photons efficiently to luminescent manganese ions. The Mn^{2+} ions are themselves hard to pump directly because their d → d transitions are forbidden on both spin and parity grounds for electrical-dipole transitions. The presence of Eu^{2+} ions in the same host allows first their pumping through f → d transitions, followed by efficient radiation-less transfer of energy to Mn^{2+} ions, causing them to fluoresce. While the green emission in this phosphor originates from Mn^{2+} ions, some emission also comes from the Eu^{2+} ions (low-intensity broad band between 425 and 500 nm seen in Figure 13.9b).

The chromaticity coordinates of phosphor emissions change with temperature. This leads to the well-known problem with white LEDs changing their color point with changes in operating current and thus

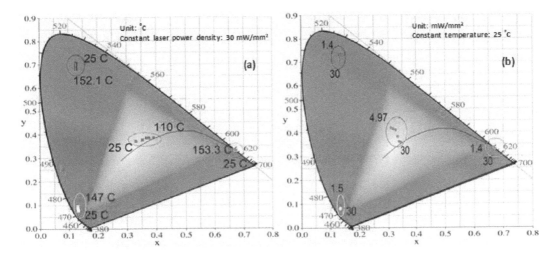

FIGURE 13.11 Changes in chromaticity due to (a) variation of phosphor temperature and (b) variation of incident laser power density. The curved black line in each diagram is the Planckian locus of chromaticity points of an ideal black body.

LED temperature. This phenomenon has been well investigated with LED light excitation of phosphors (Chiang et al., 2008). Changes in the chromaticity of trichromatic white-emission phosphor under coherent laser excitation (404 nm) at different phosphor temperatures have also been studied. In these studies, the temperature increase was deliberately caused by an external heater and was not due to heating from the LD or from the absorption of pump radiation. Figure 13.11a shows the variation of *CIE* chromaticity coordinates, due to phosphor temperature variations, for the red, green, blue, and white phosphors. In this case, the incident laser power was kept constant at 30 mW/mm². Figure 13.11b shows *CIE* chromaticity coordinate variations for these phosphors when the pump power was varied, keeping the phosphor temperature constant at 25°C. In each plot, the range of parameter causing the variation (temperature in °C or pump power density in mW/mm²) is shown alongside each group of chromaticity points. The red and blue phosphors showed the least variation, whereas the green phosphor showed more variation. It is clearly seen from Figure 13.11 that changes in *CIE* coordinates of the white phosphor due to temperature variation approximately follow the Planckian locus, whereas changes in *CIE* coordinates due to excitation power density variation do not show the same trend. These results show that one can build LD-pumped white light sources using the trichromatic phosphor with good white point chromaticities. It is also worth noting that for proper selection of monochromatic color phosphors for generating trichromatic white light using LD pumping, one should also take into account the individual red, green, and blue phosphors' luminescence relaxation dynamics. It should be mentioned here that the green phosphor's luminescence decay time is in the milli-seconds range whereas for red and blue phosphors it is typically a few orders of magnitude faster. This can be an important issue for light-dimming applications. In general, the *CIE* coordinates' variation due to temperature is expected to be much smaller than what these results show because the phosphor temperature and incident laser pump power are not likely to vary much at all. In this sense, these results are illustrations of worst-case scenario that will only be observed in real lighting devices if they malfunction for some reason, causing the phosphor temperature or incident pump power to increase catastrophically.

The aging characteristics of red, green, and blue component phosphors and their various mixtures have also been studied. Accelerated life tests have been done by periodically measuring the intensity of the main emission peak for each phosphor, as samples were kept heated at 120°C in ambient air. The results are shown in Figure 13.12, where the gradual decrease in the phosphor's luminescence efficiency over an extended period of time can be clearly seen. All phosphors show an initial period of intensity adjustment before their emission stabilizes. This is seen very clearly in Figure 13.10, where the intensity of the red

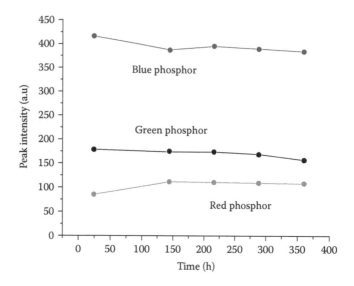

FIGURE 13.12   Changes in the peak emission intensity from red, green, and blue phosphor as a function of time. All phosphors were kept at 120°C.

phosphor shows an initial increase, while that of the green and blue phosphors show an initial decrease followed by an increase in emission intensity. None of these phosphors are hygroscopic. However, phosphor particles always carry some adsorbed water and this is gradually released over a period of time. Phosphor crystals also contain internal defects that heal over time (accelerated by heating). These phenomena may play a role in the initial adjustments that are seen. Afterward, the emission stabilizes with a small downward slope. Phosphors designed for pumping with LEDs, emitting around 460 nm, show a similar behavior. However, in the case of remote laser excitation the lifetime can be much longer, as phosphors operate at temperatures only a little above room temperature. Typical temperatures for laser-excited phosphors are in the range of 30°C–35°C (assuming the experimental arrangement itself is kept at around 23°C). At such temperatures, it is estimated that the phosphors mentioned earlier will still emit 80% of their initial intensity after 40,000 h. This estimate is based on the observed rate at which phosphor luminescence decreases over time in aging tests and takes into account the much lower temperature the phosphors will encounter during actual device operation. Note that 80% reduction from initial brightness is a metric that is standard in solid-state lighting industry for quoting luminaire lifetimes. Furthermore, this lifetime also assumes no appreciable degradation in the pump LD itself.

13.9 Luminaire Designs for LD-Based Lighting

Depending upon application requirements, various types of LD-pumped white light sources can be designed. An illustrative example of a domestic-type lamp is described here although most recent lamps have been designed for such high-intensity applications as automobile headlights. This is because LD-pumped luminaires have an immediate advantage over LED-pumped luminaires in that they are unaffected by LED droop and can thus easily achieve higher brightness levels at respectable conversion efficiencies. The laser-driven white light source described here was designed to have approximately the same size and form factor as a traditional MR-40-style tungsten halogen lamp. The device comprised a 405 nm violet laser module (actual emission at 404 nm peak wavelength) with a Sony LD, attached to an interior-aluminized plastic reflector shell. A phosphor-coated borosilicate glass disc was attached to the wide end of the reflector. Figure 13.13 shows two-dimensional (2D) and three-dimensional (3D) models of the parabolic reflector. This component had a uniform wall thickness of 2 mm and it was made through

FIGURE 13.13   Reflector design: 2D sketch (left) and revolved 3D models (right).

3D printing on a Stratys 60 printer using a hard polymer. The optimized profile of the reflector was first created as a 2D sketch. This profile was subsequently revolved using SolidEdge software to produce a suitable 3D model. The .stl file thus created was directly used by the printer to build the reflector.

The surface of the reflector had roughness on the order of 25 μm RMS and thus it was a somewhat diffuse reflector rather than a purely specular reflector. It was aluminized by placing it in an electron beam evaporator and evaporating a 500 nm thick layer of silver on its interior surface to form the reflecting surface. This type of development shows that 3D printing is a powerful tool for creating optimized reflector optics for solid-state lighting applications. The excitation source was a 100 mW-rated GaN/InGaN Sony violet LD module powered by a compact 120 V AC/5V DC in-plug switch-mode power supply. The cylindrical housing of the laser module fitted snugly into the narrow end of the reflector. The laser module had glass optics for beam collimation and beam divergence adjustments. An additional external 6 mm diameter bi-concave lens was used to expand the beam so as to fill the entire 25 mm diameter of the phosphor-coated disc. This ensured that as much of the laser light as possible was utilized for phosphor pumping without causing phosphor saturation and/or excessive pump light bleed. The phosphor disc itself was a 25 mm diameter and 2 mm thick optical grade borosilicate wafer with ground but unpolished edges. In order to uniformly coat it with phosphor, a plasma activation procedure was performed on one surface of the disc. This process made the coating surface highly hydrophilic and greatly aided uniform spread of the phosphor + silicate binder slurry. For plasma activation, discs were thoroughly cleaned with acetone and isopropyl alcohol (IPA) and then rinsed with deionized water, followed by blow drying in nitrogen. Oxygen plasma used for surface treatment was generated using March Instrument's CS-1701 reactive ion etcher (RIE) system with a parallel-plate chamber at a base pressure of 40 mTorr, capacitively coupled to an RF source at 13.56 MHz. Typically, 5 min of plasma exposure was used.

Phosphor was coated on the plasma-activated side of the disc by mixing it with a potassium silicate solution (KASIL 2132) and spreading the resulting slurry evenly on the disc surface by dropping it with a glass pipette in a radial pattern. Both single trichromatic phosphor layer and four-layer alternating (RGBR) phosphor coating schemes were tried, with the latter appearing to be slightly more beneficial in lighting characteristics and uniformity. Different ratios of phosphor to silicate binder were tried and it was found that most ratios produced coatings that were 45 ± 5 μm thick, as measured by a micrometer. This procedure produces reasonably uniform coatings that are suitable for use in LD-pumped lamp assemblies. The entire assembly was 70 mm high and weighed 32 g (excluding the power supply adaptor). Figure 13.14 shows a photograph of the phosphor plate mounted onto the reflector shell.

Geometric ray tracing simulations of the luminaire assembly have been carried out, using Lambda Research Corporation's TracePro software (version: 7.6.2). For this purpose, a geometric model was set up. It consisted of a parabolic-figured reflector with aluminized inner surface with 95% reflectance (ALCOA. BriteCoat 85 Specular), a phosphor plate modeled by a diffuse scattering coating on its reflector-facing surface, a double concave glass lens with 12 mm focal length and a cylindrical base/adapter. Laser light was simulated by a circular 1 mm diameter grid source with a Gaussian intensity distribution. Typically, 271 rays per wave were simulated and the results are shown in Figure 13.15.

FIGURE 13.14 A 3D printed reflector with phosphor plate attached.

FIGURE 13.15 TracePro modeling of LD-pumped phosphor down-conversion lamp showing beam expansion lens, reflector, and phosphor plate.

For simulation visualization, the external surface of the phosphor plate was made opaque in order to clearly see the behavior of light inside the laser-reflector-phosphor plate assembly. The expansion of the bundle of rays (red) comprising the incident light from the LD module is clearly seen on the left. The middle image shows the expanded light cone striking the inside (phosphor-coated) surface of the phosphor plate. Some light is backscattered from the plate (blue) and travels back to be reflected by the specular inner surface of the reflector. This light is then re-incident on the phosphor plate. Note that only very few rays come out of the back end of the reflector. These rays are wasted as they are absorbed by the LD module housing. Rays seen escaping the gap between the reflector and the phosphor plate is a result of deliberately having a gap between the two, during system simulation, in order to see inside the assembly. In actual practice, the phosphor plate is flush with the end of the reflector and no light escapes there. The actual behavior of the luminaire was found to be very close to the simulation results depicted here.

Yet another arrangement is shown in Figure 13.16a where LD light strikes a much smaller opaque phosphor plate and the down-converted light is completely directed outward by a spherical reflector. Although not shown, a beam expansion lens is needed here too in order to avoid saturating the phosphor coating.

Figure 13.16b shows another arrangement where down-conversion takes place in a special enclosed spherical phosphor chamber that serves as an integrating sphere. Here, the pump light and the down-converted light have a 90° angular separation, but other separations are also possible due to the symmetry of the integrating sphere.

In extremely high-power luminaires, such as powerful headlights and searchlights, multiple LDs need to be used to pump phosphors effectively. In such cases, simple mirrors can be used to combine light from multiple LDs. Alternatively, polarization beam combiners can also be used, which combine orthogonally

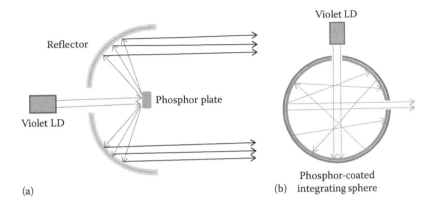

FIGURE 13.16 Alternative configurations for LD-pumped phosphor-converted white light sources. (a) Phosphor plate/mirror. (b) Phosphor in an integrating chamber.

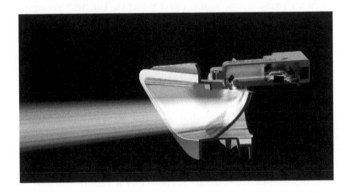

FIGURE 13.17 Cutaway diagram of BMW Laserlight. Courtesy of BMW GmbH. (Courtesy of BMW GmbH, Munich, Germany.)

polarized laser beams into a single higher power beam. This is seen in Figure 13.17, which shows a cutaway diagram of the BMW Laserlight auto headlight. Combined light from three (or more) LDs is incident on a phosphor plate and the resulting white light is directed out by a curved mirror (Hanafi and Erdl, 2015). A possible concern in this and similar high-intensity lamps is related to damage to phosphor due to the use of high incident optical flux. This problem can be mitigated by scanning the incident beam on the phosphor plate with a small electrically driven scanning mirror that produces a raster pattern on the phosphor plate. No doubt other designs will be developed in the future as this is a rapidly developing technology.

13.10 The Road Ahead

Solid-state LD-pumped white light sources are now being developed as an alternative to white LEDs. While both RGB LD sources and phosphor-converted LD-pumped sources are under development, the latter are being much more seriously pursued. This is because, just as in the case of LEDs, phosphor-pumped sources offer the great benefit of freedom from light-mixing requirements. The promise of higher efficiencies— circumventing the LED droop problem—is often cited as the most important reason to develop LD-pumped white light lamps, but these systems offer other advantages as well. These include better light quality and the potential to make much higher brightness sources than is currently possible with LED technology. This is the reason LD-pumped white light lamps have been developed for car headlight applications. For the near future, this type of LD-pumped luminaire appears set to get the most attention with more auto makers coming out with LD-based headlights, just as happened with automotive LED lighting. This development will also spill into other related areas such as search lights and high-power lighting for sports arenas and open outdoor spaces. Thus, significantly higher brightness than can be currently obtained from LED-based sources appears to be the prime driver for LD-pumped luminaire development at this time. However, further into the future, LD-pumped luminaires will also be developed for domestic/office uses, along the lines described in this chapter. These developments will require advances on several fronts. First, better, high-power, and, most importantly, long-life pump LDs will be needed. Ongoing developments in LD technology such as better device structures, multichip stacked LDs, and use of native GaN substrates will surely help in this quest. New phosphors that can be efficiently pumped at 405 nm (and future near-UV wavelengths) also need to be developed. These will likely encompass the entire range spanning broadband yellow emitters, RGB trichromatics, and single-component multi-wavelength down-converters. New optical systems that allow easy phosphor pumping and maximum light extraction also need to be developed. Finally, technologies behind lamp power supplies and LD drivers will need to be improved in order to take full advantage of the potential inherent in LDs for pumping down-conversion phosphors.

Future prospects for LD-pumped lamp development are very bright and offer numerous opportunities for both scientific research and technological developments. The next generation of solid-state light sources—a step up from LEDs—appears to be just around the corner. Time will tell how quickly this technology will mature, but it appears to be a given that within a few years, LD lamps, as opposed to LED lamps, will become increasingly more important for all types of lighting applications.

References

Chen L, Lin C, Yeh C, Liu R (2010) Light converting inorganic phosphors for white light-emitting diodes. *Materials*. 3:2172–2195.

Chiang C, Tsai M, Hon M (2008) Luminescent properties of cerium-activated garnet series phosphor: Structure and temperature effects. *Journal of Electrochemical Society*. 155:B517–B520.

Denault KA, Cantore M, Nakamura S, DenBaars SP, Seshadri R (2013) Efficient and stable laser-driven white lighting. *AIP Advances*. 3:072107 (6pp).

DenBaars SP, Feezell D, Kelchner K, Pimputkar S, Pan C-C, Yen C-C, Tanaka S et al. (2013) Development of gallium-nitride-based light-emitting diodes (LEDs) and laser diodes for energy-efficient lighting and displays. *Acta Materialia*. 61:945–951.

Hanafi A, Erdl H (2015) Lasers light the road ahead. *Compound Semiconductor*. 21:36–42.

Kim M-H, Schubert MF, Dai Q, Kim JK, Schubert EF, Piprek J, Park Y (2007) Origin of efficiency droop in GaN-based light-emitting diodes. *Applied Physical Letters*. 91:183507 (3pp).

Lei Z, Xia G, Ting L, Xiaoling G, Ming LQ, Guangdi S (2007) Color rendering and luminous efficacy of trichromatic and tetrachromatic LED-based white LEDs. *Microelectronics Journal*. 38:1–6.

Neumann A, Wierer JJ, Davis W, Ohno Y, Brueck SRJ, Tsao JY (2011) Four-color laser white illuminant demonstrating high color-rendering quality. *Optics Express*. 19:A982–A990.

Ryu H-Y, Kim D-H (2010) High-brightness phosphor-conversion white light source using InGaN blue laser diode. *Journal of the Optical Society of Korea*. 14:415–419.

Schubert EF, Kim JK, Luo H, Xi J-Q (2006) Solid-state lighting—A benevolent technology. *Reports on Progress in Physics*. 69:3069–3099.

Xu Y, Chen L, Li L, Song G, Wang Y, Zhuang W, Long Z (2008) Phosphor-conversion white light using InGaN ultraviolet laser diode. *Applied Physical Letters*. 92:021129 (3pp).

Zukaukas A, Shur M, Gaska R (2002) *Introduction to Solid State Lighting*. New York: Wiley.

14

InGaN Laser Diodes by Plasma-Assisted Molecular Beam Epitaxy

Czeslaw
Skierbiszewski
*Institute of High
Pressure Physics
Polish Academy of Sciences
and
TopGaN Ltd*

Muziol Grzegorz
*Institute of High
Pressure Physics
Polish Academy of Sciences*

Turski Henryk
*Institute of High
Pressure Physics
Polish Academy of Sciences*

Siekacz Marcin
*Institute of High
Pressure Physics
Polish Academy of Sciences
and
TopGaN Ltd*

Marta Sawicka
*Institute of High
Pressure Physics
Polish Academy of Sciences
and
TopGaN Ltd*

Abstract   The present status of nitride-based laser diodes (LDs) made by plasma-assisted molecular beam epitaxy (PAMBE) is reviewed. We demonstrate continuous-wave LDs operating in the range of 408–482 nm on (0001) *c*-plane bulk GaN substrates and ultraviolet LDs at 390 nm on $\left(20\bar{2}1\right)$ semipolar GaN substrates. The peculiarities of low-temperature GaN and InGaN growth by PAMBE are discussed. Recent improvements of InGaN growth have allowed us to demonstrate high-quality quantum wells (QWs) and excellent morphology for thick InGaN layers. The use of high N fluxes— up to 3 µm/h—during the InGaN growth was essential to push the lasing wavelengths of PAMBE LDs above 460 nm. The novel LDs design with the InGaN waveguides is demonstrated. We discuss the influence of thickness and composition of InGaN waveguides on the internal and external LDs parameters. Additionally, a solution to the problem of light leakage to GaN substrate is proposed.

14.1 Introduction

The demonstration of the first blue–violet light-emitting diodes (LEDs) and edge-emitting laser diodes (LDs) by Shuji Nakamura at Nichia and other groups in the early 1990s was a turning point for the development of a variety of optoelectronics devices based on nitrides [1]. The importance of LEDs for energy savings and impact on human life was confirmed when the Nobel Prize in 2014 was awarded to the pioneers of nitride-based devices: Isamu Akasaki, Hiroshi Amano, and Shuji Nakamura. The large potential of the nitride-based LEDs and LDs for efficient light sources spanning from ultraviolet to infrared regions, detectors as well as transistors, has been triggering continuous interest in the III-N materials. However until very recently, the key achievements in this field were made by metal–organic vapor phase epitaxy (MOVPE) technology [1,2]. The other growth technology, molecular beam epitaxy (MBE), in spite of its potential in situ characterization and monitoring tools, was lagging behind MOVPE for LEDs and LDs applications. It is believed that the reason for this is the lower quantum efficiency of MBE-grown structures. The main issue deliberated from the beginning of nitride epitaxy by MBE was the choice of nitrogen precursor [3–5]. The apparent one was ammonia. Ammonia MBE has been successful in improving the optical quality of nitride films and produced the first blue–violet nitride LDs [6]. However, the use of very high ammonia flow required to stabilize growth surface at high growth temperatures combined with the highly corrosive nature of NH_3 was the subject of serious technological challenges [3,4]. The alternative source of N for nitride growth was a plasma source, where among different types, the radio frequency (RF) plasma was the most successful and widely employed. Early experimental results showed that unlike ammonia MBE, plasma-assisted MBE (PAMBE) requires group III-rich conditions to achieve good material quality [7]. Much progress has been made in both theoretical [8,9] and experimental [10–13] understanding of the growth kinetics for such metal-rich conditions. In spite of the relatively low growth temperatures, 650°C–750°C, state-of-the-art GaN/AlGaN heterostructures with record high mobilities of two-dimensional electron gas [14–16] have been grown by PAMBE, making it the technique of choice for the production of electronic devices. The refinement of the PAMBE growth conditions on GaN/sapphire MOVPE templates allowed the demonstration of promising LED devices [17]. The turning point in low-temperature PAMBE technology was the introduction of high-quality GaN substrates, which led to truly dramatic improvements in the optical quality of PAMBE-grown structures, resulting in the room temperature high-power pulsed and cw blue–violet LDs [18–21]. In this work, we report on the progress made by PAMBE in the area of edge-emitting nitride LDs, which demonstrates the potential of this technology.

14.2 Basic Concepts of Low-Temperature Growth of Nitrides by Plasma-Assisted Molecular Beam Epitaxy

14.2.1 Experimental

The growth of LDs presented in this work was performed in Riber V90 MBE and Veeco GEN20A reactors. Both were equipped with two Veeco RF plasma sources. The N_2 flow for V90 MBE was varied from 0.8 to 2 sccm allowing for the growth rate up to 0.9 µm/h, while for GEN 20 the maximum flow rate was 10 sccm, which allows for growth rate up to 3 µm/h. The maximum growth rate cited here is obtained for both RF plasma working simultaneously. The pressure during growth was $1.5–7 \times 10^{-5}$ Torr. We used four types of (0001) GaN substrates: a bulk GaN with threading dislocation densities (TDDs) $10^3–10^5$ cm^{-2} grown by high nitrogen solution pressure (HNSP) [22] or by ammonothermal method [23]; bulk GaN fabricated by hydride vapor phase epitaxy, that is, HVPE-GaN (from Saint Gobain Crystals) with TDDs about 5×10^7 cm^{-2} and GaN/sapphire templates with TDDs >5×10^8 cm^{-2}. We also used semipolar (20–21) substrates grown by ammonothermal method [24]. The epi-ready bulk substrates were prepared by mechanical polishing and mechanochemical polishing. For the growth, 10 × 10 mm or

1 in. substrates were attached to 2 in. Si wafers with In or to 2 in. GaN/sapphire templates with Ga. For 1 in. substrate we also used metal-free mounting—where substrate was hung on special springs supplied by Veeco. The high-quality InAlGaN structures were grown on (0001) and (20–21) surfaces in the metal-rich growth regime [7,9,11–13]. The N flux was calibrated from GaN growth rate at Ga-rich conditions.

14.2.2 Growth of GaN

We will describe and investigate the growth of nitrides on vicinal substrates. For the growth on such substrates, the best quality of semiconductor layers is obtained in the so-called step-flow growth mode. In Figure 14.1, we show schematically such a situation, when crystal growth is realized through continuing advancement of atomic steps. The analysis of growth for other semiconductors like GaAs or Si indicates that the optimum growth temperature for step-flow mode equals to about half of their melting temperature, T_M [25,26]. This rule is also valid for GaN growth, where the melting point lies somewhere between 2540 K (experimentally determined for 6 GPa [27]) and 2800 K (theoretically calculated [28]). The optimum growth temperature for GaN (used in MOVPE) is in the range 1050°C–1100°C (1320–1370 K). The explanation for this phenomenon on the microscopic scale is that at $T = 0.5T_M$ atomic kinks start to become active; that is, atoms can detach from an atomic kink at a rate comparable to typical attachment rates during deposition. For the ideal Kossel crystal [29] depicted in Figure 14.1, atoms that reach the surface during growth can be adsorbed at the terraces (atom C), at atomic edges (atom B), or at atomic kinks (atom A). The binding energy for the atoms on these sites increases with decreasing number of the dangling bonds ($E_C < E_B < E_A$). Therefore, atom A has the highest binding energy among atoms A, B, and C. The step-flow growth mode takes place when all the atoms are finally incorporated at the atomic kinks (like atom A), while those that did not reach the atomic kinks are desorbed from the growth surface. For Kossel crystal, the atoms at the kink positions have a half-filled number of dangling bonds, while atoms inside the crystal have no dangling bonds. Therefore, the binding energy of the atom at the atomic kink is approximately equal to half the binding energy of an atom inside the crystal. Thus, the atoms at the kinks are expected to become mobile close to half of the melting temperature of the crystal.

The main difficulty in GaN growth in MBE is that for temperatures near $0.5T_M = 1320$ K, a high nitrogen overpressure (about 60 bars for N_2) is required to prevent GaN decomposition (see stars in Figure 14.2) [30,31]. For more chemically active nitrogen precursors, like ammonia, the situation is better, but still an overpressure in the range of 0.1–2 bar must be used—like it is realized in MOVPE systems. Such overpressures are not compatible with MBE technology, which relies on high vacuum conditions for the delivery of atoms from the effusion cells to the growing layer. This situation is even more difficult for InN growth. The expected melting point of InN is similar to that for GaN; therefore, the equilibrium nitrogen overpressure over InN at half its melting temperature is orders of magnitude higher than that for GaN (Figure 14.2). In our opinion, this is the main reason why MBE using "classical" growth conditions (i.e., N-rich growth) could not compete for many years with MOVPE in the field of optoelectronic devices containing InGaN layers.

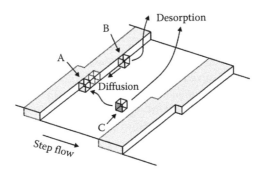

FIGURE 14.1 Schematic picture introducing step-flow growth mode on vicinal substrate.

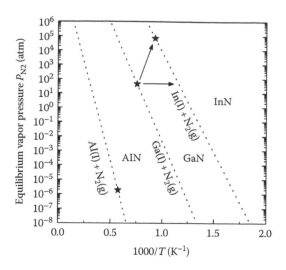

FIGURE 14.2 Equilibrium N_2 pressure over the MN(s) + M(l) systems (M = In, Ga, or Al), after Ambacher et al. [31]. Stars indicate N_2 pressure for $T = 1/2T_M$.

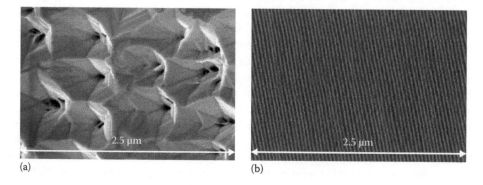

(a)   (b)

FIGURE 14.3 The surface morphology of GaN grown at (a) N-rich conditions and (b) Ga-rich conditions. Parallel atomic steps are visible on sample grown at Ga-rich conditions on low dislocation density GaN vicinal substrate.

The first breakthrough in the study of growth kinetics in PAMBE came with the finding that using Ga-rich conditions it is possible to grow relatively smooth layers at temperatures much lower than expected from considerations outlined earlier [7,10–14,16]. What is very important is that the lower growth temperature reduces by orders of magnitude the nitrogen overpressure required for high-quality epitaxy. However, such growth conditions were prone to the formation of Ga droplets on the GaN surface and high-quality material was mainly formed in the regions between the droplets. Further study of the Ga surfactant effect revealed that this problem could be avoided provided that Ga to N flux ratio was maintained in a very narrow range of values: low enough to be just below the formation of the droplets, but high enough to ensure the formation of a metallic Ga bilayer on the Ga polarity surface [10–14,16]. As an example we show in Figure 14.3 the PAMBE-grown GaN surface morphology for N-rich (rough surface) and Ga-rich (smooth surface step-flow growth) conditions, both layers grown at 710°C [32].

There are two processes that can explain why we can grow crystal much below $0.5T_M$. First is that for the metal-rich regime for PAMBE (we can call it liquid phase epitaxy–like conditions), the formation energies of the kink sites are substantially reduced, lowering the minimum growth temperature necessary to sustain the desirable step-flow growth mode [33]. As it was pointed out by Schoonmaker et al., the decomposition of GaN (and therefore the activity of atomic kinks) is controlled by a kinetic barrier that is strongly reduced when GaN is covered by liquid Ga [34]. It was also experimentally evidenced and discussed by others [35,36] that, for a given temperature, the presence of a metal layer on a GaN surface greatly accelerates the thermal decomposition of GaN in comparison to the exposed surface.

The second very important process necessary for the high-quality growth at low temperature, which has also been a subject of intensive theoretical studies, is the surface adatom diffusivity. Perhaps the most insightful were results reported by Neugebauer et al. [8], who postulated the existence of a very efficient lateral diffusion channel for nitrogen adatoms on semiconductor surface just below the thin metallic adlayer. Low diffusion barriers for this so-called adlayer enhanced lateral diffusion (AELD) facilitate high-quality step-flow epitaxy at low temperatures [8–21]. This enhanced surface mobility of nitrogen adatoms, coupled with the reduction in the kink formation energy discussed earlier, promotes step-flow growth at temperatures far below those needed for the "classical" group V-rich conditions. However, it was found that for PAMBE, in the group III metal-rich regime and for relatively low growth rates used there (up to 0.3 µm/h) and low substrate miscut angle, the spiral stepped growth is always dominant on substrates containing high density of threading dislocations [12]. Such morphological features are located at the outcrops of threading dislocations and are common for all crystals grown in the step flow regime [37]. Thus, the availability of the low-dislocation density GaN substrates turned out to be the key to further optimization of PAMBE process.

14.2.3 InGaN Growth Peculiarity

The growth of efficient high In content QWs continues to be a subject of intensive studies [38–45]. Successful growth of the InGaN layers is a central point in the nitride technology of visible emitters. The discovery by S. Nakamura that addition of small amount of In to GaN increases substantially the quantum efficiency of nitride devices (regardless of high-dislocation density) was a fundament of development of InGaN LEDs [1]. This phenomenon works for In composition up to 15%–17%, that is, for emission wavelengths between 405 and 460 nm. Above these values, the difficulties of InGaN growth prevail and decrease of quantum efficiency is observed and such effect is called the green gap [39]. There are two main reasons for low InGaN quality for high In content. First is the large lattice mismatch between GaN and InN. It is more than 10% and growth of thick InGaN layers on GaN results in strain relaxation, which drastically deteriorates optical properties of grown layers. A more severe problem is the growth thermodynamics. The melting temperature for InN is very high—close to GaN; therefore, the optimum growth temperature for InN should lie somehow between 900°C and 1050°C. However, even at temperatures below $0.5T_M$, InN decomposes very fast, so such high growth temperatures would require huge nitrogen overpressure to stabilize the surface. In practice (and this is particularly important for MBE conditions) for successful InGaN growth, understanding the mechanisms of nitrogen loss is of paramount importance.

In this section, we will describe the nitrogen loss mechanism and limitations of In incorporation into InGaN layers in PAMBE for In-rich regime. We assume that during the growth a thin In layer is present at the surface, which means in practice that we have always infinite reservoir of In for InGaN growth. This requires application of In flux, which is equal to active N flux plus part of In that would desorb plus part of In that forms thin liquid In layer. Similarly to GaN growth, the thin monolayer of liquid In at the surface enhances diffusion of adatoms at low growth temperatures [8].

The wurtzite crystal symmetry influences the growth kinetics in PAMBE. Atoms in III-N wurtzite structure on (0001) c-plane are arranged in such a way that adjacent crystal planes along c-axis are rotated by 60° with respect to each other and spaced by $c_0/2$, where c_0 is the lattice constant equal 0.5186 nm for GaN. Such crystallographic arrangement for vicinal substrates leads to the situation when two neighboring atomic step edges (named here step edge "A" and step edge "B") are *not equivalent*. The role of the exact arrangement of atoms on the surface was noted by Xie et al. [46], where nonequivalent character of two adjacent atomic steps was discussed for the low-temperature growth of wurtzite GaN on SiC substrates. Here, following Turski et al. [47], we discuss how such asymmetry influences InGaN growth.

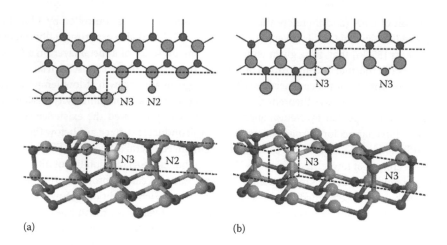

(a)   (b)

FIGURE 14.4 Top view (upper part) and perspective view [48] (lower part) of arrangement of atoms at the surface of (In)GaN in metal-rich conditions for the atomic step edge "A" (a) and "B" (b). Smaller atoms stand for nitrogen and larger atoms stand for metal (Ga or In). Nitrogen positions at atomic kinks are labeled in black, while positions at atomic step edges are labeled in gray. In both cases, nitrogen positions are marked "NX," where index X stands for the number of saturated bonds at that site.

We assume that the probability of incorporation of nitrogen atom at a particular site is governed by the number of bonds that N atom will saturate on full adsorption at this site. The N atoms can be bound to the atomic steps "A" in two places—at the atomic kink, where the number of bonds to cations is equal to three (therefore, we depicted it as N_3 in Figure 14.4a), or at the step edges, with two bonds (N_2 in Figure 14.4a). For step "B," on the other hand, N atoms attach by three bonds at both: the kink and at the edge (N_3 in Figure 14.4b). In other words, taking into account only the nearest neighbors, there is no difference in microscopic picture for step "B" when N atom is bonded to the atomic kink or to the atomic edge. Such conclusion is related directly to the microscopic arrangement of atoms in wurtzite crystal. This has important consequences for the difference in stability between atomic steps "A" and "B." For steps "B," the N atom can be incorporated into the atomic kink or atomic edge with comparable binding energies, since for both cases the number of bonds to cations equals three. Therefore, at step "B" the average bonding energy for nitrogen is much higher than at step "A." This leads to faster nucleation at the step edge "B" as well as to *instabilities* in the step-flow growth mode. As a result, the step edges "A" will advance slower than "B," so after a short growth time, the surface will be dominated by step edges "A" even for terraces that at commencing the growth had edges of purely "B" type, as is schematically shown in Figure 14.5b.

The inspiration for the InGaN growth model discussed here comes from the investigation of morphologies of InGaN layers with low In content (2%), as presented in Figure 14.6. In this figure, we show AFM images of 100 nm thick InGaN layers with about 2% of In grown by PAMBE at $T_G = 650°C$ on GaN substrate with very low miscut toward m direction [1–100]. For the growth, we used low TDDs GaN substrates fabricated by HNPS method. The AFM images reveal clearly the sequences of single atomic steps with height of

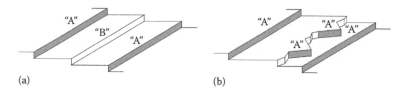

(a)   (b)

FIGURE 14.5 Schematic image of the surface of (a) vicinal substrate (surface before growth) and (b) InGaN layer grown on vicinal substrate where during growth, on step edge "B," V-shaped step edges "A" had been formed.

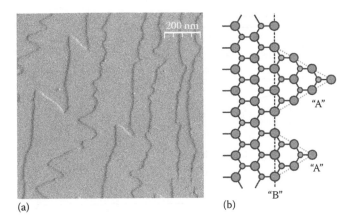

(a)   (b)

FIGURE 14.6 (a) Amplitude signal measured by AFM for $In_{0.02}Ga_{0.98}N$ layer grown on low dislocation density, free-standing GaN bulk substrates with miscut angle of approximately 0.1° oriented toward *m*-plane. (b) Schematic picture showing the evolution of the atomic step edges during InGaN growth: from initial step "B" to step "A."

$\sim c_0/2$ with interesting step morphology where every second atomic step is approximately parallel, while the steps in-between have irregular edges. Under closer inspection, these irregular steps have triangle-shaped features which edges are aligned along direction twisted by 60°. The characteristic evolution of straight step edge "B" into the "zigzag" shaped, sawlike atomic edges is explained with atomistic model in Figure 14.6b. Nitrogen atoms arriving at step edge "B" attach to it and quickly form a microscopic triangular shaped feature. Arrangement of atoms at the edges of these triangles is the same as at step edge "A," so further nitrogen atoms that appear near the edge of the step will not be incorporated at the edge but rather at the kink of that step (N_3 position in Figure 14.4a). The evolution of the atomic step edges observed experimentally (presented in Figure 14.6) is indeed dominated by the slowly advancing step edges of type "A."

Detailed studies of evolution of atomic step edges for other miscut orientations, for example, toward *a*-plane [11–20] confirmed that it is dominated by the slowly advancing step edges of type "A," regardless of the substrate orientation [47]. The arguments presented earlier point toward dependence of In incorporation into the InGaN layer on microscopic growth scenario—whether we can create a situation where only steps "A" are present (for low In content) or when both types of atomic steps are present during growth. The qualitative description of the In incorporation in InGaN layers based on the analysis of the atomic step symmetry was given by Turski et al. [47,49,50]. According to the arguments presented there, the In content x in InGaN layers grown at In-rich conditions is given by the following equation:

$$x = 1 - \frac{\Phi_{Ga}^{\searrow}}{\Phi_{N}^{\searrow} - \Phi_{N}^{\nearrow}} \tag{14.1}$$

where

$\Phi_{N}^{\searrow}$ and $\Phi_{Ga}^{\searrow}$ are the total N and Ga fluxes used for the growth
$\Phi_{N}^{\nearrow}$ is a N flux evaporated from the surface due to the InGaN decomposition

$$\Phi_{N}^{\nearrow} = \left(\frac{C_A}{\Phi_{N}^{\searrow}} + \frac{C_B}{\Phi_{Ga}^{\searrow}} \right) \cdot \left(\Phi_{N}^{\searrow} - \Phi_{Ga}^{\searrow} \right) \cdot \exp\left(\frac{E_A(x)}{kT_G} \right) \tag{14.2}$$

$\Phi_{N}^{\nearrow}$ depends on the microscopic arrangement of atoms at the growth sites present at the surface, so the balance between the edges "A" and "B" (what is represented by parameters c_A and c_B). The E_A stands for the activation energy of InGaN decomposition. This model describes well the evolution of In content as a function of the growth temperature (Figure 14.7), as well as Ga and N fluxes (Figure 14.8) using common

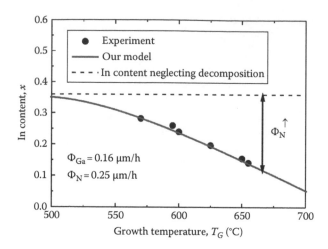

FIGURE 14.7 Indium content in InGaN layers grown at different T_G using constant fluxes: $\Phi_N^\searrow = 0.25\,\mu\text{m/h}$ and $\Phi_{Ga}^\searrow = 0.16\,\mu\text{m/h}$. Dotted line stands for the case of $\Phi_N^\nearrow = 0$, while solid line stands for $\Phi_N^\nearrow$ after our model (Equation 14.2).

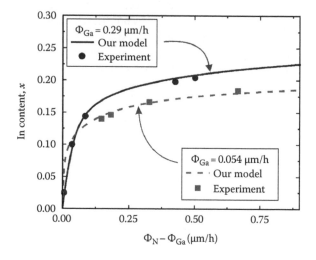

FIGURE 14.8 Indium content for InGaN layers grown using different Ga and N fluxes. Experimental data for two series of layers grown using gallium fluxes of 0.054 and 0.29 μm/h. In content predicted by our model (Equation 14.2) is plotted using dashed and solid lines, respectively.

set of fitting parameters: $C_A = 2.88 \times 10^{17}\,\dfrac{\mu\text{m}}{\text{h}}, C_B = 3.12 \times 10^{16}\,\mu\text{m}/\text{h}$. The activation energy E_A is assumed to be a function of In content x: $E_A = xE_{\text{InN}} + (1-x)E_{\text{GaN}}$, where E_{InN} and E_{GaN} are taken from the literature data of thermal decomposition of InN and GaN, respectively [47]. The important conclusion coming from the experiments and the model is that we can increase the In content (1) by reducing the growth temperature (which is known and was used already for the growth of high In content layers) or (2) by the *increase* of the active N flux at constant growth temperature, which effectively suppresses the decomposition of the InN sublattice in InGaN.

At the end, we would like to point out that the "zigzag" shape of atomic step edges for both InGaN and GaN can be observed when the growth temperature is substantially lower than 700°C. Previously, the "zigzag" shape of GaN atomic steps observed on layers grown by PAMBE was reported by Xie et al. [46].

For the case of GaN layers grown at higher growth temperature, the atomic steps have a tendency to bunch. Therefore, the straight step edges of GaN grown above 700°C observed on AFM pictures usually are double steps (~0.5 nm high). We believe that in such case, the "zigzag" shape steps do not form due to the fact that at elevated temperatures there is enhanced diffusion of nitrogen atoms along step edge "B." That would result in preventing the formation of slower edge "A" and in the end to step bunching.

14.2.4 Role of Active Nitrogen in Growth of InGaN Layers and Efficient Quantum Wells

There are two strategies for the growth of high In content InGaN layers. The first relies on lowering the growth temperature T_G, while the second is based on the increase of N flux (see the horizontal and almost vertical arrows in Figure 14.2). It is straightforward and easy in practical realization to increase the In content by lowering T_G—but as a result, the quality of grown InGaN layers is worse with the increase of In composition [51]. The most likely reason for such degradation is the tendency toward formation of In-rich clusters at lower growth temperature. As a result of large lattice mismatch between InN and GaN, such In clustering may lead to premature layer relaxation by the nucleation of dislocations in In-rich regions with all the negative consequences on optical layer quality and morphology. As demonstrated for thin $In_xGa_{1-x}N$ layers ($x < 0.25$), optimizing growth conditions through the increase of the N flux during InGaN growth allows to increase the In content without decreasing the growth temperature [51]. The advantages of such an approach are (1) reduced In content fluctuations in the InGaN layers, which decreases the density of localized states [51] and (2) improved quality of the interfaces between quantum wells and barriers [52].

In Figure 14.9, we compare the time-resolved photoluminescence (TRPL) spectra for two multiquantum wells (MQWs) samples grown with low (Figure 14.9a) and high (Figure 14.9b) N flux. Both MQWs structures have maximum photoluminescence (PL) around 520 nm, but wavelength dependence of PL decay time, $\tau_{PL}(\lambda)$ for sample (a) is qualitatively different than for sample (b). For sample (a) grown with low nitrogen flux, τ_{PL} increases from 3 ns at $\lambda = 485$ nm to 13 ns at $\lambda = 570$ nm. Such a spectral dispersion of τ_{PL} implies carriers' localization and can be attributed to indium content fluctuations [51]. In the presence of dislocations, it is expected that such localization can effectively reduce the influence of the related nonradiative recombination channels, which can be beneficial for LEDs. However, for high-performance LDs localized states are undesirable. Indeed, even though localized states will have tendency to preferentially trap electron–hole pairs, their integrated density of states is typically insufficient to obtain population inversion and thus laser action. For sample shown in Figure 14.9b grown with high N flux, the TRPL measurements show no spectral dispersion of τ_{PL}. This indicates that the degree of carrier localization in

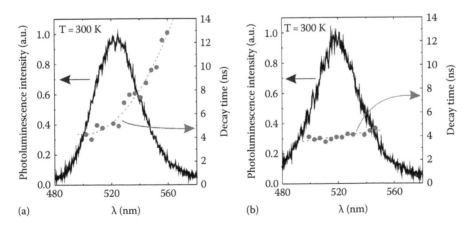

FIGURE 14.9 Photoluminescence intensity and its decay time at 520 nm for the MQWs structures grown by PAMBE using (a) low and (b) high N flux.

this sample is substantially smaller than in sample shown in Figure 14.9a grown with much lower N flux. Indeed, the spectral dispersion of τ_{PL} can be used as a fingerprint of carrier localization, giving valuable feedback for optimizing InGaN MQWs growth parameters for LD structures.

The second benefit of high N flux is the reduction of the surface roughness of InGaN layers with important consequences to QW–barrier interfaces. In Figure 14.10a and b, we present atomic force microscope (AFM) images of two 30 nm $In_{0.17}Ga_{0.83}N$ layers, both grown at the same growth temperature (650°C) but using two different N fluxes of 0.3 and 0.9 μm/h, respectively. The Ga fluxes were adjusted appropriately to obtain the same In composition. The root mean square roughness (RMS) measured on $1 \times 1 \, \mu m^2$ AFMs can be decreased from 0.52 nm for the layer grown at lower N flux to 0.39 nm for the layer grown using higher N flux. Qualitatively, for the higher N flux smoother surface with visible atomic step edges was grown, while for low N flux the surface is substantially rougher with numerous deep trenches and no clear atomic steps. For the low N flux sample the local surface orientation changes from place to place, which means that we can expect the In incorporation inhomogeneity, since In content in InGaN layers depends on substrate miscut [50]. The additional confirmation of the improved quality of InGaN layer grown at higher N flux is XRD measurements of these samples, presented in Figure 14.10c. The small surface roughness of InGaN layers obtained in PAMBE for high N fluxes allows growing QWs with smooth interfaces with direct positive impact on QW thickness homogeneity.

The combination of reduced In fluctuations and improved QW thickness uniformity should translate into narrowing of the PL line. In Figure 14.11a, we demonstrate the room-temperature PL of $In_{0.17}Ga_{0.83}N/In_{0.08}Ga_{0.92}N$ MQWs grown using two different N fluxes. The quantum barriers and wells

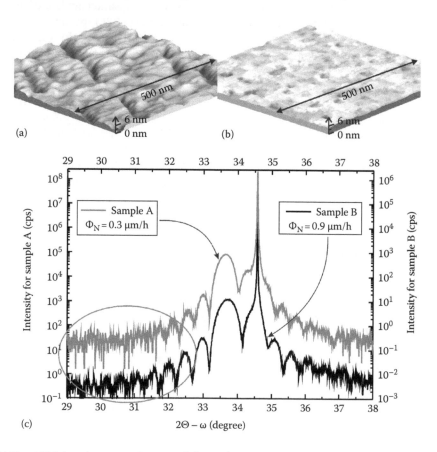

FIGURE 14.10 AFM data showing surface morphology of 30 nm $In_{0.17}Ga_{0.83}N$ layers grown using nitrogen fluxes equal to (a) 0.3 μm/h, (b) 0.9 μm/h, and (c) The XRD $2\Theta - \omega$ scan of these layers.

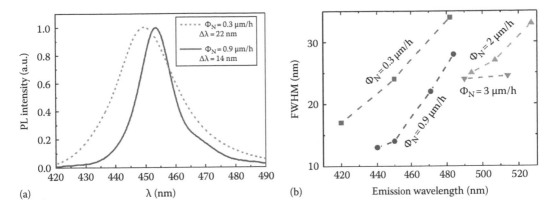

FIGURE 14.11 (a) Photoluminescence spectra of two MQWs structures grown by PAMBE using active nitrogen flux of 0.3 and 0.9 μm/h, respectively. (b) Photoluminescence full width at half maximum as a function of the emission wavelength for active nitrogen fluxes 0.3, 0.9, 2, and 3 μm/h, respectively.

thicknesses were 7 and 3 nm, respectively. The QWs were grown using the same growth conditions as for the InGaN layers shown in Figure 14.10. As can be seen in Figure 14.11a, by increasing the N flux from 0.3 to 0.9 μm/h, we reduced the PL full width at half maximum (FWHM) from 22 down to 14 nm for the wavelength λ = 450 nm. The smaller PL FWHM leads to better performance of LDs due to narrowing of the gain spectra. It is important to point out that we observe systematic narrowing of the PL lines for all QWs structures grown with higher N flux. To confirm the relation between narrowing of the PL line for QWs and increase of N flux used for the growth, we present in Figure 14.11b PL FWHM as a function of emission wavelength for QW structures grown with different N fluxes. The demonstrated pronounced narrowing of the PL lines for longer wavelengths is an important milestone on the path toward PAMBE-grown green LDs.

14.3 Evolution of Laser Diodes Grown by Plasma-Assisted Molecular Beam Epitaxy

14.3.1 First Laser Diodes by Plasma-Assisted Molecular Beam Epitaxy

The first blue–violet PAMBE LDs were grown in a custom-made VG V90 MBE reactor equipped with one Veeco RF plasma source on (0001) Ga-polarity low TDDs GaN crystals [18,19,53], using the active N flux of 0.25 μm/h (4.2 nm/min). The substrates were prepared by mechanopolishing, dry etching, and deposition of 2 μm GaN:Si buffer layer in MOVPE reactor. The GaN:Si MOVPE buffer layer was grown for (0001) GaN surface planarization, since the mechanochemical polishing of (0001) surface was not well developed at that time. Figure 14.12a shows the structure of lasers operating at 405–420 nm. The 40 nm GaN:Si buffer layer and 450 nm $Al_{0.08}Ga_{0.92}$N:Si cladding were grown under Ga-rich conditions at 720°C. The bottom $In_{0.02}$GaN waveguide, MQWs, the $In_{0.02}Al_{0.18}$GaN electron blocking layer (EBL), top $In_{0.02}$GaN waveguide, upper $In_{0.02}$GaN/$In_{0.02}Al_{0.18}$GaN superlattice cladding, and $In_{0.02}$GaN contact layer were grown under In-rich conditions at 650°C. The active region consisted of three 3 nm thick $In_{0.1}Ga_{0.9}$N QWs with 7 nm $In_{0.02}Ga_{0.98}$N barriers. The transmission electron microscopy (TEM) picture presented at Figure 14.12b confirms the high quality of PAMBE epitaxy and the sharp interfaces across entire LD. The devices were processed as ridge-waveguide, oxide-isolated lasers. The mesa structures were etched to a depth of 0.3 μm. The 20 μm-wide and 500 μm-long stripes were used as laser resonators. The oxidized Ni/Au ohmic contacts were deposited on the top surface of the devices, and Ti/Au contacts were prepared on the backside of the highly conducting *n*-GaN substrate crystal. The cleaved laser mirror facets were coated with symmetrically reflecting mirrors. Figure 14.13a shows the light–current–voltage (L–I–V) characteristics of the cw LDs with lasing threshold current density and voltage of 5.5 kA/cm² and 5.7 V, respectively.

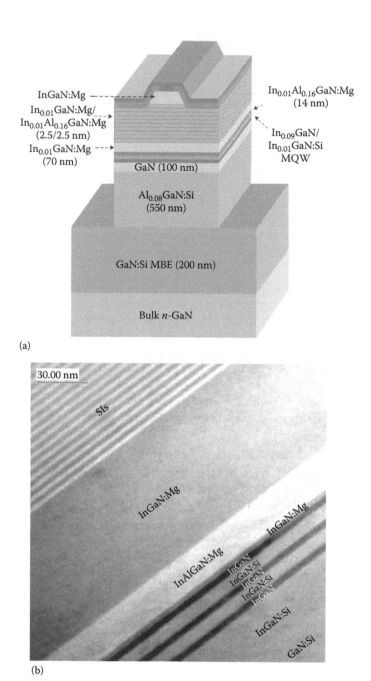

(a)

(b)

FIGURE 14.12 Structure details of first PAMBE laser diode (a) and transmission electron microscope image of the active region of a PAMBE laser diode (b). (Reprinted from *J. Cryst. Growth*, 305, Skierbiszewski, C., Siekacz, M., Perlin, P., Feduniewicz-Zmuda, A., Cywiński, G., Grzegory, I., Leszczyński, M., Wasilewski, Z.R., and Porowski, S., Role of dislocation-free GaN substrates in the growth of indium containing optoelectronic structures by plasma-assisted MBE, 346, Copyright 2007, with permission from Elsevier.)

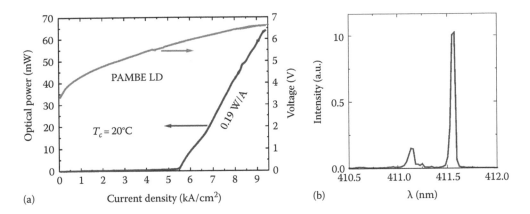

FIGURE 14.13 L–I–V characteristics of PAMBE-grown cw laser diode (a) and lasing spectrum (b). (Reprinted from *J. Cryst. Growth*, 305, Skierbiszewski, C., Siekacz, M., Perlin, P., Feduniewicz-Zmuda, A., Cywiński, G., Grzegory, I., Leszczyński, M., Wasilewski, Z.R., and Porowski, S., Role of dislocation-free GaN substrates in the growth of indium containing optoelectronic structures by plasma-assisted MBE, 346, Copyright 2007, with permission from Elsevier.)

Lasing was observed up to 60 mW of optical output power (30 mW per facet) at a wavelength of 411 nm as indicated in Figure 14.13b [19,53].

14.3.2 True-Blue Laser Diodes with InGaN Waveguides

Nitride-based LDs have gained much attention as they cover a very wide spectral range and have many possible applications, for example, laser projectors and general lightning [54,55]. Conventional nitride LDs are grown by MOVPE with GaN waveguides and AlGaN cladding layers to provide confinement of optical modes with InGaN active layer. As we demonstrated in previous chapters, the advantage of low-temperature PAMBE is growth of high-quality InGaN layers with In content up to 15%–17%. Therefore, thick InGaN with In composition in a range of 4%–10% (where limiting factor will be relaxation of the grown InGaN layer) can be used as a waveguide to enhance the optical confinement. Kim et al. showed that the use of InGaN waveguide with 3% In content allowed for the demonstration of long wavelength $\lambda = 485$ nm LDs [56]. Additionally, such approach allowed the demonstration of AlGaN-cladding-free LDs on *c*-plane [57,58]. To date, MOVPE technique is reported to be capable of growth of AlGaN-cladding-free structures on *c*-plane GaN with InGaN waveguide with 3% indium content [59]. Therefore, the use of InGaN layers seems to completely change the waveguiding properties and strain distribution in LDs, which makes such design very interesting.

The schematic of the true-blue LD design is shown in Figure 14.14a. The laser structures consist of a 700 nm $Al_{0.065}Ga_{0.935}N$:Si cladding layer, followed by 100 nm GaN:Si (Si doping level is 2×10^{18} cm^{-3}), and 80 nm undoped $In_xGa_{1-x}N$ bottom waveguide ($x = 0.04$–0.08). An active region is composed of three 2.6 nm $In_{0.17}Ga_{0.83}N$ QWs separated by 8 nm $In_xGa_{1-x}N$ quantum barriers. Then, the $In_xGa_{1-x}N$ upper waveguide is grown with thickness D changed from 5 to 80 nm. The EBL with thickness 20 or 30 nm is composed of $Al_{0.15}Ga_{0.85}N$ with a Mg doping level of 5×10^{19} cm^{-3}. This is followed by a 100 nm GaN:Mg and 400 nm $Al_{0.065}Ga_{0.935}N$:Mg to further enhance optical confinement. The LD structure is capped with 60 nm $In_{0.01}Ga_{0.99}N$:Mg acting as a contact layer.

The growth of LD structures presented in this section was performed in a customized VG V90 MBE reactor equipped with two Veeco RF plasma sources, operating at 240–450 W power for 0.8–2 sccm of N$_2$ flow. All of the LDs were grown on bulk ammono-GaN substrates with TDDs in the order of 10^4–10^5 cm^{-2}. The epi-ready substrates were prepared by mechanical polishing and mechanochemical polishing. The crystals had miscut 0.5° toward [1–100] direction. The growth of GaN and AlGaN cladding layers were

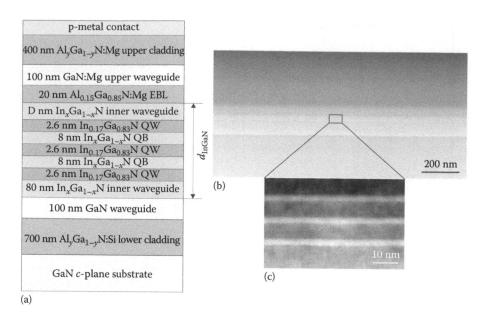

FIGURE 14.14 (a) Schematics of LD structure, (b) large-area HAADF-STEM cross-section image showing no extended defects, and (c) HAADF-STEM image of MQW region.

performed at 720°C under Ga-rich conditions, while InGaN layers were grown at 650°C under In-rich conditions. The growth of AlGaN in EBL was performed at 650°C with In as a surfactant. The growth rate for all layers apart from MQW was equal to supplied N flux of 0.35 μm/h. The MQWs were grown using 1 μm/h of N flux. The N flux was calibrated during GaN growth under Ga-rich conditions. PAMBE technique allows for a very good control of composition in grown layers, including QWs; therefore, all LDs had nearly the same emission wavelength $\lambda = 450 \pm 2$ nm. Details of InGaN growth in PAMBE can be found in References 21 and 47. The composition of InGaN waveguide was corroborated after growth with x-ray diffraction.

The fundamental question arises whether we can grow thick InGaN waveguide strained to the substrate, which is necessary for the high-quality growth of laser structure. In Figure 14.14b, we present a large-area high-angle annular dark-field scanning transmission electron microscopy (HAADF-STEM) image of LD with the upper InGaN waveguide thickness $D = 60$ nm taken in a Titan Cubed 80–300 system operating at 300 kV equipped with Cs-corrector. Cross-sectional STEM specimens were prepared by a standard method based on mechanically thinning of the samples, followed by an Ar ion milling procedure. In this LD, the total thickness of $In_{0.08}Ga_{0.92}N$ was 160 nm and we did not observe creation of any extended defects after the growth of such thick InGaN layers. Figure 14.14c shows magnification of the MQW region. Furthermore, we investigated the whole LD structure using x-ray reciprocal space mapping. We used Philips X'Pert MRD x-ray diffractometer equipped with a fourfold Ge (220) monochromator, a threefold Ge (220) analyzer, and an x-ray mirror. Figure 14.15 presents the reciprocal space map of asymmetric $\left(1\bar{1}24\right)$ reflection of LD with the 160 nm thick $In_{0.08}Ga_{0.92}N$ waveguide. As can be seen from Figure 14.15 there is no relaxation and the 160 nm $In_{0.08}Ga_{0.92}N$ waveguide is fully strained to GaN substrate. An additional and very important advantage of the InGaN waveguide is the fact that after the growth of the full LD structure the wafers did not suffer from bowing due to the opposite strain in InGaN waveguide and AlGaN claddings (the curvature radii measured on 1 in. wafers were around 40–60 m). This is beneficial for processing and will be extremely important for the growth of LD structures on 2–4 in. GaN wafers. In addition, standard approach for true-blue LDs with GaN waveguides

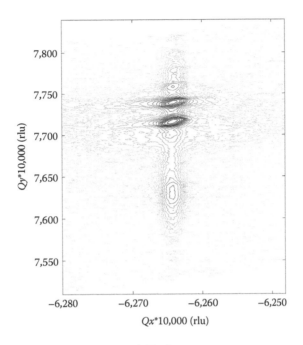

FIGURE 14.15 Reciprocal space map of asymmetric $\left(\bar{1}\,\bar{1}24\right)$ reflection of full LD structure with thick InGaN waveguides showing no relaxation. (Reprinted from *J. Cryst. Growth*, 425, Muziol, G., Siekacz, M., Turski, H., Wolny, P., Grzanka, S., Grzanka, E., Feduniewicz-Żmuda, A. et al., High power nitride laser diodes grown by plasma assisted molecular beam epitaxy, 398, Copyright 2015, with permission from Elsevier.)

requires presence of very thick AlGaN cladding, which makes necessary preparation of special substrates (patterning) for strain relaxation.

Figure 14.16 presents L–I–V characteristics of high-power true-blue LD by PAMBE, together with the emission spectra below and above lasing threshold. The mesa size was 5×1000 μm and the maximum output power was 0.5 W per facet [60]. It is important to stress that the facets of this LD were without coating, so the light was emitted equivalently in both directions.

Next, we will discuss the lifetime of the blue LDs grown by PAMBE. In Figure 14.17, the threshold current as a function of aging time is shown. The life tests were performed at the optical power set to 10 mW at room temperature. After about 4000 h, the observed rate at which the threshold current was increasing was 0.004 mA/h. The expected lifetime for these diodes, defined as the operating time for which the threshold current increased by 50% from its starting value, equals 15,000 h.

The high optical power and very long lifetime of LDs grown by PAMBE show the high potential of this technique in the production of reliable nitride-based lasers.

14.3.2.1 Influence of InGaN Waveguide LDs Design on Threshold Current, Slope Efficiency, Internal Losses, and Differential Gain

The influence of the indium composition in InGaN waveguide on properties of LDs was studied theoretically by Huang et al., who found an increase of optical confinement factor Γ with indium content [61]. An unknown factor is the possible generation of additional internal optical losses α_{int} by the application of InGaN waveguide due to alloy scattering [62]. Furthermore, InGaN layers are well known to suffer from indium fluctuations, which may further increase their absorption coefficient at lasing wavelength. Maintaining low optical losses in InGaN waveguide will be crucial to fully benefit from the enhanced Γ. In this section, the influence of indium content and design of InGaN waveguide on both Γ and α_{int} will be discussed.

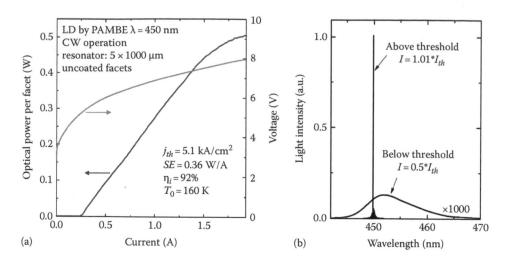

FIGURE 14.16 Light–current–voltage characteristics of high-power LD grown by PAMBE (a) and spectrum below and above lasing threshold (b). (Reprinted from *J. Cryst. Growth*, 425, Muziol, G., Siekacz, M., Turski, H., Wolny, P., Grzanka, S., Grzanka, E., Feduniewicz-Żmuda, A. et al., High power nitride laser diodes grown by plasma assisted molecular beam epitaxy, 398, Copyright 2015, with permission from Elsevier.)

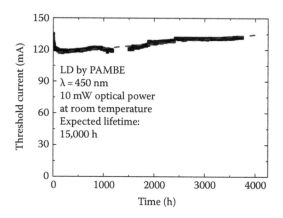

FIGURE 14.17 Threshold current as a function of time during life tests of blue LD grown by PAMBE.

The internal optical losses α_{int} in early nitride LDs were very high in the order of 40 cm^{-1} [63,64]. The commercialization demanded reduction of α_{int} to obtain low-threshold current densities and high optical power operation. The early studies on violet LDs had proven Mg-doped layers to be responsible for the generation of the great part of losses because of the necessity of high Mg doping due to the high Mg acceptor ionization energy [63]. Solution for this problem was proposed by Uchida et al., who increased the distance between MQW and Mg-doped layers (by introducing an undoped AlGaN interlayer), which reduced the overlap of the optical mode with p-type layers and therefore reduced α_{int} down to 14 cm^{-1} [65]. In the case of LDs with InGaN waveguide, there is very little literature data about α_{int}. Son et al. investigated α_{int} of LDs with In$_{0.01}$Ga$_{0.99}$N waveguide and claimed that the absorption coefficient of In$_{0.01}$Ga$_{0.99}$N is as high as $\alpha_{InGaN} = 40$ cm^{-1} for LDs operating at 405 nm, regardless of the distance between Mg-doped EBL and MQWs [66]. Therefore, the application of InGaN waveguides to LD design requires a comprehensive study of the interplay between enhanced Γ and increased internal losses. Moreover, if the absorption

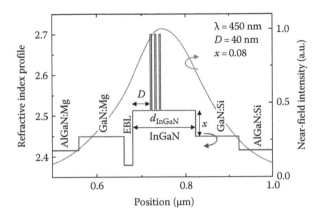

FIGURE 14.18 Refractive index profile and corresponding optical mode distribution of blue LD.

coefficient of InGaN layers would be low, they could be used as an interlayer between active region and Mg-doped region to reduce the overall α_{int} of the device.

In this section, we will study the influence of InGaN waveguide design of true-blue ($\lambda = 450$ nm) nitride LDs grown by PAMBE on the internal and external LD properties. We will explore in detail the behavior of the internal losses, differential gain, lasing threshold, and slope efficiency as a function of two parameters: (1) the In composition of In$_x$Ga$_{1-x}$N waveguide in the range of $x = 0.04$–0.08 and (2) thickness D (see Figure 14.14a) of undoped In$_x$Ga$_{1-x}$N interlayer (upper waveguide) between MQW region and the Mg-doped layers in the range of 5–80 nm. Figure 14.18 shows the refractive index profile and calculated optical mode for $x = 0.08$ and $D = 40$ nm using 1D transfer matrix method. Details of the calculations and refractive index data are given in References 67 and 68, respectively. On the one hand, an increase of the thickness D as well as an increase of In content x in waveguide should decrease the overlap of optical mode with Mg-doped region, thus reducing the internal losses of the device. On the other hand, higher In content InGaN layers may have a higher absorption coefficient and may lead to an increase of α_{int}.

Figure 14.19 presents collected threshold current densities j_{th} of LDs with different In content and different thickness of the InGaN waveguide [69]. We observed a decrease in j_{th} with increase of distance D and In composition x. For the highest composition $x = 0.08$, the j_{th} drops from 12.5 kA/cm² for $D = 5$ nm to 3.6 kA/cm² for $D = 60$ nm. We also observed a decrease of j_{th} with In composition for constant D.

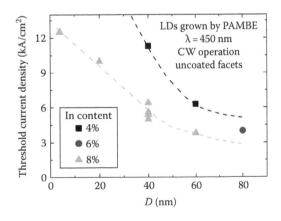

FIGURE 14.19 Threshold current density of LDs as a function of distance D between MQW and Mg-doped EBL. Dashed lines are guides to the eye. (From Muziol, G. et al., *Appl. Phys. Express*, 8, 032103, Copyright 2015, The Japan Society of Applied Physics.)

Change in the In content of InGaN waveguide from $x = 0.04$ to $x = 0.08$ resulted in a 50% reduction of j_{th}. To understand the exact influence of InGaN waveguide on LD performance, we need to correlate the change of j_{th} with the internal parameters of LD. The j_{th} is given by

$$j_{th} = \frac{\alpha_{int} + \alpha_m}{\eta_i \Gamma (dg/dj)} + \frac{j_{tr}}{\eta_i} \tag{14.3}$$

where

j_{tr} is the transparency current density

η_i is the injection efficiency

Γ is the optical confinement factor

g is the material gain of MQW

α_{int} are internal losses

$\alpha_m = \dfrac{1}{2L} \ln \dfrac{1}{R_1 R_2}$ are mirror losses, L is the resonator length, and R_1 and R_2 are reflectivities of front and rear mirrors

The term $\eta_i \Gamma \dfrac{dg}{dj}$ is the differential modal gain. The MQWs in every LD were grown with the same growth parameters; therefore, j_{tr} and $\dfrac{dg}{dj}$ should be the same for all LDs. The resonator length was 1000 μm and the mirrors were cleaved with facets left uncoated in all LDs. The resulting α_m were equal to 17 cm^{-1}. To obtain insight into internal parameters, we measured the optical gain of few LDs from the assembly shown in Figure 14.20 using the Hakki–Paoli method [70]. The optical modal gain G is given by

$$G = \eta_i \Gamma g - \alpha_{int} = \frac{1}{L} \ln \left(\frac{\sqrt{\dfrac{I_{max}}{I_{min}}} - 1}{\sqrt{\dfrac{I_{max}}{I_{min}}} + 1} \right) + \alpha_m \tag{14.4}$$

where I_{max} and I_{min} are light intensities in maximum and minimum of ripples of amplified spontaneous emission below lasing threshold [70]. Exemplary optical gain spectra of LD with $D = 80$ nm and $x = 0.06$

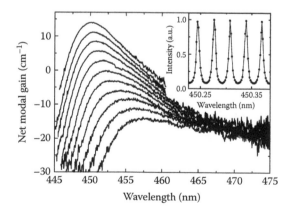

FIGURE 14.20 Net modal gain as a function of wavelength for different pumping currents for LD with $D = 80$ nm and $x = 0.06$. The long-wavelength tail is used to derive internal losses. Inset shows high-resolution amplified spontaneous emission spectrum measured at $j = 0.99 j_{th}$. (From Muziol, G. et al., *Appl. Phys. Express*, 8, 032103, Copyright 2015, The Japan Society of Applied Physics.)

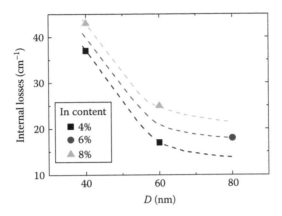

FIGURE 14.21 Dependence of optical losses on distance D for LDs with different In content in waveguide. Dashed lines are guides to the eye. (From Muziol, G. et al., *Appl. Phys. Express*, 8, 032103, Copyright 2015, The Japan Society of Applied Physics.)

are presented in Figure 14.20 for current densities from 0.67 to 4.0 kA/cm² in steps of 0.33 kA/cm², with inset showing high-resolution spectra at 4.0 kA/cm².

The Hakki–Paoli method allows obtaining insight into two internal parameters important to understand the change of j_{th}: (1) internal losses α_{int}, which are derived from the long-wavelength part of gain spectra that is not changing with current density (see Figure 14.20 where α_{int}=18 cm⁻¹); (2) differential modal gain $\eta_i \Gamma \frac{dg}{dj}$, which is defined as the derivative of maximum of modal gain G with respect to current density.

Figure 14.21 shows measured internal losses for LDs as a function of thickness D for three different compositions of InGaN waveguide. We observed a decrease of α_{int} with distance D, which we attribute to the lower overlap of optical mode with Mg-doped layers [71]. We also found an increase of α_{int} for LDs with higher In composition in waveguide, which we attribute to an increase of the absorption coefficient of InGaN α_{InGaN} with increasing In content. The measured change of α_{int} between LDs with $x = 0.04$ and 0.08 is 8 cm⁻¹. This is primarily due to the change in energy difference between absorption edge of InGaN layers and emission wavelength of LD. Smaller difference leads to higher absorption coefficient of InGaN layers. For LDs operating at $\lambda = 450$ nm with 4% and 8% of indium in InGaN waveguide, this difference is about 500 and 350 meV, respectively. Moreover, the indium fluctuations are likely to increase for higher indium content layers, causing additional states with energy gap closer to emission wavelength of LD, which would additionally raise the absorption coefficient of InGaN layers.

Knowing α_{int}, we can derive η_i from the measurement of optical power P above lasing threshold. The slope efficiency of LD is given by [72]

$$\frac{dP}{dI} = \eta_i \left(\frac{\alpha_m}{\alpha_m + \alpha_{int}} \right) \frac{h\nu}{q} \frac{\left(1 - R_1\right)\sqrt{R_2}}{\left(\sqrt{R_1} + \sqrt{R_2}\right)\left(1 - \sqrt{R_1 R_2}\right)} \tag{14.5}$$

where the last term represents the fraction of optical power emitted through the front mirror. In this study, we used uncoated facets, so the term $\left(1 - R_1\right)\sqrt{R_2} / \left[\left(\sqrt{R_1} + \sqrt{R_2}\right)\left(1 - \sqrt{R_1 R_2}\right)\right]$ reduces to $\frac{1}{2}$. Figure 14.22 presents slope efficiencies measured for LDs with the same cavity lengths $L = 1$ mm and uncoated mirrors. Slope efficiencies were measured on the same devices that were used in the Hakki–Paoli experiment. The lines in Figure 14.22 show calculated $\frac{dP}{dI}$ for two values of $\eta_i = 0.9$ and 1 with $\alpha_m = 17$ cm⁻¹. As can be seen from Figure 14.22, the injection efficiency is almost the same for all LDs and is slightly below 100%,

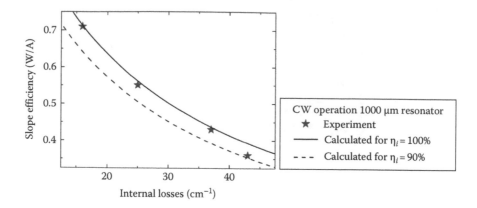

FIGURE 14.22 Measured slope efficiencies and calculated dependence of slope efficiency on internal losses for two values of injection efficiency. (From Muziol, G. et al., *Appl. Phys. Express*, 8, 032103, Copyright 2015, The Japan Society of Applied Physics.)

showing that there is no leakage of current above threshold and that the EBL is well optimized. The high slope efficiencies (up to 0.7 W/A) for LDs with uncoated facets are in agreement with the low internal losses measured by the Hakki–Paoli method.

Next, the influence of InGaN waveguide composition on the differential gain will be studied. Figure 14.23a presents the maximum net modal gain as a function of current density for three LDs with different In content in the waveguide. Additionally, the solid lines represent the linear fit used to derive the differential gain. To better illustrate the influence of change in the differential gain, the optical gain arising from the MQW $\eta_i \Gamma g$ is shown in Figure 14.23b. It can be clearly seen that the higher the InGaN waveguide composition, the higher is the differential gain. Figure 14.24 shows the dependence of differential modal gain on distance D. For constant In content in waveguide, the differential gain $\eta_i \Gamma \dfrac{dg}{dj}$ does not depend on D, which shows that moving EBL away from MQW does not affect injection efficiency that might be counterintuitive. Perhaps for LDs with D above 80 nm we would observe a drop of η_i, but in our experiments, the growth of LDs with thicker InGaN was limited by structural relaxation.

We observed an increase of $\eta_i \Gamma \dfrac{dg}{dj}$ for higher In content in waveguide. This change has to be due to an increase of Γ, because η_i is equal to nearly 1 in all LDs (see Figure 14.22) and $\dfrac{dg}{dj}$ is also constant, since MQWs in all LDs are the same. To confirm this, we performed 1D calculations for three structures with $D = 40$, 60, and 80 nm and varied the In composition in waveguide [67,69]. Figure 14.25 shows the calculated Γ. We observed only a slight increase of Γ with D and a significant increase of Γ with In content in waveguide. This is consistent with the experimental results, which show no change of differential gain for structures with different thickness D up to 80 nm and a high increase of differential gain with In content. Such a high increase of Γ with In content is due to a large increase of refractive index contrast of InGaN layers with respect to GaN at this wavelength [73]. It is worth noting here that for longer lasing wavelengths (e.g., for green-color region) this is not so well pronounced.

Figure 14.26 shows the expected change in the differential gain due to change of optical confinement Γ. The differential gain of LD with $x = 0.08$ has been taken as a reference value. The expected change of the $\eta_i \Gamma \dfrac{dg}{dj}$ between LDs with $x = 0.04$ and 0.08 is only 2.2 cm/kA, whereas the measured change is 4.4 cm/kA. The explanation of this inconsistency lies in leakage of light to the GaN substrate in case of LDs with $x = 0.04$.

In Figure 14.27a, a schematic view of the ridge-waveguide laser diode together with the output laser beam is shown. The fast axis and slow axis of the far-field profile are marked as $\Theta \perp$ and $\Theta_\parallel$, respectively.

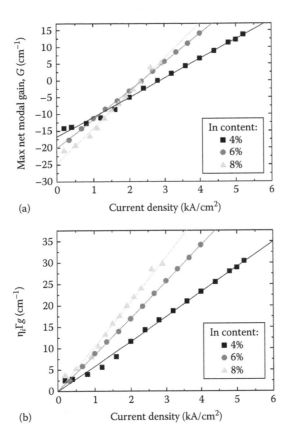

FIGURE 14.23 Maximum of net modal gain G (a), the optical gain arising from the MQW $\eta_i\Gamma g$ (b) as a function of current density for three LDs with different InGaN waveguide composition. The solid lines are linear fits and are used to derive the differential gain $\eta_i\Gamma\dfrac{dg}{dj}$.

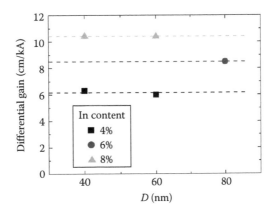

FIGURE 14.24 Dependence of differential gain on distance D for LDs with different In content in waveguide. Dashed lines are guides to the eye.

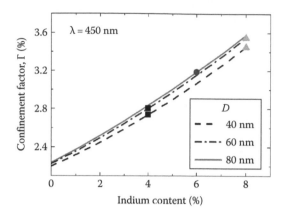

FIGURE 14.25 Calculated dependence of optical confinement factor Γ on indium content x in waveguide for LDs with different distance D.

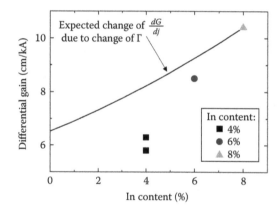

FIGURE 14.26 Change of differential gain due to change of optical confinement factor Γ. The solid line presents theoretical predictions, while squares, circles, and triangles present the measured differential gain for LDs with InGaN composition $x = 0.04$, 0.06, and 0.08, respectively.

Figure 14.27b and c shows the experimentally measured far-field patterns of LDs with $x = 0.08$ and 0.04, respectively. We found a large leakage in case of LD with $x = 0.04$, whereas for LDs with $x = 0.08$ we observed a perfect Gaussian-shaped far-field pattern with no leakage to GaN substrate. For LD with $x = 0.04$, the measured leaking part to GaN substrate was 25%, which decreases Γ and, thus, $\eta_i \Gamma \dfrac{dg}{dj}$ by 25% from the expected 8.2 to 6.2 cm/kA. This is in good agreement with the experimentally measured value.

In the next section, a detailed analysis of leakage of optical mode to GaN substrate for LDs with InGaN waveguide is given.

14.3.2.2 Leakage of the Optical Modes to the GaN Substrate

Historically, the beam quality of the first violet laser diodes, grown on sapphire substrates, was very poor [74,75]. This was due to the leakage of the optical mode to the GaN buffer layer resulting from an insufficient thickness of n-AlGaN cladding layer—a thicker n-AlGaN cladding was used to reduce the leakage [76].

Unfortunately, as the operating wavelength is increased the thickness of n-AlGaN cladding should also be increased. Strauss et al. have shown that for violet ($\lambda = 405$ nm) LDs a 2 µm thick n-AlGaN cladding has to be used to obtain Gaussian-shaped far-field pattern, but for blue ($\lambda = 440$ nm) LDs, this thickness is insufficient

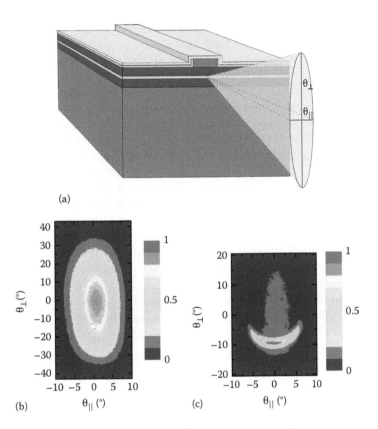

FIGURE 14.27 (a) Schematic view of the ridge-waveguide laser diode and the output laser beam, (b) Far-field pattern of LD with 8% InGaN waveguide, and (c) far-field pattern of LD with 4% InGaN waveguide.

and has to be increased up to 3 µm [54]. In case of nitride material system, there is a substantial lattice mismatch between AlN and GaN, which strongly limits the thickness of AlGaN alloy coherently strained to GaN [77]. The AlGaN alloys thicker than the critical thickness tend to crack and complicated processing solutions that allow for relaxation outside of the laser region need to be used [78]. Due to the limitation in the thickness of AlGaN cladding, a need of a different method of dealing with optical mode leakage has emerged.

In this section we propose a simple solution to the problem of leakage of optical modes to GaN substrate in nitride LDs. Instead of using a thick n-AlGaN cladding, we used an InGaN waveguide not only to reduce but to fully suppress leakage. This is achieved by implementing an InGaN waveguide with thickness and In content sufficient to increase the effective refractive index n_{eff} of the optical mode to values higher than the refractive index of GaN substrate $n_{Substrate}$ [79]. Engineering of n_{eff} can be done by changing the following parameters: (1) In content in InGaN waveguide, (2) thickness of InGaN waveguide, and (3) Al content in AlGaN cladding layers. n_{eff} also strongly depends on the operating wavelength. It is important to note that n_{eff} does not depend on the thickness of n-AlGaN cladding. In case of $n_{eff} < n_{Substrate}$, leakage of the optical mode will occur and its magnitude will depend on the thickness of the n-AlGaN cladding. However, in case of $n_{eff} > n_{Substrate}$, leakage will not occur and the thickness of the n-AlGaN cladding becomes irrelevant. This is a very desirable feature because the thickness of AlGaN cladding is strongly limited by cracking due to large lattice mismatch to GaN substrate.

To study the mode leakage in the LDs grown on sapphire substrates, a standing wave model has been used [76,80–83]. In this model, the optical field leaked through the n-AlGaN cladding into the few-microns-thick GaN buffer layer and was reflected back by the GaN/sapphire interface. The resulting far-field pattern had two symmetrical peaks at ±20° ÷ 25°. One was coming from the light traveling toward the substrate and the other one from the reflected light. The theoretical simulations were used with success to predict the dependence of the n-AlGaN cladding design on the far field. Furthermore, the simulation results were used to design an epitaxial structure with n-AlGaN thickness sufficient to reduce the leakage and therefore substantially improve the beam quality [76].

The same issue of mode leakage was observed in LDs grown on freestanding GaN substrate [78,84,85]. In the experimentally measured far-field pattern, there was observed only one asymmetrical peak at an angle of 20° ÷ 25° on the GaN substrate side. The standing wave model was used to simulate the leakage of optical modes to GaN substrate. However, the predicted far-field pattern differed from the one observed experimentally [85]. The standing wave model predicts two symmetrical peaks in the far-field pattern due to the waves traveling in both directions in the substrate. However, in case of a 100 μm thick GaN substrate, the wave that leaks to the substrate does not reflect on the n-type metal contact and does not form a standing wave. Therefore, a different mode leakage model has to be proposed to qualitatively describe mode leakage in nitride LDs grown on free-standing GaN substrate. The model of leaky modes was used to describe the leakage of light in dielectric waveguides [86,87]. Wenzel had shown that, in case of GaAs LDs grown on an infinitely thick substrate, the far-field pattern predicted by this model would have only one asymmetrical peak arising from the leakage to the substrate [88]. This is exactly what is observed in the experimentally measured far-field pattern in nitride-based LDs.

We adopted the leaky modes model to describe the leakage of the optical modes to the GaN substrate in nitride LDs. The electric field distribution is calculated by solving the standard wave equation:

$$\frac{d^2 E(x)}{dx^2} - \left[\beta^2 - k_0^2 n^2(x) \right] E(x) = 0 \tag{14.6}$$

where

$\beta = k_0 n_{eff} + \dfrac{1}{2} i \alpha_{leak}$ is the complex propagation constant

k_0 is the free space wave number

$n(x)$ is the refractive index profile

The boundary condition at the interface between the n-cladding layer and the substrate is set to be [88]

$$\frac{dE(x)}{dx} = -i \sqrt{k_0^2 n_{GaN}^2 - \beta^2} \, E(x) \tag{14.7}$$

This ensures that a wave traveling inside the waveguide in the z direction suffers a constant leakage of light out of the waveguide if $n_{eff} < n_{GaN}$. The power intensity of wave traveling inside the waveguide in the z direction is therefore being reduced at the rate of $dI(z)/dz = -\alpha_{leak} I(z)$.

The calculation of n_{eff} was made for the structure presented in Figure 14.14a. It consisted of $Al_y Ga_{1-y} N$ claddings with an identical composition y. The thickness of the n-AlGaN bottom cladding and the p-AlGaN upper cladding are set to 700 and 400 nm, respectively. The shift of the refractive index of the GaN substrate due to the plasmonic effect is taken into account [89]. The electron concentration in our substrate is equal to 1×10^{19} cm^{-3}. Therefore, the $n_{Substrate}$ is lowered by 0.1% [90].

In this section, we will discuss the impact of the InGaN waveguide and AlGaN cladding compositions on the magnitude of the leakage loss. Figure 14.28 presents the calculated leakage losses α_{leak} for three AlGaN compositions $y = 0.03$, 0.06, and 0.09 as a function of In content in the InGaN inner waveguide. In case of only GaN waveguide (InGaN composition $x = 0$), α_{leak} strongly depends on the AlGaN composition. The higher the Al content, the lower the leakage loss. This is because the intensity of the optical field in the n-AlGaN bottom cladding decreases more rapidly in case of higher Al content. Therefore, its intensity at the end of this layer is lower and causes lower radiation losses. As the In content is increased, the confinement in the InGaN waveguide increases, which also means that the intensity of the optical

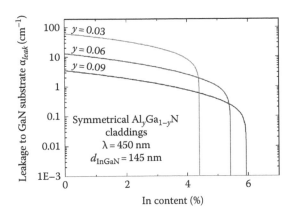

FIGURE 14.28 Leakage loss α_{leak} as a function of InGaN waveguide composition x for three AlGaN cladding compositions y.

field at the end on n-AlGaN cladding will be lower. Therefore, we observe a decrease of the leakage loss for increased InGaN composition. At high In contents, the leakage starts to decrease more rapidly and at a certain In content, dependent on the AlGaN composition, drops to zero. Above this point there is no leakage to the GaN substrate. For $y = 0.03$ the cutoff In content is $x = 0.044$. It is very interesting that above this In content there will be no leakage of optical modes in case of $y = 0.03$, but for higher AlGaN compositions the leakage will still be observed despite the fact that below this cutoff In content it was much lower. To understand this behavior it is important to look at the change of n_{eff}. Figure 14.29 presents the calculated n_{eff} as a function of In content in InGaN waveguide for three AlGaN compositions $y = 0.03, 0.06$, and 0.09. As the In content is increased the n_{eff} increases and, at a certain composition, surpasses $n_{Substrate}$ (marked as a dashed line in Figure 14.29). Above this composition, the leakage to GaN substrate will be fully suppressed. The composition at which the n_{eff} exceeds $n_{Substrate}$ depends on the Al content in AlGaN claddings. When the Al content is increased, the leakage cutoff InGaN composition also increases.

The second condition required for the realization of LDs without leakage to the substrate is the sufficient thickness of InGaN waveguide. Figure 14.30 presents the calculated n_{eff} as a function of the thickness of $In_xGa_{1-x}N$ waveguide d_{InGaN} (see Figure 14.14a) for five values of the composition $x = 0.04, 0.05, 0.06, 0.07$, and 0.08 for a constant $Al_yGa_{1-y}N$ composition of $y = 0.06$. As in case of Figure 14.29, the $n_{Substrate}$ value has been added as a dashed line. When n_{eff} surpasses the value of $n_{Substrate}$, the leakage of optical modes to substrate is fully suppressed. As can be seen, this is dependent on the thickness as well as on the

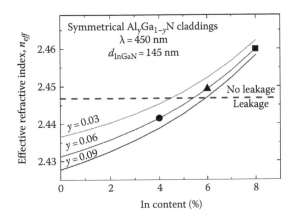

FIGURE 14.29 Dependence of n_{eff} on the InGaN composition x for three AlGaN cladding compositions y.

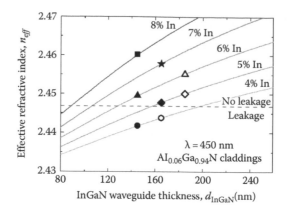

FIGURE 14.30 Dependence of n_{eff} on the InGaN waveguide thickness d_{InGaN} for five compositions x.

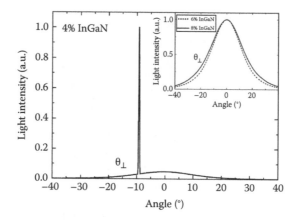

FIGURE 14.31 Calculated far-field pattern for LD with 4% In content in the waveguide. Inset shows the calculated far-field patterns for 6% and 8% In content in the waveguide.

composition of the InGaN waveguide. The lower the composition of InGaN waveguide, the higher the thickness should be. For $x = 0.04$, the thickness would have to be higher than 200 nm, and for $x = 0.08$, the thickness of 90 nm would be sufficient to fully eliminate leakage to GaN substrate.

The far-field patterns predicted by the model for three LDs are presented in Figure 14.31 for the InGaN composition of $x = 0.04$, 0.06, and 0.08 and $d_{InGaN} = 145$ nm. In the case of the 4% In content in the waveguide, there is a significant peak at 9.2° coming from the light leaking into the GaN substrate. The angle at which the light leaking to GaN substrate is seen in the far-field pattern strongly depends on the $n_{eff}/n_{Substrate}$ ratio. Therefore, in the case of an LD with 4% InGaN waveguide, the peak is observed at lower angles than in LDs with GaN waveguide. It is important to stress that the far-field pattern predicted by the model has an asymmetrical peak. Integrating the amount of light in this peak gives a value of 25% of the overall far field. This is a substantial part and significantly deteriorates the beam quality of the LD as well as the differential gain. On the other hand, the predicted far-field patterns of LDs with InGaN composition of $x = 0.06$ and 0.08 (shown as inset to Figure 14.31), due to the fact that there is no leakage in these cases, show a Gaussian beam profile with no additional peaks.

To validate the theoretical predictions, eight blue ($\lambda = 450$ nm) LDs were prepared with different thicknesses of the InGaN waveguide and the composition varied from $x = 0.04$ to 0.08 [69,79]. The thickness of n-AlGaN cladding was 700 nm and the composition was $y = 0.06$ in all LDs. This thickness is chosen small enough to allow for leakage to GaN substrate. It is too small to reduce leakage even in case of LDs operating at 405 nm and in case of longer wavelength (λ = 450 nm in this case), the leakage is strongly increased. The parameters of the thickness and composition of the InGaN waveguide of these LDs have been marked in Figures 14.29 and 14.30. As can be seen from those figures, the model predicts full elimination of leakage for structures with In content $x \geq 0.05$.

The far-field patterns of these LDs were measured to investigate the occurrence of leakage of optical mode to GaN substrate. A simple CCD camera was used to collect the far-field profiles of the LDs. Figure 14.32 presents the measured fast axis profiles of LDs. The compositions and thicknesses of InGaN waveguides are marked in Figure 14.32 for all of the LDs. It is worth noting that both LDs with InGaN composition $x = 0.04$ suffer from leakage to GaN substrate. In case of LD with $x = 0.04$ and $d_{\text{InGaN}} = 145$ nm, an enormous narrow peak at 9.5° is observed. Its magnitude is seven times larger than the maximum of Gaussian profile of the far field. This is a manifestation of a strong leakage to GaN substrate. It is so strongly pronounced because the thickness of the n-AlGaN cladding layer is insufficient to reduce it. If the

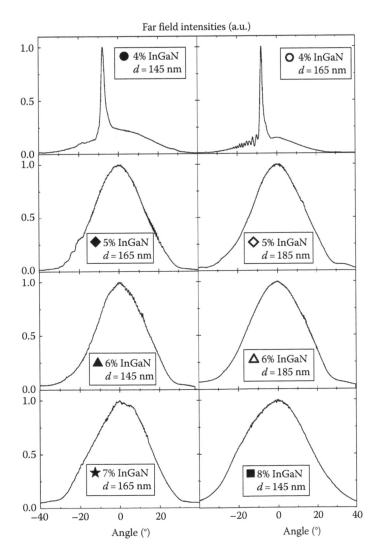

FIGURE 14.32 Experimentally measured fast axis $\Theta_\perp$ of the far-field patterns of LDs with various InGaN compositions and thicknesses. Leakage to the GaN substrate is observed in LDs with 4% InGaN waveguide.

thickness would be higher, the leakage and, therefore, the magnitude of the peak could be reduced. The amount of light in this peak is 25% of the overall far-field pattern and agrees well with the value of 21% predicted theoretically. However, the experimental peak coming from the substrate is much broader and of lower maximum intensity than the theoretical one. This difference might be due to small inhomogeneities in the electron concentration and thus the refractive index of the substrate. In the case of other LDs, the far-field profiles have Gaussian shape and there are no observable additional peaks. As it was predicted by our model, the LDs with InGaN composition $x \geq 0.05$ and sufficient thickness of the waveguide do not suffer from mode leakage despite the thickness of the n-AlGaN cladding being only 700 nm.

14.3.2.3 Aluminum-Free True-Blue Laser Diodes

The structures of LDs described in previous sections contained Al in several layers. The use of AlGaN layers in conventional LDs with GaN waveguides is a natural solution for optical mode confinement. Here—for LDs with InGaN waveguides—the interesting question arises about performance of Al-free LDs with GaN claddings and GaN EBL. Elimination of Al in nitride LDs might enhance the properties of the devices as it was in a case of arsenide LDs operating in the infrared region [91]. In particular, the change of the material system from AlGaAs/GaAs to InGaAsP/GaAs increased the lifetime of LDs operating at 808 nm by few orders of magnitude. This change was solely attributed to elimination of, highly susceptible to oxidation, Al from the device structure [92]. In case of nitride LDs, Al have been successfully eliminated from the cladding layers [57,59,93]. In these devices, the confinement of optical modes is provided by high refractive index InGaN waveguide (see Figure 14.33 for design details). However, the AlGaN-based EBL was still needed due to the high anisotropy of electron and hole mobilities. It is placed near the quantum wells to prevent electron escape from the active region. Although this layer is relatively thin (20 nm), its proximity to the active region might deteriorate the properties of the LDs.

The possible designs of nitride LDs with InGaN waveguide are shown in Figure 14.33. The LDs shown in Figure 14.33a and b differ in the optical mode confinement, while those presented in Figure 14.33b and c differ in carrier injection to active region.

The PAMBE, Al-free, true-blue nitride LDs were grown on ammono-GaN substrates. The structure consisted of 140 nm thick $In_{0.08}Ga_{0.92}N$ waveguide surrounded by GaN claddings. Inside the waveguide,

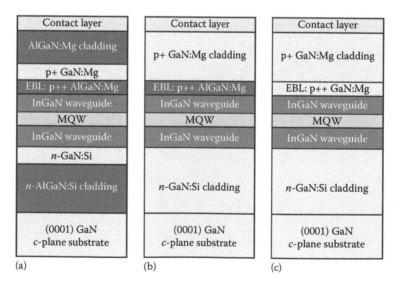

FIGURE 14.33 The different designs for nitride LDs with InGaN waveguides (a) with AlGaN claddings and AlGaN EBL, (b) with GaN claddings and AlGaN EBL, and (c) with GaN claddings and GaN EBL.

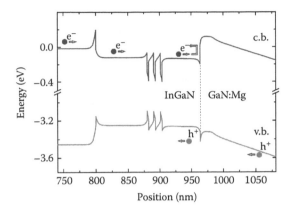

FIGURE 14.34 The conduction and valence band profiles of Al-free laser diodes at a current density of 5 kA/cm². The interface between InGaN waveguide and doped GaN:Mg cladding (marked by dashed line) plays the role of electron blocking layer by preventing the electron overflow.

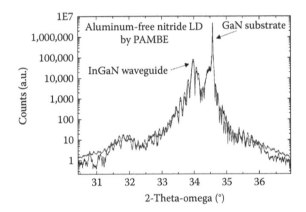

FIGURE 14.35 The 2-Theta-Omega XRD scan of the laser diode structure. The black and gray lines represent the measured and simulated curves, respectively.

three 2.6 nm thick $In_{0.17}Ga_{0.83}N$ quantum wells were separated by 8 nm thick $In_{0.08}Ga_{0.92}N$ quantum barriers. The EBL here was formed by an abrupt bandgap and doping change between the $In_{0.08}Ga_{0.92}N$ waveguide and p-GaN cladding. Figure 14.34 presents the calculated band profile of the device (using the SiLENSe simulation package [94]) at a current density of 5 kA/cm².

Figure 14.35 shows the measured and simulated 2-Theta-Omega x-ray diffraction scans showing only GaN substrate and InGaN waveguide. The L–I–V characteristics are presented in Figure 14.36. The threshold current density of 4.2 kA/cm² is comparable to LDs with AlGaN EBL and AlGaN claddings grown by PAMBE [69]. The measured life tests indicate that degradation rate is about 0.004–0.005 mA/h—which is comparable to our best LDs.

14.3.3 Role of Active Nitrogen Flux in Growth of Long-Wavelength Laser Diodes

The main barriers for long-wavelength nitride-based LDs are related to the decrease of the differential gain for LDs operating above λ = 450 nm. It is due to (1) the decrease of the internal quantum efficiency

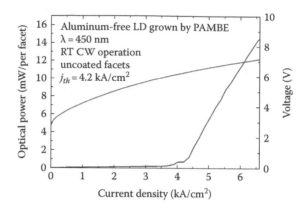

FIGURE 14.36 The L–I–V characteristics of the aluminum-free laser diode. The threshold current density and threshold voltage are 4.2 kA/cm^2 and 6.2 V, respectively.

of QWs for green emission, (2) the broadening of the EL lines (and subsequent broadening of the gain spectra), and (3) the decrease of the optical mode confinement. The first two factors are related with the quality of the InGaN layers. As we have already demonstrated, the quality of InGaN QWs can be improved by an increase of the N flux during the growth [51]. A decrease of the PL line width for long wavelength is achieved by an increase of the N flux (see Figure 14.11). Furthermore, high N flux reduces the number of localized states related to In clustering (Figure 14.9). These improvements could have an impact on the long-wavelength LDs as pointed out by Kojima et al. [95], who stated that additional density of states related to In clusters can play a role of a parasitic radiative recombination channel.

The other issue is the optical mode confinement, which is reduced for the long-wavelength LDs due to the reduction of the contrast of refractive indexes of InGaN, GaN, and AlGaN. In Figure 14.37, we show the calculated optical confinement factor as a function of wavelength for two LDs with 145 nm In$_{0.08}$GaN waveguide. The first one is an AlGaN-cladding-free design, and the second contains both bottom and upper AlGaN claddings.

The strategy we applied for the growth of QWs regions of LDs involved the combination of high growth temperature with highest accessible N flux. In Figure 14.38, we show model predictions of In content in PAMBE (according to Equation 14.1). For the constant growth temperature and constant N flux we can influence the In content by changing the Ga flux (see Figure 14.38a). The maximum value of In content is

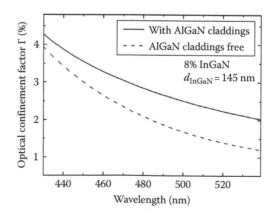

FIGURE 14.37 Dependence of optical confinement factor Γ as a function of light wavelength for LDs with AlGaN claddings (solid line) and AlGaN claddings free (dashed line).

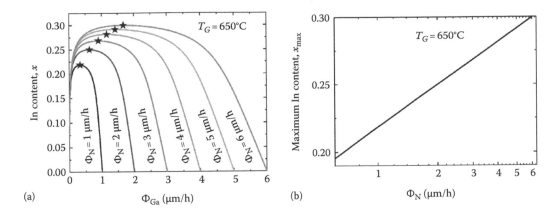

FIGURE 14.38 (a) The In content *x* as a function of Ga flux for five different N fluxes at constant growth temperature 650°C. Stars indicate maximum In content for each nitrogen flux. (b) The maximum In content at 650°C as a function of the N flux.

marked there by stars. For different N fluxes and constant growth temperature, the maximum In content obtained that way is shown in Figure 14.38b. It is worth noting that the maximum In content is a logarithmic function of the N flux, so to obtain an increase of In content by a few percent, the N flux needs to be increased by an order of magnitude.

With continuous increase of the pumping capability of our MBE systems we were able to grow InGaN with the active N flux of 3 μm/h. The longest lasing wavelength from LDs grown by PAMBE is 482 nm, which is achieved in GEN20 PAMBE system equipped with two RF plasma cells with available active N flux of 2 μm/h (Figure 14.39).

The importance of the increase in the available N flux for the growth of nitride-based LDs is illustrated in Figure 14.40, where the maximum lasing wavelength as a function of the N flux is plotted for optically pumped structures (open circles) and LDs (dots) [21]. One can see that the maximum wavelength is roughly a logarithmic function of N flux—similar to In incorporation into the InGaN layers. Even though the number of other factors influences the LDs performance (e.g., injection efficiency or waveguide design related with light losses), such evident correlation provides important guidance for future

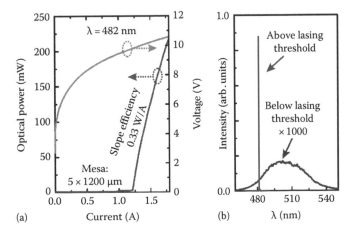

FIGURE 14.39 L–I–V characteristics of AlGaN-cladding-free cw LD operating at 482 nm (a) and spectrum below and above lasing threshold (b).

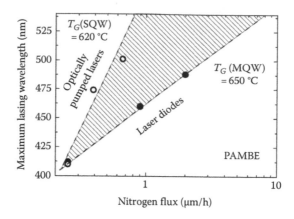

FIGURE 14.40   Maximum lasing wavelength as a function of N flux for optically pumped laser structures (circles) and LDs (dots).

development of PAMBE technology. We would like to stress here that in the commercially available PAMBE systems, the typical N flux is in the range 0.1–0.3 µm/h. In our case, we systematically increased the N flux by using two RF plasma cells, designing new plasma shower heads, and adding the pumping capacity to maintain acceptable vacuum conditions during growth. This optimization strategy was instrumental for successful fabrication of PAMBE LDs up to 482 nm and achieves optically pumped lasing in the range 410–501 nm.

14.3.4 Laser Diodes on Semipolar Crystal Orientations

While the development of polar *c*-plane (0001) nitride LDs has taken place since the early 1990s, the semi-polar and nonpolar LD research boosted up only after the breakthroughs in bulk GaN growth by HVPE. The availability of high-quality, low-defect density semipolar substrates was a key factor in the success of the non-*c*-plane laser structures. Since the year 2009, several groups working with MOVPE technology have demonstrated semipolar LDs operating at ultraviolet, blue, and green wavelengths using various GaN surfaces: $(10\bar{1}1)$ [96], $(20\bar{2}1)$ [97–99], $(11\bar{2}2)$ [100], $(30\bar{3}1)$ [101]. The PAMBE, apart from the fundamental studies reports [102,103], also demonstrated ultraviolet semipolar LDs [104,105].

The advantage of the semipolar structures over the *c*-plane counterparts lies in the reduced piezoelectric fields [106]. Additionally, in non-*c*-plane structures two different ridge orientations can be realized. The in-plane anisotropy on semipolar surfaces allows for the choice of laser resonator orientation along the higher or lower optical gain direction due to the birefringent property of GaN crystal [107]. For green LDs on semipolar $(20\bar{2}1)$ surface, the calculated gain difference is about five times in favor for the ridge direction along the *c*-plane axis, that is, $\left[\bar{1}01\bar{4}\right]$ direction when compared to perpendicular [11–20] direction. However, there are two important challenges for the semipolar devices: low critical thickness for InGaN and AlGaN growth on GaN [108] and laser mirrors fabrication. Unwanted stress relaxation can be addressed either by limited area epitaxy [109–110] or by proper structure strain engineering [100]. In the absence of easy cleavage planes, the laser mirror fabrication can be supported by advanced processing involving, for example, dry etching [99] or mechanical polishing [105]. The growth of semipolar InGaN layers by PAMBE encountered one more challenge such as low indium incorporation efficiency [111]. On the one hand, the low indium content QWs limit the emission wavelength to ultraviolet-blue region, but on the other hand, it protects from early strain relaxation.

The LDs structures were grown by PAMBE on low-dislocation density, conductive semipolar $(20\bar{2}1)$ GaN substrates. The bulk GaN crystals were fabricated by ammonothermal method [24] and sliced for the appropriate orientation. The substrate surface was mechanically and chemo-mechanically polished to obtain the root mean square (rms) surface roughness of 0.2 nm measured by AFM on 5×5 μm² area. Metal-rich conditions provided smooth surface morphologies after the LD structures growth [104]. The nominal structure consisted of the n-type 90 nm GaN:Si buffer, n-type $Al_{0.04}Ga_{0.96}$N:Si cladding layers, n-type 90 nm GaN:Si waveguide, bottom 60 nm $In_{0.02}Ga_{0.98}$N optical confinement layer (OCL), three 3 nm $In_{0.06}Ga_{0.94}$N/7 nm $In_{0.02}Ga_{0.98}$N QWs, top 8 nm $In_{0.02}Ga_{0.98}$N OCL, p-type 15 nm $In_{0.01}Al_{0.15}Ga_{0.86}$N:Mg EBL, p-type 180 nm GaN:Mg waveguide, p-type 180 nm $Al_{0.04}Ga_{0.96}$N:Mg cladding layers, and p-type 55 nm $In_{0.02}Ga_{0.98}$N:Mg contact layer. We discuss here the results on two LD structures: A and B, which differed in nominal bottom cladding $Al_{0.04}Ga_{0.96}$N:Si layer thickness being 180 and 270 nm, respectively. The processing details for these structures also were different.

The LDs were processed as ridge-waveguide, oxide-isolated lasers. Structure A was processed with 5 μm-wide and 1200 μm-long laser stripes along the $[\overline{1}2\overline{1}0]$ direction, while structure B with 5 μm-wide and 700 μm-long stripes oriented along the $[\overline{1}014]$ direction, that is, the in-plane projection of the c-axis. In the first case, the laser mirrors were cleaved, while for the second case, the mirrors were obtained by mechanical polishing. No facet coating was applied.

Figure 14.41a shows the laser L–I curve for the semipolar LD structure A operating in pulse mode operation at room temperature with duty cycle 0.1% and pulse duration of 200 ns. The emission wavelength is 388.2 nm, as shown in Figure 14.41b. Threshold current density measured for the presented device is $j_{th} = 13.2$ kA/cm², which is comparable to the MOCVD-grown semipolar LD emitting at similar wavelength [97].

Figure 14.42a presents L–I–V characteristics for the semipolar LD structure B operating at room temperature in cw mode. Threshold current density is $j_{th} = 12$ kA/cm² at voltage of $U_{th} = 8$V. Slope efficiency of $\eta = 0.4$ W/A is obtained. Single mode laser emission at 390.5 nm is measured above the threshold and shown in Figure 14.42b.

The fabricated LD structures with relatively thin AlGaN claddings, GaN waveguide, and low indium content OCL did not provide sufficient transverse modal confinement because the mode leakage was observed in a far-field projection from an LD operating in cw mode. As an alternative to AlGaN claddings, the quaternary AlInGaN cladding layers or waveguiding layers with higher In content should be introduced in order to provide a better transverse modal confinement. Despite very low In content $In_{0.02}Ga_{0.98}$N OCL used, laser mode confinement improved when compared to the structure without OCL [105].

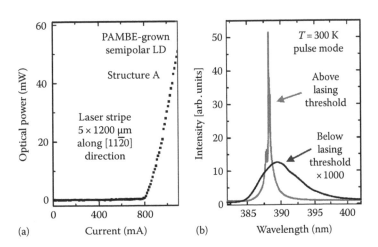

FIGURE 14.41 (a) Light–current characteristics for the semipolar LD operating in pulse mode. (b) Laser emission at 388.2 nm.

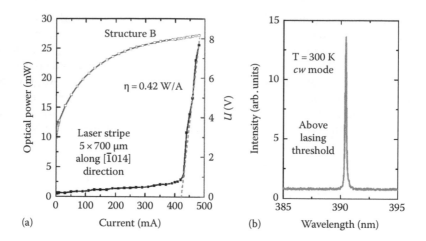

FIGURE 14.42 (a) L–I–V characteristics for the semipolar LD operating in cw mode. (b) Single mode laser emission at 390.5 nm.

14.4 Summary and Conclusions

In this work, we have shown recent progress in plasma-assisted molecular beam epitaxy (PAMBE) of nitride-based laser diodes (LDs). We point out three important milestones contributing to the progress in fabrication of LDs by PAMBE: (1) the availability of low-dislocation density bulk GaN substrates, (2) application of the low-temperature growth mechanism with thin metal layer at a surface during the growth in so-called liquid phase epitaxy–like regime, and (3) understanding the essential role of high N flux during the growth of InGaN.

We investigated in detail the growth peculiarities of GaN and InGaN on low-dislocation density GaN substrates. In particular, we described the influence of nonequivalent atomic steps in wurtzite symmetry crystals on InGaN growth. We demonstrated LDs grown on (0001) and on $\left(20\bar{2}1\right)$ GaN substrates. These LDs operate in a range of 408–482 nm for c-plane and at 388 nm for semipolar orientation. The low-dislocation density GaN substrates allowed us to fabricate continuous-wave LDs with the lifetime exceeding 15,000 h. We observed an increase of the maximum wavelength of the lasing action with increasing N flux applied during the quantum wells (QWs) growth. Lasing at 482 nm was obtained for 2 μm/h N flux. We discussed the influence of InGaN waveguide design on internal and external parameters of blue LDs grown by PAMBE. We observed a significant improvement in threshold current density with increased In content in waveguide and increased thickness of undoped InGaN interlayer between QWs and Mg-doped region. The detailed analysis of the InGaN waveguides and AlGaN claddings composition on quality of the optical beam was discussed. We demonstrated LDs containing InGaN waveguides with excellent optical beam quality.

As can be seen from these results, there are no obvious barriers for the growth of laser diodes by PAMBE. Thus, the long-term prospects for PAMBE in this area appear to be as good as those for MOVPE.

Acknowledgments

This work was supported partially by the National Centre for Research and Development Grants (INNOTECH 157829 and PBS3/A3/23/2015), the National Science Centre Grants (Nos. DEC-2013/11/N/ST7/02788, DEC-2011/03/N/ST3/02938, and DEC-2011/03/N/ST3/02950), and the Ministry of Science and Higher Education (405/E-72/STYP/8/2013). We would like to thank P. Perlin, T. Suski, P. Wisniewski, I. Grzegory, Z. R. Wasilewski, S. Porowski, M. Boćkowski, A. Feduniewicz-Zmuda,

P. Wolny, J. Smalc-Koziorowska, Jola Borysiuk, S. Grzanka, E. Grzanka, G. Nowak, J. Weyher, L. Marona, R. Kudrawiec, M. Krysko, S. Krukowski, A. Sarzyńska, A. Nowakowska-Siwińska, and W. Purgal for the support in experiments, discussions, laser processing, and testing.

References

1. S. Nakamura, S. J. Pearton, and G. Fasol, *The Blue Laser Diode: The Complete Story*, 2nd edn., Springer-Verlag, Berlin, Germany, 368pp. (2000).

2. H. Morkoc, *Nitride Semiconductors and Devices*, Springer Science & Business Media, Berlin, Germany, 489pp. (2013).

3. N. Grandjean, B. Damilano, and J. Massies, Group-III nitride quantum heterostructures grown by molecular beam epitaxy, *J. Phys. Condens. Matter* 13, 6945 (2001).

4. H. Morkoc, III-Nitride semiconductor growth by MBE: Recent issues, *J. Mater. Sci. Mater. Electron.* 12, 677 (2001).

5. T. D. Moustakas, E. Iliopoulos, A. V. Sampath, H. M. Ng, D. Doppalapudi, M. Misra, M. Korakakis, and R. Singh, Growth and device applications of III-nitrides by MBE, *J. Cryst. Growth* 227, 13 (2001).

6. M. Kauer, S. E. Hooper, V. Bousquet, K. Johnson, C. Zellweger, J. M. Barnes, J. Windle, T. M. Smeeton, and J. Heffernan, Continuous-wave operation of InGaN multiple quantum well laser diodes grown by molecular beam epitaxy, *Electron. Lett.* 41, 739 (2005).

7. H. Riechert, R. Averbeck, A. Graber, M. Schienle, U. Strauss, and H. Tews, MBE growth of (In)GaN for LED applications, *Mater. Res. Soc. Symp. Proc.* 449, 149 (1996).

8. J. Neugebauer, T. K. Zywietz, M. Scheffler, J. E. Northrup, H. Chen, and R. M. Feenstra, Adatom kinetics on and below the surface: The existence of a new diffusion channel, *Phys. Rev. Lett.* 90, 056101 (2003).

9. J. E. Northrup, J. Neugebauer, R. M. Feenstra, and A. R. Smith, Structure of GaN(0001): The laterally contracted Ga bilayer model, *Phys. Rev. B* 61, 9932 (2000).

10. C. R. Elsass, I. P. Smorchkova, B. Heying, E. Haus, P. Fini, K. Maranowski, J. P. Ibbetson et al., High mobility two-dimensional electron gas in AlGaN/GaN heterostructures grown by plasma-assisted molecular beam epitaxy, *Appl. Phys. Lett.* 74, 3528 (1999).

11. B. Heying, R. Averbeck, L. F. Chen, E. Haus, H. Riechert, and J. S. Speck, Control of GaN surface morphologies using plasma-assisted molecular beam epitaxy, *J. Appl. Phys.* 88, 1855 (2000).

12. B. Heying, E. J. Tarsa, C. R. Elsass, P. Fini, S. P. DenBaars, and J. S. Speck, Dislocation mediated surface morphology of GaN, *J. Appl. Phys.* 85, 6470 (1999).

13. C. Adelmann, J. Brault, D. Jalabert, P. Gentile, H. Mariette, G. Mula, and B. Daudin, Dynamically stable gallium surface coverages during plasma-assisted molecular-beam epitaxy of (0001) GaN, *J. Appl. Phys.* 91, 9638 (2002).

14. M. J. Manfra, K. W. Baldwin, A. M. Sergent, K. W. West, R. J. Molnar, and J. Caissie, Electron mobility exceeding 160,000 cm²/Vs in AlGaN/GaN heterostructures grown by molecular-beam epitaxy, *Appl. Phys. Lett.* 85 5394 (2004).

15. C. Skierbiszewski, K. Dybko, W. Knap, M. Siekacz, W. Krupczynski, G. Nowak M. Bockowski et al., High mobility two-dimensional electron gas in AlGaN/GaN heterostructures grown on bulk GaN by plasma assisted molecular beam epitaxy, *Appl. Phys. Lett.* 86, 102106 (2005).

16. C. Skierbiszewski, Z. R. Wasilewski, M. Siekacz, A. Feduniewicz, B. Pastuszka, I. Grzegory, M. Leszczynski, and S. Porowski, Growth optimisation of the GaN layers and GaN/AlGaN heterojunctions on bulk GaN substrates using plasma-assisted molecular beam epitaxy, *Phys. Status Solidi A* 201, 320 (2004).

17. P. Waltereit, H. Sato, C. Poblenz, D. S. Green, J. S. Brown, M. McLaurin, T. Katona et al., Blue GaN-based light-emitting diodes grown by molecular-beam epitaxy with external quantum efficiency greater than 1.5%, *Appl. Phys. Lett.* 84, 2748 (2004).

18. C. Skierbiszewski, Z. R. Wasilewski, M. Siekacz, A. Feduniewicz, P. Perlin, P. Wisniewski, J. Borysiuk, I. Grzegory, M. Leszczynski, T. Suski, and S. Porowski, Blue-violet InGaN laser diodes grown on bulk GaN substrates by plasma-assisted molecular-beam epitaxy, *Appl. Phys. Lett.* 86, 011114 (2005).

19. C. Skierbiszewski, P. Wisniewski, M. Siekacz, P. Perlin, A. Feduniewicz-Zmuda, G. Nowak, I. Grzegory, M. Leszczynski, and S. Porowski, 60 mW continuous-wave operation of InGaN laser diodes made by plasma-assisted molecular-beam epitaxy, *Appl. Phys. Lett.* 88, 221108 (2006).

20. C. Skierbiszewski, M. Siekacz, P. Perlin, A. Feduniewicz-Zmuda, G. Cywiński, I. Grzegory, M. Leszczyński, Z. R. Wasilewski, and S. Porowski, Role of dislocation-free GaN substrates in the growth of indium containing optoelectronic structures by plasma-assisted MBE, *J. Cryst. Growth* 305, 346 (2007).

21. C. Skierbiszewski, H. Turski, G. Muziol, M. Siekacz, M. Sawicka, G. Cywiński, Z. R. Wasilewski, and S. Porowski, Nitride based laser diodes grown by plasma assisted molecular beam epitaxy, *J. Phys. D Appl. Phys.* 47, 073001 (2014).

22. I. Grzegory and S. Porowski, GaN substrates for molecular beam epitaxy growth of homoepitaxial structures, *Thin Solid Films* 367, 281 (2000).

23. R. Dwiliński, R. Doradziński, J. Garczyński, L. Sierzputowski, A. Puchalski, Y. Kanbara, K. Yagi, H. Minakuchi, and H. Hayashi, Excellent crystallinity of truly bulk ammonothermal GaN, *J. Cryst. Growth* 310, 3911 (2008).

24. R. Kucharski, M. Zajac, R. Doradzinski, J. Garczynski, L. Sierzputowski, R. Kudrawiec, J. Serafinczuk, J. Misiewicz, and R. Dwilinski, Structural and optical properties of semipolar GaN substrates obtained by ammonothermal method, *Appl. Phys. Express* 3, 101001 (2010).

25. W. K. Burton, N. Cabrera, and F. C. Frank, The growth of crystals and the equilibrium structure of their surfaces, *Philos. Trans. R. Soc. London* 243, 299 (1951).

26. A. Ishizaka and Y. Murata, Crystal growth model for molecular beam epitaxy: Role of kinks on crystal growth, *J. Phys. Condens. Matter* 6, L693 (1994).

27. H. Saitoh, W. Utsumi, H. Kaneko, and K. Aoki, The phase and crystal-growth study of group-III nitrides in a 2000°C at 20 GPa region, *J. Cryst. Growth* 300, 26 (2007).

28. J. A. Van Vechten, Quantum dielectric theory of electronegativity in covalent systems. III. Pressure-temperature phase diagrams, heats of mixing, and distribution coefficients, *Phys. Rev. B* 7, 1479 (1973).

29. W. Kossel, Zur Energetik von Oberflächenvorgängen, *Annalen der Physik* 21, 457 (1934).

30. J. Karpinski and S. Porowski, High pressure thermodynamics of GaN, *J. Cryst. Growth* 66, 11 (1984).

31. O. Ambacher, Growth and applications of Group III-nitrides, *J. Phys. D Appl. Phys.* 31, 2653 (1998).

32. C. Skierbiszewski, Nitride-based light-emitting diodes and nitride-based laser diodes by plasma-assisted molecular beam epitaxy, *Lattice Engineering: Technology and Applications*, Ed: S. Wang, Pan Stanford Publishing Pvt. Ltd. Singapore, Chapter 10, pp. 355–385 (2012).

33. C. Skierbiszewski, Z. R. Wasilewski, I. Grzegory, and S. Porowski, Nitride-based laser diodes by plasma-assisted MBE—From violet to green emission, *J. Cryst. Growth* 311, 1632 (2009).

34. R. C. Schoonmaker, A. Buhl, and J. Lemley, Vaporization catalysis. The decomposition of gallium nitride[1], *J. Phys. Chem.* 69, 3455 (1965).

35. R. Groh, G. Gerey, L. Bartha, and J. I. Pankove, On the thermal decomposition of GaN in vacuum, *Phys. Status Solidi A* 26, 353 (1974).

36. N. Newman, Thermochemistry of III–N semiconductors, in: *Semiconductors and Semimetals*, J. I. Pankove and T. D. Moustakas, eds., Vol. 50, Elsevier, Chapter 4, pp. 55–101 (1997).

37. R. H. Swendsen, P. J. Kortman, D. P. Landau, and H. Muller-Krumbhaar, Spiral growth of crystals: Simulations on a stochastic model, *J. Cryst. Growth* 35, 73 (1976).

38. X. Q. Shen, K. Furuta, N. Nakamura, H. Matsuhata, M. Shimizu, and H. Okumura, Quality improvement of III-nitride epilayers and their heterostructures grown on vicinal substrates by rf-MBE, *J. Cryst. Growth* 301–302, 404 (2007).

39. S. Saito, R. Hashimoto, J. Hwang, and S. Nunoue, InGaN light-emitting diodes on c-face sapphire substrates in green gap spectral range, *Appl. Phys. Express* 6, 111004 (2013).

40. C. Adelmann, R. Langer, G. Feuillet, and B. Daudin, Indium incorporation during the growth of InGaN by molecular-beam epitaxy studied by reflection high-energy electron diffraction intensity oscillations, *Appl. Phys. Lett.* 75, 3518 (1999).

41. R. Averbeck and H. Riechert, Quantitative model for the MBE-growth of ternary nitrides, *Phys. Status Solidi A* 176, 301 (1999).

42. T. Bottcher, S. Einfeldt, V. Kirchner, S. Figge, H. Heinke, D. Hommel, H. Selke, and P. L. Ryder, Incorporation of indium during molecular beam epitaxy of InGaN, *Appl. Phys. Lett.* 73, 3232 (1998).

43. H. Chen, R. M. Feenstra, J. E. Northrup, T. Zywietz, and J. Neugebauer, Spontaneous formation of indium-rich nanostructures on InGaN(0001) surfaces, *Phys. Rev. Lett.* 85, 1902 (2000).

44. M. Siekacz, A. Feduniewicz-Zmuda, G. Cywinski, M. Krysko, I. Grzegory, S. Krukowski, K. E. Waldrip, W. Jantsch, Z. R. Wasilewski, S. Porowski, and C. Skierbiszewski, Growth of InGaN and InGaN/InGaN quantum wells by plasma-assisted molecular beam epitaxy, *J. Cryst. Growth* 310, 3983 (2008).

45. M. Siekacz, M. Ł. Szańkowska, A. Feduniewicz-Zmuda, J. Smalc-Koziorowska, G. Cywiński, S. Grzanka, Z. R. Wasilewski, I. Grzegory, B. Łucznik, S. Porowski, and C. Skierbiszewski, InGaN light emitting diodes for 415 nm–520 nm spectral range by plasma assisted MBE, *Phys. Status Solidi C* 6, S917 (2009).

46. M. H. Xie, S. M. Seutter, W. K. Zhu, L. X. Zheng, H. Wu, and S. Y. Tong, Anisotropic step-flow growth and island growth of GaN(0001) by molecular beam epitaxy, *Phys. Rev. Lett.* 82, 2749 (1999).

47. H. Turski, M. Siekacz, Z. R. Wasilewski, M. Sawicka, S. Porowski, and C. Skierbiszewski, Nonequivalent atomic step edges—Role of gallium and nitrogen atoms in the growth of InGaN layers, *J. Cryst. Growth* 367, 115 (2013).

48. Jmol. An open-source Java viewer for chemical structures in 3D. http://www.jmol.org//, Accessed August 20, 2016.

49. H. Turski, M. Siekacz, M. Sawicka, Z. R. Wasilewski, S. Porowski, and C. Skierbiszewski, Role of nonequivalent atomic step edges in the growth of InGaN by plasma-assisted molecular beam epitaxy, *Jpn. J. Appl. Phys.* 52, 08JE02 (2013).

50. H. Turski, M. Siekacz, M. Sawicka, G. Cywinski, M. Krysko, S. Grzanka, J. Smalc-Koziorowska, I. Grzegory, S. Porowski, Z. R. Wasilewski, and C. Skierbiszewski, Growth mechanism of InGaN by plasma assisted molecular beam epitaxy, *J. Vac. Sci. Technol. B* 29, 03C136 (2011).

51. M. Siekacz, M. Sawicka, H. Turski, G. Cywinski, A. Khachapuridze, P. Perlin, T. Suski et al., Optically pumped 500 nm InGaN green lasers grown by plasma-assisted molecular beam epitaxy, *J. Appl. Phys.* 110, 063110 (2011).

52. C. Skierbiszewski, M. Siekacz, H. Turski, G. Muziol, M. Sawicka, P. Perlin, Z. R. Wasilewski, and S. Porowski, MBE fabrication of III-N-based laser diodes and its development to industrial system, *J. Cryst. Growth* 378, 278 (2013).

53. C. Skierbiszewski, P. Perlin, I. Grzegory, Z. R. Wasilewski, M. Siekacz, A. Feduniewicz, P. Wisniewski et al., High power blue–violet InGaN laser diodes grown on bulk GaN substrates by plasma-assisted molecular beam epitaxy, *Semicond. Sci. Technol.* 20, 809 (2005).

54. U. Strauss, C. Eichler, C. Rumbolz, A. Lell, S. Lutgen, S. Tautz, M. Schillgalies, and S. Brüninghoff, Beam quality of blue InGaN laser for projection, *Phys. Status Solidi C* 5, 2077 (2008).

55. J. J. Wierer, Jr., J. Y. Tsao, and D. S. Sizov, The potential of III-nitride laser diodes for solid-state lighting, *Phys. Status Solidi C* 11, 674 (2014).

56. K. S. Kim, J. K. Son, S. N. Lee, Y. J. Sung, H. S. Paek, H. K. Kim, M. Y. Kim et al., Characteristics of long wavelength InGaN quantum well laser diodes, *Appl. Phys. Lett.* 92, 101103 (2008).

57. C. Skierbiszewski, M. Siekacz, H. Turski, G. Muziol, M. Sawicka, A. Feduniewicz-Żmuda, G. Cywiński et al., AlGaN-free laser diodes by plasma-assisted molecular beam epitaxy, *Appl. Phys. Express* 5, 022104 (2012).

58. C. Skierbiszewski, M. Siekacz, H. Turski, G. Muziol, M. Sawicka, P. Wolny, G. Cywiński et al., True-blue nitride laser diodes grown by plasma-assisted molecular beam epitaxy, *Appl. Phys. Express* 5, 112103 (2012).

59. J. Dorsaz, A. Castiglia, G. Cosendey, E. Feltin, M. Rossetti, M. Duelk, C. Velez, J.-F. Carlin, and N. Grandjean, AlGaN-free blue III–nitride laser diodes grown on c-plane GaN substrates, *Appl. Phys. Express* 3, 092102 (2010).

60. G. Muziol, M. Siekacz, H. Turski, P. Wolny, S. Grzanka, E. Grzanka, A. Feduniewicz-Żmuda et al., High power nitride laser diodes grown by plasma assisted molecular beam epitaxy, *J. Cryst. Growth* 425, 398 (2015).

61. C. Huang, Y. Lin, A. Tyagi, A. Chakraborty, H. Ohta, J. S. Speck, S. P. DenBaars, and S. Nakamura, Optical waveguide simulations for the optimization of InGaN-based green laser diodes, *J. Appl. Phys.* 107, 023101 (2010).

62. E. Kioupakis, P. Rinke, and C. G. Van de Walle, Determination of internal loss in nitride lasers from first principles, *Appl. Phys. Express* 3, 082101 (2010).

63. M. Kuramoto, C. Sasaoka, N. Futagawa, M. Nido, and A. A. Yamaguchi, Reduction of internal loss and threshold current in a laser diode with a ridge by selective re-growth (RiS-LD), *Phys. Status Solidi A* 192, 329 (2002).

64. U. T. Schwarz, E. Sturm, W. Wegscheider, V. Kümmler, A. Lell, and V. Härle, Optical gain, carrier-induced phase shift, and linewidth enhancement factor in InGaN quantum well lasers, *Appl. Phys. Lett.* 83, 4095 (2003).

65. S. Uchida, M. Takeya, S. Ikeda, T. Mizuno, T. Fujimoto, O. Matsumoto, T. Tojyo, and M. Ikeda, Recent progress in high-power blue-violet laser diodes, *IEEE J. Sel. Top. Quantum Electron.* 9(5), 1252 (2003).

66. J. K. Son, S. N. Lee, H. S. Paek, T. Sakong, H. K. Kim, Y. Park, H. Y. Ryu, O. H. Nam, J. S. Hwang, and Y. H. Cho, Measurement of optical loss variation on thickness of InGaN optical confinement layers of blue-violet-emitting laser diodes, *J. Appl. Phys.* 103, 103101 (2008).

67. G. Muziol, H. Turski, M. Siekacz, M. Sawicka, P. Wolny, C. Cheze, G. Cywinski, P. Perlin, and C. Skierbiszewski, Waveguide design for long wavelength InGaN based laser diodes, *Acta Phys. Pol. A* 122, 1031 (2012).

68. G. M. Laws, E. C. Larkins, I. Harrison, C. Molloy, and D. Somerford, Improved refractive index formulas for the $Al_xGa_{1-x}N$ and $In_yGa_{1-y}N$ alloys, *J. Appl. Phys.* 89, 1108 (2001).

69. G. Muziol, H. Turski, M. Siekacz, P. Wolny, S. Grzanka, E. Grzanka, P. Perlin, and C. Skierbiszewski, Enhancement of optical confinement factor by InGaN waveguide in blue laser diodes grown by plasma-assisted molecular beam epitaxy, *Appl. Phys. Express* 8, 032103 (2015).

70. B. W. Hakki and T. L. Paoli, cw degradation at 300°K of GaAs double-heterostructure junction lasers. II. Electronic gain, *J. Appl. Phys.* 44, 4113 (1973).

71. E. Kioupakis, P. Rinke, A. Schleife, F. Bechstedt, and C. G. Van de Walle, Free-carrier absorption in nitrides from first principles, *Phys. Rev. B* 81, 241201 (2010).

72. K. Petermann (Ed.), *Laser Diode Modulation and Noises*, Kluwer Academic Publishers, Dordrecht, the Netherlands, 315pp. (1988).

73. G. Muziol, H. Turski, M. Siekacz, M. Sawicka, P. Wolny, P. Perlin, and C. Skierbiszewski, Determination of gain in AlGaN cladding free nitride laser diodes, *Appl. Phys. Lett.* 103, 061102 (2013).

74. S. Nakamura, RT-CW Operation of InGaN multi-quantum-well structure laser diodes, *Mater. Sci. Eng. B* 50, 277 (1997).

75. D. Hofstetter, D. P. Bour, R. L. Thornton, and N. M. Johnson, Excitation of a higher order transverse mode in an optically pumped $In_{0.15}Ga_{0.85}N/In_{0.05}Ga_{0.95}N$ multiquantum well laser structure, *Appl. Phys. Lett.* 70, 1650 (1997).

76. T. Takeuchi, T. Detchprohm, M. Iwaya, N. Hayashi, K. Isomura, K. Kimura, H. Amano et al., Improvement of far-field pattern in nitride laser diodes, *Appl. Phys. Lett.* 75, 2960 (1999).

77. S. Einfeldt, V. Kirchner, H. Heinke, M. Dießelberg, S. Figge, K. Vogeler, and D. Hommel, Strain relaxation in AlGaN under tensile plane stress, *J. Appl. Phys.* 88, 7029 (2000).

78. M. Sarzyński, M. Kryśko, G. Targowski, R. Czernecki, A. Sarzyńska, A. Libura, W. Krupczyński, P. Perlin, and M. Leszczyński, Elimination of AlGaN epilayer cracking by spatially patterned AlN mask, *Appl. Phys. Lett.* 88, 121124 (2006).

79. G. Muziol, H. Turski, M. Siekacz, S. Grzanka, P. Perlin, and C. Skierbiszewski, Elimination of leakage of optical modes to GaN substrate in nitride laser diodes by use of a thick InGaN waveguide, *Appl. Phys. Express* 9, 092103 (2016).

80. M. J. Bergmann and H. C. Casey Jr., Optical-field calculations for lossy multiple-layer $Al_xGa_{1-x}N/In_xGa_{1-x}N$ laser diodes, *J. Appl. Phys.* 84, 1196 (1998).

81. P. G. Eliseev, G. A. Smolyakov, and M. Osinski, Ghost modes and resonant effects in AlGaN-InGaN-GaN lasers, *IEEE J. Sel. Top. Quantum Electron.* 5, 771 (1999).

82. G. Hatakoshi, M. Onomura, and M. Ishikawa, Optical, electrical and thermal analysis for GaN semiconductor lasers, *Int. J. Numer. Model.* 14, 303 (2001).

83. G. Hatakoshi, M. Onomura, S. Saito, K. Sasanuma, and K. Itaya, Analysis of device characteristics for InGaN semiconductor lasers, *Jpn. J. Appl. Phys.* 38, 1780 (1999).

84. V. Laino, F. Roemer, B. Witzigmann, C. Lauterbach, U. T. Schwarz, C. Rumbolz, M. O. Schillgalies, M. Furitsch, A. Lell, and V. Härle, Substrate modes of (Al, In)GaN semiconductor laser diodes on SiC and GaN substrates, *IEEE J. Quantum Electron.* 43, 16 (2007).

85. T. Lermer, M. Schillgalies, A. Breidenassel, D. Queren, C. Eichler, A. Avramescu, J. Muller, W. Scheibenzuber, U. Schwarz, S. Lutgen, and U. Strauss, Waveguide design of green InGaN laser diodes, *Phys. Status Solidi A* 207, 1328 (2010).

86. D. B. Hall and C. Yeh, Leaky waves in a heteroepitaxial film, *J. Appl. Phys.* 44, 2271 (1973).

87. J. Hu and C. R. Menyuk, Understanding leaky modes: Slab waveguide revisited, *Adv. Opt. Photon.* 1, 58–106 (2009).

88. H. Wenzel, Basic aspects of high-power semiconductor laser simulation, *IEEE J. Sel. Top. Quantum Electron.* 19(5), 1502913 (2013).

89. P. Perlin, K. Holc, M. Sarzyński, W. Scheibenzuber, Ł. Marona, R. Czernecki, M. Leszczyński et al., Application of a composite plasmonic substrate for the suppression of an electromagnetic mode leakage in InGaN laser diodes, *Appl. Phys. Lett.* 95, 261108 (2009).

90. P. Perlin, T. Czyszanowski, L. Marona, S. Grzanka, A. Kafar, S. Stanczyk, T. Suski et al., Highly doped GaN: A material for plasmonic claddings for blue/green InGaN laser diodes, *Proc. SPIE* 8262, 826216 (2012).

91. M. Razeghi, High-power laser diodes based on InGaAsP alloys, *Nature* 369, 631 (1994).

92. J. Diaz, H. J. Yi, M. Razeghi, and G. T. Burnham, Long-term reliability of Al-free InGaAsP/GaAs ($\lambda = 808\,nm$) lasers at high-power high-temperature operation, *Appl. Phys. Lett.* 71, 3042 (1997).

93. H. Turski, G. Muziol, P. Wolny, S. Grzanka, G. Cywinski, M. Sawicka, P. Perlin, and C. Skierbiszewski, Cyan laser diode grown by plasma-assisted molecular beam epitaxy, *Appl. Phys. Lett.* 104, 023503 (2014).

94. SiLENSe 5.4 package. http://www.str-soft.com/products/SiLENSe/, Accessed August 20, 2016.

95. K. Kojima, M. Funato, Y. Kawakami, S. Nagahama, T. Mukai, H. Braun, and U. T. Schwarz, Gain suppression phenomena observed in $In_xGa_{1-x}N$ quantum well laser diodes emitting at 470 nm, *Appl. Phys. Lett.* 89, 241127 (2006).

96. A. Tyagi, H. Zhong, R. B. Chung, D. F. Feezell, M. Saito, K. Fujito, J. S. Speck, S. P. DenBaars, and S. Nakamura, Semipolar (1011) InGaN/GaN laser diodes on bulk GaN substrates, *Jpn. J. Appl. Phys.* 46, L444 (2007).

97. D. A. Haeger, E. C. Young, R. B. Chung, F. Wu, N. A. Pfaff, M. Tsai, K. Fujito, S. P. DenBaars, J. S. Speck, S. Nakamura, and D. A. Cohen, 384 nm laser diode grown on a (20-21) semipolar relaxed AlGaN buffer layer, *Appl. Phys. Lett.* 100, 161107 (2012).

98. Y. Enya, Y. Yoshizumi, T. Kyono, K. Akita, M. Ueno, M. Adachi, T. Sumitomo, S. Tokuyama, T. Ikegami, K. Katayama, and T. Nakamura, 531 nm green lasing of InGaN based laser diodes on semi-polar {20-21} free-standing GaN substrates, *Appl. Phys. Express* 2, 082101 (2009).

99. A. Tyagi, R. M. Farrell, K. M. Kelchner, C.-Y. Huang, P. S. Hsu, D. A. Haeger, M. T. Hardy et al., AlGaN-cladding free green semipolar GaN based laser diode with a lasing wavelength of 506.4 nm, *Appl. Phys. Express* 3, 011002 (2010).

100. P. Shan Hsu, F. Wu, E. C. Young, A. E. Romanov, K. Fujito, S. P. DenBaars, J. S. Speck, and S. Nakamura, Blue and aquamarine stress-relaxed semipolar (11-22) laser diodes, *Appl. Phys. Lett.* 103, 161117 (2013).

101. P. S. Hsu, K. M. Kelchner, A. Tyagi, R. M. Farrell, D. A. Haeger, K. Fujito, H. Ohta, S. DenBaars, J. S. Speck, and S. Nakamura, InGaN/GaN blue laser diode grown on semipolar (30-31) free-standing GaN substrates, *Appl. Phys. Express* 3, 052702 (2010).

102. M. Sawicka, C. Chèze, H. Turski, J. Smalc-Koziorowska, M. Kryśko, S. Kret, T. Remmele, M. Albrecht, G. Cywiński, I. Grzegory, and C. Skierbiszewski, Growth mechanisms in semipolar and nonpolar m-plane AlGaN/GaN structures grown by PAMBE under N-rich conditions, *J. Cryst. Growth* 377, 184 (2013).

103. M. Sawicka, M. Kryśko, G. Muziol, H. Turski, M. Siekacz, P. Wolny, J. Smalc-Koziorowska, and C. Skierbiszewski, Strain relaxation in semipolar (20-21) InGaN grown by plasma assisted molecular beam epitaxy, *J. Appl. Phys.* 119, 185701 (2016).

104. M. Sawicka, G. Muziol, H. Turski, S. Grzanka, E. Grzanka, J. Smalc-Koziorowska, J. L. Weyher et al., Ultraviolet laser diodes grown on semipolar (20-21) GaN substrates by plasma-assisted molecular beam epitaxy, *Appl. Phys. Lett.* 102, 251101 (2013).

105. M. Sawicka, G. Muziol, H. Turski, A. Feduniewicz-Żmuda, M. Kryśko, S. Grzanka, E. Grzanka et al., Semipolar (20-21) GaN laser diodes operating at 388 nm grown by plasma-assisted molecular beam epitaxy, *J. Vac. Sci. Technol. B* 32, 02C115 (2014).

106. A. E. Romanov, T. J. Baker, S. Nakamura, and J. S. Speck, Strain-induced polarization in wurtzite III-nitride semipolar layers, *J. Appl. Phys.* 100, 023522 (2006).

107. W. G. Scheibenzuber, U. T. Schwarz, R. G. Veprek, B. Witzigmann, and A. Hangleiter, Calculation of optical eigenmodes and gain in semipolar and nonpolar InGaN/GaN laser diodes, *Phys. Rev. B* 80, 115320 (2009).

108. A. E. Romanov, E. C. Young, F. Wu, A. Tyagi, C. S. Gallinat, S. Nakamura, S. P. DenBaars, and J. S. Speck, Basal plane misfit dislocations and stress relaxation in III-nitride semipolar heteroepitaxy, *J. Appl. Phys.* 109, 103522 (2011).

109. M. T. Hardy, F. Wu, P. Shan Hsu, D. A. Haeger, S. Nakamura, J. S. Speck, and S. P. DenBaars, True green semipolar InGaN-based laser diodes beyond critical thickness limits using limited area epitaxy, *J. Appl. Phys.* 114, 183101 (2013).

110. L. Marona, J. Smalc-Koziorowska, E. Grzanka, M. Sarzynski, T. Suski, D. Schiavon, R. Czernecki, P. Perlin, R. Kucharski, and J. Domagala, Suppression of extended defects propagation in a laser diodes structure grown on (20-21) GaN, *Semicond. Sci. Technol.* 31, 035001 (2016).

111. M. Sawicka, C. Cheze, H. Turski, G. Muziol, S. Grzanka, C. Hauswald, O. Brandt et al., Ultraviolet light-emitting diodes grown by plasma-assisted molecular beam epitaxy on semipolar GaN (20-21) substrates, *Appl. Phys. Lett.* 102, 111107 (2013).

GaN-Based Blue and Green Laser Diodes

Jianping Liu
Chinese Academy of Sciences

Hui Yang
Chinese Academy of Science

Abstract We describe our studies on GaN-based laser diodes, including homoepitaxial growth of GaN layers on GaN substrates, laser diode layer structure design and doping optimization, epitaxial growth of InGaN quantum wells, and suppression of unintentional carbon incorporation in AlGaN:Mg to improve its conductivity. Based on these studies, we have demonstrated both blue laser diodes with output power of 1.3 W and green laser diodes with output power of 60 mW under continuous wave operation.

15.1 Introduction

Stimulus emission had been observed first time in AlGaN/GaN/InGaN separate confinement heterostructure (SCH) by Prof. Akasaki and Prof. Amano in 1995,[1] 2 years after candela-class high–brightness InGaN/AlGaN double-heterostructure (DH) blue light-emitting diodes (LEDs) were reported by Prof. Shuji Nakamura. First GaN-based laser diode (LD) was then reported by Prof. Shuji Nakamura of Nichia the following year,[2] and Nichia is a leading company in this field till date. GaN-based LDs in the early days were grown on sapphire substrate with a dislocation density of 3×10^9 cm^{-2}, which was too high to obtain reliable laser diodes. Epitaxial lateral overgrowth (ELOG) was then developed to reduce the dislocation density of overgrown GaN layers in which dislocation density could be reduced to 10^6 cm^{-2}. LDs with lifetime of several thousand hours were fabricated on ELOG GaN in 1997.[3]

GaN bulk substrates are preferred to fabricate high-performance LDs. It is difficult to obtain cleaved mirror facets employed for the cavities of GaN-based LDs grown on sapphire. Also, the thermal conductivity of the sapphire is insufficient to efficiently dissipate the heat generated by the LDs. The first GaN-based LD grown on GaN bulk substrate was also developed in Nichia in 1998 when GaN substrate became available.[4] GaN substrates can offer three benefits for the fabrication of InGaN-based LDs. First, smooth facets can be obtained by cleaving along m-plane by using GaN substrates, which help reduce the threshold current density and improve fabrication yield. Second, vertical LD structure can be fabricated by depositing n-electrode on the N-face of GaN substrates, which eliminates the current crowding effect that exists in lateral LD structures, and thus reduces operation voltage. Third, the lifetime of InGaN-based LDs can be improved by using GaN substrates with low dislocation density, since dislocation defects can seriously reduce the lifetime of InGaN-based LDs operated at a current density of more than 5 kA/cm².

By employing high-quality GaN substrates, the performance of GaN-based LDs has made great advancement in the past several years. Blue laser diodes with a threshold current density lower than 2 kA/cm² and an output power higher than 1 W have been commercialized by Nichia[5] and Osram.[6]

At Suzhou Institute of Nano-tech and Nano-bionics, we have carried out extensive studies on GaN-based laser diodes on bulk GaN substrates, including homoepitaxial growth of GaN layers on GaN substrates, LD layer structure design and doping optimization, epitaxial growth of InGaN quantum wells (QWs), and suppression of unintentional carbon incorporation in AlGaN:Mg to improve its conductivity. Based on these studies, we have demonstrated both blue LDs with an output power of 1.3 W and green LDs with an output power of 60 mW under continuous wave operation. The chapter is organized as follows. In Section 15.2, the homoepitaxial growth of GaN layers on GaN substrates is investigated. In Section 15.3, layer structure design and optimization of blue LDs are inspected. In Section 15.4, the epitaxial growth of InGaN QWs and suppression of unintentional carbon incorporation in AlGaN:Mg are examined.

15.2 Homoepitaxial Growth of GaN Layers on GaN Substrates

Most of GaN-based light-emitting diodes, laser diodes, and transistors utilize multilayer quantum heterostructures. Therefore, the control of growth surface morphology is extremely important, since rough surface will result in rough interface in the heterostructures and thus poor device performances.

Due to difficulties in achieving bulk GaN crystal, previously most of GaN layers and devices were grown on foreign substrates, such as sapphire, silicon, and SiC. Heteroepitaxial GaN layers grown on these substrates contain high density of threading dislocations (TDs) ranging from ~10⁸ to ~10⁹ cm⁻².[7] Shen et al. systematically reported the surface morphology of heteroepitaxial GaN layers,[8] which featured wavy atomic steps and varied terrace width because of pinning effect of high density of dislocations.

In contrast, much lower density of dislocations in bulk GaN substrates result into quite different surface morphologies. We observed two kinds of surface structures, that is, hillock and striation on the surface of homoepitaxial GaN layers. We have found that two key factors that affect the morphology of homoepitaxial GaN layers are substrate miscut angle and direction.[9] We found a correlation between hillocks and dislocation clusters. Moreover, we found that an increase of substrate miscut angle can suppress hillock formation. On the other hand, we found that surface striation was due to substrate miscut orientation.

GaN layers were grown on epi-ready c-plane GaN substrates by a Thomas Swan CCS metalorganic vapor phase epitaxy (MOVPE) system. Substrates A, B, and C from the same manufacturer have miscut angles of 0.1°, 0.2°, and 0.4°, respectively, while they have the same average TD density of about 4×10^7 cm⁻². Substrate D from another manufacturer has a miscut angle of 0.2°, while the TD density is about 4×10^6 cm⁻². All four substrates mentioned here were fabricated by HVPE and miscut toward $[1\bar{1}00]$ direction.

A 1 μm-thick Si doped GaN layer was deposited on these substrates under identical conditions: at a pressure of 100 torr, at a V/III ratio of 1600, and at a temperature of 1030°C. H_2 was used as the carrier gas, and trimethylgallium (TMGa) and ammonia (NH_3) were used as the Ga and the N precursors, respectively. The growth rate was about 2.5 μm/h. The Si doping level was around 3×10^{18} cm⁻³ measured by secondary ion mass spectrometry (SIMS). These grown samples are named by their substrate names.

Surface morphology was characterized by Nomarski optical microscope and atomic force microscopy (AFM). Substrate miscut angle and direction were determined from the offset angle between the incident angle ω and the diffraction angle 2θ of GaN (006) plane by x-ray diffraction (XRD). Cathodoluminescence (CL) measurement was used to evaluate dislocation density and distribution.

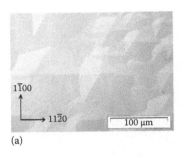
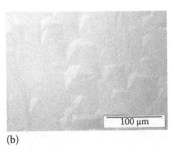
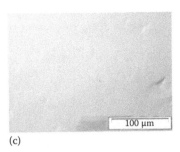

(a) (b) (c)

FIGURE 15.1 (a) Nomarski optical image of sample A, (b) sample B, and (c) sample C.

Figure 15.1 shows Nomarski optical images of GaN epilayers grown on GaN substrates A, B and C. It is found that a lot of hillocks were formed on GaN substrates with 0.1° and 0.2° miscut angle as shown in Figure 15.1a and b. Hillock density is estimated at 5×10^4 cm^{-2} from Figure 15.1a. Each pyramidal hillock was composed of six faces, with all faces of the hillock symmetrically inclined toward $\left[1\bar{1}00\right]$ directions. The diameters of the hillocks decreased as the substrate miscut angle increased. For a sample grown on substrate with a miscut angle of 0.4°, hillocks disappear and the surface morphology is mirrorlike. Since growth conditions are identical, we believe the morphology change is caused by substrate miscut angle. Hillocks have been observed by several groups [10–12] on layers grown on c-plane free-standing GaN substrates. The formation mechanism has not been clearly stated.

Spiral growth hillocks arise from the lateral growth of pinned steps, which form when a threading dislocation with a screw component intersects the free surface of a crystal.[13] However, for hillocks to be formed, spirals need to survive in a field of flowing steps originating from the miscut of the substrate. Spirals can only survive if the miscut terrace width W_{MC} meets the following condition.[14]

$$W_{MC} = \frac{c}{2\tan\theta} > \frac{4\pi\rho_c}{2},$$ (15.1)

where

θ is the miscut angle
c is the lattice parameter of c-axis
ρ_c is the critical radius of curvature of a pinned step around one single dislocation

According to this condition, a miscut angle of 0.1° should be enough to suppress hillock formation for MOVPE-grown layers where $\rho_c \gg 20$ nm.[15] Therefore, hillocks are usually absent for MOVPE-grown GaN layers. However, in our case, hillocks appear in GaN layers even with miscut angle of 0.2°. In order to clarify the reason, more studies were done as follows.

Figure 15.2 shows AFM images near the apex of a hillock in sample A. Atomic steps are clearly resolved on six faces of the pyramidal hillock, as shown in Figure 15.2a. Figure 15.2b shows a relatively high-resolution AFM image of the hillock apex. Dozens of depressions are observed at the hillock apex. These depressions form at the surface terminations of dislocations with screw components. Atomic steps are pinned by these dislocations, and spiral ramps are observed to rotate around them. By counting these suppressions in the scanned area, dislocation density is estimated to be 3×10^9 cm^{-2}, which is much higher than the average dislocation density of the substrate. It is clear that there exists a dislocation cluster near the hillock apex. This is inconsistent with what Sarzynski et al.[11] assumed—that each TD produces a hillock.

In order to learn the correlation between hillock formation and dislocation cluster, we studied dislocation distribution in samples A, B, and C by CL measurement. As shown in Figure 15.3, dislocation clusters are observed in all samples. It should be noted that in order to observe a dislocation cluster clearly, Figure 15.3b and c has a higher resolution than Figure 15.3a. Although the average dislocation density of

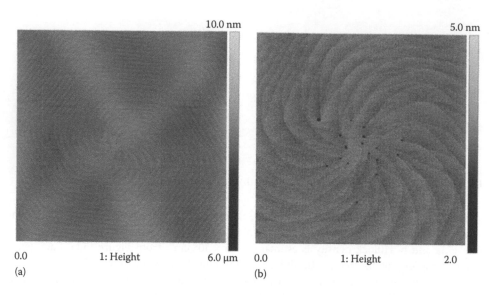

FIGURE 15.2 (a) AFM images showing the apex of a hexagonal hillock of sample A, (b) a higher-resolution image of (a).

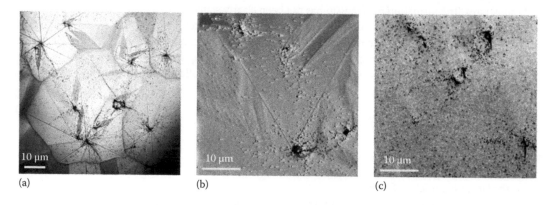

FIGURE 15.3 (a) CL image of sample A, (b) sample B, and (c) sample C.

these samples is around 4×10^7 cm^{-2}—much lower than that found on sapphire substrate—dislocation clusters containing dislocations several orders of magnitude higher than that were observed. Moreover, it is found that all the apex of hillocks are located at the regions with dislocation clusters in samples A and B, as shown in Figure 15.3a and b. Therefore, we believe that a dislocation cluster other than single dislocation results in hillock formation. Absence of hillocks in sample C implies that a miscut angle of 0.4° can suppress hillock formation even though dislocation clusters also exist in the substrate. These findings are inconsistent with the speculation by Lee et al.[10] They attributed the hillock formation that they observed to more dislocations with screw components than the edge components on GaN substrate with a low miscut angle.

Equation 15.1 indicates that a miscut angle of 0.1° can suppress hillock formation from single dislocation for MOVPE-grown layers. When there exists dislocation cluster, it is expected that strain around a dislocation cluster will be much larger than that around single dislocation. Strain field was found to affect the advancement of steps, as reported by Cabrera and Vermilyea.[16] Hannon et al.[17] have experimentally and theoretically demonstrated that the surface strain field of dislocation fundamentally influences how surface steps move. Sunagawa and Bennema[18] have observed various effects of strain associated with dislocations depending on the sizes of Burgers vector b and the concentration of dislocations. In spiral growth, both

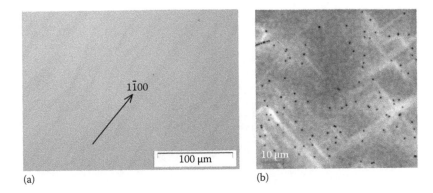

(a)   (b)

FIGURE 15.4   (a) Nomarski optical image of sample D, (b) CL image of sample D.

the advancing rate and the curvature of the spiral layers are affected by strain field. The spiral steps may advance much faster around a dislocation cluster than in the case of single dislocation, thus spiral growth can win the competition with the aid of step flow induced by substrate miscut. This may be the reason why the hillock formation condition no longer follows Equation 15.1 in our samples. In order to suppress hillocks on layers grown on GaN substrate with a dislocation cluster, a relatively larger miscut angle is needed.

Our viewpoint is further supported by the fact that hillocks are absent in sample D, which was grown on a substrate with the same miscut angle of 0.2° as sample B, as shown in Figure 15.4a. No dislocation cluster is observed by CL measurements and the dislocations were uniformly distributed, as shown in Figure 15.4b. This result verifies that spiral growth can be suppressed on substrate with a miscut angle of 0.2° without the presence of dislocation clusters.

Although hillock is absent in sample D, striation feature was observed. This surface feature has been reported by other groups,[10,19,20] but the formation mechanism has not been clearly stated. We found a correlation between striation surface and miscut direction. A miscut along $\left[11\bar{2}0\right]$ direction causes striation surface, while miscut along $\left[1\bar{0}10\right]$ gives a smooth surface. The mechanism behind it could be that the atomic steps perpendicular $\left[10\bar{1}0\right]$ direction, which appears by cutting substrate along $\left[10\bar{1}0\right]$ direction, are stable, whereas the atomic steps parallel $\left[10\bar{1}0\right]$ direction are unstable, causing striation on the surface.

By choosing appropriate miscut angle and direction, smooth surface morphology with straight and parallel atomic steps can be obtained for homoepitaxial GaN layer grown on GaN substrate with a dislocation density around 10^6 cm^{-2}, as shown in Figure 15.5a. As a comparison, the surface morphology of GaN layer grown on sapphire is shown in Figure 15.5b, which has curved atomic steps pinned by dislocations. Curved atomic steps result in variation of terrace width, which may cause fluctuation of indium composition. Figure 15.6 shows a comparison of EL spectra of blue LD structures grown on GaN substrate and sapphire in the same MOCVD run. It is noted that the FWHM of LD structure grown on GaN substrate is 2.4 nm narrower than that of LD structure grown on sapphire, which is attributed to little variation in atomic terrace width of the underlying GaN layer.

In summary, spiral growth hillocks were observed on GaN layers grown on free-standing GaN substrates with a miscut angle ≤0.2°. A close correlation was found between hillock formation and dislocation cluster by CL and AFM measurements. It is believed that hillocks originate from dislocation clusters. Absence of hillocks on a layer grown on GaN substrate free of a dislocation cluster further supports the viewpoint. A larger strain field around a dislocation cluster than around a single dislocation may be the reason for hillock formation. Hillocks disappear on GaN layers grown on free-standing GaN substrates with a miscut angle of 0.4° even though dislocation clusters still exist. On the other hand, homoepitaxial layers grown on substrates with miscut direction along $\left[1\bar{1}20\right]$ have striation on the surface morphology, which is related to the instability of atomic steps parallel to $\left[1\bar{1}20\right]$ direction.

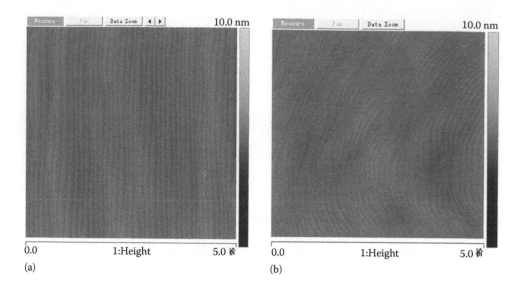

(a)   (b)

FIGURE 15.5   AFM images of GaN layers grown on (a) GaN substrate and (b) sapphire substrate.

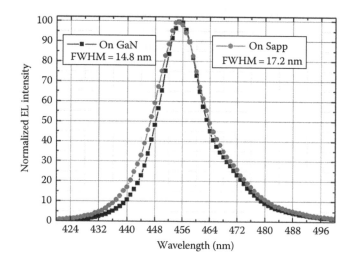

FIGURE 15.6   Electroluminescence spectra of blue LD structures grown on GaN and sapphire, respectively.

15.3 Blue Laser Diodes

GaN-based blue and green LDs, which are a desirable light source for laser display, are still under development. Although high-quality GaN substrates are available, which is essential to obtain a longer lifetime for laser diodes, difficulties of fabricating high-performance InGaN-based laser diodes still exist, including high internal loss and large strain due to lattice mismatch. The main internal loss in GaN-based LD structures comes from Mg-doped upper waveguide and upper AlGaN cladding layers. Layer structure optimization including waveguide thickness and composition as well as doping profile to reduce internal loss will be investigated by both simulations and experiments. Large tensile strain between AlGaN cladding layer and GaN substrate could result in cracking, which will be described in Section 15.3.3. On the other hand, large compressive strain between InGaN QW and GaN layer causes In segregation, especially in high-In-content green QWs, which is the main challenge in green LDs, and will be described in Section 15.4.

15.3.1 Design Consideration of LD Layer Structures

As emitting wavelength extends longer, an issue that needs to be addressed is decreased optical confinement due to reduced refractive index difference of GaN and AlGaN. InGaN waveguide (WG) layers are usually used to improve the optical confinement. Moreover, other advantages of inserting an InGaN interlayer between active region and *p*-AlGaN electron blocking layer (EBL) in GaN-based blue LDs have been reported.[21–25] Asano et al. reported that this InGaN interlayer could enhance lasing performance of LDs due to increase in optical confinement factor (OCF) and the reduction of strain.[21,22] Lee et al. reported that it could also suppress the diffusion of Mg from *p*-type layers to prevent the device from degradation.[23–25]

Here, we have calculated dependences of OCF and internal absorption loss (IAL) on location, In composition, and thickness of the InGaN layer. Based on complex refractive index, the 2D optical mode of GaN-based LDs is calculated with a finite difference beam propagation method. OCFs of each layer are calculated, and total IAL is the sum of the products of OCF and absorption coefficient of each layer. The typical values for internal loss used in simulation and experimental studies reported in the literature range from ~10 up to ~50 cm^{-1}.[26–28]

15.3.1.1 Location and Doping of the Upper Waveguide

Figure 15.7a and b schematically shows the two-dimensional (2D) diagrams of two ridge waveguide LDs. To analyze the effects of the location of the upper InGaN waveguide layer, simulation of two LDs with different structures, that is, LD I and LD II as shown in Figure 15.7a and b, are performed. For conventional GaN-based LDs, the upper waveguide layer is located between *p*-AlGaN EBL and *p*-AlGaN/GaN superlattices (SLs),[2,29] as shown in Figure 15.7b. An InGaN interlayer that acts as upper waveguide layer is inserted between active region and *p*-AlGaN EBL in the new LD structure I, as shown in Figure 15.7a.

LD I in Figure 15.7a is composed of a 400 µm *n*-type GaN layer (Si: 3×10^{18} cm^{-3}), a 1 µm *n*-type Al$_{0.16}$Ga$_{0.84}$N/GaN SLs (Si: 3×10^{18} cm^{-3}), a 40 nm *n*-type GaN layer (Si: 5×10^{17} cm^{-3}), a 80 nm undoped lower InGaN waveguide layer, the multiple quantum wells (MQWs) active region, an undoped upper InGaN waveguide layer (i.e., inserted InGaN interlayer), a 10 nm undoped GaN layer, a 20 nm *p*-type Al$_{0.2}$Ga$_{0.8}$N EBL (Mg: 3×10^{19} cm^{-3}), a 500 nm *p*-type Al$_{0.16}$Ga$_{0.84}$N/GaN SLs (Mg: 1×10^{19} cm^{-3}), and a 20 nm *p*-type GaN contact layer (Mg: 1×10^{20} cm^{-3}). The MQWs consist of two 2.5 nm undoped In$_{0.17}$Ga$_{0.83}$N well layers and three 14 nm *n*-GaN barrier layers (Si: 1×10^{17} cm^{-3}). Figure 15.7b shows the structure of LD II, which is similar to LD I except for the different location and doping of upper waveguide layer. In LD II, the upper waveguide layer is Mg doped (Mg: 5×10^{18} cm^{-3}), and it is located between *p*-AlGaN EBL and

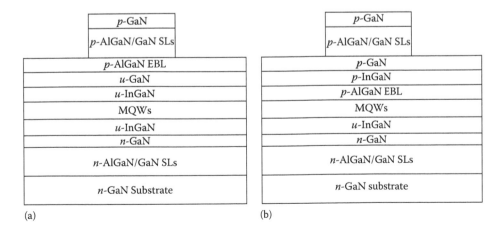

(a)   (b)

FIGURE 15.7 The schematic 2D structures of ridge waveguide LD I (a) and LD II (b), where *u*-is an abbreviation of "undoped."

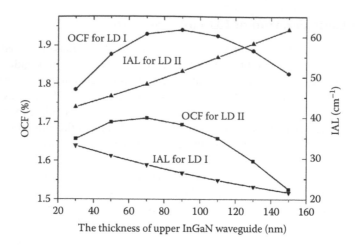

FIGURE 15.8 The dependences of OCF and IAL on the thickness of upper InGaN waveguide layer for LD I and LD II.

p-Al$_{0.16}$Ga$_{0.84}$N/GaN SLs as in most conventional GaN-based LD structures. The width and height of the ridge are 5 and 0.52 μm, respectively. The cavity length is 800 μm, and the mirror loss is supposed to be 10 cm^{-1}. The data of refractive index are taken as the same as those reported in previous studies.[30] And the absorption coefficients of Mg-doped, Si-doped, and undoped GaN are taken as 100, 30, and 10 cm^{-1}, respectively.[31] For AlGaN and InGaN, the absorption coefficients are approximately taken to be the same as GaN.

We calculated dependences of OCF and IAL of both LD I and LD II on the thickness of upper InGaN waveguide layer. In the simulation, the In composition of all InGaN waveguide layers is taken as 3%. As shown in Figure 15.8, with the same thickness of upper InGaN waveguide layer, OCF of LD I is always larger than that of LD II, and IAL is always smaller for LD I. It is understood that for LD II, AlGaN EBL pushes the optical field profile to n-type side, which reduces OCF. Due to the strong absorption of the nearest p-InGaN waveguide layer, IAL of LD II is larger than that of LD I. The OCF gets the maximum value of 1.71% when the thickness of upper InGaN waveguide is 70 nm. In this case, IAL is equal to 48.28 cm^{-1}. For LD I, at the same thickness of upper InGaN waveguide layer (70 nm), OCF is as high as 1.93%, which is 13% larger than that of LD II, and IAL decreases from 48.28 to 28.42 cm^{-1}.

15.3.1.2 The In Composition and Thickness of InGaN Interlayer

Difference of the refractive index between Al$_{0.16}$Ga$_{0.84}$N/GaN SLs and GaN is smaller with increasing wavelength, and thus, the optical confinement factor of MQWs is smaller in GaN-based blue LDs compared to violet LDs if GaN is used as waveguide layers, which has a significant influence on threshold current density of LDs.[32] To get enough higher OCF, we take InGaN layers as waveguide layers of LD I as shown in Figure 15.7a. Figure 15.9 shows the dependence of OCF on the thickness of upper u-InGaN waveguide layer at different In composition. It shows that OCF rises with increasing of In composition of InGaN with the same thickness. Figure 15.10 shows the dependence of IAL on the thickness of the upper InGaN waveguide layer at different In composition values. It shows that IAL decreases with increasing In composition of InGaN with the same thickness. Figure 15.11 shows the dependence of threshold material gain on the thickness of the upper InGaN waveguide layer at different In composition values. It shows that the threshold material gain decreases with increasing In composition and thickness of upper InGaN waveguide layer.

Taking into account the difficulty of growing high In-composition InGaN with high quality, the optimal composition of InGaN should be chosen not larger than 3%–4%, and then the optimal thickness of the upper InGaN waveguide is 70–90 nm according to the simulation result mentioned earlier. When the thickness of the InGaN waveguide layer is taken as 70 nm, for GaN, In$_{0.02}$Ga$_{0.98}$N, In$_{0.03}$Ga$_{0.97}$N, and In$_{0.04}$Ga$_{0.96}$N, the calculated OCF values are 1.49%, 1.81%, 1.93%, and 2.01%, IALs are 33.62, 30.11, 28.42, and 27.08 cm^{-1}, and threshold material gains are 2923, 2220, 1991, and 1847 cm^{-1}, respectively.

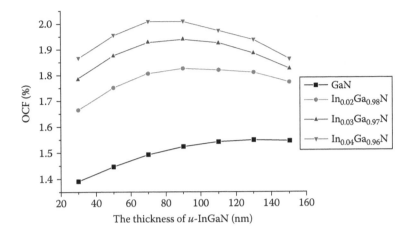

FIGURE 15.9 The dependence of OCF on the thickness of upper InGaN waveguide layer at different In compositions of InGaN.

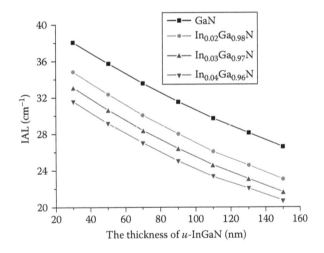

FIGURE 15.10 The dependences of IAL on thickness of upper InGaN waveguide layer at different In compositions of InGaN.

15.3.2 Suppression of Internal Loss

As discussed earlier, an effective approach to suppress internal loss is to make the Mg-doped layers away from MQW active region. We have then experimentally investigated the internal loss and the threshold current density dependent on the distance d between the last QW and Mg-doped layer. The internal loss was measured by the Hakki–Paoli method. As shown in Figure 15.12, increasing d from 20 to 80 nm, the internal loss reduces from 60 to 15 cm^{-1}, which reduces the threshold current density from 4 to 2.3 kA/cm^2. However, a further increase of d results in increasing threshold current density, which is attributed to reduced hole injection efficiency. The effective mass of a hole in GaN material system is as large as $0.8m_0$. Therefore, there exists an optimum distance *d* so as not to degrade the efficiency of hole injection into the MQWs.

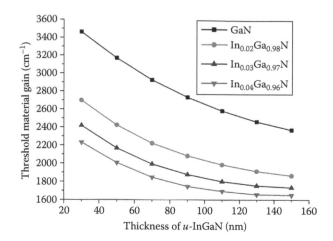

FIGURE 15.11 The dependences of threshold material gain on the thickness of upper InGaN waveguide layer at different In compositions of InGaN.

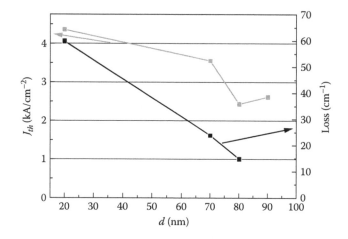

FIGURE 15.12 Threshold current density and internal loss dependent on the distance d between the last QW and Mg-doped layer.

15.3.3 Strain Control

AlGaN cladding layer thick enough to suppress optical mode leak into GaN substrate often results in cracking due to lattice mismatch with GaN substrate, as shown in Figure 15.13a. In order to eliminate cracks, an InGaN layer with compressive strain with GaN can be inserted beneath AlGaN cladding layer to compensate the strain. Figure 15.13b shows the curvature evolution during the epitaxial growth of blue LD wafer with and without InGaN insert layer. Its composition and thickness is 0.05 and 100 nm, respectively. It shows that the curvature of blue LD wafer is reduced significantly, indicating the InGaN layer effectively compensates the tensile strain of AlGaN cladding layer.

15.3.4 Device Characteristics of Blue LDs

After optimizing the layer structures and the epitaxial growth, blue LDs were fabricated in various sizes. Chip size $4 \times 400 \ \mu m^2$ was used for low threshold current LDs. Figure 15.14 shows power–current–voltage curve for a typical LD with size of $4 \times 400 \ \mu m^2$. The threshold current is 36 mA, and the threshold voltage

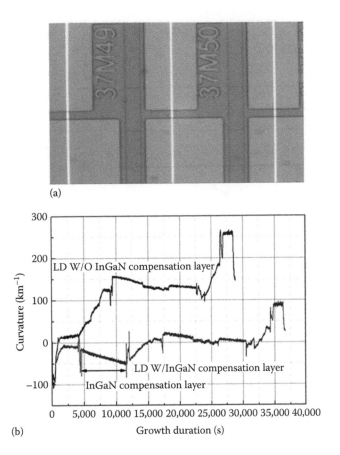

(a)

(b)

FIGURE 15.13 (a) Optical microscopy image showing cracks on an LD wafer and (b) curvature evolution of LD structures with and without InGaN strain compensation layers.

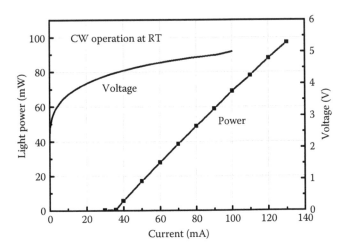

FIGURE 15.14 Power–current–voltage curves of a blue LD with chip size of 4 × 400 μm² under continuous-wave operation.

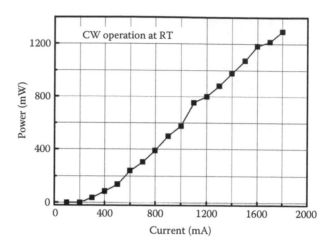

FIGURE 15.15   Power–current curve of a blue LD with chip size of 15 × 1200 μm² under continuous-wave operation.

is 4.3 V, respectively. The output power is 97 mW at 130 mA current under continuous-wave operation at room temperature, which corresponds to a typical operation current density of 8.1 kA/cm² for blue LDs. We also fabricated LD chips with a size of 15 × 1200 μm² to obtain high output power. Figure 15.15 shows the power–current curve of a high power blue InGaN LDs lasing at 450 nm under room-temperature and continuous-wave condition. The threshold current is 180 mA, corresponding to a threshold current density of 1 kA/cm². The output power is 1.3 W at the current of 1.8 A.

15.4 Green Laser Diodes

Considerable attention has been focused on InGaN green LDs during the past few years due to their potential applications in pico-projectors and laser display. Since the first breakthrough of 488 nm InGaN-based LDs achieved by Nichia Corp. in 2008,[33] tremendous progress has been made in InGaN green LDs. Green InGaN LDs with emission wavelength above 500 nm grown on *c*-plane,[34–40] $\left(1\,\bar{1}22\right)$ plane,[41] and $\left(2\bar{0}21\right)$ plane GaN[42–45] have been realized within 2 years.

As reported in those publications, the key issue for realizing green InGaN LDs is the preparation of In-rich (normally >30%) InGaN QWs with high material quality. The development of green LDs is restricted by two main challenges. One challenge is the luminescence inhomogeneity caused by the rough interface and indium composition fluctuation. Pronounced broadening of spontaneous emission spectra is often observed for green LD structures.[46] Since maximum gain reduces significantly due to inhomogeneous broadening, it is essential to suppress inhomogeneous broadening to realize green laser diodes and to improve their performance. The other challenge is the low thermal stability of In-rich InGaN QWs, which results in the degradation of InGaN QWs during *p*-layer growth and therefore low internal quantum efficiency. We employed two-temperature growth approach for InGaN QWs in green LD structures.[47] We will describe how to improve the interface sharpness and how to eliminate the thermal degradation for InGaN QWs in green LD structures in the following.

15.4.1 Improvement of QW Interface Sharpness

In a two-temperature growth process, InGaN QW tends to locally decompose during temperature ramping up to QB temperature, which results in rough interface and In composition fluctuation. In order to protect InGaN QW from decomposing, a GaN cap layer is grown at the same temperature as InGaN QW. The effect of GaN cap layer thickness will be described later.

Green LD structure samples were grown on *c*-plane sapphire substrates in a commercial Thomas Swan 6 × 2 in. Close Coupled Showerhead (CCS) metalorganic chemical vapor deposition (MOCVD) reactor. The LD structures consisted of a 30 nm nucleation layer, a 5 μm *n*-GaN layer, *n*-AlGaN/GaN SLs cladding layer, an 80 nm lower InGaN waveguide layer, two pairs of InGaN/GaN QWs as the active region, a 70 nm upper InGaN WG, a 20 nm *p*-$Al_{0.2}Ga_{0.8}N$ EBL, *p*-AlGaN/GaN SLs cladding layer, and a heavily doped p+-GaN contact layer. For active layers, a thin GaN layer (LT-cap) was grown succeeding the growth of InGaN quantum well layer at the same growth temperature to protect the QW during subsequent temperature ramping process.

EL full width at half maximum (FWHM) is an indication of the extent of inhomogeneous broadening and potential homogeneity in the active region. EL FWHMs of LD structures with different LT-cap thicknesses are plotted as a function of EL wavelengths, as shown in Figure 15.16a. Each data point represents an LD structure, while the dashed lines plotted along the lower limit of the EL FWHMs are guide lines for the eyes. It shows a tendency that the EL FWHMs of each group of samples increased with emission wavelengths, which is an indication of enhanced potential inhomogeneity as indium composition in the InGaN QWs increases.[46] However, it is noted that EL FWHMs of LDs-12 group of samples are larger than that of LDs-18 and LDs-25 groups, and they show a more pronounced increase with increasing wavelength. As we mentioned previously, the LT-cap layers act to protect InGaN QW layers from decomposition during temperature ramping up. Therefore, we believe that the LT-cap layers with nominal thickness of 1.2 nm in LDs-12 group of LD structures may not be thick enough to protect InGaN QW layers, which results into additional fluctuation of InGaN QW layer thickness and of indium composition caused by indium desorption. It is expected that this effect increases as the indium composition of InGaN QW layers further increases. As a result, the EL FWHMs of LDs-12 group of samples show a pronounced increase in wavelength. On the other hand, the guidelines for EL FWHM lower limit of LDs-18 and LDs-25 groups of samples have similar slopes, which suggests an LT-cap layer with nominal thickness of 1.8 nm and higher can fully protect the InGaN QW layer from temperature ramping.

As shown in Figure 15.16b, we compare the EL FWHMs of our optimized LD structures with FWHM value reported in other works. It should be noted that the measured conditions are different for data from different groups since only limited data are available in the literature. The current density of our measurements is 10 A/cm² under DC condition, while it is 14 A/cm² for semipolar green LEDs reported by UCSB,[48,49] 150 A/cm² for semipolar green LD structures reported by Sumitomo Electric Industries.[50] PL data are used for *c*-plane green LD structures of Osram[51] and Nichia.[52] It shows that our FWHM value for the emission wavelength of 530 nm (2.34 eV) is roughly comparable to that from the other works, except for that which was recently reported by Kyoto University and Nichia for c-plane green LD structure,[52] which is remarkably narrower than all other works although it is PL FWHM for undoped LD structures. It indicates that the potential inhomogeneity of Nichia's LD structure is significantly suppressed. We believe the capping technique used in our LD structures suppresses potential inhomogeneity more related to well thickness fluctuation. However, potential inhomogeneity caused by In composition fluctuation needs to be further reduced. In composition fluctuation may be more related to InGaN QW growth conditions such as growth rate and temperature and so on.

In order to study the effect of LT-cap thickness on the microstructures of InGaN/GaN active region, HAADF-STEM measurements were carried out. The STEM images of InGaN/GaN active layers with LT-cap layer thickness of 2.5 and 1.2 nm are shown in Figure 15.17. For InGaN QW layers covered by 2.5 nm LT-cap layer prior to temperature ramping up, continuous InGaN QWs are observed, as shown in Figure 15.17a and c. The thickness of InGaN QWs is determined to be 2.5 nm. As shown in Figure 15.17b and d, for the LD structure with LT-cap thickness of 1.2 nm, InGaN QWs turn out to be broken, and separated InGaN islands can be observed clearly. The height of these InGaN islands is about 2.5 nm, the same as the intended thickness of InGaN QW layers. It indicates that 1.2 nm thick LT-cap is insufficient to protect InGaN QW from temperature ramping. These island structures may form when temperature is ramped from 700°C to 830°C and growth is interrupted due to absence of triethylgallium (TEGa) or

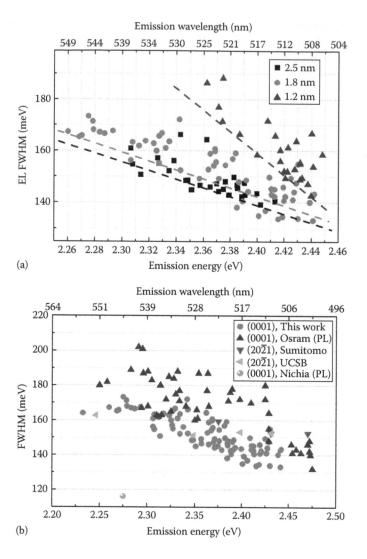

FIGURE 15.16 (a) Wavelength-dependent FWHMs of green LD structures with varying LT-cap thickness. The dashed lines are guide lines for the eyes. (b) Comparison of FWHMs of green LD structures from this and the other works. It should be noted that the measured conditions are different for data from different groups. The current density of our measurements is 10 A/cm² under DC condition, while it is 14 A/cm² for semipolar green LEDs reported by UCSB,[48,49] 150 A/cm² for semipolar green LD structures reported by Sumitomo Electric Industries.[50] PL data are used for c-plane green LD structures of Osram[51] and Nichia.[52]

trimethylindium (TMIn) in the chamber. During this stage, LT-cap layers may locally decompose due to thermal annealing. For the LD structures with 1.2 nm LT-cap layers, the decomposition of LT-cap layers may have exposed InGaN QWs partially and result into desorption of indium from InGaN QWs and fluctuation of InGaN QW thickness. In contrast, for the LD structures with 2.5 nm LT-cap layers, the LT-cap layer was thick enough to protect the InGaN QWs from the temperature ramping, leading to the homogeneous InGaN QWs. Composition analysis indicates that indium composition is highest in the center of the islands and reduces towards peripheral regions. As a result, EL emission broadens due to fluctuation in both InGaN layer thickness and indium composition.

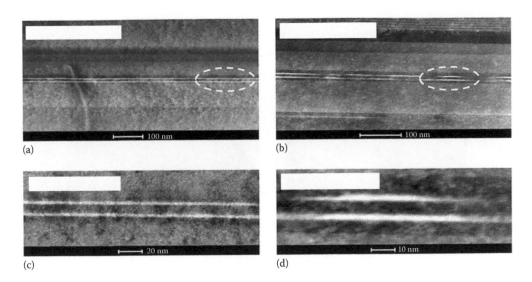

FIGURE 15.17 STEM images of InGaN/GaN active region of green LD structures with varying LT-cap thickness: (a) 2.5 nm LT-cap, (b) 1.2 nm LT-cap. (c) and (d) are magnification of the circled areas in (a) and (b), respectively.

15.4.2 Suppression of Thermal Degradation

In-rich InGaN QWs tend to degrade during the growth of p-type layers, which reduces the optical properties of InGaN QW active region. We investigated the origin of thermal degradation of green InGaN QWs by a combined use of micro-photoluminescence (micro-PL), Z-contrast scanning transmission electron microscopy (STEM), and high-resolution transmission electron microscopy (HRTEM). Based on this investigation, we then obtain green LD structures with high optical quality by suppressing local thermal decomposition of InGaN QWs.

Green LD structure samples were grown on c-plane sapphire substrates in a commercial Thomas Swan 6 × 2 in. CCS MOCVD reactor. The LD structures consisted of a 30 nm nucleation layer, a 5 μm n-GaN layer, n-AlGaN/GaN SLs cladding layer, a 80 nm lower InGaN waveguide layer, two pairs of InGaN/GaN QWs as the active region, an 70 nm upper InGaN WG, an 20 nm p-Al$_{0.2}$Ga$_{0.8}$N EBL, p-AlGaN/GaN SLs cladding layer, and a heavily doped p$^+$-GaN contact layer. For active layers, a thin GaN layer (LT-cap) was grown succeeding the growth of InGaN quantum well layer at the same growth temperature to protect the QW during the subsequent temperature ramping process. For comparison, an additional LED structure that had InGaN/GaN QWs grown under nominally the same conditions as those of LD structures but without AlGaN cladding layers was also grown.

Micro-PL imaging was performed on an inverted Nikon A1 confocal laser scanning microscope with a 405 nm laser source. Z-contrast STEM and HRTEM characterizations were performed on a Gatan Tecnai G2 F20 S-Twin transmission electron microscope. After activation of p-type layers, quick on-wafer electroluminescence (EL) tests were carried out using indium dots as electrodes at a current of 20 mA.

When we extend LD structure emission wavelength from blue to green region, it is found that the EL intensity decreases greatly (data not shown here). It is noted that the green LD epitaxial wafers appear dark color when observed with naked eyes, which is different with green LED epitaxial wafers that appear normal light-yellow color. To investigate the origin of weak emission, micro-PL imaging was performed at room temperature. Figure 15.18a shows the micro-PL image of a typical green LD of this kind (named LD-I thereafter). It has weak emission intensity with a peak wavelength at 504 nm. It is noted that the photoluminescence is very inhomogeneous. Large dark areas with diameters up to 30 μm can be observed. These dark areas correspond to the regions with very weak luminescence in the QWs. We suppose that these dark regions result from the thermal degradation of InGaN/GaN QWs during p-type layers' growth.

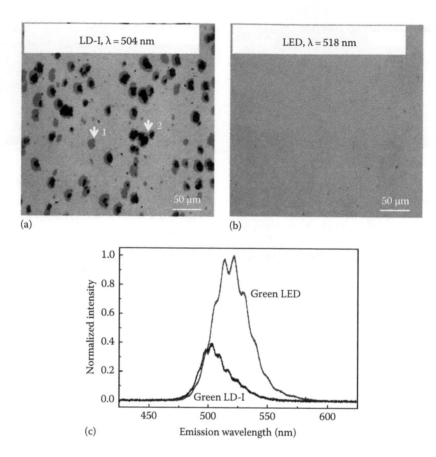

FIGURE 15.18 Micro-PL images of (a) green LD-I sample and (b) a green LED sample. (c) is the normalized on-wafer EL spectra of the LD-I sample and the LED sample. The intensity in (c) is normalized according to the peak intensity of the LED sample. The inset shows the normalized integrated EL intensity of these two samples.

Calculation by Stringfellow shows that InGaN alloy with indium content larger than 6% tend to be meta-stable at a temperature above 800°C.[53] Therefore, InGaN QWs readily decompose when indium content increases to extend to green emission. This problem is particularly severe in laser diode epitaxial growth since longer growth time and higher growth temperature are used to grow p-AlGaN/GaN SL cladding layer. It is also noted that two kinds of dark areas with different gray scales can be distinguished clearly, as marked by arrows 1 and 2 in Figure 15.18a. The different gray scales may indicate that the decomposition in the top and bottom QW does not happen simultaneously. The less dark area marked with "1" in Figure 15.18a may imply that in this region only one InGaN QW layer is degraded, but another QW layer still performs well. For the area marked with 2, both top and bottom InGaN QW layers are damaged resulting in a much weaker emission.

In order to prove that the inhomogeneous luminescence and weak EL intensity result from thermal degradation of InGaN/GaN QWs during p-type layers growth, an LED structure was grown under the same growth conditions for the active region, but the growth temperature for the p-type layer was 30°C lower and the growth time was reduced by a factor of two compared to LD-I sample. The micro-PL image of the LED structure is very homogeneous except for several tiny spots as shown in Figure 15.18b, indicating that the InGaN/GaN QWs have good quality, and thermal degradation hardly happens. EL measurement shows that the LED structure has strong emission with peak wavelength at 518 nm, and its intensity is three times higher than that of the LD-I, as shown in Figure 15.18c. It is noted that the EL peak wavelength of LD-I is 14 nm shorter than that of the LED sample although the same active layer growth conditions were used, which should result from longer growth time and higher growth temperature of *p*-type layers in LD-I. It is well known that the emission efficiency of InGaN/GaN MQWs decreases as emission wavelength increases in the green spectra range due to increased crystalline defects and quantum confinement Stark effect (QCSE) with increasing indium composition.[38,51] However, in our case the EL intensity of the LED structure is three times higher than that of LD-I even though the emission wavelength of LD-I is 14 nm shorter. We believe that the much lower EL intensity is caused by the thermal degradation of the InGaN QWs and the thermal degradation happens during *p*-type layer growth due to too high thermal budget.

We further study the microscopic structures of InGaN QWs in LD-I to confirm that thermal decomposition happens. Z-contrast STEM cross-section images from $(1\bar{1}20)$ plane are examined for LD-I. Figure 15.19a and b shows the Z-contrast STEM images for non-degraded regions and degraded regions, respectively. InGaN QW layers appear as bright parallel lines in STEM images. For non-degraded regions, two InGaN QWs appear as continuous and homogeneous bright lines, as shown in Figure 15.19a. However, the InGaN QWs in the degraded region turn out to be broken, as shown in Figure 15.19b. Bright precipitates and black voids are observed. Away from the precipitates, the contrast between InGaN well and GaN barrier in Figure 15.19b is weaker than that in Figure 15.19a, indicating a decreased composition difference between InGaN QW and GaN barrier layers. The height of the precipitates is much larger than the original QW thickness. They spread from the upper InGaN/GaN interface to the lower GaN barrier, as shown more clearly in the magnified image of Figure 15.19c.

Further investigation on the precipitate formation is performed by HRTEM. The HRTEM image in Figure 15.20a shows the atomic structure of a precipitate, which reveals it is metallic indium within the matrix of nitrides. InGaN QWs decomposition into metallic indium in green LED structures due to high thermal budget has also been reported previously.[54,55] Figure 15.20b shows the HRTEM image of an area near the boundary of a precipitate. In Figure 15.20b, stacking faults (SFs) can be observed as marked by the short arrows. The formation of In–In bonds may be the cause of the SFs, though the formation mechanism of the SFs is not clearly understood yet.[56] Several nanometers away from the precipitate, the typical ABAB stacking order of GaN can be observed. Around the precipitate, an edge dislocation is found as shown in Figure 15.20b marked by the long arrow that can be attributed to the relaxation of metallic indium.

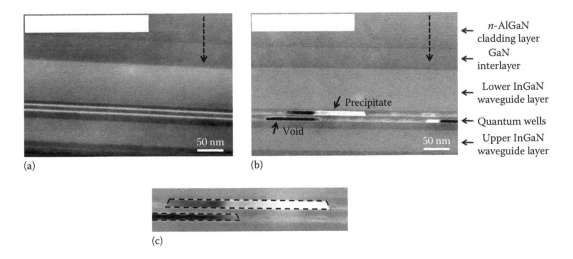

(a) (b) (c)

FIGURE 15.19 Cross-section Z-contrast STEM of LD-I (a) non-degraded region and (b) degraded region. Uniform QWs can be observed in (a), but only broken QW layers as well as precipitates and voids are detected in (b). (c) is a magnification of the precipitate and void region in (b).

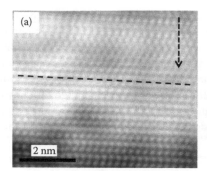
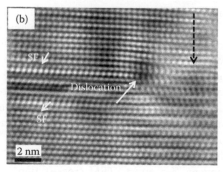

FIGURE 15.20 HRTEM taken from $(1\bar{1}20)$ plane of different regions in LD-I. (a) HRTEM of a precipitate within the matrix of nitrides, revealing the structure of metallic indium cluster. The dotted line shows the upper interface of metallic indium and GaN. (b) HRTEM of an area near the interface of GaN barrier and metallic indium cluster, showing stacking faults and a dislocation (marked by the arrows).

Considering the optical properties and growth conditions of the LED sample and LD-I sample as mentioned earlier, the observed metallic indium in HRTEM should result from the decomposition of InGaN QWs induced by the thermal budget during p-type layers growth. It is expected that InGaN decomposition initiates at local In-rich InGaN clusters.[57] It is well known that In-rich InGaN clusters exist around threading dislocations. Therefore, local InGaN decomposition and resultant dark spots in micro-PL images around threading dislocations have been reported by several groups.[38,51,54] However, in our case, the LD structures and the LED structures have the same threading dislocation density around 5×10^8 cm^{-2} because they are grown on the same n-GaN buffer layers. Therefore, threading dislocations are not the main cause triggering local InGaN decomposition in LD-I. We propose another InGaN degradation mechanism in the following text.

It is noted that the metallic indium precipitates have a truncated pyramid shape with a flat top and faceted sidewall, as shown in Figure 15.19c. Island formation with truncated pyramid shape has been found in the cases of InAs quantum dots grown on GaAs (001) substrate[58,59] and Ge dots on Si substrate.[60] Convex truncated pyramid along the island growth direction is equilibrium shaped as the result of strain relaxation.[60] Assuming that such a case is valid for metallic indium, the fact that the convex edges of the metallic cluster are toward the direction of $\left[00\bar{0}1\right]$ indicates that the growth direction of the metallic cluster is along $\left[00\bar{0}1\right]$. Considering the fact that the longer edges of metallic indium clusters coincide with the InGaN/GaN upper interface, namely, the surface of InGaN well layers, we can deduce that the formation of metallic clusters initiate at this position. For the growth of InGaN/GaN MQWs here, a cap layer is deposited right after InGaN QW growth to protect InGaN QW during temperature ramping up. Therefore, floating indium molecule at the growing surface due to In segregation may lose the chance to evaporate and result in the formation of nanoscale In-rich clusters at the upper interface of InGaN QW and GaN barrier.[61] These In-rich clusters located at InGaN QW upper surface may decompose first and act as the initiation centers of metallic indium phase. This assumption agrees with our TEM observation. Once a new metallic indium cluster appears, the strain relaxation makes the chemical potential of indium atoms in the metallic phase lower than that in normal InGaN alloy phase of QWs. It means that the metallic phase thermodynamically favors indium atoms. The difference in chemical potential drives the diffusion of indium atoms and growing of the metallic clusters, leading to a reduction of indium content in the surrounding InGaN layer. The growth uni-directivity of metallic indium phase along $\left[00\bar{0}1\right]$ GaN is possibly caused by the etching anisotropy of indium to Ga face and N face of GaN.

As mentioned earlier, suppression of local InGaN QW decomposition is critical to develop green LDs. Local InGaN decomposition is caused by thermal budget during p-type layer growth and happens in local

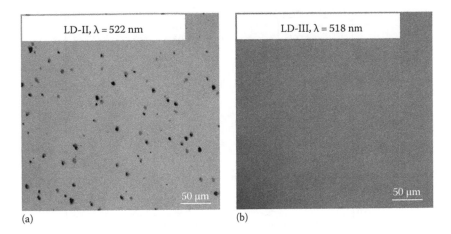

FIGURE 15.21 Micro-PL images of (a) LD-II, with a *p*-type layer growth temperature 30°C lower than that of LD-I, and (b) LD-III, with nominally identical growth parameters as LD-II except for a thinner LT-cap layer.

In-rich InGaN clusters. Therefore, reducing thermal budget and suppressing the formation of In-rich InGaN clusters on the InGaN upper surface are two approaches we have employed to improve the EL emission of green LD epitaxial wafers.

To mitigate the thermal budget in InGaN QWs, the growth temperature of *p*-type layers was lowered by 30°C to grow LD-II sample. Figure 15.21a shows the micro-PL image of LD-II. Compared to LD-I, it is noted that both the size and the density of the dark spots are reduced, indicating that the decomposition of InGaN QW is suppressed to a certain degree, which results from reduced thermal budget to InGaN QWs during *p*-type growth. However, compared to LED structure, LD-II still exhibits a lower EL intensity and worse luminescence homogeneity, which should be attributed to the longer thermal load. Therefore, the decomposition of InGaN QWs needs to be further suppressed. Reducing the growth temperature further is undesirable, since it is very difficult to obtain low resistance *p*-type AlGaN cladding layer at a lower growth temperature.

In order to fully suppress the decomposition of InGaN QWs, preventing the formation of In-rich clusters is critical. Several previous works reported that growth interruption[54] or introduction of H_2 after the growth of InGaN well layer[61,62] can remove In-rich InGaN clusters on the InGaN QW upper interface and enhance the thermal stability of InGaN well layer. However, these approaches may blueshift the emission wavelength greatly, which needs to be avoided when growing green InGaN LDs. Here, we reduce the thickness of the LT cap layer, which is employed to protect InGaN QW layer from evaporation during the subsequent temperature ramping process. A suitably thin LT-cap layer allows a slight evaporation of the InGaN surface without shortening the emission wavelength remarkably. Figure 15.21b shows the micro-PL image of LD-III, which has thinner LT cap layer compared to LD-II. It is noted that the luminescence is extremely homogeneous and the dark regions disappear totally. As a result of suppression of local InGaN decomposition and improvement of luminescence homogeneity, the EL intensity of LD-II and LD-III is greatly enhanced by 110% and 450% compared with LD-I, as shown in Figure 15.22.

In summary, we investigate the origin of thermal degradation of green InGaN QWs by means of micro-PL, Z-contrast STEM, and HRTEM. The InGaN QW decomposition is found to happen during *p*-type layer growth due to too high thermal budget and is the major cause of poor EL intensity and inhomogeneous luminescence of the epitaxial LD structures. The decomposition is suggested to initiate at In-rich InGaN clusters located at the surface of InGaN QW. Metallic indium clusters are generated as a result of the decomposition. Edge dislocation is observed around the metallic indium clusters. Reducing the thermal budget and optimizing InGaN/GaN QW growth suppress the InGaN QW decomposition. Green LD structures with homogeneous luminescence and bright EL intensity are obtained.

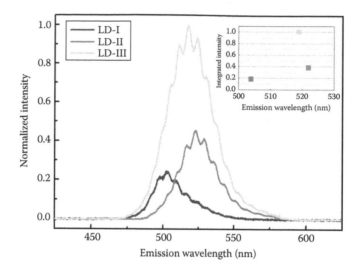

FIGURE 15.22 Comparison of on-wafer EL spectra of samples LD-I, LD-II, and LD-III at 20 mA at room temperature. (The intensity is normalized according to the peak intensity of sample LD-III). The inset shows normalized EL integral intensity of these samples.

15.4.3 Carbon Impurity in *p*-AlGaN:Mg Cladding Layer

As described earlier, in order to suppress the thermal degradation of high indium content in green InGaN/GaN QWs during the growth of *p*-type AlGaN cladding layers,[46,63,64] *p*-type cladding layers are usually grown at a temperature lower than the optimal temperatures, which is higher than 1000°C for AlGaN. However, *p*-type AlGaN grown at a low temperature often shows high resistivity due to increased defects and impurity incorporations.[65–68] Due to the chemical activity of Al, AlGaN layer grown by MOCVD usually contains higher carbon impurity concentration than GaN, especially when the growth temperature is lower than the optimal growth temperatures. Experimental study of the effect of carbon impurity on the electrical properties of AlGaN layer has not been reported.

We first studied the dependence of carbon impurity concentration in AlGaN layers on growth conditions such as growth temperature, pressure, and growth rate. Then Hall measurements were carried out to study the effect of carbon concentration on the electrical conductivity in AlGaN layers with Al content of 0.07. A correlation between carbon concentration and electrical conductivity has clearly been found. By reducing carbon concentration from 2×10^{18} to 5×10^{16} cm^{-3}, the resistivity of *p*-Al$_{0.07}$Ga$_{0.93}$N decreases from 7.4 to 2.2 $\Omega \cdot$ cm. Based on the analysis of charge neutrality equation, we found carbon concentration is close to the compensating donor concentration in the Mg-doped *p*-AlGaN, which suggests that carbon acts as the main compensating donor in the *p*-type AlGaN.

All AlGaN samples were grown on *c*-plane GaN/sapphire templates in an Aixtron 6 × 2 in. CCS MOCVD reactor, using TMGa, trimethyl-aluminum (TMAl), NH$_3$, and bis(cyclopentadienyl)magnesium (Cp$_2$Mg) as the organometallic precursors for Ga, Al, N, and Mg. H$_2$ was used as a carrier gas when growing Mg-doped AlGaN samples.

First, we investigated the effect of growth conditions on carbon impurity incorporation in Al$_{0.15}$Ga$_{0.85}$N layers. SIMS were performed on an Al$_{0.15}$Ga$_{0.85}$N sample with multiple layers, which was grown at varied temperatures, growth rates, and growth pressures. We observed that Al incorporation of AlGaN layers is increased when grown at a lower growth rate and at a lower growth pressure, and therefore, the Al/Ga ratio was changed to keep a constant Al content under different growth conditions. Figure 15.23 shows the carbon concentration dependent on growth temperature, growth pressure, and growth rate, which is consistent with reports in the literature.[66–68] Therefore, if the *p*-type AlGaN cladding layer is grown at a lower

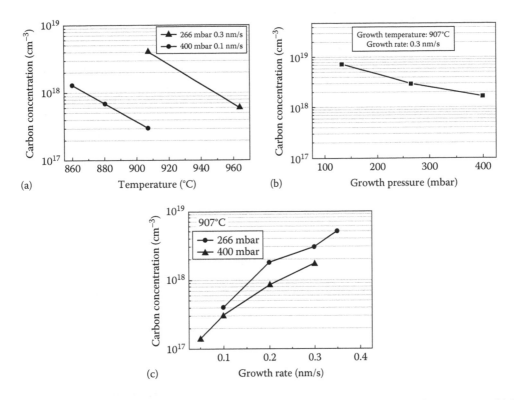

FIGURE 15.23 Dependence of the growth conditions on carbon incorporations. (a) the growth temperature, (b) the growth pressure, and (c) the growth rate.

temperature to avoid the thermal degradation of the active region, the incorporation of carbon impurity will be enhanced. For example, simply reducing the growth temperature of AlGaN:Mg to 907°C leads to an increase of carbon concentration to as high as 4×10^{18} cm^{-3}. However, by adopting a lower growth rate and a higher growth pressure, p-type AlGaN with carbon concentrations less than 1×10^{17} cm^{-3} can be achieved.

The main source of carbon impurity in MOCVD grown p-type AlGaN is the metalorganic precursors. Carbon incorporation can be reduced by the chemical reaction:

$$GaCH_3 + N - H \rightarrow GaN + CH_4$$

It is expected that this chemical reaction can be enhanced when NH_3 partial pressure is higher, which results in a lower incorporation of carbon impurity in the samples. A low growth rate means a high V/III ratio, while a high growth pressure can enhance the crack of NH_3. This may explain that low growth rate and high growth pressure is effective in reducing carbon incorporation.

In order to study the influence of carbon concentration on electrical properties of Mg-doped AlGaN, Hall measurements were carried out on a series of $Al_{0.07}Ga_{0.93}N$:Mg samples with different carbon concentrations. We decreased the Al content to 7% to obtain high-quality AlGaN:Mg samples without lattice relaxation. So the carbon concentrations in these samples were lower than those in the AlGaN samples with Al content of 15%, even under the same growth conditions. The samples consisted of a 30 nm low-temperature grown GaN nucleation layer, a 1 μm unintentionally doped GaN layer, a 4 μm Si-doped n-GaN, a 0.7 μm Mg-doped $Al_{0.07}Ga_{0.93}N$ bulk layer, and a 20 nm heavily Mg-doped p-GaN contact layer. The as-grown samples were activated in nitrogen environment at a temperature of 950°C for 3 min.

These samples were then cut to a size of 10 × 10 mm² and Pd/Pt/Au electrodes were deposited by magnetron sputtering in the four corners to obtain electrical contacts . The contacts were annealed at 550°C for 90 s in nitrogen ambient, and all of them exhibited ohmic properties. During the Hall measurements, the voltage was kept below 3 V to avoid any leakage through the underlying *n*-type GaN layer, and the magnetic field applied was 5800 G. The growth parameters, SIMS results, and room-temperature Hall results of these samples are listed in Figure 15.24 and Table 15.1. Even though the Mg-doping level is a little bit different, all of them were below the self-compensation doping level of Mg (3 × 10¹⁹ cm⁻³) in (Al)GaN.[69-71] According to the SIMS results in Figure 15.24, oxygen and silicon concentrations in all the samples are around 3 × 10¹⁶ and 2 × 10¹⁵ cm⁻³, respectively—both of which are close to the detection limit. Hydrogen concentration after activation was 2.5 × 10¹⁸ cm⁻³, which means most of the Mg–H complexes were decomposed. The SIMS measurement error is within 5%–10%.

By comparing samples A and B as shown in Table 15.1, it can be seen that decreasing the growth temperature of Al₀.₀₇Ga₀.₉₃N:Mg from 964°C to 907°C leads to significant decrease in hole concentration and thus increase of resistivity, although these two samples have almost the same Mg concentration. By increasing the growth pressure to 400 mbar and decreasing the growth rate to 0.05 nm/s, sample D has hole concentration close to that of sample A despite the fact that sample D has a lower Mg concentration. The results indicate that the electrical properties of these samples are not determined by Mg concentration. On the contrary, when the carbon concentration is higher than a certain level in the order of 10¹⁷, there is a clear tendency that the hole concentration decreases and thus the resistivity increases as the carbon concentration increases in the samples, as shown in Figure 15.25.

Carbon impurity in Al(GaN) can substitute for either nitrogen (C_N) or gallium (C_{Ga}) atoms[72-74] due to its amphoteric nature. According to the calculation reported in the reference,[74] the site preference depends on the growth condition of Al(GaN) (relative abundance of Ga and N) and on the Fermi level. In the case of AlGaN:Mg grown at N-rich conditions in this work, carbon impurity should prefer to occupy the Ga(Al) site.

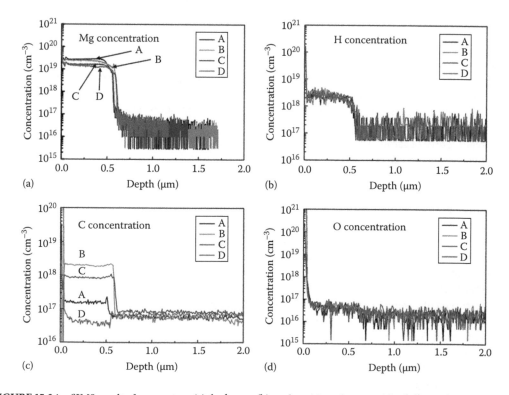

FIGURE 15.24   SIMS result of magnesium (a), hydrogen (b), carbon (c), and oxygen (d) of all samples.

TABLE 15.1 Growth Condition and Hall Result of $Al_{0.07}Ga_{0.93}N$:Mg Samples

Sample	GP[a] (mbar)	GR[b] (nm/s)	GT[c] (°C)	[C][d] (cm⁻³)	[Mg][e] (cm⁻³)	ρ[f] (Ω · cm)	p[g] (cm⁻³)
A	266	0.3	964	2×10^{17}	2.8×10^{19}	2.06	3.5×10^{17}
B	266	0.3	907	2×10^{18}	2.5×10^{19}	7.43	7.5×10^{16}
C	400	0.3	907	9×10^{17}	1.8×10^{19}	4.13	1.2×10^{17}
D	400	0.05	907	5×10^{16}	1.5×10^{19}	2.24	3.1×10^{17}

[a] GP is the growth pressure.
[b] GR is the growth rate.
[c] GT is the growth temperature.
[d] [C] is the carbon concentration.
[e] [Mg] is the Mg-doping level.
[f] ρ is the resistivity.
[g] p is the hole concentration.

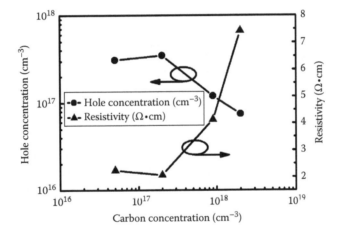

FIGURE 15.25 Hole concentration and resistivity dependent on carbon concentration of $Al_{0.07}Ga_{0.93}N$:Mg samples.

It has been widely reported the $C_{Ga(Al)}$ acts as a shallow donor.[72-76] We believe that the compensation effect of $C_{Ga(Al)}$ as a shallow donor in AlGaN:Mg explains the dependence of the hole concentration and the resistivity on the carbon concentration in our samples, as shown in Figure 15.25. We[77] then analyze the compensating donor concentration in the samples listed in Table 15.1 based on the charge neutrality equation and its relationship with the concentration of carbon impurity in the following.

The relationship of compensating donor concentration, acceptor concentration, and hole concentration in partially compensated semiconductors can be described by the charge neutrality equation[78,79]:

$$\frac{p(p+N_D)}{N_A - N_D - p} = \frac{N_V}{g} \cdot \exp\left(-\frac{E_A}{kT}\right)$$

where N_A, N_D, E_A, N_V, g, and k are acceptor concentration, compensating donor concentration, acceptor ionization energy, effective density of states in the valence band, degeneracy factor and Boltzmann constant, respectively. g is assumed to be equal to 4. N_V can be expressed by

$$N_V = \frac{2\left(2\pi m_h^* kT\right)^{3/2}}{h^3}$$

where h is the Planck constant. N_V for $Al_{0.07}Ga_{0.93}N$ is calculated to be 6.0×10^{19} cm^{-3}, by taking the value of 1.8 for the density-of-state effective mass, m_h^* / m_0, deduced from a linear extrapolation between 1.5 of GaN and 5.4 of AlN, these numbers being calculated as $m_{lh(hh)} = \left(m_{lh(hh)}^{\perp} \cdot m_{lh(hh)}^{\perp} \cdot m_{lh(hh)}^{\parallel} \right)^{1/3}$ and $m_h^* = \left(m_{hh}^{3/2} + m_{lh}^{3/2} \right)^{2/3}$ based on the reported values given in the literature.[20] The value of E_A used in our calculations was chosen to be 194 meV, because it yields an improved agreement between the extracted acceptor concentration and the Mg concentration obtained from SIMS measurement. This value also agrees with that reported by Li et al.[80]

In Figure 15.26, the colored curves were drawn based on the charge neutrality equation at room temperature with different compensating-donor concentrations N_D, which are assumed to be identical to the carbon concentrations in Mg-doped AlGaN measured by SIMS as shown in Table 15.1. The straight lines parallel to the coordinate axis are used to determine the value of acceptor concentration $[N_A]$ in each sample, corresponding to the measured hole concentration, p.

In these samples, Mg concentration is much larger than the concentration of hydrogen, so the residual hydrogen in the samples is thought to exist as the form of Mg–H complex. Thus, the effective value of acceptor concentration $[N_A]$ should be close to the SIMS-determined value of the Mg concentration minus the hydrogen concentration ([Mg]–[H]). Therefore, we use Figure 15.26 to estimate the acceptor concentration $[N_A]$ and then compare the estimated $[N_A]$ with the value of [Mg]–[H]. If these two values are in good agreement, it means our assumption that carbon is the main compensating donor in Mg-doped AlGaN should be justified. As shown in Figure 15.27, a very good agreement was actually confirmed between the estimated acceptor concentration $[N_A]$ and the measured value of [Mg]–[H]. Therefore, the decrease of hole concentration and the increase of electrical resistivity of AlGaN:Mg samples with a high carbon concentration are believed to be due to the compensation effect of carbon impurities, and carbon impurity is the main compensating donor in our Mg-doped AlGaN samples.

In conclusion, growth conditions were explored to suppress carbon impurity incorporation in AlGaN:Mg grown at low temperature. Carbon concentration was reduced from 2×10^{18} to 5×10^{16} cm^{-3} by increasing the growth pressure and decreasing the growth rate. A correlation between carbon concentration and electrical conductivity of $Al_{0.07}Ga_{0.93}N$:Mg has clearly been found. Based on the analysis of charge neutrality equation, we found that carbon concentration is close to the compensating donor concentration in the AlGaN:Mg, which suggests that carbon acts as the main compensating donor in the p-type AlGaN samples grown at low temperature.

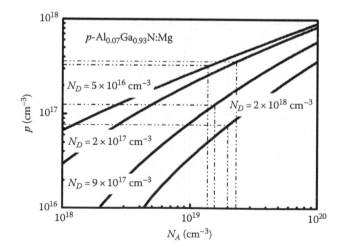

FIGURE 15.26   Hole concentration dependent on N_A with different compensating-donor levels using $N_v/g = 1.5 \times 10^{19}$ cm^{-3}, $E_A = 194$ meV, at 300 K.

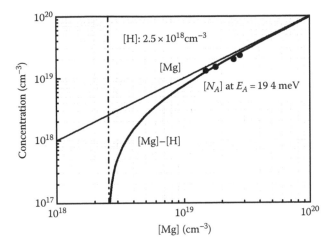

FIGURE 15.27 Comparison of $[N_A]$ and $[Mg]-[H]$, $[N_A]$ is estimated using Figure 15.26 assuming $E_A = 194$ meV, and is identical to $[Mg]-[H]$ assuming.

15.4.4 Device Characteristics of Green LDs

After the investigation mentioned earlier, green LD structure was then grown on bulk GaN substrate and fabricated into ridge waveguide LD chips. Ridge-guided lasers with stripes of 10 μm width and 800 μm length were fabricated by conventional lithography and lift-off technique. The 800 μm long cavities and mirror facets were formed by cleaving, and then rear and front facets were coated with dielectric films with reflectivity of 90% and 70%, respectively. Figure 15.28 shows the power–current–voltage curve of a typical green LD under continuous-wave operation at room temperature. It can be shown that the threshold current density and the threshold voltage are 1.8 kA/cm² and 4.4 V, respectively. The output power is 58 mW at 6 kA/cm². It lases at 508 nm as shown in Figure 15.29.

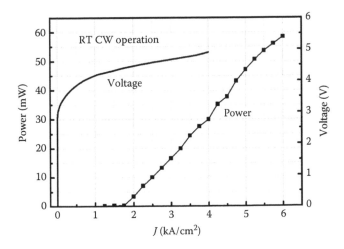

FIGURE 15.28 Power–current–voltage curve of a typical green LD under continuous-wave operation at room temperature.

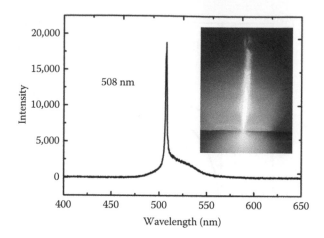

FIGURE 15.29   Lasing spectrum and photograph of the laser beam of the 508 nm LD.

15.5 Conclusion

We have carried out extensive studies on GaN-based laser diodes on bulk GaN substrates, including homoepitaxial growth of GaN layers on GaN substrates, LD layer structure design and doping optimization, epitaxial growth of InGaN QWs, and effect of unintentional carbon incorporation in AlGaN:Mg.

Appropriate miscut angle and direction of GaN substrates are critical to obtain smooth homoepitaxial GaN layers. LD layer structure design and doping optimization to reduce the overlap between optical field and *p*-layers have been carried out to reduce internal loss in LD structures. Thermal degradation of green InGaN QWs is a major challenge for green LDs. Reducing thermal budget and suppressing In segregation in InGaN/GaN QWs can suppress the InGaN QW degradation, which greatly improves the quantum efficiency of green InGaN QWs. Unintentional carbon impurity concentration increases significantly in *p*-type AlGaN grown at low temperature. Carbon impurity acts as a compensating donor in the *p*-type AlGaN, which results in increasing resistivity.

Based on these studies, we have demonstrated blue LDs with output power of 1.3 W under continuous wave operation and green LDs with output 60 mW under continuous wave operation.

Acknowledgments

This work was supported by the National Natural Science Foundation of China (Grant Nos. 61574160, 61334005), the National High Technology Research and Development Program of China (863 Program) (Grant Nos. 2013AA030502), the Science and Technology Support Project of Jiangsu Province (Grant No. BE2013007), the Strategic Priority Research Program of the Chinese Academy of Science (Grant No. XDA09020401), and Natural Science Foundation of Jiangsu province (Grant No. BK20130362).

References

1. I. Akasaki, H. Amano, S. Sota, H. Sakai, T. Tanaka, and M. Koike, *Jpn. J. Appl. Phys.* 34, L1517 (1995).
2. S. Nakamura, M. Senoh, S. Nagaham, N. Iwasa, T. Yamada, T. Matsushita, H. Kryoku, and Y. Sugimoto, *Jpn. J. Appl. Phys.* 35, L74–L76 (1996).
3. S. Nakamura, M. Senoh, S. Nagahama et al., *Appl. Phys. Lett.* 72(2), 211 (January 12, 1998).
4. S. Nakamura, M. Senoh, S. Nagahama et al., *Appl. Phys. Lett.* 72(16), 2014 (April 20, 1998).
5. A. Michiue, T. Miyoshi, T. Kozaki, T. Yanamoto, S. Nagahama, and T. Mukai, *IEICE Trans. Electron.* E92-C, 194 (2009).

6. C. Vierheilig, C. Eichler, S. Tautz, A. Lell, J. Muller, F. Kopp, J. Ristic, and U. Strauss, *Proc. SPIE* 8277, 82770K-1 (2012).

7. B. Heying, X.H. Wu, S. Keller, Y. Li, D. Kapolnek, B.P. Keller, S.P. DenBaars, and J.S. Speck, *Appl. Phys. Lett.* 68, 643 (1996).

8. X.Q. Shen, M. Shimizu, H. Okumura, F.J. Xu, B. Shen, and G.Y. Zhang, *J. Crystal. Growth* 311, 2049 (2009).

9. K. Zhou, J.P. Liu, S.M. Zhang et al., *J. Crystal Growth* 371, 7–10 (2013).

10. S.-N. Lee, H.S. Paeka, J.K. Sona, T. Sakonga, E. Yoonb, O.H. Nama, Y. Parka, *Phys. B* 376, 532 (2006).

11. M. Sarzynski, M. Leszczynski, M. Krysko, J.Z. Domagala, R. Czernecki, and T. Suski, *Cryst. Res. Technol.* 47, 321 (2012).

12. A.R.A. Zauner, J.L. Weyher, M. Plomp, V. Kirilyuk, I. Grzegory, W.J.P. van Enckevort, J.J. Schermer, P.R. Hageman, and P.K. Larsen, *J. Crystal Growth* 210, 435 (2000).

13. F.C. Frank, *Discuss. Faraday Soc.* 5, 67 (1949).

14. W.K. Burton, N. Cabrera, and F. C. Frank, *Philos. Trans. R. Soc. Lond. Ser. A* 243, 299 (1951).

15. B. Heying, E.J. Tarsa, C.R. Elsass, P. Fini, S.P. DenBaars, and J.S. Speck, *J. Appl. Phys.* 85, 6470 (1999).

16. N. Cabrera and D.A. Vermilyea, in *Growth and Perfection of Crystals*, eds. R.H. Doremus, B.W. Roberts, and V. Turnbull, John Wiley & Sons, New York (1958), pp. 393–410.

17. J.B. Hannon, V.B. Shenoy, and K.W. Schwarz, *Science* 313, 1266 (2006).

18. I. Sunagawa and P. Bennema, *J. Cryst. Growth* 53, 490 (1981).

19. J. Smalc-Koziorowska, S. Grzanka, E. Litwin-Staszewska, R. Piotrzkowski, G. Nowak, M. Leszczynski, P. Perlin, E. Talik, J. Kozubowski, and S. Krukowski, *Solid-State Electron.* 54, 701 (2010).

20. K. Tachibana, H. Nago, and S.-y. Nunoue, *Phys. Status Solidi C* 3, 1819 (2006).

21. T. Asano, M. Takeya, T. Tojyo, T. Mizuno, S. Ikeda, K. Shibuya, T. Hino, S. Uchida, and M. Ikeda, *Appl. Phys. Lett.* 80, 3497–3499 (2002).

22. T. Asano, T. Tojyo, T. Mizuno, M. Takeya, S. Ikeda, K. Shibuya, T. Hino, S. Uchida, and M. Ikeda, *IEEE J. Quantum Electron.* 39(1), 135–140 (January 2003).

23. S.N. Lee, S.Y. Cho, H.Y. Ryu et al., *Appl. Phys. Lett.* 88, 111101-1–111101-3 (2006).

24. S.N. Lee, J.K. Son, H.S. Paek, Y.J. Sung, K.S. Kim, H.K. Kim, H. Kim, T. Sakong, Y. Park, K.H. Ha, and O.H. Nam, *Appl. Phys. Lett.* 93, 091109-1–091109-3 (2008).

25. K.S. Kim, J.K. Son, S.N. Lee et al., *Appl. Phys. Lett.* 92, 101103-1–101103-3 (2008).

26. H.Y. Ryu, K.H. Ha, A.N. Lee et al., *IEEE Photon. Technol. Lett.* 19(21), 1717–1719 (November 2007).

27. W.W. Chow and M. Kneissl, *J. Appl. Phys.* 98, 114502-1–114502-6 (2005).

28. H. Zhao, R. A. Arif, Y.-K. Ee, and N. Tansu, *IEEE J. Quantum Electron.* 45(1), 66–78 (January 2009).

29. S. Nakamura, M. Senoh, S. Nagaham, N. Iwasa, T. Matsushita, and T. Mukai, *Appl. Phys. Lett.* 76, 22–24 (2000).

30. L.Q. Zhang, D.S. Jiang, J.J. Zhu, D.G. Zhao, Z.S. Liu, S.M. Zhang, and H. Yang, *J. Appl. Phys.* 105, 023104-1–023104-8 (2009).

31. M. Kuramoto, C. Sasaoka, N. Futagawa, M. Nido, and A.A. Yamaguchi, *Phys. Stat. Sol. (A)* 192, 329–334 (2002).

32. W. Gotz, N.M. Johnson, J. Walker, D.P. Bour, and R.A. Street, *Appl. Phys. Lett.* 68, 667–669 (1996).

33. T. Miyoshi, T. Yanamoto, T. Kozaki, S. Nagahama, Y. Narukawa, M. Sano, T. Yamada, and T. Mukai, *Proc. SPIE* 6984, 698414 (2008).

34. A. Avramescu, T. Lermer, J. Müller, C. Eichler, G. Bruederl, M. Sabathil, S. Lutgen, and U. Strauss. *Appl. Phys. Express* 3, 061003 (2010).

35. S. Lutgen, A. Avramescu, T. Lermer, D. Queren, J. Müller, G. Bruederl, and U. Strauss, *Phys. Status Solidi A* 207, 1318 (2010).

36. J. Müller, U. Strauß, T. Lermer, G. Brüderl, C. Eichler, A. Avramescu, and S. Lutgen, *Phys. Status Solidi A* 208, 1590 (2011).

37. D. Queren, A. Avramescu, G. Brüderl, A. Breidenassel, M. Schillgalies, S. Lutgen, and U. Strauß, *Appl. Phys. Lett.* 94, 081119 (2009).

38. U. Strauß, A. Avramescu, T. Lermer, D. Queren, A. Gomez-Iglesias, C. Eichler, J. Müller, G. Brüderl, and S. Lutgen, *Phys. Status Solidi B* 248, 652 (2011).
39. T. Miyoshi, S. Masui, T. Okada, T. Yanamoto, T. Kozaki, S. Nagahama, and T. Mukai, *Appl. Phys. Express* 2, 062201 (2009).
40. T. Miyoshi, S. Masui, T. Okada, T. Yanamoto, T. Kozaki, S. Nagahama, and T. Mukai, *Phys. Status Solidi A* 207, 1389 (2010).
41. D. Sizov, R. Bhat, A. Heberle, N. Visovsky, and C. Zah, *Appl. Phys. Lett.* 99, 041117 (2011).
42. C.Y. Huang, M.T. Hardy, K. Fujito, D.F. Feezell, J.S. Speck, S.P. DenBaars, and S. Nakamura, *Appl. Phys. Lett.* 99, 241115 (2011).
43. Y. Zhao, S. Tanaka, Q. Yan et al., *Appl. Phys. Lett.* 99, 051109 (2011).
44. S. Takagi, Y. Enya, T. Kyono et al., *Appl. Phys. Express* 5, 082102 (2012).
45. K. Yanashima, H. Nakajima, K. Tasai et al., *Appl. Phys. Express* 5, 082103 (2012).
46. D. Queren, M. Schillgalies, A. Avramescu, G. Bruderl, A. Laubsch, S. Lutgen, and U. Strauß, *J. Cryst. Growth* 311, 2933 (2009).
47. J.P. Liu, Z.C. Li, L.Q. Zhang et al., *Appl. Phys. Express* 7, 111001 (2014).
48. S. Yamamoto, Y. Zhao, C.C. Pan, R.B. Chung, K. Fujito, J. Sonoda, S.P. DenBaars, and S. Nakamura, *Appl. Phys. Express* 3, 122102 (2010).
49. R.B. Chung, Y.D. Lin, I. Koslow, N. Pfaff, H. Ohta, J. Ha, S.P. DenBaars, and S. Nakamura, *Jpn. J. Appl. Phys.* 49, 070203 (2010).
50. M. Adachi, *Jpn. J. Appl. Phys.* 53, 100207 (2014).
51. D. Queren, A. Avramescu, M. Schillgalies, M. Peter, T. Meyer, G. Brüderl, S. Lutgen, and U. Strauß, *Phys. Status Solidi C* 6, S826 (2009).
52. M. Funato, Y.S. Kim, T. Hira, A. Kaneta, Y. Kawakami, T. Miyoshi, and S. Nagahama, *Appl. Phys. Express* 6, 111002 (2013).
53. I. Ho and G.B. Stringfellow, *Appl. Phys. Lett.* 69, 2701 (1996).
54. B. Van Daele, G. Van Tendeloo, K. Jacobs, I. Moerman, and M.R. Leys, *Appl. Phys. Lett.* 85, 4379 (2004).
55. H.K. Cho, J.Y. Lee, C.S. Kim, and G.M. Yang, *J. Electron. Mater.* 30, 1348 (2001).
56. S. Tomiya, O. Goto, and M. Ikeda, *Proc. IEEE* 98, 1208 (2010).
57. J.P. Liu, Y.T. Wang, H. Yang, D.S. Jiang, U. Jahn, and K.H. Ploog, *Appl. Phys. Lett.* 84, 5449 (2004).
58. L.G. Wang, P. Kratzer, N. Moll, and M. Scheffler, *Phys. Rev. B* 62, 1897 (2000).
59. N. Moll and M. Scheffler, *Phys. Rev. B* 58, 4566 (1998).
60. Z.M. Wang, *Self-Assembled Quantum Dots*, Springer, New York (2008), p. 432.
61. Y.T. Moon, D.J. Kim, K.M. Song, C.J. Choi, S.H. Han, T.Y. Seong, and S.-J. Park, *J. Appl. Phys.* 89, 6514 (2001).
62. S. Suihkonen, O. Svensk, T. Lang, H. Lipsanen, M. A. Odnoblyudov, and V. E. Bougrov, *J. Cryst. Growth* 298, 740 (2007).
63. Z.C. Li, J.P. Liu, M.X. Feng et al., *Appl. Phys. Lett.* 103(15), 152109 (2013).
64. S. Nagahama, T. Yanamoto, M. Sano, and T. Mukai, *Jpn. J. Appl. Phys.* Part 1—Regular Papers Short Notes & Review Papers 40(5A), 3075 (2001).
65. J. Yang, D.G. Zhao, D.S. Jiang et al., *J. Appl. Phys.* 115(16), 163704 (2014).
66. D.D. Koleske, A.E. Wickenden, R.L. Henry, and M.E. Twigg, *J. Cryst. Growth* 242(1), 55 (2002).
67. G. Parish, S. Keller, S.P. Denbaars, and U.K. Mishra, *J. Electron. Mater.* 29(1), 15 (2000).
68. J.T. Chen, U. Forsberg, and E. Janzén, *Appl. Phys. Lett.* 102(19), 193506 (2013).
69. H. Obloh, K.H. Bachem, U. Kaufmann, M. Kunzer, M. Maier, A. Ramakrishnan, and P. Schlotter, *J. Cryst. Growth* 195(1), 270 (1998).
70. A. Castiglia, J.F. Carlin, and N. Grandjean, *Appl. Phys. Lett.* 98(21), 213505 (2011).
71. M. Suzuki, J. Nishio, M. Onomura, and C. Hongo, *J. Cryst. Growth* 189, 511 (1998).
72. A.F. Wright, *J.Appl. Phys.* 92(5), 2575 (2002).
73. P. Bogusławski and J. Bernholc, *Phys. Rev. B* 56(15), 9496 (1997).

74. J.L. Lyons, A. Janotti, and C.G. Van de Walle, *Appl. Phys. Lett.* 97(15), 152108 (2010).

75. C.H. Seager, A.F. Wright, J. Yu, and W. Götz, *J. Appl. Phys.* 92(11), 6553 (2002).

76. A.Q. Tian, J.P. Liu, M. Ikeda et al., *Appl. Phys. Express* 8, 051001 (2015).

77. D. Demchenko, I. Diallo, and M. Reshchikov, *Phys. Rev. Lett.* 110, 087404 (2013).

78. T. Tanaka, A. Watanabe, H. Amano, Y. Kobayashi, I. Akasaki, S. Yamazaki, and M. Koike, *Appl. Phys. Lett.* 65(5), 593 (1994).

79. P. Kozodoy, H. Xing, S.P. DenBaars, U.K. Mishra, A. Saxler, R. Perrin, S. Elhamri, and W.C. Mitchel, *J. Appl. Phys.* 87(4), 1832 (2000).

80. J. Li, T.N. Oder, M.L. Nakarmi, J.Y. Lin, and H.X. Jiang, *Appl. Phys. Lett.* 80(7), 1210 (2002).

V

Nano and Other Types of LEDs

16

Photonic Crystal Light-Emitting Diodes by Nanosphere Lithography

Kwai Hei Li
*Department of Electrical
and Electronic Engineering*

Hoi Wai Choi
*Department of Electrical
and Electronic Engineering*

Abstract The group of III–V semiconductors is emerging as highly attractive materials for a wide range of applications, particularly the gallium nitride (GaN) family of alloys. The development of nitride-based light-emitting diodes (LEDs) represented a quantum leap in the advancement of optoelectronics. The timely arrival of InGaN blue LEDs enables full-color mixing with existing red and green LEDs based on AlInGaP and GaP alloys, respectively, promoting the progress of solid-state lighting and displays. Due to total internal reflection at the GaN–ambience interface with a high refractive index contrast, low extraction efficiency is one of the major bottlenecks for LEDs. Extensive research efforts have been conducted on producing energy-efficient and highly reliable LEDs in the past decade. Among potential strategies, nanotechnology promises to offer significant boosts to device performance. Nanostructure on a scale of wavelength of light exhibits prominent effects on the propagation behavior of photons. However, the formation of well-defined nanostructure relies heavily on processing techniques. Although electron beam lithography enables precise direct writing of nanopatterns, high equipment cost and time-consuming processes make mass production impractical. On the other hand, the technique of nanosphere lithography (NSL) as adopted in the works reported in this chapter is a practical alternative approach. Uniform spheres acting as etch masks are capable of self-assembling into hexagonal closed-packed arrays. The resultant nanopillar array serves as the photonic crystals (PhCs) extracting guided light. The feature dimensions of the resultant patterns are scalable according to the diameter of nanospheres used. Such ordered closed-packed arrays are capable of promoting light extraction via the dispersion and diffraction properties of weak PhCs. To extend the functionality of sphere-patterned arrays, a dimension-adjusting procedure is developed to realize strong PhC structures. Finite spacing between individual spheres is introduced, resulting in strong air-spaced nanopillar PhCs structure with a wavelength-tunable photonic bandgap (PBG). Distinguished from typical PhCs in the form of air holes or pillars, a clover-shaped structure with a wide PBG is fabricated by dual-step NSL. The PBG structures have been exploited for suppressing lateral wave guiding and possibly redirecting a significant proportion of trapped photons for extraction. Additionally, the PhC is able to enhance light extraction in the visible and in the infrared simultaneously, so the devices emit more light while radiating more heat, a feature that is beneficial to overall efficiencies of the emitters.

16.1 Introduction

Undoubtedly, the development of nitride-based light-emitting diodes (LEDs) represented a quantum leap in the advancement of optoelectronics. Although gallium nitride (GaN) LEDs are on the way to replace light bulbs and fluorescent tubes, most of the emitted light suffers from total internal reflection (TIR) and is trapped inside the device, resulting in extremely low light extraction efficiency. Numerous approaches have been adopted to efficiently extract light from devices and prevent unwanted guiding modes, such as geometrical shaping [1–3], micro-LEDs [4–6], and surface roughening [7–10]. These methods rely on the formation of nonparallel surfaces to minimize reflections, albeit at different dimensional scales. In particular, two-dimensional (2D) photonic crystals (PhCs) [11,12], typically in the form of arrays of air-holes or pillars inscribed on the light-emitting surfaces, have been demonstrated to have the capability of redirecting photons in guided modes into free space, offering better control over directionality of light emission than random surface texturing.

As the characteristic length scales of PhC structures are of the order of the wavelength, nanopatterning techniques are involved during the fabrication of visible light devices, often increasing manufacturing costs. Thus, the development of processes feasible and suitable for mass production of nanoscale features for the III-nitride optoelectronic devices is definitely a task of high priority. The research works in this chapter are based on a novel nanostructuring technique—nanosphere lithography (NSL)—to pattern various nanostructures on nitride semiconductors. These regularly patterned structures are of great technological importance, enabling us to manipulate light at subwavelength dimensions through nanophotonic effects. Compared with other possible top-down approaches for fabricating III-nitride nanostructures such as electron beam evaporation and focused ion beam, NSL has its predominance in efficiently forming arrays across large areas. Moreover, NSL overcomes resolution issues arising from diffraction limit in optical lithography and even limitation of beam spot in electron beam lithography.

In this chapter, the main focus is placed on the fabrication and optical properties of nanosphere-patterned PhCs, leading to the enhancement of LED efficiency. Following an in-depth introduction of the NSL process, the methods of coating and the subsequent fabrication flow are highlighted. Specifically, five PhC structures fabricated by NSL combined with micro-fabrication technologies are described. The optical properties of these sphere-patterned PhCs will be discussed in detail, based on diverse geometries and light extraction mechanisms.

16.2 Strategies for Light Extraction

The growing demand for blue light LEDs has prompted for devices with maximal external quantum efficiency (EQE), which is determined by both internal quantum efficiency (IQE) and LEE. With the rapid and massive improvements in growth techniques, epitaxial structures, and crystal quality, the IQE has been greatly enhanced to more than 80% [13]. However, the extremely low extraction efficiency (<10%)

is still one of the major bottlenecks restricting the performance of LEDs [14], attributed to absorption of substrate, current spreading layer, ohmic contacts and bonding wire, as well as the main challenge of TIR, thus implying that there is still plenty of room for improving the LEE. In the following sections, the influence of TIR is discussed. Numerous approaches aiming to extract optically guided light from devices and suppress TIR are highlighted, including surface roughening, micro-LEDs, geometrical shaping, and PhCs. These methods rely on the formation of nonparallel surfaces to minimize reflections and reduce reabsorption loss, albeit at different dimensional scales.

16.2.1 Overview

Owing to high refractive index contrast at the semiconductor/air interface, the majority of photons emitted from the active region remain trapped, as depicted in Figure 16.1. The light-trapping phenomenon is known as total internal reflection (TIR), which strictly limits the LEE of LED. According to Snell's law, TIR occurs when light rays strike on the flat-top semiconductor/air interface with the incident angle greater than the critical angle: $\theta_c = \sin^{-1}(n_{air}/n_{sc})$, where n_{air} and n_{sc} are the refractive indexes of air and semiconductor, respectively. Photons outside of the escape angle are likely to be reabsorbed after multiple reflections. In particular, the refractive index is about 2.45 for III-nitride semiconductor and the light extraction angle (escape angle or escape cone) is about 23.5°. TIR resulting from the narrow escape cone prevents the photons from escaping from the semiconductor. The flat-top emission surface of conventional LEDs is found to be as low as 4% [15], while the overall extraction efficiency is strictly limited to around 12%.

A popular, cost-effective, and practical approach is to roughen the surface of the LED chip via natural chemical etching. Common roughening techniques, including photo-electrochemical etching and wet etching, are capable of developing high-density randomly oriented miniature facets/features on the LED surface [8–10], as shown in Figure 16.2. The processed surface can randomize the path of trapping and significantly increase the probability for light striking the boundary at an angle close to normal. Surface roughing techniques can possibly produce about a factor of enhancement of the light output power and effectively scatter the trapped light outside the LED devices.

When emitted light incidents upon the boundary at an angle greater than the critical angle, it suffers TIR and become laterally guided modes. The trapped photons would either be reabsorbed by active region or escape through the edges of LED. To increase the changes for light to be extracted into free space before reabsorption, micro-LEDs provide additional photon escape pathways through the peripheries of microstructures [4,6]. The interconnected microstructures are generally formed by the dry etching process so as to remove the material between the microelements, thus significantly increasing the exposed sidewalls [5]. The enhancement of light extraction is attributed to the increased surface area, especially the etched sidewall, and the charge-coupled device (CCD) images shown in Figure 16.3 clearly indicate that higher brightness is observed along the edges of microstructures.

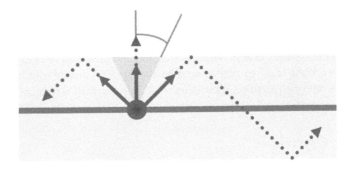

FIGURE 16.1 Depicts how the emitted light remains trapped within the device.

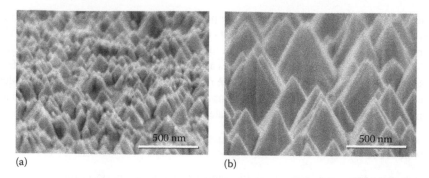

(a)   (b)

FIGURE 16.2 SEM micrographs of an n-face GaN surface etched by a KOH-based photoelectrochemical (PEC) method. (a) 2-min etching and (b) 10-min etching. (Reprinted with permission from Fujii, T., Gao, Y., Sharma, R., Hu, E.L., DenBaars, S.P., and Nakamura, S., *Appl. Phys. Lett.*, 84(6), 855. Copyright 2004, American Institute of Physics.)

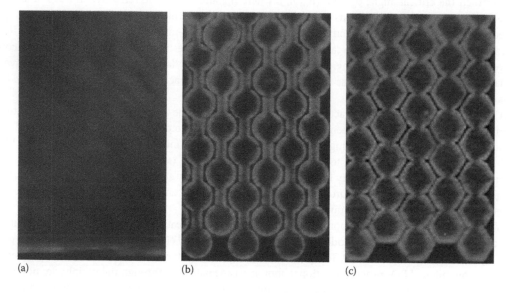

(a)   (b)   (c)

FIGURE 16.3 Optical microphotographs showing emission regions of the (a) large area, (b) micro-disk, and (c) micro-hexagon LEDs. (Reprinted with permission from Li, Z.L., Li, K.H., and Choi, H.W., *J. Appl. Phys.*, 108(11). Copyright 2010, American Institute of Physics.)

For a conventional LED with cubical geometry, a light ray reflected from one face is likely to hit another parallel facet and bounce around inside the LED chip until it is reabsorbed. One way to overcome this problem is to change the shape of the LED die by creating beveled sidewalls such that the facet pairs are no longer parallel and possibly alter the propagation direction of the reflected light [2]. Kao, C.C. et al., demonstrated a 70% enhancement in light output power for a nitride-based LED with 22° undercut side-wall LED [1]. Moreover, Wang et al. reported various polygonal LEDs shaped with laser micromachining and proved that LEDs with polygonal geometries increase light extraction compared with conventional rectangle LEDs [3] (Figure 16.4).

16.2.2 Photonic Crystals

Photonic crystals (PhCs) [11,12], with unique capabilities of being able to control and manipulate the propagation of light, have been widely adopted in diverse optoelectronic and photonic applications,

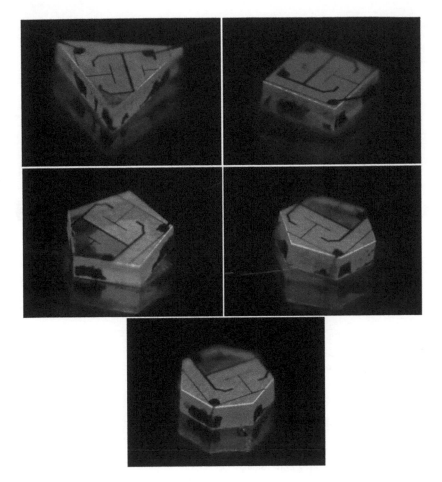

FIGURE 16.4 Optical microscopy images of polygonal LED chips. (Reprinted with permission from Wang, X.H., Lai, P.T., and Choi, H.W., *J. Appl. Phys.*, 108(2). Copyright 2010, American Institute of Physics.)

including laser resonant cavities [17], high-speed optical fiber transmission [18], and polarization filtering [19]. The incorporation of 2D PhCs onto the surfaces of nitride-based LEDs enables strong interaction of the guided modes and has also been demonstrated to effectively promote the LEE [20,21]. Such ordered periodic nanostructures are typically in the form of arrays of recessed air holes or protruding pillars, as shown in Figure 16.5. Because of their ability of manipulating spontaneous emission, the 2D PhCs can be extremely useful for extracting guided modes to air, thus enlarging the escape cone.

The light extraction behaviors of 2D PhCs can be explained by the dispersion diagrams showing normalized frequency versus in-plane wave vector. As illustrated in Figure 16.6, the green solid line in the band structure, namely, the light line, represents a dividing line between guided and leaky modes. In the presence of PhCs, band folding will occur at the Brillouin zone edges and guided modes are folded above the air light line, meaning that the guided modes can radiate out from the device. The radiative modes located at the region above the light line corresponds to leaky modes, in which the optical mode leaks energy into the surrounding air as it propagates down the waveguide. For the frequency bands that are below the light line, they are the guided modes and do not leak energy as they propagate. Therefore, the light extraction enhancement originates from the coupling of leaky modes above the light line of the band structure. PhCs can also act as 2D diffraction gratings in slabs to extract guided modes to the air and to redirect the emissions.

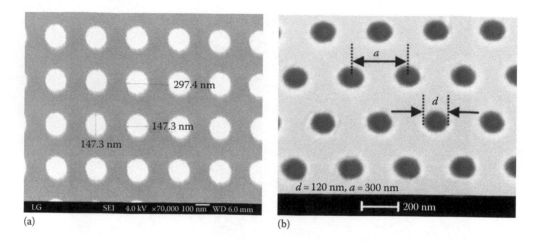

(a)   (b)

FIGURE 16.5 Cubic array of pillars by nanoimprint (a) and hexagonal air-hole array patterned by electron beam lithography (b). (Reprinted with permission from Oder, T.N., Shakya, J., Lin, J.Y., and Jiang, H.X., *Appl. Phys. Lett.*, 83(6), 1231. Copyright 2003, American Institute of Physics.)

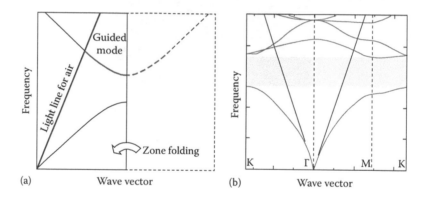

(a)   Wave vector   (b)   Wave vector

FIGURE 16.6 Band diagram illustrating the band-folding effect (a) and band structure of a well-designed PhC (b).

With a well-defined periodic arrangement and with sufficiently large refractive index contrast between GaN and the ambience, a photonic bandgap (PBG) may be established, which forbids the propagation of light within a specific range of frequencies dependent on the dimension and pitch of the array, as illustrated in Figure 16.6. The PBG can thus be exploited for suppressing lateral wave guiding and possibly redirect a significant proportion of trapped photons for extraction, overcoming one of the major limitations of nitride LEDs.

16.3 Nanosphere Lithography

As the characteristic length scale of PhCs structures is of the order of the wavelength or subwavelength, nanopatterning techniques are involved during the fabrication of short-wavelength optoelectronics devices, often increasing manufacturing costs. Unlike conventional AlInGaP emitters, which can be processed by standard microlithographic techniques, the PhCs require further dimensional shrinkage in order to fulfill constraints associated with short-wavelength blue/ultraviolet (UV) light interaction, so traditional optical patterning techniques are no longer able to offer the required resolutions due to the diffraction limit. While direct-write techniques such as electron beam lithography and focused ion beam milling are capable of producing arbitrary 2D feature accurately down to the nanometer scale, they each also have their own drawbacks. High equipment cost, time-consuming point-by-point processing, and thus low throughput make large-volume manufacturing impractical. To overcome such limitations, a high-throughput yet low-cost approach is introduced that is particularly suitable for the processing of hard nitride semiconductors: NSL.

16.3.1 Overview

Nanosphere lithography (NSL) [22] is an inexpensive and ultra-efficient nanopatterning technique with attractive abilities of producing well-ordered periodic arrays over large areas with minimal processing time. With wide ranges of commercially available sphere dimensions from hundreds of micrometers down to tens of nanometers, the spheres can self-assemble into mono- and multilayers of periodically ordered arrays and are sparsely distributed on the sample surface, as illustrated in Figure 16.7. Nanospheres have previously found uses in technologies such as catalysis, biochemical devices, cell cultures, surface-enhanced Raman spectroscopy, and sensing applications. In the field of optoelectronics, ordered periodic structures based on nanospheres may be used as PhCs to manipulate the flow of light; in particular, three-dimensional (3D) PhCs [23–24], which can interact with light in both the vertical and lateral directions. In this chapter, we shall mainly focus on lithographic processes for generating 2D monolayers of silica spheres, which serve as high etch selectivity masks for GaN materials during dry etch, with the target of producing various 2D regular nanostructure arrays on the surface of the wafer. We shall first discuss the fundamental steps in the formation of a self-assembled ordered monolayer in a hexagonal closed-packed (HCP) arrangement for pattern transfer.

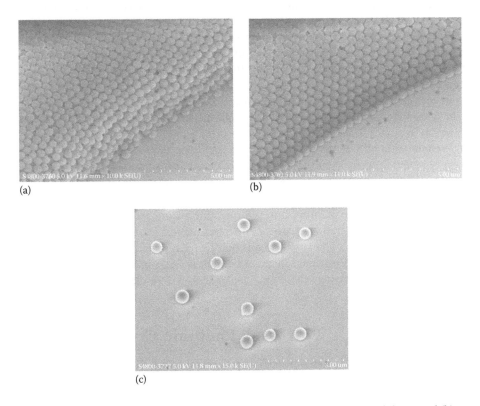

(a) (b)

(c)

FIGURE 16.7 Field-emission scanning electron microscope (FE-SEM) images of (a) multilayers and (b) monolayer of spheres and (c) sparsely distributed spheres.

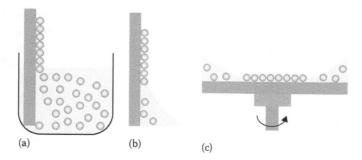

FIGURE 16.8 Schematic diagrams showing various nanosphere coating procedures, including (a) vertical deposition, (b) dip coating, and (c) spin coating.

16.3.2 Process Development

The self-assembly of an ordered HCP monolayer of spheres over large areas can be readily achieved by one of several efficient and cost-effective methods, including spin coating, vertical deposition, dip coating, as described in Figure 16.8. The successes of these strategies are dependent on various factors determining the quality of the coated monolayer. If the processing conditions are suboptimal, the spheres will simply aggregate to form undesirable clusters, or spread out loosely without any particular order. Due to the low withdrawal speed in dip coating and slow evaporation rate in vertical deposition, a long processing time (of the order of hours) may be required to establish a monolayer over centimeter-scale areas. Moreover, it is often difficult to maintain precise control of the ambient conditions, such as temperature, humidity, and pressure, which strongly affect the evaporation rate. On the other hand, the spin-coating method, which is mainly governed by rotation speed, concentration, size of spheres, and wettability of substrate, is a more reliable method with distinct advantages of higher throughput and better reproducibility. The areas of a closed-packed (CP) monolayer are typically of the order of centimeter square, while the equipment cost of spin-coating technique is low. When these parameters are fine-tuned, monolayer arrays of CP nanospheres over centimeters can be obtained within a matter of minutes.

Uniform silica nanospheres are initially diluted in deionized (DI) water to produce the optimal volume concentration of ~2%. The diluted colloidal suspension is then mixed with sodium dodecyl sulfate (SDS) at a volume ratio of 10:1. The introduction of a surfactant lowers the surface tension of the colloidal suspension and thus facilitates the spreading of nanospheres to prevent particle agglomeration or aggregation. The well-mixed colloidal suspension is then carefully dispensed onto the sample surface by mechanical micro-pipetting or other means. An optimized rotational speed is necessary to balance the centrifugal force with the solvent capillary force. During spin coating, the excess suspension is gradually flung off and spheres spread laterally, self-assembling into a monolayer of HCP array across the sample. The coated spheres then act as a lithographic mask and the pillar pattern is transferred to the wafer by inductively coupled plasma (ICP) etching. The etched sample is finally immersed in DI water under sonication to remove the spheres, leaving behind a nanopillar array. The resultant HCP nanopillar array after etching is shown in Figure 16.9.

16.4 III-Nitride LEDs with Weak Photonic Crystals

In this section, two electroluminescence devices incorporating nanosphere-patterned structure are discussed: the self-assembled array of spheres with varying dimensions serves as a hard mask to form the CP PhCs onto the indium tin oxide (ITO) film and the intermediate layer of the semiconductor. Optical properties of the nanostructures and devices are extensively studied through a range of spectroscopy techniques and simulations.

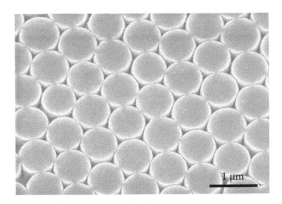

FIGURE 16.9 FE-SEM images of an HCP nanopillar array patterned by NSL.

16.4.1 InGaN LEDs with ITO PhC Current-Spreading Layer

Without a doubt, significant improvement in the LEE can be achieved by texturing the surface of LEDs. However, the choice of substrate for surface patterning is a critical issue, highly affecting the electrical characterization of devices. When the top *p*-GaN contact layer involves plasma etching, the plasma damage induced will significantly degrade electrical conduction in the device due to increased ohmic contact resistance and leakage currents, thus sacrificing overall efficiency. Plasma damage of the *p–n* junction has also been shown to affect device lifetime [16]. Instead of directly processing the GaN layer, the surface roughening of the ITO current-spreading layer has been demonstrated as an alternative method for improving the LEE without degradation of electrical characteristics.

The fabrication and characterization of PhC-on-ITO LEDs patterned by NSL are reported in this section. The self-assembled array of spheres serves as a hard mask for pattern transfer onto the ITO layer, resulting in HCP ITO pillar arrays. Although such PhC structures do not possess a PBG in the visible region, light extraction can be improved via the dispersion effect. The guided modes are effectively diffracted by the periodic refractive lattice. The performance of the ITO-PhC LEDs is evaluated, together with an investigation of the mechanisms involved. The rigorous coupled-wave analysis (RCWA) algorithm and finite-difference time-domain (FDTD) method are employed to investigate the effect of incorporating PhCs of different dimensions on the performance of LEDs.

16.4.1.1 Experimental Details

A schematic diagram illustrating the process flow of ITO-PhCs LEDs in this work is depicted in Figure 16.10. The InGaN/GaN LED wafers are grown on a *c*-plane sapphire substrate by metal–organic chemical vapor deposition (MOCVD), with embedded multiple quantum wells (MQWs) designed for emission at around 470 nm. A 200 nm thick transparent ITO coating is deposited by sputtering as the current-spreading layer, as shown in Figure 16.10a. The ITO-PhC structure is patterned by NSL, beginning with the dispensing of a colloidal suspension onto the surface of the wafer using a micro-pipette. The colloidal suspension is prepared by mixing silica spheres with mean diameters of 500, 700, and 1000 nm suspended in DI water with an anionic surfactant, SDS. The spheres self-assemble naturally and uniformly across the ITO layer with the aid of spin coating at optimized conditions. The rotation speed is varied between 140 nm and 200 rotations per minute (rpm), depending on the sphere diameter, for a duration of 10 min, resulting in the formation of a monolayer of spheres over an area of approximately 8×8 mm^2.

The self-assembled HCP array serves as an etch mask, the pattern of which is subsequently transferred to the ITO layer by ICP etching. The etch parameters are set to 500 W of coil power, 150 W of platen power at 5 mTorr of chamber pressure, using a gas comprising 15 SCCM of Cl$_2$ and 10 SCCM of Ar.

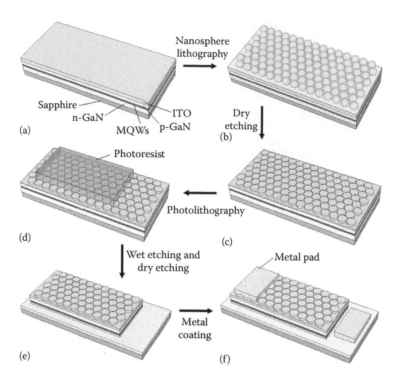

FIGURE 16.10   Schematic diagrams showing the process flow: (a) the starting wafer, (b) silica spheres spin coated onto ITO layer, (c) pattern transfer to ITO layer by ICP etching followed by sphere removal, (d) mesa definition by photolithography, (e) exposure of *n*-GaN region after dry etching, and (f) metal pads deposition by electron beam evaporation. (Reprinted with permission from Li, K.H. and Choi, H.W., *J. Appl. Phys.*, 110(5). Copyright 2011, American Institute of Physics.)

Photolithographic patterning defines 600 μm × 300 μm mesa regions, followed by dry etching to expose the *n*-GaN layer. Another photolithographic step is performed to define the contact pad regions for metallization. The *p*-pads and *n*-pads are deposited by electron beam evaporation and the wafer is subject to rapid thermal annealing at 500°C in an N_2 ambience to form ohmic contacts. For comparison, an unpatterned LED of identical dimensions is fabricated alongside. The chips are diced by UV nanosecond laser micromachining and die bonded onto transistor outline (TO) headers, followed by Al wire bonding. The surface morphologies of PhCs LEDs are imaged using field-emission scanning electron microscopy.

16.4.1.2 Surface Morphologies

Figure 16.11a shows an ordered hexagonal monolayer of nanospheres. To minimize the occurrence of defects such as dislocations and vacancies, and to avoid the formation of multiple layers, which disrupts the desired hexagonal pattern, the spin coat rotation speed must be optimized. With decreasing sphere dimensions, an increase in spin velocity (centrifugal force) is required to overcome the viscosity of the suspension. The optimized speeds for achieving high-quality monolayer array are determined to be 200, 160, and 140 rpm for nanosphere diameters of 500, 700, and 1000 nm, respectively. After dry etching, periodic HCP ITO pillar arrays are formed with triangle air gap voids between adjacent pillars exposed, as shown in Figure 16.11b. The etch depths of pillars are around 100 nm, corresponding to an etching duration of 90 s, as estimated from the scanning electron microscope (SEM) image in Figure 16.11c captured at a tilted angle of 30°. Figure 16.11d through f shows the resultant PhC structures after sphere removal with pillars diameters of 500, 700, and 1000 nm, respectively.

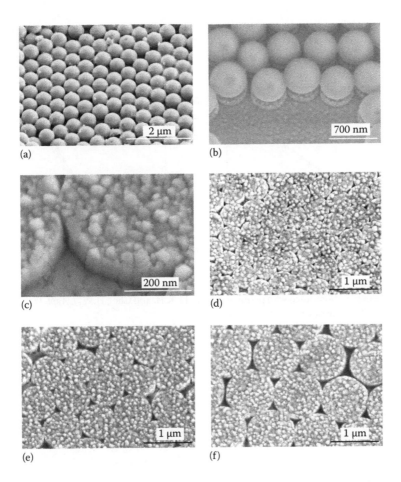

FIGURE 16.11 FE-SEM images showing (a) ordered hexagonal monolayer arrays of nanospheres on a GaN LED wafer, (b) the nanosphere-coated sample after ICP etching, (c) close-up view of the ITO nanopillars, and resultant nanopillar arrays with diameters of (d) 500 nm, (e) 700 nm, and (f) 1000 nm. (Reproduced with permission from Li, K.H. and Choi, H.W., *J. Appl. Phys.*, 110(5). Copyright 2011, American Institute of Physics.)

16.4.1.3 Device Characterizations

Electroluminescence (EL) measurements are conducted on the packaged and unencapsulated PhC and unpatterned LEDs by collecting the emitted light with a 2-inch integrating sphere optically coupled to a radiometrically calibrated spectrometer. Figure 16.12a shows plots of EL intensity versus injection current for the devices, from which it is observed that the PhCs LEDs exhibit strong enhancements in light emission over unpatterned LED. At an injection current of 100 mA, the output powers of PhCs LEDs with pillar diameters of 500, 700, and 1000 nm were enhanced by 64.6%, 39.1%, and 31.2%, respectively. The current–voltage (I–V) characteristics of the LEDs are plotted in Figure 16.12b. The forward voltages at 20 mA dc current are 3.30, 3.28, 3.26, and 3.25 V, for PhCs LEDs with diameters of 500, 700, and 1000 nm, and for the unpatterned LED respectively. The slopes of the I–V curves in the linear region (thus series resistance) are also identical. The I–V data testify to the fact that nanostructuring of the ITO layer does not degrade electrical characteristics of the LEDs, an important consideration for minimizing power consumption.

Figure 16.13a through d shows plan-view microphotographs of the PhCs LEDs operated at 5 mA, with pillar diameters of 500, 700, and 1000 nm, together with the as-grown LED. As the ITO layer is not degraded by micro-structuring, uniform emission is maintained. The PhCs LEDs also appear brighter

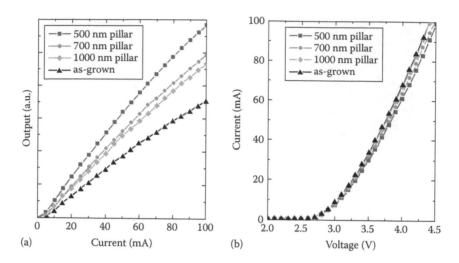

(a)   Current (mA)   (b)   Voltage (V)

FIGURE 16.12 (a) Light output power as a function of injection currents and (b) I–V characteristics of PhCs LEDs and as-grown LED samples. (Reprinted with permission from Li, K.H. and Choi, H.W., *J. Appl. Phys.*, 110(5). Copyright 2011, American Institute of Physics.)

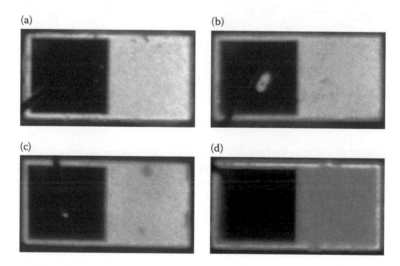

FIGURE 16.13 Optical microphotographs showing emission from ITO-PhC LEDs with pillar diameters of (a) 500 nm, (b) 700 nm, (c) 1000 nm, and (d) the as-grown LED. The devices are biased at 5 mA. (Reprinted with permission from Li, K.H. and Choi, H.W., *J. Appl. Phys.*, 110(5). Copyright 2011, American Institute of Physics.)

with decreasing pillar diameters. Compared with the PhCs LEDs, emission along the edges is significantly stronger than the planar regions from the as-grown LED since the guided photons are either reabsorbed within the active layer or escape through the sidewalls. The PhCs LEDs, on the other hand, offer enhanced emission intensity over the entire planar surface. To investigate the function of the patterned ITO layer, a reflectivity simulation is performed based on the RCWA algorithm. The defined unit cell is as shown in the inset of Figure 3.6 and a λ = 450 nm beam was incident onto the periodic array. At the ITO(n_{air} = 1.9)/air(n_{ITO} = 1.0) interface, the critical angle determined by $\sin^{-1}(n_{air}/n_{ITO})$ is ~ 31.8° such that incident light rays striking the interface at angles greater than the critical angle are totally reflected, as illustrated in Figure 16.14(a). On the other hand, the ITO film incorporating PhCs serves as a light

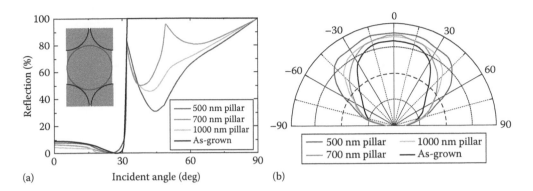

FIGURE 16.14 (a) Calculated reflection spectrum under varying incident angles and (b) angular emission patterns of PhCs LEDs and the as-grown sample. (Reprinted with permission from Li, K.H. and Choi, H.W., *J. Appl. Phys.*, 110(5). Copyright 2011, American Institute of Physics.)

extraction layer for the suppression of TIR so that more photons are capable of escaping from the devices. Angular-resolved emission patterns of the LEDs are determined by collecting EL intensities with a fiber probe at different angles while maintaining a fiber–LED separation of 50 mm. The light collected by the fiber is channeled to an optical spectrometer. The peak EL intensity at each angle is taken to plot the emission pattern, as shown in Figure 16.14b. For the as-grown LED, the intensity drops rapidly beyond ~30°, while the full-width at half-maximum (FWHM) divergence angle increases with reducing pillar diameters. The results are consistent with RCWA-simulated predictions.

16.4.1.4 Theoretical Simulations

To further evaluate the effect of PhCs patterns on LEDs, a 3D FDTD simulation is carried out. Periodic boundary conditions are applied to the x–y plane. The simulated LED structure consists of 200 nm thick ITO/150 nm thick p-GaN/40 nm thick MQWs/2000 nm GaN. The wavelength of the source is set as 450 nm and the mesh size is 20 nm. A sufficient simulation period is allowed so that the light output signal attains a steady state. Figure 16.15a shows FDTD simulated emission pattern from an unpatterned LED; photons emitted outside the critical angle are seen to be totally reflected at the flat interface. On the other hand, the incorporation of periodic PhC structures onto the ITO layer is seen to suppress lateral guiding modes and redirect the trapped photons into radiated modes, as illustrated in Figure 16.15b. The simulated time-resolved light intensity plot in Figure 16.15c shows that PhC LEDs with pillar diameters of 500, 700, and 1000 nm transmit 61.4%, 29.5%, and 20.0%, respectively, more light than the flat-top sample, correlating well with previous experimental and simulated results.

FDTD simulations are also performed to study the effects of varying pillar heights on the optical output power. The heights of pillars are varied from 0 to 160 nm, while the other device parameters remain unchanged. The computed output powers are normalized with respect to those of an unpatterned LED. From the simulated results plotted in Figure 16.16, it is apparent that taller pillars generally deliver larger optical powers. It is also observed that the rate of increase in output power for the 500 nm PhC slows down significantly after exceeding a height of ~100 nm. As for the 700 and 1000 nm PhCs, although a gradual increase in output power continues beyond pillar heights of 100 nm, a higher degree of penetration into the ITO film would degrade the lateral conductivity of the current-spreading layer and thus the electrical properties. In view of such considerations, the heights of pillar are design to be ~100 nm, which is half of the total thickness of the ITO film, in order to maximize overall device performance in terms of both optical and electrical characteristics.

2D PhCs are known to promote light extraction in LEDs via two possible mechanisms. If the PhC possesses a PBG along the plane, lateral guiding mode can effectively be eliminated over the range of

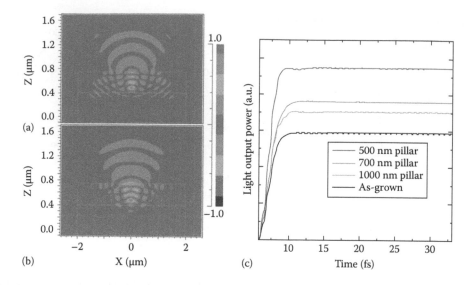

(a)

(b)

(c)

FIGURE 16.15 Comparison of FDTD simulation between the (a) as-grown flat-top and (b) PhCs samples; (c) FDTD simulation results of light output power of PhCs. (Reprinted with permission from Li, K.H. and Choi, H.W., *J. Appl. Phys.*, 110(5). Copyright 2011, American Institute of Physics.)

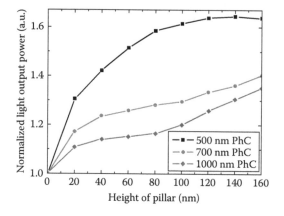

FIGURE 16.16 Plot of FDTD-computed normalized light output power as a function of pillar height for the 500, 700, and 1000 nm PhC LEDs. (Reprinted with permission from Li, K.H. and Choi, H.W., *J. Appl. Phys.*, 110(5). Copyright 2011, American Institute of Physics.)

frequencies covered by the bandgap. However, PhCs comprising CP pillar structures as fabricated by NSL do not possess PBGs, as confirmed by the simulated transverse electric (TE) and transverse magnetic (TM) band diagrams shown in Figure 16.17. Of course, it is possible to produce PBG structures using NSL. In this chapter, the fabrication of air-spaced nanopillar structures will be introduced by shrinking the patterned sphere pattern prior to pattern transfer. In this way, a PBG can be induced from such air-spaced pillar structures. In the present study, the "weak" PhCs serve to redirect emissions from guided modes into radiative modes [26]. A periodic refractive index is capable of altering the propagation behavior of photons, as described by the dispersion relation $\omega(k)$ with the light line $\omega = k_0 c$ for free-space propagation. According to Bloch's theorem, the dispersion curves of Bloch modes are folded at the Brillouin zone boundary, as evident from Figure 16.17. As a result, the waveguided modes originally located below the light line can be folded to the diffracted mode, which is located above the light line, and thus can be extracted, provided

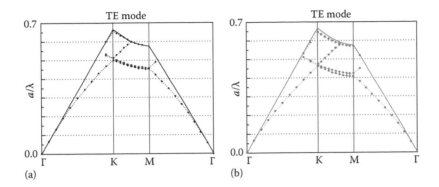

FIGURE 16.17 Simulated (a) transverse electric (TE) and (b) TE band structures for closed-packed (CP) pillar arrays. (Reprinted with permission from Li, K.H. and Choi, H.W., *J. Appl. Phys.*, 110(5). Copyright 2011, American Institute of Physics.)

the lattice constant is larger than the cutoff (Λ_{cutoff}), which is evaluated by $\Lambda_{cutoff} \approx \lambda/(n_{eff}+1)$, where n_{eff} is the effective index of the PhC layer [27]. The lattice constants of the PhCs LEDs described in this study are in the range of 500–1000 nm and thus satisfy the cutoff condition.

16.4.2 III-Nitride LED with Embedded PhCs

The integration of PhCs has been demonstrated to improve optical performance and directionality control. Although PhCs can be of the form of arrays of air holes or pillars, reported PhC LEDs mostly adopt the former configuration [20,28,29], where device processing does not involve planarizatin. Yet, periodic arrays of pillars are not just equally capable of extracting guided photons [30], but potentially can enhance IQE simultaneously due to partial strain relaxation. While this sounds attractive, electrical injection to individually isolated nanoscale emitters poses a challenge to device processing. Typically, electrical interconnection over a nonplanar surface involves planarization via gap filling [31] followed by chemical–mechanical polishing; though feasible, filling the air gaps between adjacent nanopillars reduces the refractive index contrast, disrupting optical properties of the PhC. In this work, an innovative approach of planarization through wafer regrowth is reported. A nanostructured device comprising a pillar-type PhC is planarized through the regrowth of a continuous *p*-doped contact layer. Such overgrowth techniques produce epitaxial layers with reduced threading dislocation density, ensuring high crystal quality. With regard to the formation of the nanoscale pillar array, a standard NSL process is adopted to define a large-area HCP array, serving as a weak PhC after pattern transfer.

16.4.2.1 Designing the Embedded PhC Structure

It is not uncommon to form nanostructures on the surfaces of GaN LEDs in order to improve their performances. Different types of nanostructures affect the optical properties of LEDs to different extents via different mechanisms. In the simplest case, random surface roughening promotes light extraction via scattering effects. On the other hand, nanostructures, in the form of periodic arrays of air holes or pillars of nanometer dimensions, produce a more pronounced effect through the dispersion or PBG effect. However, when the nanostructured device is packaged and encapsulated, the epoxy will fill the gaps between nanostructures, as illustrated in Figure 16.18a and b, affecting the effectiveness of the designed structures. In some cases, the nanostructures may be designed to penetrate the quantum wells (QWs) to induce strain relaxation in the light-emitting region, leading to an increase in quantum efficiencies. In such cases, the nanostructures have to be planarized before current can be injected into the nanostructured QW regions. As mentioned earlier, the filling up of air gaps between adjacent

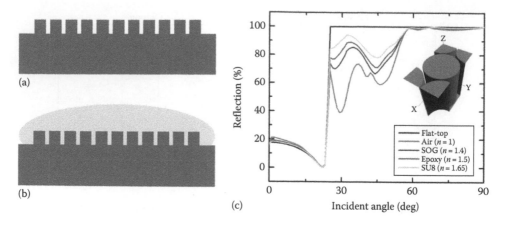

FIGURE 16.18   A nanopillar array (a) without and (b) with encapsulation. (c) Plots of reflectivities as a function of incident angle for 192 nm CP nanopillar arrays gap filled with different materials.

nanostructures reduces their effectiveness. Figure 16.18c plots the simulated reflectivities of HCP 192 nm nanopillar arrays exposed to air and a variety of common gap-filling media. As observed, the air-spaced nanopillars offer the lowest reflectivities at angles above the critical angle. The reflectivities drop when the air gaps are infiltrated with a medium of higher refractive index. Therefore, an innovative approach of planarization through wafer regrowth is employed. A nanostructured device comprising an array of nanopillars is planarized through the regrowth of a continuous p-doped contact layer. In this way, the air gap between nanopillars can be retained so that the nanostructures will not be affected by subsequent encapsulation.

16.4.2.1.1 Experimental Details

The LED wafer used in this study consists of $In_{0.2}Ga_{0.8}N/GaN$ QWs grown by MOCVD on a c-plane sapphire substrate. Silica nanospheres with mean diameters (d) of 192, 310, and 500 nm are used to assemble HCP monolayers, serving as masking layers, the patterns of which are subsequently transferred to GaN by dry etching. Details of the NSL process have been described in the previous section. A continuous p-doped GaN layer is grown over the nanopatterned surface by epitaxial lateral overgrowth (ELO) and also by MOCVD. The 300 nm thick p-type GaN ELO layer is grown at a temperature of 1070°C and at a pressure of 60 Torr. A mixture of N_2 and H_2 is used as the carrier gas, while TMGa, Cp_2Mg, and NH_3 are the Ga, dopant, and N sources, respectively. The growth temperature is subsequently decreased to 800°C while maintaining an N_2 ambience for 30 min to activate the Mg dopants. A Ni/Au (10 nm/10 nm) current-spreading layer is deposited over the p-GaN ELO layer by electron beam evaporation, followed by contact alloying at 600°C in an oxygen ambience. The LED mesa regions with areas of $400 \times 200 \ \mu m^2$ are defined by photolithography and are dry etched to expose the n-GaN layer. Another photolithographic process is employed to define the p-pad and n-pad regions for contact metallization. A bilayer of Ti/Au (40/200 nm) is electron beam evaporated, followed by annealing at 550°C in N_2 for 5 min to form ohmic contacts. The chips are diced by laser micromachining and mounted onto TO headers for wire bonding. A Spectra Physics 349 nm diode-pumped solid-state (DPSS) pulsed laser is used as an excitation source for photoluminescence (PL) studies of the PhCs. The PL signals are dispersed by an Acton SP2500A 500 mm spectrograph and detected by a PI-PIXIS TE-cooled CCD, giving spectral resolutions of ~0.3 meV.

The schematic diagram in Figure 16.19a highlights the key features of the device with an embedded PhC. Planar and cross-sectional views of the embedded nanostructure are captured by field-emission scanning electron microscope (FE-SEM; Hitachi S-4800), as shown in Figure 16.19b and c, clearly revealing that

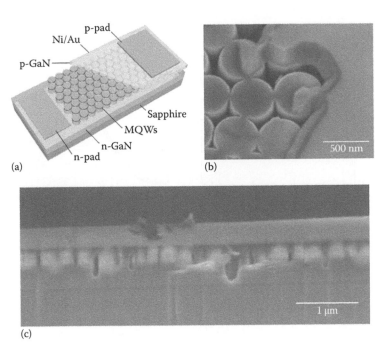

FIGURE 16.19 (a) Schematic diagram showing the structure of an LED with an embedded PhC; FE-SEM images showing (b) the edge region of the ELO regrown GaN layer covering the nanopillar array and (c) cross-sectional view of the nanopillars planarized by the ELO layer. (Reprinted with permission from Li, K.H., Zang, K.Y., Chua, S.J., and Choi, H.W., *Appl. Phys. Lett.*, 102(18). Copyright 2013, American Institute of Physics.)

the nanopillars have been planarized by the ELO layer, leaving the air gaps intact. The electrical, optical, and structural properties of the packaged devices are comprehensively characterized and complemented by FDTD simulations for the optical properties of the PhC structure as well as by finite-element strain simulations to illustrate the role of strain relaxation.

16.4.2.1.2 Optical Properties of PhCs

The enhancement ratios of the integrated PL signal for the PhCs before and after ELO, compared with the as-grown sample, are plotted in Figure 16.20a. The PL intensity is raised by as much as ~1.9 for $d = 192$ nm; the factor decreases with increasing d. Such CP pillar arrays do not possess PBGs in the visible region; instead, they operate as "weak" PhCs due to the folding of dispersion curves of Bloch modes at the Brillouin zone boundaries. The enhancement effects are weakened with the insertion of the ELO layer, attributed to absorption. The PL spectra of the PhCs after regrowth, plotted in Figure 16.20b, reveal spectral blueshifts as d decreases, tentatively attributed to strain relaxation.

The optical characteristics of the PhC with and without the ELO layer are further studied through 3D FDTD simulations. The QWs are modeled as light sources inserted within each pillar, as illustrated in the diagrams depicted in Figure 16.21a and b. In the simulation, the propagation of waves from the sources is detected and analyzed with a wide-field planar monitor placed on the top of the device models. The simulated data are plotted in Figure 16.21c, also summarized in its inset. Consistent with the measured PL data, insertion of the ELO layer does reduce light extraction probabilities but light output remains above as-grown levels. It is also noted that the simulated enhancement factors are, on average, 15% higher than the measured values.

The transmission characteristics of the PhC may be diminished by the presence of defects within the pillar array, a common feature of self-assembly techniques, including NSL. Dimensional nonuniformities among spheres give rise to point defects, as illustrated in the inset of Figure 16.22a. The impact of

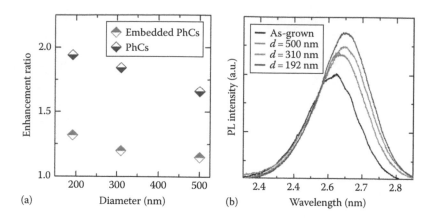

FIGURE 16.20 (a) Enhancement ratios of the integrated PL signals for the PhCs before and after ELO growth; (b) PL spectra of the PhC after regrowth, together with that of the as-grown sample. (Reprinted with permission from Li, K.H., Zang, K.Y., Chua, S.J., and Choi, H.W., *Appl. Phys. Lett.*, 102(18). Copyright 2013, American Institute of Physics.)

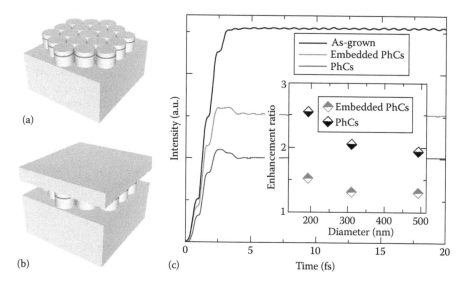

FIGURE 16.21 Models of the (a) nanopillar PhC and (b) PhC after ELO regrowth used in 3D FDTD simulations. (c) Simulated results showing the light output intensities over time for the PhC LEDs before and after ELO regrowth, as well as for the as-grown LED. The data are summarized in the inset of the figure. (Reprinted with permission from Li, K.H., Zang, K.Y., Chua, S.J., and Choi, H.W., *Appl. Phys. Lett.*, 102(18). Copyright 2013, American Institute of Physics.)

imperfections in the PhC on its optical properties is investigated through additional simulations, conducted using modified models containing pillars with diameter deviations (σ) of ±2.5%, ±5%, and ±7.5% from the nominal value of 192 nm. The simulated results, as shown in Figure 16.22a, show that imperfect PhCs extract as much as 20% less light compared with perfect arrays. Since the diameters of the nanospheres used in this work are known to have deviations of up to ±7.5%, the observation of lower-than-predicted PL intensities is justified.

Micro-Raman spectroscopy is employed to detect strain in the nanopillars. The 325 nm line of a He–Cd laser, with shallow penetration in GaN, is chosen for Raman excitation so that the signals originate from the near-surface region. The spectral resolution of the setup is ~0.2 cm⁻¹. Raman spectra of the PhCs after regrowth showing the E_2 (high) phonon mode, corresponding to lattice vibrations perpendicular to the surface (which is known to be sensitive to strain), is plotted in Figure 16.22b. The degree of stress can be estimated using $\Delta\omega_{E2} = K_R\sigma$, where σ is the in-plane biaxial stress, K_R (= 4.2 cm^{-1} GPa^{-1}) is the proportionality factor, and $\Delta\omega_{E2}$ is the shift of the E_2 (high) phonon peak with respect to strain-free GaN, obtained as 567.5 cm^{-1} from a free-standing GaN wafer. The typical GaN epilayer exhibits large stresses due to mismatches in both the lattice constants and the thermal expansion coefficients of GaN and sapphire. By patterning the GaN surface into nanopillars, a Raman shift ($\Delta\omega_{E2}$) of 0.8 ± 0.2 cm^{-1} is observed for d = 192 nm, corresponding to a residual stress of 0.35 ± 0.05 GPa. Strain relaxation occurs mainly in the vicinity of surfaces as the surface atoms are less constrained by surrounding materials. Pillars of small diameters have a high surface area to volume ratio, thereby increasing the extent of strain relaxation. Raman measurements are also conducted on an identical set of nanopillar structures before ELO regrowth; the results are summarized in Figure 16.22c. A slight reduction in redshift is observed from samples after ELO regrowth, signifying that the ELO cap layer is lightly strained (albeit much less than the as-grown). After all, ELO is a growth technique originally developed for achieving high-quality epilayers.

With the growth of a cap layer bridging isolated pillars, it becomes possible to inject currents into devices with embedded PhCs. Figure 16.23a plots EL output powers of LEDs versus injection currents.

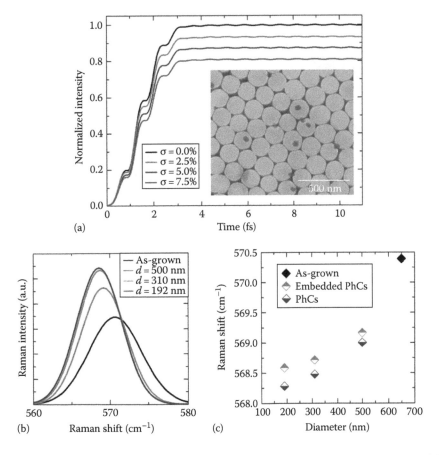

FIGURE 16.22 (a) Simulated light output versus time for the imperfect PhCs with various dimensional deviations; an FE-SEM image of the d = 192 nm nanopillar is shown in the inset, whereby defects caused by dimensional nonuniformity (σ) among nanospheres are seen; (b) Raman spectra for the PhC (after ELO) and the as-grown sample showing the E_2 (high) phonon mode; (c) Plot of the Raman E_2 (high) phonon mode frequency for the nanopillar PhCs of varying diameters. (Reprinted with permission from Li, K.H., Zang, K.Y., Chua, S.J., and Choi, H.W., *Appl. Phys. Lett.*, 102(18). Copyright 2013, American Institute of Physics.)

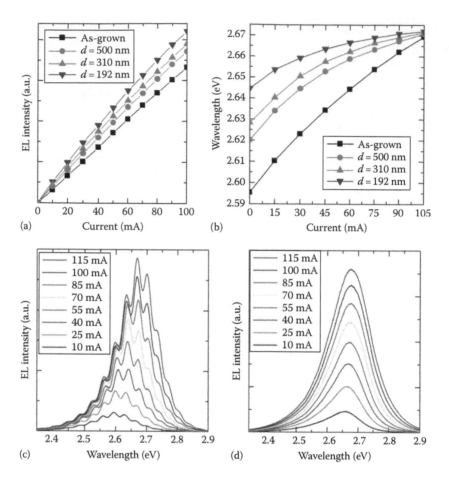

FIGURE 16.23   (a) Plot of light output power as a function of injection current. (b) Plot of peak emission wavelength as a function of injection current. Room-temperature EL spectra of (c) an as-grown LED and (d) an LED with embedded PhC (d = 192 nm) operated at currents between 10 and 115 mA. (Reprinted with permission from Li, K.H., Zang, K.Y., Chua, S.J., and Choi, H.W., *Appl. Phys. Lett.*, 102(18). Copyright 2013, American Institute of Physics.)

LEDs with embedded PhCs emit as much as 20% more light over the unpatterned counterparts. Since the majority of photons emitted from QWs are trapped within the device due to TIR, they would eventually be reabsorbed or escape from the sidewalls. On the other hand, the embedded PhC is capable of altering the directionalities of the propagating photons. Thus, the overall enhancement of light extraction can be attributed to diffraction and scattering caused by the periodic change in the refractive index of the PhC.

The emission center wavelength of an InGaN LED not only depends on the bandgap energy of the QW, but is also strongly affected by the large piezoelectric field arising from the built-in strain. The triangular-shaped potential well caused by the piezoelectric field and spontaneous polarization tilts the band alignment separating the electron and hole wavefunctions, causing spectral redshifts and reductions in radiative recombination rates. To determine dimensional effects on the emission spectrum, the LEDs are driven in pulsed mode (1 μs, 1 kHz) to minimize self-heating effects. The emission spectra of the as-grown LED in Figure 16.23c indicate significant blueshifts of ~74 meV as the bias currents increase from 10 to 115 mA. With increasing currents, the injected carriers partially screen the polarization field; such screening reshapes the potential function back to a rectangular profile, leading to the observed spectral blueshift. The filling of localized states in the well and barrier layers may yet be another factor leading to the reduction in effective carrier separation. However, a gradual saturation of localized states would lead to a reduction in quantum efficiency caused by enhanced capture rates by nonradiative recombination centers. The L–I data in Figure 16.23a indicate that the LEDs maintain nearly linear increases in light emission at higher injection currents. In view of this observation, the spectral blueshifts are mainly attributed to the screening of the polarization field. The emission spectra for the PhC LED ($d = 192$ nm) are plotted in Figure 16.23d; it is observed that the peak wavelength is nearly invariant of driving currents, signifying that the strain effect is significantly diminished. The spectra are also characterized by the lack of interference fringes as compared with the as-grown LED spectra, indicating that Febry–Pérot oscillations have been effectively suppressed by the embedded PhC. A plot of peak wavelength with respect to bias currents for the LEDs is shown in Figure 16.23b. As d shrinks from 500 to 192 nm, the extent of spectral blueshift reduces from 50 to 27 meV over the current range of 10–115 mA, compared with a 74 meV spectral shift of the as-grown LED over the same current range. Such observations of diminishing blueshifts clearly attest to the role of strain relaxation through nanostructuring.

A 3D simulation has also been performed by employing a finite-element approach to map strain distributions along the QW layers, based on a model containing four pairs of 192 nm disk-shaped QWs, as illustrated in Figure 16.24a. A cross-sectional view of in-plane strain distribution along the various layers is depicted in Figure 16.24b, revealing that a higher degree of relaxation occurs toward the disk edges, while the central region remains strongly strained. Similar degrees of relaxation occur at the edges independent of pillar diameters. Figure 16.24c shows a high-resolution close-up strain map of the sidewall region; the extent of compressive strain gradually increases over a distance of ~20 nm from the edge before attaining its maximum value. Consequently, the observed blueshift of spectral peaks suggests that a significant proportion of EL signals originate from the strain-relaxed region surrounding the nanopillars.

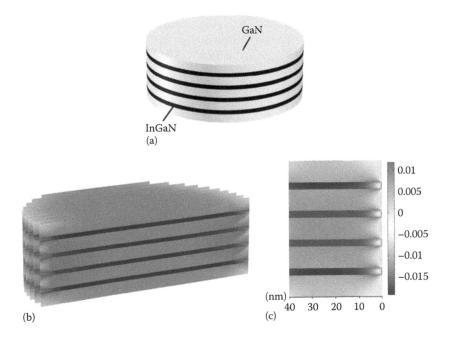

FIGURE 16.24 (a) The disk model containing four pairs of InGaN/GaN QWs as used in the 3D finite-element strain simulations. (b) Simulated results showing the strain profile along the planes of the QWs. (c) An enlarged view showing strain distributions near the edge of the disk. The scale bar on the right represents the value of the in-plane strain. (Reproduced with permission from Li, K.H., Zang, K.Y., Chua, S.J., and Choi, H.W., *Appl. Phys. Lett.*, 102(18). Copyright 2013, American Institute of Physics.)

In summary, an InGaN LED with an embedded PhC is demonstrated. The PhC is fabricated by etching through an NSL pattern, followed by ELO regrowth. The optical performances of the PhC LEDs have been studied and compared with FDTD simulation results. The periodic ordered nanopillar structure not only promotes light extraction, but also partially suppresses the piezoelectric field through strain relaxation of the InGaN/GaN QWs, supported with μ-Raman spectroscopy spectra. The LEDs with embedded PhCs emit as much as 20% more light than the as-grown LED, as well as exhibiting emission wavelengths nearly invariant of injection currents.

16.5 III-Nitride LEDs with Strong PhCs

To extend the function of a sphere-patterned array, a dimension-adjusting procedure is employed to overcome the restrictions of CP patterning and realize PBG structures. With a well-defined periodic arrangement and with sufficiently large refractive index contrast between GaN and the ambience, a wave-length-tunable PBG in the visible region can be opened up, which forbids the propagation of light within a specific range of frequencies and is extremely useful in molding the flow of emitted light from an LED. Two non-closed-packed (NCP) periodic patterns, namely, air-spaced and clover-shaped PBG structures, are highlighted in this section.

16.5.1 Air-Spaced GaN Nanopillar PBG Structures

Conventionally, PhCs are patterned as ordered arrays of recessed air holes or protruding pillars. NSL is a practical alternative approach toward large-scale nanofabrication and with the capability of forming 2D and 3D PhCs. Uniform spheres are capable of self-assembling into hexagonal arrays over large areas, but the ability of spheres to spontaneously form CP structures acts also as a limitation: PhCs require alternating layers of different materials with defined and constant separations, and a close-packed structure is obviously not favorable for achieving this. Establishing finite spacing between individual spheres is a critical step toward realizing PBG structures. Formation of PBG structures inhibits all wave vectors within the PBG [33] and thus promotes light extraction by diffracting waveguided modes out with the semiconductor, as illustrated in Figure 16.25. An air-spaced nanopillar PhC structure with a wavelength-tunable PBG is proposed. Prior to pattern transfer to the wafer, silica nanospheres are shrunk by a selective dry etch process. As a result, spacing is induced between spheres on the plane without altering the sites of spheres; therefore, the packing of the modified array remains largely regular. This dimension-adjusting procedure overcomes the restrictions of CP patterning to achieve low-cost, high-efficiency, and PBG-tunable nanopillar arrays, which has been applied to InGaN LEDs for realizing the enhancement of light extraction.

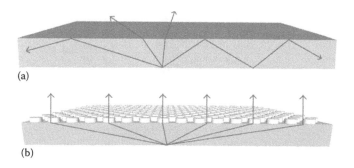

(a)

(b)

FIGURE 16.25   (a) A large fraction of the light is trapped within the semiconductor layer due to total internal reflection (TIR). (b) PBG structures promote light extraction by diffracting waveguided modes out of the layer. (Reprinted with permission from Li, K.H. and Choi, H.W., *J. Appl. Phys.*, 109(2). Copyright 2011, American Institute of Physics.)

16.5.1.1 Experimental Details

Figure 16.26 illustrates the proposed fabrication process of air-spaced nanopillar arrays by NSL. The hybrid NSL process was performed on III-nitride LED wafers consisting of InGaN/GaN MQWs grown by MOCVD on a *c*-plane sapphire substrate. Uniform silica nanospheres, with mean diameters of 192 nm, suspended in DI water, were mixed with SDS at a volume ratio of 10:1. SDS acted as a surfactant to reduce the surface tension of water and thus facilitated the spreading of particles to prevent clustering. Around 1.5 μL of well-mixed colloidal suspension was then dispensed onto a sample surface by mechanical micro-pipetting. The nanospheres spread laterally upon spin coating at 1000 rpm for 5 min, self-assembling into a monolayer HCP array across the sample. At this time, the pattern can be transferred to the LED wafer to form CP nanopillars, as demonstrated in the previous sections. In this work, the nanosphere-coated wafer was subjected to reactive ion etching using CHF_3-based plasmas at low radio frequency (RF) power prior to pattern transfer. This choice of etchant gas ensured that the silica spheres are selectively etched, without affecting the GaN substrate. During the etching process, the dimensions of the silica nanospheres are reduced. The RF power was maintained low in order to avoid overheating (causing distortion of sphere geometry) and translation of spheres (destroying the orderliness of packing). Due to dimensional reduction, an air gap is induced between spheres, as illustrated in Figure 16.26. The shrunk nanospheres then served as an etch mask and the pattern was subsequently transferred to the GaN wafer by ICP etching using Cl_2 to form an air-spaced nitride nanopillar array. The ICP and platen powers were maintained at 300 and 100 W, respectively, while the chamber pressure was fixed at 5 mTorr. The etch depth was approximately 350 nm after etching for 75 s. The spacing between nanopillars was exposed as imaged by FE-SEM in Figure 16.27b. The nanospheres were subsequently removed by sonication in DI water, leaving behind the nitride air-spaced nanopillar array.

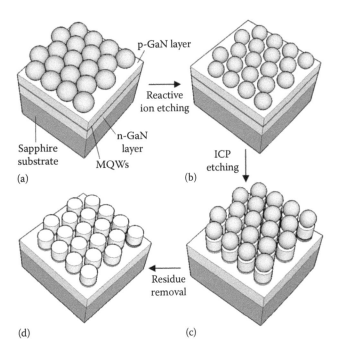

FIGURE 16.26 Schematic diagrams illustrating the fabrication flow. (a) Silica nanospheres are coated onto the surface of a GaN wafer, forming CP arrays. (b) Shrinkage of spheres using reactive ion etching (RIE). (c) Pattern transfer to GaN using ICP etching. (d) Silica residue removal. (Reprinted with permission from Li, K.H. and Choi, H.W., *J. Appl. Phys.*, 109(2). Copyright 2011, American Institute of Physics.)

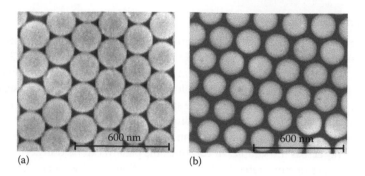

FIGURE 16.27 FE-SEM images showing (a) the original ordered CP nanopillar array and (b) the air-spaced nano-pillars. (Reprinted with permission from Li, K.H. and Choi, H.W., *J. Appl. Phys.*, 109(2). Copyright 2011, American Institute of Physics.)

Optical transmittance measurements were conducted to verify the existence and position of a PBG. The incident beam from a high-power broadband solid-state plasma light source (Thorlabs HPLS-30-03) was collected by an optical fiber and focused onto the samples in planar direction. The transmitted beam was collected and channeled to an optical spectrometer via another fiber. The optical properties of the PhC structures were further evaluated by time-integrated PL at room temperature. A DPSS UV laser at 349 nm was used as an excitation source (120 µs, 1 kHz) while the PL signal was coupled to a spectrometer comprising an Acton SP2500A 500 mm spectrograph and a Princeton Instrument PIXIS open-electrode CCD via an optical fiber bundle, which offers optical resolutions of better than 0.1 nm.

16.5.1.2 Designing the PhC Structure

To design and predict the existence of a PBG in a nitride nanopillar array, band diagrams were computed using Rsoft BandSOLVE, which employs the plane wave expansion (PWE) algorithm for band computations. The supercell technique was utilized during 3D PWE simulations. We begin with the simulation of a CP nanopillar array, the structure obtained naturally with NSL. In this case, the periodicity of the array is considered to be the diameter of a single nanosphere. The positions of the PBG (a/λ) for the TE modes were computed as a function of sphere diameter (between 100 and 900 nm) and plotted in Figure 16.28, since emissions from InGaN/GaN MQWs are dominated by TE modes [35]. The plot shows that the TE-PBGs are mainly located in the UV and infrared (IR) regions of the spectrum. For visible InGaN/GaN LED applications, the PBG should obviously be located within the visible spectrum to achieve any beneficial effects. Of course, even in the absence of a PBG, the CP nanopillar structure can still be used to increase light extraction via dispersive and geometrical effects [36], albeit with reduced effectiveness.

In spite of the said limitations, a PBG can still be introduced into the visible spectral region by modifying the NSL process to enlarge the physical gaps between nanopillars to produced air-spaced nanopillar arrays, as described in the section on experimental details. To demonstrate this concept, our experiments and computations were carried out using nanospheres with initial nominal diameters of 192 nm with variations of ±2 nm. The positions of the TE bandgaps for shrunk nanopillar arrays of diameters 120–170 nm, with a fixed pitch of 192 nm, were computed and are shown in Figure 16.29. By reducing the diameters of spheres while maintaining their pitch, the positions of the TE-PBG shift accordingly. As we are interested in the visible region, the range of frequencies (a/λ) of interest lie between 0.39 and 0.45, corresponding to wavelengths between 426 and 492 nm. From the simulation results in Figure 16.29, it can be deduced that the diameters of spheres should be reduced in the range of 126–162 nm, coinciding with the range of InGaN/GaN MQWs emission centered at 455 nm for the material used in our experiments.

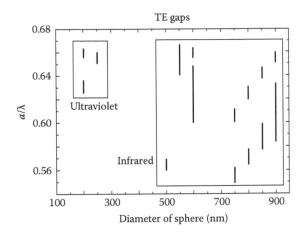

FIGURE 16.28 Plot of simulated TE photonic bandgap as a function of diameter of CP spheres. (Reprinted with permission from Li, K.H. and Choi, H.W., *J. Appl. Phys.*, 109(2). Copyright 2011, American Institute of Physics.)

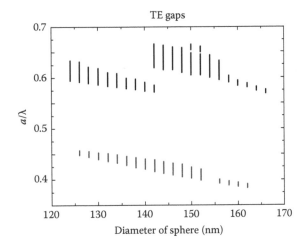

FIGURE 16.29 Plot of simulated TE photonic bandgap as a function of diameter of sphere (with a pitch of 192 nm). (Reprinted with permission from Li, K.H. and Choi, H.W., *J. Appl. Phys.*, 109(2). Copyright 2011, American Institute of Physics.)

16.5.1.3 Fabrication and Optical Characterization of the PhC Structure

The proposed air-spaced nanopillar structure was fabricated using a modified NSL process. The self-assembled spheres, which acted as a sacrificial masking layer, were etched for dimensional reduction. During this first etching step, the silica spheres are etched without pattern transfer to the wafer. As the dry etch process is directional, the rate of etching in the vertical direction is faster than in the lateral plane. Gradually, the incident ions trim the diameters of the spheres, opening up air gaps between spheres. However, under prolonged etching, the geometry of spheres may become irregular and adjacent spheres tend to aggregate, resulting in disruption of order in the arrays, due to physical bombardment of ions and accumulation of heat. Figure 16.30a and b shows FE-SEM images of nanosphere arrays after etching at 300W and 70W of RF power, respectively. To produce nanopillar arrays of desired order and uniformity, moderate etch conditions are required to minimize defect formation. Based on our calculations, four sets

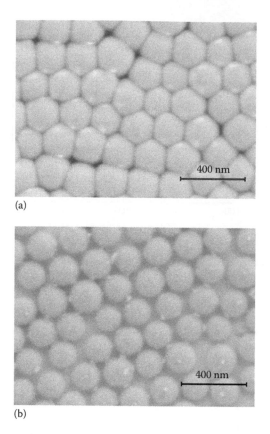

(a)

(b)

FIGURE 16.30 FE-SEM images showing nanosphere arrays after dry etching at (a) 300 W and (b) 70 W RF power. (Reprinted with permission from Li, K.H. and Choi, H.W., *J. Appl. Phys.*, 109(2). Copyright 2011, American Institute of Physics.)

of samples were developed with incremental etch durations. After pattern transfer to the wafer and residue removal, four samples with air-spaced nanopillar structures were produced. The FE-SEM images in Figure 16.31a (i–iv) illustrate nanopillar arrays etched for durations of 14, 12, 10, and 8 min, respectively; their diameters are roughly equal to 130, 140, 150, and 160 nm, respectively.

Plan-view microphotographs in Figure 16.31b (i–iv) illustrate physical color changes observed from the sample surface in the normal direction, changing from purplish blue to greenish blue with decreasing pillar diameter. Light spots are attributed to larger area of defects that are unpatterned. To verify the existence and position of the PBG, an optical transmission measurement in the planar direction was conducted. Figure 16.31c shows the measured transmission spectra. Four distinct transmission minima were observed at the wavelengths of 490.59, 471.49, 444.51, and 431.96 nm, respectively, with decreasing sphere diameters of 160, 150, 140, and 130 nm. As the PBG structure forbids lateral propagation at the range of wavelengths within the bandgap, the propagation of the incident beam along the plane is restricted, giving rise to reduced transmission at those wavelengths. For light with wavelengths beyond the bandgap region, laterally propagating photons do not experience PBG confinement effects. The transmission spectra, which indicate the PBG positions, correlated well with the PWE stimulated results in Figure 16.29.

16.5.1.4 Enhancement of PL Intensity in PhC Structures

The optical effects of incorporating these PhC structures onto LED wafers were evaluated by PL measurement. Figure 16.32 shows measured PL spectra from the four nanopillar arrays, together with the PL spectra from an as-grown sample. Interference fringes in the spectrum of the as-grown sample indicate

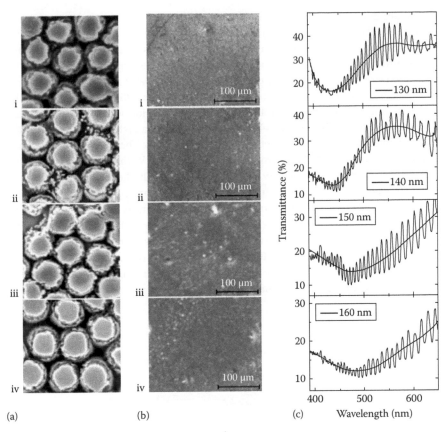

FIGURE 16.31 (a) FE-SEM images showing air-spaced nanopillar arrays that have been RIE etched for durations of (i) 8 min, (ii) 10 min, (iii) 12 min, and (iv) 14 min; (b) optical microphotographs of nanopillar arrays with diameters of (i) 160 nm, (ii) 150 nm, (iii) 140 nm, (iv) 130 nm, and (c) measured reflection spectra from the respective nanopillar arrays. (Reprinted with permission from Li, K.H. and Choi, H.W., *J. Appl. Phys.*, 109(2). Copyright 2011, American Institute of Physics.)

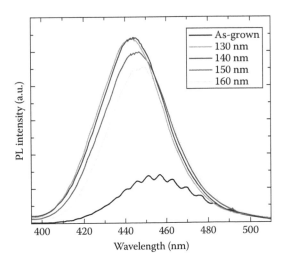

FIGURE 16.32 Measured photoluminescence spectra from air-spaced nanopillar arrays, compared with an as-grown sample. (Reprinted with permission from Li, K.H. and Choi, H.W., *J. Appl. Phys.*, 109(2). Copyright 2011, American Institute of Physics.)

Fabry–Pérot modes due to vertical optical confinement. The majority of the photons generated by the MQWs remain trapped within the wafer [37], forming standing waves, which are subsequently reabsorbed. Such oscillations were clearly suppressed in the nanopillar samples with diameters of 130, 140, 150, and 160 nm due to diffraction of the guided modes by the PhC, together with the observation of significant PL intensity enhancements, demonstrating that PhCs can indeed play a remarkable role in manipulating spontaneous emission by suppressing unwanted optical modes via the PBG. Compared with the as-grown sample with peak emission wavelength at 455 nm, the peak emission wavelengths from the nanopillar samples were centered at 443.79, 444.05, 446.84, and 448.46 nm, a systematic blueshift with respect to their diameters, which can be attributed to the position of the PBG. The spectral shifts were consequential of the overlap between the MQW emission band and the PBG. With decreasing pillar diameters, the position of the PBG was shifted from 490.59 to 431.96 nm. As a result, the spectra contents in the shorter-wavelength region were enhanced to a greater extent, giving rise to an apparent spectral blueshift. For the 140 nm diameter nanopillar array, a fourfold increase in the PL intensity was observed since the position of the PBG coincided with the emission wavelength, as evidenced through the computed TE and TM band structures in Figure 16.33. The TE bandgap (the TE mode dominates in InGaN/GaN MQW LEDs) occurs in the frequency range, a/λ, of 0.4211–0.4402 within the light line of air, corresponding to wavelengths between 436.17 and 455.95 nm. It indicates that the PBG band corresponding to the 140 nm nanopillar array overlaps optimally with the MQW emission band, correlating well with PL measurements.

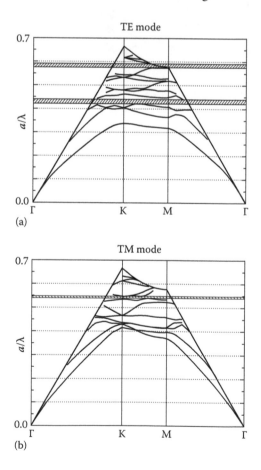

FIGURE 16.33 Computed (a) TE and (b) TM band structures for an air-spaced nanopillar array with a pillar diameter and a pitch of 140 and 192 nm, respectively. (Reprinted with permission from Li, K.H. and Choi, H.W., *J. Appl. Phys.*, 109(2). Copyright 2011, American Institute of Physics.)

16.5.1.5 Effects of Disordering in Nanopillar Arrays

With the technique of NSL, the formation of CP hexagonal arrays relies on the packing and dimensions of self-assembling microspheres. To achieve a CP monolayer array by spin coating, the sphere diffusion rate and concentration of the suspension play important roles in determining the coverage area. The former factor involves a balance between centrifugal forces controlled by rotation speed and surface tension forces, which can be reduced by adding the surfactant SDS. Nevertheless, the presence of defects, including point defects and line defects, and the nonuniformity of sphere diameters are inevitable. Point and line defects are naturally and randomly formed during the self-assembly process. These defects are then transferred to the GaN wafer during etching, affecting the optical properties of the PhC, which is investigated and reported in this section.

The nanospheres were shrunk with a CHF_3-based etch recipe that targets SiO_2. Figure 16.34 plots the diameters of nanospheres as a function of etch duration. The etching process also induces dimensional nonuniformity among spheres. Initially, the 192 nm diameter spheres self-assemble into a monolayer with high uniformity (±2 nm). Once the etch duration exceeds the seventh minute, the sphere shrinks rapidly. The rate of shrinkage increases further after the 25 min. Nevertheless, with increasing etching duration, the variation in diameters between spheres gradually enlarges. When the diameters of spheres are below one-third of their original values, the variation becomes significant and causes poor uniformity (±8 nm), as shown in Figure 16.35. Therefore, there is a limit on the extent of shrinkage that can be tolerated. When the pattern is transferred to the GaN wafer, nanopillar arrays with poor uniformity of diameters are unfavorable for establishing a well-defined PBG.

To illustrate the effect of nonuniformity qualitatively, FDTD simulations were carried out to predict the field distribution when a continuous wave at 440 nm was emitted from the center-bottom position from arrays of nanopillars with different packing orders. Figure 16.36a shows the simulation result for an ideal air-spaced nanopillar array with diameter/pitch/height of 140/192/350 nm, whereby the lateral propagation of light is obviously suppressed; based on the simulated band diagram, a PBG is indeed predicted between ~436 and 455 nm, correlating well with the FDTD simulated results. However, the presence of defects and nonuniformities disrupts the orderliness of the PhC. As a result, losses and scattering effects are superimposed upon the PBG, the degree of which depends on the extent of defects. A simulation was then performed on a disordered array of nanopillars, whereby randomly selected pillars with a height of 350 nm have been shifted from their equilibrium positions. At the same time, the diameters of nanopillars in the array range between 130 and 150 nm; such a "defective" array bears

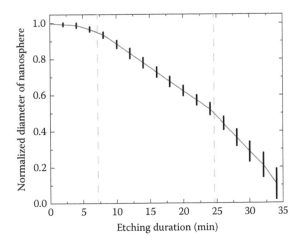

FIGURE 16.34 Plot of normalized diameter of nanosphere versus etch duration. (Reprinted with permission from Li, K.H. and Choi, H.W., *J. Appl. Phys.*, 109(2). Copyright 2011, American Institute of Physics.)

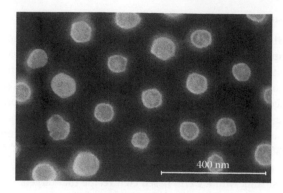

FIGURE 16.35   FE-SEM image demonstrating poor dimensional uniformity of spheres due to excessive etching. (Reprinted with permission from Li, K.H. and Choi, H.W., *J. Appl. Phys.*, 109(2). Copyright 2011, American Institute of Physics.)

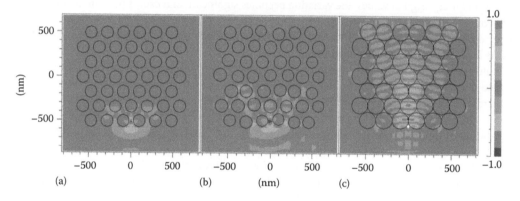

FIGURE 16.36   Computed FDTD results for (a) a perfect PhC structure comprising regular air-spaced nanopillars, (b) a PhC structure with defects, and (c) a CP nanopillar array. (Reprinted with permission from Li, K.H. and Choi, H.W., *J. Appl. Phys.*, 109(2). Copyright 2011, American Institute of Physics.)

close resemblance to an actual fabricated array under worst-case conditions. The results of this simulation are shown in Figure 16.36b. The introduced defects interrupt the periodic order, causing some degree of scattering. The incident source is now partially reflected and partially transmitted by the array due to the presence of leakage modes. This also explains why the measured transmission dips are Gaussian in profile, instead of being abruptly sharp. To complete the picture, an FDTD simulation was also performed on a CP 192 nm nanopillar array, as illustrated in Figure 16.36c. The wave is allowed to propagate freely through the array in the absence of a PBG. Comparing the three presented scenarios, one can conclude that the air-spaced nanopillar structure proposed in this work behaves as a leaky PhC, comprising a superposition of defect leakage modes upon PBG confinement modes. In spite of imperfections, its effectiveness in enhancing light extraction from GaN materials is well demonstrated, especially in consideration of the ease and cost of the self-assembled nanoscale patterning process.

In summary, the fabrication of an ordered hexagonal array of air-spaced nanopillar on GaN wafers by NSL has been demonstrated. Employing a dual-step dry etch process, the dimensions of the nanopillars in an array can be adjusted without altering their pitch; at the same time, PBG properties of the self-assembled nanostructures can also be modified. The positions of the PBG were identified by optical transmission measurement, correlating well with the prediction of PWE stimulations. A maximum fourfold increase in the PL intensity was observed compared to an as-grown sample, depending on the overlap between the PBG position and the emission band. Despite the presence of leakage modes due to defects, we have demonstrated the effectiveness of the self-assembled PBG structures in suppressing unwanted guiding modes and promoting the LEE.

16.5.2 Tunable Clover-Shaped GaN PBG Structures Patterned by Dual-Step NSL

For the standard NSL process, the self-assembled sphere array then serves as an etch mask for pattern transfer to form periodic arrays of recessed air holes or protruding pillars. However, CP nanostructures do not provide a suitable periodic variation in the refractive index in the lateral direction, so a PBG corresponding to a TE-dominated emission from InGaN/GaN MQWs does not exist. From the previous section, the air-spaced PBG structures in the visible spectral regions have been demonstrated. The work being presented in this section adopts a dual-step NSL for generating CP clover-shaped PhC structures. It can also be extended to pattern NCP clover-shaped structures through an additional dimension-adjusting process. By adjusting the air spacing, the PBG position can be tuned to spectrally overlap with the emission band of InGaN/GaN MQWs. Transmission measurement are carried out to locate the position of the PBG, which is found to correlate well with 3D FDTD simulated results. The effects of having a PBG on light extraction and recombination decay lifetime are also discussed.

16.5.2.1 Process Flow

Figure 16.37 illustrates the process flow for fabricating a clover-shaped PhC structure. The MOCVD-grown InGaN/GaN MQW LED on a *c*-plane sapphire substrate emits at a center wavelength of ~450 nm and an FWHM of ~44 nm. Silica spheres with a mean radius of 132 nm are initially diluted in DI water to produce the optimized solid concentration of 2%. Around 5 μL of a diluted colloidal suspension mixed with SDS at a volume ratio of 10:1 is dispensed and dispersed uniformly across the sample by spin coating. The SDS surfactant

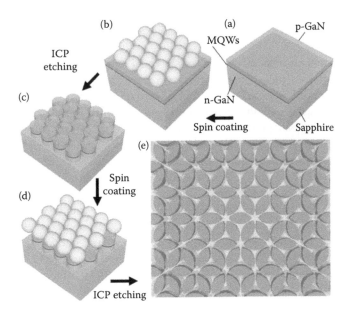

FIGURE 16.37 Schematic diagram depicting the process flow: (a) the starting LED wafer, (b) silica spheres coated onto the wafer surface by spin coating, (c) pattern transfer to GaN by ICP etching, (d) second monolayer of sphere array coated on top of pillar array, and (e) clover-shaped pattern formed after ICP etching. (Reprinted with permission from Li, K.H., Ma, Z.T., and Choi, H.W., *Appl. Phys. Lett.*, 100(14). Copyright 2012, American Institute of Physics.)

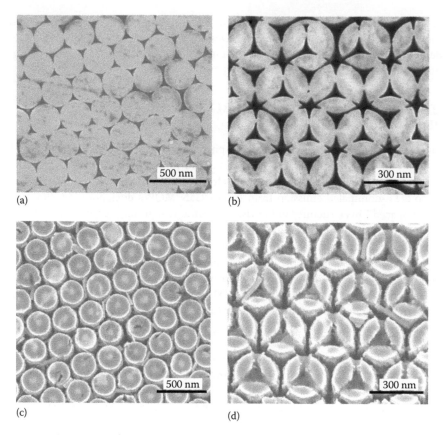

FIGURE 16.38 FE-SEM images showing (a) the ordered hexagonal CP pillar array, (b) the CP clover-shaped PhC, (c) the NCP pillar array, and (d) the NCP clover-shaped PhC. (Reprinted with permission from Li, K.H., Ma, Z.T., and Choi, H.W., *Appl. Phys. Lett.*, 100(14). Copyright 2012, American Institute of Physics.)

reduces water tension and prevents spheres from aggregating into clusters, thus forming a monolayer of spheres. The ordered hexagonal pillar pattern is transferred to GaN by ICP etching using Cl_2/He gas mixtures. The coil and platen powers are maintained at 500 and 135 W, respectively, while the chamber pressure is held constant at 5 mTorr. The spheres are then removed via sonication in DI water. The etched sample is subsequently subjected to another NSL process. During spin coating, the spheres spontaneously occupy locations at the triangular voids between adjacent pillars. After another dry etch process, the HCP clover-shaped PhC is formed. The surface morphologies of the resultant structures are imaged by FE-SEM. The FE-SEM images in Figure 16.38 show the resultant CP and NCP clover-shaped structure, respectively.

Time-integrated PL at room temperature is conducted to characterize the optical properties of the PhC structure. The 349 nm excitation laser beam is focused onto the sample, while the PL signal is fiber collected to a spectrometer. Time-resolved PL spectroscopy is conducted at room temperature using a picosecond laser (Passat Compiler) as an excitation source, whose wavelength is 266 nm with an 8 ps pulse width and a 100 Hz repetition rate. The PL signal is bandpass filtered and collected via a 40× UV objective, subsequently detected by a high-speed photodetector (Thorlabs SVC-FC, <150 psrise time), whose electrical signal is read on a 4 GHz digital real-time sampling oscilloscope (Agilent DSO9404A, 85 ps rise time).

16.5.2.2 PBG Calculation

The band structures are computed by modified 3D FDTD simulations to predict the PBG position of the clover-shaped PhC structure. The unit cell transformation technique is employed [39], offering compatibility with any geometry, including the clover-shaped structure in this study. Since the TE mode is dominant in the emission of InGaN/GaN MQWs, the TE band structures with varying ratios of pillar radius to pitch (r/a) are computed for both pillar and clover-shaped arrays, as plotted in Figure 16.39a and b, respectively. For the pillar arrays, the calculated bandgaps are discretely distributed throughout the plot. Although tuning of PBG can be achieved by modifying r/a ratios, the bandwidths are relatively narrow, varying from 8.98 nm at $r/a = 0.48$ (CP) to 20.55 nm at $r/a = 0.41$, for effective coupling with QW emission. A detailed study on air-spaced pillar arrays has been reported in the previous section. On the other hand, the CP clover-shaped structure, with an r/a ratio of 0.48, possesses a PBG centered at 517.72 nm and a bandwidth of 12.18 nm. This PBG, being located at the green spectral region, does not coincide with the

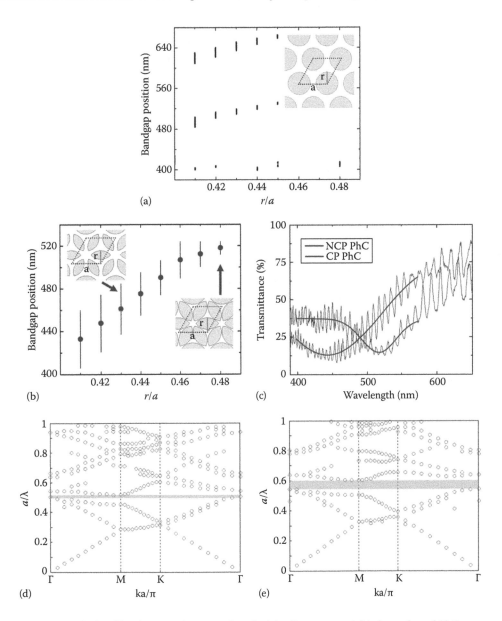

FIGURE 16.39 Calculated bandgap as a function of r/a for (a) pillar arrays and (b) clover-shaped PhCs structures, (c) measured optical reflection spectrum from clover-shaped PhCs, simulated band diagram of (d) CP and (e) NCP clover-shaped structure, predicting PBG centered at ~517.72 and ~460.75 nm, respectively. (Reprinted with permission from Li, K.H., Ma, Z.T., and Choi, H.W., *Appl. Phys. Lett.*, 100(14). Copyright 2012, American Institute of Physics.)

blue-light emission from the InGaN/GaN MQWs of the wafers used in this study. However, the computed results also indicate that the PBG shifts gradually toward shorter wavelengths with decreasing r/a ratios, together with a broadening of the bandwidth to 53.87 nm at r/a of 0.41.

To achieve a reduction in r/a, a selective dry etch step [34] can be inserted into the process flow to reduce the sphere radius r prior to pattern transfer onto GaN, ensuring that the centroids of the spheres do not shift so that the pitch a remains unchanged. Compared with the CP pillar array, spacing between pillars has been established with diameters of pillars reducing to ~230 nm. The sphere shrinkage process is repeated for the second sphere coating. The resultant NCP clover-shaped structure, as illustrated in Figure 16.38D, has an r/a ratio of ~0.43, and is predicted to have a complete TE-PBG in the blue spectral region according to Figure 16.39b.

To experimentally determine the position of the PBG, an optical transmission measurement is conducted in the near-planar direction. The incident beam from a broadband solid-state plasma light source is focused onto the sample while the transmitted signal is measured by an optical spectrometer. Two pronounced dips with center wavelengths at 517.97 and 447.85 nm are clearly observed in Figure 16.39c from the transmission spectra of the CP and NCP clover-shaped structures, corresponding to their respective bandgap positions. Transmission within the bandgap region does not fall to zero, attributed to the outcoupling of light induced by random disorder within the PhCs. Such disorders originate from point and line defects, formed during the coating process, mainly arising from nonuniformities of sphere sizes and geometrical irregularities. At wavelengths beyond the range of frequencies covered by the PBG, a relatively high transmission is maintained. The calculated band structure for the NCP clover-shaped structure shown in Figure 16.39e also confirms the presence of a PBG at the frequency range a/λ of 0.545–0.604, corresponding to wavelengths of 437.09–484.40 nm. Figure 16.39d shows the narrower PBG located at the green-light spectral region of 511.63–523.81 nm (a/λ = 0.504–0.516) for the CP array. The measured transmittance data correlate well with simulated results.

The optical enhancement of PhCs incorporation is assessed by conducting an angular-resolved PL measurement on the structures. The integrated PL intensity is collected at 1° intervals over an angular range of 90° at room temperature. Owing to the high refractive index contrast at the GaN/air planar interface, a light beam striking the interface at incident angles beyond the critical angle determined by $\sin^{-1}(n_{air}/n_{GaN})$ ≈ 24.6° will remain confined due to TIR and sequentially lost due to the reabsorption by the active layer. Hence, the rapid drop in the PL intensity at angles beyond ~30° with respect to the normal, giving a narrow escape cone. According to the angular PL plot in Figure 16.40a, the FWHM of emission divergence increases from 128.72° in the unpatterned sample to 138.26° in the NCP PhC and 137.18° in the CP PhC; the clover-shaped PhCs can indeed expand the escape cone of light. Although the PBG position of the NCP clover-shaped structure does not correspond to the emission spectrum, the nanotextured surface can still increase the probability of light escaping from the wafer via surface scattering, diminishing the losses caused by TIR. Compared with the unpatterned sample, a threefold increase in the PL intensity is observed for the NCP structure. The result indicates that proper designing of a PhC is crucial for maximizing light extraction. The approach described here offers the capability of bandgap tuning, allowing optimal overlap between the PBG and emission wavelengths.

To further investigate the emission behavior of the PhCs, the PL spectra at each angle are measured; the individual normalized spectra are combined to generate an angular-resolved emission pattern. For the as-grown sample, sharp Fabry–Pérot interference fringes are observed, as shown in Figure 16.40b, resulting from multiple reflections at the GaN/sapphire and the air/GaN interfaces, both with high refractive index contrasts. The NCP PhC structure maintains uniform emission intensity with faint fringes, as shown in Figure 16.40c, signifying that the optical confinement effect has been weakened. The improvement is attributed to the Brillouin zone folding. Embedding such "weak" PhCs on top of the GaN wafer causes the dispersion curves of Bloch modes to become folded at the Brillouin zone boundary, allowing phase matching to the radiation modes that lie above this cutoff frequency. The PhCs can further Bragg scatter the light emitted from the active region to avoid Fabry–Pérot oscillations. The CP clover-shaped structure, on the other hand, acts as a "strong" PhCs via a different mechanism—the PBG effect—which significantly alters the properties of

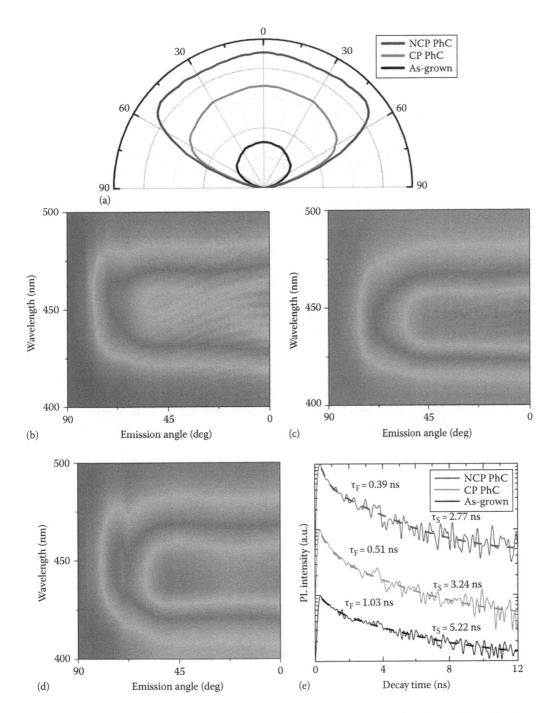

FIGURE 16.40 (a) Angular PL plot showing integrated PL intensity versus azimuthal angle. Angular-resolved emission patterns for (b) as-grown, (c) CP, and (d) non-close-packed (NCP) clover-shaped PhCs. (e) Time-resolved PL decay profiles measured at room temperature. These decay profiles are fitted by double-exponential decay curves. (Reprinted with permission from Li, K.H., Ma, Z.T., and Choi, H.W., *Appl. Phys. Lett.*, 100(14). Copyright 2012, American Institute of Physics.)

light propagation. The fringe-free spectrum obtained in Figure 16.40d indicates that the lateral propagation of Bloch guided modes is prohibited by the PBG, thereby the light generated from MQWs will couple directly to radiation modes. Additionally, the spectrum from the clover-shaped structures exhibits a spectral blueshift of ~6 nm compared with the as-grown sample, attributed to partial strain relaxation of the MQW.

A time-resolved PL measurement is then carried out to examine the carrier dynamics. The PL decay rate can generally be expressed as the sum of radiative and nonradiative recombination rates. The experimental data are fitted into two exponential decay profiles, as shown in Figure 16.40e. The fast decay component τ_F, which is strongly dependent upon thermally activated nonradiative recombinations at room temperature, is shortened to 0.39 and 0.51 ns for the NCP and CP PhCs, respectively, from 1.03 ns for the as-grown sample, attributed to higher surface recombination velocity as a result of increased etched sidewalls. On the other hand, the clover-shaped nanostructures indeed enhance carrier localization and reduce the quantum-confined stark effect, thereby accelerating the radiative recombination rates, as evident from the slow decay component, τ_S.

In summary, the formation of CP and NCP clover-shaped PhCs has been demonstrated by dual-step NSL. The PBG determined from measured optical transmission spectra agree well with the band structure as calculated by modified 3D FDTD simulations. A threefold enhancement of the PL intensity is observed from the NCP clover-shaped PhCs, which has been optimized for the MQW emission band. Shortened PL decay lifetimes observed at room temperature from PhCs structures suggest nanocavities effects.

16.6 Enhanced Visible and IR Emissions in PhC Thin-Film LEDs

While such solid-state emitters are generally more energy efficient than their predecessors, their performances are often limited by heat-related side effects at high-current operations, including brightness recession, color shift, and lifetime shortening. The thermal performances of GaN-on-sapphire LEDs are limited by the poor thermal conductivity of sapphire. Integration with cooling components such as heat sinks or even peltier coolers improves heat dissipation but makes the manufacturing process more tedious and costly. The laser liftoff (LLO) technique [40], being capable of separating the epitaxial GaN film from its sapphire substrate, allows both faces of the semiconductor film to be accessible for contacting, thus enabling vertical conduction.

While there have been vast numbers of papers reporting on PhC effects at visible and UV wavelengths, its role with regard to IR wavelengths for InGaN LEDs has not been considered. Understanding the optical properties of the PhC in the IR region is meaningful even though the devices emit in the visible range as the temperatures of the chips are directly affected by IR radiation, a means of heat loss particularly at elevated temperatures. In this section, the optical and thermal effects of a thin-film InGaN LED incorporating a 2D PhC patterned by NSL are investigated. The thin-film platform is ideal for such studies as it allows the devices to be driven harder. Apart from the established effects of improved electrical and optical (at visible wavelengths) performances, the PhC is found to enhance the extraction of IR light simultaneously, contributing to improved heat dissipation via radiation at high-power operations.

16.6.1 Experimental Details

The wafers used in this study are grown on a c-plane sapphire substrate by MOCVD. The epitaxial structure consists of 3 μm of undoped GaN, 2.5 μm of Si-doped GaN, 10 periods of InGaN/GaN QW with center emission wavelength of ~470 nm and capped with a 0.25 μm Mg-doped contact GaN layer. Around 200 nm of ITO is deposited on top of p-GaN as a current-spreading layer. The sapphire surface is polished to optical smoothness and the wafer is diced into samples of 5 × 5 mm² by laser micro-machining, which are then mounted p-face downward onto Cu submounts. The collimated beam from a 266 nm Nd:YAG laser (Continuum Surelite) is uniformly irradiated through the sapphire surfaces of the samples for LLO, during which GaN decomposes into gaseous N_2 and Ga droplets at the interface, after which the entire sapphire substrate can be mechanically removed. The detached films are then immersed into dilute HCl for dissolving

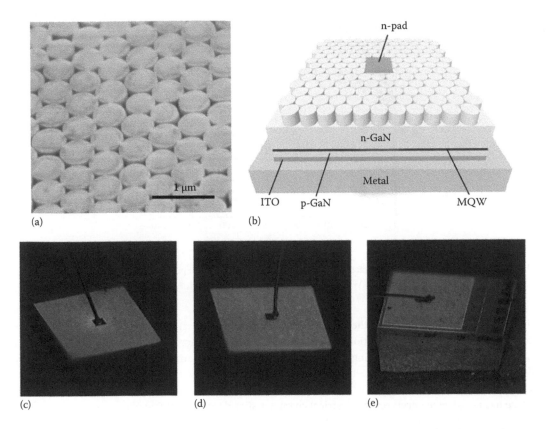

FIGURE 16.41 (a) FE-SEM image showing the HCP nanopillar array fabricated by NSL. (b) Schematic diagram (not to scale) of the vertical PhC thin-film LED. Microphotographs of the (c) V-PhCLED, (d) V-LED, and (e) L-LED. (Reprinted with permission from Cheung, Y.F., Li, K.H., Hui, R.S.Y., and Choi, H.W., *Appl. Phys.*, 105(7). Copyright 2014, American Institute of Physics.)

the residual droplets. To expose the *n*-GaN contact layer, the undoped GaN layer is removed by ICP etching using a gaseous mixture of BCl_3 and He, at a pressure of 5 mTorr with platen and coil RF power of 100 W and 400 W, respectively. A monolayer of HCP silica nanospheres with nominal diameters of 500 nm is spin coated onto the exposed *n*-GaN surface. The coated samples are then subjected to ICP etching for pattern transfer. After removal of spheres through sonication, a nanopillar array with a diameter of 500 nm and a height of 600 nm is formed, as shown in the SEM image in Figure 16.41a. The device mesa is then photo-lithographically defined so that the unmasked regions are etched by ICP. After a second photo-lithographic definition, Ti/Al/Ti/Au electrodes are deposited by electron beam evaporation and subsequently annealed in an N_2 ambience to form ohmic contacts. The schematic diagram of a vertical PhC thin-film LED (denoted V-PhCLED thereafter) is depicted in Figure 16.41b. For comparison, a thin-film LED without PhC (V-LED) and a conventional laterally conducting GaN-on-sapphire LED (L-LED) are also fabricated. All three devices, with identical mesa areas of 600 × 600 μm², are wire bonded but unencapsulated for evaluation. Microphotographs of the three emitting devices are shown in Figure 16.41c through e.

16.6.2 Device Characterizations

The light output intensity–current–voltage (L–I–V) characteristics of the packaged devices are plotted in Figure 16.42. At an injection current of 20 mA, the forward voltages of the L-LED, V-LED, and V-PhCLED, as extracted from the plot, are 4.0, 3.0, and 3.2 V, respectively. Expectedly, the vertical configuration allows

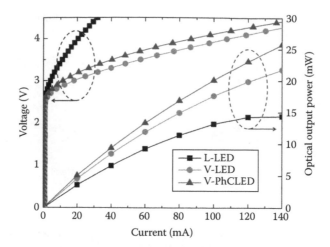

FIGURE 16.42 Plots of L–I–V characteristics of the packaged devices. (Reprinted with permission from Cheung, Y.F., Li, K.H., Hui, R.S.Y., and Choi, H.W., *Appl. Phys.*, 105(7). Copyright 2014, American Institute of Physics.)

higher currents to flow through the equally sized junctions at the same bias voltage due to significant shortening of the current conduction pathway from hundreds of microns in a laterally conducting LED to a few microns, thereby reducing resistances. Likewise, the dynamic resistances of the V-LED and V-PhCLED of 8.1 and 7.7 Ω, respectively, as determined from the slopes of the I–V curves in the linear regions, are significantly lower than that of the L-LED of 43.8 Ω. Notably, the formation of the PhC on the *n*-face has no adverse effects on the electrical characteristics despite involving plasma processes [42]; the same would not have been true for *p*-GaN patterning.

EL measurements are conducted to characterize the optical performance of the LEDs. The light emitted from the packaged devices is collected within a 4 in. integrating sphere coupled to a radiometrically calibrated spectrometer via an optical fiber. At lower currents, the optical power of the L-LED rises almost linearly with increasing currents, as shown in the L–I curves in Figure 16.42, before thermal effects set in. With further increases in currents, the rate of increase in output power declines and the optical output eventually saturates at currents above 120 mA. For the thin-film devices, the L–I relations remain nearly linear over the measured current range, having been spared of heat-induced degradations. At 20 mA, the V-LED and V-PhCLED emit ~28% and ~42% more light than the L-LED, respectively. The 28% enhancement of the V-LED over the L-LED can be attributed to a combination of reduced absorption due to the removal of the light-absorbing sapphire and direct extraction of light from GaN to air without having to pass through the light-absorbing ITO layer. Thermal effects are assumed to be negligible at this low current. The additional 14% offered by the V-PhCLED over the V-LED is a result of "weak" PhC effects on light extraction. Note that the CP nanopillar array does not possess a PBG; instead, it causes band folding at the Brillouin zone edge so that light from guided modes is coupled into leaky modes above the light line, thus promoting light extraction. The band-folding capability of the PhC is evident from the computed TE band structure shown in Figure 16.43a (TE mode dominates emissions from InGaN/GaN MQWs). An FDTD simulation is also conducted to study the interaction of light with the PhC, based on 3D models constructed according to the fabricated structure, injected with a continuous-wave dipole at $\lambda = 470$ nm (being λ_{peak} of the LEDs). The simulated plots of electric-field intensity distribution for an unpatterned GaN/air interface and the PhC are illustrated in Figure 16.43b and c, respectively. Quantitatively, the simulation predicts that the PhC increases light transmission by ~11.2% compared with a flat interface, consistent with experimental results.

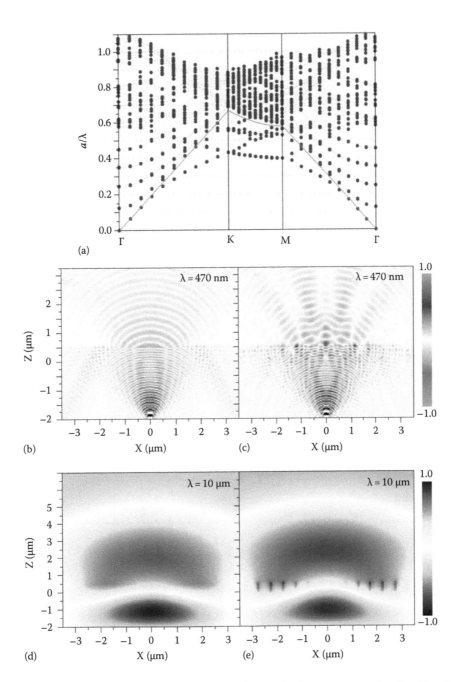

FIGURE 16.43 (a) TE band structure of the PhC computed using the plane wave expansion algorithm. Simulation radiation profiles of continuous waves at wavelengths of 470 nm and 10 μm propagating through the flat-top [(b) and (d)] and PhC interfaces [(c) and (e)]. (Reprinted with permission from Cheung, Y.F., Li, K.H., Hui, R.S.Y., and Choi, H.W., *Appl. Phys.*, 105(7). Copyright 2014, American Institute of Physics.)

16.6.3 Optical and Thermal Analyses

At higher currents, light extraction effects remain unchanged, while thermal effects begin to kick in. In the L-LED, the reducing rate of increase in output power is mainly due to the drop in IQE at elevated temperatures. At a current of 140 mA, the V-LED emits 51% more optical power than the L-LED, of which 28% is attributed to temperature-independent light extraction effects. The thermal effect on the L-LED is associated with the sapphire substrate with high thermal resistance. One would then expect thermal effects to be identical in both vertical LEDs with identical thermal conduction pathways, but this is found not to be the case. As observed from Figure 16.42, the L–I characteristics of the V-LED appear to deviate from linearity at high currents to a greater extent than of the V-PhCLED. For a clearer view, the light output enhancement ratios of the V-PhCLED over the V-LED at different currents are plotted in Figure 16.44a; the ratio increases from 11% to 18% as the current is raised from 20 to 140 mA. Noting that light extraction is temperature independent, the observation suggests that the temperature of V-LED is higher than that of the V-PhCLED.

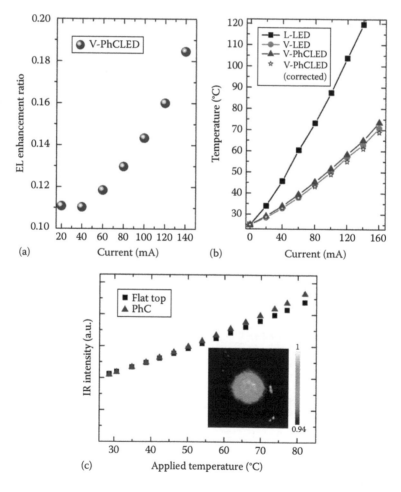

FIGURE 16.44 (a) EL enhancement ratios of the V-PhCLED over the V-LED, as calculated from the EL data of Figure 16.42. (b) The variations in surface temperatures versus current as measured by LWIR camera. (c) IR intensities as a function of applied temperature. The inset shows the LWIR image illustrating intensity contrast between the PhC (circular central region) and the flat-top GaN surface (outside the central region). The scale bar on the right represents the normalized LWIR intensity. (Reprinted with permission from Cheung, Y.F., Li, K.H., Hui, R.S.Y., and Choi, H.W., *Appl. Phys.*, 105(7). Copyright 2014, American Institute of Physics.)

The surface temperatures of the LEDs are measured by IR thermometry using a long-wave infrared (LWIR) camera (FLIR SC645), adopting emissivity values for GaN and ITO of 0.62 and 0.265, respectively, which have been experimentally determined by comparison with a reference heat source. Multiple data points are taken across the LEDs from the LWIR map, from which an average temperature is evaluated and the standard error of measurement is 0.89%. Figure 16.44b plots the average surface temperatures of the three devices as a function of current. At 20 mA, the measured temperatures of all LEDs are nearly identical and are close to the ambient temperature. Hence, the earlier-made assumption of negligible thermal effect experienced by the devices at this low current is justified. Subsequently, the temperatures of the thin-film LEDs rise at much slower rates than of the L-LED. At 140 mA, whereby light output saturates, the L-LED attains a surface temperature of 120°C, or 57°C higher than that of the V-LED of 63°C, which can simply be attributed to poor thermal conductivity of sapphire, exacerbated by higher dynamic resistance. In terms of optical power, the V-LED emits 51% more light than the L-LED, of which 23% is related to changes in IQE, which is temperature dependent. According to the temperature dependence of IQE in InGaN/GaN LEDs, as reported in Reference 43, a reduction in temperature by 57°C gives rise to an increase in IQE by ~29% based on linear interpolation of the given data, which is in good agreement with our measured data.

The temperature data in Figure 16.44b also indicate that the V-PhCLED attains higher temperatures than the V-LED over the entire measurement range, which is in stark contradiction to the information conveyed by the optical data. Since the V-PhCLED emits more light with similar I–V characteristics compared with the V-LED, its chip temperature should also be inferred as being lower. This intriguing observation can be explained by interpreting the LWIR data in terms of LWIR signal strength rather than absolute temperature, which has been evaluated based on the assumption of equal emissivities. Therefore, the LWIR data should be interpreted as increased emission of IR signal from the V-PhCLED over the V-LED, rather than increased temperatures. In other words, the PhC structure on the surface of the V-PhCLED enhances LWIR radiation generated from devices.

To further investigate this phenomenon, an identical PhC structure is selectively patterned in the central region of a 5×5 mm^2 n-doped GaN sample, leaving the surrounding area flat top. The sample is then placed on a hot plate for a sufficiently long period to ensure that the entire body reaches thermal equilibrium. The LWIR intensities are evaluated from the temperature readings using Planck's law. The captured LWIR map of the sample is shown in the inset of Figure 16.44c, which reflects the variation in LWIR intensities detected across the sample. Even though the sample attains thermal equilibrium, the LWIR intensities emitted from the PhC and flat-top regions of the sample differ. Figure 16.44c plots the LWIR intensities at the center of the PhC and at the flat-top region of the sample as a function of applied temperature. When the applied temperature is close to the ambient temperature, the detected LWIR intensities from both points are nearly the same. With an increase in the applied temperature, the PhC region emits more IR radiation than the surrounding flat-top surface, with an enhancement factor of 6.9% at 86°C. Unfortunately, as indicated by the band structure diagram shown in Figure 16.43a, the band-folding effect on guided modes is only valid up to near-IR wavelengths of ~1.25 μm and thus may not contribute to light extraction in the mid-IR range (5–30 μm). The relation $\Lambda_{cutoff} \approx \lambda/(n_{eff}+1)$ also indicates that a 2D PhC with a lattice constant of 500 nm is unlikely to introduce any band-folding effects at wavelengths >1.6 μm. Therefore, Λ_{cutoff}, the enhancement of IR intensity, is mainly attributed to an enlarged exposed surface area provided by the nanopillar PhC for heat exchange with the ambient air, thereby promoting the emission of LWIR radiation. To verify this postulation, a similar FDTD simulation is carried out, this time at LWIR wavelengths of 7.5–13 μm. The simulated plots of electric-field intensity distribution for an unpatterned GaN/air interface and the PhC at 10 μm are illustrated in Figure 16.43d and e, respectively. Quantitatively, the PhC enhances IR radiation by 4.2%–7.4% between 7.5 and 13 μm, consistent with the experimental data obtained from LWIR thermography. Although higher LWIR intensities are detected from the PhC region, the temperatures of the PhC and flat-top regions are actually identical. In other words, the emissivity of the PhC region has been enhanced and is determined to be 0.70 (compared with 0.62 of flat-top GaN). Using the true value of emissivity for the PhC, the corrected temperature of the V-PhCLED is evaluated and replotted in Figure 16.44b. Evidently, the V-PhCLEDs have the lowest surface temperatures among the devices, consistent with the fact that these devices also emit the most light. Therefore, the PhC enhances light extraction in the visible and IR regions simultaneously, so devices emit more light while radiating more heat at higher currents, a feature that is beneficial to the overall efficiencies of the emitters.

16.7 Conclusion

In this chapter, NSL has been demonstrated as a practical alternative approach toward large-scale nano-fabrication. Silica spheres acting as etch masks are self-assembled into monolayers and transfer patterns to the substrate after dry etching. A wide range of dimensions of silica spheres are used to generate two types of well-ordered periodic structures, namely, weak and strong PhCs. Both PhCs nanostructures enable a strong interaction between the guided modes introduced by TIR so as to realize the enhancement of light extraction from LEDs. Weak PhCs composed of CP pillars can extract light from LEDs via dispersion behavior and the diffraction property of PhCs, while strong PhCs patterned by modified NSL are capable of suppressing undesired optical guiding modes via PBG and of promoting light extraction. The optical performances and mechanisms of weak and strong PhCs are also thoroughly investigated. Additionally, devices with PhCs can benefit from enhanced IR extraction due to increased surface areas, further reducing chip temperatures to give higher IQEs. The superior thermal properties have also been demonstrated and verified by IR thermometric imaging. Ordered hexagonal nanostructures patterned by NSL open an opportunity and become a breakthrough to realize the mass production of PhC structures on light emitters.

References

1. C. C. Kao, H. C. Kuo, H. W. Huang, J. T. Chu, Y. C. Peng, Y. L. Hsieh, C. Y. Luo, S. C. Wang, C. C. Yu, and C. F. Lin, *IEEE Photon Technol* 17 (1), 19 (2005).
2. C. E. Lee, Y. C. Lee, H. C. Kuo, M. R. Tsai, T. C. Lu, and S. Cwang, *Semicond Sci Technol* 23 (2), 025015 (2008).
3. X. H. Wang, P. T. Lai, and H. W. Choi, *J Appl Phys* 108 (2), 023110 (2010).
4. S. X. Jin, J. Li, J. Y. Lin, and H. X. Jiang, *Appl Phys Lett* 77 (20), 3236 (2000).
5. H. W. Choi and S. J. Chua, *J Vac Sci Technol B* 24 (2), 800 (2006).
6. H. W. Choi, M. D. Dawson, P. R. Edwards, and R. W. Martin, *Appl Phys Lett* 83 (22), 4483 (2003).
7. T. Fujii, Y. Gao, R. Sharma, E. L. Hu, S. P. DenBaars, and S. Nakamura, *Appl Phys Lett* 84 (6), 855 (2004).
8. Y. Gao, T. Fujii, R. Sharma, K. Fujito, S. P. Denbaars, S. Nakamura, and E. L. Hu, *Jpn J Appl Phys* 243 (5A), L637 (2004).
9. M. S. Minsky, M. White, and E. L. Hu, *Appl Phys Lett* 68 (11), 1531 (1996).
10. C. C. Yang, R. H. Horng, C. E. Lee, W. Y. Lin, K. F. Pan, Y. Y. Su, and D. S. Wuul, *Jpn J Appl Phys* 144 (4B), 2525 (2005).
11. T. N. Oder, J. Shakya, J. Y. Lin, and H. X. Jiang, *Appl Phys Lett* 83 (6), 1231 (2003).
12. S. H. Kim, K. D. Lee, J. Y. Kim, M. K. Kwon, and S. J. Park, *Nanotechnology* 18 (5), (2007).
13. T. Akasaka, H. Gotoh, T. Saito, and T. Makimoto, *Appl Phys Lett* 85 (15), 3089 (2004).
14. A. David, T. Fujii, R. Sharma, K. McGroddy, S. Nakamura, S. P. DenBaars, E. L. Hu, C. Weisbuch, and H. Benisty, *Appl Phys Lett* 88 (6), 061124 (2006).
15. S. D. Lester, F. A. Ponce, M. G. Craford, and D. A. Steigerwald, *Appl Phys Lett* 66 (10), 1249 (1995).
16. Z. L. Li, K. H. Li, and H. W. Choi, *J Appl Phys* 108 (11), 114511 (2010).
17. H. Matsubara, S. Yoshimoto, H. Saito, J. L. Yue, Y. Tanaka, and S. Noda, *Science* 319 (5862), 445 (2008).
18. J. C. Knight, *Nature* 424 (6950), 847 (2003).

19. C. F. Lai, J. Y. Chi, H. H. Yen, H. C. Kuo, C. H. Chao, H. T. Hsueh, J. F. T. Wang, C. Y. Huang, and W. Y. Yeh, *Appl Phys Lett* 92 (24), 243118 (2008).

20. J. J. Wierer, M. R. Krames, J. E. Epler, N. F. Gardner, M. G. Craford, J. R. Wendt, J. A. Simmons, and M. M. Sigalas, *Appl Phys Lett* 84 (19), 3885 (2004).

21. D. H. Kim, C. O. Cho, Y. G. Roh, H. Jeon, Y. S. Park, J. Cho, J. S. Im et al., *Appl Phys Lett* 87 (20), 203508 (2005).

22. C. L. Haynes and R. P. Van Duyne, *J Phys Chem B* 105 (24), 5599 (2001).

23. Q. Zhang, K. H. Li, and H. W. Choi, *2012 12th IEEE Conference on Nanotechnology (IEEE-Nano)* (2012).

24. Q. Zhang, K. H. Li, and H. W. Choi, *IEEE Photon Tech L* 24 (18), 1642 (2012).

25. K. H. Li and H. W. Choi, *J Appl Phys* 110 (5), 053104 (2011).

26. C. Wiesmann, K. Bergenek, N. Linder, and U. T. Schwarz, *Laser Photon Rev* 3 (3), 262 (2009).

27. Y. J. Lee, S. H. Kim, J. Huh, G. H. Kim, Y. H. Lee, S. H. Cho, Y. C. Kim, and Y. R. Do, *Appl Phys Lett* 82 (21), 3779 (2003).

28. T. N. Oder, K. H. Kim, J. Y. Lin, and H. X. Jiang, *Appl Phys Lett* 84 (4), 466 (2004).

29. J. J. Wierer, A. David, and M. M. Megens, *Nat Photon* 3 (3), 163 (2009).

30. S. H. Fan, P. R. Villeneuve, J. D. Joannopoulos, and E. F. Schubert, *Phys Rev Lett* 78 (17), 3294 (1997).

31. H. M. Kim, Y. H. Cho, H. Lee, S. I. Kim, S. R. Ryu, D. Y. Kim, T. W. Kang, and K. S. Chung, *Nano Lett* 4 (6), 1059 (2004).

32. K. H. Li, K. Y. Zang, S. J. Chua, and H. W. Choi, *Appl Phys Lett* 102 (18), 181117 (2013).

33. E. Yablonovitch, *Phys Rev Lett* 58 (20), 2059 (1987).

34. K. H. Li and H. W. Choi, *J Appl Phys* 109 (2), 023107 (2011).

35. P. G. Eliseev, G. A. Smolyakov, and M. Osinski, *IEEE J Sel Top Quant* 5 (3), 771 (1999).

36. M. Boroditsky, T. F. Krauss, R. Coccioli, R. Vrijen, R. Bhat, and E. Yablonovitch, *Appl Phys Lett* 75 (8), 1036 (1999).

37. H. X. Jiang and J. Y. Lin, *Crit Rev Solid State* 28 (2), 131 (2003).

38. K. H. Li, Z. T. Ma, and H. W. Choi, *Appl Phys Lett* 100 (14), (2012).

39. Z. T. Ma and K. Ogusu, *Opt Commun* 282 (7), 1322 (2009).

40. W. S. Wong, T. Sands, and N. W. Cheung, *Appl Phys Lett* 72 (5), 599 (1998).

41. Y. F. Cheung, K. H. Li, R. S. Y. Hui, and H. W. Choi, *Appl Phys Lett* 105 (7), 071104 (2014).

42. S. J. Chua, H. W. Choi, J. Zhang, and P. Li, *Phys Rev B* 64 (20), 205302 (2001).

43. W. W. Chow, *Opt Express* 22 (2), 1413 (2014).

17

ZnO-Based LEDs

Hao Long

*School of Electronics and
Information Engineering
South-Central University
for Nationalities*

Abstract Zinc oxide (ZnO) is one of the II–VI semiconductor materials with a wide direct bandgap of 3.37 eV and a large exciton binding energy of 60 mV. In addition, ZnO is environmentally friendly and presents high thermal stability, low cost, and abundant availability. Because of these advantages, ZnO is considered as one of the most promising materials for generating high-efficiency short-wavelength excitonic luminescence and fabricating semiconductor lasers with a low threshold. In this chapter, properties of ZnO are introduced and recent progress in ZnO-based light-emitting diode (LED) research is reviewed. The ZnO thin-film LEDs and ZnO nanostructure LEDs are discussed in detail. With regard to ZnO thin-film LEDs, though ZnO-based homojunction LEDs have been continuously researched, the lack of reproducibility of high-quality *p*-type ZnO is still the bottleneck hindering their further development. Then, the advantages of ZnO are explored and exploited by forming ZnO-based heterojunctions, such as metal–insulator–semiconductor (MIS) heterojunction, *n–n* heterojunction, *p–n* heterojunction, and other complicated structures, including multiple heterojunction LEDs, and quantum well (QW) and multiple quantum well (MQW) LEDs. Meanwhile, various kinds of novel ZnO-based LEDs with nanoscale structures have been fabricated recently. They show exciting and satisfactory performance. With regard to ZnO-based LEDs, we have much more work to do. We believe they will emit bright short-wavelength light, even bright short-wavelength laser, in future.

17.1 Introduction

There has been a great deal of interest in zinc oxide (ZnO) semiconductor materials in recent decades, as seen from a large number of publications. ZnO is not new to the semiconductor field, and the study of it dates back to 1935 [1]. It is now a well-known material that has many applications in such diverse fields as abrasives, brake linings, cosmetics, dental cements, lubricants, paints, phosphors, rubber products, and other devices [2–14]. Also, it has many applications in the optoelectronics field owing to its direct wide bandgap (E_g = 3.37 eV at 300 K) and large exciton binding energy (60 meV), which should pave the way for exciton-based lasing action with very low threshold currents even above room temperature. As we know, gallium nitride (GaN), another wide-bandgap semiconductor with a bandgap of 3.4 eV at 300 K, has been widely used for the production of green, blue-ultraviolet (UV), and white light-emitting diodes (LEDs) and laser devices. Actually, some optoelectronic applications of ZnO overlap with those of GaN.

However, comparing with GaN, the advantages of ZnO, such as simpler crystal-growth technology, larger exciton binding energy, environmental friendliness, higher thermal stability, lower cost, and abundant availability, make it suitable for the next generation of optoelectronic devices, and many research groups have been making efforts on the ZnO material and ZnO-based optoelectronic devices.

Previous reviews of ZnO focused mainly on material processing, doping, and transport properties, while in this chapter, we pay attention to the recent progress in ZnO-based LED research. The organization of this chapter is as follows: First, properties of ZnO, including its basic characterization and conductivity type, are presented in Section 17.2. Followed by ZnO-based LEDs, in Section 17.3, we will discuss ZnO thin-film LEDs and ZnO nanostructure LEDs in detail. Finally, a summary of this chapter and future outlook of ZnO-based LEDs are presented in Section 17.4.

17.2 Properties of ZnO

17.2.1 Basic Characterization

ZnO is a wide-gap semiconductor. The E_g at 0 K is 3.441 ± 0.003 eV and E_g at 300 K is 3.365 ± 0.005 eV. It belongs to the group of II^b–VI compound semiconductors. It crystallizes almost exclusively in the hexagonal wurtzite-type structure (point group C_{6v} or 6 mm, space group C_{6v}^4 or P6$_3$mc). Compared with similar II^b–VI (e.g., ZnS) or III–V (e.g., GaN) semiconductors, it has a relatively strong polar binding and relatively large exciton binding energy of 59.5 ± 0.5 meV. Due to the light mass of oxygen, the upmost LO phonon has a relatively high energy of 72 meV. The lattice dimensions of ZnO are found to be $a_0 = 3.24265$ ± 0.0001, $c_0 = 5.1948 ± 0.0003$, and axial ratio $c_0 a_0 = 1.60200 ± 0.0001$, all at 18°C [1]. The density of ZnO is 5.6 g/cm^3 corresponding to 4.2×10^{22} ZnO molecules/cm^3. These data can be found, for example, in a data collection [15,16] and references therein.

Calculations on the defect energetics and electronic structure in ZnO have been reviewed by Wang et al. [17]. The calculations consistently indicate that oxygen vacancies (V_O), zinc vacancies (V_{Zn}), and oxygen interstitials (O_i) create deep energy levels in the bandgap, and among these, V_O has the lowest formation energy [18]. Reliable results concerning the charge transition levels can be obtained by aligning the electronic band structure through an external potential, which yields the +2/0 charge transition level of the V_O in ZnO to be 2.2–2.4 eV from the valence band maximum (VBM) [19]. Figure 17.1 shows the schematic illustration of the excitation and emission processes of ZnO [17]. Also, Alvi et al. [20] reported the schematic band diagram of the deep-level emissions in ZnO, as shown in Figure 17.2, based on the full-potential linear muffin-tin orbital method and the reported data [21–32].

17.2.2 Conductivity Type

As mentioned earlier, ZnO shows a strong potential for various short-wavelength optoelectronic device applications. In order to attain the potential offered by ZnO, both high-quality n- and p-type ZnO are indispensable [34]. However, difficulty in bipolar carrier doping (both n- and p-types) is a major obstacle, as seen in other wide-bandgap semiconductors such as GaN and II–VI compound semiconductors, including ZnS, ZnSe, and ZnTe [35–40]. Except for ZnTe, in which p-type doping can be easily done, most wide-bandgap inorganic semiconductors, such as ZnO, GaN, ZnS, and ZnSe, are easily doped to n-type, while p-type doping is very difficult.

ZnO with a wurtzite structure is naturally an n-type semiconductor because of a deviation from stoichiometry due to the presence of intrinsic defects such as O vacancies (V_O) and Zn interstitials (Zn_i) [41]. Also, doping ZnO n-type is easy with group III elements on Zn place [42]. The main obstacle to the development of ZnO material and ZnO-based devices is the lack of reproducible and low-resistivity p-type ZnO. Many research groups have focused on the fabrication of p-ZnO and the result is that group I elements (Li, Na, K) on Zn place from deep acceptors result in high-resistivity material but not efficient p-type conductivity. Group V elements (N, P, As, Sb) on O place result in p-type conductivity, sometimes

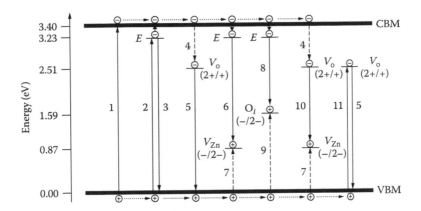

FIGURE 17.1 Schematic illustration of the excitation and emission processes of ZnO. (1) Band-to-band excitation; (2) and (3) adsorption and emission by excitons E; (4), (7), and (9) trapping of charge carriers by V_O, V_{Zn}, and O_i; (5) radiative recombination of deeply trapped electrons at V_O with holes from the valence band; (6) and (8) radiative recombination of shallowly trapped electrons with deeply trapped holes at V_{Zn} and O_i; (10) donor–acceptor transition involving V_O and V_{Zn}; (11) subband excitation. The defect energy levels for V_O, V_{Zn}, and O_i are provided in Reference 33, and the energy level for the excitons is determined according to the 385 nm UV emission. (Reprinted with permission from Wang, M., Zhou, Y., Zhang, Y., Jung Kim, E., Hong Hahn, S., and Gie Seong, S., Near-infrared photoluminescence from ZnO, *Appl. Phys. Lett.*, 100(10), 101906, Copyright 2012, American Institute of Physics.)

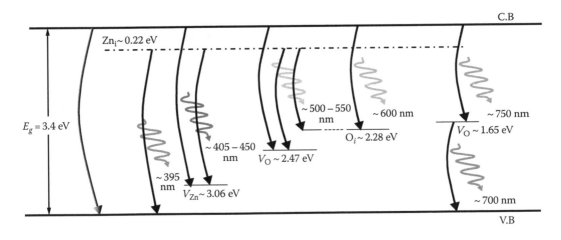

FIGURE 17.2 Schematic band diagram of the deep level emissions in ZnO based on the full-potential linear muffin-tin orbital method. (From Alvi, N.H. et al., *Nanoscale Res. Lett.*, 6(1), 130, 2011.)

also in *n*-type; meanwhile, some authors also claim about observing *p*-type conductivity after incorporation of Ag, Cu, C, and other elements [16,43].

Despite all the progress that has been made and the reports of *p*-type conductivity in ZnO films using various growth methods and various group V dopant elements, a reliable and reproducible high-quality *p*-type conductivity has not yet been achieved for ZnO. Therefore, it remains the most pivotal topic in ZnO material research. To overcome this bottleneck, a clear understanding of physical processes in ZnO is necessary. Till now, some of the basic properties of ZnO are still unclear. For example, the nature of the residual *n*-type conductivity in undoped ZnO films, whether due to impurities of some native defect or defects, is still under some degree of debate [8].

17.3 ZnO-Based LED

There are many researches on LED involving both homojunctions based on ZnO and its alloys and heterojunctions such as metal–insulator–semiconductor (MIS) heterojunction, *n–n* heterojunction, *p–n* heterojunction, and other complicated structures (multiple heterojunction LEDs, quantum well [QW] and multiple quantum well [MQW] LEDs, and so on) [44–64]. Also, various kinds of novel ZnO-based LEDs with nanoscale structures have been fabricated and they show exciting and satisfactory performance.

17.3.1 ZnO LED Based on Thin Films

17.3.1.1 ZnO Homojunction Thin-Film LED

As early as 2000, Aoki et al. [65] fabricated a ZnO homojunction diode by using a laser doping technique to form a phosphorous-doped *p*-type ZnO layer on an *n*-type ZnO substrate. The current–voltage characteristics showed a diode characteristic between the *p*-layer and the *n*-type substrate. However, light emission can be observed by forward current injection only at 110 K or even lower temperatures. In spite of this, the research results raised the hopes for the development of ZnO-based homojunction LEDs.

Subsequently, more and more people began to study this field. Tsukazaki et al. [66,67] reported a repeated temperature modulation technique as a reliable and reproducible way to fabricate *p*-type ZnO and continued fabricating a typical homostructural *p–i–n* junction [66]. Excellent rectification was obtained from current–voltage characteristics. The threshold voltage was about 7 V, which was higher than the bandgap of ZnO. It was mainly caused by the high resistivity of the *p*-type ZnO layer. The electroluminescence (EL) spectrum showed a redshift comparing the exciton emission of 3.2 eV from *i*-ZnO. The authors thought it was partly due to the low hole concentration in *p*-ZnO, that is, electron injection from *i*-ZnO to *p*-ZnO overcame hole injection from *p*-ZnO to *i*-ZnO. From this result, we can see that optimizing the growth process for increasing the hole concentration of *p*-ZnO is very important to the development of ZnO-based homojunction LEDs.

In 2006, Lim et al. [68] also reported a ZnO-based homojunction LED. Two $Mg_{0.1}Zn_{0.9}O$ epitaxial layers, as the carrier confinement layers, were deposited by co-sputtering MgO and ZnO targets on Ga-doped *n*-type ZnO, and then P-doped *p*-type ZnO was grown on it. The structure schematic diagram is shown in Figure 17.3b. This LED showed clear rectification with a threshold voltage of 3.2 V and emitted 380 nm UV light at room temperature. Figure 17.3a shows the band diagram. The energy barrier for electrons (ΔE_C) is larger than that for holes (ΔE_V). Figure 17.3c shows EL spectra of the ZnO LED with $Mg_{0.1}Zn_{0.9}O$ layers operated at forward currents of 20 and 40 mA, and the comparison of EL spectra of the *p–n* homojunction ZnO LED and the ZnO LED with $Mg_{0.1}Zn_{0.9}O$ layers operated at a forward current of 40 mA is shown in Figure 17.3d. From the results, we can see that the intensity of near-band edge (NBE) emission was increased further and the deep-level emission was greatly suppressed by using $Mg_{0.1}Zn_{0.9}O$ layers as energy barrier layers to confine the carriers to the high-quality *n*-type ZnO, and the advantages of ZnO have been fully taken. This research demonstrated the importance of carrier-confined effect in ZnO-based homojunction LEDs, and Section 17.3.1.2 will show more applications of the electron (hole)-blocking layer (EBL) in ZnO-based heterojunction LEDs.

In the same year, Xu et al. [69] reported a ZnO homojunction LED by metal–organic chemical vapor deposition (MOCVD). *p*-type ZnO thin films with a hole concentration of 10^{16}–10^{17} cm^{-3} and a mobility of 1–10 cm^2/V·s were grown using nitric oxide (NO) plasma. Room-temperature photoluminescence (PL) spectra revealed nitrogen-related emissions. EL at room temperature was demonstrated with band-to-band emission at I = 40 mA and defect-related emissions in the blue-yellow spectrum range. In the next year, Wei et al. [70] fabricated ZnO *p–n* homojunction junction LEDs on *c*-plane Al_2O_3 substrates by plasma-assisted molecular beam epitaxy. The *p*-type ZnO layers, doping with the help of a gas mixture of N_2 and O_2 by which the double-donor doping of $N_{2(O)}$ can be avoided significantly, had a high hole density and carrier mobility. The LEDs showed a very good rectification characteristic with a low threshold

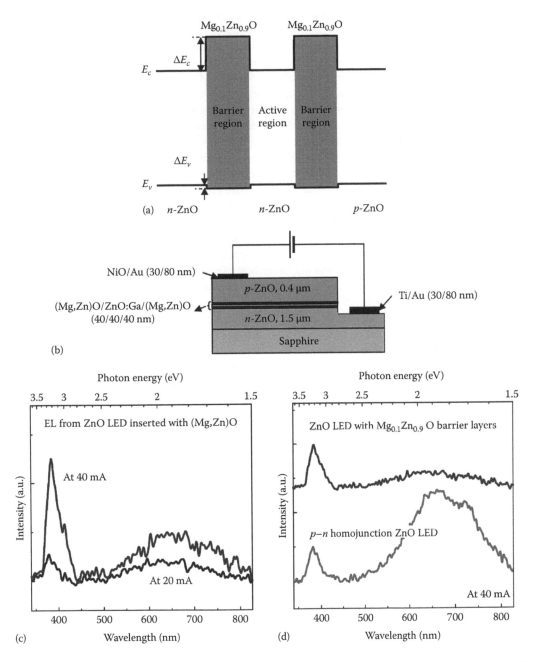

FIGURE 17.3 (a) Band diagram of the ZnO LED with $Mg_{0.1}Zn_{0.9}O$ layers. (b) Schematic diagram of a ZnO LED containing $Mg_{0.1}Zn_{0.9}O$ barrier layers for carrier confinement in *n*-type ZnO. (c) EL spectra of the ZnO LED with $Mg_{0.1}Zn_{0.9}O$ layers operated at forward currents of 20 and 40 mA. (d) Comparison of EL spectra of the *p–n* homojunction ZnO LED and the ZnO LED with $Mg_{0.1}Zn_{0.9}O$ layers, operated at a forward current of 40 mA. (From Lim, J.H., Kang, C.K., Kim, K.K., Park, I.K., Hwang, D.K., and Park, S.J.: UV electroluminescence emission from ZnO light-emitting diodes grown by high-temperature radiofrequency sputtering. *Adv. Mater.* 2006. 18(20). 2720–2724. Copyright Wiley-VCH Verlag GmbH & Co. KGaA. Reproduced with permission.)

voltage of 4.0 V even at a temperature above 300 K. Comparing previous works, the ZnO homojunction LED in this paper showed EL emission at an even higher temperature. Authors thought the blue-violet EL emission was attributed to the donor–acceptor pair recombination at the *p*-type layer of the LED.

In addition to the research discussed earlier, with the development of *p*-type ZnO in the twenty-first century, more and more researchers began to focus on the ZnO-based homojunction LEDs. In 2008, Kong et al. [71] and Chu et al. [72] both fabricated ZnO homojunction LEDs on Si substrate by molecular beam epitaxy (MBE). Kong et al. [71] reported a Sb-doped *p*-type ZnO/Ga-doped *n*-type ZnO homojunction on Si (100) substrate. Considering the large lattice mismatch between Si and ZnO, a thin magnesium oxide (MgO) buffer layer was deposited at 350°C on Si substrate. We can use this kind of buffer layer in other semiconductor electronic devices and integrate them in silicon planar technology. Current–voltage (*I–V*) and capacitance–voltage (*C–V*) measurements showed good rectification behavior, indicating that reliable *p*-type ZnO was formed. EL experiments demonstrated dominant UV emissions and insignificant deep-level-related yellow/green band emissions at different drive currents from 60 to 100 mA at room temperature, as shown in Figure 17.4. NBE emission at 3.2 eV started to appear when the current was 60 mA. And then the intensity of this emission increased as the injection current increased from 60 to 100 mA. This research discussed the influence on device performance of the heat effect in detail. As exhibited in Figure 17.4, the intensity of this emission peak increased evidently from 60 to 80 mA, while it changed less significantly from 80 to 100 mA, which was due to the heat effect as a result of the increasing current passing through the diode. Moreover, while injection current increased from 60 to 100 mA, the UV emission peak also slightly redshifted from 385 to 393 nm. This was typical in a radiative recombination process for direct bandgap material because heat induced by increased injection current decreased its effective bandgap. Despite progress, the output power of this LED was estimated to be only 1 nW at a drive current of 100 mA. To improve the efficiency, ZnO homojunctions with higher crystal quality are in progress. Zhao et al. [73] also used the Sb-doped *p*-type ZnO with a hole concentration of 1.27×10^{17} cm^{-3} to fabricate homojunction LEDs with *n*-ZnO. However, the material preparation method was not MBE but MOCVD. Trimethylantimony (TMSb) was used as the Sb-doping source. The *I–V* characteristics of the device

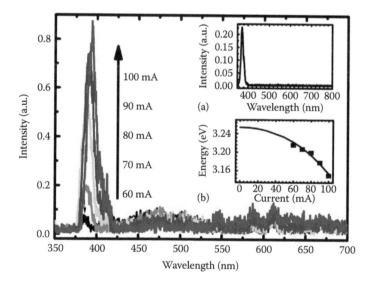

FIGURE 17.4 EL spectra of *p–n* ZnO diode at room temperature, with increasing injection current from 60 to 100 mA. (a) The inset is PL spectrum of the same device at room temperature; (b) is the fitting using Varshni's equation for near-band emission energy at different temperatures. Here, $\alpha = 8.2 \times 10^{-4}$, $\beta = 1060$. (Reproduced with permission from Kong, J., Chu, S., Olmedo, M., Li, L., Yang, Z., and Liu, J., Dominant ultraviolet light emissions in packed ZnO columnar homojunction diodes, *Appl. Phys. Lett.*, 93(13), 132113, Copyright 2008, American Institute of Physics.)

exhibited a desirable rectifying behavior with a turn-on voltage of 3.3V and a reverse breakdown voltage of greater than 5.0 V. Distinct EL with UV and visible emissions was detected from this device under forward bias at room temperature. The results indicated that this technique may pave the way for further industrialization of ZnO LEDs and laser diodes (LDs).

With the in-depth study of ZnO nanostructure, especially *p*-type ZnO nanostructure, many research groups used ZnO nanoscale homojunction to fabricate LEDs, and the results were also satisfactory. Recent papers of homojunction LEDs based on ZnO nanostructure can be found in References 74–77, and we will discuss the ZnO LEDs based on nanostructure in Section 17.3.2.

17.3.1.2 ZnO Heterojunction Thin-Film LED

While *p*-type ZnO is difficult to attain, the advantages of ZnO are being explored and exploited by forming ZnO-based heterojunctions, such as MIS heterojunction, *n–n* heterojunction, *p–n* heterojunction, and other complicated structures, such as multiple heterojunction LEDs, and QW and MQW LEDs.

17.3.1.2.1 MIS Heterojunction LEDs

Ever since the 1970s, there has been some research [78–81] about ZnO-based MIS LEDs. In 1974, Minami et al. [79] reported the ZnO MIS diode with an SiO layer as the insulating layer and an Au layer as the metal layer. As shown in Figure 17.5, the EL emission spectrum formed a broadband extending from 2.0 to 3.3 eV and consisted of an intense peak at about 3.13 eV and another broad peak at about 2.3 eV. As the temperature decreased from 300 to 100 K, the intensity of the UV peak increased about several tens of times and the emission spectrum was blueshifted, according to the increase of the bandgap of the bulk material.

Also in 1970s, Shimizu et al. [81] reported on the ZnO MIS LED made of ZnO single crystals with a green band. The single crystals were epitaxially grown on sapphire substrates by chemical vapor deposition (CVD). The surface of *n*-type ZnO was oxidized and inverted into *i*-type by dipping it in 30% solution of H_2O_2 at room temperature for hours. An Au layer of 0.5 mm diameter was then evaporated on the *i*-type surface. Figure 17.6 shows the EL and PL spectra of the ZnO MIS LED obtained at 82 and 290 K under forward-bias condition of 50 mA. The EL spectra coincided fairly well with the PL spectra, suggesting

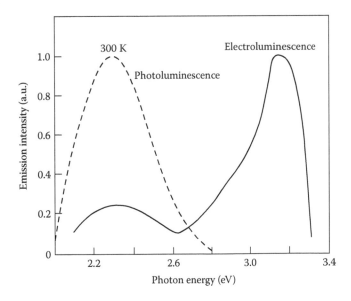

FIGURE 17.5 EL and PL spectra of ZnO at 300 K. The pulsed current (400 Hz, 10 μs) was about 2.6 A. (Reproduced with permission from Minami, T. et al., *Jpn. J. Appl. Phys.*, 13(9), 1475, Copyright 1974, The Japan Society of Applied Physics.)

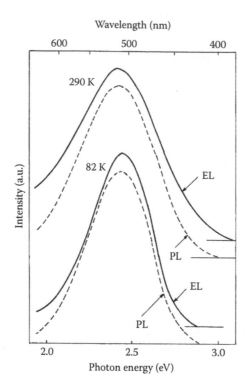

FIGURE 17.6   EL and PL spectra of the ZnO MIS LED obtained at 82 K and 290 K under forward-bias condition of 50 mA. (Reproduced with permission from Shimizu, A. et al., *Jpn. J. Appl. Phys.*, 17(8), 1435, Copyright 1978, The Japan Society of Applied Physics.)

that the emission mechanism was similar in both cases. The emission peak was located at about 2.43 eV at room temperature and shifted toward the higher energy with decreasing temperature. The temperature coefficient was about -1×10^{-4} eV/K, which was about one tenth of that of the energy gap of ZnO.

ZnO-based MIS LEDs have been in continuous development. In 2006, Wang et al. [82] also used ZnO single-crystal substrates to fabricate ZnO MIS LEDs. N+ ion implantation at moderate doses (10^{13}–10^{14} cm^{-2}) into nominally undoped ($n \sim 10^{17}$ cm^{-3}) bulk single-crystal ZnO substrates followed by annealing in the temperature range of 600°C–950°C was used to fabricate diodes that showed visible yellow luminescence at 300 K and NBE EL of about 390 nm at 120 K under forward-bias conditions. The I–V behavior of the diodes was characteristic of MIS devices and suggested that the implantation created a more resistive region in the n-ZnO in which holes were created by impact ionization during biasing, similar to the case of EL in ZnO varistors. The authors believed that future work on acceptor implantation should be focused on achieving p-type conductivity in ZnO so that true injection LEDs could be realized.

From the research discussed earlier, we can see that early works mostly used ZnO single-crystal material as the substrate and active layer. However, the preparation of ZnO single-crystal material is complex and difficult, and the LEDs reported earlier show weak luminescence intensity. In recent years, with the development of ZnO thin-film materials, more and more researchers have begun to focus on MIS LEDs based on ZnO thin films.

In 2006, Chen et al. [83] reported a ZnO-based MIS LED with a structure of Au/SiO$_x$/ZnO on the n^+ silicon substrate fabricated by the reactive direct current sputtering and electron beam evaporation. The room-temperature PL spectrum for the sputtered ZnO film excited by an He–Cd laser operating at 325 nm is shown as curve a in Figure 17.7, which reveals the UV emission peaking at about 382 nm, characteristic

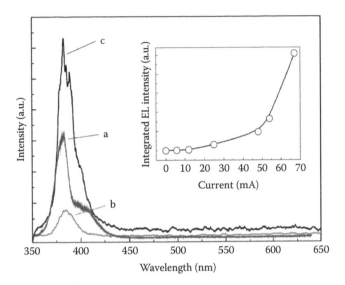

FIGURE 17.7 Room-temperature PL spectrum of the sputtered ZnO film (curve a) and EL spectra of the ZnO-based MIS structure with an ~100 nm thick SiO$_x$ layer at different injection currents (curve b: 25 mA at 8.0 V, curve c: 66 mA at 12.0 V). The inset shows the integrated intensity of UV EL peaking at about 383 nm as a function of injection current. (Reproduced with permission from Chen, P., Ma, X., and Yang, D., Fairly pure ultraviolet electroluminescence from ZnO-based light-emitting devices, *Appl. Phys. Lett.*, 89(11), 111112–111113, Copyright 2006, American Institute of Physics.)

of NBE emission of ZnO with a full width at half maximum (FWHM) of 15 nm and the defect-related emission shouldering at about 400 nm. Figure 17.7 also shows the typical EL spectra obtained at the forward currents of 25 (curve b) and 66 mA (curve c), respectively. The integrated intensity of the UV EL peaking at 383 nm as a function of injection current is shown in the inset of Figure 17.7. From this, we can see that the UV EL intensity increased significantly with the current over ~50 mA. From the experimental results and schematic energy band diagrams, the authors thought that the thickness of the SiO$_x$ layer was a key factor in achieving UV EL from the ZnO-based MIS structure.

In the next year, Hwang et al. [84] fabricated a Au/*i*-ZnO/*n*-ZnO MIS LED using a radio frequency magnetron sputtering method on a *c*-plane sapphire substrate, which showed a good diode characteristic and EL emission peaks at 380 and 550 nm from the *i*-ZnO grown on *n*-ZnO. The ZnO MIS diode showed a threshold voltage of 8.9 V. The high threshold voltage was a typical electrical characteristic of the MIS structure since the MIS structure used an insulating semiconductor barrier layer where holes can be produced by impact ionization at a high bias voltage and they tunneled into semiconductor layers. In 2009, Huang et al. [85] improved the structure of ZnO-based MIS LEDs and reduced the threshold voltage of emission to 2.0 V. The authors fabricated ZnO-based MIS LED devices with the structure of Au/*i*-HfO$_2$/*n*-ZnO on *n*$^+$-GaN/sapphire substrates using a radio frequency magnetron sputtering system. Fairly pure UV EL emission around 372 nm with a line width of 7.8 nm was observed. The MIS LED device with a proper thickness of the insulator HfO$_2$ layer had a low threshold voltage of 2 V. The reason for low threshold voltage may lie in the n$^+$-GaN/sapphire substrate (it was as the ideal substrate to grow ZnO layer because of their similarity in the crystalline structure and closely matched lattice constant, and because it provided sufficient electrons to the ZnO layer when the device applied a forward bias) and the high-*k* insulator HfO$_2$ layer (it was a high-*k* dielectric material compared with SiO$_2$, *i*-ZnO, and some of other insulator materials; that is, under the same circumstance, equal thickness, forward bias, etc., the MIS structure with HfO$_2$ as the insulator layer could confine more electrons in the semiconductor–insulator interface). Even though further research work need to be done to improve the quality of the UV LED, the authors thought that this study demonstrated the possibility of fabricating ZnO-based UV LED devices that can be driven by two ordinary dry batteries.

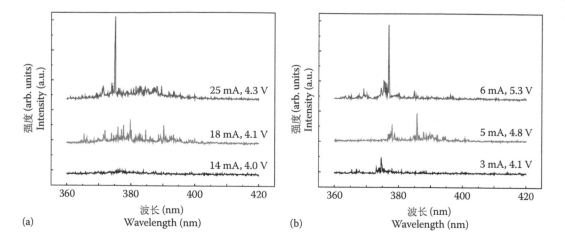

FIGURE 17.8 Room-temperature EL spectra of Au/SiO₂/ZnO MIS random-lasing devices based on (a) reactive sputtering and (b) pulse laser deposition. The horizontal axis shows the wavelength range (350–430 nm), and the vertical axis shows the EL intensity. (Reproduced with permission from Xu Yun, L.Y.-P. et al., *Acta Phys. Sin.*, 62(8), 84207, Copyright 2013, Acta Physica Sinica.)

In the recent years, we have noticed that many works have obtained the random lasing from ZnO-based MIS LEDs [86–90]. The random lasing is particularly attractive compared with cavity lasing due to its board angular distribution, which is suitable for lighting and display applications, as well as due to its easy realization [91]. Since the discovery of optically pumped random lasing in ZnO polycrystalline films [92], much attention has been focused on the investigation of ZnO-based random lasing. Compared with optically pumped random lasing, electrically pumped lasing is more desirable [86]. Xu et al. [93] prepared ZnO films on silicon substrates to fabricate the MIS random-lasing devices with a structure of Au/SiO₂/ZnO by two methods: reactive sputtering and pulsed laser deposition. Electrically pumped random laser actions of the two MIS structured devices based on the sputtered and pulse laser–deposited ZnO films were comparatively investigated, as shown in Figure 17.8. It was found that the device fabricated using the pulse laser–deposited ZnO film possessed a much lower threshold current for random lasing and higher output optical power. The authors considered that the pulse laser–deposited ZnO film had much fewer defects, leading to remarkably lower optical loss during multiple scattering within such a ZnO film.

17.3.1.2.2 *n–n Heterojunction LED*

As mentioned earlier, ZnO is a direct wide-bandgap semiconductor with a large exciton binding energy at room temperature and has been extensively considered as the most suitable material for fabricating short-wavelength light sources. However, the development of ZnO-based LEDs is hindered by the lack of stable and reproducible *p*-type ZnO [94]. For realizing the EL emission from ZnO, some effort has been made on heterostructures with *n*-type ZnO and other *n*-type materials.

Among these *n*-type materials, *n*-type GaN is a good candidate for LED application because *n*-type GaN with high carrier mobility is easier to realize and has been widely used in industrial production. Li et al. [95] reported a simple *n*-ZnO/*n*-GaN isotype heterostructure that was fabricated by radio frequency magnetron sputtering of *n*-ZnO on an *n*-GaN substrate. An MgO layer was inserted between *n*-ZnO and *n*-GaN to block the electrons. Intensive UV emissions of these simple isotype heterojunction LEDs were observed. The EL characteristics of *n*-ZnO/*n*-GaN and *n*-ZnO/*i*-MgO/*n*-GaN LEDs were investigated. The EL spectra of both heterojunction devices are shown in Figure 17.9. The EL spectra of *n*-ZnO/*n*-GaN heterojunction LED are dominated by UV emission peaked at around 368 nm under various voltages. With the increase in forward-bias voltage from 3.0 to 6.5 V, the UV EL emission showed a significant increase in its intensity. The room-temperature EL spectra of the devices with and without *i*-MgO layer

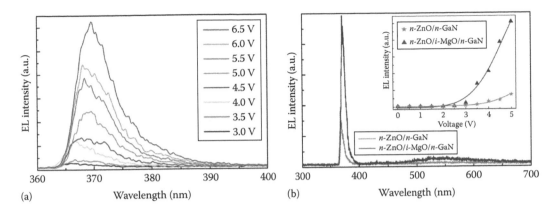

FIGURE 17.9 Room-temperature EL spectra of *n*-ZnO/*n*-GaN and *n*-ZnO/*i*-MgO/*n*-GaN heterojunction LEDs. (a) EL spectra under different bias voltages for *n*-ZnO/*n*-GaN heterojunction LEDs. (b) EL spectra of both heterojunction LEDs under the same bias voltage of 3.5 V. The inset shows the intensity of UV emission peaks for *n*-ZnO/*n*-GaN and *n*-ZnO/*i*-MgO/*n*-GaN heterojunction LEDs as a function of bias voltage. (Reproduced with permission from Li, S., Fang, G., Long, H., Mo, X., Huang, H., Dong, B., and Zhao, X., Enhancement of ultraviolet electroluminescence based on n-ZnO/n-GaN isotype heterojunction with low threshold voltage, *Appl. Phys. Lett.*, 96(20), 201111, Copyright 2010, American Institute of Physics.)

under the same forward-bias voltage of 3.5 V are shown in Figure 17.9b. The dominant sharp UV emission peaks are both at ~368 nm. The intensity of the UV emission peak with the *i*-MgO layer is much stronger than that without the *i*-MgO layer. The FWHM of both UV emission peaks is only 7 nm. What is more, in the meantime, a quite weak visible band was detected in both the devices. The inset of Figure 17.9b illustrates the intensity of UV emission peaks for both heterojunction LEDs as a function of bias voltage. With the insertion of an EBL of *i*-MgO, the UV emission intensity has been much enhanced under the same voltage, and the threshold voltage drops down from 3.5 to 2.5 V. Such *n*-ZnO/*i*-MgO/*n*-GaN isotype heterojunction LEDs are promising for the development of short-wavelength and high-performance optoelectronic devices.

Li et al. [96] continued studying the ZnO-based *n–n* heterojunction LED and reported the *n*-ZnO:Al(AZO)/*i*-layer/*n*-GaN LEDs in 2012. Figure 17.10a shows the schematic diagram of the fabricated *n*-AZO/*i*-layer/*n*-GaN heterojunction device. Figure 17.10b shows the x-ray diffraction (XRD) pattern of a sputtered ZnO film on *n*-GaN substrate. The high-quality *n*-GaN substrate has extremely strong (002) diffraction peak in the XRD pattern. There was a pronounced peak, at around 34.41°, indexed as the hexagonal ZnO (002), definitely indicating that the ZnO film was hexagonal-wurtzite structured with highly *c*-axis preferred orientation [97]. Under various voltages, the EL spectra of both AZO/*i*-ZnO/*n*-GaN and AZO/*i*-MgO/*n*-GaN heterojunctions are dominated by UV emission peaks, while the visible emissions were greatly suppressed. The UV emission of AZO/*i*-MgO/*n*-GaN heterojunction was stronger than that of AZO/*i*-ZnO/*n*-GaN at the same voltage. The threshold voltage of AZO/*i*-MgO/*n*-GaN heterostructured device was as low as 2.3 V. This study demonstrated the possibility of fabrication of AZO/*i*-MgO/*n*-GaN UV EL devices that can be driven by two ordinary dry batteries. Such AZO/*i*-MgO/*n*-GaN isotype heterojunction is promising for the development of short-wavelength and high-performance optoelectronic devices.

17.3.1.2.3 *p–n Heterojunction LED*

Besides ZnO-based *n–n* heterojunction LED, for realizing the EL emission from ZnO, much more efforts have been made on heterostructures with *n*-type ZnO and other *p*-type materials, such as *p*-GaN [98–106], *p*-AlGaN [107], *p*-NiO [108], *p*-Si [109–112], *p*-SrCu$_2$O$_2$ [113], *p*-CuGaS$_2$ [114], poly 3,4-ethylenedioxythiophene:poly styrenesulfonate (PEDOT:PSS) [115].

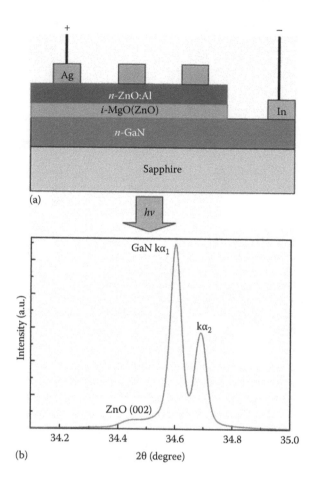

FIGURE 17.10 (a) Schematic diagram of the fabricated *n*-AZO/*i*-layer/*n*-GaN heterojunction device and (b) XRD pattern of a sputtered ZnO film on *n*-GaN substrate. (Reproduced from Li, S., Fang, G., Long, H., Wang, H., Huang, H., Mo, X., and Zhao, X., Low-threshold pure UV electroluminescence from n-ZnO:Al/i-layer/n-GaN heterojunction, 1642–1645, Copyright 2012, with permission from Elsevier.)

Rogers et al. [105] fabricated *n*-ZnO/*p*-GaN:Mg heterojunction LEDs on *c*-Al$_2$O$_3$ substrate using pulsed laser deposition for ZnO. The LEDs showed *I*–*V* characteristics, confirming a rectifying diode behavior and a room-temperature EL peak at about 375 nm. Yang et al. [72] also reported a heterojunction LED with *n*-ZnO/*p*-GaN structure. The unintentionally doped *n*-type ZnO layer with an electron concentration of ~10^{18} cm^{-3} was prepared not by pulsed laser deposition but by the MOCVD technique. A distinct blue-violet EL with a dominant emission peak centered at ~415 nm was observed at room temperature from the heterojunction structure under forward-bias conditions.

Alivov et al. [107] changed GaN into AlGaN to fabricate ZnO/AlGaN LEDs, taking advantage of the fact that AlGaN is also well matched to ZnO, has a large bandgap, and can be doped as *p*-type. Energy band diagrams built using the Anderson model [116] showed a much smaller barrier for holes than that which existed for electrons. Under forward bias, the device produced intense UV EL with a peak emission near 389 nm and an FWHM of 26 nm. No other emission bands were observed in the EL spectrum. The stability of device performance at high temperatures indicated possible applications in harsh environments. The authors believed that, with further development, it may even be possible to fabricate UV LDs, with low, "excitonic" thresholds and high-temperature capability, by exploiting the thermal stability of the ZnO exciton.

Ohta et al. [113] used another p-type material, p-SrCu$_2$O$_2$, with n-ZnO to fabricate ZnO-based p–n heterojunction. Multilayered films prepared by a pulsed laser deposition technique were processed by conventional photolithography with the aid of reactive ion etching to fabricate the LED device. A relatively sharp emission band centered at about 382 nm could be assigned as the transition associated with electron–hole plasma in ZnO. Figure 17.11a shows the UV emission spectra of the p-SrCu$_2$O$_2$/n-ZnO p–n junction LED for currents of 10, 11, 14, and 15 mA, respectively. Figure 17.11b shows LED emission intensity as a function of injection current. The turn-on voltage of the LED device was approximately 3 V, and the external efficiency was estimated to be less than 10^{-3}%. Improvements in interface uniformity may lead to improved efficiency. Chichibu et al. [114] used p-CuGaS$_2$, instead of p-SrCu$_2$O$_2$, to fabricate ZnO-based p–n junction LEDs. Greenish-white EL was observed from p-type (001) CuGaS$_2$ epilayers grown by metal–organic vapor-phase epitaxy (MOVPE) due to electron injection from the n-type preferentially (0001)-oriented polycrystalline ZnO films deposited by the surface damage–free helicon wave–excited plasma sputtering method. The EL spectra exhibited emission peaks and bands between 1.6 and 2.5 eV. Since the spectral line shape resembled that of the PL from the identical CuGaS$_2$ epilayers, the EL was assigned to originate not from n-ZnO, but from p-CuGaS$_2$. The authors believed that further improvements in the qualities, increase in the layer thickness, and proper control of the hole concentration in p-CuGaS$_2$ epilayers are necessary to improve LED performances.

The p-type materials in ZnO-based p–n heterojunction LEDs discussed earlier were inorganic. Considering the good p-type conductivity and high flexibility of organic materials, combining n-ZnO with p-type polymers may lead to new flexible, possibly elastic devices. Usually, the n-ZnO in the inorganic/organic hybrid devices is based not on thin films, but on nanostructures, which are suitable for flexible devices. Nadarajah et al. [115] reported a novel LED structure that used vertically oriented ZnO nanowires grown on a flexible transparent substrate and embedded in a polymeric matrix. The design scheme is shown in Figure 17.12. Single-crystal inorganic nanowires were the optically active component in this device structure, while the polymeric environment provided a robust, yet flexible support matrix. This flexible hybrid LED generated a broad emission spectrum covering most of the visible range and reaching to the near UV. The optically active components in this flexible and potentially stretchable device were inorganic nanowires, making this structure an interesting alternative to all-organic electronic and photonic devices. The authors considered that a monolithic p–n junction

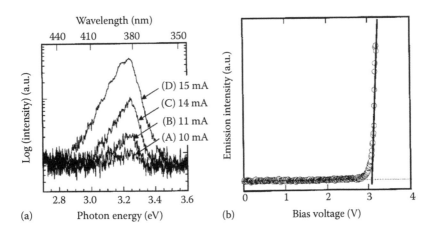

FIGURE 17.11 (a) UV emission spectra of the p-SrCu$_2$O$_2$/n-ZnO p–n junction LED for several currents. Electric currents were (A) 10 mA, (B) 11 mA, (C) 14 mA, and (D) 15 mA. (b) LED emission intensity as a function of injection current. (Reproduced with permission from Ohta, H., Kawamura, K.-I., Orita, M., Hirano, M., Sarukura, N., and Hosono, H., Current injection emission from a transparent p–n junction composed of p-SrCu2O2/n-ZnO, *Appl. Phys. Lett.*, 77(4), 475–477, Copyright 2000, American Institute of Physics.)

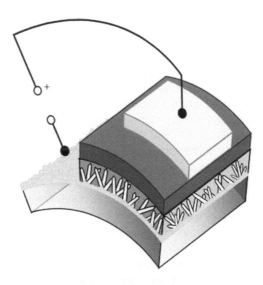

FIGURE 17.12 Design scheme for a flexible LED structure consisting of vertically oriented single-crystal nanow-ires grown on a polymeric indium tin oxide (ITO)-coated substrate. The top contact consisted of *p*-type polymer and an evaporated Au layer. Light was emitted through the transparent polymer. (Reprinted with permission from Nadarajah, A., Word, R.C., Meiss, J., and Konenkamp, R., Flexible inorganic nanowire light-emitting diode, *Nano Lett.*, 8(2), 534–537. Copyright 2008 American Chemical Society.)

and higher nanowire bulk doping appeared as the next development goals for improved efficiency in these devices. More work on the inorganic/organic hybrid heterojunction ZnO LEDs will be discussed in Section 17.3.2.2.

Among all the *p*-type materials discussed earlier in ZnO-based *p–n* junction LEDs, including inorganic and organic materials, *p*-GaN has no doubt got the most attention for its similar physical properties to ZnO. GaN and ZnO are both direct wide-bandgap semiconductors and have been extensively considered as the most suitable materials for fabricating short-wavelength light sources [94,117–119]. However, for the similar bandgap of ZnO (3.37 eV) and GaN (3.40 eV), it is usually difficult to identify whether the NBE emission comes from ZnO or GaN. Furthermore, because of lower hole mobility and carrier concentration in *p*-GaN, the EL usually emits mainly from the *p*-GaN instead of *n*-ZnO. To obtain the EL emission from *n*-ZnO, EBLs, such as MgO [99,101,120,121], SiO_2 [98], AlN [104], HfO_2 [100], Ga_2O_3 [101,122], and ZnS [123], were introduced between GaN and ZnO.

In 2007, Chen et al. [98] compared ZnO-based LED samples with and without a SiO_2 current-blocking layer and obtained the 394 nm UV light emission from low-temperature-sputtered n-ZnO/SiO_2 thin films on top of the *p*-GaN heterostructure. With a SiO_2 layer, EL spectrum showed a sharp emission peak at 394 nm, which was attributed to the recombination of accumulated carriers between n-ZnO/SiO_2 and *p*-GaN/SiO_2 junctions. As for the sample without a SiO_2 layer, a broadband ranging from 400 to 800 nm was observed, which was due to Mg^+ deep-level transition in the GaN along with defects in the ZnO layers. The authors further extended the thickness of SiO_2 layers, and the quantum efficiency, calculated from the EL spectra at 20 mA, of samples with 3, 6, and 9 nm SiO_2 was 7.6%, 24.5%, and 57.2%, respectively, higher than that of an n-ZnO/*p*-GaN device. In this experiment, they also fabricated a sample with a 12 nm SiO_2 layer. But unfortunately, the required bias voltage was too large to be of practical measurement. From the results, we can see that SiO_2 can be used as the EBL for ZnO/GaN *p–n* junction LEDs and a suitable thickness is necessary.

With regard to the large conduction band offset (CBO) between ZnO and MgO and comparatively smaller valence band offset (VBO) between MgO and GaN, many researchers paid much attention to the

MgO EBL in ZnO/GaN heterojunction LEDs. Jiao et al. [120] prepared the ZnO/GaN heterojunction and the *p–i–n* junction with the *n*-ZnO and *i*-MgO layers. For the *n*-ZnO/*i*-ZnO/*p*-GaN heterojunction, it was concluded that the emission emerged from the *i*-ZnO layer due to the confinement of electrons and holes. More importantly, *n*-ZnO/*i*-MgO/*p*-GaN *p–i–n* heterojunction was fabricated by MgO, which showed a greater ability of electron blocking comparing with *i*-ZnO, and then UV emission at 382 nm originated from ZnO was achieved. The authors discussed the conclusion supported by the energy band diagram of this heterojunction built using the Anderson model in detail. The electron affinities χ_{ZnO}, χ_{MgO}, and χ_{GaN} were taken as 4.35 eV [124], 0.80 eV [125], and 4.20 eV [126], respectively. The E_g of ZnO, MgO, and GaN was 3.37, 7.70 [127], and 3.40 eV, respectively. The energy barrier for electrons ΔE_C and holes ΔE_V in the interface of ZnO/MgO and GaN/MgO, respectively, was

$$\Delta E_C = \chi_{ZnO} - \chi_{MgO} = 4.35 - 0.80 = 3.55 \text{ eV} \tag{17.1}$$

$$\Delta E_V = E_{g\ MgO} + \chi_{MgO} - E_{g\ GaN} - \chi_{GaN} = 7.70 + 0.80 - 3.40 - 4.20 = 0.90 \text{ eV} \tag{17.2}$$

The results showed that the energy barrier for electrons was much higher than that for holes, which indicated that the existence of MgO could block effectively the electrons injected from the ZnO/GaN region. On the contrary, for a simple ZnO/GaN *p–n* heterojunction, the energy barrier for electrons ΔE_C and holes ΔE_V in the interface of ZnO/GaN, respectively, was

$$\Delta E_C = \chi_{ZnO} - \chi_{GaN} = 4.35 - 4.20 = 0.15 \text{ eV} \tag{17.3}$$

$$\Delta E_V = E_{g\ ZnO} + \chi_{ZnO} - E_{g\ GaN} - \chi_{GaN} = 3.37 + 4.35 - 3.40 - 4.20 = 0.12 \text{ eV} \tag{17.4}$$

The barrier height for electrons and that for holes were almost the same. Therefore, it was very difficult to realize the emission of ZnO. In 2012, Zhu et al. [121] reported *n*-ZnO/*i*-MgO/*p*$^+$-GaN heterojunction LEDs fabricated by radio frequency magnetron sputtering, and a UV emission from the NBE exciton recombination of ZnO was achieved. Effects of the insulator MgO layer on the EL performance of *n*-ZnO/*i*-MgO/*p*$^+$-GaN LEDs were investigated, and the authors obtained the same result.

Another Zhu et al. [99] also introduced MgO as the EBL into ZnO/MgO heterojunction LEDs. However, a novel phenomenon was found. For the fabrication of the heterojunction diodes, undoped ZnO and MgO layers were deposited onto commercially available GaN/Al$_2$O$_3$ (0001) templates using a VG V80H plasma-assisted molecular-beam epitaxy system. Compared with the previous work, not conventional EL emission but continuous-current-driven laser operating at room temperature was realized in ZnO. The threshold of this diode was about 0.8 mA, the smallest threshold for semiconductor LDs operating in the blue/UV-light spectrum range at that time. The reason for the ultralow threshold may lie in the microcavities formed in the ZnO columns as well as in the large exciton binding energy and high optical gain of the ZnO small-sized structures. The *p*-GaN used in this paper served as a hole source for the ZnO, and the authors believed that similar results may be attainable by extending the hole source to other *p*-type materials with proper conduction and valence-band offsets with ZnO. Also, this random lasing was achieved by Long et al. [108] in the same year. They fabricated UV LEDs based on ZnO/NiO heterojunctions on commercially available *n*$^+$-GaN/sapphire substrates using a radio frequency magnetron sputtering system. Here, they changed *p*-GaN into *p*-NiO because NiO was cheaper and simpler to fabricate and was a natural *p*-type direct-gap semiconductor with larger bandgap energy of 3.6 eV and a smaller electron affinity of 1.8 eV [128]. NBE emission of ZnO peaking at ~370 nm with an FWHM of ~7 nm was achieved at room temperature when the *n*$^+$-GaN/*n*-ZnO/*p*-NiO devices were under sufficient forward bias. With the help of an EBL of *i*-Mg$_{1-x}$Zn$_x$O ($0 < x < 1$) inserted between the ZnO and NiO layers (as shown in Figure 17.13), the emission intensity was much enhanced and the threshold current dropped down to

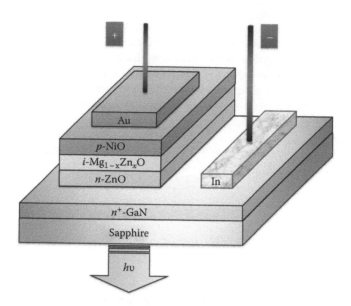

FIGURE 17.13 Schematic diagram of n^+-GaN/n-ZnO/i-Mg$_{1-x}$Zn$_x$O/p-NiO LED structure. (Reproduced with permission from Long, H., Fang, G.J., Huang, H.H., Mo, X.M., Xia, W., Dong, B.Z., Meng, X.Q., and Zhao, X.Z., Ultraviolet electroluminescence from ZnO/NiO-based heterojunction light-emitting diodes, *Appl. Phys. Lett.*, 95(1), 013509, Copyright 2009, American Institute of Physics.)

~23 from ~70 mA. The i-Mg$_{1-x}$Zn$_x$O film was used as an EBL. The effect of the i-Mg$_{1-x}$Zn$_x$O layer on the EL characteristics can be explained by the energy band structure. The effective barrier in the conduction band offset between ZnO and i-Mg$_{1-x}$Zn$_x$O was greatly enhanced due to band bending so that electrons will be confined in the n-ZnO layer. On the other hand, because of the reduced effective barrier in the vicinity of valence band offset, holes in the p-NiO layer can easily tunnel through it and enter into the n-ZnO layer. Therefore, under the same injection current, there could be many more electron–hole pairs recombining in the n-ZnO layer with the EBL of i-Mg$_{1-x}$Zn$_x$O and the emission intensity could also be enhanced. This explanation was much similar to ZnO/GaN-based p–i–n heterojunction LEDs with MgO EBLs.

In 2010, You et al. [104] put forward a new EBL for ZnO/GaN-based heterojunction LEDs. They fabricated n-ZnO/p-GaN heterojunction LEDs with and without a sandwiched AlN EBL. AlN, with the widest direct bandgap (6.2 eV) among group III nitrides, has outstanding physical and chemical properties, including high thermal conductivity, high stability, and good lattice match with ZnO and GaN. Therefore, it was possible to adopt AlN to improve the performance of n-ZnO/p-GaN LEDs. The EL spectra of the n-ZnO/p-GaN and n-ZnO/AlN/p-GaN heterojunction diodes under various currents are shown in Figure 17.14a and b, respectively. As shown in the figure, emission from n-ZnO/AlN/p-GaN was much stronger than that from n-ZnO/p-GaN even if the injected current in the n-ZnO/AlN/p-GaN was lower. Using the Anderson model, the CBO and VBO of ZnO/GaN were determined to be 0.15 and 0.12 eV, respectively, as mentioned earlier. On the other hand, the CBO for ZnO/AlN was experimentally determined by x-ray photoelectron spectroscopy (XPS) to be 3.29 eV, while the VBO of 0.94 eV for AlN/GaN can be derived by the transitivity rule [129]. Thus, the energy barrier for electrons was much higher than that for holes, indicating that the existence of AlN EBL can effectively block electron injection from ZnO to GaN in the n-ZnO/AlN/p-GaN LEDs, whereas holes in the GaN layer can tunnel through the barrier and enter the ZnO layer due to the lower VBO and band bending under forward bias. Hence, a recombination of carriers took place in the ZnO layer and the emission intensity was remarkably enhanced due to the lower density of interface defects and accumulated electrons. Besides the energy band structure, the authors also studied the heterojunction interfacial layer in detail, which was also

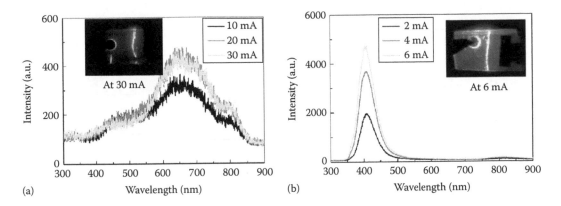

FIGURE 17.14 Room-temperature EL spectra of (a) *n*-ZnO/*p*-GaN LED and (b) *n*-ZnO/AlN/*p*-GaN LED under various currents and their EL images (inset). (Reproduced with permission from You, J.B., Zhang, X.W., Zhang, S.G., Wang, J.X., Yin, Z.G., Tan, H.R., Zhang, W.J. et al., Improved electroluminescence from n-ZnO/AlN/p-GaN heterojunction light-emitting diodes, *Appl. Phys. Lett.*, 96(20), 201102–201103, Copyright 2010, American Institute of Physics.)

important to device performance and had not been reported much at that time. XPS depth profiles of the heterojunction and XPS core-level spectra at the interface were acquired. From the experimental results, in addition to the Ga–N bonding, it was found that the Ga–O bonding also existed, indicating that the GaO_x interfacial layer formed at the interface of the ZnO/GaN. Also, the GaO_x interfacial layer has been identified by high-resolution transmission electron microscopy in other references [130,131]. The GaO_x interfacial layer may be responsible for the low-intensity EL from the ZnO/GaN device by providing a high density of interface states acting as the nonradiative recombination centers [132]. After inserting a thin AlN layer, the diffusion of O/N atoms was prevented by this intermediate layer and formation of the GaO_x interfacial layer was suppressed. Therefore, the performance of the *n*-ZnO/AlN/*p*-GaN LEDs was improved significantly.

However, from their research results, Zhang et al. [101] found that Ga_2O_3 did not degrade the performance of ZnO/GaN-based heterojunction LEDs, but was good for the improvement of EL performance. They fabricated *n*-ZnO:Ga/*p*-GaN heterojunction LEDs with different interfacial layers by pulsed laser deposition. All the devices demonstrated nonlinear rectifying behavior. Due to the formation of Ga_2O_3 interfacial layers, *n*-ZnO:Ga/*p*-GaN exhibited strong UV emission centered at 382 nm and blue emission centered at 423 nm. Compared with an *n*-ZnO:Ga/MgO/*p*-GaN LED, the turn-on voltage of an *n*-ZnO:Ga/*p*-GaN LED with a Ga_2O_3 interfacial layer dropped down to 7.6 V and the UV emission intensity was enhanced. The mechanism of radiative recombination in heterojunction LEDs under forward bias can be understood by examining the energy band structure. To construct the band diagrams, the E_g and electron affinities (χ) of each material should be known. The E_g and χ of ZnO, GaN, and MgO have been presented earlier. The E_g and χ of Ga_2O_3 are 4.9 and 2.5 eV, respectively [133,134]. Figure 17.15 shows the energy band diagrams of LEDs derived from the Anderson model. Without the intermediate layer, ΔE_C and ΔE_V of ZnO/GaN are determined to be 0.15 and 0.12 eV, respectively, which have been mentioned earlier. For the *n*-ZnO:Ga/*p*-GaN heterojunction with an intermediate layer, ΔE_C for Ga_2O_3/ZnO and MgO/ZnO can be calculated to 1.85 and 3.55 eV, and ΔE_V for Ga_2O_3/GaN and MgO/GaN are equal to 0.20 and 0.90 eV, respectively. Since $ZnGa_2O_4$, forming with Ga_2O_3 and confirmed by XRD, is a binary compound oxide consisting of ZnO and Ga_2O_3 with a bandgap energy of 4.4–4.7 eV, it can be treated as Ga_2O_3 in the energy band calculation [135]. For the ZnO/Ga_2O_3/GaN LED, although ΔE_C for the Ga_2O_3/ZnO interface (1.85 eV) has been improved compared with ΔE_C for the GaN/ZnO interface (0.15 eV), electron injection from ZnO:Ga to *p*-GaN could overcome hole injection from *p*-GaN to *n*-ZnO due to the larger electron concentration and mobility in ZnO:Ga. As a result, emissions from the *p*-GaN layers can be detected when the forward-bias voltage is small. When the forward-bias voltage is large enough,

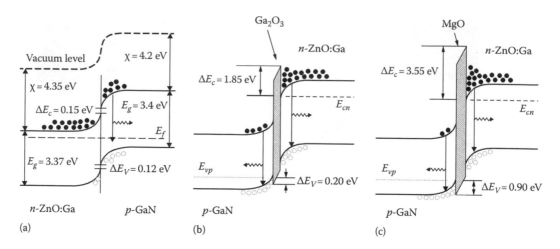

FIGURE 17.15 Energy band diagrams of the *n*-ZnO:Ga/*p*-GaN heterojunction LEDs (a) without interfacial layers, and with (b) Ga$_2$O$_3$ and (c) MgO interlayer, under forward bias. (From Zhang, L.C., Li, Q.S., Shang, L., Zhang, Z.J., Huang, R.Z., and Zhao, F.Z., Electroluminescence from n-ZnO : Ga/p-GaN heterojunction light-emitting diodes with different interfacial layers, *J. Phys. D: Appl. Phys.*, 45, 485103, Copyright 2012, IOP Publishing. Reproduced with permission. All rights reserved.)

the carriers may possess higher energy. Thus, the holes in the valence band of *p*-GaN have a higher possibility of being driven directly across the Ga$_2$O$_3$/GaN interface due to the much smaller value of ΔE_V for the Ga$_2$O$_3$/GaN interface (0.20 eV). As a result, holes can be recombined easily in *n*-ZnO:Ga and can lead to the band-edge emission of ZnO. That is why the emission peak of this LED of ZnO/Ga$_2$O$_3$/GaN shifted from blue to UV when the forward-bias voltage was increased. The authors thought that although the formed Ga$_2$O$_3$ interlayer is naturally oxidized during the deposition of ZnO:Ga, it can be an effective electron-blocking and hole-transporting layer for *n*-ZnO/*p*-GaN LED devices. Around the same time, Liu et al. [122] also used Ga$_2$O$_3$ as the EBL in the LED of ZnO/Ga$_2$O$_3$/*p*-GaN heterojunction by MOCVD. The heterojunction LED exhibited a strong UV EL emission at ~392 nm. This work showed that Ga$_2$O$_3$ can effectively block carriers for radiative recombination in ZnO thin films. The performance of this device was dependent upon efficient holes' injection from *p*-GaN to ZnO. The authors believed that the Ga$_2$O$_3$ thin film was indeed a good candidate for dielectric EBL for ZnO-based LEDs. They also expected that the light intensity can be enhanced remarkably by improving the crystal quality of ZnO and optimizing the design of ZnO/Ga$_2$O$_3$ QWs.

The same year, based on the previous research of EBLs, such as MgO, AlN, etc., Huang et al. [100] reported another suitable EBL material, HfO$_2$, for ZnO/GaN-based heterojunction LEDs. As we know, the EL of the *n*-ZnO/*p*-GaN heterojunction LEDs is emitted mainly from the *p*-GaN layer instead of the *n*-ZnO layer, because electron injection from *n*-ZnO would prevail over hole injection from *p*-GaN due to the higher carrier concentration in *n*-ZnO. As mentioned earlier, some researchers employed wide-bandgap materials such as MgO or AlN as the intermediate layer between the ZnO and GaN to block electron injection from ZnO due to the huge ΔE_C for MgO/ZnO ($\Delta E_C = 3.55$ eV) and AlN/ZnO ($\Delta E_C = 3.29$ eV) interfaces, respectively, thus suppressing the emission from GaN. However, these two structures also showed high ΔE_V for MgO/GaN ($\Delta E_V = 0.90$ eV) and AlN/GaN ($\Delta E_V = 0.94$ eV) interfaces, which block hole injection from *p*-GaN. In this work, the authors introduced the high-*k* (with a dielectric constant of 25–30) and modest bandgap (5.8–6.0 eV) insulator HfO$_2$ as the intermediate layer. Cook et al. [136] calculated the ΔE_V of the HfO$_2$/GaN interface as 0.3 eV, which is much smaller than that of the MgO/GaN and AlN/GaN interfaces. Besides, the ΔE_C of HfO$_2$/ZnO is around 2.29 eV [137], which in combination with the high-*k* nature of HfO$_2$ should make it a good electron-blocking and hole-transporting layer for the *n*-ZnO/*p*-GaN LED devices. The experimental results validated this view. The authors fabricated samples with and without HfO_2. The EL spectra of the n-ZnO/p-GaN device without the HfO_2 EBL displayed broad and weak emissions centered at ~435 nm (originating from the transitions between conduction band or donors and Mg-related acceptors) and ~523 nm (originating from defects emission in ZnO), respectively. However, for devices with the intermediate HfO_2 layer, a much stronger violet emission centered at around 415 nm dominates all the EL spectra. The output light powers of the devices with the HfO_2 layer are 2–4 times larger than those of the n-ZnO/p-GaN heterojunction device under the same injection currents. On the other hand, a color shift from violet to cold white was realized by fabricating devices with different preparation conditions of HfO_2 by varying the Ar/O_2 flow ratio during the radio frequency deposition for comparison, without any phosphors or sacrificing the efficiency of luminescence, which was ascribed to the control of deep-level emission bands in ZnO. The authors believed that this work was helpful to expand the applications of n-ZnO/p-GaN heterojunction LEDs.

In 2003, Zhang et al. [123] continued making efforts on the EBL of ZnO/GaN-based heterojunction LEDs and introduced ZnS material as the EBL. ZnS, with a modest bandgap (3.68 eV) and large exciton binding energy of 40 meV, has similar fundamental physical properties as ZnO and GaN, including crystal structures, lattice constants, melting points, and so on [138,139]. The n-ZnO/ZnS/p-GaN LED was fabricated by pulsed laser deposition. In contrast, the n-ZnO/p-GaN LED without EBL and the n-ZnO/AlN/p-GaN LED with AlN EBL were also fabricated. The thicknesses of the ZnS and AlN layer were both 20 nm. A broad blue-violet emission band centered at 430 nm was observed in the EL spectra of the n-ZnO/p-GaN heterojunction LED under various currents ranging from 0.4 to 2.0 mA. Comparing the EL spectra with the PL spectra, it was considered that the broad blue-violet emission band originates from the p-GaN layer. The reason has been discussed several times earlier. With the injection currents increasing from 0.4 to 2.0 mA, all the EL spectra of the n-ZnO/ZnS/p-GaN and n-ZnO/AlN/p-GaN heterojunction diodes displayed typical near-UV emission peaks at around 381 nm. As the injection currents increased from 0.4 to 2.0 mA, the FWHM of n-ZnO/ZnS/p-GaN narrowed from 68.9 to 31.8 nm, while the FWHM of n-ZnO/AlN/p-GaN narrowed from 33.9 to 27.5 nm. Compared with the EL spectra of the n-ZnO/AlN/p-GaN heterojunction LED, the EL spectra of the n-ZnO/ZnS/p-GaN heterojunction LED showed a much broader emission band. The EL-integrated intensities of the n-ZnO/ZnS/p-GaN LED were obviously both higher than those of the n-ZnO/AlN/p-GaN LED at the same input power, which indicated that ZnS films can effectively improve the EL efficiency of n-ZnO/p-GaN heterojunction diodes. In order to better understand the origin of the EL emission, a Gaussian function was exploited to simulate the EL spectra of the n-ZnO/ZnS/p-GaN and n-ZnO/AlN/p-GaN heterojunction diodes with injection currents of 2.0 mA, as shown in Figure 17.16. The radiative recombination processes are illustrated with energy band diagram in the insets of Figure 17.16. As the electron barrier height at the ZnS/ZnO interface ($\Delta E_C = 0.45$ eV) is smaller than that of the AlN/ZnO interface ($\Delta E_C = 3.75$eV), some of the electrons would inject from n-ZnO to p-GaN and the blue emission has been observed merely in n-ZnO/ZnS/p-GaN diode. The hole barrier height of the ZnS/ZnO interface (0.14 eV) was almost the same as that of the GaN/ZnO interface (0.13 eV), which was far smaller than the hole barrier height of the AlN/ZnO interface (0.92 eV). As the much smaller hole barrier height could make hole injection from GaN to ZnO easier, the possibility of radiative recombination in ZnO was enhanced in the n-ZnO/ZnS/p-GaN diode, which was the very reason that the EL intensity of the n-ZnO/ZnS/p-GaN diode was higher than that of the n-ZnO/AlN/p-GaN diode. The present work provided a feasible method for improving the EL performance of ZnO/p-GaN heterojunction LEDs.

All the ZnO-based p–n heterojunction LEDs discussed earlier are formed by n-type ZnO with other p-type materials. With the development of p-type ZnO, several research groups fabricated ZnO-based p–n heterojunctions by combining p-ZnO with other n-type materials and studied their p-type conductivity and luminescent properties.

Mandalapu et al. [140] reported NBE emissions from Sb-doped p-ZnO/n-Si heterojunction LEDs. Heterojunction LEDs were fabricated by making top Ohmic contacts on a Sb-doped p-type ZnO film with low specific contact resistivity and back Ohmic contacts on an n-type Si substrate. NBE and deep-level emissions were observed from the LED devices at both low temperatures and room temperature, which is

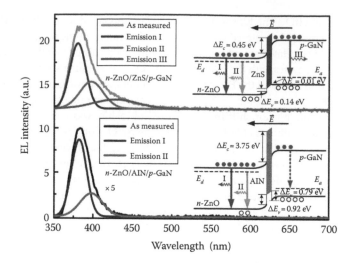

FIGURE 17.16 Gaussian deconvolution for EL spectra of the *n*-ZnO/ZnS/*p*-GaN and *n*-ZnO/AlN/*p*-GaN heterojunction diodes. Insets show the energy band diagram of heterojunction LEDs. (Reproduced from Zhang, L.C., Li, Q.S., Shang, L., Wang, F.F., Qu, C., and Zhao, F.Z., Improvement of UV electroluminescence of n-ZnO/p-GaN heterojunction LED by ZnS interlayer, *Opt. Express*, 21(14), 16578–16583. Copyright 2013 With permission of Optical Society of America.)

due to band-to-band and band-to-deep-level radiative recombinations in ZnO, respectively. The EL emissions precisely matched those of the PL spectra from Sb-doped *p*-type ZnO, indicating that the ZnO layer acted as the active region for the radiative recombinations of electrons and holes in the diode operation. The authors believed that the demonstration of UV emission from Sb-doped *p*-type ZnO films made Sb doping of *p*-ZnO promising for LED applications. Hwang et al. [141] also reported a *p*-ZnO based *p*–*n* heterojunction LED with a structure of *p*-ZnO/*n*-GaN. The LED structure consisted of a P-doped *p*-ZnO film with a hole concentration of 6.68×10^{17} cm^{-3} and an Si-doped *n*-GaN film with an electron concentration of 1.1×10^{18} cm^{-3}. The *I–V* curve of the LED showed a threshold voltage of 5.4 V and an EL emission of 409 nm at room temperature. The EL emission peak at 409 nm was attributed to the bandgap of *p*-ZnO, which was reduced as a result of the band offset at the interface of *p*-ZnO and *n*-GaN.

17.3.1.2.4 Multiple Heterostructures and QW LED

To further improve the performance of ZnO-based LEDs, many researchers designed novel LEDs with more complicated structures compared with conventional MIS, *n*–*n*, and *p*–*n* structures, such as multiple heterojunction LEDs, and QW and MQW LEDs. Satisfactory EL emission of ZnO-based LEDs can be expected.

The present worldwide research activities on ZnO, its alloys $Zn_{1-x}A_xO$ (with A = Cd, Mg, or Be), and nanostructures or quantum structures based on them are driven by various hopes. The predominant one is possibly the hope to obtain a material for optoelectronics that covers the spectral range from the green (A = Cd) over the blue to the near UV (A = Mg or Be), especially to obtain LEDs or LDs in these spectral ranges.

Ohashi et al. [142] have successfully realized high-quality red-bandgap $Zn_{1-x}Cd_xO$ films by remote plasma-enhanced metal–organic chemical vapor deposition (RPE-MOCVD) and fabricated a ZnO-based double heterojunction diode with a structure of *n*-ZnO cap layer, *n*-$Mg_{0.12}Zn_{0.88}O$ cladding layer, *n*-$Zn_{1-x}Cd_xO$ emission layer, and *p*-$Mg_{0.12}Zn_{0.88}O$:N barrier layer on *p*-4H–SiC substrates. By using the in-situ thermal annealing process, the compositional homogeneity of the $Zn_{1-x}Cd_xO$ films has been markedly improved. Proper annealing in vacuum provided an FWHM in PL of 126 meV, which was almost comparable to that of typical ZnO films. Figure 17.17 shows emission images under current injection, from which it is found that the emissions of blue and red were coming from different places. When the

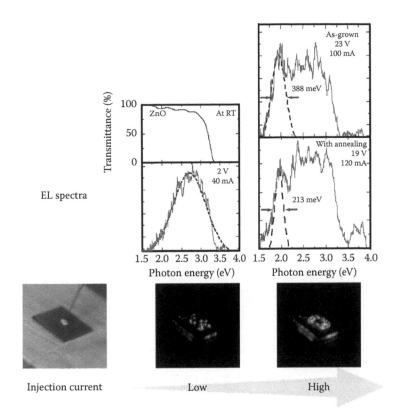

FIGURE 17.17 EL spectra and EL images from ZnO-based double heterojunctions under current injection. (Reproduced with permission from Ohashi, T. et al., *Jpn. J. Appl. Phys.*, 47(4), 2961. Copyright 2008, The Japan Society of Applied Physics.)

injection current level was as low as 40 mA, blue emissions coming from the *p*-4H–SiC substrate were observed. When a higher current of around 120 mA was injected, the red emission appeared from the as-grown and annealing samples. For both cases of EL spectra under high injection excitation, curve fitting using data of blue emission from the SiC substrate and the PL emission energy of the emitting layer was carried out. As can be seen from the results, the authors obtained an improved EL FWHM of 213 meV by utilizing the annealing technique and succeeded in the reduction of EL FWHM by 175 meV compared with that of the as-grown emission layer. The authors thought that the blue emission coming from the SiC substrate was caused by the remote junction due to the lack of adequate *p*-type conductivity of the N-doped $Mg_{0.12}Zn_{0.88}O$ layer. Although the reproducibility of proper *p*-$Mg_yZn_{1-y}O$:N was considered to be a problem in comparison with that of *p*-ZnO:N, an improvement was in progress and a suppression of the desirable higher-energy emission could be highly expected.

From the research discussed earlier, we can see that the ZnO-based symmetry double heterostructures (e.g., MgZnO/ZnO/MgZnO [102,143,144] and SiO_2/ZnO/SiO_2 [145] structures) have been often reported for the enhancement of radiative recombination in the active ZnO layer because of their better carrier confinement. Shi et al. [146] reported an asymmetric double heterostructure of MgZnO/ZnO/SiO_2/Si. The SiO_2 layer was used as the EBL, and a fashioned asymmetric waveguide mechanism leads to an improved light extraction efficiency. In 2014, Long et al. [147] designed and reported a new improved asymmetric double heterostructure of Ta_2O_5/ZnO/HfO_2 that was utilized in the ZnO/GaN-based heterojunction LEDs. The device structure is illustrated in the inset of Figure 17.18. Good Ohmic contacts were achieved for both electrodes, as shown in the inset of Figure 17.18. The LED with the structure *n*-ZnO/Ta_2O_5/*i*-ZnO/HfO_2/*p*-GaN was labeled as LED 3. As a contrast, LED 1 with the structure *n*-ZnO/*i*-ZnO/*p*-GaN

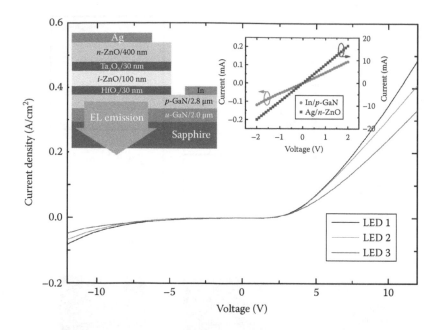

FIGURE 17.18 The *I–V* characteristic of LEDs in dark at room temperature. The insets show the schematic diagram of LED 3 with the structure *n*-ZnO/Ta$_2$O$_5$/*i*-ZnO/HfO$_2$/*p*-GaN and the *I–V* curves of the In/*p*-GaN and the Ag/*n*-ZnO contacts, respectively. (From Long, H., Li, S., Mo, X., Wang, H., Chen, Z., Feng, Z.C., and Fang, G., Enhanced electroluminescence using Ta$_2$O$_5$/ZnO/HfO$_2$ asymmetric double heterostructure in ZnO/GaN-based light emitting diodes, *Opt. Express*, 22(S3), A833–A841. Copyright 2014 With permission of Optical Society of America.)

and LED 2 with the structure *n*-ZnO/*i*-ZnO/HfO$_2$/*p*-GaN were also fabricated. The *I–V* characteristic of the devices in dark at room temperature is illustrated in Figure 17.18. A rectifying diode-like behavior was visible from the curves. The results showed that LED 3 had the best EL performance. The main reason was that the large conduction (*p* side) and valence (*n* side) band offsets at the interfaces of the active layer provide significant suppression of both electron and hole leakage from the active region [148,149]. HfO$_2$ is one of the most suitable materials for the EBL in *n*-ZnO/*p*-GaN LEDs because of its large ΔE_C (2.29 eV) for the ZnO/HfO$_2$ interface and small ΔE_V (0.30 eV) for the HfO$_2$/GaN interface, which means that electrons can be well blocked from ZnO to GaN, and meanwhile, holes from GaN to ZnO are rarely affected, thus effectively enhancing the possibility of radiative recombination in ZnO. On the other hand, the ΔE_C for the *n*-ZnO/Ta$_2$O$_5$ interface and the ΔE_V for the Ta$_2$O$_5$/*i*-ZnO interface are calculated to be 0.60 eV and 0.68 eV, respectively. Comparing with the conventional MgZnO/ZnO/MgZnO and SiO$_2$/ZnO/SiO$_2$ symmetric double heterostructures used in ZnO-based LEDs, Ta$_2$O$_5$, as a barrier layer in contact with the *n*-type electron injection layer and the ZnO active layer, has a much lower conduction band minimum (CBM), which is very close to that of ZnO. As a result, the Ta$_2$O$_5$ HBL blocks hole injection from *i*-ZnO into *n*-ZnO while relatively affecting electron injection from *n*-ZnO into *i*-ZnO to a lesser degree. This is why the EL intensity of LED 3 is greatest.

Furthermore, the stability of this asymmetric double heterostructure LED has also been measured. The emission intensity of the band at 633 nm of LED 3 was recorded every 120 s intermittently with a continuous injection current of 2.00 mA. The emission intensity as a function of the driving time is shown in Figure 17.19a. The EL spectra of LED 3 before and after running for 24 h are shown in Figure 17.19b. After stopping the 24-h stability test for 2 h, the EL spectrum was also tested. The inset of Figure 17.19a shows the result of long-time stability test recorded every 10 h intermittently with a continuous injection current of 2.00 mA. The LED lost ~30% of the EL intensity at 160 hours of operation. The ZnO-based LED lifetime in this work was longer than that demonstrated in previous reports [102,103,150,151].

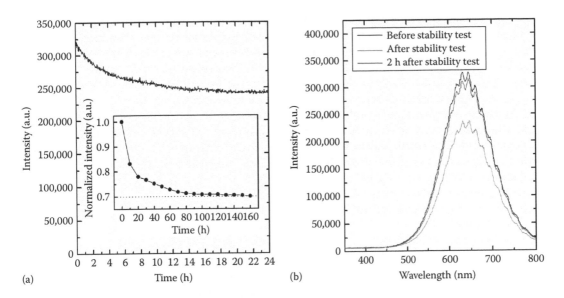

FIGURE 17.19 (a) The emission intensity of the band at 633 nm of LED 3 as a function of the 24-h driving time recorded every 120 s intermittently with a continuous injection current of 2.00 mA. The inset shows the result of long-time stability test recorded every 10 h intermittently with a continuous injection current of 2.00 mA. (b) The EL spectra of LED 3 before and after running 24 h and after stopping the continuous work for 2 h. (Reproduced from Long, H., Li, S., Mo, X., Wang, H., Chen, Z., Feng, Z.C., and Fang, G., Enhanced electroluminescence using Ta_2O_5/ZnO/HfO_2 asymmetric double heterostructure in ZnO/GaN-based light emitting diodes, *Opt. Express*, 22(S3), A833–A841. Copyright 2014 With permission of Optical Society of America.)

Considering the known advantages of QW and MQW used in GaN-based LEDs, many researchers attempted to fabricate ZnO QW and MQW structures and use them in ZnO-based LEDs, hoping to improve the quantum efficiency and emission intensity of ZnO-based LEDs, reduce the threshold, and obtain lasing.

Mares et al. [152] reported on the first demonstration, at the time, of a hybrid GaN/CdZnO LED, grown using plasma-enhanced MBE. They employed a hybrid LED structure consisting of a 300 nm p-type GaN spreading layer ($p \sim 7 \times 10^{17}$ cm^{-3}), an unintentionally doped $Cd_{0.12}Zn_{0.88}O$ single QW, a 200 nm n-type $Mg_{0.12}Zn_{0.88}O$ confinement layer doped with gallium ($n \sim 1 \times 10^{17}$ cm^{-3}), and a 15 nm n^+-ZnO contact layer ($n = 5 \times 10^{18}$ cm^{-3}). The $Cd_{0.12}Zn_{0.88}O$ QW thickness was expected to be ~3 nm. However, as discussed next, the growth of strained $Cd_{0.12}Zn_{0.88}O$ on p-GaN may have resulted in a smaller QW layer thickness. The LED structures and die were characterized by both PL and EL measurements. PL measurements showed strong luminescence centered at 430 nm with an FWHM of ~90 nm. No measurable low-energy defect luminescence was observed. Consistent with the PL spectra, no significant defect luminescence band in the EL emission was observed, indicating high-quality growth. EL emission redshifted from 390 to 410 nm as the forward current was increased from 20 to 40 mA. This observation was qualitatively explained in terms of competing intrawell optical transitions determined through simulations involving both ground and excited electron states in the QW. Domination of the intrawell emission, despite there being only an extremely thin active region in the LED structure, provided evidence for the advantage of using CdZnO to provide efficient light emission from hybrid LEDs. The measured PL spectrum differed significantly from that of the EL emission. Unlike the EL spectra, the PL spectrum appeared to be due to spatially indirect optical transitions between the electrons confined in the QW and the holes accumulated at the GaN/CdZnO interface.

In 2013, Horng-Shyang et al. [153] fabricated and characterized a vertical light-emitting diode (VLED) with CdZnO/n-ZnO QWs and the n^+-ZnO capping layer grown with molecular beam epitaxy and the *p*-GaN layer grown with MOCVD. Its performances were compared with those of a lateral LED based on the same epitaxial structure to show the significantly lower device resistance, smaller leakage current, weaker output intensity saturation, relatively lower defect emission, and stronger emissions from the *p*-GaN and *n*-ZnO layers in the VLED. However, there was still some improvement that could be done on the device structure. For example, adding an EBL, such as an AlGaN or MgZnO layer, between the *p*-GaN and the first *n*-ZnO barrier layer could reduce the emission from the *p*-GaN layer, because it could decrease carriers located in the layers outside the QWs for recombination. Also, a higher Cd content in growing the QW structure for increasing quantum confinement can help in reducing the emissions from the barrier and carrier transport layers. In the next year, Pandey et al. [154] presented an in-depth analysis of a $Cd_{0.4}Zn_{0.6}O/ZnO$ MQW LED using commercial simulation software and experimentally optimized growth conditions of *n*-type ZnO on Si(001) substrate by the dual-ion-beam sputtering deposition (DIBSD) system. The theoretical study revealed that an internal quantum efficiency of about 93.5% can be achieved at room temperature from the device, which emitted at 510 nm with a turn-on voltage of 3 V.

As we know that by alloying with CdO, the bandgap of ZnO can be shrunk from 3.3 to 1.8 eV, making it useful to fabricate visible optoelectronic devices. ZnO-based wide-bandgap oxide alloys have attracted researchers' attention as they can be used to achieve UV emissions or to produce ZnO-based MQWs as barrier layers. Ohtomo et al. [155] proposed MgZnO for such ZnO alloys. The bandgap of $Mg_xZn_{1-x}O$, where *x* is the atomic fraction, would be increased to 3.99 eV at room temperature as the content of Mg was increased upward to $x = 0.33$. Superlattices were fabricated by employing ZnO and ZnMgO as alternate layers [156], and ZnO/ZnMgO MQWs have also been fabricated and used in UV random LDs. Long et al. [157] prepared 10 ZnO/ZnMgO QWs on an *n*-GaN wafer, which was used as the substrate because GaN has the same wurtzite structure as that of ZnO with a small in-plane lattice mismatch of 1.8%. Then, a *p*-type NiO film was deposited on MQW layers. NiO was used as a hole injection layer because NiO is a cheap and natural *p*-type direct bandgap semiconductor with a large bandgap energy of 3.6 eV and a small electron affinity of 1.8 eV. The EL spectrum consists of a clear spontaneous emission centered at ~370 nm from NBE recombination in ZnO with an FWHM of ~7 nm. As the forward-bias voltage increased to 5.0 V, the UV emission was remarkably enhanced, and some sharp peaks, with the FWHMs less than 0.4 nm, emerged based on the normal emission spectrum background. These discrete sharp peaks appearing in the UV band essentially resulted from a random lasing action in the ZnO layers with a forward bias. The mechanism of random lasing has been discussed in the paper. Because of the structure *p*-NiO/MQWs(ZnO/ZnMgO)/*n*-GaN, the injected electrons and holes quickly form excitons as the diode is forward biased and become localized around the ZnO QWs. Each well has a width of only several nanometers, which is comparable with the de Broglie wavelength of electrons. At enough forward bias, carriers injected into the junction result in exciton–exciton inelastic collisions, which dissociate the localized excitons to form carrier population inversion conditions for stimulated emissions. Carrier population inversion is a critical condition required for lasing in diode lasers [158]. Besides, the device had a very low threshold current density of 4.7 A/cm² and extremely weak visible emission. It is anticipated that this low threshold and fairly pure UV random lasing will be beneficial to future applications.

Considering the large lattice matching between ZnO and MgO (ZnO is hexagonal with *a*-axis lattice constant being 3.25 Å, and MgO is cubic with *a*-axis lattice constant being 4.22 Å), Ryu et al. [159] deposited BeZnO films, a wide-bandgap oxide alloy, on sapphire substrates by a hybrid beam deposition growth method. The *a*-axis lattice constant values for ZnO and BeO are 3.25 and 2.70 Å, respectively [160]. The value of the bandgap of BeZnO can be efficiently engineered to vary from the ZnO bandgap (3.4 eV) to that of BeO (10.6 eV). The authors further improved lattice matching between ZnO and BeZnO by adding an appropriate amount of Mg into a BeZnO alloy. This work showed that the bandgap of ZnO can be tuned to higher values by using BeZnO alloys and that ZnO and BeZnO alloys are useful for designing and developing QWs and superlattices. ZnO/BeZnO heterostructures may be useful in developing new ZnO-based optoelectronic devices.

17.3.2 ZnO LED Based on Nanostructure

As mentioned earlier, for realizing the EL emission from ZnO, numerous efforts have been made on heterojunction structures with n-type ZnO and other p-type materials. However, compared to homojunctions, the $p-n$ heterojunction diode suffers lower carrier injection efficiency because of the large band offset at the interface of the junction [102]. Nanostructures have been introduced to overcome the problem by increasing the carrier injection rate through the nanojunction [161–163]. Furthermore, one-dimensional ZnO nanowires and nanorods have attracted increasing attention due to their unique physical properties arising from ultra-high specific surface area and quantum confinement, which shows promising potentials in extensive applications [161,164–175]. Next, ZnO LEDs based on the ZnO nanowire/nanorod array, organic/inorganic hybrid nanostructure, and single ZnO nanowire [176–193] are reviewed.

17.3.2.1 ZnO Nanowire/Nanorod Array LED

In 2008, Sun's group [194] reported stable and repeatable UV and red EL from ZnO nanorod array LEDs. Vertically aligned ZnO nanorods with tip diameters ranging from 100 to 400 nm and a length of ~5 μm were uniformly grown on a (111) n-type Si substrate by vapor-phase transport method. The as-grown ZnO nanorod arrays were subsequently implanted using a VARIAN (E-220) ion implanter with 50 and 180 keV As$^+$ ions at a dosage of 1×10^{14} or 1×10^{15} cm^{-2} perpendicular to the aligned nanorods. The EL spectrum of the ZnO nanorod diode array implanted with 10^{14} cm^{-2} As$^+$ ions was dominated by a strong UV band centered at ~380 nm, with a weak broad red band peaking at ~630 nm. The high luminescence intensity in UV emission bands and low input power of amplified spontaneous emission indicated the high efficiency of the nanostructure-based devices. In the next year, Sun et al. [195] continued their work on ZnO nanowire array LEDs and reported the production of a UV-emitting ZnO rod $p-n$ homojunction LED with an ion-implanted P-doped p-type ZnO. The as-grown ZnO rod arrays were implanted with P$^+$ ions with 50 keV (device I) and 100 keV (device II) at a dosage of 1×10^{14} cm^{-2} perpendicular to the aligned rods. The spacing between a- and c-ZnO:P planes of the ZnO:P was expected to be smaller than that of intrinsic ZnO nanowires (i.e., 5.629 and 5.207 Å) [196]. From low-temperature photoluminescence (LTPL) results, the intensity ratio of donor-bound to acceptor-bound exciton peaks of device II was higher than that of device I, suggesting that more donor defects (such as V_O) were induced in device II during implantation with higher ion energy. The EL spectra of devices I and II were measured at various injection currents. Strong UV emission was observed from both devices under forward bias, which corresponded to NBE emission of ZnO. The authors believed that their approach may provide a relatively simple and low-cost process for producing UV LEDs with some commercial implications.

Chen et al. [76] also made efforts on the ZnO nanowire array LED based on n-type ZnO and P-doped p-type ZnO. Catalyst-free $p-n$ homojunction ZnO nanowire arrays in which P and Zn served as p- and n-type dopants, respectively, were synthesized by a controlled in-situ doping process for fabricating efficient UV LEDs. The morphological and structural characterization of $p-n$ ZnO nanowire arrays, such as the scanning electron microscope (SEM) image, the transmission electron microscope (TEM) image, the selected area electron diffraction (SAED) pattern with the [0001] growth direction and the XRD spectrum, were carefully studied in the work. Further, the EL spectra from the $p-n$ ZnO nanowire arrays distinctively exhibited the short-wavelength emission at 342 nm, and the blueshift from 342 to 325 nm was observed as the operating voltage further increased. The authors believed that the ZnO nanowire $p-n$ homojunctions comprising p-type segment with high electron concentration were promising building blocks for short-wavelength lighting devices and photoelectronics.

Besides, Sb-doped ZnO has also been used in ZnO nanowire (nanorod) array LEDs, or even laser devices. Chu et al. [197] achieved the fabrication of electrically pumped ZnO nanowire diode lasers using p-type Sb-doped ZnO nanowires and an n-type ZnO film. Fabry–Perot (FP) type UV lasing was demonstrated at room temperature with good stability. The laser device consisted of an n-type ZnO thin film on a c-sapphire substrate, p-type vertically aligned ZnO nanowires, ITO contact, and Au/Ti contact. The structure schematic is shown in Figure 17.20a. Figure 17.20b through d shows the photo-image

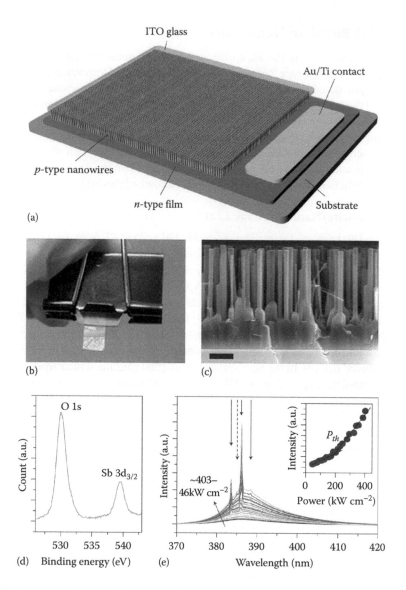

FIGURE 17.20 Structure and material properties of a ZnO nanowire/film laser device. (a) Schematic of the laser device, which consists of an *n*-type ZnO thin film on a *c*-sapphire substrate, *p*-type vertically aligned ZnO nanowires, ITO contact, and Au/Ti contact. (b) Photo-image of the device. (c) Side-view SEM image of the device structure showing the ZnO thin film and nanowires; scale bar, 1 μm. (d) XPS spectrum of the Sb-doped ZnO nanowires array. (e) Room-temperature optically pumped lasing spectra from 46 to 403 kW cm^{-2} with average steps of ~20 kW cm^{-2}. Solid arrows denote equidistant lasing peaks, and a spacing of 2.4 nm is extracted. Inset: integrated spectra intensity as a function of pumping power density. Solid lines represent threshold P_{th} (~180 kW cm^{-2}). (Reprinted by permission from Macmillan Publishers Ltd. *Nat. Nanotechnol.*, Chu, S., Wang, G., Zhou, W., Lin, Y., Chernyak, Y., Zhao, J., Kong, J., Li, L., Ren, J., and Liu, J., Electrically pumped waveguide lasing from ZnO nanowires, 6(8), 506–510, Copyright 2011.)

of the device, side-view SEM image of the device structure (scale bar, 1 μm), and XPS spectrum of the Sb-doped ZnO nanowires array, respectively. Figure 17.20e is the room-temperature optically pumped lasing spectra from 46 to 403 kW cm^{-2} with average steps of ~20 kW cm^{-2}. Solid arrows denoted equidistant lasing peaks, and a spacing of 2.4 nm was extracted. Inset: integrated spectra intensity as a function of pumping power density. Solid lines represent threshold P_{th} (~180 kW cm^{-2}). Future work is needed to further optimize laser performance. Also, heterojunction nanowire diode structures may be used to achieve stronger power output.

Besides As-, P-, and Sb-doped *p*-type ZnO, N–In codoped *p*-type ZnO has also been used in ZnO nanowire (nanorod) array LEDs. Wu's group [198] reported that ZnO homojunction LEDs based on ZnO nanowires were fabricated on Si (100) substrates. An N–In codoped *p*-type ZnO film grown by ultrasonic spray pyrolysis and an unintentionally doped *n*-type ZnO nanowire quasi-array grown by an easy low-temperature hydrothermal method were employed to form the homojunction diode. EL emissions of as-fabricated ZnO homojunction LEDs were observed under various forward biases. Typical EL spectra of the diodes showed evident UV peaks at 386 nm and broad greenish bands centered at ~540 nm. It was concluded that the EL emission was contributed by the ZnO nanowires. The intensities of both UV and green emissions grew nearly linearly with the current.

The researches discussed earlier are all based on the ZnO nanowire (nanorod) array LEDs with homojunctions. However, the fabrication of ZnO homojunction devices is difficult because of the instability of the *p*-type material. This motivated a search for other *p*-type materials that have been mentioned in Section 17.3.1.2. Similar to ZnO thin-film LEDs, many research groups have paid much attention to ZnO heterojunction nanowire (nanorod) array LEDs. Lupan et al. [199] have successfully grown high-quality epitaxial ZnO nanowires on *p*-type GaN epitaxial layers supported on sapphire by an electrochemical technique. The room-temperature PL measurements showed only a UV emission at 382 nm due to exciton recombination. The heterojunction was subsequently used to construct an LED structure. A narrow UV emission peak centered at 397 nm was measured at room temperature above an applied forward bias of 4.4 V. The EL steeply increased with the applied voltage and the observed strong emission was highly stable and reproducible. The results clearly stated the remarkable effectiveness of electrodeposited epitaxial ZnO as an active layer in solid-state UV LEDs. Shi et al. [200] also fabricated a ZnO heterojunction nanowire array LED based on *n*-ZnO nanowires and *p*-GaN. However, there was a single-crystal film between ZnO nanowires and GaN, which was used as the emitting layer. ZnO nanowires were used as the electron injection layer and light extraction layer. A highly efficient UV emission was observed under forward bias. And the improved EL performance was confirmed in comparison with the *n*-ZnO single-crystal films/*p*-GaN structure. Moreover, the suitability of the diode to practical LED applications was confirmed by the long-time aging test in harsh environments. It is reasonably believed that this route opens possibilities for the design and development of highly efficient UV LEDs, in which the advantages of ZnO single-crystal films and nanostructures can be integrated together. In the same year, Dong et al. [201] also reported the production of a ZnO nanorod array/*p*-GaN LED, but the preparation method of ZnO nanorods was different, and they used graphene as the current-spreading layer. Highly ordered ZnO nanorod arrays on the *p*-GaN substrates were fabricated by a hydrothermal process using a TiO_2 ring template derived from a polystyrene (PS) microsphere self-assembled monolayer. Herein, the diameters of ZnO nanorods were tuned by varying the solution concentration during hydrothermal growth and reactive ion etching (RIE) of PS microspheres. To grow the ZnO nanorod arrays, the aqueous solutions of $Zn(NO_3)_2 \cdot 6H_2O$ and hexamethylenetetramine (HMTA) of identical concentration were used. By choosing an optimal diameter of 170 nm, the light emission of the ZnO nanorod array/*p*-GaN heterojunction LEDs was enhanced further. This work has great potential applications in solid-state lighting, high-performance optoelectronic devices, and so on. Also based on ZnO nanorod array and *p*-GaN, Long et al. [102] reported a MgZnO/ZnO nanorods/MgZnO double heterojunction structure on a *p*-GaN/*u*-GaN/sapphire substrate and studied its room-temperature EL spectra. Because of the double heterostructure, LEDs showed better visible EL performance than that of LEDs with ordinary *p–i–n* structure. By replacing the ZnO film in this double heterostructure with ZnO nanorod arrays, which showed large carrier injection rate, relaxed strain to the substrate, and led to good light extraction, a strong UV EL emission around 380 nm was achieved. The ZnO nanorods–based double heterostructured LED exhibits superior stability with an intensity degradation of less than 3% over 8 h. In future, one can expect that the leakage current would be further reduced by optimizing the preparation process and EL performance could be further improved by updating the double heterostructure to ZnO/MgZnO MQWs in the ZnO nanorods–based LED.

p-Si is another suitable material for the *n*-ZnO nanorod array LED and is compatible with silicon technology. Lee et al. [202] reported the observation of blue-white emission in the vertical ZnO nanorod array contact LEDs. The proposed configuration of the heterojunction contact diode allowed the creation of self-organized ZnO/Si *p–n* nanodiodes of high density (~10^9 cm^{-2}) and high quality due to the structural perfection of the *p*$^+$-Si wafer and *n*$^-$-ZnO nanorod tips as well as provided high injection current and light emission from vertical nanorod array contact LEDs needed for solid-state lighting that is compatible with silicon technology.

Besides *p*-GaN and *p*-Si, *p*-type 6H–SiC has also been used in ZnO nanorod array LED. Hassan et al. [203] fabricated high-quality vertically aligned ZnO nanorods on *p*-type 6H–SiC substrate by microwave-assisted chemical bath deposition. A novel seed material, poly(vinyl alcohol)-Zn(OH)$_2$ nanocomposite, was used to seed the 6H–SiC substrate. The optical properties were examined by PL spectroscopy, which showed a high-intensity UV peak compared with visible defect peaks. The *I–V* property proved the good rectifier characteristic of the *n*-type ZnO nanorod/*p*-type SiC diode heterojunction. The EL spectrum of the hetero-epitaxial ZnO/6H–SiC LEDs yields a broadband (400–800 nm) centered at 520 nm, which was related to the combination of ZnO defect states and SiC defect states, which means that the EL emission happened on the two sides of the heterojunction. The EL emission of the heterojunction LED was sufficiently high to be seen with the naked eye. The EL property of LEDs makes the ZnO/SiC structure a promising heterojunction in the fabrication of white LEDs. Hassan's group [204] has also worked on the ZnO nanorod/*p*-GaN LED, which showed high-brightness UV EL emission under both forward and reverse bias.

17.3.2.2 ZnO Organic/Inorganic Hybrid Nanostructure LED

Nanoscale hybrid materials containing organic as well as inorganic components have attracted considerable attention as they promise new properties that may not easily be available using conventional materials. For example, hybrid structure devices can combine the high flexibility of polymers with the structural and chemical stability of inorganic nanostructures and are compatible with low-cost polymer printing techniques [115,205,206]. Sometimes, they show a higher light extraction efficiency than thin-film structures.

Sun et al. [207] presented a heterostructure LED made of *N,N*′-di(naphth-2-yl)-*N,N*′-diphenyl-benzidine (NPB) and ZnO nanorods fabricated by aqueous thermal deposition. A schematic diagram of the inorganic/organic heterostructure LED is shown in Figure 17.21. An excitonic UV emission at 342 nm, which corresponded to the blue-shifted ZnO band edge emission caused by ZnO conduction band filling by electron accumulation at the ZnO/NPB interface, was observed in the EL spectrum of the heterostructure LED. The heterostructure ZnO-based LED was fabricated at low temperature, indicating a low-cost process to fabricate UV LEDs, which shows great potential applications.

PEDOT:PSS, as another *p*-type polymer, has been used frequently in solar cells and other optoelectronic devices [208–216]. Li et al. [217] reported an *n*-ZnO/*p*-PEDOT:PSS hybrid LED with ZnO nanorods embedded in the PEDOT:PSS film. The embedded ZnO nanorods increased the junction area and played a key role in the emission process. With an increase in voltage, the intensity of the UV emission increased

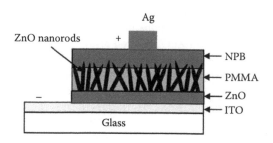

FIGURE 17.21 Schematic diagram of the inorganic/organic heterostructure LED. (Reproduced with permission from Sun, X.W., Huang, J.Z., Wang, J.X., and Xu, Z., A ZnO nanorod inorganic/organic heterostructure light-emitting diode emitting at 342 nm, *Nano Lett.*, 8(4), 1219–1223. Copyright 2008, American Chemical Society.)

faster than that of the visible emissions. As a result, a strong UV peak was exhibited when the forward-bias voltage was higher than 8V. These results indicated that the ZnO/polymer heterojunction with embedded ZnO nanorods was a promising LED structure.

From recent reports, we can see that the external quantum efficiency of UV LEDs based on ZnO nano-structures is usually low. One of the most important reasons is the lack of efficient methods to achieve a balance between electron-contributed current and hole-contributed current that reduces nonradiative recombination at the interface. Yang et al. [206] demonstrated that the piezo-phototronic effect can largely enhance the efficiency of a hybridized inorganic/organic LED made of a ZnO nanowire/p-polymer structure by trimming the electron current to match the hole current and increasing the localized hole density near the interface through a carrier channel created by piezoelectric polarization charges on the ZnO side. Figure 17.22a and b shows the change in relative injection current I_ε/I_0 under different strains and the change in relative light intensity P_ε/P_0 and external efficiency η_ε/η_0 under different strains, respectively. Figure 17.22d presents the charge-coupled device (CCD) images recorded from the emitting end of a packaged single-wire LED under different applied strains; the dashed line represents the position of ZnO nanowire/p-polymer core-shell structure, at a scale bar of 10 μm. From the results, it can be seen that the external efficiency of the hybrid LED was enhanced by at least a factor of 2 after applying a proper strain, reaching 5.92%. This study offers a new concept for increasing the efficiency of organic LEDs. The approach pioneered in the study can be applied to other optoelectronic devices and may bring about significant performance improvement and energy saving.

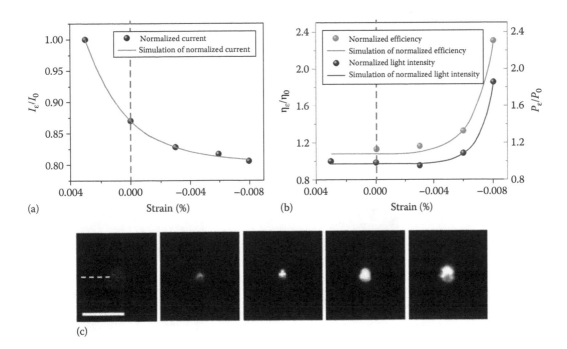

(a) (b) (c)

FIGURE 17.22 Enhancement of emission light intensity and conversion efficiency of a ZnO nanowire/p-polymer core–shell UV LED with the c-axis pointing away from p-polymer under different strains at 25 V biasing voltage. (a) Change in relative injection current I_ε/I_0 under different strains. (b) Change in relative light intensity P_ε/P_0 and external efficiency η_ε/η_0 under different strains. (c) CCD images recorded from the emitting end of a packaged single-wire LED under different applied strains; the dashed line represents the position of the ZnO nanowire/p-polymer core–shell structure; scale bar 10 μm. (Reproduced with permission from Yang, Q., Liu, Y., Pan, C., Chen, J., Wen, X., and Wang, Z.L., Largely enhanced efficiency in ZnO nanowire/p-polymer hybridized inorganic/organic ultraviolet light-emitting diode by piezo-phototronic effect, *Nano Lett.*, 13(2), 607–613. Copyright 2013, American Chemical Society.)

17.3.2.3 Single ZnO Nanowire LED

With the development of nanoscience and nanotechnology, much more research groups have started paying attention to fabricating optoelectronic devices based on a single nanowire for the requirement for the integration down to nanoscale and better controllability over parameters compared with the average information of nonuniform samples [66,196,218].

For these, in 2010, Bie et al. fabricated photovoltaic devices and UV LEDs based on single ZnO nanowire/p-type GaN heterojunctions and first reported the photovoltaic effects of single n-ZnO nanowire/p-GaN devices. The ZnO nanowire lay with a length of about 15 μm on the GaN surface and about 8 μm on the Al$_2$O$_3$ surface. The device not only can output a large power of ~80 nW and a large current density of ~4250 mA cm^{-2} (detailed characterizations showed that the V_{oc} was sensitive to both temperature and incident power), but also demonstrated good UV EL, which indicated the high quality of the heterojunction.

Besides p-GaN, p-AlGaN is also a suitable material for the single ZnO nanowire LED. Tang et al. [196] fabricated n-ZnO/p-AlGaN heterojunction nanowires by a green and effortless processing of CVD, which was different from the methods discussed earlier. Figure 17.23a shows the TEM image of a single n-type ZnO/p-type AlGaN heterojunction nanowire, and the inset shows the energy dispersive spectroscopy (EDS) of the heterojunction nanowire. Figure 17.23b shows the high-magnification TEM image of the as-fabricated product showing p-AlGaN, n-ZnO segments, and the interface (insets: the upper left and lower right for the SAED and high-resolution TEM of the interface region marked by the white box). Figure 17.23c exhibits the elemental mapping images obtained by the electron energy loss spectroscopy (EELS) of the heterojunctioned nanowire. Studies on single diode–like heterojunction nanowires showed excellent UV emission centered at about 390 nm under a drive current of 4 μA for application in optoelectronic devices and LEDs, which indicated the high-quality structure of the heterojunction of n-type ZnO and p-type AlGaN. Simultaneously, the UV EL measurements reveal that this device shows good stability due to the good ohmic contacts characteristics of the electrodes deposited on the nanodevices.

In addition, by using PEDOT:PSS and single ZnO nanowires, Zhang et al. [219] reported the observation of electrically driven UV lasing behavior by electron injection into a single ZnO micro/nanowire in organic/inorganic p–n heterostructures. At an excitation injection of ~1 A/cm^2, the EL spectrum showed a near-UV lasing action. Through experimental evidence and theory analysis, the authors attributed this

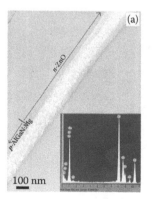
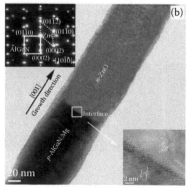
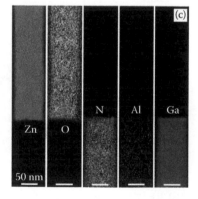

FIGURE 17.23　(a) TEM image of a single n-type ZnO/p-type AlGaN heterojunction nanowire (inset: EDS of the heterojunction nanowire). (b) A high-magnification TEM image of the as-fabricated product showing p-AlGaN, n-ZnO segments, and the interface (insets: the upper left and lower right for SAED and high-resolution TEM of the interface region marked by the white box). (c) The elemental mapping images obtained by EELS reveal heterojunctioned nanowires. (Reproduced with permission from Tang, X., Li, G., and Zhou, S., Ultraviolet electroluminescence of light-emitting diodes based on single n-ZnO/p-AlGaN heterojunction nanowires, *Nano Lett.*, 13(11), 5046–5050. Copyright 2013, American Chemical Society.)

lasing to whispering-gallery mode (WGM), which resulted from trajectories that traversed a polygonal cross-section near the edges of the cylinder. The reduction of some modes was associated with the surface states or defects, and the resonance below the intrinsic exciton emission of ZnO is related to electron accumulation at the polymer/ZnO interface. This study supported the application of ZnO micro/nanostructures in optoelectronic devices and was a step toward the realization of electrically pumped UV lasers in an organic/inorganic system.

For the WGM lasing in ZnO micro/nanowires, Xu's group [220–229] has done a lot of research. Besides ZnO micro/nanowires, the WGM resonant phenomena have been investigated in ZnO nanodisks [230] and nanonails [231]. However, it is hard to find distinct WGM modes. Xu's group [223] demonstrated the WGM UV laser at room temperature from an individual hexagonal ZnO microcavity, which showed a high quality factor of 1990, low threshold of 330 kW/cm^2, and distinct mode structure. A series of high quality factor lasing modes were combined together when two or more ZnO microrods were excited simultaneously. The authors continued working on the ZnO micro/nanowire WGM laser [228]. In the next year, they used the dodecagonal ZnO microrods to investigate the WGM lasing emission at room temperature. The typical exciton lasing was obtained in a microrod with $D = 6.35$ μm. In the smaller ZnO microrod $D = 3.37$ μm, exciton emission band emerged first under the low excitation power density of 320 kW/cm^2, and when the excitation power density was over the threshold of 380 kW/cm^2, the electron hole plasma lasing was generated. The authors also fabricated the ZnO twin-rods microstructure by the hydrothermal method, wherein the microstructure was employed as a WGM lasing microcavity [224]. The formation of twin-rods microstructure was attributed to the surface energy difference on the two polar surfaces of the ZnO nuclei. Under excitation of femtosecond laser pulses at a wavelength 800 nm, the effective three-photon absorption (3PA)–induced lasing emission was observed in the hexagonal WGM microcavity. The branches of the twin rods lased with different modes, and the larger branch lased with a lower threshold and a higher quality factor, which made the lasing modes from the smaller branch depressed when both branches of the twin rods were pumped simultaneously. Xu's group [225] continued to observe multiphoton absorption (MPA) optical WGMs in the UV–visible region, which corresponds to the near-bandgap exciton recombination, defect emission and second-harmonic generation in a ZnO hexagonal microrod. Based on the total internal reflection, the hexagonal microrod provides an effective optical path to strengthen the interaction between excitation light and the ZnO gain matter; thus, the 3PA-induced UV stimulated emission, defect emission, and second-harmonic generation resonance phenomenon can be effectively generated in the WGM cavity and the present distinct mode structures. Under the excitation of 1200 femtosecond pulse laser, the 3PA-induced UV lasing has a threshold of 0.98 TW/cm^2 and a quality factor of 440 for the microrod with a diameter of 5.6 μm. Under the excitation of 1240 nm femtosecond pulse laser, the four-photon absorption (4PA)–induced stimulated emission was realized in a bigger cavity with a diameter of 7.6 μm, which increased the optical gain path more effectively as compared with the smaller cavity. The 4PA-induced UV lasing has a lasing threshold of 1.30 TW/cm^2 and a quality factor of 1120. All the resonant mode numbers were indexed according to the plane wave theory. These significant WGM modes in a ZnO microcavity might find potential applications in frequency upconversion lasers and photomedical therapy. The work by Xu's group showed perfect lasing modes and wide lasing wavelength, which are useful in designing multiwavelength UV laser devices working at room temperature.

17.4 Summary and Outlook

A comprehensive review of LED applications of ZnO is presented. ZnO as a material has been the subject of varying degrees of research effort over the past decades. This effort has recently been intensified for a better understanding of ZnO's physical properties as well as for developing high-quality ZnO films and crystals for device applications. ZnO has some unique properties and some advantages, such as wide direct bandgap, large exciton binding energy, simple crystal-growth technology, environmentally friendliness, high thermal stability, low cost, and abundant availability, and is suitable for the production of white and short-wavelength light-emitting devices and lasing devices. For these reasons, many research groups have paid much attention to ZnO-based LEDs. There are many reports on ZnO-based thin-film and nanostructure LEDs involving both homojunctions and heterojunctions. Though ZnO-based homojunction LEDs have been continuously researched, the lack of reproducibility of high-quality p-type ZnO is still the bottleneck that hinders their further development. Then, the advantages of ZnO are explored and exploited by forming ZnO-based heterojunctions, such as MIS heterojunction, *n–n* heterojunction (taking *n*-ZnO with other *n*-type materials), and *p–n* heterojunction (taking *n*-ZnO with other *p*-type materials). To further improve the performance of ZnO-based LEDs, many researchers designed novel LEDs with more complicated structures compared with conventional MIS, *n–n*, and *p–n* structures, such as multiple heterojunction LEDs, and QW and MQW LEDs. Also, various kinds of novel ZnO-based LEDs with nanoscale structures have been fabricated recently and show exciting and satisfactory performance. With regard to ZnO-based LEDs, we still have much more work to do. We believe they will emit bright short-wavelength light, even bright short-wavelength laser, in future.

References

1. C. W. Bunn, The lattice-dimensions of zinc oxide, *Proc. Phys. Soc.* **47**(5):835–842 (1935).
2. H. Gleiter, Nanostructured materials: Basic concepts and microstructure, *Acta Mater.* 48(1):1–29 (2000).
3. G. K. Tapan, Application of zinc oxide varistors, *J. Am. Ceram. Soc.* **73**(7):1817–1840 (1990).
4. H. Yan, R. He, J. Pham, and P. Yang, Morphogenesis of one-dimensional ZnO nano- and microcrystals, *Adv. Mater.* **15**(5):402–405 (2003).
5. K. Keem, J. Kang, C. Yoon, D. Y. Jeong, B. M. Moon, and S. Kim, Enhanced performance of ZnO nanowire field effect transistors by H_2 annealing, *Jpn. J. Appl. Phys.* **46**:6230–6232 (2007).
6. D. C. Look, D. C. Reynolds, C. W. Litton, R. L. Jones, D. B. Eason, and G. Cantwell, Characterization of homoepitaxial p-type ZnO grown by molecular beam epitaxy, *Appl. Phys. Lett.* **81**:1830–1832 (2002).
7. D. C. Look, D. C. Reynolds, J. R. Sizelove, R. L. Jones, C. W. Litton, G. Cantwell, and W. C. Harsch, Electrical properties of bulk ZnO, *Solid State Commun.* **105**(6):399–401 (1998).
8. Ü. Özgür, Y. I. Alivov, C. Liu, A. Teke, M. A. Reshchikov, S. Doğan, V. Avrutin, S. J. Cho, and H. Morkoç, A comprehensive review of ZnO materials and devices, *J. Appl. Phys.* **98**(4):041301 (2005).
9. E. A. Meulenkamp, Synthesis and growth of ZnO nanoparticles, *J. Phys. Chem. B* **102**(29):5566–5572 (1998).
10. H. Pan, J. B. Yi, L. Shen, R. Q. Wu, J. H. Yang, J. Y. Lin, Y. P. Feng, J. Ding, L. H. Van, and J. H. Yin, Room-temperature ferromagnetism in carbon-doped ZnO, *Phys. Rev. Lett.* **99**(12):127201 (2007).
11. Z. K. Tang, G. K. L. Wong, P. Yu, M. Kawasaki, A. Ohtomo, H. Koinuma, and Y. Segawa, Room-temperature ultraviolet laser emission from self-assembled ZnO microcrystallite thin films, *Appl. Phys. Lett.* **72**(25):3270–3272 (1998).
12. C. C. Yong, S. Lee, H. P. Ji, W. K. Kim, H. H. Nahm, C. H. Park, and S. Y. Jeong, Hydrogen-induced anomalous Hall effect in Co-doped ZnO, *New J. Phys.* **16**(7):1137–1143 (2014).
13. X. J. Yue, T. S. Hong, X. Wei, K. Cai, and X. Xing, High performance micro CO sensors based on $ZnO–SnO_2$ composite nanofibers with anti-humidity characteristics, *Chin. Phys. Lett.* **29**(12):120702–120705 (2012).
14. Q. Zhang, C. S. Deneau, X. Zhou, and G. Cao, ZnO nanostructures for dye-sensitized solar cells, *Adv. Mater.* **21**(41):4087–4108 (2009).
15. O. Madelung, M. Schulz, and H. Weiss, *Landolt-Bornstein Group III: Condensed Matter*, Springer-Verlag, Berlin, Germany, 242pp, 1983.
16. C. Klingshirn, J. Fallert, H. Zhou, J. Sartor, C. Thiele, F. Maier-Flaig, D. Schneider, and H. Kalt, 65 years of ZnO research–old and very recent results, *Phys. Status Solidi B* **247**(6):1424–1447 (2010).

17. M. Wang, Y. Zhou, Y. Zhang, E. Jung Kim, S. Hong Hahn, and S. Gie Seong, Near-infrared photoluminescence from ZnO, *Appl. Phys. Lett.* **100**(10):101906 (2012).

18. J. Anderson and G. V. d. W. Chris, Fundamentals of zinc oxide as a semiconductor, *Rep. Prog. Phys.* **72**(12):126501 (2009).

19. A. Alkauskas and A. Pasquarello, Band-edge problem in the theoretical determination of defect energy levels: The O vacancy in ZnO as a benchmark case, *Phys. Rev. B* **84**(12):125206 (2011).

20. N. H. Alvi, K. Ul Hasan, O. Nur, and M. Willander, The origin of the red emission in n-ZnO nanotubes/p-GaN white light emitting diodes, *Nanoscale Res. Lett.* **6**(1):130 (2011).

21. A. B. Djurišić, Y. H. Leung, K. H. Tam, Y. F. Hsu, L. Ding, W. K. Ge, Y. C. Zhong et al., Defect emissions in ZnO nanostructures, *Nanotechnology* **18**(9):095702 (2007).

22. K. H. Tam, C. K. Cheung, Y. H. Leung, A. B. Djurišić, C. C. Ling, C. D. Beling, S. Fung et al., Defects in ZnO nanorods prepared by a hydrothermal method, *J. Phys. Chem. B* **110**(42):20865–20871 (2006).

23. S.-H. Jeong, B.-S. Kim, and B.-T. Lee, Photoluminescence dependence of ZnO films grown on Si(100) by radio-frequency magnetron sputtering on the growth ambient, *Appl. Phys. Lett.* **82**(16):2625–2627 (2003).

24. F. K. Shan, G. X. Liu, W. J. Lee, and B. C. Shin, The role of oxygen vacancies in epitaxial-deposited ZnO thin films, *J. Appl. Phys.* **101**(5):053106 (2007).

25. A. B. Djurišić, Y. H. Leung, K. H. Tam, L. Ding, W. K. Ge, H. Y. Chen, and S. Gwo, Green, yellow, and orange defect emission from ZnO nanostructures: Influence of excitation wavelength, *Appl. Phys. Lett.* **88**(10):103107 (2006).

26. P. F. Carcia, R. S. McLean, M. H. Reilly, and G. Nunes, Transparent ZnO thin-film transistor fabricated by rf magnetron sputtering, *Appl. Phys. Lett.* **82**(7):1117–1119 (2003).

27. Q. X. Zhao, P. Klason, M. Willander, H. M. Zhong, W. Lu, and J. H. Yang, Deep-level emissions influenced by O and Zn implantations in ZnO, *Appl. Phys. Lett.* **87**(21):211912 (2005).

28. S. Yamauchi, Y. Goto, and T. Hariu, Photoluminescence studies of undoped and nitrogen-doped ZnO layers grown by plasma-assisted epitaxy, *J. Cryst. Growth* **260**(1–2):1–6 (2004).

29. B. Cao, W. Cai, and H. Zeng, Temperature-dependent shifts of three emission bands for ZnO nanoneedle arrays, *Appl. Phys. Lett.* **88**(16):161101 (2006).

30. X. Q. Wei, B. Y. Man, M. Liu, C. S. Xue, H. Z. Zhuang, and C. Yang, Blue luminescent centers and microstructural evaluation by XPS and Raman in ZnO thin films annealed in vacuum, N_2 and O_2, *Phys. B* **388**(1–2):145–152 (2007).

31. B. Lin, Z. Fu, and Y. Jia, Green luminescent center in undoped zinc oxide films deposited on silicon substrates, *Appl. Phys. Lett.* **79**(7):943–945 (2001).

32. C. H. Ahn, Y. Y. Kim, D. C. Kim, S. K. Mohanta, and H. K. Cho, Erratum: A comparative analysis of deep level emission in ZnO layers deposited by various methods, *J. Appl. Phys.* **105**(8):089902 (2009).

33. A. Janotti and C. G. Van de Walle, Native point defects in ZnO, *Phys. Rev. B* **76**(16):165202 (2007).

34. H. Morkoç and Ü. Özgür, *Zinc Oxide: Fundamentals, Materials and Device Technology*, Wiley, Dublin, Ireland, 488pp, 2008.

35. G. F. Neumark, Achievement of well conducting wide band-gap semiconductors: Role of solubility and of nonequilibrium impurity incorporation, *Phys. Rev. Lett.* **62**(15):1800–1803 (1989).

36. D. J. Chadi, Doping in ZnSe, ZnTe, MgSe, and MgTe wide-band-gap semiconductors, *Phys. Rev. Lett.* **72**(4):534–537 (1994).

37. S. B. Zhang, S. H. Wei, and A. Zunger, Microscopic origin of the phenomenological equilibrium "doping limit rule" in n-type III-V semiconductors, *Phys. Rev. Lett.* **84**(6):1232–1235 (2000).

38. M. Tadatsugu, S. Hirotoshi, N. Hidehito, and T. Shinzo, Group III impurity doped zinc oxide thin films prepared by RF magnetron sputtering, *Jpn. J. Appl. Phys.* **24**(10):L781–L784 (1985).

39. S. B. Zhang, S.-H. Wei, and A. Zunger, A phenomenological model for systematization and prediction of doping limits in II–VI and I–III–VI$_2$ compounds, *J. Appl. Phys.* **83**(6):3192–3196 (1998).

40. D. B. Laks, C. G. Van de Walle, G. F. Neumark, and S. T. Pantelides, Acceptor doping in ZnSe versus ZnTe, *Appl. Phys. Lett.* **63**(10):1375–1377 (1993).

41. M. Henini, *Molecular Beam Epitaxy: From Research to Mass Production*, Elsevier Science, Amsterdam, the Netherlands, 744pp, 2012.
42. M. Göppert, F. Gehbauer, M. Hetterich, J. Münzel, D. Queck, and C. Klingshirn, Infrared-optical properties of undoped and gallium doped ZnO, *J. Lumin.* **72–74**(6):430–431 (1997).
43. D. C. Look and B. Claflin, p-Type doping and devices based on ZnO, *Phys. Status Solidi B* **241**(3):624–630 (2004).
44. S. S. Chen, X. H. Pan, H. P. He, W. Chen, J. Y. Huang, B. Lu, and Z. Z. Ye, Enhanced internal quantum efficiency in non-polar ZnO/Zn$_{0.81}$Mg$_{0.19}$O multiple quantum wells by Pt surface plasmons coupling, *Opt. Lett.* **40**(15):3639–3642 (2015).
45. C. J. Fan, Y. Lei, Z. Liu, R. X. Wang, Y. L. Lei, G. Q. Li, Z. H. Xiong, and X. H. Yang, High-efficiency phosphorescent hybrid organic-inorganic light-emitting diodes using a solution-processed small-molecule emissive layer, *ACS Appl. Mater. Inter.* **7**(37):20769–20778 (2015).
46. M. Guziewicz, R. Schifano, E. Przezdziecka, J. Z. Domagala, W. Jung, T. A. Krajewski, and E. Guziewicz, n-ZnO/p-4H-SiC diode: Structural, electrical, and photoresponse characteristics, *Appl. Phys. Lett.* **107**(10):101105 (2015).
47. H. H. Huang, C. C. Dun, W. X. Huang, Y. Cui, J. W. Xu, Q. K. Jiang, C. W. Xu, H. Zhang, S. C. Wen, and D. L. Carroll, Solution-processed yellow-white light-emitting diodes based on mixed-solvent dispersed luminescent ZnO nanocrystals, *Appl. Phys. Lett.* **106**(26):263506 (2015).
48. S. Iwan, J. L. Zhao, S. T. Tan, S. Bambang, M. Hikam, H. M. Fan, and X. W. Sun, Ion-dependent electroluminescence from trivalent rare-earth doped n-ZnO/p-Si heterostructured light-emitting diodes, *Mater. Sci. Semicond. Process.* **30**(30):263–266 (2015).
49. S. Jeong and H. Kim, Enhanced performance characteristics of n-ZnO/p-GaN heterojunction light-emitting diodes by forming excellent Ohmic contact to p-GaN, *Mater. Sci. Semicond. Process.* **39**:771–774 (2015).
50. S. Jeong and H. Kim, Carrier transport analysis of n-ZnO:Al/p-GaN:Mg heterojunction light-emitting diodes, *J. Vac. Sci. Technol. B* **33**(2):021205 (2015).
51. S. Jeong, K. K. Kim, and H. Kim, Growth and fabrication of reliable n-ZnO/p-GaN heterojunction light-emitting diodes, *J Nanoelectron. Optoe.* **10**(2):260–264 (2015).
52. G. B. Lin, B. Pandit, Y. Park, J. K. Kim, Y. R. Ryu, E. F. Schubert, and J. Cho, Effect of a p-type ZnO insertion layer on the external quantum efficiency of GaInN light-emitting diodes, *Appl. Phys. Express* **8**(9):092102(2015).
53. L. L. Liu, X. Z. Tan, D. D. Teng, M. Y. Wu, and G. Wang, Simultaneously enhancing the angular-color uniformity, luminous efficiency, and reliability of white light-emitting diodes by ZnO@SiO$_2$ modified silicones, *IEEE Trans. Compon. Pack. Man. B* **5**(5):599–605 (2015).
54. P. Mao, M. S. Xu, J. Chen, B. Xie, F. Q. Song, M. Han, and G. H. Wang, Dual enhancement of light extraction efficiency of flip-chip light-emitting diodes with multiple beveled SiC surface and porous ZnO nanoparticle layer coating, *Nanotechnology* **26**(18):185201 (2015).
55. M. M. Morshed, Z. Zuo, J. Huang, and J. L. Liu, Ultraviolet random lasing from Mg$_{0.12}$Zn$_{0.88}$O:N/ZnO:Ga single-heterostructure diode, *Appl. Phys. A* **118**(3):817–821 (2015).
56. C. X. Shan, J. S. Liu, Y. J. Lu, B. H. Li, F. C. C. Ling, and D. Z. Shen, p-Type doping of MgZnO films and their applications in optoelectronic devices, *Opt. Lett.* **40**(13):3041–3044 (2015).
57. V. P. Sirkeli, O. Yilmazoglu, F. Kuppers, and H. L. Hartnagel, Effect of p-NiO and n-ZnSe interlayers on the efficiency of p-GaN/n-ZnO light-emitting diode structures, *Semicond. Sci. Technol.* **30**(6):065005 (2015).
58. R. Udayabhaskar and B. Karthikeyan, Enhanced fluorescence and local vibrational mode in near-white-light-emitting ZnO:Mg nanorods system, *J. Am. Ceram. Soc.* **98**(6):1807–1811 (2015).
59. S. Verma, S. K. Pandey, S. K. Pandey, and S. Mukherjee, Theoretical simulation of Hybrid II-O/III-N green light-emitting diode with MgZnO/InGaN/MgZnO heterojunction, *Mater. Sci. Semicond. Process.* **31**:340–350 (2015).

60. C. X. Wang, C. Y. Lv, C. Zhu, Z. F. Gao, D. S. Li, X. Y. Ma, and D. R. Yang, Electrically pumped random lasing with an onset voltage of sub-3 V from ZnO-based light-emitting devices featuring nanometer-thick MoO₃ interlayers, *Nanoscale* 7(20):9164–9168 (2015).

61. H. Wang, Y. Zhao, C. Wu, X. Dong, B. L. Zhang, G. G. Wu, Y. Ma, and G. T. Du, Ultraviolet electroluminescence from n-ZnO/NiO/p-GaN light-emitting diode fabricated by MOCVD, *J. Lumin.* 158:6–10 (2015).

62. H. N. Wang, H. Long, Z. Chen, X. M. Mo, S. Z. Li, Z. Y. Zhong, and G. J. Fang, Fabrication and characterization of alternating-current-driven ZnO-based ultraviolet light-emitting diodes, *Electron. Mater. Lett.* 11(4):664–669 (2015).

63. H. Zhang, C. X. Xia, X. M. Tan, T. X. Wang, and S. Y. Wei, Effects of polar and nonpolar on band structures in ultrathin ZnO/GaN type-II superlattices, *Solid State Commun.* 221:14–17 (2015).

64. Y. Zhao, H. Wang, C. Wu, W. C. Li, F. B. Gao, G. G. Wu, B. L. Zhang, and G. T. Du, Study on the electroluminescence properties of diodes based on n-ZnO/p-NiO/p-Si heterojunction, *Opt. Commun.* 336:1–4 (2015).

65. T. Aoki, Y. Hatanaka, and D. C. Look, ZnO diode fabricated by excimer-laser doping, *Appl. Phys. Lett.* 76(22):3257–3258 (2000).

66. A. Tsukazaki, A. Ohtomo, T. Onuma, M. Ohtani, T. Makino, M. Sumiya, K. Ohtani et al., Repeated temperature modulation epitaxy for p-type doping and light-emitting diode based on ZnO, *Nat. Mater.* 4(1):42–46 (2005).

67. A. Tsukazaki, M. Kubota, A. Ohtomo, T. Onuma, K. Ohtani, H. Ohno, S. F. Chichibu, and M. Kawasaki, Blue light-emitting diode based on ZnO, *Jpn. J. Appl. Phys.* 44(21):L643–L645 (2005).

68. J. H. Lim, C. K. Kang, K. K. Kim, I. K. Park, D. K. Hwang, and S. J. Park, UV electroluminescence emission from ZnO light-emitting diodes grown by high-temperature radiofrequency sputtering, *Adv. Mater.* 18(20):2720–2724 (2006).

69. W. Z. Xu, Z. Z. Ye, Y. J. Zeng, L. P. Zhu, B. H. Zhao, L. Jiang, J. G. Lu, H. P. He, and S. B. Zhang, ZnO light-emitting diode grown by plasma-assisted metal organic chemical vapor deposition, *Appl. Phys. Lett.* 88(17):173506 (2006).

70. Z. P. Wei, Y. M. Lu, D. Z. Shen, Z. Z. Zhang, B. Yao, B. H. Li, J. Y. Zhang, D. X. Zhao, X. W. Fan, and Z. K. Tang, Room temperature p-n ZnO blue-violet light-emitting diodes, *Appl. Phys. Lett.* 90(4):042113 (2007).

71. J. Kong, S. Chu, M. Olmedo, L. Li, Z. Yang, and J. Liu, Dominant ultraviolet light emissions in packed ZnO columnar homojunction diodes, *Appl. Phys. Lett.* 93(13):132113 (2008).

72. S. Chu, J. H. Lim, L. J. Mandalapu, Z. Yang, L. Li, and J. L. Liu, Sb-doped p-ZnO/Ga-doped n-ZnO homojunction ultraviolet light emitting diodes, *Appl. Phys. Lett.* 92(15):152103 (2008).

73. J. Z. Zhao, H. W. Liang, J. C. Sun, J. M. Bian, Q. J. Feng, L. Z. Hu, H. Q. Zhang, X. P. Liang, Y. M. Luo, and G. T. Du, Electroluminescence from n-ZnO/p-ZnO: Sb homojunction light emitting diode on sapphire substrate with metal–organic precursors doped p-type ZnO layer grown by MOCVD technology, *J. Phys. D: Appl. Phys.* 41(19):195110 (2008).

74. P. C. Tao, Q. J. Feng, J. Y. Jiang, H. F. Zhao, R. Z. Xu, S. Liu, M. K. Li, J. C. Sun, and Z. Song, Electroluminescence from ZnO nanowires homojunction LED grown on Si substrate by simple chemical vapor deposition, *Chem. Phys. Lett.* 522(10):92–95 (2012).

75. J. C. Sun, J. M. Bian, Y. Wang, Y. X. Wang, Y. Gong, Y. Li, K. C. Liu et al., Fabrication of a homojunction light emitting diode with ZnO-nanorods/ZnO:As-film structure, *Electrochem. Solid State Lett.* 15(5):H164–H166 (2012).

76. M. T. Chen, M. P. Lu, Y. J. Wu, J. H. Song, C. Y. Lee, M. Y. Lu, Y. C. Chang, L. J. Chou, Z. L. Wang, and L. J. Chen, Near UV LEDs made with in situ doped p-n homojunction ZnO nanowire arrays, *Nano Lett.* 10(11):4387–4393 (2010).

77. X. Fang, J. H. Li, D. X. Zhao, D. Z. Shen, B. H. Li, and X. H. Wang, Phosphorus-doped p-type ZnO nanorods and ZnO nanorod p-n homojunction LED fabricated by hydrothermal method, *J. Phys. Chem. C* 113(50):21208–21212 (2009).

78. B. W. Thomas and D. Walsh, Metal-insulator-semiconductor electroluminescent diodes in single-crystal zinc oxide, *Electron. Lett.* **9**(16):362–363 (1973).

79. T. Minami, M. Tanigawa, M. Yamanishi, and T. Kawamura, Observation of ultraviolet-luminescence from the ZnO MIS diodes, *Jpn. J. Appl. Phys.* **13**(9):1475–1476 (1974).

80. T. Minami, S. Takata, M. Yamanishi, and T. Kawamura, Metal-semiconductor electroluminescent diodes in ZnO single crystal, *Jpn. J. Appl. Phys.* **18**(8):1617–1618 (1979).

81. A. Shimizu, M. Kanbara, M. Hada, and M. Kasuga, ZnO green light emitting diode, *Jpn. J. Appl. Phys.* **17**(8):1435–1436 (1978).

82. H.-T. Wang, B. S. Kang, J.-J. Chen, T. Anderson, S. Jang, F. Ren, H. S. Kim, Y. J. Li, D. P. Norton, and S. J. Pearton, Band-edge electroluminescence from N⁺-implanted bulk ZnO, *Appl. Phys. Lett.* **88**(10):102107 (2006).

83. P. Chen, X. Ma, and D. Yang, Fairly pure ultraviolet electroluminescence from ZnO-based light-emitting devices, *Appl. Phys. Lett.* **89**(11):111112–111113 (2006).

84. D.-K. Hwang, M.-S. Oh, J.-H. Lim, Y.-S. Choi, and S.-J. Park, ZnO-based light-emitting metal-insulator-semiconductor diodes, *Appl. Phys. Lett.* **91**(12):121113–121123 (2007).

85. H. H. Huang, G. J. Fang, X. M. Mo, H. Long, L. Y. Yuan, B. Z. Dong, X. Q. Meng, and X. Z. Zhao, ZnO-based fairly pure ultraviolet light-emitting diodes with a low operation voltage, *IEEE Electron Device Lett.* **30**(10):1063–1065 (2009).

86. Y. Fang, Y. Wang, X. Ding, R. Lu, L. Gu, and J. Sha, Electrically pumped random lasing from FTO/porous insulator/n-ZnO/p⁺-Si devices, *Opt. Express* **21**(9):10483–10489 (2013).

87. Y. P. Li, C. X. Wang, L. Jin, X. Y. Ma, and D. R. Yang, Electrically pumped random lasing in ZnO-based metal-insulator-semiconductor structured devices: Effect of ZnO film thickness, *J. Appl. Phys.* **113**(21):213103 (2013).

88. K. W. Wu, P. Ding, Y. F. Lu, X. H. Pan, H. P. He, J. Y. Huang, B. H. Zhao, C. Chen, L. X. Chen, and Z. Z. Ye, Electrically pumped ultraviolet lasing from ZnO in metal-insulator-semi devices, *Appl. Phys. A* **111**(3):689–694 (2013).

89. Y. Xu, Y. P. Li, L. Jin, X. Y. Ma, and D. R. Yang, Low-threshold electrically pumped ultraviolet random lasing from ZnO film prepared by pulsed laser deposition, *Acta Phys. Sin.* **62**(8):786–790 (2013).

90. Y. Tian, X. Y. Ma, L. L. Xiang, M. V. Ryzhkov, A. A. Borodkin, S. I. Rumyantsev, and D. R. Yang, Optically and electrically pumped random lasing from ZnO films annealed at different temperatures, *Opt. Commun.* **285**(24):5323–5326 (2012).

91. H. Zhu, C.-X. Shan, J.-Y. Zhang, Z.-Z. Zhang, B.-H. Li, D.-X. Zhao, B. Yao, D.-Z. Shen, X.-W. Fan, Z.-K. Tang, X. Hou, and K.-L. Choy, Low-threshold electrically pumped random lasers, *Adv. Mater.* **22**(16):1877–1881 (2010).

92. H. Cao, Y. G. Zhao, H. C. Ong, S. T. Ho, J. Y. Dai, J. Y. Wu, and R. P. H. Chang, Ultraviolet lasing in resonators formed by scattering in semiconductor polycrystalline films, *Appl. Phys. Lett.* **73**(25):3656–3658 (1998).

93. L. Y.-P. Xu Yun, Jin Lu, Ma Xiang-Yang, Yang De-Ren, Low-threshold electrically pumped ultraviolet random lasing from ZnO film prepared by pulsed laser deposition, *Acta Phys. Sin.* **62**(8):786–790 (2013).

94. Y. S. Choi, J. W. Kang, D. K. Hwang, and S. J. Park, Recent advances in ZnO-based light-emitting diodes, *IEEE Tran. Electr. Dev.* **57**(1):26–41 (2010).

95. S. Li, G. Fang, H. Long, X. Mo, H. Huang, B. Dong, and X. Zhao, Enhancement of ultraviolet electroluminescence based on n-ZnO/n-GaN isotype heterojunction with low threshold voltage, *Appl. Phys. Lett.* **96**(20):201111 (2010).

96. S. Li, G. Fang, H. Long, H. Wang, H. Huang, X. Mo, and X. Zhao, Low-threshold pure UV electroluminescence from n-ZnO:Al/i-layer/n-GaN heterojunction, *J. Lumin.* **132**(7):1642–1645 (2012).

97. H.-C. Chen, M.-J. Chen, M.-K. Wu, W.-C. Li, H.-L. Tsai, J.-R. Yang, H. Kuan, and M. Shiojiri, UV electroluminescence and structure of n-ZnO/p-GaN heterojunction LEDs grown by atomic layer deposition, *IEEE J. Quantum Electron.* **46**(2):265–271 (2010).

98. C. P. Chen, M. Y. Ke, C. C. Liu, Y. J. Chang, F. H. Yang, and J. J. Huang, Observation of 394 nm electroluminescence from low-temperature sputtered n-ZnO/SiO$_2$ thin films on top of the p-GaN heterostructure, *Appl. Phys. Lett.* **91**(9):091107 (2007).

99. H. Zhu, C. X. Shan, B. Yao, B. H. Li, J. Y. Zhang, Z. Z. Zhang, D. X. Zhao et al., Ultralow-threshold laser realized in zinc oxide, *Adv. Mater.* **21**:1613–1617 (2009).

100. H. H. Huang, G. J. Fang, Y. Li, S. Z. Li, X. M. Mo, H. Long, H. N. Wang, D. L. Carroll, and X. Z. Zhao, Improved and color tunable electroluminescence from n-ZnO/HfO$_2$/p-GaN heterojunction light emitting diodes, *Appl. Phys. Lett.* **100**(23):233502 (2012).

101. L. C. Zhang, Q. S. Li, L. Shang, Z. J. Zhang, R. Z. Huang, and F. Z. Zhao, Electroluminescence from n-ZnO: Ga/p-GaN heterojunction light-emitting diodes with different interfacial layers, *J. Phys. D: Appl. Phys.* **45**(48):485103 (2012).

102. H. Long, S. Z. Li, X. M. Mo, H. N. Wang, H. H. Huang, Z. Chen, Y. P. Liu, and G. J. Fang, Electroluminescence from ZnO-nanorod-based double heterostructured light-emitting diodes, *Appl. Phys. Lett.* **103**(12):123504 (2013).

103. Z. F. Shi, Y. T. Zhang, J. X. Zhang, H. Wang, B. Wu, X. P. Cai, X. J. Cui et al., High-performance ultraviolet-blue light-emitting diodes based on an n-ZnO nanowall networks/p-GaN heterojunction, *Appl. Phys. Lett.* **103**(2):021109 (2013).

104. J. B. You, X. W. Zhang, S. G. Zhang, J. X. Wang, Z. G. Yin, H. R. Tan, W. J. Zhang et al., Improved electroluminescence from n-ZnO/AlN/p-GaN heterojunction light-emitting diodes, *Appl. Phys. Lett.* **96**(20):201102–201103 (2010).

105. D. J. Rogers, F. Hosseini Teherani, A. Yasan, K. Minder, P. Kung, and M. Razeghi, Electroluminescence at 375 nm from a ZnO/GaN:Mg/c-Al$_2$O$_3$ heterojunction light emitting diode, *Appl. Phys. Lett.* **88**(14):141918 (2006).

106. T. P. Yang, H. C. Zhu, J. M. Bian, J. C. Sun, X. Dong, B. L. Zhang, H. W. Liang, X. P. Li, Y. G. Cui, and G. T. Du, Room temperature electroluminescence from the n-ZnO/p-GaN heterojunction device grown by MOCVD, *Mater. Res. Bull.* **43**(12):3614–3620 (2008).

107. Y. I. Alivov, E. V. Kalinina, A. E. Cherenkov, D. C. Look, B. M. Ataev, A. K. Omaev, M. V. Chukichev, and D. M. Bagnall, Fabrication and characterization of n-ZnO/p-AlGaN heterojunction light-emitting diodes on 6H-SiC substrates, *Appl. Phys. Lett.* **83**(23):4719–4721 (2003).

108. H. Long, G. J. Fang, H. H. Huang, X. M. Mo, W. Xia, B. Z. Dong, X. Q. Meng, and X. Z. Zhao, Ultraviolet electroluminescence from ZnO/NiO-based heterojunction light-emitting diodes, *Appl. Phys. Lett.* **95**(1):013509 (2009).

109. H. H. Huang, G. J. Fang, X. M. Mo, H. Long, H. N. Wang, S. Z. Li, Y. Li, Y. P. Zhang, C. X. Pan, and D. L. Carroll, Improved and orange emission from an n-ZnO/p-Si heterojunction light emitting device with NiO as the intermediate layer, *Appl. Phys. Lett.* **101**(22):223504 (2012).

110. Y. Yang, Y. P. Li, L. L. Xiang, X. Y. Ma, and D. R. Yang, Low-voltage driven ~1.54 µm electroluminescence from erbium-doped ZnO/p$^+$-Si heterostructured devices: Energy transfer from ZnO host to erbium ions, *Appl. Phys. Lett.* **102**(18):181111 (2013).

111. J. Ahn, H. Park, M. A. Mastro, J. K. Hite, C. R. Eddy, and J. Kim, Nanostructured n-ZnO/thin film p-silicon heterojunction light-emitting diodes, *Opt. Express* **19**(27):26006–26010 (2011).

112. Y. F. Chan, W. Su, C. X. Zhang, Z. L. Wu, Y. Tang, X. Q. Sun, and H. J. Xu, Electroluminescence from ZnO-nanofilm/Si-micropillar heterostructure arrays, *Opt. Express* **20**(22):24280–24287 (2012).

113. H. Ohta, K.-I. Kawamura, M. Orita, M. Hirano, N. Sarukura, and H. Hosono, Current injection emission from a transparent p–n junction composed of p-SrCu$_2$O$_2$/n-ZnO, *Appl. Phys. Lett.* **77**(4):475–477 (2000).

114. S. F. Chichibu, T. Ohmori, N. Shibata, T. Koyama, and T. Onuma, Greenish-white electroluminescence from p-type CuGaS$_2$ heterojunction diodes using n-type ZnO as an electron injector, *Appl. Phys. Lett.* **85**(19):4403–4405 (2004).

115. A. Nadarajah, R. C. Word, J. Meiss, and R. Konenkamp, Flexible inorganic nanowire light-emitting diode, *Nano Lett.* **8**(2):534–537 (2008).

116. A. G. Milnes and D. L. Feucht, *Heterojunctions and Metal-Semiconductor Junctions*, Academic Press, Salt Lake City, UT, 428pp, 1972.

117. S. Pimputkar, J. S. Speck, S. P. DenBaars, and S. Nakamura, Prospects for LED lighting, *Nat. Photon.* **3**(4):180–182 (2009).

118. Z. C. Feng, *Handbook of Zinc Oxide and Related Materials: Volume One, Materials*, Taylor & Francis, Boca Raton, FL, 446pp, 2012.

119. Z. C. Feng, *III-Nitride Devices and Nanoengineering*, Imperial College Press, London, U.K., 462pp, 2008.

120. S. J. Jiao, Y. M. Lu, D. Z. Shen, Z. Z. Zhang, B. H. Li, J. Y. Zhang, B. Yao, Y. C. Liu, and X. W. Fan, Ultraviolet electroluminescence of ZnO based heterojunction light emitting diode, *Phys. Status Solidi C* **3**(4):972–975 (2006).

121. G. Y. Zhu, J. T. Li, Z. L. Shi, Y. Lin, G. F. Chen, T. Ding, Z. S. Tian, and C. X. Xu, Ultraviolet electroluminescence from n-ZnO/i-MgO/p$^+$-GaN heterojunction light-emitting diodes fabricated by RF-magnetron sputtering, *Appl. Phys. B* **109**(2):195–199 (2012).

122. Y. D. Liu, H. W. Liang, X. C. Xia, R. S. Shen, Y. Liu, J. M. Bian, and G. T. Du, Introducing Ga$_2$O$_3$ thin films as novel electron blocking layer to ZnO/p-GaN heterojunction LED, *Appl. Phys. B* **109**(4):605–609 (2012).

123. L. C. Zhang, Q. S. Li, L. Shang, F. F. Wang, C. Qu, and F. Z. Zhao, Improvement of UV electroluminescence of n-ZnO/p-GaN heterojunction LED by ZnS interlayer, *Opt. Express* **21**(14):16578–16583 (2013).

124. J. A. Aranovich, D. Golmayo, A. L. Fahrenbruch, and R. H. Bube, Photovoltaic properties of ZnO/CdTe heterojunctions prepared by spray pyrolysis, *J. Appl. Phys.* **51**(8):4260–4268 (1980).

125. J. Yamashita, Oxygen band in magnesium oxide, *Phys. Rev.* **111**(3):733–735 (1958).

126. D. Qiao, L. S. Yu, S. S. Lau, J. M. Redwing, J. Y. Lin, and H. X. Jiang, Dependence of Ni/AlGaN Schottky barrier height on Al mole fraction, *J. Appl. Phys.* **87**(2):801–804 (2000).

127. D. M. Roessler and W. C. Walker, Electronic spectrum and ultraviolet optical properties of crystalline MgO, *Phys. Rev.* **159**(3):733–738 (1967).

128. M. D. Irwin, D. B. Buchholz, A. W. Hains, R. P. H. Chang, and T. J. Marks, p-Type semiconducting nickel oxide as an efficiency-enhancing anode interfacial layer in polymer bulk-heterojunction solar cells, *Proc. Natl. Acad. Sci. USA* **105**(8):2783–2787 (2008).

129. T. D. Veal, P. D. C. King, S. A. Hatfield, L. R. Bailey, C. F. McConville, B. Martel, J. C. Moreno, E. Frayssinet, F. Semond, and J. Zúñiga-Pérez, Valence band offset of the ZnO/AlN heterojunction determined by x-ray photoemission spectroscopy, *Appl. Phys. Lett.* **93**(20):202108 (2008).

130. J. Y. Lee, J. H. Lee, H. Seung Kim, C.-H. Lee, H.-S. Ahn, H. K. Cho, Y. Y. Kim, B. H. Kong, and H. S. Lee, A study on the origin of emission of the annealed n-ZnO/p-GaN heterostructure LED, *Thin Solid Films* **517**(17):5157–5160 (2009).

131. C. Hsing-Chao, C. Miin-Jang, W. Mong-Kai, L. Wei-Chih, T. Hung-Ling, Y. Jer-Ren, K. Hon, and M. Shiojiri, UV electroluminescence and structure of n-ZnO/p-GaN heterojunction LEDs grown by atomic layer deposition, *IEEE J. Quantum Electron.* **46**(2):265–271 (2010).

132. S. Lee and D. Y. Kim, Characteristics of ZnO/GaN heterostructure formed on GaN substrate by sputtering deposition of ZnO, *Mater. Sci. Eng. B* **137**(1–3):80–84 (2007).

133. S. Gowtham, M. Deshpande, A. Costales, and R. Pandey, Structural, energetic, electronic, bonding, and vibrational properties of Ga$_3$O, Ga$_3$O$_2$, Ga$_3$O$_3$, Ga$_2$O$_3$, and GaO$_3$ Clusters, *J. Phys. Chem. B* **109**(31):14836–14844 (2005).

134. H. Y. Yang, S. F. Yu, H. K. Liang, S. P. Lau, S. S. Pramana, C. Ferraris, C. W. Cheng, and H. J. Fan, Ultraviolet electroluminescence from randomly assembled n-SnO$_2$ nanowires/p-GaN:Mg Hetero junction, *ACS Appl. Mater. Inter.* **2**(4):1191–1194 (2010).

135. X. Chen, G. Fei, J. Yan, Y. Zhu, and L. De Zhang, Synthesis of ZnGa$_2$O$_4$ hierarchical nanostructure by Au catalysts induced thermal evaporation, *Nanoscale Res. Lett.* **5**(9):1387–1392 (2010).

136. T. E. Cook, C. C. Fulton, W. J. Mecouch, R. F. Davis, G. Lucovsky, and R. J. Nemanich, Band offset measurements of the GaN (0001)/HfO$_2$ interface, *J. Appl. Phys.* **94**(11):7155–7158 (2003).

137. Q. Chen, M. Yang, Y. P. Feng, J. W. Chai, Z. Zhang, J. S. Pan, and S. J. Wang, Band offsets of HfO$_2$/ZnO interface: In situ X-ray photoelectron spectroscopy measurement and ab initio calculation, *Appl. Phys. Lett.* **95**(16):162104 (2009).

138. K. M. Yeung, S. G. Lu, C. L. Mak, and K. H. Wong, Structural and optical properties of ZnS:Mn films grown by pulsed laser deposition, *MRS Proc. Libr.* **667**:G5.2.1–G5.2.6 (2001).

139. M.-Y. Lu, J. Song, M.-P. Lu, C.-Y. Lee, L.-J. Chen, and Z. L. Wang, ZnO–ZnS heterojunction and ZnS nanowire arrays for electricity generation, *ACS Nano* **3**(2):357–362 (2009).

140. L. J. Mandalapu, Z. Yang, S. Chu, and J. L. Liu, Ultraviolet emission from Sb-doped p-type ZnO based heterojunction light-emitting diodes, *Appl. Phys. Lett.* **92**(12):122101 (2008).

141. D.-K. Hwang, S.-H. Kang, J.-H. Lim, E.-J. Yang, J.-Y. Oh, J.-H. Yang, and S.-J. Park, p-ZnO/n-GaN heterostructure ZnO light-emitting diodes, *Appl. Phys. Lett.* **86**(22):222101 (2005).

142. T. Ohashi, K. Yamamoto, A. Nakamura, and J. Temmyo, Red emission from ZnO-based double heterojunction diode, *Jpn. J. Appl. Phys.* **47**(4):2961–2964 (2008).

143. P.-C. Wu, H.-Y. Lee, and C.-T. Lee, Enhanced light emission of double heterostructured MgZnO/ZnO/MgZnO in ultraviolet blind light-emitting diodes deposited by vapor cooling condensation system, *Appl. Phys. Lett.* **100**(13):131116–131123 (2012).

144. S. Chu, J. Zhao, Z. Zuo, J. Kong, L. Li, and J. Liu, Enhanced output power using MgZnO/ZnO/MgZnO double heterostructure in ZnO homojunction light emitting diode, *J. Appl. Phys.* **109**(12):123110 (2011).

145. M.-Y. Ke, T.-C. Lu, S.-C. Yang, C.-P. Chen, Y.-W. Cheng, L.-Y. Chen, C.-Y. Chen, J.-H. He, and J. Huang, UV light emission from GZO/ZnO/GaN heterojunction diodes with carrier confinement layers, *Opt. Express* **17**(25):22912–22917 (2009).

146. S. Zhi-Feng, Z. Yuan-Tao, X. Xiao-Chuan, Z. Wang, W. Hui, Z. Long, D. Xin, Z. Bao-Lin, and D. Guo-Tong, Electrically driven ultraviolet random lasing from an n-MgZnO/i-ZnO/SiO$_2$/p-Si asymmetric double heterojunction, *Nanoscale* **5**(11):5080–5085 (2013).

147. H. Long, S. Li, X. Mo, H. Wang, Z. Chen, Z. C. Feng, and G. Fang, Enhanced electroluminescence using Ta$_2$O$_5$/ZnO/HfO$_2$ asymmetric double heterostructure in ZnO/GaN-based light emitting diodes, *Opt. Express* **22**(S3):A833–A841 (2014).

148. S. V. Ivanov, V. A. Solov'ev, K. D. Moiseev, I. V. Sedova, Y. V. Terent'ev, A. A. Toropov, B. Y. Meltzer, M. P. Mikhailova, Y. P. Yakovlev, and P. S. Kop'ev, Room-temperature midinfrared electroluminescence from asymmetric AlSbAs/InAs/CdMgSe heterostructures grown by molecular beam epitaxy, *Appl. Phys. Lett.* **78**(12):1655–1657 (2001).

149. S. V. Ivanov, V. A. Kaygorodov, S. V. Sorokin, B. Y. Meltser, V. A. Solov'ev, Y. V. Terent'ev, O. G. Lyublinskaya et al., A 2.78-µm laser diode based on hybrid AlGaAsSb/InAs/CdMgSe double heterostructure grown by molecular-beam epitaxy, *Appl. Phys. Lett.* **82**(21):3782–3784 (2003).

150. J. S. Liu, C. X. Shan, H. Shen, B. H. Li, Z. Z. Zhang, L. Liu, L. G. Zhang, and D. Z. Shen, ZnO light-emitting devices with a lifetime of 6.8 hours, *Appl. Phys. Lett.* **101**(1):011106 (2012).

151. X. Y. Liu, C. X. Shan, C. Jiao, S. P. Wang, H. F. Zhao, and D. Z. Shen, Pure ultraviolet emission from ZnO nanowire-based p-n heterostructures, *Opt. Lett.* **39**(3):422–425 (2014).

152. J. W. Mares, M. Falanga, A. V. Thompson, A. Osinsky, J. Q. Xie, B. Hertog, A. Dabiran, P. P. Chow, S. Karpov, and W. V. Schoenfeld, Hybrid CdZnO/GaN quantum-well light emitting diodes, *J. Appl. Phys.* **104**(9):093107 (2008).

153. C. Horng-Shyang, T. Shao-Ying, L. Che-Hao, C. Chih-Yen, H. Chieh, Y. Yu-Feng, C. Hao-Tsung, K. Yean-Woei, and Y. Chih-Chung, Vertical CdZnO/ZnO quantum-well light-emitting diode, *IEEE Photonic Technol. Lett.* **25**(3):317–319 (2013).

154. S. K. Pandey, S. K. Pandey, V. Awasthi, and S. Mukherjee, Design and growth optimization by dual ion beam sputtered ZnO-based high-efficiency multiple quantum well green light emitting diode, *Nanosci. Nanotech. Lett.* **6**(2):146–152 (2014).

155. A. Ohtomo, M. Kawasaki, T. Koida, K. Masubuchi, H. Koinuma, Y. Sakurai, Y. Yoshida, T. Yasuda, and Y. Segawa, $Mg_xZn_{1-x}O$ as a II–VI widegap semiconductor alloy, *Appl. Phys. Lett.* **72**(19):2466–2468 (1998).

156. A. Ohtomo, M. Kawasaki, I. Ohkubo, H. Koinuma, T. Yasuda, and Y. Segawa, Structure and optical properties of $ZnO/Mg_{0.2}Zn_{0.8}O$ superlattices, *Appl. Phys. Lett.* **75**(7):980–982 (1999).

157. H. Long, G. J. Fang, S. Z. Li, X. M. Mo, H. N. Wang, H. H. Huang, Q. K. Jiang, J. B. Wang, and X. Z. Zhao, A ZnO/ZnMgO multiple-quantum-well ultraviolet random laser diode, *IEEE Electron Device Lett.* **32**(1):54–56 (2011).

158. S. Chu, M. Olmedo, Z. Yang, J. Kong, and J. Liu, Electrically pumped ultraviolet ZnO diode lasers on Si, *Appl. Phys. Lett.* **93**(18):181106 (2008).

159. Y. R. Ryu, T. S. Lee, J. A. Lubguban, A. B. Corman, H. W. White, J. H. Leem, M. S. Han, Y. S. Park, C. J. Youn, and W. J. Kim, Wide-band gap oxide alloy: BeZnO, *Appl. Phys. Lett.* **88**(5):052103 (2006).

160. O. Madelung, *Semiconductors: Data Handbook*, Springer, Berlin, Germany, 691pp, 2004.

161. M.-C. Jeong, B.-Y. Oh, M.-H. Ham, and J.-M. Myoung, Electroluminescence from ZnO nanowires in n-ZnO film/ZnO nanowire array/p-GaN film heterojunction light-emitting diodes, *Appl. Phys. Lett.* **88**(20):202105 (2006).

162. J. J. Dong, X. W. Zhang, Z. G. Yin, J. X. Wang, S. G. Zhang, F. T. Si, H. L. Gao, and X. Liu, Ultraviolet electroluminescence from ordered ZnO nanorod array/p-GaN light emitting diodes, *Appl. Phys. Lett.* **100**(17):171109 (2012).

163. W. I. Park and G. C. Yi, Electroluminescence in n-ZnO nanorod arrays vertically grown on p-GaN, *Adv. Mater.* **16**(1):87–90 (2004).

164. L. E. Greene, M. Law, D. H. Tan, M. Montano, J. Goldberger, G. Somorjai, and P. Yang, General route to vertical ZnO nanowire arrays using textured ZnO seeds, *Nano Lett.* **5**(7):1231–1236 (2005).

165. N. Kouklin, Cu-doped ZnO nanowires for efficient and multispectral photodetection applications, *Adv. Mater.* **20**(11):2190–2194 (2008).

166. Z. Liang, A. S. Susha, A. Yu, and F. Caruso, Nanotubes prepared by layer-by-layer coating of porous membrane templates, *Adv. Mater.* **15**(21):1849–1853 (2003).

167. D. F. Liu, Y. J. Xiang, X. C. Wu, Z. X. Zhang, L. F. Liu, L. Song, X. W. Zhao, S. D. Luo, W. J. Ma, and J. Shen, Periodic ZnO nanorod arrays defined by polystyrene microsphere self-assembled monolayers, *Nano Lett.* **6**(10):2375–2378 (2006).

168. M. R. Parra and F. Z. Haque, Optical study and ruthenizer (II) N3 dye-sensitized solar cell application of ZnO nanorod-arrays synthesized by combine two-step process, *Opt. Spectrosc.* **119**(4):672–681 (2015).

169. W. Riedel, Y. Fu, Ü. Aksüngera, J. Kavalakkatt, C. H. Fischer, M. C. Lux-Steiner, and S. Gledhill, ZnO and ZnS nanodots deposited by spray methods: A versatile tool for nucleation control in electrochemical ZnO nanorod array growth, *Thin Solid Films* **589**:327–330 (2015).

170. S. Saito, M. Miyayama, K. Koumoto, and H. Yanagida, Gas sensing characteristics of porous ZnO and Pt/ZnO ceramics, *J. Am. Ceram. Soc.* **68**(1):40–43 (2006).

171. S. S. Soni, M. J. Henderson, B. Jean-François, and G. Alain, Visible-light photocatalysis in titania-based mesoporous thin films, *Adv. Mater.* **20**(8):1493–1498 (2008).

172. D. Wang and C. Song, Controllable synthesis of ZnO nanorod and prism arrays in a large area, *J. Phys. Chem. B* **109**(26):12697–12700 (2005).

173. M. Willander, O. Nur, Q. X. Zhao, L. L. Yang, M. Lorenz, B. Q. Cao, J. Pérez Zúñiga, C. Czekalla, G. Zimmermann, and M. Grundmann, Zinc oxide nanorod based photonic devices: Recent progress in growth, light emitting diodes and lasers, *Nanotechnology* **20**(33):375–380 (2009).

174. P. Yan, D. Zhang, K. Cheng, Y. Wang, K. Ye, D. Cao, B. Wang, G. Wang, and Q. Li, Preparation of Au nanoparticles modified TiO_2/C core/shell nanowire array and its catalytic performance for $NaBH_4$ oxidation, *J. Electroanal. Chem.* **745**:56–60 (2015).

175. Y. Zhang, K. Yu, D. Jiang, Z. Zhu, H. Geng, and L. Luo, Zinc oxide nanorod and nanowire for humidity sensor, *Appl. Surf. Sci.* **242**(1–2):212–217 (2005).

176. J. C. D. Faria, A. J. Campbell, and M. A. McLachlan, ZnO nanorod arrays as electron injection layers for efficient organic light emitting diodes, *Adv. Funct. Mater.* **25**(29):4657–4663 (2015).

177. Q. J. Feng, H. W. Liang, Y. Y. Mei, J. Y. Liu, C. C. Ling, P. C. Tao, D. Z. Pan, and Y. Q. Yang, ZnO single microwire homojunction light emitting diode grown by electric field assisted chemical vapor deposition, *J. Mater. Chem. C* **3**(18):4678–4682 (2015).

178. H. Jeong, R. Salas-Montiel, and M. S. Jeong, Optimal length of ZnO nanorods for improving the light-extraction efficiency of blue InGaN light-emitting diodes, *Opt. Express* **23**(18):23195–23207 (2015).

179. X. Y. Li, M. X. Chen, R. M. Yu, T. P. Zhang, D. S. Song, R. R. Liang, Q. L. Zhang et al., Enhancing light emission of ZnO-nanofilm/Si-micropillar heterostructure arrays by piezo-phototronic effect, *Adv. Mater.* **27**(30):4447–4453 (2015).

180. S. S. Lo, L. Yang, and C. P. Chiu, ZnO/poly(N-vinylcarbazole) coaxial nanocables for white-light emissions, *J. Mater. Chem. C* **3**(3):686–692 (2015).

181. Y. J. Lu, C. X. Shan, M. M. Jiang, G. C. Hu, N. Zhang, S. P. Wang, B. H. Lia, and D. Z. Shen, Random lasing realized in n-ZnO/p-MgZnO core-shell nanowire heterostructures, *Crystengcomm* **17**(21):3917–3922 (2015).

182. P. Mao, A. K. Mahapatra, J. Chen, M. R. Chen, G. H. Wang, and M. Han, Fabrication of polystyrene/ZnO micronano hierarchical structure applied for light extraction of light-emitting devices, *ACS Appl. Mater. Inter.* **7**(34):19179–19188 (2015).

183. G. C. Park, S. M. Hwang, S. M. Lee, J. H. Choi, K. M. Song, H. Y. Kim, H. S. Kim et al., Hydrothermally grown in-doped ZnO nanorods on p-GaN films for color-tunable heterojunction light-emitting-diodes, *Sci. Rep.* **5**:10410 (2015).

184. T. Pauporte, O. Lupan, J. Zhang, T. Tugsuz, I. Ciofini, F. Labat, and B. Viana, Low-temperature preparation of Ag-doped ZnO nanowire arrays, DFT study, and application to light-emitting diode, *ACS Appl. Mater. Inter.* **7**(22):11871–11880 (2015).

185. L. S. Vikas, C. K. Sruthi, and M. K. Jayaraj, Defect-assisted tuning of electroluminescence from p-GaN/n-ZnO nanorod heterojunction, *Bull. Mater. Sci.* **38**(4):901–907 (2015).

186. Y. T. Xu, L. Xu, J. Dai, Y. Ma, X. W. Chu, Y. T. Zhang, G. T. Du, B. L. Zhang, and J. Z. Yin, Ultraviolet-enhanced light emitting diode employing individual ZnO microwire with SiO_2 barrier layers, *Appl. Phys. Lett.* **106**(21):212105 (2015).

187. X. Yang, C. X. Shan, M. M. Jiang, J. M. Qin, G. C. Hu, S. P. Wang, H. A. Ma, X. P. Jia, and D. Z. Shen, Intense electroluminescence from ZnO nanowires, *J. Mater. Chem. C* **3**(20):5292–5296 (2015).

188. Z. P. Yang, Z. H. Xie, C. C. Lin, and Y. J. Lee, Slanted n-ZnO nanorod arrays/p-GaN light-emitting diodes with strong ultraviolet emissions, *Opt. Mater. Express* **5**(2):399–407 (2015).

189. H. Zhou, P. B. Gui, Q. H. Yu, J. Mei, H. Wang, and G. J. Fang, Self-powered, visible-blind ultraviolet photodetector based on n-ZnO nanorods/i-MgO/p-GaN structure light-emitting diodes, *J. Mater. Chem. C* **3**(5):990–994 (2015).

190. J. H. Jeon, P. J. Choi, S. J. Oh, Y. J. Kang, J. Y. Kim, and M. K. Kwon, Improvement of the light extraction efficiency of InGaN/GaN blue light emitting diodes using ZnO nanostructures, *J. Nanosci. Nanotech* **15**(7):5215–5219 (2015).

191. X. L. Ren, X. H. Zhang, N. S. Liu, L. Wen, L. W. Ding, Z. W. Ma, J. Su, L. Y. Li, J. B. Han, and Y. H. Gao, White light-emitting diode from Sb-doped p-ZnO nanowire arrays/n-GaN film, *Adv. Funct. Mater.* **25**(14):2182–2188 (2015).

192. C. F. Wang, R. R. Ba, K. Zhao, T. P. Zhang, L. Dong, and C. F. Pan, Enhanced emission intensity of vertical aligned flexible ZnO nanowire/p-polymer hybridized LED array by piezo-phototronic effect, *Nano Energy* **14**:364–371 (2015).

193. Z. F. Shi, Y. T. Zhang, X. J. Cui, S. W. Zhuang, B. Wu, J. Y. Jiang, X. W. Chu, X. Dong, B. L. Zhang, and G. T. Du, Epitaxial growth of vertically aligned ZnO nanowires for bidirectional direct-current driven light-emitting diodes applications, *Crystengcomm* **17**(1):40–49 (2015).

194. Y. Yang, X. W. Sun, B. K. Tay, G. F. You, S. T. Tan, and K. L. Teo, A p-n homojunction ZnO nanorod light-emitting diode formed by As ion implantation, *Appl. Phys. Lett.* **93**(25):253107–253113 (2008).

195. X. W. Sun, B. Ling, J. L. Zhao, S. T. Tan, Y. Yang, Y. Q. Shen, Z. L. Dong, and X. C. Li, Ultraviolet emission from a ZnO rod homojunction light-emitting diode, *Appl. Phys. Lett.* **95**(13):133124 (2009).

196. X. Tang, G. Li, and S. Zhou, Ultraviolet electroluminescence of light-emitting diodes based on single n-ZnO/p-AlGaN heterojunction nanowires, *Nano Lett.* **13**(11):5046–5050 (2013).

197. S. Chu, G. Wang, W. Zhou, Y. Lin, L. Chernyak, J. Zhao, J. Kong, L. Li, J. Ren, and J. Liu, Electrically pumped waveguide lasing from ZnO nanowires, *Nat. Nanotechnol.* **6**(8):506–510 (2011).

198. H. Sun, Q. Zhang, J. Zhang, T. Deng, and J. Wu, Electroluminescence from ZnO nanowires with a p-ZnO film/n-ZnO nanowire homojunction, *Appl. Phys. B* **90**(3–4):543–546 (2008).

199. O. Lupan, T. Pauporté, and B. Viana, Low-voltage UV-electroluminescence from ZnO-nanowire array/p-GaN light-emitting diodes, *Adv. Mater.* **22**(30):3298–3302 (2010).

200. Z. Shi, Y. Zhang, X. Cui, B. Wu, S. Zhuang, F. Yang, X. Yang, B. Zhang, and G. Du, Improvement of electroluminescence performance by integration of ZnO nanowires and single-crystalline films on ZnO/GaN heterojunction, *Appl. Phys. Lett.* **104**(13):131109 (2014).

201. J.-J. Dong, H.-Y. Hao, J. Xing, Z.-J. Fan, and Z.-L. Zhang, Electroluminescence of ordered ZnO nanorod array/p-GaN light-emitting diodes with graphene current spreading layer, *Nanoscale Res. Lett.* **9**(1):1–6 (2014).

202. S. W. Lee, H. D. Cho, G. Panin, and T. W. Kang, Vertical ZnO nanorod/Si contact light-emitting diode, *Appl. Phys. Lett.* **98**(9):093110–093113 (2011).

203. J. J. Hassan, M. A. Mahdi, A. Ramizy, H. Abu Hassan, and Z. Hassan, Fabrication and characterization of ZnO nanorods/p-6H–SiC heterojunction LED by microwave-assisted chemical bath deposition, *Superlattices Microstruct.* **53**:31–38 (2013).

204. J. J. Hassan, M. A. Mahdi, Y. Yusof, H. Abu-Hassan, Z. Hassan, H. A. Al-Attar, and A. P. Monkman, Fabrication of ZnO nanorod/p-GaN high-brightness UV LED by microwave-assisted chemical bath deposition with Zn(OH)$_2$–PVA nanocomposites as seed layer, *Opt. Mater.* **35**(5):1035–1041 (2013).

205. C. Sanchez, B. Lebeau, F. Chaput, and J. P. Boilot, Optical properties of functional hybrid organic–inorganic nanocomposites, *Adv. Mater.* **15**(23):1969–1994 (2003).

206. Q. Yang, Y. Liu, C. Pan, J. Chen, X. Wen, and Z. L. Wang, Largely enhanced efficiency in ZnO nanowire/p-polymer hybridized inorganic/organic ultraviolet light-emitting diode by piezo-phototronic effect, *Nano Lett.* **13**(2):607–613 (2013).

207. X. W. Sun, J. Z. Huang, J. X. Wang, and Z. Xu, A ZnO nanorod inorganic/organic heterostructure light-emitting diode emitting at 342 nm, *Nano Lett.* **8**(4):1219–1223 (2008).

208. E. Arici, N. S. Sariciftci, and D. Meissner, Hybrid solar cells based on nanoparticles of CuInS$_2$ in organic matrices, *Adv. Funct. Mater.* **13**(2):165–171 (2003).

209. L. C. Chen, J. C. Chen, C. C. Chen, and C. G. Wu, Fabrication and properties of high-efficiency perovskite/PCBM organic solar cells, *Nanoscale Res. Lett.* **10**(1):1020 (2015).

210. X. Gong, Y. Jiang, M. Li, H. Liu, and H. Ma, Hybrid tapered silicon nanowire/PEDOT:PSS solar cells, *Rsc Advances* **5**(14):10310–10317 (2015).

211. F. C. Krebs, Polymer solar cell modules prepared using roll-to-roll methods: Knife-over-edge coating, slot-die coating and screen printing, *Sol. Energ. Mat. Sol. Cells* **93**(4):465–475 (2009).

212. F. C. Krebs, Roll-to-roll fabrication of monolithic large-area polymer solar cells free from indium-tin-oxide, *Sol. Energ. Mat. Sol. Cells* **93**(9):1636–1641 (2009).

213. J. Ouyang, C. W. Chu, F. C. Chen, Q. Xu, and Y. Yang, High-conductivity poly(3,4-ethylenedioxythiophene):poly(styrene sulfonate) film and its application in polymer optoelectronic devices, *Adv. Funct. Mater.* **15**(2):203–208 (2005).

214. M. Pietsch, S. Jäckle, and S. Christiansen, Interface investigation of planar hybrid n-Si/PEDOT:PSS solar cells with open circuit voltages up to 645 mV and efficiencies of 12.6 %, *Appl. Phys. A* **115**(4):1109–1113 (2014).

215. H. Spanggaard and F. C. Krebs, A brief history of the development of organic and polymeric photovoltaics, *Sol. Energ. Mat. Sol. Cells* **83**(2):125–146 (2004).

216. J. M. Topple, S. M. McAfee, G. C. Welch, and I. G. Hill, Pivotal factors in solution-processed, non-fullerene, all small-molecule organic solar cell device optimization, *Org. Electron.* **27**:197–201 (2015).

217. L. Duan, P. Wang, F. Wei, W. Zhang, R. Yao, and H. Xia, Electroluminescence of ZnO nanorods embedded in a polymer film, *Solid State Commun.* **200**:14–16 (2014).

218. P.-J. Li, Z.-M. Liao, X.-Z. Zhang, X.-J. Zhang, H.-C. Zhu, J.-Y. Gao, K. Laurent, Y. Leprince-Wang, N. Wang, and D.-P. Yu, Electrical and photoresponse properties of an intramolecular p-n homojunction in single phosphorus-doped ZnO nanowires, *Nano Lett.* **9**(7):2513–2518 (2009).

219. Q. Zhang, J. Qi, X. Li, F. Yi, Z. Wang, and Y. Zhang, Electrically pumped lasing from single ZnO micro/nanowire and poly(3,4-ethylenedioxythiophene):poly(styrenexulfonate) hybrid heterostructures, *Appl. Phys. Lett.* **101**(4):043119 (2012).

220. J. Dai, Y. Ji, C. X. Xu, X. W. Sun, K. S. Leck, and Z. G. Ju, White light emission from CdTe quantum dots decorated n-ZnO nanorods/p-GaN light-emitting diodes, *Appl. Phys. Lett.* **99**(6):063112 (2011).

221. J. Dai, C. X. Xu, and X. W. Sun, ZnO-Microrod/p-GaN heterostructured whispering-gallery-mode microlaser diodes, *Adv. Mater.* **23**(35):4115–4119 (2011).

222. G. Y. Zhu, C. X. Xu, Y. Lin, Z. L. Shi, J. T. Li, T. Ding, Z. S. Tian, and G. F. Chen, Ultraviolet electroluminescence from horizontal ZnO microrods/GaN heterojunction light-emitting diode array, *Appl. Phys. Lett.* **101**(4):041110–041114 (2012).

223. J. Dai, C. X. Xu, R. Ding, K. Zheng, Z. L. Shi, C. G. Lv, and Y. P. Cui, Combined whispering gallery mode laser from hexagonal ZnO microcavities, *Appl. Phys. Lett.* **95**(19):191117 (2009).

224. J. Dai, C. X. Xu, Z. L. Shi, R. Ding, J. Y. Guo, Z. H. Li, B. X. Gu, and P. Wu, Three-photon absorption induced whispering gallery mode lasing in ZnO twin-rods microstructure, *Opt. Mater.* **33**(3):288–291 (2011).

225. J. Dai, C. X. Xu, L. X. Sun, Z. H. Chen, J. Y. Guo, and Z. H. Li, Multiphoton absorption-induced optical whispering-gallery modes in ZnO microcavities at room temperature, *J. Phys. D: Appl. Phys.* **44**(2):025404 (2011).

226. J. Dai, C. X. Xu, and X. W. Sun, Single-photon and three-photon absorption induced whispering-gallery mode lasing in ZnO micronails, *Opt. Commun.* **284**(16–17):4018–4021 (2011).

227. J. Dai, C. X. Xu, X. W. Sun, and X. H. Zhang, Exciton-polariton microphotoluminescence and lasing from ZnO whispering-gallery mode microcavities, *Appl. Phys. Lett.* **98**(16):161110 (2011).

228. J. Dai, C. X. Xu, P. Wu, J. Y. Guo, Z. H. Li, and Z. L. Shi, Exciton and electron-hole plasma lasing in ZnO dodecagonal whispering-gallery-mode microcavities at room temperature, *Appl. Phys. Lett.* **97**(1):011101 (2010).

229. J. Dai, C. X. Xu, K. Zheng, C. G. Lv, and Y. P. Cui, Whispering gallery-mode lasing in ZnO microrods at room temperature, *Appl. Phys. Lett.* **95**(24):241110 (2009).

230. D. Yu, Y. Chen, B. Li, X. Chen, M. Zhang, F. Zhao, and S. Ren, Structural and lasing characteristics of ultrathin hexagonal ZnO nanodisks grown vertically on silicon-on-insulator substrates, *Appl. Phys. Lett.* **91**(9):091116 (2007).

231. D. Wang, H. W. Seo, C.-C. Tin, M. J. Bozack, J. R. Williams, M. Park, and Y. Tzeng, Lasing in whispering gallery mode in ZnO nanonails, *J. Appl. Phys.* **99**(9):093112 (2006).

18

Natural Light-Style Organic Light-Emitting Diodes

Jwo-Huei Jou
National Tsing-Hua University

Meenu Singh
National Tsing-Hua University

Yi-Fang Tsai
National Tsing-Hua University

Abstract Natural light from the sun is a primary sustainable element for life on Earth. Sunlight is a diffused and continuous visible radiation with varying chromaticity from dawn to dusk, also changing with the weather and the region. This radiation, which is of the highest quality, reproduces the true and natural colors of objects but, unfortunately, cannot be retained for illumination after dusk. Five thousand years ago, the ancient Egyptians introduced candles to provide lighting at night. These hydrocarbon-burning lighting sources, while emitting a high-quality radiation that is almost free of blue hazard, are unfortunately energy inefficienct, a major obstacle. Lighting sources powered by electricity emerged 200 years ago and developed continuously in terms of their energy efficiency. However, most of the emission spectra of these lighting sources are not diffused and soft, and their chromaticity is not perfectly similar to that of blackbody radiation. Modern lighting measures, such as compact fluorescent tubes and light-emitting diodes, have a large share of the illumination market, but barely satisfy our desire for the natural lighting given by the sun. By using organic light-emitting diode (OLED) technology, chromaticity with a varying color temperature similar to that of sunlight can be obtained. In this book, we comprehensively discuss the design and development of natural light-style OLEDs such as a sunlight-style OLED, low color-temperature OLED, candlelight-style OLED, sunset-style OLED, and pseudo-style natural light OLED. We also discuss the status of and issues with current light-quality matrices and introduce a new metric with which to measure the quality of light, such as its resemblance to the spectra of sunlight and natural light. The discussion we present is intended to provide a path leading to healthy illumination and to open up a new era in solid-state lighting in order to trigger a "Lighting Renaissance."

18.1 Introduction

Natural light is daylight, the color temperature of which varies according to the time of day. There are various sources of natural light, for example, from the sun, the light from the stars, fires, insects, and so on. Light from the moon is basically the reflected light of the sun; even so, the moon can provide significant luminance, especially when it is full. Besides these sources, certain other living things—such as glow worms, certain kinds of fish, and mushrooms—generate light via light-producing cells; this is known as bioluminescence.[1-8] Sunlight and the light from wood or oil fires provide light that covers a smooth and continuous spectrum over the entire visible range, perfectly matching that of blackbody radiation at the same corresponding color temperature. Natural light, like sunlight, is a basic component required to sustain all plants, animals, and people on earth.[9] It comprises a smooth and continuous spectrum, which, at its highest quality, produces sufficient light to reproduce the actual color of objects. The color temperature of sunlight varies according to the time of day, location, and weather states. It ranges from 1500 to 3000 K at dawn or sunset to 5500 K at noon on a sunny day. While the sky has a color temperature of around 10,000 K most of the daytime in clear weather, it can drop to 6,500 K on a cloudy day.[10,11]

The color-temperature transition of natural light influences all living things and causes a gradually shift in all the physiological and psychological functions of living things.[12-21] For instance, brighter natural light is responsible for the secretion of cortisol, which makes people more awake, better able to concentrate, and more productive at work.[14,17,18] It is also responsible for the suppression of melatonin secretion. On the other hand, dimmer light levels with a lower color temperature permit the increased production of melatonin, leading to a pleasant sensation of relaxation. Importantly, the lack of melatonin due to frequent exposure to intense light at night can increase the risk of cancer.[12,13,15,16,19-21] Thus, natural light provides a pleasant radiation for people's well-being.

Today, we have many artificial lighting sources that simulate light components ranging from near ultraviolet (UV) to infrared (IR). These sources can be divided into two categories, namely, (1) hydrocarbons burning and (2) electrical. Light sources that burn hydrocarbons include oil lamps, gas lamps, candles, and charcoal lights. Electrical lighting sources are based on a variety of lighting mechanisms and comprise incandescent bulbs,[22-25] gas-discharge lamps (mercury lamps, sodium lamps, and metal-halide lamps),[26,27] fluorescent lamps,[27,28] light-emitting diodes (LEDs),[29-34] and organic light-emitting diodes (OLEDs).[35,36]

Lighting sources that burn hydrocarbons—such as candles, which were invented 5000 years ago—are energy inefficient and provide a flickering, glaring light. They also increase air pollution, are a fire hazard, and emit greenhouse gas into the ambience atmosphere. On the other hand, lighting sources powered by electricity are energy efficient; nevertheless, they do not provide good quality light and not pollution free. Typical electrical lighting sources, such as fluorescent tubes and incandescent bulbs, have critical issues in terms of efficiency, light quality, and environmental friendliness,[37,38] although alternative systems, such as LEDs, are energy efficient in comparison with incandescent lamps and their use is widespread in the display and lighting market.[30-34] Even so, electrically powered light may cause certain hazards, such as increasing the risk of various cancers that result from the suppressed secretion of melatonin, an oncostatic hormone,[12,13,15,16,19-21] not only producing irreversible retinal damage to people's eyes,[39] but also degrading certain pigments in famous artworks that form part of people's heritage.[40,41] Furthermore, these electrical lighting sources have certain evident issues, for example, mercury discharge, UV radiation, blue-light hazard, high heat levels, high color temperature, glaring light, a low color rendering index (CRI), and nonuniform color chromaticity.[42–45] Current illumination devices, both those that are electrically powered and those that burn hydrocarbons, satisfy basic lighting requirements, but they do not satisfy our need for a light source that reflects the behavior of natural light, the color temperature of which has many governing factors.[10,11] This user-friendly lighting requirement means that OLEDs are a favorable contender in solid-state lighting. Basically, as an electrical lighting source, considering all illumination technologies, OLEDs demonstrate excellent characteristics. They are an environmentally friendly, high-quality lighting source that is free of UV, mercury discharge, glare, spiking emission, blue-light hazard, and heating surface.[46–53] Light quality has a marked impact on human psychology and physiology, and OLEDs provide high-quality light with a high CRI,[46] high spectrum resemblance (SRI),[47,48] uniform color chromaticity,[49–51] and a color-temperature span covering that of natural light.[49–53] OLEDs exhibit advantages such as self-emission, a wide beam angle (>160°),[54] fast response time (<1 μs),[55] a high color contrast ratio,[56] and low power consumption[57] and are fully dimmable.[58] Also, they show an efficiency of greater than 90 lm/W, approaching that of a fluorescent tube.[59] In contrast to other lighting source, such as LEDs, OLEDs are a planar lighting source and provide pleasant diffused light over a large area. Besides these features, OLED luminaires possess special properties such as being ultra-thin, lightweight, transparent, printable, and flexible (they can be rolled and are wearable and bendable), which could lead to new design opportunities. These characteristics, coupled with their tailorable spectrum, enable OLEDs to produce a variety of emissions that mimic different states of natural light.

This chapter will present the results of a comprehensive study on various OLEDs capable of simulating various states of natural light, such as sunlight, candlelight, low color temperature light, sunset, and pseudo-natural light, and will discuss the design and fabrication processes. There will also be discussion of new light-quality metrics, issues in current light-quality metrics, commercialization of the world's first candlelight-style OLED, and a critic review on the need for a "Lighting Renaissance."

18.2 Some Natural Light-Style OLEDs

18.2.1 Sunlight-Style Color-Temperature Tunable OLEDs

It is essential to have illumination that mimics sunlight, a source that provides emission with daylight chromaticity and a wide color-temperature span. However, no current illumination device provides lighting capable of offering the variation in color temperature found with natural light. Current lighting sources include electrically powered point-type incandescent bulbs, mercury lamps, and LEDs, and line-type cool-white and warm-white fluorescent tubes, together with traditional lighting devices that burn hydrocarbons, such as torches and candles. For example, the color temperature is 1900 K for candles, 2500 K for incandescent bulbs, 3000 K for warm-white fluorescent tubes, and 5000 K for cold-white fluorescent tubes (Figure 18.1).

Therefore, devising a light source with color-temperature tunability would be highly valuable. Some efforts have been made with regard to LEDs, and partial color-temperature matching with a natural light locus has been achieved. No attention was given to devising a color-temperature tunable lighting source in OLED technology until 2009.

In 2009, Jou and colleagues invented the world's first electrically powered sunlight-style color-temperature tunable OLED that yielded a wide color-temperature span and contributed a noteworthy incentive to OLED technology in general lighting.[49] As shown in the study, a single OLED device could exhibit various chromaticities with a color temperature that ranged between 2300 and 7900 K, which fully covers the entire daylight spectrum at different times and in a variety of regions.[10,11] The device included fluorescent emitters such as an 8 nm blue layer of 2-(*N,N*-diphenyl-amino)-6-[4-(*N,N*-diphenylamino)styryl]naphthalene (DPASN), a 2 nm green layer bis[(p-isopropylohenyl)(p-tolyl)amino]—10-10′-phenanthracene (BPTAPA) 0.05 wt% doped in DPASN, a 5 nm red layer with 4-(dicyanomethylene)-2-tertbutyl-6-(1,1,7,7-tetramethyljulolidin-4-yl-vinyl)-4H-pyran (DCJTB) doped in DPASN, a 32 nm electron-transporting layer of 1,3,5-tris(N-phenylbenzimidazol-2-yl)benzene (TPBi), and a 0.8 nm electron-injection layer of lithium fluoride (LiF). A 3 nm hole modulation

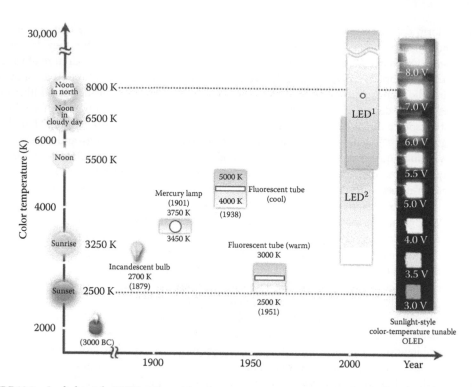

FIGURE 18.1 Sunlight-style OLED with a wide color-temperature span, compared with the color temperature and/or color-temperature span exhibited by candles, as well as typical electrically powered lighting devices. (Reprinted with permission from Jou, J.H. et al., Sunlight-style color temperature tunable organic light-emitting diode, *Appl. Phys. Lett.*, 95, 013307 Copyright 2009, American Institute of Physics, AIP Publishing LLC.)

layer (HML) TPBi was introduced between the green and red emissive layers, which resulted in a broad color-temperature span in comparison with that without the HML (Figure 18.2a).

The resultant OLED had a sunlight-style emission with a power efficiency of 7.0 lm/W at a brightness of 100 cd/m^2. Variation in red dopant DCJTB concentration effectively controlled the emission track and, at 0.8 wt%, the device exhibited an emission that closely matched with a daylight locus (Figure 18.2).

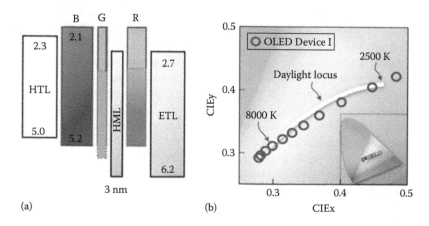

FIGURE 18.2 (a) Device architecture of the sunlight-style OLED and (b) the resultant emission track with color temperature between 2500 and 8000 K. (Reprinted with permission from Jou, J.H. et al., Sunlight-style color temperature tunable organic light-emitting diode, *Appl. Phys. Lett.*, 95, 013307 Copyright 2009, American Institute of Physics, AIP Publishing LLC.)

The device showed a maximum luminance of 5900 cd/m² at 9 V. Initially, at 3 V, it emitted a red spectrum, which turned to pure white at 5.5 V, and further converted to bluish white at 9 V, as shown in the electroluminescence (EL) spectra in Figure 18.3. The corresponding EL spectra showed a suppression in the red emission, peaking at 590 nm, and a simultaneous rising blue peak at 460 nm. More electrons passed through the interface of the green and blue emissive layers as the applied voltage was increased, which caused the recombination zone to shift toward the blue emissive layer.

Further, in 2011, Jou and colleagues also reported a high-efficiency sunlight-style OLED with a wide color-temperature span.[50] They investigated the effect of the thickness of HML from 1.5 to 6 nm on the chromaticity span of the device, which was found to increase gradually as the thickness increased. At 3 nm, the entire emission span was sufficiently wide to cover that of daylight and, at 4.5 nm, the device could show either a hypsochromic or bathochromic shift in the resultant chromaticity, depending on the applied voltage. Moreover, when a second HML was employed between the green and blue emissive layers (Figure 18.4), the device showed a sunlight-style emission with a much wider color-temperature span, that is, from 2,400 to 18,000 K. The corresponding power efficiency, at 1000 cd/m², for example, was increased to 9.0 lm/W in contrast to the 7.0 lm/W for the single HML device, which also showed a smaller color-temperature span, that is, from 2300 to 7900 K (Figure 18.4).

Subsequently, the efficiency of the sunlight-style OLED was significantly enhanced from 2.2 to 30 lm/W at 100 cd/m² as the electro-fluorescent emitters were replaced by their phosphorescent counterparts. Moreover, the phosphorescent OLED device also demonstrated a relatively high efficiency, even at 10,000 cd/m².

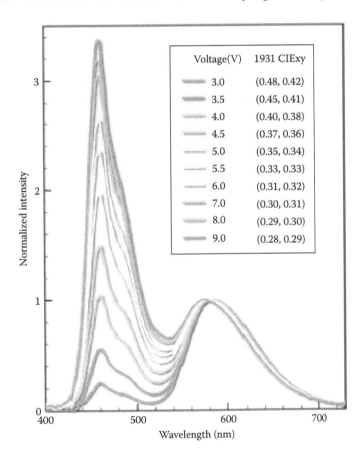

FIGURE 18.3 Electroluminescence spectra of the sunlight-style OLED at various applied voltages. (Reprinted with permission from Jou, J.H. et al., Sunlight-style color temperature tunable organic light-emitting diode, *Appl. Phys. Lett.*, 95, 013307 Copyright 2009, American Institute of Physics, AIP Publishing LLC.)

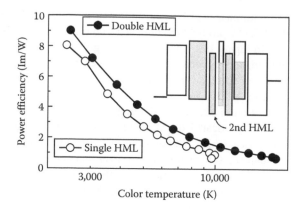

FIGURE 18.4 Power efficiency with respect to color temperature for the OLED with a double HML compared with that of the single HML counterpart. (Reproduced from Jou. J.H. et al., *J. Photon. Energy*, 1(1), 011021. Copyright 2011, Society of Photo Optical Instrumentation Engineers.)

This may provide an energy efficient alternative to incandescent bulbs, whose power efficiency is 10–15 lm/W. As to the lower color-temperature range, it may be suitable for lighting at night to minimize blue hazard.

Recently, Jou and colleagues reported a sunlight-style OLED device with a CRI ranging from 74 to 84.4 and a color temperature ranging from 2300 to 9300 K without any carrier modulation layer (CML).[60] The resultant higher color-temperature span, 7000 K, had been achieved by using a tailored device architecture. The color-temperature span can be significantly affected by the doping concentrations of emitters and the incorporation of a carrier modulation layer. For example, the device with a 2 wt% 5,6,11,12-tetra-phenylnaphthacene (rubrene) yellow dopant, a 0.3 wt% red dopant (WPRD931, a proprietary material from Wan Hsiang OLED Ltd.), and a 10 wt% blue dopant (EB515, a proprietary material from e-Ray Optoelectronics Technology Co. Ltd.) exhibited a color temperature from 2460 to 9340 K.[60] As the concentration of the yellow dopant was increased from 2 to 4 wt%, the red dopant from 0.3 to 0.5 wt%, and the blue dopant from 10 to 15 wt%, the device showed an emission with a slight blueshift and the color temperature covered the entire daylight locus. The resultant device exhibited an orange-red emission spectrum at 4.0 V with a Commission Internationale de l'Eclairage (CIE) of (0.45, 0.39) and turned to bluish white (0.29, 0.32) at 9.5 V (Figure 18.5a). Besides having a wide color-temperature span, from 2300 to 9300 K, it

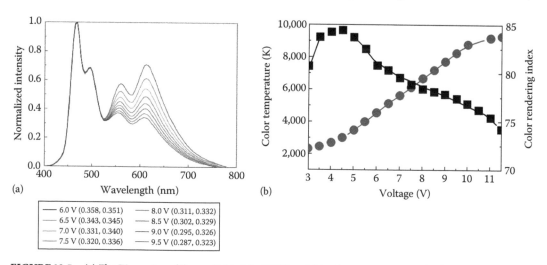

6.0 V (0.358, 0.351)	8.0 V (0.311, 0.332)
6.5 V (0.343, 0.345)	8.5 V (0.302, 0.329)
7.0 V (0.331, 0.340)	9.0 V (0.295, 0.326)
7.5 V (0.320, 0.336)	9.5 V (0.287, 0.323)

FIGURE 18.5 (a) The EL spectra of the sunlight-style OLED and (b) color temperature and color rendering index at various voltages. (From Liao, S.Y. et al., *Int. J. Photo Energy*, 2014, 480829. Copyright 2014.)

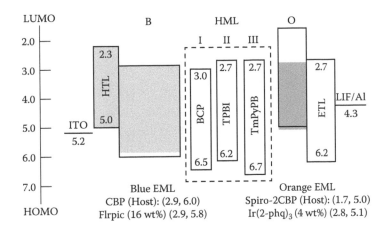

FIGURE 18.6 Energy-level diagram of the device with different carrier modulation layers. (From Jou, J.H. et al., Highly efficient color temperature tunable organic light-emitting diodes, *J. Mater. Chem.*, 22, 8117–8120. Copyright 2012 Reproduced by permission of the Royal Society of Chemistry.)

also showed a significantly high CRI, which ranged from 74 to 84.4 for various voltages from 3.0 to 11.5 V (Figure 18.5). When a 3 nm TPBi layer was employed as the CML, the device showed a color temperature from 2900 to 4890 K. However, as the thickness of CML was increased to 5 nm, the device showed a much smaller color-temperature span of 1700 K.

Moreover, Jou and colleagues also reported a high-efficiency OLED device with a color-temperature span from 1700 to 5200 K with a current efficiency of 23.7 cd/A at 1000 cd/m^2.[61] They fabricated a novel OLED device architecture by using a blue phosphorescent emitter bis(3,5-difluoro-2-(2-pyridyl)-phenyl-(2-carboxypyridyl)iridium(III) (FIrpic) and an orange-red emitter Ir(2- phq)$_3$ with two host materials, 4, 4′-N, N′-dicarbazole-biphenyl (CBP) and 2,7-bis(9-carbazolyl)-9,9-spirobifluorene (Spiro-2CBP), as shown in Figure 18.6.

As shown in Figure 18.6, three electron-transporting materials—namely, 2,9-dimethyl-4-7-diphenyl-1-10-phenanthroline (BCP), TPBi, and1,3,5-tri(m-pyrid-3-yl-phenyl)benzene (TmPyPB)—were introduced as carrier modulation layers in low color-temperature OLED devices. The resultant device exhibited the largest color-temperature span, 3700 K, when TmPyPB was employed. This may be attributed to the resulting highest hole-injection barrier (0.5 eV) and its high triplet energy (2.8 eV),[62] which, respectively, effectively modulated the flux of holes and confined the triplet excitons within the blue emissive layer. In addition, TmPyPB also showed a much higher electron mobility than that of its BCP and TPBi[63,64] counterparts, leading more excitons to generate in the blue emissive layer as the operational voltage was increased from 4 to 8 V (Figure 18.7).

Most recently, Jou and colleagues reported an OLED device with a tunable color temperature from 1580 to 2600 K.[51] The resultant device exhibited a color temperature between that of sunset (2500 K) and that of candlelight (1900 K). The color temperature of the OLED device can be further tuned from 5200 to 2360 K by varying the emissive layer thickness from 8 to 2 nm.

18.2.2 Low Color-Temperature OLEDs

Numerous medical studies reported that white light sources with a high color temperature emitted an intense blue emission, which may cause certain serious health issues, such as irreversible retinal damage,[39] physiological disorders,[12–17] and an increasing risk of various types of cancers resulting from the suppression of melatonin secretion.[21] Therefore, devising a new lighting source with a low color temperature or free of blue emission is as important as realizing a high-efficiency lighting device.

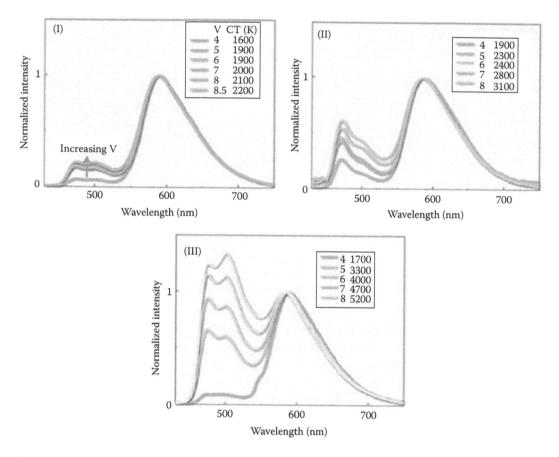

FIGURE 18.7 EL spectra comparison of the devices with BCP (I), TPBi (II), and TmPyPB (III). (From Jou, J.H. et al., Highly efficient color temperature tunable organic light-emitting diodes, *J. Mater. Chem.*, 22, 8117–8120. Copyright 2012 Reproduced by permission of the Royal Society of Chemistry.)

Low color-temperature OLEDs can be realized by maximizing the long-wavelength (red) emission and minimizing its short-wavelength (blue) counterparts. In 2011, Jou and colleagues demonstrated an OLED device with a color temperature of 1940 K and an external quantum efficiency (EQE) of 20.8% at 1000 cd/m^2 by employing a blend nano CML between the phosphorescent sky blue and orange-red emissive layers.[65] Figure 18.8 shows the architecture of the device with corresponding energy levels and the plausible distribution of charge carriers (holes and electrons) in the low color-temperature OLED devices with nano CMLs.

The resulted OLED initially showed an EQE of 17.2% with a color temperature of 1860 K at 1000 cd/m^2 without the CML (Device I). The incorporation of a 3 nm CML of 4,4″-di(triphenylsilyl)-p-terphenyl (BSB) between the blue and orange-red emissive layers caused a sharp reduction in the EQE to 5.1% and increased the color temperature to 2780 K. As shown in Figure 18.8, the BSB interlayer possessed a 0.5 eV barrier-to-hole injection; hence, a large number of holes would be blocked in the blue emissive layer, leading to an increasing color temperature. However, a 3 nm mixed CML with a 2:1 weight ratio of BSB: Spiro2-CBP markedly improved the performance of the device. The resultant device (Device III) exhibited a color temperature of 1940 K with a power efficiency of 29.1 lm/W, and an EQE of 20.8%, at 1000 cd/m^2. This noteworthy enhancement in device efficiency resulted as a blend CML was incorporated. This enhanced efficiency was due to the blend CML improving the distribution of the charge carriers in the available recombination zones and balanced the carrier injection.[65]

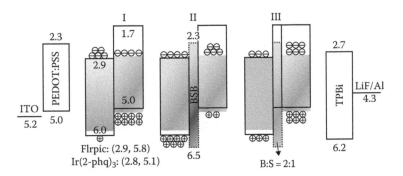

FIGURE 18.8 Schematic illustration of low color-temperature OLEDs without an interlayer (Device I), with a neat BSB interlayer (Device II), and with a blend interlayer (Device III). The ratio of BSB-to-Spiro-2CBP in the blend layer of Device III was 2:1. (From Jou, J.H. et al., High-efficiency, low color temperature organic light-emitting diode with a blend interlayer, *J. Mater. Chem.*, 21, 17850–17854. Copyright 2011, Reproduced by permission of the Royal Society of Chemistry.)

Jou and colleagues reported two further devices: an orange OLED device with a power efficiency of 38.8 lm/W and a color temperature of 2280 K at 100 cd/m² and a white OLED device with a power efficiency of 34.9 lm/W and a color temperature of 2860 K.[66] The devices consisted of an emissive layer containing a 0.6 wt% tris(2-phenylquinoline)iridium(III) (Ir-(2-phq)3) orange-red dye, a 0.2 wt% bis[5-methyl-7-tri-fluoromethyl-5H-benzo(c)(1,5)naphthyridin-6one]iridium(picolinate) (CF3BNO) green dye, and a 14 wt% FIrpic blue dye in a host CBP. Two other host materials—namely, 3,5-Di(9H-carbazol-9-yl)tetraphenylsilane (SimCP2) and 4,4′,4″-tri(*N*-carbazolyl)triphenylamine (TCTA)—were also used for comparison. The CBP composing device showed a current efficiency of 55.9 cd/A, which was higher than those of the simCP2 (26.8 cd/A) and TCTA (21.5 cd/A) (Figure 18.9a). The reason why the CBP device showed better efficiency is due to its having the lowest LUMO level (2.9 eV) of these hosts, which is very favorable to electron injection (Figure 18.9). Its ambipolar-transporting property also favors a balance carrier injection.

The CBP device exhibited a warm-white emission with a color temperature ranging between 2690 and 3410 K when the concentration of the green dopant was increased from 0.1 to 0.4 wt%. For example, by doping with a 0.2 wt% of the green dye, the resultant device showed an efficiency of 34.9 lm/W (55.9 cd/A) at 100 cd/m² with a color temperature of 2860K and a CIE of (0.46, 0.45). Further, by decreasing the concentration of the blue dye from 14 to 8 wt%, the resultant device exhibited an orange emission with an improved efficiency of 38.8 lm/W (66.5 cd/A) at 100 cd/m² with a color temperature of 2280 K and a CIE of (0.50, 0.44).

Additionally, Jou and colleagues fabricated an OLED with an ultra-low color temperature of 1773 K and a high CRI of 87 by using fluorescent electroluminescent materials.[67] The reported devices consisted of a single emissive layer with a yellow dye, Spiro-fluorene-dibenzosuberan[d](1,4-bis(4-(*N*,*N*-diphenylamine)-phenyl)-quinoxaline) (Spiro-QDPAP), and a red dye, ER55, doped into three different blue light-emitting hosts, namely, 2,7-bis{2[phenyl(m-tolyl)amino]-9,9-dimethyl-fluorene-7-yl}-9,9-dimethyl-fluorene (MDP3FL), 1-butyl-9,10-naphthalene-anthracene (BANE), and 2-(*N*,*N*-diphenyl-amino)-6-[4-(*N*,*N*-diphenylamine)styryl]naphthalene (DPASN). All the devices were doped with a 5 wt% of the yellow dye and had a red dye concentration varying from 0.1 to 0.7 wt%. The device using the host MDP3FL showed the highest power efficiency, 11.9 lm/W, and had an EQE of 6.4%, and a CRI of 87, which are higher levels than those of its BANE and DPASN counterparts. The high efficiency of the MDP3FL device may be attributed to a balance carrier injection, as the energy barriers for holes and electrons to the MDP3FL are 0.5 and 0.3 eV, respectively, whereas they are 0.8 and 0.1 eV for the BANE and 0.3 and 0.6 eV for the DPASN.[67] The high CRI may be related to five emission bands, including one band from the blue emitting host, one band from the yellow dopant, and three bands from the red dopant. Notably, the reported device showed a much lower color temperature than that of the candle, which may

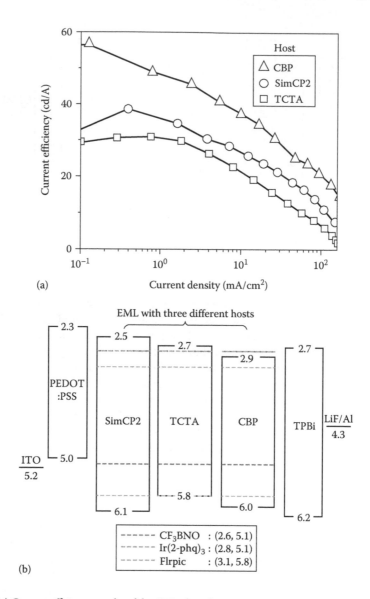

FIGURE 18.9 (a) Current efficiency results of the CBP white device compared with those of the SimCP2 and TCTA counterparts and (b) energy-level diagram of the white OLEDs using three different hosts: SimCP2, TCTA, and CBP. (Reprinted from *Org. Electron.*, 13, Jou, J.H. et al., High-efficiency low color temperature organic light-emitting diodes with solution-processed emissive layer, 899–904. Copyright 2012, with permission from Elsevier B.V.)

be attributed to a stronger red emission peak occurring beyond 610 nm. These plausibly physiologically and psychologically friendly characteristics show the reported OLED devices to be an ideal lighting source to use at night to safeguard people's health.

18.2.3 Candlelight-Style OLEDs

Candles create a romantic ambience and a general feeling of well-being. A candle shows a spectrum that is almost free of blue emission and is capable of creating a pleasant sensation that may originate from naturally occurring melatonin secretion. Therefore, it is essential to devise a lighting source with color temperature and color coordinates closely matching those of candlelight. Currently, all illumination devices have a color temperature much greater than 2500 K (an incandescent bulb). They may be suitable for lighting in the daytime or at work, but are apparently not suitable for lighting at night. Hence, it is as important to devise a new light source that is user-friendly, that is, with a low color temperature or free of blue hazard, as it is to achieve high efficiency. Little attention had been paid to user-friendly lighting till Jou and colleagues reported a low color-temperature OLED.[65] In 2013, Jou and colleagues invented the world's first candlelight-style OLED with a color temperature of 2000 K.[52] The device exhibited a yellowish-orange emission with a CIE (0.52, 0.43) which closely matched with the CIE (0.52, 0.42) of the candle being studied. The reported device consisted of two emissive layers. The first emissive layer was composed of a 20 wt% of blue dye (Flrpic) in a host TCTA, and the second emissive layer a 12.5 wt% of green dye (Ir(ppy)$_3$), a 5 wt% of yellow dye (PO-01), and a 1 wt% of red dye (Ir(piq)2acac) in a host TPBi (Figure 18.10). As the thickness of the blue emissive layer decreased from 15 to 5 nm, and the thickness of the orange emissive layer increased from 5 to 15 nm, the device showed a decrement in color temperature from 3000 to 2361 K at 1000 cd/m^2. By employing a blend nano CML of TCTA and TPBi between the emissive layers, the device exhibited an improved CRI from 64 to 79 with a color temperature of 2300 K. Furthermore, by reducing the yellow dopant concentration from 5 to 3 wt%, the device showed a very high CRI of 93 with a color temperature of 2150 K and a power efficiency of 19.2 lm/W.

The reported candlelight-style OLED showed a power efficiency 300 times that of the candles. The luminance spectrum of the device matched over 80% with that of the candle. As shown in Figure 18.11a, both spectra exhibited the same trend in emission intensity below 620 nm, that is, they increase with the increase of wavelength. Besides having a high CRI, the resultant OLED yielded no emission near the infrared region and, hence, it provided a glow that was physically cold but psychologically warm. According to Figure 18.11b, the luminous spectrum of the candlelight-style OLED showed double emission peaks at 560 and 608 nm, whose combined emission closely matched with that of the candle emission spectrum peaking at 580 nm.

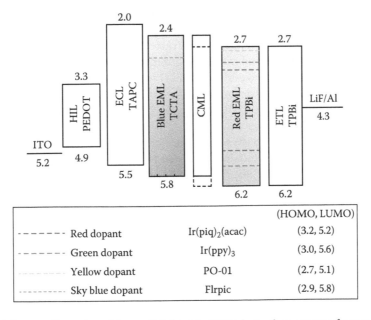

FIGURE 18.10 Schematic illustration of the candlelight-style OLED device that composes four candlelight emission complementary dyes, namely, red, yellow, green, and sky blue, dispersed in two emissive layers. (From Jou, J.H. et al.: Candle light-style organic light-emitting diodes. *Adv. Funct. Mater.* 2013. 23. 2750–2757. Copyright Wiley VCH Verlag GmbH & Co. KGaA, Weinheim. Reproduced with permission.)

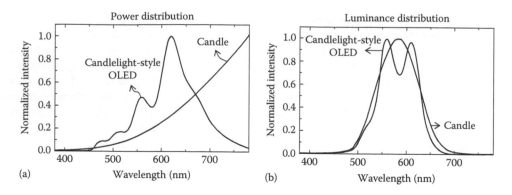

FIGURE 18.11 Comparison of the (a) spectral power and (b) luminosity distributions of the candlelight-style OLED with the candle. (From Jou, J.H. et al.: Candlelight-style organic light-emitting diodes. *Adv. Funct. Mater.* 2013. 23. 2750–2757. Copyright Wiley VCH Verlag GmbH & Co. KGaA, Weinheim. Reproduced with permission.)

Recently, they developed a blue hazard-free, low color-temperature candlelight-style OLED device.[53] It showed a power efficiency of 85.4 lm/W, a color temperature of 2279 K, and an SRI of 79 at 100 cd/m². In comparison with all other electrically powered source, the reported OLED device exhibited the lowest color temperature and was free of flickering, glare, and UV/IR emission. The device does not constitute a fire or burning hazard and, unlike candles, does not generate unpleasant smoke.

18.2.4 Sunset-Style OLEDs

The light at sunset is composed of a near continuous spectrum with the strongest emission being red and the weakest blue. In terms of color temperature, natural light at sunset exhibits the lowest daytime color temperature. The low color temperature of the light at sunset has an appealing chromaticity, attracting considerable attention and stirring the emotions of numerous poets. However, current lighting sources are not able to recreate this delightful sunset hue indoors. Therefore, it would be beneficial to design a lighting source with a low color temperature and with color chromaticity tunability that covers the chromaticity of sunset. In 2013, Jou and colleagues reported an OLED with chromaticity tunable between sunset hue (2500 K) and candlelight (2000 K).[51] The reported device exhibited a color-temperature tunable from 1600 to 2600 K with a CIE (0.58, 0.40) to (0.52, 0.43). It showed a maximum CRI of 80, a power efficiency of 23 lm/W, and luminance of 31,770 cd/m².

As shown in Figure 18.12, the reported OLED device comprised three different emissive layers. The first emissive layer, a sky blue emission, consisted of a 20 wt% sky blue emitter (Flrpic) doped in a host bis-4-(*N*-carbazolyl)phenyl)phenylphosphine oxide (BCPO) and the second emissive layer, a green emission, consisted of a 10 wt% green emitter (Ir(ppy)₃) doped in a host TCTA. The third emissive layer, an orange emission, was obtained by doping a 3 wt% red emitter (Ir(piq)2(acac)), a 5 wt% yellow emitter (PO-01), and a 12.5 wt% green emitter (Ir(ppy)3) in a host TPBi. An additional electron confining layer of Spiro-2CBP and a CML of TPBi were employed between the sky blue and the green emissive layers. The relative thickness of the emissive layers affected the color-temperature span. For example, as the thickness of the sky blue layer was decreased from 8 to 6 nm and the thickness of the orange emissive layer was increased from 5 to 7 nm with a fixed 2 nm green emissive layer, the color-temperature span shifted from 1860–2370 K to 2120–3960 K. Moreover, the device showed a much lower and narrower color-temperature span, from 1820 to 2430 K, as the thickness of the sky blue emissive layer was decreased to 5 nm and the thickness of the orange emissive layer was increased to 8 nm. Further, as the thickness of the green layer was increased from 2 to 2.7 nm and the thickness of the orange emissive layer was increased from 8 to 9.3 nm with a fixed 8 nm sky blue

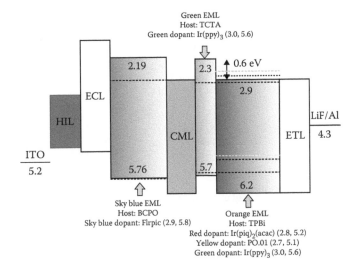

FIGURE 18.12 Schematic illustration of the structure of the device, in terms of energy levels, of the color-temperature tunable, low color-temperature OLED devices composing four blackbody radiation complementary emitters dispersed in three emissive layers with a nano-scale CML. (Reprinted from *Org. Electron.*, 14, Jou, J.H. et al., OLEDs with chromaticity tunable between dusk-hue and candlelight, 47–54. Copyright 2013, with permission from Elsevier B.V.)

emissive layer, the resultant OLED device exhibited an ultra-low color-temperature span ranging from 1560 to 2600 K with a chromaticity covering the range from sunset to candlelight. The incorporation of a 3 nm TPBi as a CML was used to regulate the injection of the hole, which resulted in a very high CRI of 91 at 10,000 cd/m². An electron confining layer of 1 nm thickness of spiro-2CBP with a 4 nm CML enhanced the efficiencies of the device and the CRI from 88 to 92.[51]

Recently, Jou and colleagues reported an OLED with a color temperature ranging between 1500 and 3000 K, and that covered the color temperature of sunset.[47] It showed a very high CRI of 92 with an 87% spectrum resemblance with sunset hue. The reported device consisted of two emissive layers separated by a CML. The first emissive layer with an orange emission was obtained by doping a 1 wt% of red emitter (Ir(piq)2(acac)), a 5 wt% of yellow emitter (PO-01), and a 5 wt% of green emitter (Ir(ppy)₃) in a host CBP. The second emissive layer yielded a mild bluish-green emission obtained by doping a 1 wt% of bis[(p-isopropy lohenyl)(p-tolyl)-amino]-10–10-phenanthrene (BPTAPA) in mixed hosts of 3,7-bis(4-*N*,*N*-diphenylamino)-5,5-spirofluorenyl-5H-dibenzo[a,d]cycloheptene (Ph2N-STIF-NPh2) and 2,7-bis-2[phenyl(m-tolyl)-amino]-9,9-dimethyl-fluorene-7-yl-9,9-dimethyl-fluorene (MDP3FL) with a 1:1 weight ratio. Initially, the reported device with a single orange emissive layer showed a CRI of 66 with a sunlight spectrum resemblance (SSR) of 78%. As the blue emissive layer was introduced, the device exhibited a CRI of 74 and the color temperature was increased from 1500 to 1700 K. Further, as a 4 nm CML layer of CBP was employed between the orange and the bluish-green emissive layers, the device showed an SSR of 84% and CRI of 86, with a color temperature of 2046 K. While a 2 nm TPBi instead of CBP was used as a CML, the device exhibited a much higher SSR of 87% and CRI of 92, with a color temperature of 2746 K.[47]

The resultant OLED exhibited a chromaticity similar to that of sunset with a color temperature covering that of sunset-light at different times before sunset, for example, a color temperature of 2800 K 30 min before the sunset and a color temperature of 1500 K 3 min before sunset.

During the period from 108 to 1 min before sunset, the light changes color from yellowish-white, to orange-yellow, orange, and to reddish-orange (Figure 18.13a). The locus of the sunset hue deviated from the blackbody radiation locus because of strong red absorption by the ozonosphere. The sunset-light-style OLED device with a color temperature of 2760 K compared with that of the sunset hue at

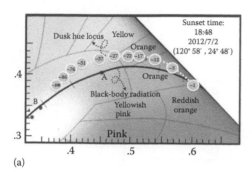 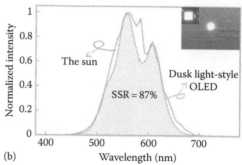

(a)   (b)

FIGURE 18.13   (a) Experimentally measured chromaticity of the sunset hue within the last 108 min before sunset and (b) direct comparison of the sunset hue-style OLED (2760 K) with the sunset hue at 2600 K in terms of spectral luminosity distribution. (Reprinted with permission from Jou, J.H. et al., Artificial dusk-light based on organic light-emitting diodes, *ACS Photon.*, 1, 27–31. Copyright 2014 American Chemical Society.)

2600 K (Figure 18.13b). It showed that the two lights have an 87% similarity in their luminance spectra, with the major emissions peaking at around 580 nm, where the luminance spectrum was obtained by convoluting the experimentally determined power spectra with the luminosity function.

18.2.5 Pseudo-Natural Light OLEDs

Natural light from the sun is a very pleasing high-quality light with a time-dependent chromaticity. It has a profound effect on people in many ways, including psychology, physiology, and aesthetics. However, the emission spectra of current illumination sources are quite different than those of diffused and soft natural light. The quality of light becomes questionable for the current sources of illumination. Therefore, devising a pseudo-natural light is of great value, assuming that natural light has the best light quality.

Recently, Jou and colleagues reported a pseudo- (1) sunny daylight with a color temperature of 5800 K, (2) dawn light with a color temperature of 3600 K, and (3) sunset hue with a color temperature of 2100 K, based on state-of-the-art solid-state lighting technologies, that is, LED and OLED.[48] As a very high SRI and a very high CRI with a wide range of color temperature are required to harvest the all-weather pseudo-natural light levels for general illumination and a wide range of colors for displays, Jou and colleagues proposed the design approach of taking blackbody radiation as the reference light. Initially, six different color emitters—namely deep red, orange-red, yellow, green, sky blue, and deep blue—were used for the simulation. The OLED device exhibited a 92%–96% similarity with sunlight from dawn to sunset, a color temperature tunability from 1100 K to infinity, and a color gamut of 80% in terms of the National Television System Committee (NTSC) requirements (Figure 18.14). This combination of six LED chips (deep red, orange-red, yellow, green, sky blue, and deep blue) exhibited an 83% similarity with sunlight, a color-temperature tunable from 1000 K to infinity, and a much wider range of 121% of NTSC requirements (Figure 18.14).

Moreover, they paid specific attention to the LED and OLED technologies by considering the effects of bandwidth and band number on the light quality of the all-weather pseudo-natural lights.[48] For example, a two-band pseudo-sunny daylight LED showed an SRI of 58 and a poor CRI of 25. However, the two-band pseudo-sunny daylight OLED showed a SRI of 90 and a CRI of 62. The poorer light-quality performance was also observed in the two-band pseudo-dawn and pseudo-sunset LEDs when compared with their OLED counterparts. The comparatively poor light quality in the LEDs may be attributed to the intrinsically narrowband nature of the emitters. This explains why LED lamps invariably need to incorporate one or multiple broadband phosphors. Furthermore, as the number of bands was increased from 2 to 4 in the pseudo-sunny daylight LED, the value of the SRI increased from 58 to 73 and the value of the CRI increased from 25 to 62. The same trend was observed in the OLED counterpart, that is, its SRI increased

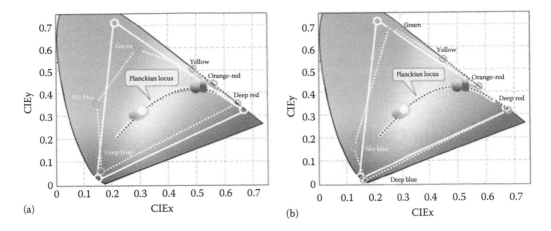

(a)

(b)

FIGURE 18.14 The employment of multiple (a) organic and (b) inorganic emitters with colors varying from deep red to deep blue enables the generation of all-weather pseudo-natural lights for illumination and a wide color range for displays. (From Jou, J.H. et al.: Pseudo-natural light for displays and lighting. *Adv. Opt. Mater.* 2015. 3. 95–102. Copyright Wiley VCH Verlag GmbH & Co. KGaA, Weinheim. Reproduced with permission.)

from 90 to 96 and its CRI increased from 62 to 89. Therefore, it is clear that high-quality pseudo-natural lights can be fabricated by utilizing high band number, natural light complementary organic emitters. This is because almost all organic electroluminescent materials can emit any color throughout the entire visible range, and their spectra are broad and diffused.

Besides simulation, Jou and colleagues also fabricated all-weather pseudo-natural light OLED devices with emissions mimicking natural light levels from dawn to sunset. The devices were composed of six different emitters dispersed in three different emissive layers. The comparable results of the simulated and experimental pseudo-natural lights are shown in Figure 18.15.

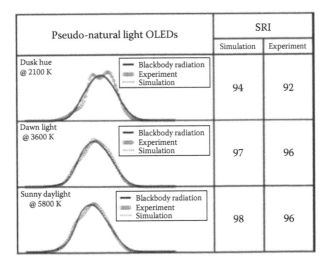

Pseudo-natural light OLEDs		SRI	
		Simulation	Experiment
Dusk hue @ 2100 K	— Blackbody radiation ▥▥ Experiment ····· Simulation	94	92
Dawn light @ 3600 K	— Blackbody radiation ▥▥ Experiment ····· Simulation	97	96
Sunny daylight @ 5800 K	— Blackbody radiation ▥▥ Experiment ····· Simulation	98	96

FIGURE 18.15 The luminous spectra obtained from simulation and experiment for the pseudo-sunlight at noon, dawn, and sunset using an OLED comprising all six different emitters. As can be seen, very high SRIs are obtainable experimentally by dispersing emitters in three different emission layers. (From Jou, J.H. et al.: Pseudo-natural light for displays and lighting. *Adv. Opt. Mater.* 2015. 3. 95–102. Copyright Wiley VCH Verlag GmbH & Co. KGaA, Weinheim. Reproduced with permission.)

As illustrated in Figure 18.15, the pseudo-dawn and pseudo-sunny daylight lights with a color temperature of 3600 and 5800 K exhibited a very high SRI of 96, which closely matched with the simulated values. Experimentally, the pseudo-sunset hue light with a low color temperature of 2100 K showed a very high SRI of 92. The luminous spectra obtained from the simulation of and experimentation with the pseudo-dawn, pseudo-sunny daylight, and pseudo-sunset hue lights exhibited a high consistency with the blackbody luminance spectrum.

18.3 New Light-Quality Metrics

18.3.1 Issues with Current Metrics

The quality of the light is one of the critical parameters of any source of illumination. The impact of a light source on the appearance of an object's color was first reported in 1974 by the Commission Internationale de l'Eclairage (CIE).[68,69] Nowadays, several light metrics have been proposed in industries and by academics to calculate the ability of a source illumination to reflect the actual color of an object.[70–76] These metrics evaluate the quality of light from two perspectives, namely, objective measure and subjective measure.[77,78] Metrics such as the CIE CRI provide the color shift of chromaticity with respect to a reference illuminant, offering an objective measure of the light quality of a test source. Other metrics, such as the flattery index R_f[70] color preference index (CPI),[71] gamut area index (GAI),[79,80] color quality scale (CQS),[81] and memory color quality metric,[82,83] have emphasized the subjective measure, such as color discrimination, preference, and naturalness of light quality.

Many alternative light-quality metrics have been developed since the CRI was introduced. However, the CRI has been considered the most useful tool with which to measure the color rendering of a lighting source within the lighting community. Over time, certain shortcomings have been noticed in CRI values, especially for the color rendering of a white light source. For example, a fluorescent tube with spiky emission spectra exhibits a CRI of 98. Moreover, a low-pressure sodium lamp shows a CRI of –12. The value for the poorest quality light should be zero instead of a negative one, as this does not provide precise information about the quality of light. To overcome these limitations, researchers are focusing on developing a more appropriate light-quality index. For example, Judd et al.[70] proposed a flattery index R_f, which is a reference-based metric similar to the CRI except that in the calculation of R_f an average of the chromaticity difference of the sample in test light and reference light is considered. Thornton et al.[71] reported a metric CPI, based on preferred chromaticity shift. Both R_f and CPI have a limitation of correlated color temperature (CCT) greater than 3500 K. Rea et al.[79] introduced a new metrics, GAI, which is used in conjunction with the CRI for a prediction of how well one can discriminate between subtle differences in hue, whereas CRI only indicates how natural as well as how vivid an object appears under the test light.[84] Xu et al.[74,75] reported a light-quality index of color rendering capacity, using the 1960 CIE chromaticity color space, in which they provide a maximum chromaticity in three dimensions at various points in the luminosity range.[74,75] Davis et al.[81] developed an index of color quality scale in which they derived a new calculation by extending the use of the CIE CRI, taking a new set of test color samples of low to high chromatic saturation, and using the CIELAB as the color space. This advancement improves several facets of color quality such as color rendering, chromatic discrimination, and observer preference.[85]

The dilemma arises in all lighting indexes from the CCT constraint, that is, in all metrics calculations a fixed CCT value above 2800 K was used. For example, a typical reference illuminant called D_{65} shows only a fixed color temperature of 6503 K, which would, again, skew the calculation results, especially when the color temperature for the light sources being tested is very different from that of D_{65}. Moreover, there is no reference light source that is a blackbody radiator that can emit a wide and continuous series of colors with different color temperatures. Another important fact is that an unreal daylight spectrum with a color temperature greater than 5000 K is used as the reference light source for mathematical modeling analysis. Again, this is a misleading approach for the measurement of light quality. Thus, currently no metric of light quality is available that can calculate the illumination color chromaticity from the respective color temperature perspective. In 2013 and 2014, with health and productivity in mind, the Department of

Energy emphasized a need for research into the development of an appropriate light-quality index before 2020 to give accurate descriptions of the color of light.[86,87]

18.3.2 Sunlight Spectrum Resemblance

Ironically, any quantification will be inappropriate if the CRI is adopted because inappropriate artifacts were introduced when setting the criteria for CRI calculation due to the selection being limited to specific samples and an improper reference light source. To solve this problem, in 2011 Jou and colleagues introduced a new light-quality index, SSR, which quantifies the resemblance of the power spectrum of any given light source to the sunlight spectrum at the same color temperature.[50] The values assigned to SSR range from 0 to 100, being from perfect dissimilarity to perfect resemblance. The CRI and SSR of an OLED device were compared with those of several conventional illumination sources (Figure 18.16).

As shown in Figure 18.16, all sources were studied at the same correlated color temperature of daylight. The incandescent bulb, with a power efficiency of 15 lm/W, exhibited a CRI of 98% and 72% SSR at a color temperature of 2500 K. The fluorescent tube, with a power efficiency of 90 lm/W, showed a CRI of 82 and a very low 16% SSR at a color temperature of 3900 K. The sodium lamp showed a monochromatic light with a CRI of 0 and only 2% SSR at a color temperature of 1800 K. The LED showed a CRI of 75% and 49% SSR at a color temperature of 5700 K. The sunlight-style OLED under examination exhibited a very high CRI of 97% and 63% SSR at a color temperature of 4500 K.[50]

However, the SSR index does not completely solve the issues of light quality as, in the SSR calculations, equal weight is assigned to all the different visible levels of light with different wavelengths. When the value of the SSR approaches 100, this would lead to a less energy-efficient lighting design. This is because some of the power would be wasted generating light frequencies to which the human eye is less sensitive, the deep blue and deep red frequencies, in order to achieve the ultimate quality. Moreover, the reference sunlight spectrum cannot be considered as universal, that is, the power spectrum of sunlight frequently varies in accordance with the time of day, weather, and/or latitude and is not at all smooth, especially in the long-wavelength frequencies at low color temperatures.

18.3.3 Spectrum Resemblance Index

The earlier reported SSR index did not weight the contributions of the different colors according to the photopic function, which is the reason why the different contributions of the luminosity of different colors are not distinguishable. To resolve this problem, in 2014 Jou and colleagues again proposed a universal index for light quality that would be easy to apply, namely, the SRI, which quantifies a given lighting source on the basis of lumen rather than its power spectrum.[88] The SRI is defined as the percentage similarity, in terms of its luminance spectrum, between a given light source and the reference source, blackbody radiation, based on the same color or color temperature. It is defined as

$$\text{SRI} \equiv \frac{\int L(\lambda, T)\,dy}{\int L_{BR}(\lambda, T)\,dy} \times 100\% \tag{18.1}$$

where

$L_{BR}(\lambda, T)$ is the luminance spectrum of the blackbody radiation

$L(\lambda, T)$ is the overlapping area between the luminance spectrum of the studied lightsource and its corresponding blackbody radiation[88]

The value of the SRI to be determined for any given lighting source may fall between 0 and 100, which reveals a range from total dissimilarity to perfect similarity. The luminance spectrum, $L(\lambda, T)$, can be obtained by combining the power spectrum, $I(\lambda, T)$, with the luminosity function, or the photopic function if under well-lit condition, $V(\lambda)$ (Figure 18.17).[89–91]

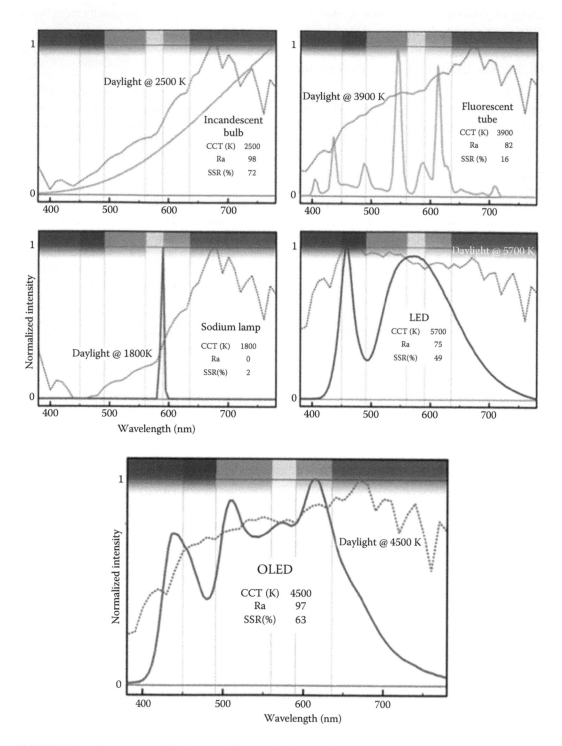

FIGURE 18.16 Comparison of the CRI and SSR of an OLED with those of an incandescent bulb, fluorescent tube, sodium lamp, and LED. (Reproduced from Jou, J.H. et al., *J. Photon. Energy*, 1, 011021. Copyright 2011, Society of Photo Optical Instrumentation Engineers.)

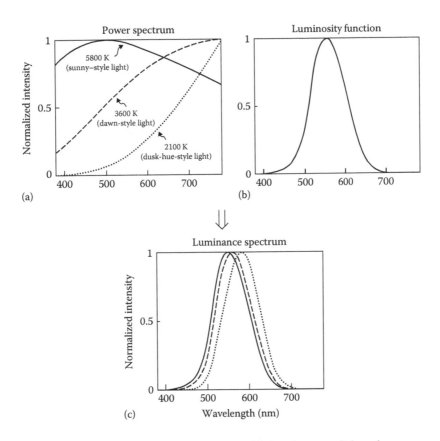

FIGURE 18.17 Exemplifications of the spectra of three natural lights, that is, sunlight at dawn, noon, and sunset, from the perspectives of a machine and human eyes, where the luminance or lumen spectra (c) are obtained by weighting the machine detected power spectra (a) with the luminosity or photopic function (b). (Reprinted with permission from Jou, J.H. et al., A universal, easy-to-apply light-quality index based on natural light spectrum resemblance, *Appl. Phys. Lett.*, 104, 203304 Copyright 2014, American Institute of Physics Publishing LLC; Schnapf, J.L., Kraft, T.W., and Baylor, D.A., Spectral sensitivity of human cone photoreceptors, *Nature*, 325, 439–441 Copyright 1987, American Institute of Physics Publishing LLC; Sharpe, L.T., Stockman, A., Jagla, W., and Jägle, H., A luminous efficiency function, $V^*(\lambda)$, for daylight adaptation, *J. Vis.*, 5, 948–968 Copyright 2015, American Institute of Physics Publishing LLC; Wald, G., Human vision and the spectrum, *Science*, 101, 653–658 Copyright 1945, American Institute of Physics Publishing LLC.)

Human eyes sense light differently at varying levels of brightness and varying wavelengths. The photopic function quantitatively depicts different values of lumen that different lights can contribute from the perspective of human eyes in well-lit conditions.[92] For example, the visual sensitivity of human eyes to a slightly yellowish-green light at 555 nm is 17 times stronger than that to 460 nm of a blue light and 4 times that to 630 nm of a red light. The SRI shows that the pursuit of a better quality of light can be achieved by emitting more light at or around the 555 nm yellowish-green light region, where human eyes perceive the highest visual sensitivity. In contrast to the SSR index, there is no wastage of energy in the deep blue and deep red wavelengths. However, the natural light from the sun would also, during the daytime, emit light in the strong deep blue and violet wavelengths. But human eyes are less sensitive to these colors, which would only contribute slightly, if at all, to the luminous intensity.

Jou and colleagues also studied the power spectrum, luminance spectrum, CRI, and SRI of different street lamps and residential illumination sources.[48] A correlated study between the CRI and SRI for all illumination sources was undertaken and is illustrated in Figure 18.18.

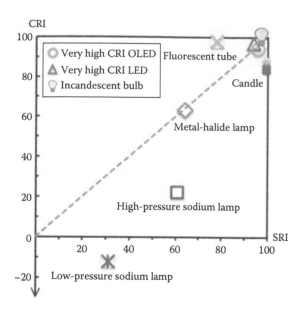

FIGURE 18.18 Correlation between the SRI and CRI light-quality indexes for the lighting measures under study. (From Jou, J.H. et al.: Pseudo-natural light for displays and lighting. *Adv. Opt. Mater.* 2015. 3. 95–102. Copyright Wiley VCH Verlag GmbH & Co. KGaA, Weinheim. Reproduced with permission.)

They found an unintelligible negative value of CRI (−12) but a positive index, 31 SRI, for a low-pressure sodium street lamp. A high-pressure sodium lamp and a metal-halide lamp showed a CRI of 22 and 63, respectively, while they exhibited nearly equal values of SRI, 61 and 64, respectively. Moreover, candlelight showed a high SRI of 99 and an incandescent bulb an SRI of 97, whereas an incandescent bulb showed a perfect rendering with a CRI of 100 in contrast to candlelight with a CRI of 83. It is also noticeable that a fluorescent lamp showed spiky emission spectra, which do not afford a perfect resemblance to the corresponding natural light, while it exhibited a CRI of 98 and a SRI of 78. So, it is not a justification for any lighting source with a spiky emission to be assigned a nearly perfect light-quality index, such as 98 CRI in this case. These incomprehensible values of CRI were observed because of the color-temperature difference between the reference source and the test sources. In case of SRI, all illumination devices were tested with the same color and color temperature as the reference light.

18.4 Design of Natural Light-Style OLEDs

18.4.1 Effect of Bandwidth

Fluorescent tubes are frequently used for general illumination, but their output is much less similar to natural light. For example, most of the emission spectra of fluorescent tubes consist of a sharp blue emission but show perfect light quality (CRI 98). It is difficult to have a lighting source that can be considered perfect if it produces spiky emission spectra, as spectra with sharp peaks are not observed in natural light. However, the emission spectra become broad and diffused to a certain wavelength span with the employment of wide-band phosphors,[93-95] which has markedly enhanced the similarity of the emission spectrum to that of natural light. In contrast, OLEDs are capable of reaching a very high spectral similarity to natural light. OLEDs are a plane-type lighting source, which provides diffused and continuous emission for lighting purposes. As organic electroluminescent emitters are able to emit a wide-band emission that covers the full visible range, they provide a high level of visual comfort to human eyes. These specific characteristics of OLEDs make them a potential general lighting source in the illumination market. It is clear that, for natural light-style lighting, both bandwidth and band number are crucial parameters.

In 2011, Jou and colleagues reported a very high CRI of 98 white OLED device by using adequate white light complementary emitters.[46] They fabricated a five-spectrum OLED with double white emissive layers and an HML. All the devices had a 125 nm ITO as the anode, a 25 nm PEDOT:PSS as the hole-transporting layer (HTL), a 32 nm TPBi as the ETL, a 0.7 nm LiF as the electron-injection layer, and an Al for the cathode. The first device comprised a phosphorescent emissive layer of a TCTA host doped with a 14 wt% blue dye Flrpic, 5 wt% green dye Ir(ppy)$_3$, and a 1.0 wt% red dye Ir(piq)$_2$(acac). The second device comprised a fluorescent host MDP3FL doped with 1% fluorescent yellow dye rubrene. The resulted phosphorescent OLED exhibited an emission spectrum ranging from 420 to 700 nm with three broad peaks in the blue, green, and red regions of the respective emitters. Both phosphorescent and fluorescent emissive layers were combined in a single device with respective thicknesses of 20 and 8 nm. The resultant emission spectrum covered the entire visible range with a broad blue emission peaking at 420 nm and the other yellow emission at 550 nm. By employing a 2.5 nm HTL layer of TPBi between the emissive layers, the device exhibited a five-band spectrum in the visible range with a very high CRI of 98 and a fairly high efficiency of 8.3 lm/W at 100 cd/m^2 (Figure 18.19).

Later in 2012, Jou and colleagues fabricated an OLED device that exhibited an ultra-low color temperature of 1773 K and a high CRI of 87.[67] The reason behind the ultra-low color temperature was the strong emission peak in the deep red region at 610 nm, while a high CRI resulted due to five wide-band emissions from deep blue to deep red. In 2014, Jou and colleagues explained the effect of bandwidth and the band number of the emitters employed on the quality of light in terms of SRI.[88] An LED with a narrow bandwidth showed much lower CRI and SRI values in comparison with a broadband OLED. For example, a two narrowband LED exhibited a very low CRI of 25 and an SRI of 58, and a two broadband OLED showed a comparatively high CRI of 47 and an SRI of 88.

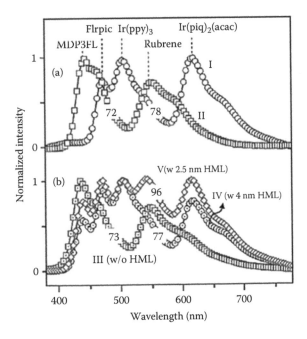

FIGURE 18.19 The resultant EL spectra of (a) the single white emissive layer containing OLEDs, Devices I and II and (b) the double white emissive layers containing OLEDs, Devices III–V. A five-wavelength emission with 96 CRI at 1000 nits was observed when a 2.5 nm HML was employed. (Reprinted from *Org. Electron.*, 12, Jou, J.H. et al., Efficient very high color rendering index organic light-emitting diode, 865–868. Copyright 2011, with permission from Elsevier B.V.)

Furthermore, in 2015 they also proposed a theoretical study on all-weather pseudo-natural light for general illumination and a wide color range for display applications.[48] By considering six broadband organic emitters, a color range of 80% in terms of NTSC requirements was achieved; six narrowband LED chips of the same colors showed a color gamut of 121%. Moreover, the light quality of a broadband OLED is much better than that of a narrowband LED. They discussed the effects of the band number and bandwidth on the pseudo-sunny daylight at 5800 K, dawn light at 3600 K, and sunset hue at 2100 K, which were fabricated by using LED chips and organic emitter-based OLEDs. In addition to the mimicking of daylight at dawn, noon, or sunset, OLEDs can easily achieve a high-quality pseudo-natural light due to their intrinsically broadband nature. Specifically, their SRI increased from 82 to 98 and their CRI increased from 39 to 95 as the band numbers employed were increased from 2 to 6. In case of LEDs, the SRI increased from 58 to 83 and the CRI increased from 25 to 86. The comparatively lower quality of light can be a consequence of the intrinsically narrowband nature of LEDs, which explains why LED lamps inevitably need to incorporate one or multiple broadband phosphors.

18.4.2 Effect of Band Number

Like bandwidth, the band number of emitters also influenced design and emitted the spectra of the natural light OLEDs. As the number of bands increase in the emission spectrum, colors from different bands go to saturation and the light quality is improved. It enhances the resemblance of the artificial light to the corresponding natural light. Jou and colleagues comprehensively studied the band number and bandwidth and their influence on light quality.[48,88] They also included issues of current metrics and developed a universal light-quality index, the SRI, that was easy to apply. They concluded that when more band numbers from the deep blue to deep red color range are introduced in the devices (LEDs and OLEDs), this resulted in an increased range of colors in terms of NTSC percentage (as shown in Figure 18.14). It also improved the light quality in terms of the CRI and the SRI. This improvement in color range was much more detectable in LEDs compared with OLEDs. The OLED emission spectrum showed considerable resemblance to natural light and better CRI and SRI values than LEDs because of the inherently broad and diffused emission spectrum of organic emitters (Figure 18.20).

Considering a white natural light with a color temperature of 5800 K, for example, it takes OLED only 4 bands to generate a pseudo-natural white light with a greater than 95% natural light similarity. In contrary, using six narrowband LEDs can only obtain an 83% similarity.[88]

18.4.3 Effect of Carrier Modulation Layer

Nowadays, OLEDs have already reached the efficiency of fluorescent tubes.[59] To realize OLEDs for lighting applications, a higher quality of light and better efficiency at elevated luminance are still required. Various approaches have been reported to achieve more efficient OLEDs,[96–116] including the employment of nano CMLs.[51,52,65] Several of them are especially effective in improving device efficiency at a high level of applied luminance. Among these, the nano CML approach is considered to be the most favorable due to its numerous excellent characteristics, such as carrier regulation function, ability to confine the excitons within a specific emissive layer, and choice of materials for the CML. In recent years, the effect of nano CMLs on the efficiency and luminance of OLED devices[65,116–118] have been reported using different terminologies, such as carrier-regulating layer, spacer, interlayer, mixed/blend interlayer, HML, and buffer layer.

Jou and colleagues reported various natural light-style OLED devices using different types of material as a nano CML, such as host materials, electron-transporting materials, hole-transporting materials, and a mixed or blend interlayer of hole and electron-transporting materials, as shown in Table 18.1. It has also been found that the nano CML played a crucial role in realizing a tunable color temperature and color

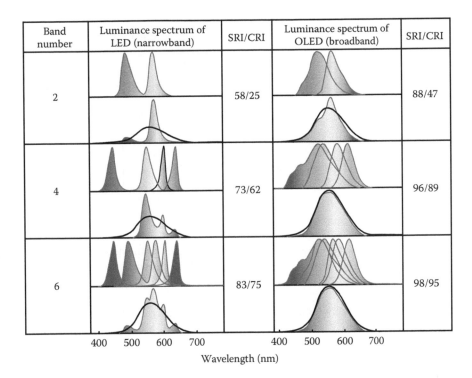

FIGURE 18.20 Effects of band number on natural light similarity for the lighting sources based on narrowband LEDs and broadband OLEDs. (Reprinted with permission from Jou, J.H. et al., A universal, easy-to-apply light-quality index based on natural light spectrum resemblance, *Appl. Phys. Lett.*, 104, 203304. Copyright 2014, American Institute of Physics.)

TABLE 18.1 Photophysical and Electrochemical Properties of Carrier Modulation Materials

Material	E_T (eV)	E_g (eV)	HOMO (eV)	LUMO (eV)	μ_h cm²·V/s	μ_e cm²·V/s	Reference
Spiro-2CBP	—	3.38	−5.03	−1.65	1×10^{-3}	—	[65]
TCTA	2.79	3.4	−5.7	−2.3	3×10^{-3}	1.0×10^{-8}	[51,52]
TPBi	2.73	3.5	−6.2	−2.7	—	3.0×10^{-5}	[50–53]
BCP	2.6	3.5	−6.1	−2.6	—	4.6×10^{-5}	[61]
BSB	2.76	4.2	−6.5	−2.3	—	—	[65]
TmPyPb	2.8	4.0	−6.7	−2.7	—	1×10^{-3}	[61]
BmPyPb	2.69	4.05	−6.67	−2.62	—	1×10^{-4}	[48,63]

temperature span.[49,51,61] In 2009, Jou and colleagues developed a sunlight-style color-temperature tunable OLED by employing a nano layer of TPBi as a CML.[49] As shown in Figure 18.2a, the position of the CML was effective in modulating the carrier injection to the emissive layer and managing the recombination zone, which resulted tunable chromaticity.

Additionally, they reported the effect of the thickness of the CML layer on the color-temperature span of a sunlight OLED, as shown in Figure 18.4.[50] Initially, a nano single HTL, TPBi, was used at varied thicknesses from 0 to 6 nm and it was found that there exists a chromic shift turning point in the vicinity of 4.5 nm. The emission spectra showed a hypsochromic shift at thicknesses lower than 4.5 nm, while a bathochromic shift was observed if more than 4.5 nm was employed (Figure 18.21).

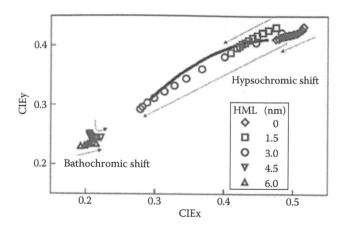

FIGURE 18.21 The effect of the thickness of the hole-modulating layer on the device emission tracks, in terms of CIE coordinates. (From Jou, J.H. et al., *J. Photon. Energy*, 1, 011021. Copyright 2011, Society of Photo Optical Instrumentation Engineers.)

Jou and colleagues reported a high-efficiency OLED device with a color-temperature tunable between 1700 and 5200 K using TPBi, BCP, and TmPyPB layers as the HML, as shown in Figure 18.6.[61] With the use of a nano HML, the recombination zone is easily shifted by varying the applied voltage.[49,119] Recently, Jou and colleagues also reported an OLED device with a color-temperature tunable from 1580 to 2600 K using a TPBi layer as a CML. The resultant device exhibited a color-temperature tunable between sunset hue (2500 K) and candlelight (1900 K). The color temperature of the OLED device was further tuned from 5200 to 2360 K by simply adjusting the emissive layer thickness ratio.[51] Furthermore, several reports also explained the effect of a mixed nano CML on efficiency and luminance of an OLED device.[52,53,65] Jou and colleagues reported a high efficiency and low color temperature of 1940 K OLED using a blend interlayer of BSB:Spiro-2CBP (Figure 18.7).[65] Most recently, they reported a candlelight-style OLED and a blue hazard free, high-efficiency candlelight-style OLED device employing a mixed CML of TCTA and TPBi.[52,53]

Moreover, a CML can effectively influence the device efficiency roll-off, lifetime, CRI, and tuning of the emission spectrum. In order to suppress the efficiency roll-off, numerous approaches have been used, for example, light-emitting materials with a lower excited state lifetime,[120] preventing the aggregation of emitter molecules,[121–124] choosing high-mobility carrier-transporting materials,[125,126] triplet managers,[127–129] and widening the recombination zone.[130–134] Various techniques have been developed to realize a wide recombination zone, such as using double emissive layers, a mixed-hosts emissive layer, graded emissive layer architecture, and a nano CML between two emissive layers. Among them, the nano CML is one of the most favorable techniques, especially for hybrid white OLEDs[135] as this prevents triplet–triplet annihilation and broadens the recombination zone. Additionally, the lifetime of a device, especially at high levels of brightness, remains a critical issue that limits the commercialization of OLED technology in lighting applications. Extensive improvement in the lifetime of an OLED was observed by the reduction of the Joule heating, which occurred due to the excessive injection of unbalanced carriers.[118] In order to obtain OLEDs with long lifetimes as a result of reduced Joule heating, it becomes imperative to extract high luminance and high efficiency at a lower current density.[136] Incorporation of a mixed CML between the fluorescent and phosphorescent emissive layers has effectively controlled exciton quenching and singlet–triplet annihilation (STA) by limiting the singlet–triplet excitons.[137,138] The use of a CML as a buffer layer also improves the efficiency and operational lifetime of OLEDs.[139,140] Using electron-transporting nano CMLs, such as Alq3, BAlq, TPBi, and Bepp2, successfully managed excitons in a wide recombination zone and enhanced the performance of the device.[141]

A nano CML therefore plays a significant role in achieving a very high CRI value in full-wavelength OLED.[46,65,142] In 2011, Jou and colleagues reported a very high CRI of 93 for a hybrid white OLED by introducing a 1.5 nm TCTA layer as the CML.[65] Moreover, Jou and colleagues reported, at 100 cd/m², for example, a CRI of 98 with an efficiency of 8.3 lm/W employing a 2.5 nm TPBi layer as a CML between the fluorescent and the phosphorescent white emissive layers.[46] In 2013, they also reported a physiologically beneficial OLED device with a high or very high CRI. For example, they fabricated an OLED device with chromaticity tunable between sunset hue and candlelight with a high CRI of 88 employing a nano layer of TPBi as CML.[51] Thus, incorporation of a nano CML within the emissive layer may enhance the performance of a device in many respects, such as chromaticity and color-temperature tunability, efficiency, lifetime, and a natural light-style emission with a high CRI.

18.5 Device Fabrication

The OLED device fabrication process includes substrate cleaning, preparation of the emissive layer solution, and layer deposition. Generally, a glass slide with a 125 nm thickness of ITO pattern was used as the substrate. A 1:3 solution of liquid detergent in deionized water was used to remove greasy layers and dust particles from the substrate. After rinsing the slide for up to 5 min with deionized water, it was blow-dried with nitrogen gas. The dried glass slide was dipped in acetone solution for surface cleaning and ultra-sonicated for 10 min at 45°C. It was then transferred to a 2-propenol solution and ultra-sonicated again for 10 min at 60°C. After that, the slide was put in a UV-O₃ slot for UV/ozone treatment for 10 min to remove organic residues and make the surface more adhesive improving its wetting properties.

OLED devices can be fabricated by solution process and thermal evaporation of organic materials. For the solution process, fabrication of OLED devices initially involves the spin coating of an aqueous solution of PEDOT: PSS at 4000 rpm for 20 s to form a 32 nm hole-injection layer on a pre-cleaned ITO anode. Before depositing the subsequent emissive layer, the solution was prepared by dissolving the host and guest molecules in an appropriate solvent at a specific temperature for 0.5 h while stirring. Then, a 20 nm emissive layer of a particular solution was deposited at 2500 rpm for 20 s. The whole process was carried out in a nitrogen gas atmosphere. A 32 nm electron-transporting layer of TPBi, a 1 nm injection layer of LiF, and a 100 nm aluminum cathode were deposited using the thermal deposition method in a high vacuum of 10^{-4} torr at the rates of 0.3, 0.1, and 10 Å/s, respectively. The thicknesses of the respective layers were adjusted according to the total OLED device thickness. For a thermal evaporated OLED device, all the layers, such as the HIL, HTL, EML, and ETL, were deposited by using the thermal evaporation method at their respective rates and thicknesses in a high vacuum chamber. The fabricated OLED device was encapsulated with a glass lid by using epoxy resin in a nitrogen glove box to protect it from moisture and oxygen.

The thermal evaporation method appears to be a successful approach with which to achieve high-efficiency with advantages such as immaculate deposition and precise control of the layer thickness. However, this approach becomes extremely expensive due to the considerable wastage (70%–80%) of organic materials in the chamber itself. The solution process, however, proved to be more efficient as it uses methods such as coating or printing, which are highly cost-effective, feasible for mass manufacturing, and promising for large area displays.[143] In contrast, the solution process still remains a major challenge due to problems such as the solubility of some organic materials in various organic solvents, undesired blending of two organic layers during subsequent coatings, and the resultant morphological and compositional defects in organic layers.[144]

The characterizations of the OLED devices, such as the current density–voltage and luminance (J–V–L), were measured by means of a Keithley 2400 electrometer using a Minolta CS-100A luminance meter. The emission spectrum and CIE color chromatic coordinates were measured using PR-655 spectroradiometer. The emission area of devices varied from 9 to 25 mm² depending on the ITO pattern, and the luminance was measured in the forward direction only.

18.6 The First Commercially Available Candlelight-Style OLED

Commercialization of the candlelight OLED began at the end of 2014. WiseChip Semiconductor Inc. has been fabricating the candlelight-emitting panels using a technology developed by Jou and colleagues (Figure 18.22).[145] WiseChip provide panels 10×10 cm² in size. The first candlelight panels were installed in a beehive-like lampshade made of rotten wood and were installed as street lights for the Atayal, an aboriginal tribe living in a village called Smangus, in Taiwan. It is noteworthy that this Atayal tribe demands general illumination sources that are both beneficial to people and eco-friendly.

18.7 Changing Lighting History

18.7.1 The Blind Leading the Blind?

White light from artificial lighting sources possesses a strong blue emission spectral content that exhibits short wavelengths and high color temperature. Many medical studies reported that high color-temperature light affects human well-being, that is, it causes considerable health hazards, such as the suppression of melatonin secretion, sleep disorders, disruption of the circadian rhythm, and the risk of various types of cancer.[12-20] Recently, Stevens et al. reported that the use of electric light at night can be considered an important cause of breast cancer, which results from the suppression of the oncostatic hormone, melatonin.[21] A survey on 147 communities in Israel on the use of low and high levels of artificial light at night showed some shocking results regarding breast cancer in female subjects. It found that women in communities that were brightly lit at night showed a 73% higher risk of breast cancer than those in areas with lower light levels.[146] It is because short-wavelength light or high color-temperature light affects hormonal balance in the human body. Additionally, the International Energy Agency reported that blue LEDs and cold-white LEDs producing light of a high color temperature (4000–5000 K) may damage the light-sensitive areas in eyes and cause blindness much more rapidly.[39] The International Dark-Sky Association reported that high color-temperature bright light disrupts the behavior of diverse nocturnal creatures and destroys the ecosystem. The Association considers that

FIGURE 18.22 Photographic image of the candlelight OLED panel. (From Jou, J.H. et al., *J. Soc. Inf. Display*, 23, 2015.)

light pollution from bright light sources is a serious threat for five species in particular, namely, fireflies, tree frogs, monarch butterflies, sea turtles, and Atlantic salmon. From a health perspective, this glaring light is not safe to use at night.[43] It is also noticeable in urban areas that bright outdoor lighting causes considerable light pollution, affecting the view of the night sky.[43] Specifically, light pollution from high color-temperature white light has become a considerable issue from the year 2006 to till now. In addition, it has also been found that high color-temperature light slowly bleaches paint colors, which has become a problem regarding the preservation of historic oil paintings. In museums, for example, the light from LEDs was thought to be the cause of discoloration of famous, priceless oil paintings by Van Gogh and Cézanne reported in 2013.[40,41]

Ironically, currently available electrically powered luminaires have a color temperature higher than 2300 K. For example, energy-saving fluorescent tubes with a color temperature of 6500 K are frequently used for indoor and outdoor lighting. Warm- and cold-white LED luminaires with a color temperature of 3000–5000 K have completely saturated the illumination market, while incandescent bulbs with a color temperature of 2300 K are frequently used in developing regions of the world. In the past years, no attention was given to issues of the color temperature of lighting fixtures. Nobody was encouraged to devise a low color temperature, electrically powered source of illumination. In particular, no lighting device with a color temperature lower than 2500 K has been encouraged or has been awarded Energy Star certification on the grounds of its specifications.[147] The world has focused solely on developing energy-saving lighting sources and, also finding the solution for an "energy-efficient, cost-effective white LED." While we congratulate the 2014 Nobel Laureates in Physics for their innovative achievement, we cannot ignore the side issue of the health hazards caused by blue and white LEDs. In recent years, many studies have revealed the notorious effect of high color-temperature bright white light on circadian rhythm and the risk of cancer.[21,146]

"Think again! We truly need a better and/or brighter light for a brighter future"

18.7.2 The Renaissance of Lighting

Make the world delightful with healthy sources of illumination!

Do not fill it with **Cancers!**

As mentioned in many medical studies, we should take greater care of the effect of light on the health of Earth's inhabitants before pursuing any other innovation in illumination technology. To achieve this, we should shift the color and color temperature of illumination sources to be not only comfortable to human eyes but also compatible with the biological clocks of all Earth's mammals. In this direction, Prof. Jwo-Huei Jou's contribution to developing natural light-style general lighting is noteworthy. His invention, candlelight-style OLEDs,[52] is considered a good source of illumination source in a medical study on cancer (*CA: A Cancer Journal of Clinicians*, impact factor 2015 144.8).[21] This invention is focused on low color temperature, diffused and continuous candlelight-style natural lighting. People benefit greatly from candlelight, which creates a physiologically warm sensation. Scientifically, it is proven that the brain works more actively in orange-red light rather than in blue and green light.[148] In April 2015, Prof. Jou and colleagues developed a candlelight-style OLED for general illumination that is free of blue hazard[53] and also created a light-quality index that allows the direct comparison of a given lighting source to its corresponding natural light. Prof. Jou has begun laying the foundations of a "Lighting Renaissance" in illumination technology.

18.8 Conclusion

In this chapter, we began with discussion of the influence of and necessity for healthy illumination both for people and for the environment. We explained the development of various natural light-style OLEDs such as the sunlight-style OLED, candlelight-style OLED, and pseudo-natural light-style OLED. All the OLED devices were fabricated using well-engineered device architecture to facilitate balanced carrier injection from the electrodes. The resultant devices were characterized in terms of color chromaticity, color temperature, device efficacy, EQE, and light-quality metrics (the CRI and SRI).

In the device design section, we described in detail how bandwidth, band number, and the CML can enhance the performance of a device. Bandwidth and band number are important factors in lighting devices in order to maintain natural light-style illumination of a high quality. The CML can be used to improve the efficiency of a device by managing the recombination zone in the emissive layer.

We included a section on light-quality metrics because clear, factual information about quality is essential for natural light-style continuous and diffused illumination. We discussed a universal, luminance-based light-quality index that is easy to apply, the SRI. It provides a percentage value regarding the resemblance of an emission to natural light without any CCT constraint.

This chapter mentioned the commercialization of candlelight-style OLEDs and spoke of a lighting renaissance with the hope of healthy illumination in the future.

References

1. Desjardin, D.E., Capelari, M., Stevani, C.V. Bioluminescent mycena species from São Paulo, Brazil, *Mycologia* 99 (2007): 317–331.
2. Hastings, J.W., Wilson, T. *Bioluminescence Living Lights: Lights for Living*, Harvard University Press, Cambridge, MA (2013).
3. Haddock, S.H.D., Moline, M.A., Case, J.F. Bioluminescence in the sea, *Annual Review of Marine Science* 2 (2010): 443–493.
4. Widder, E.A. Bioluminescence in the Ocean: Origins of biological, chemical, and ecological diversity, *Science* 328 (2010): 704–708.
5. Hastings, J.W. The gonyaulax clock at 50: Translational control of circadian expression, *Cold Spring Harbor Symposia on Quantitative Biology* 72 (2007): 141–144.
6. Visick, K.L., Ruby, E.G. Vibrio Fischeri and its host: It takes two to Tango, *Current Opinion in Microbiology* 9 (2006): 632–638.
7. Bechara, E.J.H. Bioluminescence: A fungal nightlight with an internal timer, *Current Biology* 25(7) (2015): R283–R285.
8. Oliveira, A.G., Stevani, C.V., Waldenmaier, H.E., Viviani, V., Emerson, J.M., Loros, J.J., Dunlap, J.C. Circadian control sheds light on fungal bioluminescence, *Current Biology* 25(7) (2015): 964–968.
9. Liberman, J. *Light: Medicine of the Future: How We Can Use it to Heal Ourselves Now*, Bear and Company, Inc., Santa Fe, NM (1991).
10. Judd, D.B., MacAdam, D.L., Wyszecki, G. Spectral distribution of typical daylight as a function of correlated color temperature, *Journal of the Optical Society of America* 54 (1964): 1031–1040.
11. Das, S.R., Sastri, V.D.P. Spectral distribution and color of tropical daylight, *Journal of the Optical Society of America* 55 (1965): 319–323.
12. Brainard, G.C., Richardson, B.A., King, T.S., Reiter, R.J. The influence of different light spectra on the suppression of pineal melatonin content in the Syrian–Hamster, *Brain Research* 294 (1984): 333–339.
13. Lockley, S.W., Brainard, G.C., Czeisler, C.A. High sensitivity of the human circadian melatonin rhythm to resetting by short wavelength light, *Journal of Clinical Endocrinology & Metabolism* 88 (2003): 4502–4505.
14. Mills, P.R., Tomkins, S.C., Schlangen, L.J. The effect of high correlated color temperature office lighting on employee wellbeing and work performance, *Journal Circadian Rhythms* 5 (2007): 2.
15. Pauley, S.M. Lighting for the human circadian clock: Recent research indicates that lighting has become a public health issue, *Medical Hypotheses* 63 (2004): 588–596.
16. Sato, M., Sakaguchi, T., Morita, T. The effects of exposure in the morning to light of different color temperatures on the behavior of core temperature and melatonin secretion in humans, *Biological Rhythm Research* 36 (2005): 287–292.

17. Scheer, F.A.J.L., van Doornen, L.J.P., Buijs, R.M. Light and diurnal cycleaffect human heart rate: Possible role for the circadian pacemaker, *Journal of Biological Rhythms* 14 (1999): 202–212.
18. Bommel, W.J.M.V. Non-visual biological effect of lighting and the practical meaning for lighting for work, *Applied Ergonomics* 37 (2006): 461–466.
19. Küller, L.W.R. Melatonin, cortisol, EEG, ECG and subjective comfort in healthy humans: Impact of two fluorescent lamp types at two light intensities, *Lighting Research and Technology* 25 (1993): 71–80.
20. Hatonen, T., Johansson, A.A., Mustanoja, S., Laakso, M.L. Suppression of melatonin by 2000-lux light in humans with closed eyelids, *Biological Psychiatry* 46 (1999): 827–831.
21. Stevens, R. G., Brainard, G. C., Blask, D. E., Lockley, S. W., Motta, M. E. Breast cancer and circadian disruption from electric lighting in the modern world, *CA: A Cancer Journal for Clinicians* 64(3) (2014): 207–218.
22. Woodward, H. Electric lights, U.S. Patent 0,181,613 (1876).
23. Edison, T.A. Improvement in electric lights, U.S. Patent 0,214,636 (1879).
24. Edison, T.A. Electric Lamp, U.S. Patent 0,223,898 (1880).
25. MacIsaac, D., Kanner, G., Anderson, G. Basic physics of the incandescent lamp (Lightbulb), *The Physics Teacher* 37(12) (1999): 520–525.
26. Giannuzzi, A. et al. Halogen and Incandescence Lamps, in: Paola, S., Luca, M., Alessandro, F. (Eds.), *Sustainable Indoor Lighting*, Springer-Verlag, London, U.K. (2015), pp. 87–106. DOI: 10.1007/978-1-4471-6633-7.
27. Reggiani, A. et al. Fluorescent Lamp and Discharge Lamp, in: Paola, S., Luca, M., Alessandro, F. (Eds.), *Sustainable Indoor Lighting*, Springer-Verlag, London, U.K. (2015), pp. 107–125. DOI: 10.1007/978-1-4471-6633-7.
28. Bouwknegt, A., Compact fluorescent lamps, *Journal of Illuminating Engineering Society* 14 (4) (1982): 43–46.
29. Koike, J., Kojima, T., Toyonaga, R., Kagami, A., Hase, T., Inaho, S. New tricolor phosphors for gas discharge display, *Journal of the Electrochemical Society* 126(6) (1979): 1008–1010.
30. Huang, C.H., Chen, T.M. A novel single-composition trichromatic white-light ca3y(gao)3(bo3)4:ce3+,mn2+,tb3+ phosphor for UV-light emitting diodes, *The Journal of Physical Chemistry C* 115 (2011): 2349–2355.
31. Huang, C.H., Chen, T.M. Novel yellow-emitting sr8mgln(po4)7:eu2+ (ln = y, la) phosphors for applications in white LEDs with excellent color rendering index, *Inorganic Chemistry* 50 (2011): 5725–5730.
32. Kuo, T.W., Huang, C.H., Chen, T.M. Novel yellowish-orange Sr8Al12O24S2:Eu2+ phosphor for application in blue light-emitting diode based white LED, *Optics Express* 18 (2010): A231–A236.
33. Liu, W.R., Huang, C.H., Wu, C.P., Chiu, Y.C., Yeh, Y.T., Chen, T.M. High efficiency and high color purity blue-emitting NaSrBO$_3$:Ce3+ phosphor for near-UV light-emitting diodes, *Journal of Materials Chemistry* 21 (2011): 6869–6874.
34. Liu, W.R., Yeh, C.W., Huang, C.H., Lin, C.C., Chiu, Y.C., Yeh, Y.T., Liu, R. S. (Ba,Sr)Y$_2$Si$_2$Al$_2$O$_2$N$_5$[thin space (1/6-em)]:[thin space (1/6-em)]Eu^{2+}: A novel near-ultraviolet converting green phosphor for white light-emitting diodes, *Journal of Materials Chemistry* 21 (2011): 3740–3744.
35. Service, R.F. Electronics: Organic LEDs look forward to a bright, white future, *Science* 310 (2005): 1762–1763.
36. So, F., Kido, J., Burrows, P. Organic light-emitting devices for solid-state lighting, *Materials Research Bulletin* 33 (2008): 663.
37. Lim, S.R., Kang, D., Ogunseitan, O.A., Schoenung, J.M. Potential environmental impacts from the metals in incandescent, compact fluorescent lamp (CFL), and light-emitting diode (LED) bulbs, *Environmental Science & Technology* 47 (2013): 1040–1047.
38. Eckelman, M.J., Anastas, P.T., Zimmerman, J.B. Spatial assessment of net mercury emissions from the use of fluorescent bulbs, *Environmental Science & Technology* 42(22) (2008): 8564–8570.

39. Godley, B.F., Shamsi, F.A., Liang, F.Q., Jarrett, S.G., Davies, S., Boulton, M. Blue light induces mitochondrial DNA damage and free radical production in epithelial cells, *Journal of Biological Chemistry* 280 (2005): 21061.

40. Monico, L. et al. Degradation process of lead chromate in paintings by Vincent van Gogh studied by means of spectromicroscopic methods. 3. synthesis, characterization, and detection of different crystal forms of the chrome yellow pigment, *Analytical Chemistry* 85(2) (2013): 851–859.

41. Monico, L. et al. Degradation process of lead chromate in paintings by Vincent van Gogh Studied by means of spectromicroscopic methods. 4. Artificial aging of model samples of co-precipitates of lead chromate and lead sulfate, *Analytical Chemistry* 85(2) (2013): 860–867.

42. Lim, S.R., Kang, D., Ogunseitan, O.A., Schoenung, J.M. Potential environmental impacts of light-emitting diodes (LEDs): Metallic resources, toxicity, and hazardous waste classification, *Environmental Science & Technology* 45(1) (2011): 320–327.

43. Parks, B. The IDA smart urban lighting initiative, *International Dark-Sky Association Annual Report: Nightscape* 92 (2014): 2–3.

44. Yamada, M. et al. InGaN-based near-ultraviolet and blue-light-emitting diodes with high external quantum efficiency using a patterned sapphire substrate and a mesh electrode, *Japanese Journal of Applied Physics* 41 (2002): L1431.

45. West, K.E. et al. Blue light from light-emitting diodes elicits a dose-dependent suppression of melatonin in humans, *Journal of Applied Physiology* 110(3) (2011): 619–626.

46. Jou, J.H. et al. Efficient very-high color rendering index organic light-emitting diode, *Organic Electronics* 12 (2011): 865–868.

47. Jou, J.H. et al. Artificial dusk-light based on organic light emitting diodes, *ACS Photonics* 1(1) (2014): 27–31.

48. Jou, J.H. et al. Pseudo-natural light for displays and lighting, *Advanced Optical Materials* 3 (2015): 95–102.

49. Jou, J.H. et al. Sunlight-style color-temperature tunable organic light-emitting diode, *Applied Physics Letters* 95 (2009): 013307.

50. Jou, J.H. Shen, S.M., Wu, J.H., Peng, S.H., Wang, S.C. Sunlight-style organic light-emitting diodes, *Journal of Photonics for Energy* 1(1) (2011): 011021.

51. Jou, J.H. et al. OLEDs with chromaticity tunable between dusk-hue and candle-light, *Organic Electronics* 14 (2013): 47–54.

52. Jou, J.H. et al. Candle light-style organic light-emitting diodes, *Advanced Functional Materials* 23(21) (2013): 2750–2757.

53. Jou, J.H. et al. Enabling a blue-hazard free general lighting based on candle light-style OLED, *Optics Express* 23(11) (2015): A576.

54. Choy, W.C.H., Ho, C.Y. Improving the viewing angle properties of microcavity OLEDs by using dispersive gratings, *Optics Express* 15 (2007): 13288–13294.

55. Ichikawa, M., Amagai, J., Horiba, Y., Koyama, T., Taniguchi, Y. Dynamic turn-on behavior of organic light-emitting devices with different work function cathode metals under fast pulse excitation, *Journal of Applied Physics* 94 (2003): 7796.

56. Xie, Z.Y., Hung, L.S. High-contrast organic light-emitting diodes, *Applied Physics Letters* 84 (2004): 1207.

57. Lee, J.Y., Kwon, J.H., Chung, H.Y. High efficiency and low power consumption in active matrix organic light emitting diodes, *Organic Electronics* 4 (2003): 143–148.

58. Kraus, A., Benter, N., Boerner, H. OLED technology and its possible use in automotive applications, SAE Technical Paper 01 (2007): 1230.

59. Reineke, S. et al. White organic light emitting diodes with fluorescent tube efficiency, *Nature* 459 (2009): 234–238.

60. Liao, S.Y. et al. Organic light emitting diode with color tunable between bluish-white daylight and orange-white dusk hue, *International Journal of Photoenergy* 2014 (2014): 480829.

61. Jou, J.H. et al. Highly efficient color-temperature tunable organic light-emitting diodes, *Journal of Materials Chemistry* 22 (2012): 8117–8120.

62. Su, S.J., Chiba, T., Takeda, T., Kido, J. Pyridine-containing triphenylbenzene derivatives with high electron mobility for highly efficient phosphorescent OLEDs, *Advanced Materials* 20 (2008): 2125–2130.

63. Su, S.J., Gonmori, E., Sasabe, H., Kido, J. Highly efficient organic blue-and white-light-emitting devices having a carrier- and exciton-confining structure for reduced efficiency roll-off, *Advanced Materials* 20 (2008): 4189–4194.

64. Li, Y., Fung, M.K., Xie, Z., Lee, S.T., Hung, L.S., Shi, J. An efficient pure blue organic light-emitting device with low driving voltages, *Advanced Materials* 14 (2002): 1317–1321.

65. Jou, J.H. et al. High-efficiency, low color-temperature organic light-emitting diode with a blend interlayer, *Journal of Materials Chemistry*, 21 (2011): 17850–17854.

66. Jou, J.H. et al. High-efficiency low color temperature organic light emitting diodes with solution-processed emissive layer, *Organic Electronics* 13 (2012): 899–904.

67. Jou, J.H. et al. Organic light-emitting diode-based plausibly physiologically-friendly low color temperature night light, *Organic Electronics* 13 (2012): 1349–1355.

68. CIE, *Method of Measuring and Specifying Color Rendering Properties of Light Sources*, CIE, Vienna, Austria (1974).

69. CIE, *Method of Measuring and Specifying Color Rendering Properties of Light Sources*, CIE, Vienna, Austria (1995).

70. Judd, D.B. A flattery index for artificial illuminants, *Illuminating Engineering* 62 (1967): 593–598.

71. Thornton, W.A. A validation of the colorpreference index, *Journal of Illuminating Engineering Society* 4 (1974): 48–52.

72. Thornton, W.A. Color-discrimination index, *Journal of the Optical Society of America* 62 (1972): 191–194.

73. Fotios, S.A. The perception of light sources of different colour properties, PhD thesis, UMIST, Manchester, U.K. (1997).

74. Xu, H. Colour rendering capacity of illumination, *Journal of Illuminating Engineering Society* 13 (1984): 270–276.

75. Xu, H. Colour rendering capacity and luminous efficiency of a spectrum, *Lighting Research & Technology* 25 (1993): 131–132.

76. Pointer, M.R. Measuring colour rendering–A new approach, *Lighting Research & Technology* 18 (1986): 175–184.

77. Smet, K., Ryckaert, W.R., Pointer, M.R., Deconinck, G., Hanselaer, P. Correlation between color quality metric predictions and visual appreciation of light sources, *Optics Express* 19(9) (2011): 8151–8166.

78. Smet, K.A.G., Ryckaert, W.R., Pointer, M.R., Deconinck, G., Hanselaer, P. Memory colours and colour quality evaluation of conventional and solid-state lamps, *Optics Express* 18(25) (2010): 26229–26244.

79. Rea, M., Deng, L., Wolsey, R. Lighting Answers: Light Sources and Color, Rensselaer Polytechnic Institute; National Lighting Product Information Program, Troy, NY (2004).

80. Nova, J.P.F., Rea, M.S. A two-metric proposal to specify the color-rendering properties of light sources for retail lighting, *Proceedings of SPIE Tenth International Conference of Solid-State Lighting*, San Diego, CA (2010): 77840V.

81. Davis, W., Ohno, Y. Color quality scale, *Optical Engineering* 49(3) (2010): 033602.

82. Smet, K., Ryckaert, W.R., Pointer, M.R., Deconinck, G., Hanselaer, P. Color appearance rating of familiar real objects, *Color Research & Application* 36(3) (2011): 192–200.

83. Smet, K.A.G., Ryckaert, W.R., Pointer, M.R., Deconinck, G., Hanselaer, P. A memory colour quality metric for white light sources, *Energy and Buildings* 49 (2012): 216–225.

84. Rea, M.S., Nova, J.P.F., Color rendering: A tale of two metrics, *Color Research & Application* 33 (2008): 192–202.

85. Davis, W., Ohno, Y. Toward an improved color rendering metric, *Proceedings of SPIE* 5941 (2005): 59411G.

86. US Department of Energy, Solid-State Lighting Research and Development: Multi-Year Program Plan, Office of Energy Efficiency and Renewable Energy U.S. Department of Energy, Washington, DC, p. 52 (2013).

87. US Department of Energy, Solid-State Lighting Research and Development: Multi-Year Program Plan, Office of Energy Efficiency and Renewable Energy U.S. Department of Energy, Washington, DC, p. 66 (2014).

88. Jou, J.H. et al. A universal, easy-to-apply light-quality index based on natural light spectrum resemblance, *Applied Physics Letters* 104 (2014): 203304.

89. Schnapf, J.L., Kraft, T.W., Baylor, D.A. Spectral sensitivity of human cone photoreceptors, *Nature* 325 (1987): 439–441.

90. Sharpe, L.T., Stockman, A., Jagla, W., Jägle, H. A luminous efficiency function, $V^*(\lambda)$, for daylight adaptation, *Journal of Vision* 5 (2015): 948–968.

91. Wald, G. Human vision and the spectrum, *Science* 101 (1945): 653–658.

92. Kaiser, P.K. Sensation luminance: A new name to distinguish CIE luminance from luminance dependent on an individual's spectral sensitivity, *Vision Research* 28 (1988): 455–456.

93. Schubert, E.F., Kim, J.K. Solid-state light sources getting smart, *Science* 308 (2005): 1274–1278.

94. Pimputkar, S., Speck, J.S., DenBaars, S.P., Nakamura, S. Prospects for LED lighting, *Nature Photonics* 3 (2009): 180–182.

95. Xie, R.J., Hirosaki, N., Kimura, N., Sakuma, K., Mitomo, M. 2-phosphor-converted white light-emitting diodes using oxynitride/nitride phosphors, *Applied Physics Letter* 90 (2007): 191101.

96. Zhou, X., Blochwitz, J., Pfeiffer, M., Nollau, A., Fritz, T., Leo, K. Enhanced hole injection into amorphous hole-transport layers of organic light-emitting diodes using controlled p-type doping, *Advanced Functional Materials* 11 (2001): 310–314.

97. Huang, J., Pfeiffer, M., Werner, A., Blochwitz, J., Leo, K., Liu, S. Low-voltage organic electroluminescent devices using p-i-n structures, *Applied Physics Letter* 80 (2002): 139–141.

98. Pfeiffer, M., Forrest, S.R., Leo, K., Thompson, M.E. Electrophosphorescent p-i-n organic light-emitting devices for very-high-efficiency flat-panel displays, *Advanced Materials* 14 (2002): 1633–1636.

99. D'Andrade, B.W., Forrest, S.R., Chwang, A.B. Operational stability of electrophosphorescent devices containing p and n doped transport layers, *Applied Physics Letter* 83 (2003): 3858–3860.

100. He, G., Schneider, O., Qin, D., Zhou, X., Pfeiffer, M., Leo, K. Very high-efficiency and low voltage phosphorescent organic light-emitting diodes based on a p-i-n junction, *Journal of Applied Physics* 95 (2004): 5773–5777.

101. Choudhury, K.R., Yoon, J., So, F. LiF as an n-dopant in tris(8-hydroxyquinoline) aluminum thin films, *Advanced Materials* 20 (2008): 1456–1461.

102. Yook, K.S., Jeon, S.O., Min, S.Y., Lee, J.Y., Yang, H.J., Noh, T., Kang, S.K., Lee, T.W. Highly efficient p-i-n and Tandem organic light emitting devices using an air-stable and low temperature-evaporable metal azide as an n-dopant, *Advanced Functional Materials* 20 (2010): 1797–1802.

103. Wang, Q., Tao, Y., Qiao, X., Chen, J., Ma, D., Yang, C., Qin, J. High-performance, phosphorescent, top-emitting organic light-emitting diodes with p-i-n homo junctions, *Advanced Functional Materials* 21 (2011): 1681–1686.

104. Hung, L.S., Tang, C.W., Mason, M.G. Enhanced electron injection in organic electroluminescence devices using an Al/LiF electrode, *Applied Physics Letter* 70 (1997): 152.

105. Shaheen, S.E., Jabbour, G.E., Morrell, M.M., Kawabe, Y., Kippelen, B., Peyghambarian, N. Bright blue organic light-emitting diode with improved color purity using a LiF/Al cathode, *Journal of Applied Physics* 84 (1998): 2324–2327.

106. Jou, J.H., Chiang, P.H., Lin, Y.P., Chang, C.Y., Lai, C.L. Hole-transporting-layer-free high-efficiency fluorescent blue organic light-emitting diodes, *Applied Physics Letter* 91 (2007): 043504.

107. Jou, J.H., Wang, Y.S., Lin, C.H., Shen, S.M., Chen, P.C., Tang, M.C., Wei, Y., Tsai, F.Y., Chen, C.T. Nearly non-roll-off high efficiency fluorescent yellow organic light-emitting diodes, *Journal of Materials Chemistry* 21 (2011): 12613–12618.

108. Jou, J.H., Chen, P.C., Tang, M.C., Wang, Y.S., Lin, C.H., Chen, S.H., Chen, C.C., Wang, C.C., Chen, C.T. Organic light-emitting diodes with rollup character, *Journal of Photonics for Energy* 2 (2012): 021208.

109. Li, H.Y. et al. Highly efficient green phosphorescent OLEDs based on a novel iridium complex, *Journal of Materials Chemistry* 1 (2013): 560–565.

110. Jou, J.H. et al. Using light-emitting dyes as a co-host to markedly improve efficiency roll-off in phosphorescent yellow organic light emitting diodes, *Journal of Materials Chemistry C* 1 (2013): 394–400.

111. Holmes, R.J., Forrest, S.R., Tung, Y.J., Kwong, R.C., Brown, J.J., Garon, S., Thompson, M.E. Blue organic electrophosphorescence using exothermic host–guest energy transfer, *Applied Physics Letter* 82 (2003): 2422–2424.

112. Tsai, M.H. et al. 3-(9-Carbazolyl)carbazoles and 3,6-di(9-carbazolyl) carbazoles as effective host materials for efficient blue organic electrophosphorescence, *Advanced Materials* 19 (2007): 862–866.

113. Jou, J.H., Lin, Y.P., Hsu, M.F., Wu, M.H., Lu, P. High efficiency deep-blue organic light-emitting diode with a blue dye in low-polarity host, *Applied Physics Letter* 92 (2008): 193314.

114. Choulis, S.A., Choong, V.E., Patwardhan, A., Mathai, M.K., So, F. Interface modification to improve hole-injection properties in organic electronic devices, *Advanced Functional Materials* 16 (2006): 1075–1080.

115. Hughes, G., Bryce, M.R. Electron-transporting materials for organic electroluminescent and electrophosphorescent devices, *Journal of Materials Chemistry* 15 (2005): 94–107.

116. Cheng, C.H., Chou, H.H. A highly efficient universal bipolar host for blue, green, and red phosphorescent OLEDs, *Advanced Materials* 22 (2010): 2468–2471.

117. Jou, J.H. et al. High-efficiency host free deep-blue organic light-emitting diode with double carrier regulating layers, *Organic Electronics* 13 (2012): 2893–2897.

118. Sun, Y., Giebink, N.C., Kanno, H., Ma, B., Thompson, M.E., Forrest, S.R. Management of singlet and triplet excitons for efficient white organic light-emitting devices, *Nature* 440 (2006): 908–912.

119. Choy, W.C.H., Niu, J.H., Chen, X.W., Li, W.L., Chui, P.C. Effects of carrier barrier on voltage controllable color tunable OLEDs, *Applied Physics A: Materials Science & Processing* 89 (2007): 667–671.

120. Ulbricht, C., Beyer, B., Friebe, C., Winter, A., Schubert, U.S. Recent developments in the application of phosphorescent iridium (III) complex systems, *Advanced Materials* 21 (2009): 4418–4441.

121. Jou, J.H. et al. Highly efficient ultra-deep blue organic light-emitting diodes with a wet- and dry-process feasible cyanofluorene acetylene based emitter, *Journal of Materials Chemistry C* 3 (2015): 2182–2194.

122. Kawamura, Y., Brooks, J., Brown, J.J., Sasabe, H., Adachi, C. Intermolecular interaction and a concentration-quenching mechanism of phosphorescent Ir(III) complexes in a solid film, *Physical Review Letters* 96 (2006): 17404.

123. Reineke, S., Rosenow, T.C., Lüssem, B., Leo, K. Improved high-brightness efficiency of phosphorescent organic LEDs comprising emitter molecules with small permanent dipole moments, *Advanced Materials* 22 (2010): 3189–3193.

124. Park, N.G., Choi, G.C., Lee, Y.H., Kim, Y.S. Theoretical studies on the ground and excited states of blue phosphorescent cyclometalated Ir(III) complexes having ancillary ligand, *Current Applied Physics* 6 (2006): 620–626.

125. Sasabe, H. et al. Wide-energy-gap electron-transport materials containing 3,5-dipyridylphenyl moieties for an ultra-high efficiency blue organic light-emitting device, *Chemistry of Materials* 20 (2008): 5951–5953.

126. Koene, B.E., Loy, D.E., Thompson, M.E. Asymmetric triaryldiamines as thermally stable hole transporting layers for organic light-emitting devices, *Chemistry of Materials* 10 (1998): 2235–2250.

127. Huang, Q., Evmenenko, G.A., Dutta, P., Lee, P., Armstrong, N.R., Marks, T.J. Covalently bound hole-injecting nanostructures. Systematics of molecular architecture, thickness, saturation, and electron-blocking characteristics on organic light-emitting diode luminance, turn-on voltage, and quantum efficiency, *Journal of the American Chemical Society* 127 (2005): 10227–10242.

128. Zhang, Y., Whited, M., Thompson, M.E., Forrest, S.R. Singlet-triplet quenching in high intensity fluorescent organic light emitting diodes, *Chemical Physics Letters* 495 (2010): 161–165.

129. Zhang, Y., Slootsky, M., Forrest, S.R. Enhanced efficiency in high-brightness fluorescent organic light emitting diodes through triplet management, *Applied Physics Letter* 99 (2011): 223303.

130. Giebink, N., Forrest, S.R. Quantum efficiency roll-off at high brightness in fluorescent and phosphorescent organic light emitting diodes, *Physical Review B* 77 (2008): 235215.

131. Van Mensfoort, S.L.M. et al. Measuring the light emission profile in organic light-emitting diodes with nanometer spatial resolution, *Nature Photonics* 4 (2010): 329–335.

132. Lee, J., Lee, J.I., Lee, J.Y., Chu, H.Y. Stable efficiency roll-off in blue phosphorescent organic light-emitting diodes by host layer engineering, *Organic Electronics* 10 (2009): 1529–1533.

133. Chin, B.D. Enhancement of efficiency and stability of phosphorescent OLEDs based on heterostructured light-emitting layers, *Journal of Physics D: Applied Physics* 44 (2011): 115103.

134. Hudson, Z.M., Wang, Z., Helander, M.G., Lu, Z.H., Wang, S. N-heterocyclic carbazole-based hosts for simplified single-layer phosphorescent OLEDs with high efficiencies, *Advanced Materials* 24 (2012): 2922–2928.

135. Reineke, S., Schwartz, G., Leo, K. Reduced efficiency roll-off in phosphorescent organic light emitting diodes by suppression of triplet-triplet annihilation, *Applied Physics Letter* 91 (2007): 123508.

136. Chen, P. et al. Influence of interlayer on the performance of stacked white organic light-emitting devices, *Applied Physics Letter* 95 (2009): 123307.

137. Yook, K.S., Lee, J.Y. Effect of the interlayer composition on the lifetime and color change of hybrid white organic light-emitting diodes, *Journal of Industrial and Engineering Chemistry* 17 (2011): 642–644.

138. Lee, S.J. et al. Effect of broad recombination zone in multiple quantum well structures on lifetime and efficiency of blue organic light-emitting diodes, *Japanese Journal of Applied Physics* 53 (2014): 101601.

139. VanSlyke, S.A., Chen, C.H., Tang, C.W. Organic electroluminescent devices with improved stability, *Applied Physics Letter* 69 (1996): 2160.

140. Aziz, H., Luo, Y., Xu, G., Popovic, Z.D. Improving the stability of organic light-emitting devices by using a thin Mg anode buffer layer, *Applied Physics Letter* 89 (2006): 103515.

141. Liu, B. et al. Efficient hybrid white organic light-emitting diodes with extremely long lifetime: The effect of n-type interlayer, *Scientific Reports* 4 (2014): 7198.

142. Jou, J.H., Kumar, S., Singh, M. Chen, Y.H., Chen, C.C., Lee, M.T. Carrier modulation layer-enhanced organic light-emitting diodes, *Molecules* 20 (2015): 13005–13030.

143. Jou, J.H. et al. Solution-processable, high-molecule-based trifluoromethyl-iridium complex for extraordinarily high efficiency blue-green organic light-emitting diode, *Chemistry of Materials* 21 (2009): 2565–2567.

144. Baumann, T., Rudat, B. Manufacturing of OLEDschallengesand solutions, Dipl. Chem. Daniel Volz 222908587 (2012). http://www.cynora.com/images/download/.

145. Jou, J.H. et al. Frontline technology-OLEDs with candle-like emission, *Journal of the Society for Information Display* 31(6) (November/December 2015): 23–27. (a) http://www.oled-info.com/wisechips-candle-light-oleds-installed-street-lights-aboriginal-korean-village, (b) http://www.oled-info.com/researchers-taiwan-urge-consumers-and-governments-watch-out-white-led-lighting.

146. Kloog, I., Haim, A., Stevens, R.G., Barchanade, M., Portnov, B.A. Light at night co-distributes with incident breast but not lung cancer in the female population of Israel, *Chronobiology International* 25 (2008): 65–81.

147. ANSI *Specifications for the Chromaticity of Solid State Lighting Products*, American National Standard for Electric Lamps ANSI C78.377-2015, published by National Electrical Manufacturers Association, 1300 North 17th Street, Suite 900, Rosslyn, VA 22209.

148. Neuroscience: Orange light boosts brain power, *Nature* 507 (2014): 276.

VI

Novel Technologies and Developments

IV

19

III-Nitride Semiconductor LEDs Grown on Silicon and Stress Control of GaN Epitaxial Layers

Baijun Zhang
School of Electronics and Information Technology

Yang Liu
School of Electronics and Information Technology

Abstract Gallium nitride (GaN)-based semiconductor materials and devices grown on silicon substrates are attractive due to their lower cost, larger wafer size, and better thermal conductivity in comparison with other substrates. The large wafer size of Si substrates can simply lower the fabrication process cost of GaN-based optoelectronic devices and electronic power devices. GaN-based LEDs grown on Si substrate suffer from low output efficiency due to the absorptive Si substrate. For III-nitride materials grown on silicon, the high stress that originates from the thermal mismatch between the epilayer and the substrate during the cooling down of the growth temperature is still a main problem. This chapter reviews certain key issues regarding the fabrication of optoelectronic devices and stress control in material growth.

19.1 Introduction

Since the mid-2000s, great progress, both academically and commercially, has been made in the fabrication of GaN-based materials and devices on silicon, such as light-emitting diodes (LEDs) in solid-state lighting (SSL) applications, high electron mobility transistors (HEMTs) in applications to amplify radio frequencies (RFs), and in power electronic devices in power switching applications. The fabrication of GaN-based devices on Si has certain advantages, such as the low cost of the substrate, the availability of a larger wafer size for large-scale production, the compatibility of device processing with conventional Si-based integrated circuits, which facilitate device fabrication at a low cost. However, the biggest obstacle to reaching these targets is the realization of the growth of high-quality GaN on silicon, due to the considerable mismatch of the lattice constant and thermal expansion coefficient (TEC) between these two heteromaterials. Stress control technologies are important for the growth of a GaN epitaxial layer on a silicon substrate. GaN-based LEDs grown on silicon substrates are attractive due to their advantages over those grown on sapphire substrates. However, GaN-based LEDs grown on Si substrate still suffer from issues besides that of material quality, such as a high working voltage requirement, a high series resistance, and low output efficiency. These issues are caused by the large band offset at the silicon substrate and the aluminum nitride (AlN) interface seeding layer, the high resistance of the stress compensation layers, and the light absorption of the Si substrate. Great progress has been made in GaN-based LEDs on silicon in recent years. This chapter introduces the progress and development of GaN-based materials and LEDs grown on Si substrate.

19.2 III-Nitride Semiconductor LEDs Grown on Silicon

19.2.1 Basic Structure

Guha et al. [1], in 1998, presented the first GaN-based LED grown on Si. The cracks and defects on the LED's surface limited its performance, but it showed a high turn-on voltage. In 1999, Tran et al. [2] reported GaN/InGaN multiple quantum well (MQW) blue LEDs on Si. The LEDs exhibited electroluminescence (EL) data, but the operating voltage of 20 mA current was still too high, and the surface of the LED had cracks. Using a HT-AlN buffer layer and an AlGaN interlayer, Ishikawa et al. [3] obtained a crack-free GaN thin film on Si substrate in 1999. The first crack-free GaN-based LED structure grown on a 2 in. flat silicon substrate was reported by Zhang et al. [4] in 2001. $Al_{0.27}Ga_{0.73}N/AlN$ (380/120 nm) interlayers were adopted to balance the stress. Two kinds of n-type electrode were fabricated for lateral and vertical current injection. The structure of III-nitride LEDs on silicon substrate is shown in Figure 19.1. There is an obvious difference in the operating voltage between the lateral and vertical conducting configuration due to the considerable band discontinuity at the AlN–Si interface and the presence of highly resistant AlN and AlGaN layers.

One of major advantages of GaN-based LEDs on silicon is vertical current injection, but a vertical LED on silicon usually has a high series resistance and a high operating voltage, as described earlier. An AlN layer was used to prevent GaN melt-back etching and the formation of SiN_x, and an AlGaN buffer layer was used to compensate for tensile stress [3,5,6]. They are permanent layers. The simplest method with which to weaken the effect of band discontinuity at the AlN–Si interface is to grow a thin AlN film to serve as a seeding layer. The carriers can travel through the interface when the AlN is only a few nanometers thick. Zhang and Egawa et al. [7,8] reported vertical GaN-based LEDs on silicon with a low series resistance and a low operating voltage achieved by applying a thin $AlN/Al_{0.3}Ga_{0.7}N$ buffer layer. The $I-V$ characteristics of vertical LEDs on Si can be easily improved by using a thin AlN buffer layer. However, this brings the risk of melt-back etching and degradation of the crystalline epilayer.

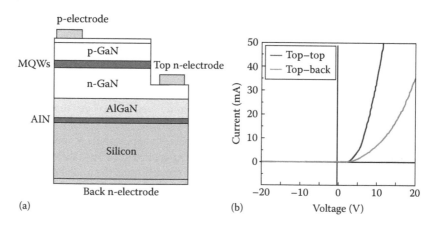

FIGURE 19.1 Structure of III-nitride LEDs on (a) silicon substrate and (b) $I-V$ characteristics.

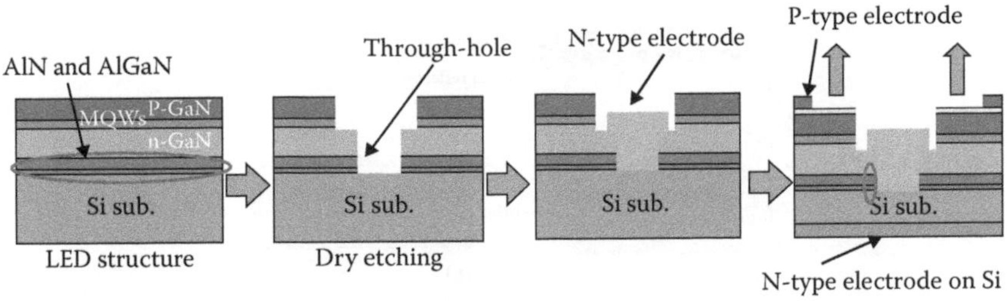

FIGURE 19.2 Process flow: (a) III-nitride LEDs structure on silicon substrate, (b) To form through holes by two steps dry etching, (c) To form N-type electrode, and fill the holes with metal, and (d) To form the transparent electrode, p-type electrode, and n-type electrode on silicon substrate.

On the other hand, in order to grow a thick high-quality crystalline epilayer, stress engineering layers are necessary to balance the tensile stress, for example, an AlGaN buffer layer and an LT-AlN interlayer. However, the layers containing Al inevitably increase the operating voltage and the series resistance. Wei et al. [9] designed a method to obtain vertical LEDs on Si by using a through-hole structure. The through-holes were formed between the n-GaN layer and the Si substrate. The fabrication processes for an LED with through-holes are shown in Figure 19.2. The through-holes were filled with metal to connect the n-GaN layer and Si substrate. The resistance induced by the large band offset at the AlN–Si interface and the high resistive interlayer was reduced by the metal-filled through-holes. The series resistance and the operating voltage were reduced, and the light output increased by 29%.

19.2.2 Nitride Thin-Film LEDs Transferred from Native Silicon Substrate

A structure with through-holes can effectively reduce the series resistance and the operating voltage. However, the problem of light absorption by the silicon substrate cannot be resolved. The thin-film transfer technique is a major approach with which to resolve the issues of light absorption, electrical conduction, and thermal dissipation. This technique was initially adopted to separate the nitride films from a sapphire substrate using the laser liftoff method and to transfer it onto a carrier with good electrical and thermal conductivity [10–13]. In comparison with a sapphire substrate, a silicon substrate can be easily removed by wet chemical etching. Metal, silicon, and thin-film epoxy resin were each adopted as a carrier for transferring a thin-film nitride LED from its native silicon substrate.

19.2.2.1 Transferred Onto Metal Carriers

As a carrier, metal is an attractive material due to its good electrical and thermal conductivity. There are two kinds of technologies for transferring GaN-based thin-film LEDs onto metal carriers: wafer bonding and metal electroplating. In 2005, crack-free GaN-based thin-film LED structures were first transferred from a native silicon substrate onto a copper carrier by Zhang et al. [14] by the adoption of wafer bonding technology. In order to avoid the generation of cracks, they proposed a selective liftoff (SLO) technique. The relaxation of stress in the epilayer could be minimized by using this technique (Figure 19.3). After the removal of the Si substrate, no obvious deterioration was observed in the crystalline quality of the structure of the LED, and the performance of the LED transferred on copper was improved significantly. The LED was fabricated with the *n*-GaN side facing upward and a vertically conductive configuration and had a low series resistance and operating voltage. An enhancement of 49% in the optical power of the LED on copper was obtained compared with the LED on the Si substrate (Figure 19.4).

Wafer bonding technology was an easy method with which to achieve thin-film LED transference; however, the concave surface of the GaN-based LED on Si, induced by the lattice and thermal mismatch, affects the wafer bonding process. The electroplating technique is not affected by the concave nature of the wafer. In 2010, Wong et al. [15] reported a double-flip approach for transferring LEDs from the epitaxial

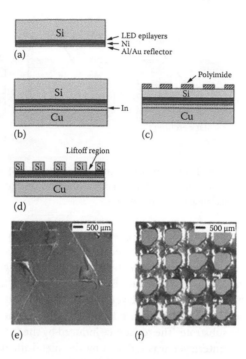

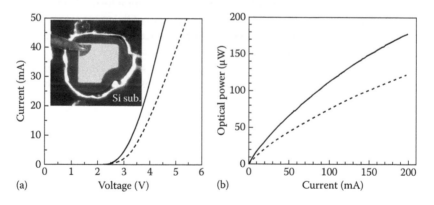

FIGURE 19.3   The SLO process flow for transferring the LED onto a Cu carrier: (a) deposit p-type ohmic contact and metal reflector, (b) bond the LED structure onto the Cu carrier, (c) thin substrate, (d) selective removal of Si substrate, (e) silicon etched fully, and (f) etched selectively. (Reprinted with permission from Zhang, B., Egawa, T., Ishikawa, H., Liu, Y., and Jimbo, T., Thin-film InGaN multiple-quantum-well light-emitting diodes transferred from Si (111) substrate onto copper carrier by selective lift-off, *Appl. Phys. Lett.*, 86, 071113, Copyright 2005, American Institute of Physics.)

FIGURE 19.4   (a) *I–V* and (b) *L–I* characteristics of the LEDs on Si carrier (dashed line) and on Cu carrier (solid line). The inset in Figure 19.4a shows a green emission from the LED transferred onto the Cu carrier. (Reprinted with permission from Zhang, B., Egawa, T., Ishikawa, H., Liu, Y., and Jimbo, T., Thin-film InGaN multiple-quantum-well light-emitting diodes transferred from Si (111) substrate onto copper carrier by selective lift-off, *Appl. Phys. Lett.*, 86, 071113, Copyright 2005, American Institute of Physics.)

substrate onto a copper carrier. The copper carrier was formed by electroplating. A lateral conductive configuration thin-film LED was transferred onto the copper carrier. The light output of the LEDs on copper increased by about 70% compared with the LEDs on silicon. In 2011, Lau et al. [16] transferred GaN-based blue LEDs onto a copper carrier using the identical process. The light output of the LEDs on copper was 80% greater. The advantage of this technique is that there is no need for further fabrication after the process, which is helpful in improving the yield and reliability of device manufacture. An omnidirectional reflector (ODR) consisting of 5-pair TiO_2/SiO_2 DBR and an Al mirror were inserted between the LED structure and the electroplating copper carrier during the thin-film transferring process to enhance the light output [17]. The reflectivity of the TiO_2/SiO_2-Al ODR was as high as 97% in the blue spectral range. The light output of ODR-based LEDs on a copper carrier was further enhanced. The thermal dissipation capacity of the two LEDs on a Si and a copper carrier was investigated by infrared (IR) microscopy. The thermal distribution at an injecting current of 300 mA after 3 min is shown in Figure 19.5. Compared with the conventional LED on Si, the ODR-based LED on a copper carrier shows a significantly lower temperature due to the better thermal conductivity of copper (401 W/(mK)).

A vertical conductive configuration thin-film LED structure with through-holes was transferred from a 2 in. Si substrate onto a copper carrier by Luo et al. [18] in 2012. The copper carrier was formed by electroplating (Figure 19.6). The large band offset at the interface of the AlN/Si and the series resistances induced by the AlN buffer layer and other Al composition interlayer were shorted by the metal-filled through-holes. The vertical through-hole structure LED showed low operating voltage and a small series resistance. The operating voltage at 350 mA and the series resistances of the LED were reduced from 5.6 to 5.1 V and from 7 to 4 Ω, respectively. The light output was improved by 75%, which was mainly due to the removal of the light absorptive substrate and the good thermal conductivity of the copper carrier (Figure 19.7). A slight blueshift in the EL spectra of the LED on copper was observed. Luo et al. believed it was due to a partial relaxation of the stress in the thin films after the LED structure transferred from the silicon substrate onto the copper carrier.

A two-inch crack-free InGaN MQWs LEDs wafer transferred from its native silicon substrate onto an electroplating copper carrier was reported by Chen et al. [19] in 2012 (Figure 19.8). After the LED structures were transferred onto copper, no obvious deterioration was found in the MQWs structure. The wavelength of the light emitted from the LED on copper showed a blueshift as compared with the conventional device (Figure 19.9a), which was due to the generation of partial stress relaxation. Stress relaxation in the LED on copper weakened the quantum confined Stark effect (QCSE) in the MQWs, which enhanced the internal quantum efficiency (IQE). The LED has an *n*-GaN facing upward. An embedded electrode structure was adopted to eliminate the electrode-shading effect. Combining this with through-hole technology, the LED had a vertical conductive configuration. The LED on copper showed an enhancement of 122% in output at 350 mA as compared with the conventional LEDs on silicon (Figure 19.9b). The forward voltage at 350 mA current was 4.75 V, which is slightly higher than that of a conventional LED on Si. Chen et al. suggest that the increase in forward voltage may be due to a poor interface contact between the metal reflector and the p-GaN. The LED on copper and on Si showed the same series resistance of about 5 Ω.

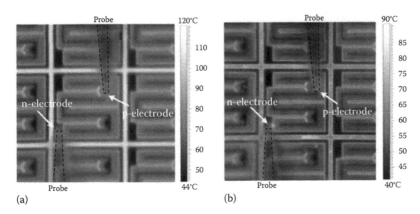

FIGURE 19.5 IR camera images of (a) the conventional LED on Si and (b) the ODR-based LED on copper at the injecting current of 300 mA after 3 min. (From Yang, Y., Hu, G., Xiang, P., Liu, M., Chen, W., Han, X., Hu, G. et al., Enhancement of GaN-based light-emitting diodes transferred from Si (111) substrate onto electroplating Cu submount with TiO2/SiO2-Al omnidirectional reflector, *J. Display Technol.*, 10, 1064, Copyright 2014 IEEE.)

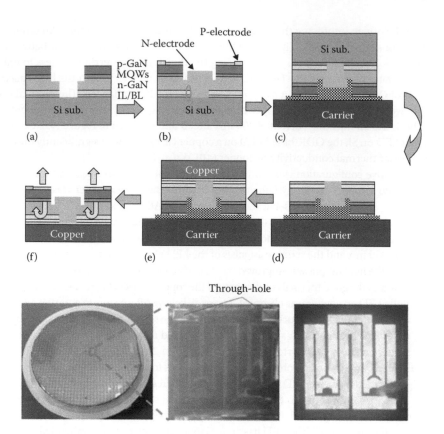

FIGURE 19.6 Devices process flow of through-hole LED on copper substrate: (a) form the through-holes and the LED mesas by inductively coupled plasma reactive ion etching (ICP-RIE), (b) deposit transparent conductive layer (TCL) and form p- and n-type electrodes, (c) bond the devices to a temporary substrate using acrylate adhesive, (d) remove the epitaxial substrate by chemical etching, (e) deposit the metal reflector onto the exposed AlN buffer layer surface and electroplate a copper layer 100 μm thick, and (f) separate the devices from the temporary substrate. A crack-free 2 in. LED wafer was successfully transferred from the silicon substrate onto the copper. The photographic image of a 1 × 1 mm² LED chip with through-holes and the light emission generated by injecting current from the p-electrode to the copper carrier are shown in this figure. (After Luo, R. et al., *Jpn. J. Appl. Phys.*, 51, 012101, 2012, Reproduced with permission from the Japan Society of Applied Physics.)

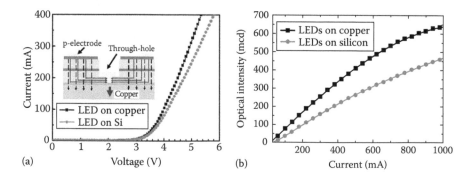

FIGURE 19.7 (a) *I–V* and (b) *L–I* characteristics of the LEDs on a silicon carrier and on a copper carrier. Inset shows the schematic of current spreading in the through-hole LED on copper. (After Luo, R. et al., *Jpn. J. Appl. Phys.*, 51, 012101, 2012, Reproduced with permission from the Japan Society of Applied Physics.)

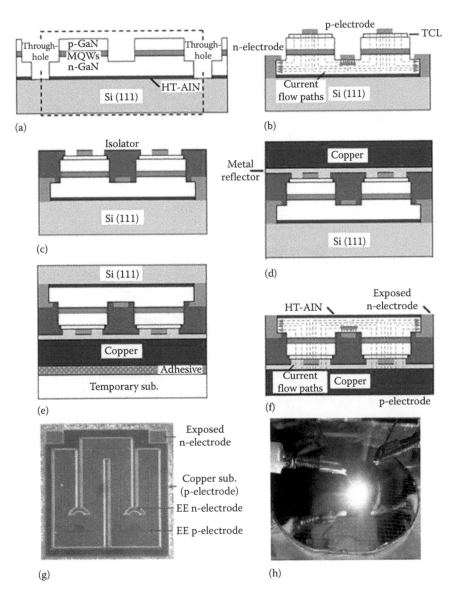

FIGURE 19.8 Fabrication process of LEDs on copper carrier: (a) etch through-holes and the mesa by ICP-RIE, (b) deposit TCL and *p/n* electrodes, (c) isolate n-electrode and p-electrode, (d) deposit metal reflector onto exposed TCL surface and electroplate a copper layer 100 μm thick, (e) bond the wafer to a temporary substrate using acrylate adhesive and thin silicon substrate, (f) remove Si substrate by wet chemical etching and separate from temporary substrate, (g) photograph of 1 mm × 1 mm LED on copper, and (h) light emission image of LEDs on 2 in. electroplated copper carrier at 350 mA current injection. Current flow paths of conventional LEDs on a Si carrier and LEDs on a copper carrier are shown in (b) and (f), respectively. (Reprinted with permission from Chen, T., Wang, Y., Xiang, P., Luo, R., Liu, M., Yang, W., Ren, Y. et al., Crack-free InGaN multiple quantum wells light-emitting diodes structures transferred from Si (111) substrate onto electroplating copper submount with embedded electrodes, *Appl. Phys. Lett.*, 100, 241112, Copyright 2012, American Institute of Physics.)

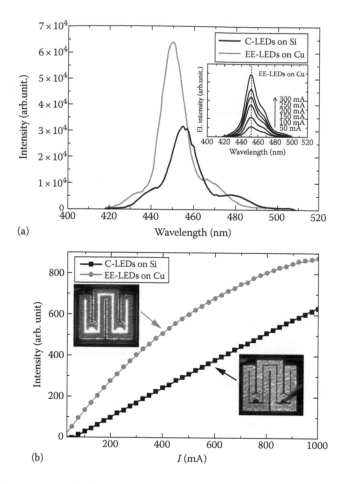

FIGURE 19.9 (a) EL spectra emitting from the C-LED and the EE-LED. The EL spectra of the EE-LED at various injecting currents are shown in the inset, no wavelength peak shift was observed with the injecting current increasing from 50 mA to 300 mA, (b) light output versus current (*L–I*) characteristics of C-LED and EE-LED. Light emission images of the LED on Si and on copper at 5 mA current injection are shown in the insets. (Reprinted with permission from Chen, T., Wang, Y., Xiang, P., Luo, R., Liu, M., Yang, W., Ren, Y. et al., Crack-free InGaN multiple quantum wells light-emitting diodes structures transferred from Si (111) substrate onto electroplating copper submount with embedded electrodes, *Appl. Phys. Lett.*, 100, 241112, Copyright 2012, American Institute of Physics.)

Crack-free LEDs transferred onto an electroplated copper carrier with embedded wide *p*-electrodes were reported by Liu et al. [20]. The widened embedded *p*-electrode covered almost the whole TCL, which improved the current spreading property and the uniformity of luminescence. The working voltage and series resistance were thereby reduced. The light output of an embedded wide *p*-electrode LEDs on copper was enhanced by 147% at a driving current of 350 mA, in comparison with conventional LEDs on Si.

19.2.2.2 Transfer Onto a Nonnative Silicon Substrate

Xiong et al. [21,22] transferred the GaN-based LED structure from the epitaxial silicon substrate onto a new silicon carrier using chemical etching method and wafer bonding technique. The wafer bonding was realized with a thick metal gold-to-gold bond at high temperature. The metal gold was deposited on the side of the LED structure and on the new silicon carrier side. After the removal of the epitaxial Si substrate by wet chemical etching, the exposed high resistive layers, including the AlN seeding layer and high resistive layers, were removed by ICP. The LEDs had a vertical conducting configuration with a

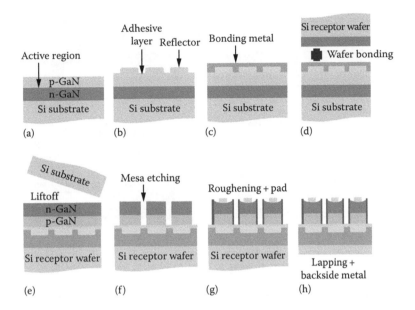

FIGURE 19.10 Fabrication processes for the vertical LED. (a) GaN-based LED grown on Si (111) substrate, (b) ITO and reflection metal deposition, (c) barrier/bonding metal deposition, (d) wafer bonding to Si (100) receptor substrate, (e) removing the Si(111) substrate, (f) mesa etching, (g) n-GaN roughening and formation of n-electrode, and (h) p-contact metallization after thinning the Si. (After Lee, S.-J. et al., *Appl. Phys. Exp.*, 4, 066501, 2011, Reproduced with permission from the *Japan Society of Applied Physics*.)

small series resistance. The output emitted from the vertical structure LED showed obvious enhancement compared with the lateral structure LED. A very similar fabrication process for a vertical GaN-based LED was reported by Lee et al. [23] (Figure 19.10).

In comparison with a copper carrier, a silicon carrier is a better option for the transfer of thin-film LEDs due to it being a better thermal match. However, a thick gold layer at the bonding interface is required in the process, which increases the cost of fabrication. On the other hand, the concave surface of a GaN-based LED on Si, induced by the lattice and thermal mismatch, affects the wafer bonding process. Compared with a silicon receptor, copper showed better thermal properties. Moreover, the copper carrier can be easily formed by electroplating. The electroplating technique will not be affected by the wafer being concave. However, copper has a high thermal mismatch to GaN, with a thermal expansion coefficient approximately three times higher, leading to thermal stresses during operation, which could affect the lifetime of the device, especially in the case of devices with a large area.

19.2.2.3 Transferred Onto a Thin Epoxy Resin Carrier

To transfer the LED, a solution of HNA (HF:HNO3:CH3COOH=1:1:1) was often used to remove the Si substrate. HNA is a very strong corrosive solution. The main problem in using this solution is how to protect the electrodes deposited on the chip and the new carrier (metal or silicon) during the wet chemical etching process. Liao et al. [24] transferred a crack-free InGaN/GaN MQWs LED structure from a 2 in. silicon wafer onto a thin epoxy resin carrier. Due to its good mechanical and anticorrosive properties, not only can the epoxy resin layer protect the electrodes deposited on the chip easily while etching the silicon with HNA solution, but also can serve as a carrier after the removal of the substrate. The fabrication process of this thin-film LED structure is shown in Figure 19.11. The metal deposition after the removal of the Si served as a metal reflector and as an electrical and thermal conductive layer, making a significant improvement in the performance of vertical thin-film LEDs transferred to an epoxy resin carrier. Photographs of a 2 in. thin-film LED wafer transferred onto an epoxy resin carrier and its light emission are shown in Figure 19.12.

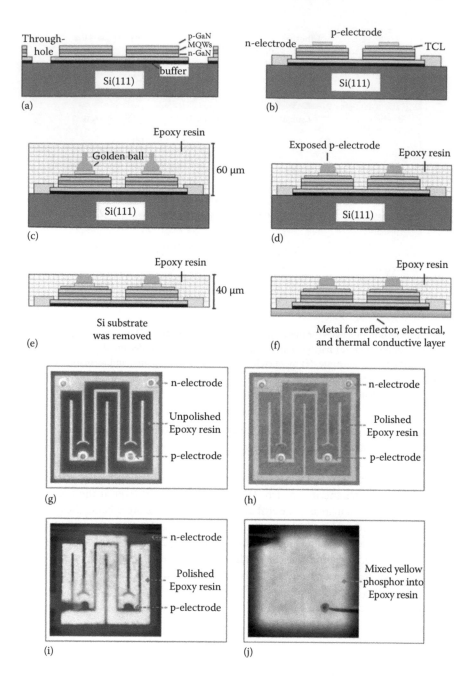

FIGURE 19.11 Fabrication process of the thin-film LEDs transferred to an epoxy carrier, (a) form the LED mesas and through-holes by ICP-RIE dry etching, (b) deposit TCL and p/n electrodes, (c) bond golden balls to the electrode pads, spin coat the epoxy layer onto the front surface of the device and harden at 135°C for 8 h, (d) polish the epoxy layer till the bonded golden balls are exposed, (e) remove Si substrate by wet chemical etching, (f) deposit the metal reflector (Ti/Al/Ni/Au) onto the exposed AlN buffer layer surface after the substrate has been removed, (g) photograph of the unpolished epoxy-coated LEDs on Si (chip size: 1 × 1 mm²), (h) photograph of the TF-LEDs on EP (chip size: 1 × 1 mm²), (i) light-emission image of TF-LEDs on EP at an injecting current of 5 mA, (j) light emission image of the thin-film white LEDs on an epoxy carrier mixed with yellow phosphor at an injecting current of 5 mA. (From Liao, Q., Yang, Y., Chen, W., Han, X., Chen, J., Zang, W., Luo, H. et al., Fabrication and properties of thin-film InGaN/GaN multiple quantum well light-emitting diodes transferred from Si (111) substrate onto a thin epoxy resin carrier, *J. Display Technol.*, 12, 1602 (2016).)

FIGURE 19.12 (a) Photograph of the 2 in. double-light emitting LEDs on an epoxy carrier viewed from the top, the remaining edge of the Si substrate acts as a framework to support the thin-film LED wafer, (b) light emission image from the double sides at 20 mA in a bright environment, a mirror was used to observe the light emission from the back surface, (c) photograph of the 2 in. double-light emitting LEDs on an epoxy carrier taken from beneath, (d) light emission image taken at 20 mA in a dark environment, (e) photograph of the TF-LEDs on EP after metal reflector deposition, and (f) light emission image taken at an injecting current of 20 mA and in a bright environment. (From Liao, Q., Yang, Y., Chen, W., Han, X., Chen, J., Zang, W., Luo, H. et al., Fabrication and properties of thin-film InGaN/ GaN multiple quantum well light-emitting diodes transferred from Si (111) substrate onto a thin epoxy resin carrier, *J. Display Technol.*, 12, 1602 (2016).)

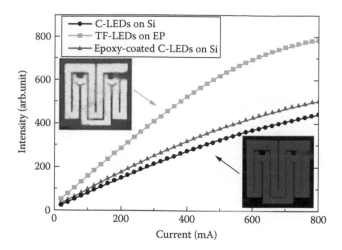

FIGURE 19.13 *L–I* characteristics of the C-LEDs on Si and TF-LEDs on EP. The inset shows the light emission images of both devices at an injecting current of 5 mA. (From Liao, Q., Yang, Y., Chen, W., Han, X., Chen, J., Zang, W., Luo, H. et al., Fabrication and properties of thin-film InGaN/GaN multiple quantum well light-emitting diodes transferred from Si (111) substrate onto a thin epoxy resin carrier, *J. Display Technol.*, 12, 1602 (2016).)

The operating voltage at 350 mA decreases from 4.95 to 4.5 V and the series resistance decreases from 3.1 to 2.6 Ω. The light output showed an increase of 92% after the LEDs were transferred to the epoxy resin carrier (Figure 19.13).

Using this fabrication process, wafer-level thin-film white LED chips can be directly fabricated by transferring an InGaN/GaN MQWs LED structure from its native Si substrate onto a thin epoxy carrier mixed with yellow phosphor. The light emission image of the thin-film white LED chips is shown in Figure 19.14. The thin-film white LED is a genuine plane light source and shows very good uniformity. Also, its color rendering properties can be easily controlled by adjusting the ratio of epoxy to yellow phosphor.

19.2.3 GaN-Based LEDs with III-Nitride-Distributed Bragg Reflector Grown on a Silicon Substrate

To solve the problem of light absorption by the silicon substrate, growing a nitride-distributed Bragg reflector (DBR) structure under the active layer is a convenient approach in comparison with the thin-film transferring technology that has been described. A greater number of DBR pairs are required to obtain high reflectivity and wide-stop bandwidth for a nitride DBR structure due to the low index contrast for GaN/AlN material systems. However, the critical thickness limits the number of DBR pairs; the thicker the epilayer, the stronger the tensile stress that would be induced. Cracks are generated when the thickness is greater than the critical thickness during cooling down to room temperature. A full GaInN LEDs structure with AlGaN/AlN DBRs grown on Si was demonstrated by Ishikawa et al. [25,26]. In comparison with the LED without the DBR structure, the LED with three pairs of DBRs showed an approximately two fold increment in light output. However, the LEDs with DBR structures showed high series resistances and high operating voltages. The output of the LED with five pairs of DBRs was drastically reduced due to the presence of cracks (Figure 19.15).

The considerable lattice mismatch and the low index contrast in systems made with AlGaN/AlN material do not permit the construction of a crack-free LED on silicon with a high reflective DBR structure. The lattice matched Ga N/AlInN system is able to suppress the generation of cracks and improves the smoothness of the interface, which affords a good opportunity to obtain a crack-free DBR structure [27,28]. An LED structure with a 14.5-pair lattice-matched GaN/AlInN DBR was fabricated on Si

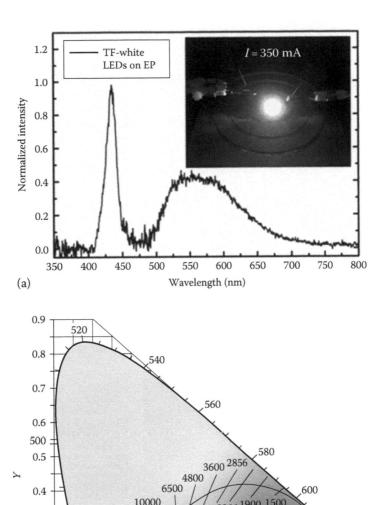

FIGURE 19.14 (a) EL spectra of the TF-white LEDs on EP. The insets show the light emission image taken at an injecting current of 350 mA. (b) CIE chromaticity diagram for the EL spectra of the TF-white LEDs on EP. (From Liao, Q., Yang, Y., Chen, W., Han, X., Chen, J., Zang, W., Luo, H. et al., Fabrication and properties of thin-film InGaN/GaN multiple quantum well light-emitting diodes transferred from Si (111) substrate onto a thin epoxy resin carrier, *J. Display Technol.*, 12, 1602 (2016).)

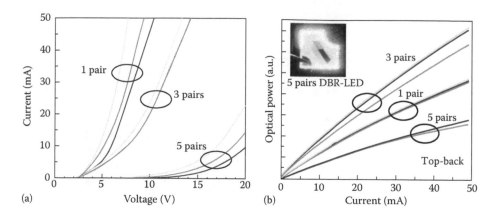

FIGURE 19.15   (a) *I–V* and (b) *L–I* characteristics of the LED with 1 pair, 3 pairs, and 5 pairs of DBRs. Inset shows the light emission image of the LED with 5 pairs of DBRs.

substrate by Ishikawa et al. [29]. The light output of the LED with the DBR structure was enhanced by 3.6 times in comparison with the conventional LED. However, the vertical conducting LED with the DBR structure usually shows a high series resistance and a high operating voltage. Crack-free InGaN/GaN MQWs LEDs with 5-pair AlN/GaN DBRs were grown on Si (111) substrate by using a linear composition-ally graded AlGaN layer to compensate for the tensile stress [30,31]. By using the through-hole structure,

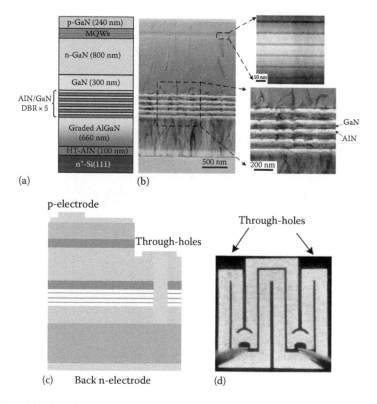

FIGURE 19.16   (a and b) The schematic structure and the cross-sectional TEM images of the LED with 5-pair AlN/GaN DBRs on Si substrate, (c) schematic view of the vertical conducting DBR-based LED with through-holes, (d) light emission image at an injecting current of 100 mA. (After Yang, Y., *Appl. Phys. Exp.*, 7, 042102, 2014, Reproduced with permission from the Japan Society of Applied Physics.)

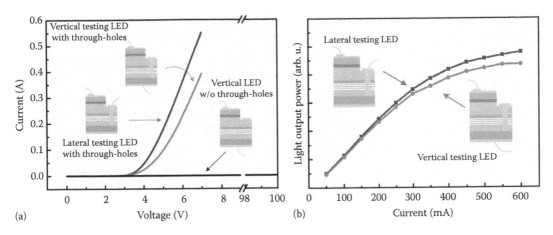

FIGURE 19.17 (a and b) *I–V* and *L–I* characteristics of the lateral and the vertical testing of DBR-based LEDs with through-holes. (After Yang, Y., *Appl. Phys. Exp.*, 7, 042102, 2014, Reproduced with permission from the Japan Society of Applied Physics.)

vertical conducting LEDs with AlN/GaN DBRs on Si (111) substrate were achieved. The through-holes were formed from the *n*-GaN layer to the Si substrate and filled with metal, which connected the *n*-GaN layer and the Si substrate. The insulating AlN, high-resistive AlGaN layers, nitride DBR, and the large band offset at the AlN/Si interface were shorted by the metal-filled through-holes. The schematic structure and the cross-sectional transmission electron microscopy (TEM) images of the LED with 5-pair AlN/GaN DBRs on Si substrate are shown in Figure 19.16. In comparison with the DBR-based LED without through-holes, the DBR-based LEDs with through-holes showed significantly better electrical and optical properties (Figure 19.17).

19.3 Stress Control of GaN Epi-Materials Grown on Silicon

GaN-on-Si has attracted intense interest since the mid-2000s, and the interest from industry is still increasing dramatically. Unfortunately, high stress in GaN-based materials originates from the thermal mismatch between the epilayer and Si substrate during cooling down, which results in considerable bowing of the wafer and the generation of cracks. The stress issue becomes more serious with the increasing thickness of the epilayer and the wafer size. Many stress control methods are used to solve these problems [32–37], such as inserting low-temperature AlN (LT-AlN), an AlGaN buffer, superlattice structures, impurity doping, and so on. Further achievements with the growth of a thick epilayer (>3 μm) and GaN-based materials grown on a large silicon wafer (> 6 in.) have been made.

19.3.1 Stress Engineering Layers

19.3.1.1 AlN Seeding Layer

AlN is the most common seeding layer for the growth of high-quality GaN on Si substrate [38–45]. An AlN seeding layer can prevent gallium and ammonia from reacting with the silicon to form an Si_xN_y amorphous layer and also melt-back etching at high temperatures. In addition, due to the low level of lattice mismatch between AlN and GaN, compressive stress can be generated on epitaxial GaN grown on AlN to counterbalance the tensile stress. Low-temperature (LT) AlN [43] is an effective method by which to reduce the stress. However, AlN grown at a temperature below 760°C shows very poor coverage of the Si substrate, which impedes effective nucleation [42]. An HT-AlN seeding layer has commonly been adopted for the growth of high crystalline quality GaN-based material on silicon due to the high surface mobility of Al atoms at high temperatures. A medium-temperature/high-temperature (MT/HT) bilayer AlN buffer for the growth of GaN on Si (111) in a metal–organic chemical vapor deposition (MOCVD) system has been reported by Xiang et al. [46]. They investigated the influence of the growth temperature of the first AlN layer. Significant improvements of the GaN quality and stress control were demonstrated using the MT/HT bilayer AlN when compared with a single HT-AlN and a LT/HT bilayer AlN. Dislocations and stresses are mainly concentrated in the bottom MT-AlN layer and are reduced greatly in the subsequent HT-AlN layer. The reduction of dislocations and stresses in the buffer layer eventually enhances the GaN film growth. In the sample with the LT/HT AlN layer, owing to the poor crystalline quality of the AlN nucleation layer grown at 700°C, considerable dislocations were still propagated into the subsequent HT-AlN layer and into the GaN film. Dislocation and stress of the samples with different AlN seeding layers were investigated by TEM and Raman spectra measurements, as shown in Figures 19.18 and 19.19.

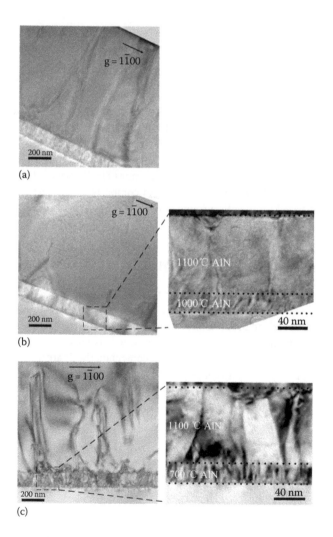

FIGURE 19.18 Cross-sectional TEM images of GaN grown on Si: (a) with a single AlN buffer grown at 1100°C, (b) with a bilayer AlN buffer grown at 1000°C and 1100°C, and (c) with a bilayer AlN buffer grown at 700°C and 1100°C. (After Xiang, P. et al., *Jpn. J. Appl. Phys.*, 52, 08JB18, 2013, Reproduced with permission from the Japan Society of Applied Physics.)

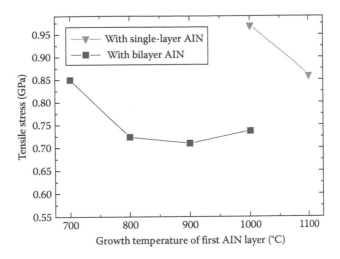

FIGURE 19.19 Tensile stress of GaN films obtained from Raman shift. (After Xiang, P. et al., *Jpn. J. Appl. Phys.*, 52, 08JB18, 2013, Reproduced with permission from the Japan Society of Applied Physics.)

19.3.1.2 Low-Temperature AlN (LT-AlN) Interlayer

An LT-AlN interlayer was first applied to reduce the dislocation density in the GaN and control the stress to yield crack-free AlGaN on sapphire. It was then introduced into the system for the growth of GaN on Si. Dadgar et al. [47] found that the LT-AlN layers can drastically reduce the pure screw dislocations and the edge dislocations (Figure 19.20). In particular, most screw dislocations were obstructed by the first AlN interlayer. The LT-AlN inserted in the middle of the thick GaN layer showed no significant reduction of the dislocations [48].

An LT-AlN interlayer is a common way to control the stresses in GaN on Si. This method was originally adopted by Amano et al. [49] when they grew GaN on a sapphire substrate. By introducing an LT-AlN interlayer, the cracking of GaN on Si was decreased [50,51]. A single LT-AlN interlayer can effectively counterbalance the tensile stress in GaN film 0.7–1 μm thick, meaning that a thick GaN film can be obtained by introducing multilayers of LT-AlN. Dadgar et al. [52], using fourfold LT-AlN interlayers, obtained GaN film thicker than 5.4 μm with a crack-free surface on a 6 in. Si substrate.

On the basis of previous reports, the stress behavior of LT-AlN itself and its effect on the overlying GaN layer as a function of the growth conditions for LT-AlN was investigated, such as the growth temperature, thickness and numbers of LT-AlN interlayer [53]. Krost et al. [54] studied the effect of the LT-AlN interlayer growth temperature on the stress in the GaN layer. The stress in the epilayer shows an obvious variation to the growth temperature. When the growth temperature increases above 1000°C, the ability of the LT-AlN to adjust stress decreased dramatically. Krost et al. explained the mechanism in terms of the decoupling effect between the GaN layers and the LT-AlN interlayer, which is dependent on the AlN growth temperature. The phenomenon of the influence of AlN growth temperature on stress was also observed by Raghavan et al. [55]. The stress transition temperature falls between 800°C and 900°C. Raghavan et al. believe it is due to the crystalline state changing from an epitaxial orientation at high temperature (above 900°C) to a defective polycrystalline state at low temperature (600°C). Meanwhile, the tensile stress in the GaN epilayer is also strongly dependent on the thickness of the AlN interlayer. Krost et al. [54] reported that the optimized thickness of AlN layer was approximately 12 nm for a GaN growth 1.3 μm thick. When the thickness was less than 12 nm, the stress compensation effect was weakened linearly with the decreasing thickness of the AlN layer, with a compensation ratio of about 0.06 ± 0.01 GPa/nm. No obvious stress adjustment effect was found by increasing the thickness of the AlN above the optimum.

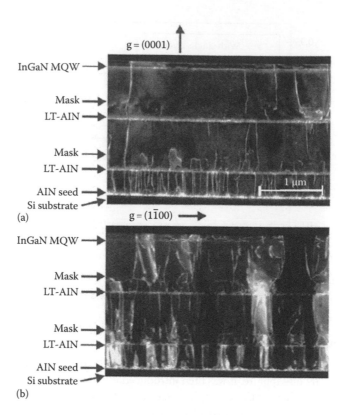

FIGURE 19.20   Cross-sectional TEM images of an LED structure with LT-AlN and Si_xN_y in situ masks: (a) $g = (0001)$ shows the screw dislocation component and (b) $g = (1\bar{1}00)$ shows the edge dislocation component. (Reprinted from *J. Cryst. Growth*, 248, Dadgar, A., Poschenrieder, M., Bläsing, J., Contreras, O., Bertram, F., Riemann, T., Reiher, A. et al., MOVPE growth of GaN on Si (111) substrates, 556, Copyright 2003, with permission from Elsevier.)

19.3.1.3 AlGaN Layer

The intrinsic stress in a GaN layer grown on HT-AlN/Si was compressive at the commencement of the growth process. The stress gradually relaxed as the thickness of the GaN increased. The compressive stress was unable to counterbalance the high tensile stress when the thickness was above the critical thickness (1.0 μm, in this case). The insertion of AlGaN layers can retain the compressive stress and prevent the generation of cracks in the GaN layer. The Al composition in AlGaN layer could be fixed [4,56,57], linearly graded [45,58–60], or step graded [61–64]. In 1999, Ishikawa et al. [57] obtained a high-quality GaN layer on a Si substrate by using a 250 nm-thick composition fixed AlGaN buffer layer on an HT-AlN seeding layer. A crack-free GaN-based LED structure was also successfully grown on 2 in. silicon substrate using an AlN/AlGaN (120/380 nm) buffer layer structure [4].

Step-graded AlGaN buffer layers were also introduced [35,62] to prevent the relaxation of compressive stress generated at the HT-AlN/Si (111) interface. Cheng et al. [62] grew a GaN layer on an HT-AlN seeding layer and on a step-graded AlGaN buffer layer. They observed that 1.0 μm thick GaN grown on HT-AlN showed a cracked surface, while no cracks appeared on the GaN surface when it was grown on a step-graded AlGaN buffer layer. The insertion of step-graded AlGaN layers can retain the compressive stress in the GaN layer, which can prevent the generation of cracks generation in the GaN layer. The stress relaxation in the step-graded AlGaN layers is also dependent on layer thickness. By optimizing the growth of the AlGaN buffer layer, the entire layer stack, measuring 2 μm thick, is kept under compressive stress with a convex bowing radius of 119 m. In this way, a crack-free GaN-based LED grown on a 6 in. Si was achieved by Zhu et al. [64].

Since a high-temperature AlN (HT-AlN) seeding layer is the most common method for GaN deposited on Si substrates, the compositionally graded AlGaN buffer layer is an effective method by which to compensate the tensile stress. Raghavan et al. [59,60] observed an increase in stress during the growth of the graded AlGaN layers. The increase in thickness of the graded AlGaN buffer layer will enhance the thickness of the subsequent GaN layer, which grows under a compressive stress. It means that the thicker the graded AlGaN buffer layer, the more thermal tensile stress can be counterbalanced in the GaN. The critical thickness of the GaN-based epilayer can be increased. The linearly graded AlGaN buffer was used for the growth of GaN [45,58,59] and an LED structure [60] on a silicon substrate.

In comparison with the Al composition fixed and step-graded AlGaN layers, a linear grade in the composition of the AlGaN buffer layer can produce a linear change in the magnitude of the incremental stress in the film, which can more effectively compensate the tensile stresses. The linearly graded AlGaN reported previously was grown by using linearly graded flow rates of TMAl and TMGa. The considerable gas-phase parasitic reaction between ammonia (NH_3) and TMAl has a strong impact on the Al composition and the growth rate of AlGaN, which leads to difficulties in achieving high-Al-composition AlGaN growth [65–67]. Therefore, the growth rate of high-Al-composition AlGaN is much slower than that of low-Al-composition AlGaN, resulting in Al-rich AlGaN being thinner than Ga-rich AlGaN [68]. The Al composition of AlGaN is not a case of linear gradient versus thickness when the flow rates of TMAl and TMGa are simply changed linearly. Yang et al. [31] varied linear and nonlinear functions of TMAl flow rates during the growth of compositionally graded AlGaN, while keeping the same linear flow rate of TMGa for all conditions. The nonlinear function flow rates compensate the parasitic reaction of TMAl and NH_3 by intentionally slowing the change rate of the TMAl flow rate in the high-Al-composition region. As a result, virtually linear compositionally graded AlGaN was obtained with the convex parabolic TMAl flow rate. This optimized buffer resulted in a better quality GaN crystalline layer and higher compressive stress (Figures 19.21 and 19.22). Using the linear compositionally graded AlGaN buffer layer, a 3 μm crack-free GaN film with low dislocation density was grown on the silicon substrate. The residual stresses in the epilayer increased with the thickness (Figure 19.23).

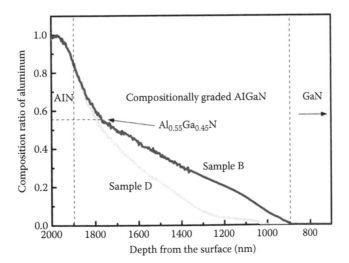

FIGURE 19.21 Composition ratio of aluminum versus thickness for samples *B* and *D*, measured by SIMS. Functions of TMAl flow rate of samples *B* and *D*: $B = 100 - 95x^2$ (sccm), $D = 95(1 - x)^2 + 5$ (sccm). (Reprinted from *J. Cryst. Growth*, 376, Yang, Y., Xiang, P., Liu, M., Chen, W., He, Z., Han, X., Ni, Y., Yang, F., Yao, Y., Wu, Z., Liu, Y., and Zhang, B., Effect of compositionally graded AlGaN buffer layer grown by different functions of trimethylaluminum flow rates on the properties of GaN on Si (111) substrates, 23, Copyright 2013, with permission from Elsevier.)

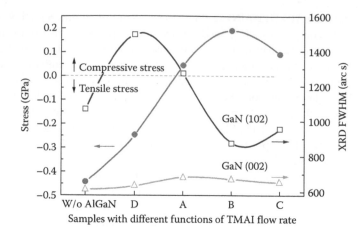

FIGURE 19.22 Biaxial stress of GaN measured by Raman shift (solid circle) and FWHMs of GaN (002) and (102) XRD rocking curves (open triangle and quadrangle) with different functions of TMAl flow rate: $A = 100 - 95x$ (sccm), $B = 100 - 95x^2$ (sccm), $C = 100 - 95x^3$ (sccm), and $D = 95(1 - x)^2 + 5$ (sccm), respectively, where x ($0 \leq x \leq 1$) is the normalized growth time of the AlGaN buffer layer. (Reprinted from *J. Cryst. Growth*, 376, Yang, Y., Xiang, P., Liu, M., Chen, W., He, Z., Han, X., Ni, Y., Yang, F., Yao, Y., Wu, Z., Liu, Y., and Zhang, B., Effect of compositionally graded AlGaN buffer layer grown by different functions of trimethylaluminum flow rates on the properties of GaN on Si (111) substrates, 23, Copyright 2013, with permission from Elsevier.)

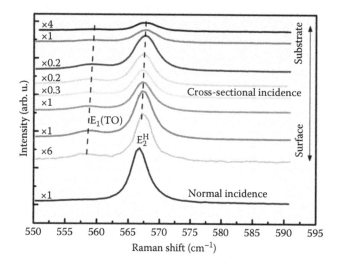

FIGURE 19.23 Raman spectra of the 3 μm crack-free GaN film, with the laser incidence scanning over the cross-section of the film, in comparison with that of normal incidence. (Reprinted from *J. Cryst. Growth*, 376, Yang, Y., Xiang, P., Liu, M., Chen, W., He, Z., Han, X., Ni, Y., Yang, F., Yao, Y., Wu, Z., Liu, Y., and Zhang, B., Effect of compositionally graded AlGaN buffer layer grown by different functions of trimethylaluminum flow rates on the properties of GaN on Si (111) substrates, 23, Copyright 2013, with permission from Elsevier.)

19.3.1.4 Superlattices Layer

The AlN/GaN superlattices (SLs) layer technique is one of the most widely used methods and was first introduced to the GaN-on-Si system by Feltin et al. [69]. Subsequently, LEDs were successfully fabricated in 2002 on a GaN-on-Si system using the SLs technique [70]. The influence of the number of pairs and the thickness of AlN/GaN SLs on the quality of the GaN on Si, such as dislocation density and surface roughness, were studied by Sanken Electric [36]. Using the AlN/GaN SLs-based high-quality GaN-on-Si template, AlGaN/GaN HFETs have been successfully fabricated in recent years and have given excellent performances [36,71–77], including a high-current operation [76], high breakdown voltage up to ~1800 V with a very thick nitride epilayer [73,75], large-diameter Si substrate with high uniformity [36,74], very high mobility with atomic surface flatness [72], and so on. Most of these achievements can be attributed to the SLs technique with its precisely controlled growth parameters.

Although great progress has been made in GaN growth on a Si substrate with an AlN/GaN SLs structure and despite the fact that it is now the main structure adopted by current industry, there is little available explanation of the mechanism regarding the issue of stress engineering in the GaN-on-Si (111) system. Commonly, the tensile stress of GaN on a Si substrate, which was generated during the cooling process, was believed to be counterbalanced [78] by the compressive stress provided from the AlN/GaN SLs structure. In fact, the stress of GaN/AlN SLs has always been studied separately on the sapphire substrate system [79]. However, this observation is unconvincing and cannot be directly applied to the much more complicated GaN-on-Si (111) system with SLs structure due to the totally mismatched stress levels of the GaN on the GaN/Si and on sapphire system, due to the difference of lattice mismatch and thermal expansion mismatch in these two systems. Ni et al. [80] investigated the stress in GaN on Si with SLs structure to understand the internal stress engineering mechanism.

Two series samples were grown on silicon substrate. In series A, AlN/GaN SLs with varying periods (4, 8, 30, 40) were grown, while, in series B, 30 periods AlN/GaN SLs with a different relative AlN thickness (from ~8% to ~30%) were grown by varying the AlN or GaN thicknesses. The relative AlN thickness was described as AlN% = $d_{AlN}/(d_{AlN} + d_{GaN}) \times 100\%$, where d_{AlN} and d_{GaN} denote the thickness of AlN and GaN in the SLs, respectively. Next, a ~1000 nm unintentionally doped GaN layer was grown on the SLs layer, followed by an AlGaN barrier layer of ~25% Al composition (Figure 19.24).

The stress of GaN on Si in an AlN/GaN SLs structure was investigated by measuring the micro-Raman scattering at room temperature. The GaN/Si system with AlN/GaN SLs shows a specific Raman spectrum for the GaN E_2^H mode. The stress of the GaN is very different in the top GaN layer than it is in the AlN/GaN SLs. The stress in the top GaN layer is dependent on the periods of the AlN/GaN SLs and the relative AlN thickness of the SLs. The micro-Raman scattering measurements of series A are shown in Figure 19.25 in which A_1 (LO), E_2 (high) modes of GaN, the modes in the silicon substrate as well as the E_2 (high) mode of AlN can be observed. Meanwhile, unlike the E_2 mode peak of GaN observed in the GaN-on-Si system with other stress engineering layers' schemes (such as step-graded/linearly graded AlGaN [31,63] or LT-AlN [81]), a satellite peak of E_2 can be found along with the E_2 (high) mode of GaN in all the

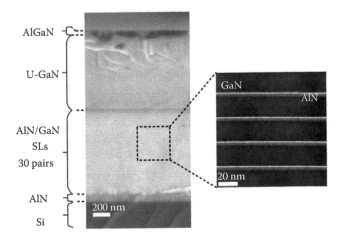

FIGURE 19.24 Cross-sectional scanning electron microscope (SEM) image of the samples investigated with AlN/ GaN SLs. The inset shows the enlarged TEM image of the AlN/GaN SLs. (After Ni, Y. et al., *Jpn. J. Appl. Phys.*, 54, 015505, 2015, Reproduced with permission from the Japan Society of Applied Physics.)

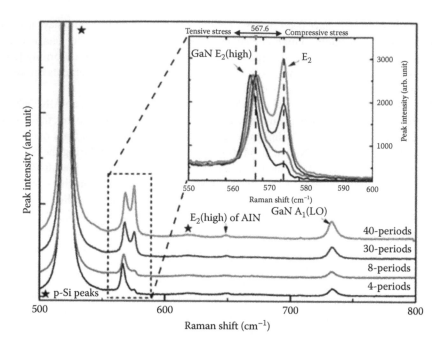

FIGURE 19.25 Raman spectrum obtained to find the E_2 (high) phonon peak shift and the satellite peak mode E2 phonon line with respect to SLs periods. Inset: Enlarged view of the GaN E_2 (high) phonon peak shift and the satellite peak mode E_2. (After Ni, Y. et al., *Jpn. J. Appl. Phys.*, 54, 015505, 2015, Reproduced with permission from the Japan Society of Applied Physics.)

investigated samples with SLs (see Figure 19.25). Although this satellite peak has always been observed by another group in their research of GaN-on-Si with SLs structure [82], its origin has not been explained. It is obvious from the inset of Figure 19.25 that the satellite peak E_2 phonon line frequency (575.7 1/cm) does not depend on the numbers of SLs periods. Nevertheless, the intensity of the abnormal satellite peak E_2 phonon line increases with the numbers of SLs periods, and this intensity can be even larger than the GaN E_2 (high) when the SLs periods equal 40. This phenomenon clearly shows that the satellite peak E_2 has a relation with the AlN/GaN SLs.

Micro-Raman scattering measurements were also carried out on samples in series B with different relative AlN thicknesses in SLs. As shown in detail in the inset of Figure 19.26, the peak of the satellite peak E_2 shifts to a lower frequency toward the GaN E_2 (high), while the relative AlN thickness in AlN/GaN SLs decreases. Actually, a similar movement of the satellite peak, along with the E_2 (high) mode, was observed in the study of AlGaN alloy material [81–83] by varying the Al composition, which was also accompanied by a major shift of the A_1(LO) mode (from ~750 to ~900 1/cm) [82,83]. In the AlN/GaN SLs system, formation of AlGaN alloy may possibly occur during the switching between the TMAl flow and the TMGa flow, and this formation can be judged by observing the shift of A_1(LO). Gleize et al. [79] used the lesser shift properties of the A_1(LO) mode in the Raman spectra of GaN/AlN SLs on AlN/Sapphire template system to exclude the possibility of the formation AlGaN in the AlN/GaN interface (alloy effects). In series B samples, this frequency shift of the A_1 (LO) peak was observed to be very small (1.9 1/cm) by varying the relative AlN thickness in SLs (see Figure 19.26) [78]. So, the possibility of AlGaN alloying effects can be definitely ruled out.

It can be seen in Figure 19.26 that, when the AlN relative thickness drops to 8%, the GaN E_2 (high) peak and the satellite peak E_2 almost overlapped so that we can observe a tendency for these two peaks to merge to into one peak if the AlN relative thickness continues to decrease down to 0%. This indicates that the satellite peak has the same origin as its "mother" peak, which originates from the

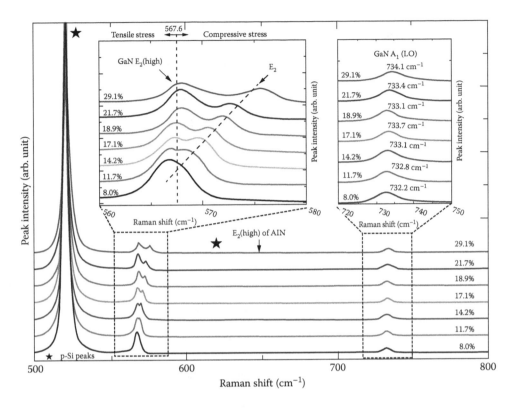

FIGURE 19.26 Raman spectra obtained to find the E_2 (high) phonon peak shift with respect to the relative AlN thickness in SLs. Inset: Partial enlarged view of Raman spectrum to observe the phonon peak shift. (After Ni, Y. et al., *Jpn. J. Appl. Phys.*, 54, 015505, 2015, Reproduced with permission from the Japan Society of Applied Physics.)

GaN material. Combined with the previous result that the satellite peak is related to the SLs presented in Figure 19.25, this satellite peak may come from the GaN layers in the AlN/GaN SLs, and the GaN E_2 peak (mother peak) comes from the top GaN layer of the epi-structure. The origin of the satellite peak was confirmed by comparing the micro-Raman scattering spectra of two 60 pairs of AlN/GaN SLs samples with and without a top GaN layer (see Figure 19.27). There is only one strong E_2 peak of GaN for the sample without a top GaN layer on SLs. But, in the sample with a top GaN layer on SLs, two E_2 peaks can be clearly observed. This result confirmed that the main GaN mode peak E_2 (high) is from the top GaN layer, while the satellite E_2 peak comes from the GaN layers in the SLs structure and not the top GaN layer. Thus, these two peaks both come from the E_2 (high) Raman scattering of GaN. The GaN E_2 (high) peak splits into two peaks when two kinds of GaN layer with differing stresses exist in the epi-system.

The stress states in GaN layers with different SLs periods were extracted from Figure 19.25 and illustrated in Figure 19.28b. By varying the SLs periods with a fixed relative AlN thickness of SLs, the compressive stress in SLs remains the same, while only the stress in the top GaN layer changes from tensile to compressive. This proves that the stress in the GaN can be controlled by varying the numbers of SLs periods.

On the other hand, the stresses in the GaN layers extracted from Figure 19.26 are illustrated in Figure 19.28a. It was found that the top GaN stress changes from tensile to compressive according to the increase of the relative AlN thickness in the SLs. It is also shown that the GaN in SLs is always compressively stressed, which is linearly dependent on the relative AlN thickness in SLs. Furthermore, the change of the stress in GaN in SLs is greater than that in the top GaN layer.

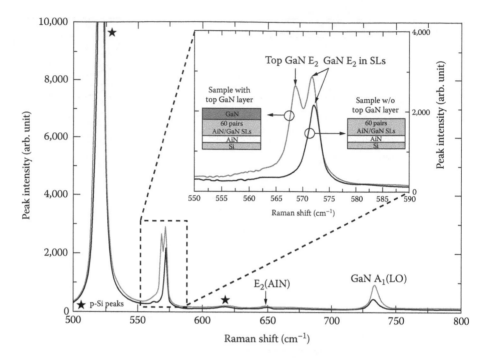

FIGURE 19.27 Micro-Raman scattering spectra of two samples with and without a top GaN layer. Inset: Partial enlarged view of the E_2 peak difference between the samples with and without a top GaN layer. (After Ni, Y. et al., *Jpn. J. Appl. Phys.*, 54, 015505, 2015, Reproduced with permission from the Japan Society of Applied Physics.)

Generally, after the growth of the GaN layer of the first SL period on the HT-AlN/Si (see Figure 19.29), the thin AlN of the first SL period grows coherently on the GaN below in a fully stressed state. If the AlN is thicker, it may become partially relaxed. If the GaN of the next SL period grows with the new in-plane lattice parameter (a_{AlN}) provided by the AlN below, it will be under compressive stress due to the fact that a_{AlN} < a_{GaN}. Obviously, a thicker AlN in SLs will have a lattice parameter a_{AlN2} more close to a relaxed one, a_{AlN}, than a thinner one, a_{AlN1} (see Figure 19.29, a_{AlN1} > a_{AlN2} > a_{AlN}), which can introduce more compressive stress in the subsequently grown GaN in each SL period. Therefore, a thicker GaN in each SL can be affected by the compressive stress provided by AlN, which results in the shift of the E_2 peak of GaN in SLs (satellite peak) to a higher frequency. Furthermore, the top GaN grown on the SLs will obtain a higher counterbalance compressive stress from AlN/GaN SLs with a thicker AlN layer in each period, which leads to the shift of the top GaN E_2 (high) peak to a higher frequency. On the basis of this discussion, the stress in GaN in SLs can only be affected by the thickness of AlN in SLs and is independent of the number of SL periods. Therefore, it can be understood that, by varying the number of SLs periods with a fixed relative AlN thickness, the compressive stress that accumulates in the GaN of each period will not change, such that the E_2 peak of GaN in SLs will always remain the same (see Figure 19.28b). A larger number of SL periods mean that many more GaN layers exist in the SLs, which will result in a higher intensity of the Raman scattering signal (see the inset of Figure 19.25). Furthermore, increasing the number of SLs periods can provide a greater counterbalancing force to compensate for the tensile stress in the top GaN during cooling, which can result in the frequency shift of the top GaN E_2 peak. This phenomenon can be explained in terms of the geometry-related stiffness coefficient KSLs of the AlN/GaN SLs, which can directly affect the stress in the top GaN layer. The stiffness coefficient (SFC) of an epitaxial film, K, can be described as [84,85]

$$K = A \times \frac{E}{L} \tag{19.1}$$

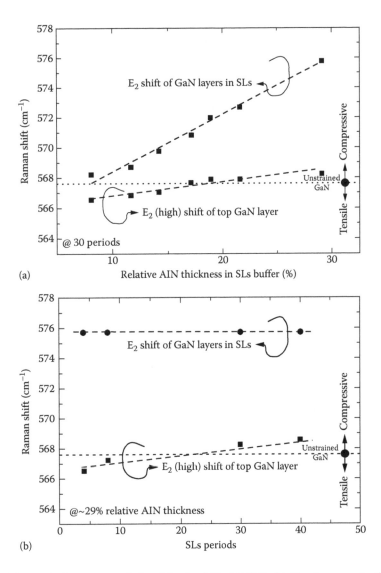

(a)

(b)

FIGURE 19.28 The residual stress was calculated for the shift of GaN E$_2$ (high) using $\sigma_{xx} = \Delta\omega/k_y$ for (a) various relative AlN thicknesses in SLs (b) and different SLs periods. (After Ni, Y. et al., *Jpn. J. Appl. Phys.*, 54, 015505, 2015, Reproduced with permission from the Japan Society of Applied Physics.)

where A, E, and L denote the cross-sectional area, the (tensile) elastic modulus (or Young's modulus), and the length of the film, respectively. Thus, the SFC of GaN/AlN SL can be described as

$$K_{SLs} = A \times \frac{E_{SLs}}{L} = (n \cdot t \cdot W) \times \frac{E_{SLs}}{L} \qquad (19.2)$$

where n, t, W, and E_{SLs} denote the number of SL periods, the thickness of one SL period, and the width and elastic modulus of SLs, respectively (see Figure 19.29). The elastic modulus of SLs, E_{SLs}, is determined by the elastic modulus of AlN and GaN in SLs. According to Vegard's law [86,87], the elastic modulus of SLs, E_{SLs}, can be described as

$$E_{SLs} = \text{AlN\%} \cdot E_{AlN} + (1 - \text{AlN\%}) \cdot E_{GaN} \qquad (19.3)$$

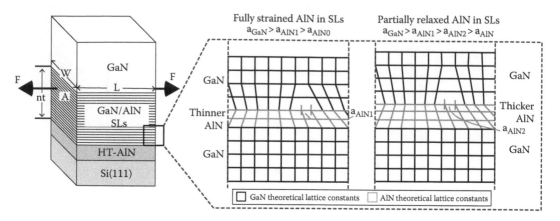

FIGURE 19.29 Comparison of lattice mismatch superlattices grown with different AlN thicknesses. (After Ni, Y. et al., *Jpn. J. Appl. Phys.*, 54, 015505, 2015, Reproduced with permission from the Japan Society of Applied Physics.)

where AlN%, E_{AlN}, and E_{GaN} denote the relative AlN thickness, the elastic modulus of AlN, and the elastic modulus of GaN, respectively. Thus, SFC can be further described as

$$K_{SLs} = (n \cdot t \cdot W) \times \frac{\left(AlN\% \cdot E_{AlN} + (1 - AlN\%) \cdot E_{GaN} \right)}{L} \qquad (19.4)$$

As we can see from Equation 19.4, by changing the number of SL periods n, the SFC of SLs can be changed, which means that the counterbalancing force provided by the SLs to the top GaN layer can also be changed. Thus, it can be easily understood that the stress in the top GaN layer can be effectively adjusted by the SFC of the SLs, which is related to both the number of periods, n, and the relative AlN thickness, AlN%, of SLs.

19.3.2 Growth on Patterned Si Substrate

The growth of GaN on the patterned Si substrate can reduce the curvature of the entire wafer and partially reduce the stress in epilayers by selective growth. In 2000, Yang et al. [88] grew a GaN-based device on SiO_2 masked Si substrate by selective area growth. However, cracks were still present if there was no stress-engineering structure. In 2001, Dadgar et al. [40] presented a crack-free GaN based LED on 100 μm × 100 μm fields patterned Si substrates with total 3.6 μm thick epilayer, in which the pattern was achieved with a grid Si_xN_y mask on a thin pre-deposited GaN layer. They proposed that the patterned Si substrates could strongly reduce stress in the GaN layer. Honda et al. [41] also obtained GaN without any cracks on an SiO_2 grid patterned Si substrate by selective epitaxy (Figure 19.30). They indicated that GaN grown on the opened windows exhibited no cracks if the window size was less than 400 × 400 μm² and that the best quality of GaN with a thickness of 1.5 μm was obtained in the 200 × 200 μm² window patterns. However, this method suffered from the issue of a thick ridge appearing at the pattern edge, which may cause certain problems in the device process presented in Figure 19.30.

Besides using a dielectric grid mask, a patterned Si substrate can also be formed with an etched trench grid. Zhang et al. [89] reported crack-free InGaN/GaN blue LED structures with GaN buffer layers 2 μm thick were grown on 340 × 340 μm² square islands patterned Si substrate, which was separated by 20 μm wide and 3 μm deep trenches. The GaN-based films were separated at the edge of the square island because the depth of the trench is greater than the thickness of epilayers. In this way, the tensile stress caused by the thermal expansion mismatch can be released in the entire wafer. Increasing the width ratio of the island and trench can reduce the growth edge effect.

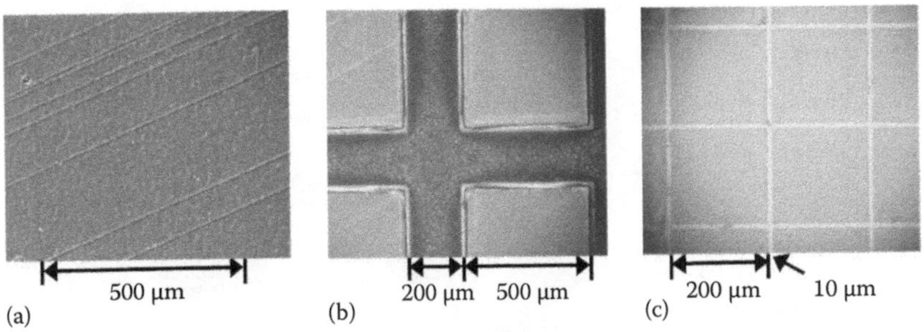

FIGURE 19.30 Microscope pictures of GaN grown on Si: (a) substrate without pattern, (b) substrate with 500 μm wide square windows separated by a 200 μm wide SiO$_2$ mask, (c) substrate with 200 μm square windows separated by a 10 μm wide SiO$_2$ mask. The crack density is reduced as the window size decreases. Ridge growth was observed in (b) and (c); no cracks were found in (c). (Reprinted with permission from Honda, Y., Kuroiwa, Y., Yamaguchi, M., and Sawaki, N., Growth of GaN free from cracks on a (111) Si substrate by selective metalorganic vapor-phase epitaxy, *Appl. Phys. Lett.*, 80, 222, Copyright 2002, American Institute of Physics.)

19.3.3 Impurity Doping–Induced Stress

19.3.3.1 N-Type Doping (Si Doping)

Si doped n-type GaN is an essential layer for certain semiconductor devices, such as LEDs and laser diodes. An additional tensile stress will be introduced due to the Si doping. The stress was found to increase with the Si-doping concentration [54,90]. Heavy Si doping can cause crack generation. Si δ-doping was reported to be an effective approach for reducing the tensile stress [54,91] and the screw dislocation density [92]. The electrical properties of Si periodic δ-doped GaN grown on sapphire substrate were investigated [93]. By inserting Si δ-doped layers, the electrostatic discharge properties of LED were significantly improved due to the enhanced current spreading property and material quality [94]. For GaN on a Si substrate, the tensile stress reducing effect of Si periodic δ-doping was reported by Wang et al. [95]; however, it was not possible to obtain electrical properties from the Si periodic δ-doped GaN due to cracks in the GaN film. In the study by Schenk and coworkers [96], however, the tensile stress in the GaN film estimated from the wafer bow was reduced, there was a ~15 times drop in electron density and ~three times increase in resistivity presented in Si periodic δ-doped samples compared with uniformly doped samples with the same doping ratio, which seems deviant and is undesirable in practical applications. Moreover, if this electrical deterioration does exist, the Si-doping level has to be enlarged to achieve the same electron density, which will increase the tensile stress. On the other hand, to achieve such a low electron density, one can simply lower the Si-doping level to reduce the tensile stress. Thus, only with comparable or even higher electron density is the stress comparison between Si periodic δ-doped GaN and Si uniformly doped GaN meaningful.

To investigate the effects of periodic δ-doping closely, especially on the stress, Xiang et al. [97] prepared two kinds of sample: Si periodic δ-doped GaN (Figure 19.31) and Si uniformly doped GaN with varying doping levels on Si (111) substrate. The electrical properties, the stress states, the crystalline qualities, and the optical properties were investigated. Compared with uniformly doped samples, only a slight decline of the electron density and conductivity was observed in the Si periodic δ-doped GaN with the same doping ratio (Figure 19.32). Moreover, the tensile stress in the periodically δ-doped GaN films was significantly reduced compared with that of uniformly doped samples, whether with the same doping ratio or with a similar electron density (Figure 19.33).

19.3.3.2 P-Type Doping (Mg Doping)

In addition to the lattice mismatch and thermal mismatch between GaN and the substrate, dopants such as Si and Mg will introduce additional stress into the epilayer. Tensile stress induced by Si has been discussed.

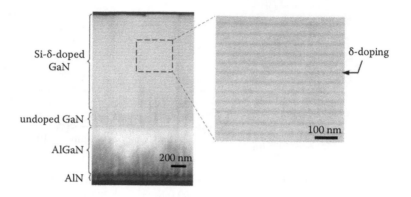

FIGURE 19.31 Cross-sectional SEM images of sample C3. (Reprinted with permission from *J. Cryst. Growth*, 387, P. Xiang, Y. Yang, M. Liu, W. Chen, X. Han, Y. Lin, et al., Influences of periodic Si delta-doping on the characteristics of n-GaN grown on Si (111) substrate, 106, Copyright 2014, with permission from Elsevier.)

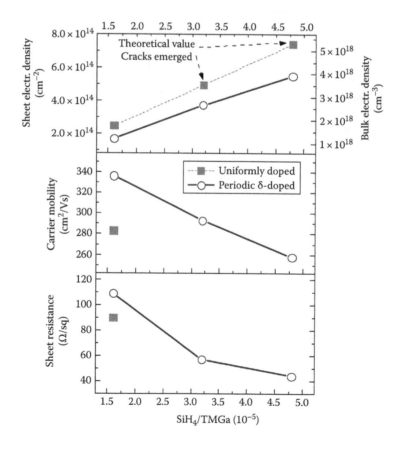

FIGURE 19.32 Electrical properties of Si uniformly doped and Si periodically δ-doped samples. (Reprinted with permission from *J. Cryst. Growth*, 387, P. Xiang, Y. Yang, M. Liu, W. Chen, X. Han, Y. Lin, et al., Influences of periodic Si delta-doping on the characteristics of n-GaN grown on Si (111) substrate, 106, Copyright 2014, with permission from Elsevier.)

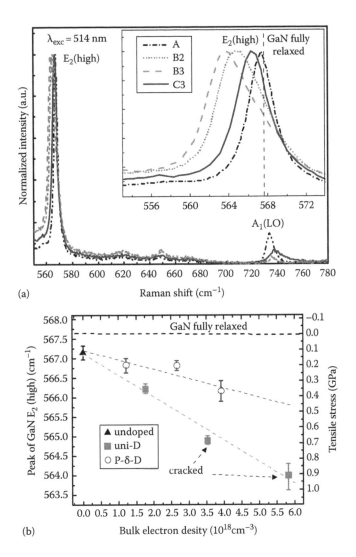

FIGURE 19.33 (a) Raman spectra of sample A (undoped), B2, B3 (Si uniformly doped), and C3 (Si periodically δ-doped), (b) tensile stress of undoped, Si uniformly doped, and Si periodically δ-doped samples estimated from the shift of GaN E_2 (high) phonon. (Reprinted with permission from *J. Cryst. Growth*, 387, P. Xiang, Y. Yang, M. Liu, W. Chen, X. Han, Y. Lin, et al., Influences of periodic Si delta-doping on the characteristics of n-GaN grown on Si (111) substrate, 106, Copyright 2014, with permission from Elsevier.)

Periodic δ-doping was found to be an effective way to reduce the tensile stress [97–99]. As for Mg doping, most research has been focused on the electrical properties, only a few reports discuss the stress. In fact, just like periodic Si δ-doping, periodic Mg δ-doping not only improves the crystalline quality, but also can control the stress effectively. Three series of samples were grown on a silicon substrate: sample A (undoped), sample B (Mg uniformly doped with difference concentrations), and sample C (periodically Mg δ-doped with difference concentrations). As shown in Figure 19.34, samples C show better crystalline quality in comparison with samples B.

Micro-Raman spectra of the three series of samples measured at room temperature are shown in Figure 19.35a. The tensile stresses estimated from the shifts of the GaN E_2 (high) mode of all the samples are shown in Figure 19.35b. For the undoped GaN samples, the tensile stress is 0.09 GPa, implying

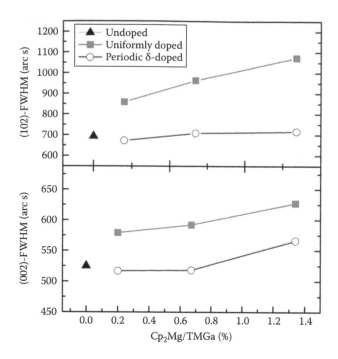

FIGURE 19.34 FWHMs of GaN(002) and GaN(102) ω-scans of undoped, Mg uniformly doped, and Mg periodically δ-doped GaN.

successful stress control with the linearly graded AlGaN buffer [31]. For Mg uniformly doped GaN, when the doping ratio is 0.20%, the additional tensile stress is very small, which agrees with the in situ curvature measurement results reported by Krost et al. [54]. When further increasing the doping ratio to 0.67% and 1.34%, an obvious redshift can be observed, indicating considerable tensile stress, 0.43 and 0.64 GPa, respectively. Also, cracks appeared on sample B3 under high tensile stress. Broader full width at half maximum FWHMs of the E_2 (high) peak also were shown in the spectra of samples B2 and B3, implying increased disorder in the GaN introduced by Mg doping. The Raman shift demonstrates that tensile stress is introduced by Mg doping in GaN and increases with the increase of the Mg doping ratio. For Mg periodically δ-doped GaN, little frequency shift of the GaN E_2 (high) phonon mode can be observed in all samples, even at a high Mg-doping ratio, which demonstrates that periodic Mg δ-doping can effectively control stress.

In addition, the periodically Mg δ-doped GaN also exhibits superior electrical properties to those of Mg uniformly doped GaN, as shown in Table 19.1. For Mg uniformly doped GaN, 3.1×10^{17} 1/cm^3 to 3.6×10^{16} 1/cm^3, when the Cp$_2$Mg /TMGa ratio increases from 0.20% to 0.67% due to the self-compensation effect [100,101]. For periodically Mg δ-doped GaN, the hole densities of samples C1 and C2 are 4.8×10^{17} and 4.1×10^{17} 1/cm^3, respectively, which are obviously greater than those of Mg uniformly doped samples, indicating reduced Mg activation energy and less self-compensation.

19.3.3.3 High Resistive Doping (C-Doping)

In order to achieve high-performance GaN-based power switching devices, a high resistive GaN on Si (111) epitaxial template with high breakdown voltage (BV) and low leakage current is required [76]. Two aspects have been the main focus for reaching this goal. One is to grow a thicker epitaxial layer [74,102,103] and the other is to boost the epitaxial layer's resistance [104–109]. Introducing carbon into the GaN (C-doping) to compensate for unintentional background donor impurities [104–109] is an effective solution by which to boost the epitaxial layer's resistance. Specifically, by introducing carbon impurities as well as varying the

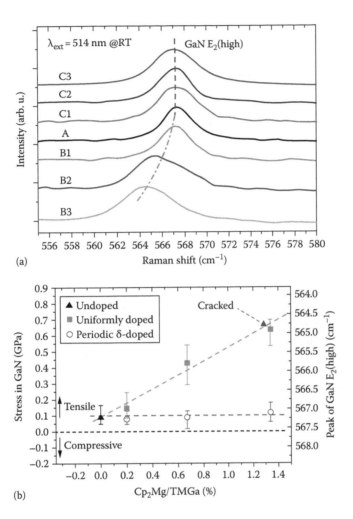

FIGURE 19.35 (a) Raman spectra around GaN E$_2$ (high) mode of all samples; (b) tensile stresses of all samples estimated from the shifts of GaN E$_2$ (high) mode. (In Figure 19.2b, spectra in three different areas were measured for each sample.)

TABLE 19.1 Electrical Properties of Mg Uniformly Doped (B Series) and Mg Periodic δ-Doped GaN (C Series)

Sample	Cp$_2$Mg/TMGa	Hole Concentration (1/cm³)	Mobility (cm²/V·s)	Sheet Resistivity (Ω/sq)
B1	0.20%	3.1×10^{17}	13.3	1.7×10^4
B2	0.67%	3.6×10^{16}	4.0	4.9×10^5
C1	0.20%	4.8×10^{17}	9.6	1.6×10^4
C2	0.67%	4.1×10^{17}	12.4	1.4×10^4
C3	1.34%	3.2×10^{17}	6.9	3.1×10^4

growth temperature, pressure, and/or V/III ratio [105,110], the background unintentionally doped donors in the GaN buffer can be partially compensated, which can form highly resistive buffer layers [109,111]. Nonetheless, the stress in the GaN-on-Si system is especially sensitive to adjustment of the growth parameters during the carbon-doping process [106,112] and is difficult to control compared with the other substrate system (sapphire or SiC) [71], which can easily lead to a rough surface, and bowing or cracking of the wafer [49,80].

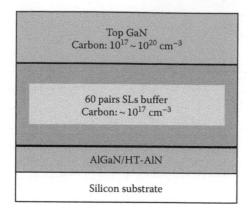

FIGURE 19.36   Schematic of the GaN-on-silicon structure with different concentrations of carbon doping.

From this aspect, the stress control issue associated with carbon doping is becoming very important and worthy of consideration during the growth processes. In order to reveal the correlation between the carbon-doping concentration and stress, GaN on silicon structures with different concentrations of carbon doping was grown by MOCVD. A schematic of the GaN-on-silicon structure with different carbon-doping concentrations is shown in Figure 19.36. By adjusting the chamber pressure and V/III ratio of the GaN during growth, the carbon concentration in the SLs is approximately 2.5×10^{17} 1/cm^3, while the carbon concentrations in the top GaN buffer vary from approximately 10^{17} to 10^{20} 1/cm^3.

The Raman shift for samples was extracted from the E$_2$ (high) phonon peak and plotted against the carbon concentrations (see Figure 19.37). It can also be seen that the GaN in the SLs is always compressively stressed and almost independent of the carbon concentration of the top GaN layer, while, for the top GaN layer, the stress first changes linearly from tensile to compressive according to the increase in carbon concentration. However, at relatively higher carbon concentrations, the carbon-introduced stress in GaN becomes tensile. It is necessary to make a detailed study of the carbon-doping mechanism on stress in a GaN-on-Si (111) system.

A possible reason for stress being dependent on the carbon-doping concentration is believed to be attributable to the incorporation site in the GaN lattice. It is well known that carbon, as a substitute atom in GaN, may exist in either a nitrogen position (C_N) or a gallium position (C_{Ga}) (see Figure 19.12) [113–116]. When the carbon atoms exist in the nitrogen position, the larger covalent radius of carbon ($r_C = 0.077$ nm) compared with that of nitrogen ($r_N = 0.074$ nm) leads to an expansion of the lattice parameters. Then, compressive stress will generate relaxation of the lattice distortion of the GaN. On the other hand, tensile stress will be introduced when the carbon replaces gallium due to the radius of carbon atoms being smaller than that of a gallium atom ($r_{Ga} = 0.126$ nm). This speculation is further confirmed according to the isotropic elastic theory [117]; the stress in the GaN lattice induced by impurities being introduced during doping can be expressed by the equation

$$\varepsilon = \frac{c - c_0}{c_o} = \left[1 - \left(\frac{r_d}{r_h} \right)^3 \right] \cdot C / 3n$$

Here, ε, r_d, r_h, n, C, c, and c_0 denote the stress in the GaN, the radius of dopant atoms (here, the carbon atom), the radius of host atoms, the number of lattice sites of the host matrix, the dopant concentration, the stressed and unstressed lattice constant, respectively.

When the carbon atoms are mainly intended to replace the N atom in GaN (see Figure 19.38a), due to the radius of carbon atoms (dopant atoms) at $r_s = r_c = 0.077$ nm being larger than the radius

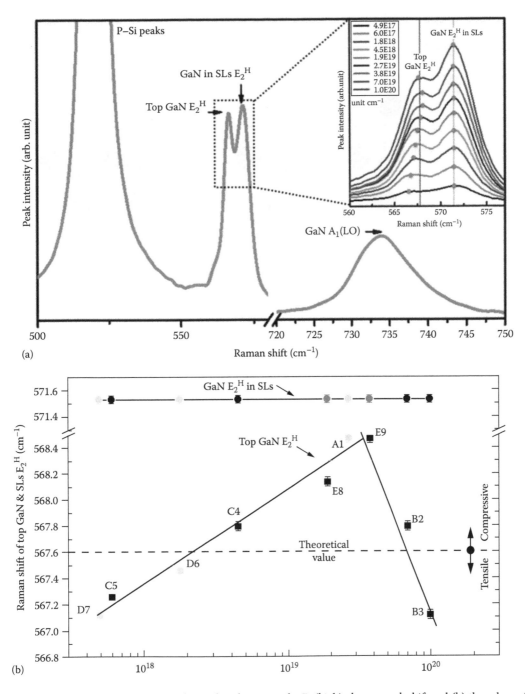

FIGURE 19.37 (a) Raman spectrum obtained to determine the E_2 (high) phonon peak shift and (b) the values of the top GaN E_2 (high) of the investigated samples were extracted and compared with the theoretical value of GaN.

of nitrogen atoms (host atoms) at $r_h = r_N = 0.074$ nm, according to the previous equation, the stress ε in GaN is less than zero, which means that, with the increase of carbon, it will introduce more compressive stress into the GaN. As shown in Figure 19.11, the stress in the top GaN buffer layer changed from tensile stress to compressive stress, regardless of whether the V/III ratio, growth pressure, or temperature was lowered.

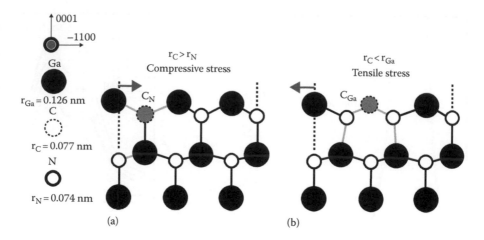

FIGURE 19.38 The relationship between stress and the carbon incorporation site in a GaN lattice, (a) N site and (b) Ga site.

On the other hand, when the carbon atoms replacing the Ga become deep enough and dominant (see Figure 19.38 b), due to the radius of carbon atoms (dopant atoms) $r_s = r_c = 0.077$ nm being smaller than the radius of gallium atom (host atoms) $r_h = r_{Ga} = 0.126$ nm, according to the previous equation, the stress ε in the GaN is greater than zero, which means that, with the further increase of carbon, it will begin to introduce more tensile stress into the GaN.

Generally, the incorporation site of carbon in GaN, whether it is introduced in the nitrogen position (C_N) or the gallium position (C_{Ga}), is determined by the formation energy required to obtain the smallest total free energy [113–116]. On the basis of first principles calculation [114], for the unintentional doping or very low carbon concentration doping sample, the existence of the n-type impurity (such as N vacancy, oxygen and silicon) causes the Fermi level of the GaN to be near the conduction band. Then, the formation energy of C_N is lower than that of C_{Ga} and the carbon atoms are intended in the main to replace the N atoms in the GaN. However, with the increase in carbon concentration, more electrons are compensated for in the GaN buffer, so the Fermi level continues to move toward to the valence band. At the same time, there is a gradual reduction in the formation energy of the C_{Ga}, which means more C_{Ga} may be introduced to the GaN. Therefore, the stress in the top GaN layer is shown to be carbon-doping concentration dependent. Carbon doping could strongly influence the stress in the top GaN buffer. So, in order to achieve a high resistance buffer GaN/Si template, careful consideration should be given to the stress state of the GaN-on-Si system when introducing carbon.

19.4 Summary

The most recent progress and development of GaN-based optoelectronic devices on a Si substrate was reviewed, in which the key issues regarding both material growth and device fabrication were summarized in detail. Due to the requirements of the solid-state lighting and power switching market, and with regard to the compatibility of a GaN/Si system with modern Si-based integrated circuit platforms, the material growth will tend to develop along the direction of increasing the wafer size and epilayer thickness. For LED applications, the development of GaN-based LEDs on Si substrate faces strong competition from the conventional LEDs on a sapphire substrate. We believe GaN-based materials and devices on large-scale Si substrates are the most likely candidates for future applications.

References

1. G. Supratik and N.A. Bojarczuk, Ultraviolet and violet GaN light emitting diodes on silicon, *Appl Phys Lett*, 72, 415 (1998).
2. C.A. Tran, A. Osinski, R.F. Karlicek, and I. Berishev, Growth of InGaN/GaN multiple-quantum-well blue light-emitting diodes on silicon by metalorganic vapor phase epitaxy, *Appl Phys Lett*, 75, 1494 (1999).
3. I. Hiroyasu, G.-Y. Zhao, and N. Naoyuki, GaN on Si substrate with AlGaN/AlN intermediate layer, *Jpn J Appl Phys*, 38, L492 (1999).
4. B.J. Zhang, T. Egawa, H. Ishikawa, N. Nishikawa, T. Jimbo, and M. Umeno, InGaN multiple-quantum-well light emitting diodes on Si (111) substrates, *Phys Stat Sol (a)*, 188, 151 (2001).
5. A. Dadgar, A. Strittmatter, J. Bläsing, et al. Metalorganic chemical vapor phase epitaxy of gallium-nitride on silicon, *Phys Stat Sol (c)*, 0, 1583 (2003).
6. H.-M. Wang, J.-P. Zhang, C.-Q. Chen, Q. Fareed, J.-W. Yang, and M. Asif Khan, AlN/AlGaN super-lattices as dislocation filter for low-threading-dislocation thick AlGaN layers on sapphire, *Appl Phys Lett*, 81, 604 (2002).
7. Takashi Egawa, Baijun Zhang, and Hiroyasu Ishikawa, High performance of InGaN LEDs on (111) silicon substrates grown by MOCVD, *IEEE Electron Device Lett*, 26, 169 (2005).
8. Baijun Zhang, Takashi Egawa, Hiroyasu Ishikawa, Yang Liu and Takashi Jimbo, Thin-film InGaN multiple-quantum-well light-emitting diodes transferred from Si (111) substrate onto copper carrier by selective lift-off. *Appl Phys Lett*, 86, 071113 (2005).
9. Jingting Wei, Baijun Zhang, Gang Wang, Bingfeng Fan, Yang Liu, Wentao Rao, Zhicong Huang, Weimin Yang, Tufu Chen and Takashi Egawa, Vertical GaN-based light-emitting diodes structure on Si (111) substrate with through-holes. *Jpn J Appl Phys*, 49, 072104 (2010).
10. M.K. Kelly, R.P. Vaudo, V.M. Phanse, L. Görgens, O. Ambacher, and M. Stutzmann, Large free-standing GaN substrate by hydride vapor phase epitaxy and laser-induced liffoff, *Jpn J Appl Phys*, 38, L217 (1999).
11. M.K. Kelly, O. Ambacher, R. Dimitrov et al., Optical process for lift off of Group III-nitride films, *Phys Stat Sol (a)*, 159, R3 (1997).
12. W.S. Wong, D.S. Timothy, N.W. Cheung, and N.M. Johnson, $In_xGa_{1-x}N$ light emitting diodes on Si substrates fabricated by Pd–In metal bonding and laser lift-off, *Appl Phys Lett* 77, 2822 (2000).
13. B.S. Tan, S. Yuan, and X.J. Kang, Performance enhancement of InGaN light-emitting diodes by laser lift-off and transfer from sapphire to copper substrate, *Appl Phys Lett*, 84, 2757 (2004).
14. B. Zhang, T. Egawa, H. Ishikawa, L. Yang, and T. Jimbo, Thin-film InGaN multiple-quantum-well light-emitting diodes transferred from Si (111) substrate onto copper carrier by selective lift-off, *Appl Phys Lett* 86, 071113 (2005)
15. K.M. Wong, X. Zou, P. Chen, K.M. Lau, Transfer of GaN-based light-emitting diodes from silicon growth substrate to copper, *IEEE Electron Device Lett*, 31, 132 (2010).
16. K.M. Lau, K.M. Wong, X. Zou, and P. Chen, Performance improvement of GaN-based light-emitting diodes grown on patterned Si substrate transferred to copper, *Opt Express* 19, A956 (2011).
17. Y. Yang, G. Hu, P. Xiang, M. Liu, W. Chen, X. Han et al., Enhancement of GaN-based light-emitting diodes transferred from Si (111) substrate onto electroplating Cu submount with TiO2/SiO2-Al omnidirectional reflector, *J Display Technol* 10, 1064 (2014).
18. R. Luo, W. Rao, T. Chen, P. Xiang, M. Liu, W. Yang et al., Vertical InGaN multiple quantum wells light-emitting diodes structures transferred from Si (111) substrate onto electroplating copper sub-mount with through-holes, *Jpn J Appl Phys* 51, 012101 (2012).
19. T. Chen, Y. Wang, P. Xiang, R. Luo, M. Liu, W. Yang et al., Crack-free InGaN multiple quantum wells light-emitting diodes structures transferred from Si (111) substrate onto electroplating copper submount with embedded electrodes, *Appl Phys Lett* 100, 241112 (2012).

20. M. Liu, Y. Wang, Y. Yang, X. Lin, P. Xiang, W. Chen et al., Performance improvement of GaN-based light-emitting diodes transferred from Si (111) substrate onto electroplating Cu submount with embedded wide p-electrodes, *Chin Phys B* 24, 038503 (2015).

21. C. Xiong, F. Jiang, W. Fang, L. Wang, H. Liu, and C. Mo, Different properties of GaN-based LED grown on Si (111) and transferred onto new substrate, *Sci China Ser E* 49, 313 (2006).

22. C. Xiong, F. Jiang, W. Fang, L. Wang, C. Mo, and H. Liu, The characteristics of GaN-based blue LED on Si substrate, *J Lumin* 122, 185 (2007).

23. S.-J. Lee, K.H. Kim, J.-W. Ju, T. Jeong, C.-R. Lee, and J.H. Baek, High-brightness GaN-based light-emitting diodes on si using wafer bonding technology, *Appl Phys Exp* 4, 066501 (2011).

24. Q. Liao, Y. Yang, W. Chen, X. Han, J. Chen, W. Zang et al., Fabrication and properties of thin-film InGaN/GaN multiple quantum well light-emitting diodes transferred from Si (111) substrate onto a thin epoxy resin carrier, *J Display Technol* 12, 1602 (2016).

25. H. Ishikawa, K. Asano, B. Zhang, T. Egawa, and T. Jimbo, Improved characteristics of GaN-based light-emitting diodes by distributed Bragg reflector grown on Si, *Phys Stat Sol (a)* 201, 2653 (2004).

26. H. Ishikawa, B. Zhang, K. Asano, T. Egawa, and T. Jimbo, Characterization of GaInN light-emitting diodes with distributed Bragg reflector grown on Si, *J Cryst Growth* 272, 322 (2004).

27. W.S. Wong, T. Sands, N.W. Cheung, M. Kneissl, D.P. Bour, P. Mei, L.T. Romano, and N.M. Johnson, Fabrication of thin-film InGaN light-emitting diode membranes by laser lift-off, *Appl Phys Lett*, 75, 1360 (1999).

28. A. Krost, C. Berger, J. Bläsing, A. Franke, T. Hempel, A. Dadgar, and J. Christen, Strain evaluation in AlInN/GaN Bragg mirrors by in situ curvature measurements and ex situ x-ray grazing incidence and transmission scattering, *Appl Phys Lett* 97, 181105 (2010).

29. H. Ishikawa, T. Jimbo, T. Egawa, GaInN light emitting diodes with AlInN/GaN distributed Bragg reflector on Si, *Phys Stat Sol (c)*, 5, 2086 (2008).

30. Y. Yang, Y. Lin, P. Xiang, M. Liu, W. Chen, X. Han et al., Vertical-conducting InGaN/GaN multiple quantum wells LEDs with AlN/GaN distributed Bragg reflectors on Si(111) substrate, *Appl Phys Exp* 7, 042102 (2014).

31. Y. Yang, P. Xiang, M. Liu, W. Chen, Z. He, X. Han et al., Effect of compositionally graded AlGaN buffer layer grown by different functions of trimethylaluminum flow rates on the properties of GaN on Si (111) substrates, *J Cryst Growth* 376, 23 (2013).

32. M.A. Mastro, Jr., C.R. Eddy, D.K. Gaskill, N.D. Bassim, J. Casey, A. Rosenberg, R.T. Holm, R.L. Henry, and M.E. Twigg, MOCVD growth of thick AlN and AlGaN superlattice structures on Si substrates, *J Cryst Growth* 287, 610 (2006).

33. A. Dadgar, M. Poschenrieder, J. Bläsing, K. Fehse, A. Diez, and A. Krost, Thick, crack-free blue light-emitting diodes on Si(111) using low-temperature AlN interlayers and in situ SixNy masking *Appl Phys Lett* 80, 3670 (2002).

34. H. Marchand, L. Zhao, N. Zhang, B. Moran, R. Coffie, U.K. Mishra, J.S. Speck, S.P. DenBaars, and J.A. Freitas, Metalorganic chemical vapor deposition of GaN on Si (111): Stress control and application to field-effect transistors, *J Appl Phys* 89, 7846 (2001).

35. M.-H. Kim, Y.-G. Do, H. Chol Kang, D.Y. Noh, and S.-J. Park, Effects of step-graded $Al_xGa_{1-x}N$ interlayer on properties of GaN grown on Si (111) using ultrahigh vacuum chemical vapor deposition, *Appl Phys Lett* 79, 2713 (2001).

36. T. Sugahara, J.-S. Lee, and K. Ohtsuka, Role of AlN/GaN multilayer in crack-free GaN layer growth on 5φ Si(111) substrate, *Jpn J Appl Phys* 43, L1595 (2004).

37. F. Eric, B. Beaumont, M. Laügt, P. de Mierry, P. Vennéguès, H. Lahrèche, M. Leroux and P. Gibart, Stress control in GaN grown on silicon (111) by metal organic vapor phase epitaxy, *Appl Phys Lett* 79, 3230 (2001).

38. S. Nakamura, M. Senoh, and T. Mukai, P-GaN/N-InGaN/N-GaN double-heterostructure blue-light-emitting diodes, *Jpn J Appl Phys* 32, L8 (1993).

39. S. Raghavan, X. Weng, E. Dickey, and J.M. Redwing. Effect of AlN interlayers on growth stress in GaN layers deposited on (111) Si, *Appl Phys Lett* 87, 2101 (2005).

40. A. Dadgar, A. Alam, T. Riemann, J. Bläsing, A. Diez, M. Poschenrieder, M. Straßburg, M. Heuken, J. Christen, and A. Krost, Crack-free InGaN/GaN light emitters on Si (111), *Phys Stat Sol (a)* 188, 155 (2001).

41. Y. Honda, Y. Kuroiwa, M. Yamaguchi, and N. Sawaki, Growth of GaN free from cracks on a (111) Si substrate by selective metalorganic vapor-phase epitaxy, *Appl Phys Lett* 80, 222 (2002).

42. S. Zamir, B. Meyler, E. Zototoyabko, and J. Salzman, The effect of AlN buffer layer on GaN grown on (111)-oriented Si substrates by MOCVD, *J Cryst Growth* 218, 181 (2000).

43. M. Poschenrieder, F. Schulze, J. Bläsing, A. Dadgar, A. Diez, J. Christen, and A. Krost, Bright blue to orange photoluminescence emission from high-quality InGaN/GaN multiple-quantum-wells on Si (111) substrates, *Appl Phys Lett* 81, 1591 (2002).

44. C. Kisielowski, J. Krüger, S. Ruvimov, T. Suski, J.W. Ager III, E. Jones, Z. Liliental-Weber, M. Rubin, and E.R. Weber, Strain-related phenomena in GaN thin films, *Phys Rev B* 54, 17745 (1996).

45. J. Li, J.Y. Lin, and H.X. Jiang, Growth of III-nitride photonic structures on large area silicon substrates, *Appl Phys Lett* 88, 171909 (2006).

46. P. Xiang, M. Liu, Y. Yang, W. Chen, Z. He, K.K. Leung, C. Surya, X. Han, Z. Wu, Y. Liu, and B. Zhang, Improving the quality of GaN on Si(111) substrate with a medium-temperature/high-temperature bilayer AlN buffer, *Jpn J Appl Phys* 52, 08JB18 (2013).

47. A. Dadgar, M. Poschenrieder, J. Bläsing, O. Contreras, F. Bertram, T. Riemann et al., MOVPE growth of GaN on Si (111) substrates, *J Cryst Growth*, 248, 556 (2003).

48. A. Krost and A. Dadgar, GaN-based devices on Si, *Phys Stat Sol A*, 194, 361 (2002).

49. H. Amano, M. Iwaya, T. Kashima, M. Katsuragawa, I. Akasaki, J. Han, S. Hearne, J.A. Floro, E. Chason, and J. Figie, Stress and defect control in GaN using low temperature interlayers, *Jpn J Appl Phys* 37, L1540 (1998).

50. A. Krost, A. Dadgar, GaN-based optoelectronics on silicon substrates, *Mater Sci Eng B* 93, 77 (2002).

51. A. Dadgar, J. Bläsing, A. Diez, A. Alam, M. Heuken, and A. Krost, Metalorganic chemical vapor phase epitaxy of crack-free GaN on Si (111) exceeding 1 μm in thickness, *Jpn J Appl Phys* 39, L1183 (2000).

52. A. Dadgar, C. Hums, A. Diez, J. Bläsing, and A. Krost, Growth of blue GaN LED structures on 150-mm Si (111), *J Cryst Growth* 297, 279 (2006).

53. J. Bläsing, A. Reiher, A. Dadgar, A. Diez, and A. Krost, The origin of stress reduction by low-temperature AlN interlayers, *Appl Phys Lett* 81, 2722 (2002).

54. A. Krost, A. Dadgar, G. Strassburger, and R. Clos, GaN-based epitaxy on silicon: Stress measurements, *Phys Stat Sol A* 200, 26 (2003).

55. S. Raghavan and J.M. Redwing, In situ stress measurements during the MOCVD growth of AlN buffer layers on (111) Si substrates, *J Cryst Growth* 261, 294 (2004).

56. Z. He, Y. Ni, F. Yang, J. Wei, Y. Yao, Z. Shen et al., Investigations of leakage current properties in semi-insulating GaN grown on Si(111) substrate with low-temperature AlN interlayers, *J Phys D: Appl Phys* 47, 045103 (2013).

57. H. Ishikawa, G.Y. Zhao, N. Nakada, T. Egawa, T. Soga, T. Jimbo, and M. Umeno, High-quality GaN on Si substrate using AlGaN/AlN intermediate layer, *Phys Stat Sol (a)*, 176, 599 (1999).

58. A. Able, W. Wegscheider, K. Engl, and J. Zweck, Growth of crack free GaN on Si (111) with graded AlGaN buffer layers, *J Cryst Growth* 276, 415 (2005).

59. S. Raghavan and J.M. Redwing, Growth stresses and cracking in GaN films on (111) Si grown by metalorganic chemical vapor deposition. II. Graded AlGaN buffer layers, *J Appl Phys* 98, 023514 (2005).

60. S. Raghavan, X. Weng, E. Dickey, and J.M. Redwing, Correlation of growth stress and structural evolution during metalorganic chemical vapor deposition of GaN on (111) Si, *Appl Phys Lett* 88, 041904 (2006).

61. D. Zhu, C. McAleese, K.K. McLaughlin, M. Häberlen, C.O. Salcianu, E.J. Thrush et al., GaN-based LEDs grown on 6-inch diameter Si (111) substrates by MOVPE, *SPIE OPTO: Integrated Optoelectronic Devices*, International Society for Optics and Photonics, 7231, 723118 (2009).

62. K. Cheng, M. Leys, S. Degroote, B. Van Daele, S. Boeykens, J. Derluyn, M. Germain, G. Van Tendeloo, J. Engelen, and G. Borghs, Flat GaN epitaxial layers grown on Si (111) by metalorganic vapor phase epitaxy using step-graded AlGaN intermediate layers, *J Electron Mater* 35, 592 (2006).

63. C.C. Huang, S.J. Chang, R.W. Chuang, J.C. Lin, Y.C. Cheng, and W.J. Lin, GaN grown on Si (111) with step-graded AlGaN intermediate layers, *Appl Surf Sci* 256, 6367 (2010).

64. D. Zhu, C. McAleese, M. Häberlen, C. Salcianu, T. Thrush, M. Kappers et al., InGaN/GaN LEDs grown on Si (111): Dependence of device performance on threading dislocation density and emission wavelength, *Phys Stat Sol (c)* 7, 2168 (2010).

65. S.C. Choi, J.H. Kim, J.Y. Choi, K.J. Lee, K.Y. Lim, and G.M. Yang, Al concentration control of epitaxial AlGaN alloys and interface control of GaN/AlGaN quantum well structures, *J Appl Phys* 87, 172 (2000).

66. S. Kim, J. Seo, K. Lee, H. Lee, K. Park, Y. Kim, C.-S. Kim, Growth of AlGaN epilayers related gas-phase reactions using TPIS-MOCVD, *J CrystGrowth* 245, 247 (2002).

67. G.S. Huang, H.H. Yao, H.C. Kuo, and S.C. Wang, Effect of growth conditions on the Al composition and quality of AlGaN film, *Mater Sci Eng, B* 136, 29 (2007).

68. R.F. Xiang, Y.Y. Fang, J.N. Dai, L. Zhang, C.Y. Su, Z.H. Wu, C.H. Yu, H. Xiong, C.Q. Chen, Y. Hao, High quality GaN epilayers grown on Si (111) with thin nonlinearly composition-graded AlxGa$_{1-x}$N interlayers via metal-organic chemical vapor deposition, *J Alloys Compd* 509, 2227 (2011).

69. E. Feltin, B. Beaumont, M. Laügt, P. de Mierry, P. Vennéguès, M. Leroux, and P. Gibart, Crack-free thick GaN layers on silicon (111) by metalorganic vapor phase epitaxy, *Phys Stat Sol (a)* 188, 531 (2001).

70. T. Egawa, T. Moku, H. Ishikawa, K. Ohtsuka, and T. Jimbo, Improved characteristics of blue and green InGaN-based light-emitting diodes on Si grown by metalorganic chemical vapor deposition, *Jpn J Appl Phys* 41, L663 (2002).

71. D. Christy, T. Egawa, Y. Yano, H. Tokunaga, H. Shimamura, Y. Yamaoka, A. Ubukata, T. Tabuchi, and K. Matsumoto, Uniform growth of AlGaN/GaN high electron mobility transistors on 200 mm silicon (111) substrate , *Appl Phys Express* 6, 026501 (2013).

72. A.F. Wilson, A. Wakejima, and T. Egawa, Step-stress reliability studies on AlGaN/GaN high electron mobility transistors on silicon with buffer thickness dependence *Appl Phys Express* 6, 056501 (2013).

73. S.L. Selvaraj, A. Watanabe, and T. Egawa, Enhanced mobility for MOCVD grown AlGaN/GaN HEMTs on Si substrate, *69th Annual IEEE Device Research Conference (DRC)*, 221 (2011).

74. S.L. Selvaraj, T. Suzue, and T. Egawa, Breakdown enhancement of AlGaN/GaN HEMTs on 4-in silicon by improving the GaN quality on thick buffer layers, *IEEE Electron Device Lett* 30, 587 (2009).

75. A. Ubukata, K. Ikenaga, N. Akutsua, A. Yamaguchi, K. Matsumoto, T. Yamazaki, and T. Egawa, GaN growth on 150-mm-diameter (1 1 1) Si substrates, *J Cryst Growth* 298, 198 (2007).

76. S. Arulkumaran, T. Egawa, S. Matsui, and H. Ishikawa, Enhancement of breakdown voltage by AlN buffer layer thickness in AlGaNGaN high-electron-mobility transistors on 4 in. diameter silicon, *Appl Phys Lett* 86, 1 (2005).

77. S. Iwakami, M. Yanagihara, O. Machida, E. Chino, N. Kaneko, H. Goto, and K. Ohtsuka, AlGaN/GaN heterostructure field-effect transistors (HFETs) on Si substrates for large-current operation, *J Appl Phys* 43, L831 (2004).

78. T. Li, M. Mastro, and A. Dadgar, *III–V Compound Semiconductors: Integration with Silicon-Based Microelectronics*, CRC Press, London, U.K., 576pp, 2010.

79. J. Gleize, F. Demangeot, J. Frandon, M.A. Renucci, F. Widmann, and B. Daudin, Phonons in a strained hexagonal GaN–AlN superlattice, *Appl Phys Lett* 74, 703 (1999).

80. Y. Ni, Z. He, D. Zhou, Y. Yao, F. Yang, G. Zhou et al., Effect of AlN/GaN superlattice buffer on the strain state in GaN-on-Si(111) system, *Jpn J Appl Phys* 54, 015505 (2015).

81. G. Cong, Y. Lu, W. Peng, X. Liu, X. Wang, and Z. Wang, Design of the low-temperature AlN interlayer for GaN grown on Si (111) substrate, *J Cryst Growth* 276, 381 (2005).

82. A.F. Wilson, A. Wakejima, and T. Egawa, Influence of GaN stress on threshold voltage shift in AlGaN/GaN high-electron-mobility transistors on Si under off-state electrical bias, *Appl Phys Express.* 6, 086504 (2013).

83. V.Y. Davydov, I.N. Goncharuk, A.N. Smirnov, A.E. Nikolaev, W.V. Lundin, A.S. Usikov et al., Composition dependence of optical phonon energies and Raman line broadening in hexagonal AlxGa$_{1-x}$N alloys, *Phys Rev B: Condens Matter* 65, 125203 (2002).

84. D.R. Askeland and P.P. Phulé, *The Science and Engineering of Materials*, Cengage Learning, Stanford, CT, 198pp, 2006.

85. W.C. Young, R.J. Roark, and R.G. Budynas, *Roark's Formulas for Stress and Strain*, McGraw-Hill, New York, 110pp, 1989.

86. Z. Zeng, N. Liu, Q. Zeng, D. Yang, Q. Shaoxing, Y. Cui, and W.L. Mao, Elastic moduli of polycrystalline Li 15 Si 4 produced in lithium ion batteries, *J. Power Sources* 242, 732 (2013).

87. A.R. Denton and N.W. Ashcroft, Vegard's law, *Phys. Rev. A* 43, 3161 (1991).

88. J.W. Yang, A. Lunev, and G. Simin et al., Selective area deposited blue GaN–InGaN multiple-quantum well light emitting diodes over silicon substrates, *Appl Phys Lett* 76, 273 (2000).

89. B. Zhang, H. Liang, Y. Wang, Z. Feng, K.W. Ng, K.M. Lau, High-performance III-nitride blue LEDs grown and fabricated on patterned Si substrates, *J Cryst Growth* 298, 725 (2007).

90. L.T. Romano, C.G. Vande Walle, J.W. Ager, W. Götz, and R.S. Kem, Effect of Si doping on strain, cracking, and microstructure in GaN thin films grown by metalorganic chemical vapor deposition, *J Appl Phys* 87, 7745 (2000).

91. A. Dadgar, M. Poschenrieder, A. Reiher, J. Bläsing, J. Christen, A. Krtschil, T. Finger, T. Hempel, A. Diez, and A. Krost, Reduction of stress at the initial stages of GaN growth on Si (111), *Appl Phys Lett* 82, 28 (2003).

92. O. Contreras, F.A. Ponce, J. Christen, A. Dadgar, A. Krost, Dislocation annihilation by silicon delta-doping in GaN epitaxy on Si, *Appl Phys Lett* 81, 4712 (2002).

93. Z. Zheng, Z. Chen, Y. Chen, S. Huang, B. Fan, Y. Xian, W. Jia, Z. Wu, G. Wang, and H. Jiang, Analysis and modeling of the experimentally observed anomalous mobility properties of periodically Si-delta-doped GaN layers, *Appl Phys Lett* 100, 212102 (2012).

94. Z. Zheng, Z. Chen, Y. Xian, B. Fan, S. Huang, W. Jia, Z. Wu, G. Wang, and H. Jiang, Enhanced electrostatic discharge properties of nitride-based light-emitting diodes with inserting Si-delta-doped layers, *Appl Phys Lett* 99, 111109 (2011).

95. L.S. Wang, K.Y. Zang, S. Tripathy, and S.J. Chua, Effects of periodic delta-doping on the properties of GaN: Si films grown on Si (111) substrates, *Appl Phys Lett* 85, 5881 (2004).

96. H.P. David Schenk, A. Bavard, E. Frayssinet, X. Song, F. Cayre, H. Ghouli, M. Lijadi, L. Naïm, M. Kennard, Y. Cordier, D. Rondi, and D. Alquier, Delta-doping of epitaxial GaN layers on large diameter Si (111) substrates, *Appl Phys Express* 5, 025504 (2012).

97. P. Xiang, Y. Yang, M. Liu, W. Chen, X. Han, Y. Lin et al., Influences of periodic Si delta-doping on the characteristics of n-GaN grown on Si (111) substrate, *J Cryst Growth* 387, 106 (2014).

98. M. Gherasimova, G. Cui, Z. Ren, J. Su, X.-L. Wang, J. Han, K. Higashimine, and N. Otsuka, Heteroepitaxial evolution of AlN on GaN grown by metal-organic chemical vapor deposition, *J Appl Phys* 95, 2921 (2004).

99. H.P.D. Schenk, A. Bavard, E. Frayssinet, X. Song, F. Cayrel, H. Ghouli et al., Delta-doping of epitaxial GaN layers on large diameter Si (111) substrates, *Appl Phys Express* 5, 025504 (2012).

100. M. Wegscheider, C. Simbrunner, T. Li, R. Jakieła, A. Navarro-Quezada, M. Quast, H. Sitter, and A. Bonanni, Periodic Mg distribution in GaN: δ-Mg and the effect of annealing on structural and optical properties, *Appl Surf Sci* 255, 731 (2008).

101. L. Eckey, U. Von Gfug, J. Holst, A. Hoffmann, B. Schineller, K. Heime, M. Heuken, O. Schön, and R. Beccard, Compensation effects in Mg-doped GaN epilayers, *J Cryst Growth* 189, 523 (1998).

102. N. Yi-Qiang, Z.-Y. He, Z. Jian, Y. Yao, F. Yang, X. Peng, Z. Bai-Jun, and L. Yang, Electrical properties of MOCVD-grown GaN on Si (111) substrates with low-temperature AlN interlayers, *Chin Phys B* 22, 88104 (2013).

103. A. Dadgar, T. Hempel, J. Bläsing, O. Schulz, S. Fritze, J. Christen, and A. Krost, Improving GaN-on-silicon properties for GaN device epitaxy, *Phys Status Solidi A* 8, 1503 (2011).

104. J.B. Webb, H. Tang, S. Rolfe, and J.A. Bardwell, Semi-insulating C-doped GaN and high-mobility AlGaN/GaN heterostructures grown by ammonia molecular beam epitaxy, *Appl Phys Lett* 75, 953 (1999).

105. X.G. He, D.G. Zhao, D.S. Jiang, Z.S. Liu, P. Chen, L.C. Le et al., Control of residual carbon concentration in GaN high electron mobility transistor and realization of high-resistance GaN grown by metal-organic chemical vapor deposition, *Thin Solid Films* 564, 135 (2014).

106. J. Selvaraj, S. Lawrence Selvaraj, and T. Egawa, Effect of GaN buffer layer growth pressure on the device characteristics of AlGaN/GaN high-electron-mobility transistors on Si, *J Appl Phys* 48, 121002 (2009).

107. S.A. Chevtchenko, E. Cho, F. Brunner, E. Bahat-Treidel, and J. Würfl, Off-state breakdown and dispersion optimization in AlGaN/GaN heterojunction field-effect transistors utilizing carbon doped buffer, *Appl Phys Lett* 100, 223502 (2012).

108. G. Verzellesi, L. Morassi, G. Meneghesso, M. Meneghini, E. Zanoni, G. Pozzovivo, S. Lavanga, T. Detzel, O. Haberlen, and G. Curatola, Influence of buffer carbon doping on pulse and AC behavior of insulated-gate field-plated power AlGaN/GaN HEMTs, *IEEE Electron Device Lett* 35, 443 (2014).

109. S. Kato, Y. Satoh, H. Sasaki, I. Masayuki, and S. Yoshida, IC-doped GaN buffer layers with high breakdown voltages for high-power operation AlGaN/GaN HFETs on 4-in Si substrates by MOVPE, *J Cryst Growth* 298, 831 (2007).

110. D.D. Koleske, A.E. Wickenden, R.L. Henry, and M.E. Twigg, Influence of MOVPE growth conditions on carbon and silicon concentrations in GaN, *J Cryst Growth* 242, 55 (2002).

111. N. Ikeda, Y. Niiyama, H. Kambayashi, Y. Sato, T. Nomura, S. Kato, and S. Yoshida, GaN power transistors on Si substrates for switching applications, *Proc IEEE* 98, 1151 (2010).

112. P.R. Hageman, S. Haffouz, A. Grzegorczk, V. Kirilyuk, and P.K. Larsen, Growth of GaN epilayers on Si(111) substrates using multiple buffer layers, *Symp Proc* 693, 13 (2001).

113. J.L. Lyons, A. Janotti, and C.G. Van de Walle, Carbon impurities and the yellow luminescence in GaN, *Appl Phys Lett* 97, 152108 (2010).

114. J.L. Lyons, A. Janotti, C.G. Van de Walle, Effects of carbon on the electrical and optical properties of InN, GaN, and AlN, *Phys Rev B* 89, 331 (2014).

115. A.F. Wright, Substitutional and interstitial carbon in wurtzite GaN, *J Appl Phys* 92, 2575 (2002).

116. M.A. Reshchikov and H. Morkoc, Luminescence properties of defects in GaN, *J Appl Phys* 97, 061301 (2005).

117. C. Kisielowski, J. Krüger, S. Ruvimov, T. Suski, E. Jones Z. Liliental-Weber et al., Strain-related phenomena in GaN thin films, *Phys Rev B* 54, 17745 (1996).

20

Hole Accelerator for III-Nitride Light-Emitting Diodes

Zi-Hui Zhang
*School of Electronics and
Information Engineering,
Hebei University of
Technology, Tianjin, China*

Yonghui Zhang
*School of Electronics and
Information Engineering,
Hebei University of
Technology, Tianjin, China*

Xiao Wei Sun
*Department of Electrical
and Electronic Engineering,
Southern University of
Science and Technology,
Guangdong, China*

Wengang Bi
*School of Electronics and
Information Engineering,
Hebei University of
Technology, Tianjin, China*

Abstract   In this chapter, we propose a hole accelerator, which is made of a polarization mismatched p-electron-blocking layer (EBL)/p-GaN/p-Al$_x$Ga$_{1-x}$N heterojunction. By setting III-nitride-based blue light-emitting diodes (LEDs) as examples, the effectiveness of the hole accelerator with different designs (i.e., the AlN composition in the p-Al$_x$Ga$_{1-x}$N layer and the thickness for the p-GaN layer and p-Al$_x$Ga$_{1-x}$N layer) on the hole injection is probed. According to our findings, the energy that the holes obtain does not monotonically increase as the AlN incorporation in the p-Al$_x$Ga$_{1-x}$N layer increases. Meanwhile, for $x > 15\%$ in our case, the energy that the holes gain increases and then reaches a saturation level as increasing the p-GaN layer or p-Al$_x$Ga$_{1-x}$N layer thickness. Therefore, the hole injection efficiency and device efficiency are very sensitive to the p-EBL/p-GaN/p-Al$_x$Ga$_{1-x}$N design, and the hole accelerator can effectively increase the hole injection if properly designed.

20.1 Introduction

As the energy-saving and healthy light source, III-nitride-based light-emitting diodes (LEDs) have found their applications in general lighting, visible light communication, automobiles and LCD backlights, etc. [1,2]. However, there is still room to improve the external quantum efficiency (EQE) for the LEDs, which is strongly influenced by various factors [3,4] such as the poor hole injection [3].

The poor hole injection is reflected by the nonuniform hole distribution within the multiple quantum wells (MQWs), such that the holes are heavily accumulated in the quantum wells close to the p-GaN side while those quantum wells close to the n-GaN side are not able to capture sufficient holes [5–9]. To solve this issue, different remedies have been suggested. One direction is to optimize the MQW region so that the holes can experience less barrier height. For example, modifying the quantum barriers by replacing the conventional GaN quantum barriers with InGaN material will facilitate the hole injection [10–12]. Meanwhile, by properly thinning the thickness of the quantum barriers, the hole distribution can be homogenized [13,14]. It is also advisable to enhance the hole transport by varying the quantum well thickness [15], for example, gradually thickening the quantum wells from the n-GaN side to the p-GaN side along the [0001] orientation [16].

Another factor influencing the hole injection is caused by the widely adopted p-type electron-blocking layer (p-EBL) for the III-nitride-based LEDs after growing the quantum well. The purpose of this p-EBL is to reduce the electron leakage [4,17]. However, the valence band discontinuity between the p-EBL and the p-GaN layer inevitably blocks the hole injection [18]. Therefore, tremendous efforts have been devoted to reduce the hole-blocking effect caused by the p-EBL [19–21]. Park and coworkers propose the Al-composition-graded AlGaN/GaN superlattice EBL to reduce the blocking effect [22]. Kang et al. find that the valence band barrier height in the p-EBL for the holes can be decreased if the p-EBL partially consists of the AlGaN/GaN super-lattice for the green InGaN/GaN LEDs [23]. Kuo et al. and their collaborators demonstrate the advantage of using the staircase AlGaN p-EBL in promoting the hole injection [24,25].

The poor hole injection is also caused by both the low hole concentration and the low hole mobility in the p-GaN layer [26,27]. Although the free hole concentration for the p-GaN layer can be improved if annealed in the ambient of N_2 [26], the activation ratio for the Mg dopants is still less than 1% at room temperature [28], which is far below the Si-doping efficiency for the n-GaN layer. To increase the hole concentration for the [000-1]-oriented LED structures, Simon and coworkers, by properly using the typical polarization effect for the polar III-nitride material [29], suggest adopting three-dimensional hole gas (3DHG) [30]. Later on, the 3DHG is realized for the [0001]-oriented LED devices as has been reported by several groups [31–34]. Most recently, we have proposed an easy way to increase the hole concentration in the p-GaN layer by using a hole modulator [35], which is achieved by doping the last quantum barrier with Mg dopants, such that the stronger polarization-induced electric field in the quantum barriers can increase the ionization ratio of the Mg dopants [30]. More importantly, the built-in electric field can deplete the holes donated by the Mg-doped last quantum barrier and store them in the p-GaN layer, which will further enhance the hole concentration for the p-GaN layer.

Another convenient way to increase the hole injection efficiency is to energize the holes [36]. According to the Fermi–Dirac distribution, the probability of the thermionic emission for the holes is formulated by $P_h = \int_{E \geq \max\{0,(\phi_{EBL}-E_k)\}}^{+\infty} F(E) \cdot P(E) dE / \int_0^{+\infty} F(E) \cdot P(E) dE$ [37], where $P(E)$ represents the valence band density of states in the p-GaN layer, ϕ_{EBL} denotes the effective valence band barrier height for the p-EBL, and E_k means the hole energy. Only those holes with energy higher than $\max\{0,(\phi_{EBL}-E_k)\}$ can jump over the p-EBL and be injected into the quantum wells. Therefore, P_h can be enhanced if $(\phi_{EBL}-E_k)$ is smaller, while a smaller $(\phi_{EBL}-E_k)$ can be obtained by reducing ϕ_{EBL} or increasing E_k. A reduced ϕ_{EBL} can be obtained by decreasing the AlN composition for the p-type EBL, which may however give rise to a worse electron blocking for the device [17]. Hence, one promising way to reduce $(\phi_{EBL}-E_k)$ is to enable holes to possess higher E_k. Note that $E_k = \frac{1}{2} m_h^* V^2$ with $V = \mu_p \cdot E_{field}$. Here, V, μ_p, and m_h^* are defined as the drift velocity, the mobility, and the effective mass for holes, respectively, and E_{field} is the electric field, by which the holes are accelerated/decelerated. Thus, the hole energy can be modified by manipulating the mobility (μ_p) and the electric field (E_{field}). The hole mobility can be increased by reducing the doping level for the p-GaN layer. However, considering the low doping efficiency (<1% at room temperature) for the p-GaN layer, it is not advisable to increase the hole mobility by reducing the Mg-doping concentration for the p-GaN layer [38]. Therefore, we have proposed a hole accelerator that uses the piezo-polarization effect to control the electric field (E_{field}) profiles and thus the hole energy [36,39].

This chapter is structured as follows: Section 20.2 defines the hole accelerator and describes the studied device architectures and the physical parameters. Section 20.3 presents the justification of the physical parameters used in this work by numerically reproducing the experimentally measured results. Note that the main purpose of the hole accelerator is to manipulate the hole energy, but the hole accelerator is also able to affect both the hole energy and the current spreading. Meanwhile, if not properly designed, the hole accelerator may simultaneously induce the hole-blocking effect, and thus Sections 20.4 through 20.6 are designed to systematically analyze the hole accelerators by discussing important architectural factors and numerically conducting parametric investigations on optimizing the hole accelerator to improve the hole injection efficiency and thus the optical output power for the LEDs. Section 20.7 discusses the conclusions and suggests the future outlooks.

20.2 Device Architectures and Physical Parameters

The holes can be energized by the hole accelerator, which is a polarized junction (see definition in Figure 20.1a). The effectiveness of the hole accelerator in increasing the hole energy and improving the hole injection efficiency is probed on an InGaN/GaN blue LED. The schematic device architecture for the studied LED with the

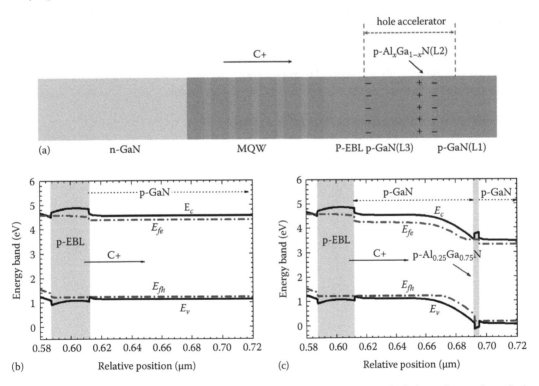

FIGURE 20.1 (a) Schematic device architecture for the InGaN/GaN LED device with a hole accelerator (layer thickness not in scale), along with which we also show the polarization-induced sheet charges at L1/L2, L2/L3, and p-EBL/L3 interfaces, respectively, (b) energy band diagram for the biased reference LED device that has the AlN composition of 0 in the L2 layer, and (c) energy band diagram for the biased LED that has the hole accelerator (L2 of 3 nm thick p-Al$_{0.25}$Ga$_{0.75}$N layer and L3 of 80 nm thick p-GaN layer). (Reprinted with permission from Zhang, Z.-H., Liu, W., Tan, S.T., Ji, Y., Wang, L., Zhu, B., Zhang, Y. et al., A hole accelerator for InGaN/GaN light-emitting diodes, *Appl. Phys. Lett.*, 105, 153503. Copyright 2014, American Institute of Physics.)

hole accelerator is shown in Figure 20.1, which comprises a 4 μm unintentionally doped n-type GaN (u-GaN) layer, a 2 μm n-GaN layer ($n = 5 \times 10^{18}$ cm^{-3}), and five-period 3 nm-In$_{0.15}$Ga$_{0.85}$N/12 nm-GaN multiple quantum wells (MQWs). The thickness for the p-Al$_{0.25}$Ga$_{0.75}$N EBL is 25 nm and that for the p-GaN (L1) is 120 nm in thickness. In order to reveal the sensitivity of the hole injection on different hole accelerator structures, we vary the AlN compositions for the p-Al$_x$Ga$_{1-x}$N (L2) layer, the thicknesses for the p-GaN (L3) layer, and the p-Al$_x$Ga$_{1-x}$N (L2) layer. The effective hole concentration in the p-type layers is assumed to be 3×10^{17} cm^{-3}. The mesa area is 350 × 350 μm^2. Note that the AlN composition for the layer L2 is set to 0 for the reference device, that is, a reference device does not have the hole accelerator. The energy band diagrams at the vicinity for the hole accelerator are shown in Figure 20.1c, in which layer L2 is a 3 nm thick p-Al$_{0.25}$Ga$_{0.75}$ N layer and layer L3 is a 80 nm thick p-GaN layer. For comparison, we also show the energy band diagram of the p-GaN layer for the reference device in Figure 20.1b. Unlike the p-GaN layer for the reference LED, the polarized p-GaN(L3)/ p-Al$_{0.25}$Ga$_{0.75}$N(L2)/p-GaN(L1) junction creates the polarization-induced electric field that strongly deforms the energy band, such that the holes are able to receive the additional energy when traveling from layer p-GaN(L1) to layer p-GaN(L3). As has been discussed previously, this can help the holes to more smoothly climb over the p-EBL through thermionic emission.

The physical parameters that we use during the numerical calculations are as follows. The band offset ratios for the InGaN/GaN and the AlGaN/GaN heterojunctions are 70/30 and 50/50 [40], respectively. The Auger recombination coefficient and the Shockley–Read–Hall recombination lifetime are set to 1×10^{-30} cm^6/s [41] and 43 ns [11,42], respectively. In order to reflect the polarization effect in these polarization mismatched heterojunctions, we have assumed 40% polarization level considering the strain release due to the dislocation generation [29]. We also set a constant hole mobility of 5.0 cm^2/V-s for both the p-GaN and the p-AlGaN layers with the hole concentration of 3×10^{17}cm^{-3}, since neglecting the dependence of the hole mobility on the electric field will not change our conclusions in this work. Furthermore, in order to better numerically reproduce the experimentally measured optical output power, we have set the light extraction efficiency for the studied LEDs to 78% in our models.

20.3 Physical Parameter Justification

We first measured the optical output power in terms of the injection current density for the reference LED, and then we reproduced the experimental results by using advanced physical models of semiconductor devices (APSYS) [5]. Both the measured and the calculated results are presented in Figure 20.2. Clearly, we see that

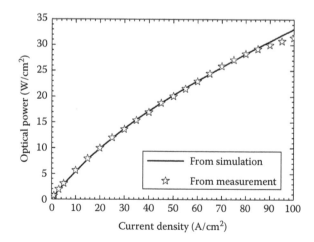

FIGURE 20.2 Experimentally measured and numerically calculated optical power in terms of the injection current for the reference InGaN/GaN LED without the hole accelerator. (Reprinted with permission from Zhang, Z.-H., Zhang, Y., Bi, W., Geng, C., Xu, S., Demir, H.V., and Sun, X.W., On the hole accelerator for III-nitride light-emitting diodes, *Appl. Phys. Lett.*, 108, 151105. Copyright 2016, American Institute of Physics.)

the numerically calculated optical power agrees well with the experimentally measured one, thus validating the physical parameters that we set during the numerical calculations.

20.4 Impact of the AlN Composition on the LED Performance

In order to illustrate the impact of the AlN composition for the p-Al_xGa_{1-x}N layer (L2) on the hole energy and the hole injection efficiency, we design hole accelerators with different AlN compositions for the p-Al_xGa_{1-x}N layer (L2) and fixed thickness for the p-Al_xGa_{1-x}N layer (L2) and the p-GaN layer (L3), which is 80 and 3 nm, respectively. The AlN compositions are chosen to be 5%, 10%, 15%, 20%, 25%, and 30%, respectively, in this work.

We first calculate and present the electric field profiles in the various hole accelerators with different designs as shown in Figure 20.3a. For a fair comparison, Figure 20.3a also includes

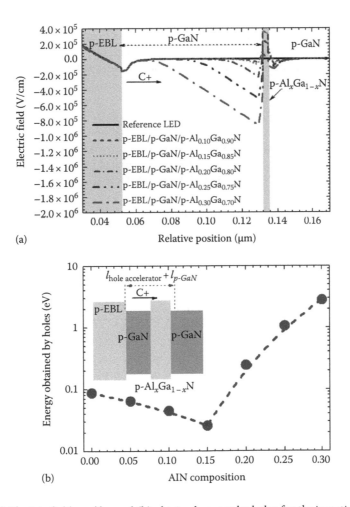

(a)

(b)

FIGURE 20.3 (a) Electric field profiles and (b) obtained energy by holes for the investigated LED devices with various AlN compositions in the p-Al_xGa_{1-x}N layer (L2). Data collected at 86A/cm². (Reprinted with permission from Zhang, Z.-H., Zhang, Y., Bi, W., Geng, C., Xu, S., Demir, H.V., and Sun, X.W., On the hole accelerator for III-nitride light-emitting diodes, *Appl. Phys. Lett.*, 108, 151105. Copyright 2016, American Institute of Physics.)

the electric field profile in the p-GaN layer for the reference LED (i.e., the AlN composition is 0% for the p-Al$_x$Ga$_{1-x}$N layer (L2)). The direction for the electric field is defined as pointing along the C+ orientation. Clearly, we can see that the polarization charge profiles in the p-GaN layer (L3) for the hole accelerators enable the acceleration effect for the holes and favor the hole injection. Meanwhile, the magnitude of the electric field in the p-GaN layer (L3) increases as more AlN composition is contained in the p-Al$_x$Ga$_{1-x}$N layer (L2), and this is ascribed to the higher polarization-induced interface charge density at the p-Al$_x$Ga$_{1-x}$N(L2)/p-GaN(L3) interface when the AlN composition increases [29]. According to Figure 20.3a, due to the negative polarization-induced interface charges at the p-GaN(L1)/p-Al$_x$Ga$_{1-x}$N(L2) interface, a portion of the layer p-GaN (L1) also possesses the electric field that favors the hole acceleration. Note that the reversed polarization-induced electric field inside the p-Al$_x$Ga$_{1-x}$N layer (L2) may reduce the drift velocity for the holes.

In order to show the impact of the polarization-induced electric field in the layers p-GaN(L1), p-Al$_x$Ga$_{1-x}$N(L2), and p-GaN(L3), we calculate the net work conducted on the holes by following $W = e\int_0^l E_{field}\ dx$, in which we have properly adjusted the integration step (dx) by optimizing the mesh distributions in the studied devices. The integration range starts from the p-EBL/p-GaN(L3) interface to the relative position of 0.17 μm (see Figure 20.3a), since beyond this range the electric field profiles are identical for all the devices. The corresponding values for the net work that the holes obtain are shown in Figure 20.3b. We can see that the holes lose more energy as the AlN composition in the p-Al$_x$Ga$_{1-x}$N layer (L2) increases from 0.0% to 15%. The reduced energy in terms of the increasing AlN composition is caused by the p-Al$_x$Ga$_{1-x}$N layer (L2), which decelerates the hole transport as has been mentioned previously. However, Figure 20.3a also reveals that the polarization-induced electric field in the p-GaN layer (L3) becomes stronger when the AlN composition for the p-Al$_x$Ga$_{1-x}$N layer (L2) exceeds 15%, and hence the holes can gain more energy as more AlN is incorporated (see Figure 20.3b), that is, the energy increase for holes from the p-GaN layer (L1) and the p-GaN layer (L3) overwhelms the energy loss from the p-Al$_x$Ga$_{1-x}$N layer (L2).

We then probe the correlation between the hole energy and the hole injection efficiency for the LEDs with variously structured hole accelerators (see Figure 20.4a1 through a5). Here, we define that the first quantum well (QW1) is the one closest to the n-GaN side while the fifth quantum well (QW5) is closest to the p-GaN side. Because the holes are not able to gain sufficient energy (see Figure 20.3b), the hole concentration in the quantum wells cannot be significantly increased for the LEDs with the p-EBL/p-GaN/p-Al$_{0.10}$Ga$_{0.90}$N and the p-EBL/p-GaN/p-Al$_{0.15}$Ga$_{0.85}$N structures as compared to the reference LED. However, thanks to the larger hole energy supplied by the p-EBL/p-GaN/p-Al$_{0.20}$Ga$_{0.80}$N, p-EBL/p-GaN/p-Al$_{0.25}$Ga$_{0.75}$N, and p-EBL/p-GaN/p-Al$_{0.30}$Ga$_{0.70}$N structures, the hole concentration in the quantum wells is becoming higher with the increasing AlN composition in the p-Al$_x$Ga$_{1-x}$N layer (L2). Meanwhile, we also show the EQE and the optical power as the function of the injection current density for the different LEDs in Figure 20.4b. Being consistent with the hole concentration profiles in Figure 20.4a, the EQE and the optical power for the LEDs with the p-EBL/p-GaN/p-Al$_{0.10}$Ga$_{0.90}$N and the p-EBL/p-GaN/p-Al$_{0.15}$Ga$_{0.85}$N structured hole accelerators are not remarkably improved, while the EQE and the optical power can be significantly improved when the AlN composition is higher than 15% for the p-Al$_x$Ga$_{1-x}$N layer (L2). Furthermore, the enhanced hole injection also reduces the efficiency droop level for the LEDs with the p-EBL/p-GaN/p-Al$_{0.20}$Ga$_{0.80}$N, the p-EBL/p-GaN/p-Al$_{0.25}$Ga$_{0.75}$N, and the p-EBL/p-GaN/p-Al$_{0.30}$Ga$_{0.70}$N structures. Note that although the holes are not able to obtain sufficient energy (see Figure 20.3b) for the LEDs with the p-EBL/p-GaN/p-Al$_{0.10}$Ga$_{0.90}$N and the p-EBL/p-GaN/p-Al$_{0.15}$Ga$_{0.85}$N structured hole accelerators, the hole concentration and the EQE are kept at the same level as the reference LED and this is ascribed to the better current spreading effect induced by the energy band discontinuity in the p-type hole injection layer [43,44].

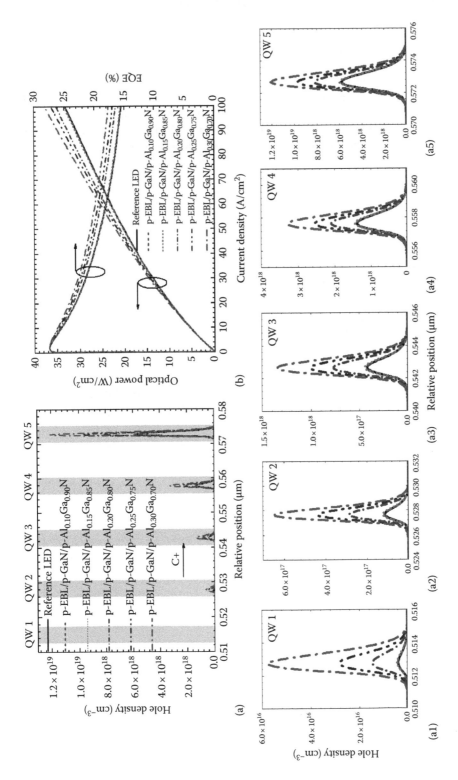

FIGURE 20.4 (a) Hole concentration profiles in the MQW region, the zoomed-in hole concentration profiles in the respective quantum well are shown in Figure 20.4a1 through a5, data collected at 86 A/cm²; (b) numerically calculated optical power and EQE for the investigated LED devices. Figure 20.4a is replotted in linear scale. (Reprinted with permission from Zhang, Z.-H., Zhang, Y., Bi, W., Geng, C., Xu, S., Demir, H.V., and Sun, X.W., On the hole accelerator for III-nitride light-emitting diodes, *Appl. Phys. Lett.*, 108, 151105. Copyright 2016, American Institute of Physics.)

20.5 Impact of the p-GaN Layer (L3) Thickness on the LED Performance

It has been mentioned in Section 20.4 that the polarization-induced electric field in the p-GaN layer (L3) contributes the most to energize the holes, making the thickness of layer L3 very essential. In this section, we first investigate the electric field profiles in the p-GaN layer (L1), p-Al$_x$Ga$_{1-x}$N layer (L2), and p-GaN layer (L3) for the hole accelerator structures with different thicknesses for the p-GaN layer (L3), that is, p-EBL/30 nm-p-GaN/p-Al$_{0.30}$Ga$_{0.70}$N, p-EBL/50 nm-p-GaN/p-Al$_{0.30}$Ga$_{0.70}$N, p-EBL/80 nm-p-GaN/p-Al$_{0.30}$Ga$_{0.70}$N, p-EBL/100 nm-p-GaN/p-Al$_{0.30}$Ga$_{0.70}$N, and p-EBL/130 nm-p-GaN/p-Al$_{0.30}$Ga$_{0.70}$N as shown in Figure 20.5a. Here, the thickness of the p-Al$_{0.30}$Ga$_{0.70}$N layer (L2) was set to 3 nm, and the

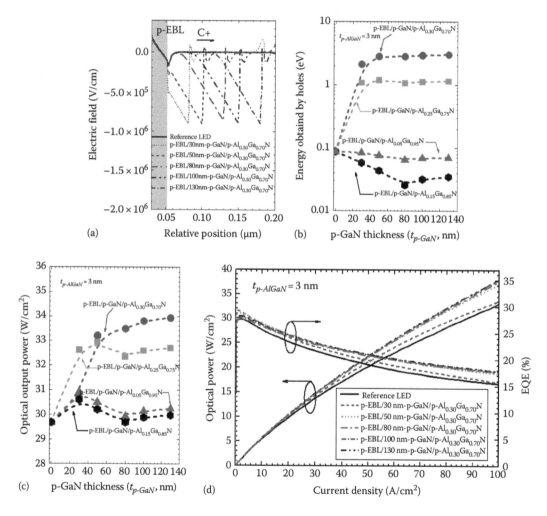

FIGURE 20.5 (a) Numerically calculated electric field profiles with various p-GaN (L3) layer thicknesses, (b) summarized net work done to the nonequilibrium holes in terms of the p-GaN (L3) layer thickness, (c) optical output power in terms of the p-GaN (L3) layer thickness, and (d) numerically calculated optical power and EQE in terms of the current density. (b) and (c) also employ p-Al$_{0.05}$Ga$_{0.95}$N (L2), p-Al$_{0.15}$Ga$_{0.85}$N (L2), p-Al$_{0.25}$Ga$_{0.75}$N (L2), and p-Al$_{0.30}$Ga$_{0.70}$N (L2) layers. Data collected at the current density level of 86 A/cm². (Reprinted with permission from Zhang, Z.-H., Zhang, Y., Bi, W., Geng, C., Xu, S., Demir, H.V., and Sun, X.W., On the hole accelerator for III-nitride light-emitting diodes, *Appl. Phys. Lett.*, 108, 151105. Copyright 2016, American Institute of Physics.)

electric field in the p-GaN layer for the reference LED is also presented for comparison. We can see that the p-GaN layer (L3) is completely polarized by the electric field for the hole accelerators with 30 nm thick p-GaN layer (L3) and 50 nm thick p-GaN layer (L3), while the p-GaN layer (L3) is only partially polarized by the electric field especially when the p-GaN layer (L3) thickness is 100 and 130 nm, respectively, for example, in the relative position ranging between 0.05 and 0.11 μm, the electric field intensity for the hole accelerator with the 130 nm thick p-GaN (L3) layer is identical to that in the reference LED device.

We then calculate the net work conducted on the holes as the function of the p-GaN layer (L3) thickness (see Figure 20.5b). In Figure 20.5b, we also adopt p-$Al_{0.30}Ga_{0.70}N$, p-$Al_{0.25}Ga_{0.75}N$, p-$Al_{0.15}Ga_{0.85}N$, and p-$Al_{0.05}Ga_{0.95}N$ as different L2 layers. According to Figure 20.5b, for the p-EBL/p-GaN/p-$Al_{0.30}Ga_{0.70}N$ and the p-EBL/p-GaN/p-$Al_{0.25}Ga_{0.75}N$ structures, the holes initially can obtain more energy from the hole accelerator with the increasing p-GaN layer (L3) thickness, which is consistent with the electric field profiles in Figure 20.5a. However, the obtained hole energy reaches saturation when the thickness for the p-GaN layer (L3) exceeds 50 nm. This is because when the thickness of the p-GaN layer (L3) exceeds 50 nm, it is only partially polarized by the electric field according to Figure 20.5a, while the other unpolarized part possesses the electric field of the same magnitude as the reference LED, which makes little contribution to energize the holes. Figure 20.5b also shows that the holes will obtain less energy from the p-EBL/p-GaN/p-$Al_{0.05}Ga_{0.95}N$ and p-EBL/p-GaN/p-$Al_{0.15}Ga_{0.85}N$ structures regardless of the thickness of the p-GaN layer (L3), mainly due to the fact that a small AlN composition in the p-AlGaN layer (L2) is not able to yield very strong polarization-induced electric field in the p-GaN layer (L3). Therefore, the increased p-GaN layer (L3) thickness for the p-EBL/p-GaN/p-$Al_{0.05}Ga_{0.95}N$ and the p-EBL/p-GaN/p-$Al_{0.15}Ga_{0.85}N$ structures will not help to further increase the hole energy as demonstrated in Figure 20.5b.

It is also worthy of investigating the impact of the p-GaN layer (L3) thickness on the optical power and the EQE for the studied LEDs. Figure 20.5c collects the optical power at the current density of 86 A/cm² for the hole accelerators with various p-GaN layer (L3) thicknesses and different AlN compositions for the p-$Al_xGa_{1-x}N$ layer (L2). Figure 20.5d presents the optical power and the EQE at different injection levels, which demonstrates the improved optical power and the reduced efficiency droop for LEDs with the hole accelerators incorporated. We can see from Figure 20.5c that the optical power for the LEDs with the p-EBL/p-GaN/p-$Al_{0.30}Ga_{0.70}N$ (also see Figure 20.5d) and the p-EBL/p-GaN/p-$Al_{0.25}Ga_{0.75}N$ designs initially increases with the p-GaN layer (L3) thickness and then becomes less thickness dependent (the threshold thickness is 50 nm), which can be well interpreted by Figure 20.5b, such that the more energized holes increase the injection efficiency and improve the optical power. However, when the hole energy reaches the saturation level at the 50 nm thickness, any further increase of the thickness for the p-GaN layer (L3) will strongly affect neither the hole energy nor the optical power. Interestingly to note, the optical power for the LED with the p-EBL/p-GaN/p-$Al_{0.25}Ga_{0.75}N$ structure is higher than that for the LED with the p-EBL/p-GaN/p-$Al_{0.30}Ga_{0.70}N$ structure when the p-GaN layer (L3) is thinner than 40 nm, and this is caused by the stronger hole-blocking effect taking place at the p-GaN/p-$Al_{0.30}Ga_{0.70}N$ interface. Note that although the holes do not receive sufficient energy from the p-EBL/p-GaN/p-$Al_{0.05}Ga_{0.95}N$ and the p-EBL/p-GaN/p-$Al_{0.15}Ga_{0.85}N$ structures (see Figure 20.5b), the p-GaN/p-$Al_xGa_{1-x}N$ heterojunction gives rise to the improved current spreading effect [43,44], and the optical power is therefore still higher than that for the reference LED within the tested p-GaN layer (L3) thickness.

20.6 Impact of the p-$Al_xGa_{1-x}N$ Layer (L2) Thickness on the LED Performance

Unlike the p-GaN layer (L3), the p-$Al_xGa_{1-x}N$ layer (L2) possesses the polarization-induced electric field that will decelerate the drift velocity for the holes. Meanwhile, the p-$Al_xGa_{1-x}N$ layer (L2) also strongly influences the polarization level in the p-GaN layer (L3). As a result, the thickness and the AlN composition for the p-Al$_x$Ga$_{1-x}$N layer (L2) are vitally important for designing effective hole accelerators. In this section, we first study the dependence of the electric field profiles on the p-EBL/p-GaN/2 nm-p-Al$_{0.30}$Ga$_{0.70}$N, the p-EBL/p-GaN/3 nm-p-Al$_{0.30}$Ga$_{0.70}$N, the p-EBL/p-GaN/5 nm-p-Al$_{0.30}$Ga$_{0.70}$N, and the p-EBL/p-GaN/10 nm-p-Al$_{0.30}$Ga$_{0.70}$N structures as shown in Figure 20.6a. Here, the p-GaN layer (L3) thickness is set to 80 nm. The electric field profile in the p-GaN region for the reference LED is also illustrated in Figure 20.6a for comparison. We observe that the electric field intensity in the p-GaN (L3) layer increases when the p-Al$_x$Ga$_{1-x}$N (L2) layer thickness increases from 2 to 3 nm. Nevertheless, the electric field then becomes less affected even when the p-Al$_x$Ga$_{1-x}$N (L2) layer is further thickened to 10 nm.

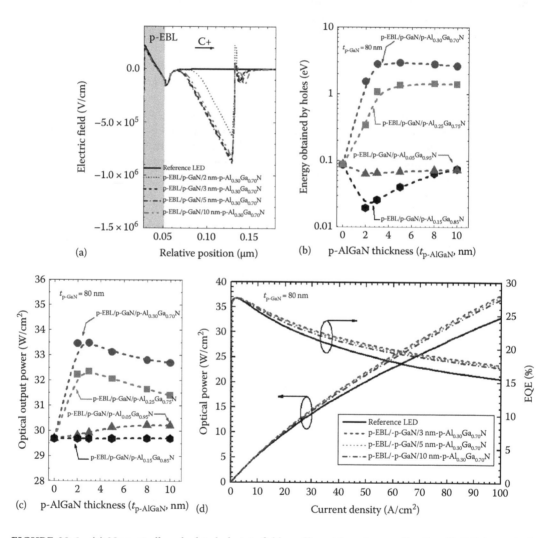

FIGURE 20.6 (a) Numerically calculated electric field profiles with various p-Al$_{0.30}$Ga$_{0.70}$N (L2) layer thicknesses, (b) summarized net work done to the nonequilibrium holes in terms of the p-Al$_x$Ga$_{1-x}$N (L2) layer thickness, (c) optical output power in terms of the p-Al$_x$Ga$_{1-x}$N (L2) layer thickness, and (d) numerically calculated optical power and EQE in terms of the current density. Figure 20.6b and c also employs p-Al$_{0.05}$Ga$_{0.95}$N (L2), p-Al$_{0.15}$Ga$_{0.85}$N (L2), p-Al$_{0.25}$Ga$_{0.75}$N (L2), and p-Al$_{0.30}$Ga$_{0.70}$N (L2) layers. Data collected at the current density of 86 A/cm^2. (Reprinted with permission from Zhang, Z.-H., Zhang, Y., Bi, W., Geng, C., Xu, S., Demir, H.V., and Sun, X.W., On the hole accelerator for III-nitride light-emitting diodes, *Appl. Phys. Lett.*, 108, 151105. Copyright 2016, American Institute of Physics.)

We then calculate the net work applied to the holes (see Figure 20.6b) by following $W = e \int_0^l E_{field} \, dx$.

For the p-EBL/p-GaN/p-Al$_{0.25}$Ga$_{0.75}$N and p-EBL/p-GaN/p-Al$_{0.30}$Ga$_{0.70}$N architectures, the energy that the holes obtain increases with the increasing p-Al$_x$Ga$_{1-x}$N layer (L2) thickness and then reaches saturation at the thickness of 3 nm. In the meanwhile, the p-EBL/p-GaN/p-Al$_{0.30}$Ga$_{0.70}$N structure can provide more energy for the holes than the p-EBL/p-GaN/p-Al$_{0.25}$Ga$_{0.75}$N structure since the higher AlN composition for the p-Al$_x$Ga$_{1-x}$N layer (L2) can induce the stronger polarization-induced electric field in the p-GaN layer (L3). On the other hand, the p-EBL/p-GaN/p-Al$_{0.05}$Ga$_{0.95}$N and the p-EBL/p-GaN/p-Al$_{0.15}$Ga$_{0.85}$N architectures supply smaller energy for the holes, and this arises from the weaker polarization-induced electric field in the p-GaN layer (L3) when the AlN composition in the p-Al$_x$Ga$_{1-x}$N layer (L2) is low.

We also present the dependence of the optical output power on the p-Al$_x$Ga$_{1-x}$N layer (L2) thickness in Figure 20.6c. Figure 20.6d demonstrates the improved optical power and the reduced efficiency droop for LEDs that have hole accelerators. Specifically, we can see that the optical output power in terms of the p-Al$_x$Ga$_{1-x}$N layer (L2) thickness is strongly determined by the hole energy in Figure 20.6b. For example, the LEDs with the p-EBL/p-GaN/p-Al$_{0.25}$Ga$_{0.75}$N and the p-EBL/p-GaN/p-Al$_{0.30}$Ga$_{0.70}$N architectures have more optical power when the p-Al$_x$Ga$_{1-x}$N layer (L2) thickness is further increased to 3 nm. However, when the thickness for the p-Al$_x$Ga$_{1-x}$N layer (L2) is beyond 3 nm, the optical power drops though the holes can still keep the high energy from the hole accelerator. We attribute the optical power decrease to the stronger hole-blocking effect caused by the valence band offset between the p-GaN layer (L3) and the p-Al$_x$Ga$_{1-x}$N layer (L2) when the p-Al$_x$Ga$_{1-x}$N layer (L2) is thicker than 3 nm. Meanwhile, although the holes are not able to obtain sufficiently high energy from the p-EBL/p-GaN/p-Al$_{0.05}$Ga$_{0.95}$N and the p-EBL/p-GaN/p-Al$_{0.15}$Ga$_{0.85}$N architectures, the optical power is not negatively influenced, which is ascribed to the suppressed current crowding effect by the energy band discontinuity for the p-GaN/p-Al$_x$Ga$_{1-x}$N heterojunctions [43,44].

20.7 Conclusions and Future Outlooks

In conclusion, we have proposed and demonstrated a promising hole accelerator, which is made of any polarization mismatched heterojunction to energize holes. By incorporating the hole accelerator, the holes can more effectively overcome the blocking effect caused by the p-EBL and the enhanced hole injection efficiency can be simultaneously obtained, resulting in an improved optical performance for LEDs. Besides, we have also investigated the sensitivity of the hole injection on the hole accelerators with different designs. Considering the different polarization features for the p-GaN layer (L3) and the p-Al$_x$Ga$_{1-x}$N layer (L2), the polarization-induced electric field in the p-GaN layer (L3) can energize the holes, while that in the p-Al$_x$Ga$_{1-x}$N layer (L2) can decelerate holes. Furthermore, we have also reported the additional advantage for the hole accelerator such that it is able to better spread the current. Nevertheless, the energy discontinuity for the hole accelerator may also give rise to the hole-blocking effect. Therefore, the p-GaN layer (L3) thickness, the p-Al$_x$Ga$_{1-x}$N layer (L2) thickness, and the AlN composition for the p-Al$_x$Ga$_{1-x}$N layer (L2) are essentially important, and the hole accelerator has to be properly designed for modifying the hole energy, enhancing the hole injection capability, improving the optical power, and reducing the efficiency droop.

In this work, we have utilized the p-EBL/p-GaN(L3)/p-Al$_x$Ga$_{1-x}$N(L2) architecture as an example to form the hole accelerator. However, the hole accelerator can be designed in various methods, such that, if properly designed, the p-EBL/p-In$_x$Ga$_{1-x}$N(L3)/p-GaN(L2) structure can also effectively increase the hole energy and improve the hole injection efficiency. Without using the p-EBL, the hole accelerator can also be designed as (p-Al$_x$Ga$_{1-x}$N/p-In$_y$Ga$_{1-y}$N/p-Al$_z$Ga$_{1-z}$N)$_n$ architecture (here n denotes the number for the heterojunction).

In addition, although the hole accelerator and its effectiveness presented in this work are for the III-nitride blue LEDs, we believe the hole accelerator is also useful and promising for deep ultraviolet LEDs (DUV LEDs), for which the EQE and the optical power are also significantly hindered by the quite low hole injection efficiency [45].

Acknowledgment

This work is supported by Natural Science Foundation of China (Project No. 51502074), Natural Science of Foundation of Tianjin (Project No. 16JCYBJC16200), and Technology Foundation for Selected Overseas Chinese Scholar Ministry of Human Resources and Social Security of the People's Republic of China (Project No. CG2016008001).

References

1. S. Pimputkar, J. S. Speck, S. P. Denbaars, and S. Nakamura, Prospects for LED lighting, *Nat. Photon.* 3, 180 (2009).
2. S. T. Tan, X. W. Sun, H. V. Demir, and S. P. DenBaars, Advances in the LED materials and architectures for energy-saving solid-state lighting toward lighting revolution, *IEEE Photon J.* 4, 613 (2012).
3. G. Verzellesi, D. Saguatti, M. Meneghini, F. Bertazzi, M. Goano, G. Meneghesso, and E. Zanoni, Efficiency droop in InGaN/GaN blue light-emitting diodes: Physical mechanisms and remedies, *J. Appl. Phys.* 114, 071101 (2013).
4. J. Cho, E. F. Schubert, and J. K. Kim, Efficiency droop in light-emitting diodes: Challenges and countermeasures, *Laser Photon Rev.* 7, 408 (2013).
5. Z.-H. Zhang, S. T. Tan, Z. Ju, W. Liu, Y. Ji, Z. Kyaw, Y. Dikme, X. W. Sun, and H. V. Demir, On the effect of step-doped quantum barriers in InGaN/GaN light emitting diodes, *J. Display Technol.* 9, 226 (2013).
6. Z.-H. Zhang, S. T. Tan, Y. Ji, W. Liu, Z. Ju, Z. Kyaw, X. W. Sun, and H. V. Demir, A PN-type quantum barrier for InGaN/GaN light emitting diodes, *Opt. Express* 21, 15676 (2013).
7. Y. K. Kuo, T. H. Wang, J. Y. Chang, and M. C. Tsai, Advantages of InGaN light-emitting diodes with GaN-InGaN-GaN barriers, *Appl. Phys. Lett.* 99, 091107 (2011).
8. Y. Li, F. Yun, X. Su, S. Liu, W. Ding, and X. Hou, Deep hole injection assisted by large V-shape pits in InGaN/GaN multiple-quantum-wells blue light-emitting diodes, *J. Appl. Phys.* 116, 123101 (2014).
9. Y. Ji, Z.-H. Zhang, S. T. Tan, Z. G. Ju, Z. Kyaw, N. Hasanov, W. Liu, X. W. Sun, and H. V. Demir, Enhanced hole transport in InGaN/GaN multiple quantum well light-emitting diodes with a p-type doped quantum barrier, *Opt. Lett.* 38, 202 (2013).
10. Y.-K. Kuo, J.-Y. Chang, M.-C. Tsai, and S.-H. Yen, Advantages of blue InGaN multiple-quantum well light-emitting diodes with InGaN barriers, *Appl. Phys. Lett.* 95, 011116 (2009).
11. Z.-H. Zhang, W. Liu, Z. Ju, S. T. Tan, Y. Ji, Z. Kyaw, X. Zhang, L. Wang, X. W. Sun, and H. V. Demir, Self-screening of the quantum confined Stark effect by the polarization induced bulk charges in the quantum barriers, *Appl. Phys. Lett.* 104, 243501 (2014).
12. K. Zhou, M. Ikeda, J. Liu, S. Zhang, D. Li, L. Zhang, J. Cai, H. Wang, H. B. Wang, and H. Yang, Remarkably reduced efficiency droop by using staircase thin InGaN quantum barriers in InGaN based blue light emitting diodes, *Appl. Phys. Lett.* 105, 173510 (2014).
13. Z. G. Ju, W. Liu, Z.-H. Zhang, S. T. Tan, Y. Ji, Z. Kyaw, X. L. Zhang et al., Improved hole distribution in InGaN/GaN light-emitting diodes with graded thickness quantum barriers, *Appl. Phys. Lett.* 102, 243504 (2013).
14. J. R. Chen, T. C. Lu, H. C. Kuo, K. L. Fang, K. F. Huang, C. W. Kuo, C. J. Chang, C. T. Kuo, and S. C. Wang, Study of InGaN-GaN light-emitting diodes with different last barrier thicknesses, *IEEE Photon Technol. Lett.* 22, 860 (2010).
15. Y.-L. Li, Y.-R. Huang, and Y.-H. Lai, Efficiency droop behaviors of InGaN/GaN multiple-quantum-well light-emitting diodes with varying quantum well thickness, *Appl. Phys. Lett.* 91, 181113 (2007).
16. C. H. Wang, S. P. Chang, W. T. Chang, J. C. Li, Y. S. Lu, Z. Y. Li, H. C. Yang, H. C. Kuo, T. C. Lu, and S. C. Wang, Efficiency droop alleviation in InGaN/GaN light-emitting diodes by graded-thickness multiple quantum wells, *Appl. Phys. Lett.* 97, 181101 (2010).

17. C. S. Xia, Z. M. Simon Li, and Y. Sheng, On the importance of AlGaN electron blocking layer design for GaN-based light-emitting diodes, *Appl. Phys. Lett.* 103, 233505 (2013).

18. S.-H. Han, D.-Y. Lee, S.-J. Lee, C.-Y. Cho, M.-K. Kwon, S. P. Lee, D. Y. Noh, D.-J. Kim, Y. C. Kim, and S.-J. Park, Effect of electron blocking layer on efficiency droop in InGaN/GaN multiple quantum well light-emitting diodes, *Appl. Phys. Lett.* 94, 231123 (2009).

19. C. S. Xia, Z. M. S. Li, W. Lu, Z. H. Zhang, Y. Sheng, W. D. Hu, and L. W. Cheng, Efficiency enhancement of blue InGaN/GaN light-emitting diodes with an AlGaN-GaN-AlGaN electron blocking layer, *J. Appl. Phys.* 111, 094503 (2012).

20. Y. Y. Lin, R. W. Chuang, S. J. Chang, S. G. Li, Z. Y. Jiao, T. K. Ko, S. J. Hon, and C. H. Liu, GaN-Based LEDs with a chirped multiquantum barrier structure, *IEEE Photon Technol. Lett.* 24, 1600 (2012).

21. Z.-H. Zhang, Z. Ju, W. Liu, S. T. Tan, Y. Ji, Z. Kyaw, X. Zhang, N. Hasanov, X. W. Sun, and H. V. Demir, Improving hole injection efficiency by manipulating the hole transport mechanism through p-type electron blocking layer engineering, *Opt. Lett.* 39, 2483 (2014).

22. J. H. Park, D. Y. Kim, S. Hwang, D. Meyaard, E. F. Schubert, Y. D. Han, J. W. Choi, J. Cho, and J. K. Kim, Enhanced overall efficiency of GaInN-based light-emitting diodes with reduced efficiency droop by Al-composition-graded AlGaN/GaN superlattice electron blocking layer, *Appl. Phys. Lett.* 103, 061104 (2013).

23. J. Kang, H. Li, Z. Li, Z. Liu, P. Ma, X. Yi, and G. Wang, Enhancing the performance of green GaN-based light-emitting diodes with graded superlattice AlGaN/GaN inserting layer, *Appl. Phys. Lett.* 103, 102104 (2013).

24. Y.-K. Kuo, J.-Y. Chang, and M.-C. Tsai, Enhancement in hole-injection efficiency of blue InGaN light-emitting diodes from reduced polarization by some specific designs for the electron blocking layer, *Opt. Lett.* 35, 3285 (2010).

25. B. C. Lin, K. J. Chen, C. H. Wang, C. H. Chiu, Y. P. Lan, C. C. Lin, P. T. Lee, M. H. Shih, Y. K. Kuo, and H. C. Kuo, Hole injection and electron overflow improvement in InGaN/GaN light-emitting diodes by a tapered AlGaN electron blocking layer, *Opt. Express* 22, 463 (2014).

26. S. Nakamura, T. Mukai, M. Senoh, and N. Iwasa, Thermal annealig effects on p-type Mg-doped GaN films, *Jpn. J. Appl. Phy.* 31, L139 (1992).

27. A. Hiroshi, K. Masahiro, H. Kazumasa, and A. Isamu, P-type conduction in Mg-doped GaN treated with low-energy electron beam irradiation (LEEBI), *Jpn. J. Appl. Phys.* 28, L2112 (1989).

28. M. Lachab, D. H. Youn, R. S. Qhalid Fareed, T. Wang, and S. Sakai, Characterization of Mg-doped GaN grown by metalorganic chemical vapor deposition, *Solid-State Electron.* 44, 1669 (2000).

29. V. Fiorentini, F. Bernardini, and O. Ambacher, Evidence for nonlinear macroscopic polarization in III–V nitride alloy heterostructures, *Appl. Phys. Lett.* 80, 1204 (2002).

30. J. Simon, V. Protasenko, C. Lian, H. Xing, and D. Jena, Polarization-induced hole doping in wide-band-gap uniaxial semiconductor heterostructures, *Science,* 327, 60 (2010).

31. L. Zhang, K. Ding, J. C. Yan, J. X. Wang, Y. P. Zeng, T. B. Wei, Y. Y. Li, B. J. Sun, R. F. Duan, and J. M. Li, Three-dimensional hole gas induced by polarization in (0001)-oriented metal-face III-nitride structure, *Appl. Phys. Lett.* 97, 062103 (2010).

32. Z.-H. Zhang, S. Tiam Tan, Z. Kyaw, W. Liu, Y. Ji, Z. Ju, X. Zhang, X. W. Sun, and H. V. Demir, p-doping-free InGaN/GaN light-emitting diode driven by three-dimensional hole gas, *Appl. Phys. Lett.* 103, 263501 (2013).

33. S. Li, M. Ware, J. Wu, P. Minor, Z. Wang, Z. Wu, Y. Jiang, and G. J. Salamo, Polarization induced pn-junction without dopant in graded AlGaN coherently strained on GaN, *Appl. Phys. Lett.* 101, 122103 (2012).

34. S. Li, T. Zhang, J. Wu, Y. Yang, Z. Wang, Z. Wu, Z. Chen, and Y. Jiang, Polarization induced hole doping in graded $Al_xGa_{1-x}N$ (x = 0.7~1) layer grown by molecular beam epitaxy, *Appl. Phys. Lett.* 102, 062108 (2013).

35. Z. -H. Zhang, Z. Kyaw, W. Liu, Y. Ji, L. C. Wang, S. T. Tan, X. W. Sun, and H. V. Demir, A hole modulator for InGaN/GaN light-emitting diodes, *Appl. Phys. Lett.* 106, 063501 (2015).

36. Z.-H. Zhang, W. Liu, S. T. Tan, Y. Ji, L. Wang, B. Zhu, Y. Zhang et al., A hole accelerator for InGaN/GaN light-emitting diodes, *Appl. Phys. Lett.* 105, 153503 (2014).

37. S. M. Sze, *Physics of Semiconductor Devices*, 2nd edn., John Wiley & Sons, Inc., New York, 1981.

38. D.-H. Youn, M. Lachab, M. Hao, T. Sugahara, H. Takenaha, Y. Naoi, and S. Sakai, Investigation on the p-type activation mechanism in Mg-doped GaN films grown by metalorganic chemical vapor deposition, *Jpn. J. Appl. Phys.* 38, 631 (1999).

39. Z.-H. Zhang, Y. Zhang, W. Bi, C. Geng, S. Xu, H. V. Demir, and X. W. Sun, On the hole accelerator for III-nitride light-emitting diodes, *Appl. Phys. Lett.* 108, 151105 (2016).

40. J. Piprek, Efficiency droop in nitride-based light-emitting diodes, *Phys. Stat. Solidi A* 207, 2217 (2010).

41. Y. C. Shen, G. O. Mueller, S. Watanabe, N. F. Gardner, A. Munkholm, and M. R. Krames, Auger recombination in InGaN measured by photoluminescence, *Appl. Phys. Lett.* 91, 141101 (2007).

42. Z.-H. Zhang, Y. Ji, W. Liu, S. Tiam Tan, Z. Kyaw, Z. Ju, X. Zhang et al., On the origin of the electron blocking effect by an n-type AlGaN electron blocking layer, *Appl. Phys. Lett.* 104, 073511 (2014).

43. Z.-H. Zhang, S. T. Tan, W. Liu, Z. Ju, K. Zheng, Z. Kyaw, Y. Ji, N. Hasanov, X. W. Sun, and H. V. Demir, Improved InGaN/GaN light-emitting diodes with a p-GaN/n-GaN/p-GaN/n-GaN/p-GaN current-spreading layer, *Opt. Express* 21, 4958 (2013).

44. Y.-J. Liu, C.-H. Yen, L.-Y. Chen, T.-H. Tsai, T.-Y. Tsai, and W.-C. Liu, On a GaN-based light-emitting diode with a p-GaN/i-InGaN superlattice structure, *IEEE Electron Device Lett.* 30, 1149 (2009).

45. A. Khan, K. Balakrishnan, and T. Katona, Ultraviolet light-emitting diodes based on group three nitrides, *Nat. Photon.* 2, 77 (2008).

21

Metalorganic Chemical Vapour Deposition (MOCVD) Growth of GaN on Foundry Compatible 200 mm Si

Li Zhang
Singapore-MIT Alliance for Research and Technology

Kenneth E. Lee
Singapore-MIT Alliance for Research and Technology

Eugene A. Fitzgerald
Singapore-MIT Alliance for Research and Technology Department of Materials Science and Engineering

Soo Jin Chua
Singapore-MIT Alliance for Research and Technology Department of Electrical and Computer Engineering

Abstract   GaN and its related group III nitrides serve as the base materials for many important applications, such as light-emitting diodes (LEDs), laser diodes (LDs), and high electron mobility transistors (HEMTs). Silicon, as a standard material for traditional Si-based very-large-scale integrated (VLSI) circuits, is becoming a preferred substrate for GaN heteroepitaxy. There are two major advantages of the GaN-on-Si platform. The first advantage is the possibility for the GaN devices to be compatible with the automated processing equipment developed for Si VLSI to reduce the cost of device fabrication. The second advantage is the potential of integrating GaN-based electronics and photonics with Si-based electronics via wafer bonding technology. In this chapter, a foundry-compatible GaN-on-Si platform is developed on 200 mm diameter 725 μm thick Si wafers to explore the full advantages of GaN-on-Si.

21.1 Introduction

21.1.1 Motivation for GaN-on-Silicon Platform

As native GaN substrates grown by hydride vapour phase epitaxy (HVPE), high nitrogen pressure solution (HNPS) and ammonothermal methods are limited in size (<50 mm) and of high cost [1], GaN-based light-emitting diodes (LEDs) and high electron mobility transistors (HEMTs) are mainly heteroepitaxially grown on foreign substrates such as sapphire, silicon (Si), and silicon carbide (SiC). The growth of GaN on Si substrates by MOCVD has attracted intense interest in the last decade [2]. First, Si substrates are available in large sizes (up to 450 mm diameter) at low cost and with high quality. Sapphire and 6H–SiC are limited to 150 mm substrates at the moment. The more significant advantage of using Si substrates is the inherent compatibility with standard Si processing equipment. By employing existing fully automated 200 mm Si production lines, GaN-on-Si technology could significantly reduce the fabrication cost of GaN-based LEDs and HEMTs. What is more important is that there is the potential of integrating GaN-based electronic and optoelectronic devices with traditional Si-based very-large-scale integrated (VLSI) circuits.

In order to reduce the cost of GaN-based devices, the size of substrates on which GaN-based devices are grown has steadily increased from 50 to 100, 150, even 200 mm. Similar to the Si industry, the cost of epitaxy and processing can be reduced with the increase in size of the substrates because the ratio of area of the edge exclusion region to the usable area is diminished [3]. As sapphire and 6H–SiC substrates are limited in size due to the difficulty of producing large-area high-quality wafers, Si is therefore a promising alternative substrate. Moreover, a large reduction in cost is anticipated for the processing of devices on large wafers as the processing cost per wafer increases sublinearly. This is particularly true for the GaN-on-Si platform with GaN devices that are compatible with the automated processing equipment developed for Si VLSI.

One of the most important potential benefits of growing GaN on large-diameter Si is the possibility of integrating GaN-based electronic and optoelectronic devices with Si-based electronics. As the scaling of critical dimensions in Si-based complementary metal–oxide–semiconductor (CMOS) transistors is reaching physical limits, traditional CMOS platforms could benefit from the higher mobility and direct bandgap of the GaN material system. GaN-based HEMTs offer higher operation frequency and power density compared to Si-based CMOS. GaN-based optical devices can serve as light sources, optical amplifiers, and detectors in hybrid GaN + Si integrated circuits. However, the mismatch in substrate orientation is a key hurdle for integrating CMOS and GaN. While wurtzite GaN can be best grown on Si (111), the Si orientation used in CMOS foundry processing is Si (001). Although there has been much effort placed on growing single-crystal GaN on CMOS-compatible Si (001) substrates, the material is still far from device quality [4].

A more realizable approach of monolithic integration of GaN with Si-CMOS is to use wafer transfer technology as illustrated in Figure 21.1. First, a 200 mm Si (100) handle wafer is bonded to a partially processed Si-CMOS layer (i.e., source-drain and gate-stack completed, but prior to formation of metal interconnects) on a standard silicon-on-insulator (SOI) substrate via SiO_2–SiO_2 hydrophilic bonding. The original Si substrate in the SOI wafer is removed via mechanical grinding and chemical etching. The CMOS layer with handle wafer is then bonded to a GaN-based device layer that was grown on a 200 mm diameter 725 μm thick Si (111) substrate. After the Si handle wafer is removed, windows are opened (in predetermined regions) in the Si-CMOS layer to expose the GaN layer for GaN-based device fabrication. CMOS-compatible W-plugs are used to bring the GaN device contacts to the same plane as the CMOS devices. Finally, standard CMOS multilayer metallization processes are applied to connect the Si-CMOS and GaN devices to form novel monolithically integrated GaN + Si-CMOS circuits. A 200 mm GaN + Si-CMOS engineered wafer is shown in Figure 21.2 and the cross-sectional transmission electron microscopy (TEM) image of the engineered wafer is shown in Figure 21.3.

One of the key requirements of the integration scheme described earlier is the availability of high-quality 200 mm GaN-on-Si device wafers. This chapter therefore covers the growth of high-quality GaN material on 200 mm diameter Si (111) substrates. Integration of optoelectronics (LEDs) and electronics (HEMTs) is attempted on Si substrate within the GaN-based material system to pave the way for the ultimate GaN + Si-CMOS integration.

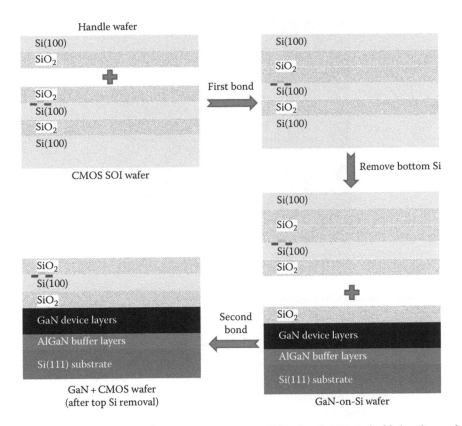

FIGURE 21.1 Schematic illustration of processing steps in monolithic Si and GaN via double bonding and transfer method.

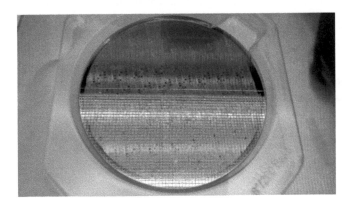

FIGURE 21.2 Optical image of 200 mm diameter Si-CMOS/GaN integrated wafer.

21.1.2 Problems with Developing Foundry-Compatible GaN-on-Si Platform

Due to the intrinsic differences between the two materials, the growth of GaN on Si is not straightforward, and there are many problems associated with growing GaN on Si substrates. The main problems are discussed in detail in this section.

In the growth of GaN on sapphire, Nakamura et al. employed a low-temperature GaN buffer to obtain high-quality GaN films [5]. However, direct growth of GaN on Si leads to the formation of deep voids in the substrate

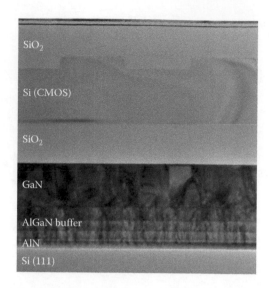

SiO$_2$

Si (CMOS)

SiO$_2$

GaN

AlGaN buffer

AlN

Si (111)

FIGURE 21.3   Cross-sectional TEM of Si-CMOS/GaN integrated wafer.

due to the eutectic reaction between Ga and Si at typical GaN growth temperatures (~1000°C). This problem in GaN-on-Si growth is often referred to as "melt-back etching." The direct contact between Ga precursor atoms and the Si substrate surface has to be avoided in all stages of the growth to prevent melt-back etching [3].

The large lattice mismatch (+20.4% relative to GaN) leads to a large density of dislocations in the GaN films on Si [6]. Dislocations can act as nonradiative recombination centers to reduce the internal quantum efficiency in LEDs [7], and they can also act as diffusion paths for impurities in HEMTs, leading to device failure [8]. Besides the large lattice mismatch, GaN and Si are also of different crystal structures. Si has a cubic diamond lattice structure, whereas GaN has a hexagonal wurtzite crystal structure.

The large coefficient of thermal expansion (CTE) mismatch (−53.5% relative to GaN) is a major problem associated with the growth of GaN on Si. At elevated growth temperatures during its growth, the GaN film is in a relaxed state. However, since the CTE of the Si substrate (2.6×10^{-6} K^{-1}) is smaller than that of GaN (5.59×10^{-6} K^{-1}), the GaN film contracts more than Si during cooling. The resulting tensile thermal stress will cause extensive cracking of the GaN film. Wafer cracking greatly affects device performance, reliability, and yield [9].

A problem directly associated with CTE mismatch between GaN and Si is the wafer bow caused by the residual tensile stress after cooling. The curvature of a GaN-on-Si wafer can be obtained according to Stoney's equation:

$$\kappa = \frac{6\sigma_f\left(1-\upsilon_s\right)h_f}{E_s h_s^2} \tag{21.1}$$

where
  κ is the curvature of the wafer
  σ_f is the film stress
  E_s is the Young's modulus of the substrate
  υ_s is the Poisson ratio of the substrate
  h_f and h_s are the thicknesses of the film and substrate, respectively

Thus, for GaN-on-Si wafers, the curvature is inversely proportional to the square of the substrate thickness, assuming that the GaN exerts the same amount of stress on the substrates in each case. In order to minimize wafer curvature changes during the growth of GaN on 200 mm Si substrates and reduce the final curvature of the wafer at room temperature, thick (≥1000 μm) 200 mm Si (111) substrates are usually chosen for GaN-on-Si applications [10–12]. However, the automated processing facilities developed for Si VLSI not only put a stringent requirement on wafer bow (<50 μm) but also require the thickness of 200 mm diameter Si substrates to be 725 μm. Bow is the measured difference in height between the center and edge of the wafer. The bow can be obtained from wafer curvature κ according to Equation 21.2 where δ is the bow of the wafer and *r* is the radius of the wafer. Therefore, while all 200 mm Si substrates used for GaN-on-Si have to be 725 μm thick to be truly Si foundry compatible, this is very challenging since the curvature of the 200 mm diameter Si wafer would be almost doubled if the same GaN epilayer was grown on a 725 μm thick Si substrate instead of a 1000 μm substrate:

$$\delta = \frac{1}{\kappa}\left(1 - \cos r\kappa\right) \tag{21.2}$$

Due to the large lattice mismatch and stacking sequence mismatch between Si and Al(Ga)N, GaN hetero-epitaxy on Si substrates usually results in device layers with high threading dislocation densities (TDD), which will deleteriously affect device performance and cause device reliability issues. A high TDD also makes strain engineering more difficult. The primary requirement of any dislocation reduction technique in the GaN-on-Si platform is to reduce TDD using a relatively thin layer thickness, because a large of amount of tensile stress will be built up after cooling down to room temperature if thick layers are employed. This stringent requirement in total layer thickness limits the use of many well-known TDD-reduction techniques that have been successfully practiced in GaN-on-sapphire growth such as epitaxial lateral overgrowth (ELOG) [13] and patterned sapphire substrates (PSS) [14].

21.2 Literature Review of GaN-on-Si Platform

Despite the difficulties presented in the previous section, both LED and HEMT structures have been realized in GaN-on-Si at 200 mm substrate size, albeit with substrate thicknesses of 1 mm or greater [10,11,15]. In this chapter, existing techniques and some unsolved problems in GaN-on-Si growth from literature are briefly presented.

21.2.1 Si Substrate Surface Treatment

Usually, a silicon substrate has a native oxide layer that has to be removed prior to the growth of the AlN buffer layer [16]. This can be achieved by *ex situ* chemical etching such as using Piranha solution ($H_2SO_4:H_2O_2$) to remove any organic contamination and HF to remove the native oxide layer [15]. On the other hand, the same effect can be achieved by *in situ* annealing in an H_2 environment. The need to use high-purity chemicals to achieve consistent results makes *ex situ* cleaning methods relatively challenging [3].

21.2.2 AlN Nucleation Layer

In most cases, GaN-on-Si epitaxial growth begins with the deposition of an AlN nucleation layer due to its ability to prevent destructive melt-back etching between Ga and Si, and its having a smaller CTE mismatch with the Si substrate [17]. The AlN layer acts as a diffusion barrier against the reaction between Si and Ga that often leads to poor GaN surface morphology. Additionally, when AlN is used as a buffer, the strain between Si and GaN will change from tensile to compressive, leading to crack reduction in the GaN epilayers.

Another approach is to use *ex situ* deposited buffer layers on the Si substrate before the growth of GaN. The *ex situ* deposited buffer layers can be AlAs [18], 3C–SiC [19], Al_2O_3 [20], or rare earth oxide Gd_2O_3 [21] to prevent melt-back etching between GaN and Si. Additionally, *ex situ* deposited buffer layers have similar functions to compliant substrates, that is, to mediate the large difference in lattice parameters and CTE between Si and GaN, so tensile stress generated upon cooling can be at least partially accommodated by the buffers. However, despite the extra complexity and higher cost, GaN/*ex situ* deposited buffer layers/Si have yet to show superior performance or higher crystal quality compared to GaN grown directly with *in situ* deposited strain compensation buffers. AlN nucleation layers and *in situ* strain compensation buffers such as AlGaN layers do not require complex external preparation or deposition steps and are free from any possible contamination arising from external wafer preparation, thus making them the most popular approach to solve the melt-back and GaN cracking issues in GaN-on-Si.

21.2.2.1 AlN/Si Interface SiN$_x$

At the moment, the two most important questions about the AlN/Si interface regard the existence of amorphous SiN$_x$ and the AlN interfacial structure. The growth of AlN on Si (111) substrate is more complicated than the case of AlN on sapphire or SiC substrates due to the possible existence of an amorphous SiN$_x$ interlayer between AlN and Si. In early studies of GaN growth on Si substrates, it was thought that avoiding the formation of amorphous SiN$_x$ by pre-deposition of Al via pre-flowing TMAl would improve the AlN film quality [16]. It was found that the initial deposition of Al resulted in a very rapid transition to a 2D growth mode of AlN. The rapid transition is essential for the subsequent growth of a high-quality GaN layer. However, more recent studies showed that despite the Al pre-deposition, a discontinuous amorphous interlayer of SiN$_x$ existed at the interface and the low-temperature AlN (LT-AlN; 735°C) was directly grown on the Si substrate via openings in the SiN$_x$ [22,23]. Another method of Si substrate surface treatment is Si nitridation by pre-flowing NH$_3$ prior to the growth of LT-AlN. The effect of nitridation duration on strain, TDD, lateral coherence lengths, and vertical coherence lengths has been discussed extensively in References 24 and 25. Usually, there exists an optimal nitridation duration. A continuous 2 nm of SiN$_x$ layer was found to be present in AlN/Si with NH$_3$ nitridation [24,25].

One possible formation mechanism of the SiN$_x$ at the AlN/Si interface is through the interdiffusion of AlN and Si at high growth temperature (>1000°C). This may explain the presence of sharp interfaces being observed only for AlN films grown at low temperature [16]. It is quite clear that the discontinuous amorphous SiN$_x$ layer could either be formed during the NH$_3$ nitridation step or be formed rapidly during the interdiffusion of AlN and Si across the interface during high-temperature (HT) AlN growth at 1080°C [26].

Another possible formation mechanism of the observed amorphous layer SiN$_x$ is the epitaxial deposition of β-Si$_3$N$_4$ with NH$_3$ nitridation of Si substrate. β-Si$_3$N$_4$ that is one of the most common phases of Si$_3$N$_4$ has hexagonal crystal structure with in-plane lattice parameter $\alpha_{\beta\text{-Si3N4}} = 0.761$ nm [27]. Ultrathin or monolayer crystalline β-Si$_3$N$_4$ has been epitaxially grown on Si (111) substrate by N$_2$ plasma, H$_2$ + N$_2$ mixture, and NH$_3$ [28–31]. Morita found that at 1000°C in H$_2$ + N$_2$ mixture, the growth of monolayer SiN is a self-limiting reaction due to the competition between film formation and H$_2$ etching [29]. In molecular beam epitaxy (MBE) GaN/Si, β-Si$_3$N$_4$ can be observed in the transition from the Si (111) 1 × 1 surface reconstruction to the β-Si$_3$N$_4$ 8 × 8 surface reconstruction. High-quality GaN can be grown with a β-Si$_3$N$_4$ buffer in the MBE environment [32]. It is likely that the epitaxial coherence relationship is transferred from Si to AlN through an ultrathin crystalline β-Si$_3$N$_4$ layer that is subsequently destroyed by the interdiffusion of AlN and Si at high temperature [33].

21.2.2.2 AlN/Si Interfacial Structures

The epitaxial relation of AlN on Si is $(0001)_{\text{AlN}}||(111)_{\text{Si}}$, $\langle 1\bar{1}00\rangle_{\text{AlN}}||\langle 11\bar{2}\rangle_{\text{Si}}$, and $\langle 11\bar{2}0\rangle_{\text{AlN}}||\langle \bar{1}10\rangle_{\text{Si}}$ [16]. Due to the large in-plane lattice mismatch (+23.4%), the critical thickness of epitaxial wurtzite AlN on Si (111) is only one monolayer [22,34,35]. The epitaxial growth of wurtzite AlN on Si (111) substrate occurs by matching four $(220)_{\text{Si}}$ planes with five $\left(2\bar{1}\bar{1}0\right)_{\text{AlN}}$ planes to reduce the 23.4% tensile strain to less than 1% residual strain in the first monolayer of the epitaxy.

In both the wurtzite III-nitride and the diamond cubic Si structure, the tetrahedral coordination leads to double (0001) or {111} close-packed atomic layers with atoms situated with their respective tetrahedral bonding orientations. For the analysis of structural transformations involving multiple possible interfacial structures, it is helpful to employ the "cells of nonidentical displacements" (cnids) concept. Dimitrakopulos et al. have listed the 12 different possible heterointerface structures or cnids, of AlN/Si; six of the structures are for Al polarity and the other six are for N polarity [36] as illustrated in Figure 21.4.

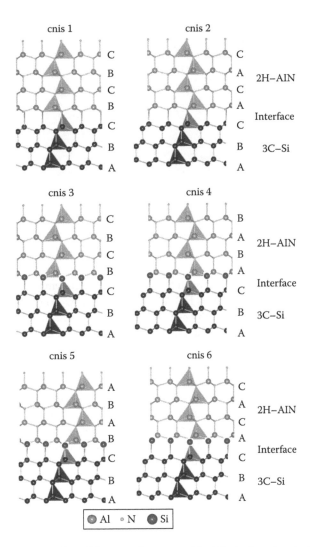

FIGURE 21.4 Interfacial structural models or cnids presented by Dimitrakopulos et al. [36]. The corresponding stacking sequences have been indicated as well as bonding tetrahedral. The dashed lines indicate the interfaces between AlN and Si. Si atoms are below the dashed lines. The bigger spheres above the dashed lines represent Al atoms, whereas the smaller ones represent N atoms.

We have systematically studied high-resolution TEM (HRTEM) images of the AlN/Si interface in the literature. In the high-angle annular dark field (HAADF) image in Radtke et al., HRTEM analysis of AlN/Si [22,23] and HRTEM images in Zang et al. [16] show that the interfacial model of AlN/Si follows cnid2 whereas HRTEM images from Liu et al. suggest that the interfacial model of AlN/Si follows cnid1 [37].

21.2.3 Strain Engineering Method

21.2.3.1 Strain Engineering Methods in GaN-on-Si

Strain is always present in heteroepitaxy either from differences in lattice constants and thermal expansion coefficients, island coalescence, or dislocation movement. For GaN-on-Si, strain can be generated by all these sources and is universally tensile. Wafer bow and epilayer cracking are one of the most problematic issues of GaN structures grown on Si substrates [3]. Due to the large tensile growth stress and thermal

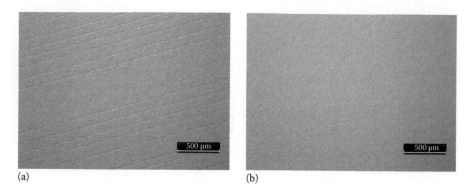

(a)   (b)

FIGURE 21.5 Nomarski microscopy (a) showing cracks on 600 nm thick GaN/240 nm HT-AlN/30 nm LT-AlN Si substrate caused by the large tensile stress introduced upon cooling from the growth temperature to room temperature and (b) crack-free 1200 nm GaN/800 nm three-step-graded AlGaN/240 nm HT-AlN/30 nm LT-AlN/Si.

stress introduced upon cooling from the growth temperature to room temperature, the maximum thickness of crack-free GaN grown directly on a HT-AlN nucleation layer is less than 600 nm, as shown in Figure 21.5a. However, when an 800 nm three-step-graded AlGaN *in situ* strain compensation layer is introduced, it is possible to obtain 1200 nm of crack-free GaN, as shown in Figure 21.5b.

For most optoelectronic and electronic device structures, 600 nm is insufficient. Furthermore, in GaN heteroepitaxy, the TDD decreases as the epilayer thickness increases [38]. Several methods have been proposed and tested to address the issue of tensile stress and associated cracking. Laterally confined epitaxy (LCE) on patterned substrates can guide the cracks in the masked or etched parts of the Si substrate to relieve the tensile stress upon cooling down from the growth temperature. This can be achieved by isolating the epitaxial region with *ex situ* deposited and patterned SiO_2 or Si_3N_4 growth masks into areas smaller than the average crack spacing [39]. Usually, the III-nitride deposited on top of the growth mask has a slow growth rate and is polycrystalline. However, LCE is typically employed in combination with *in situ* strain compensation buffer layers to grow high-quality GaN layers because employing LCE alone does not guarantee high-quality device layers.

In situ strain compensation buffers are always based on a simple concept—the introduction of compressive stress during growth to compensate for the tensile thermal stress generated upon cooling. AlN has a smaller in-plane lattice constant compared to GaN. Thus, in principle, when growing GaN on AlN, depending on relaxation, three basic compressive strain configurations can exist as shown in Figure 21.6. In the

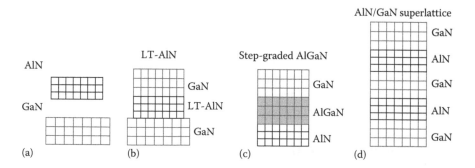

FIGURE 21.6 Different possibilities of lattice-mismatched growth in the GaN-AlN material system. (a) Relaxed AlN and GaN are depicted separately. (b) AlN grows relaxed on the underlying GaN layer, but the subsequently grown GaN layer grows pseudomorphically on AlN. (c) Graded transition from AlN to GaN leads to compressive strain in the GaN layer, if it is grown pseudomorphically or only partially relaxed. (d) Superlattice layers, both AlN and GaN, grow pseudomorphically with an equilibrium lattice constant, which will enable subsequent compressive growth of GaN.

case described in Figure 21.6b, LT-AlN grows relaxed on the underlying GaN layer, but the subsequently grown GaN layer is grown compressively strained. Low temperature (LT-AlN is typically grown at 950°C as compared to HT-AlN at 1100°C) is necessary for the growth of relaxed AlN on GaN without generating extra TDs at the GaN/AlN interface, with the further requirement that the LT-AlN is kept to ~10 nm thickness. Dadgar et al. first employed the LT-AlN interlayer technique to produce crack-free GaN/Si [9]. The advantage of using LT-AlN interlayers is that it can be inserted repeatedly to achieve a very thick GaN layer on Si [40].

An AlGaN intermediate or graded layer between the AlN seeding and GaN buffer layer is probably the oldest method to obtain crack-free GaN on Si [17]. If the material quality of the AlN seeding layer is sufficiently high, such a transition can induce a compressive stress on the subsequently grown GaN layer. A compositionally step-graded buffer is illustrated in Figure 21.6c. When a higher Ga-content $Al_xGa_{1-x}N$ layer is pseudomorphically grown with a layer thickness below the critical thickness (i.e., the thickness at which strain relaxation occurs), it grows with the same lattice parameter as the underlying lower Ga-content $Al_yGa_{1-y}N$ ($x < y$) layer, resulting in the build-up of compressive strain in the latter grown layers (including the final GaN layer).

Superlattices are well known in III–V epitaxy to enhance dislocation reaction by strain field–induced dislocation bending [41]. The same effects can also be achieved for III-nitrides where AlN/GaN superlattices are typically applied for strain compensation [42]. As shown in Figure 21.6d, thin layers of GaN and AlN layers undergo pseudomorphic growth, resulting in an effective lattice constant that is the compositionally weighted average of the GaN and AlN lattice constants. Growing AlN/GaN superlattices requires single layer thicknesses less than the critical thickness.

All strain engineering methods are of little effect if the TDD of the AlN nucleation layer is high (>10^{10} cm^{-2}). Strain fields inherently present when applying these methods can lead to dislocation climb and bending, which reduces the amount of compressive stress built up. For low layer qualities with high dislocation densities, the probability of stress relaxation via dislocation climb and annihilation processes is significant and it is observed that for such cases the strain engineering techniques will not enable the growth of thick crack-free GaN layers [6]. That is why understanding and optimizing the AlN/Si interface and heterostructure is one of the topmost priorities.

21.2.3.2 Stress Relaxation in III-Nitride

In traditional diamond cubic semiconductors, stress relaxation in mismatched heterostructures occurs through the formation and glide of misfit dislocations (MDs) at the heterointerface. This relaxation mechanism is driven by the shear stress caused by the biaxial stress perpendicular to the [001] growth direction on the <111> glide planes [43]. The Burgers vectors of threading dislocations (TDs) in the cubic system are always inclined with respect to the [001] growth direction. However, in wurtzite GaN, the pure edge TDs having Burgers vector a in the basal plane usually have a perpendicular (0001) line direction to the basal plane. The gliding planes for pure edge TDs are the prismatic m-planes that are also perpendicular to the basal plane. The biaxial stress from the lattice mismatch in the basal plane cannot induce a shear stress in the gliding planes for pure edge dislocations in GaN. However, pure edge TDs can bend from the <0001> direction to cause an effective climb to relax the stress in a growing layer [44]. Furthermore, the mixed dislocations can glide on the $\{11\bar{2}2\}$ planes to effect stress relaxation in the basal plane [38]. This secondary slip system has been found to be the main slip system in compressive (AlGaN/AlN [45]) and tensile (InGaN/GaN [46]) heterostructures in III-nitride materials. However, complete relaxation of nitride epilayers by glide in the secondary slip system is often kinetically inhibited [47]. The TDs present in AlGaN/AlN heterostructures enable a dislocation bending mechanism that acts to relax the compressive strain instead of generating MDs [44]. The dislocation bending relaxation mechanism ensures that the compressive strain from the heteroepitaxial interface can be maintained in epilayers up to hundreds of nanometers or even a few microns thick if the TDD is low (<10^{10} cm^{-2}) [47]. The dislocation bending was first theoretically modeled by Romanov and Speck [48]. In their model, the edge dislocations in the overlayer bend to project along one of the three $\langle 1\,\overline{1}00\rangle$ directions in the basal plane interface. The in-plane relaxation of the III-nitrides caused by the TD bending is given by

$$\varepsilon^{avg} = \frac{bL\rho}{4} \qquad\qquad (21.3)$$

where

$\varepsilon^{avg} = \Delta a / a$ is the average relaxation of the compressively stressed overlayer

b is the magnitude of the Burgers vector of the edge dislocation (here, b is equal to the in-plane lattice constant, a)

ρ (cm^{-2}) is the mixed and pure-edge TDD

$L = h \tan\theta^{avg}$ is the projected dislocation length over the basal plane

h is the layer thickness

θ^{avg} is the average bending angle of TDs with respect to the c-axis

21.2.4 Dislocation Reduction

As discussed in Section 21.2.3, a high TDD hinders the effect of strain engineering, which generally results in the need to grow thicker films. However, a cardinal requirement for GaN-on-Si epitaxy is to keep the total epilayer thickness low, to avoid the bow and cracking issues described earlier. This section covers a new dislocation reduction method that does not require any additional increase in the total film thickness besides SiN$_x$ *in situ* masking.

21.2.4.1 SiN$_x$ *In Situ* Masking

SiN$_x$ *in situ* masking has been used extensively in GaN-on-Si epitaxy as an *in situ* dislocation reduction interlayer technique. It was first applied to GaN-on-sapphire growths by Lahreche [49]. SiN$_x$ *in situ* masking is sometimes referred to as Si delta doping or silane burst, because silane (Si-doping precursor for GaN growth) is turned on while TMGa (Ga precursor for GaN growth) is turned off briefly during GaN growth. NH$_3$ (N precursor for GaN growth), silane, and carrier gas (H$_2$) are injected into the MOCVD chamber at the GaN growth temperature (1020°C). Markurt et al. discovered the atomic structure of a Si delta-doped layer to be a SiGaN$_3$ monolayer that acted as an antisurfactant that inhibited further GaN growth [50]. The duration of SiN$_x$ *in situ* masking, that is, the duration that TMGa is turned off and silane is turned on, determines the coverage of the monolayer SiGaN$_3$ antisurfactant. When TMGa flow is restarted, GaN islands regrow from the regions of the GaN surface that are not covered with SiGaN$_3$ and consist of facets that facilitate the bending and annihilation of TD to form half loops [51].

21.2.4.2 Migration-Enhanced AlN Buffer

As discussed earlier, GaN on Si (111) epitaxial growth typically begins with the deposition of a LT-AlN nucleation layer due to its ability to prevent destructive melt-back etching between Ga and Si and AlN's smaller CTE mismatch with the Si substrate. It is well established that in order to improve GaN device performance, a high-quality AlN nucleation layer must be employed. The AlN/Si interface must be optimized, and interfacial defects caused by lattice mismatch and stacking mismatch must be minimized. To start with, the surface morphology of the Si (111) substrate needs to be optimized for the AlN growth by forming single bilayer substrate steps. Additionally, the quality of the AlN layer is strongly affected by the nitridation duration of the Si substrate [24], V/III ratio [11], growth temperature [52], and layer thickness [53]. The results in the literature suggest that the optimizations of the growth parameters are strongly system dependent, and the range of optimized growth parameters can be large. In this book, we will focus on dislocation reduction methods that can be applied across different reactors.

Zhu et al. point out that the surface smoothness of the HT-AlN layer also plays an important role in reducing the crystal misorientation and defect density of the GaN grown on top [3]. Interestingly, achieving a smooth AlN template is also one of the top priorities in the MOCVD growth of AlN templates on sapphire substrates for UV and deep-UV emitting optical devices. Various dislocation reduction techniques applied to the AlN layers grown directly on sapphire have been proposed and applied to obtain high-quality AlN templates on sapphire. A very popular method to obtain high-quality AlN templates on sapphire is to grow the AlN template with the migration-enhanced epitaxy (MEE) method, which involves alternately pulsed flows of TMAl and NH_3 [54]. There are many versions of the MEE AlN buffer such as the alternating supply of NH_3 [55] and the NH_3 pulsed-flow methods [56] in which only the NH_3 flow is pulsed, and the flow-modulated method in which both NH_3 and TMAl are injected together for a short period of time (1 s) and alternately pulsed for a longer period of time [57,58]. It should be noted that the growth conditions reported in the literature vary significantly. One example is the growth temperature, which has been reported with values in the range of 1070°C–1250°C [54–58]. Compared with conventional growth methods, the AlN buffers produced by the MEE method universally show smoother surfaces with reduced TDDs due to the enhanced AlN lateral growth rate. In GaN-on-Si growth, Cagnon et al. [59] briefly studied the effect of the MEE AlN buffer in combination with ion implantation in the AlN/Si structure on the growth of GaN. In terms of the TDD, however, it is difficult to conclude if there is any improvement when the MEE AlN buffer is compared to the conventional AlN buffer in his paper.

We have reviewed existing techniques for GaN growth on Si in this section. In the next section, we will describe the improvements that we have made to current techniques, as well as propose new methods of growing foundry-compatible GaN-on-Si wafers.

21.3 III-Nitride MOCVD Growth on 200 mm Si

21.3.1 Introduction

In this section, a detailed discussion about the growth of III-nitride material on foundry-compatible 200 mm Si substrates is presented. Topics covered include the effect of Si_2H_6 substrate treatment, the AlN/Si interface, crystallographic tilting introduced by substrate surface steps, strain engineering, dislocation reduction with migration-enhanced epitaxy of AlN buffer, and Si substrate engineering for the growth of GaN-on-Si.

21.3.2 Si_2H_6 Treatment on Si Substrate

Usually, a silicon substrate has a native oxide layer that has to be removed prior to the growth of the AlN buffer layer [16]. This can be achieved by *in situ* annealing of the Si substrate in an H_2 environment in the MOCVD system as described in Section 21.2.1. However, a mixture of Si_2H_6 and H_2 was found to be an effective method to remove the native oxide layer in a study on Si homoepitaxy on Si substrate via MBE [60]. In our MOCVD system, 100 ppm Si_2H_6 is available as an n-type doping source. In this section, we discuss the effect of Si_2H_6 surface treatment prior to growth in the GaN-on-Si heteroepitaxial system.

21.3.2.1 Experimental Details

The epitaxial growths of the GaN-on-Si wafers were performed in an AIXTRON CRIUS® Close-Coupled-Showerhead (CCS) MOCVD reactor on 725 μm thick 200 mm diameter Si substrates. First, the Si substrate was *in situ* annealed in an H_2 ambient to remove native oxide at 1030°C as described in Section 21.2.2. Additionally, Si_2H_6 was added to the H_2 environment as pretreatment for Si substrates at the same temperature. Then, a 25 nm low-temperature AlN nucleation layer was grown at 980°C with both group III and group V precursors flowing simultaneously with a V/III ratio of 250. The temperature was increased to 1070°C for HT-AlN buffer growth with continuous AlN growth during the temperature ramping (15 nm of

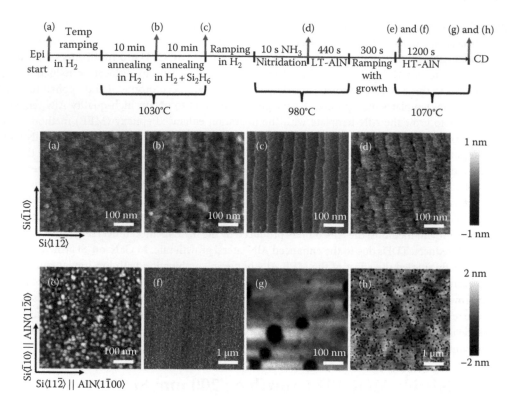

FIGURE 21.7 AFM scans of a Si substrate surface, (a) as-received, (b) after 600 s H_2 annealing, (c) after an additional 600 s $H_2 + Si_2H_6$ anneal, and (d) after subsequent 10 s NH_3 nitridation. AFM scans of heteroepitaxial wurtzite AlN/Si, (e) and (f) after ~9 nm AlN film thickness, and (g) and (h) after ~270 nm film thickness. The scale bars are 100 nm for (a–e) and (g) and 1 μm for (f) and (h).

AlN deposition took place). The carrier gas is always H_2. The growth details for different samples for atomic-force microscopy (AFM) measurement are described in Figure 21.7.

21.3.2.2 Effect of Si_2H_6 Treatment on Si Substrate Morphology

As-received Si (111) substrates (e.g., see Figure 21.7a) have essentially flat surfaces (i.e., have a moderate RMS value of ~0.23 nm at 20 × 20 μm²) without obvious step-terrace features. The Si surface after annealing under H_2 at 1030°C for 600 s (see Figure 21.7b) shows an increased surface RMS of 0.34 nm at 20 × 20 μm² and steplike features begin to appear in the direction parallel to Si⟨$\bar{1}$10⟩. Annealing under H_2 is typically used in MOCVD to remove organic impurities and native oxide layers on the surface of the substrate before LT-AlN growth. However, the island-like protrusions in the top-right and bottom-left corners of Figure 21.7b are usually ascribed to the formation of SiC due to the reaction of the Si surface with carbon contamination present on the wafer surface or from the reactor environment (coatings on the shower head and wafer susceptor).

Carbon contamination, which has been a long-standing concern for growers of thin films on Si substrates, originates from carbon-bearing gas molecules that adsorb on the silicon surface as a result of exposure to the atmosphere [61]. This contamination often takes the form of epitaxial SiC particles, which grow after the decomposition of adsorbed carbon-bearing molecules and the subsequent reaction of the freed carbon with the Si substrate at temperatures between 800°C and 1100°C in an H_2 environment [62]. The direct evidence of the formation of SiC particles (island-like protrusions) on Si substrates in a ultra-high vacuum (UHV) environment has been obtained by reflection high-energy electron diffractometry (RHEED) [63], but RHEED is not possible in the MOCVD environment. As a

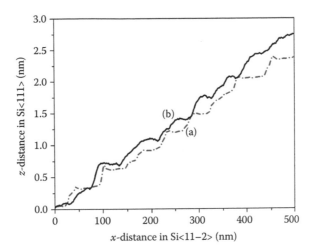

FIGURE 21.8 (a) AFM height profiles of Si (111) substrate after H_2 + Si_2H_6 annealing and (b) after NH_3 nitridation.

result, we attribute the island-like protrusions observed from AFM to be SiC particles based on the observations reported in the literature [62,64].

A further step of annealing under a mixture of Si_2H_6 + H_2 (see Figure 21.7c) promoted well-defined steps with average step heights of 0.3 nm, which is consistent with the height of the Si bilayer (0.31 nm) along the $\langle 111 \rangle$ direction or surface normal (see Figure 21.8). The surface steps have a short-range order with an average terrace spacing of 62.5 nm. The miscut measured from the AFM analysis is 0.28°, which agrees well with the miscut (0.29°) obtained from high-resolution x-ray diffractometry (HRXRD) measurements of the same sample.

We found that the formation of the well-defined bilayer steps only occurred after the additional annealing in Si_2H_6 + H_2. This observation is most likely due to (1) epitaxial growth of Si by cracking Si_2H_6 that in turn fills the voids in the terraces and reconstructs surface steps according to the original miscut and/or (2) Si_2H_6-induced enhancement in thermal cleaning of the native oxide, leading to a recovery of the native Si surface [60]. The SiC islands were also completely removed by further annealing in Si_2H_6 + H_2.

As discussed later in Section 21.3.3.3, a nitridatio step under NH_3 + H_2 ambient for 5–15 s yields the best GaN crystal quality (see Section 21.3.3.3). After nitridation, while the Si bilayer steps that were formed during the Si_2H_6 anneal still remain (see Figure 21.8), pits can now be seen on the terraces (see Figure 21.7d). The surface step configurations of the substrates prior to the nucleation of the epilayer have significant influence on the crystallographic tilt between the substrate and the epilayer, as well as the defects generated by substrate step edges. After these pretreatment steps, it is clearly seen that the Si substrate before the AlN growth consisted of vicinal atomic planes with regularly aligned bilayer steps.

21.3.2.3 Effect of Si_2H_6 Treatment on AlN Morphology

In general, the large lattice mismatch presented in AlN/Si heterostructures, together with the high sticking coefficient and low surface diffusivity of Al adatoms compared with Ga adatoms in GaN growth, leads to island nucleation at the Si (111) surface [65]. AlN tends to grow in 3D island growth mode while GaN tends to follow a 2D step-flow growth mode. A general observation obtained from the growth of $Al_xGa_{1-x}N$ ternary alloy is that the lateral growth rate of the film increases with the decrease in the Al content [66]. This is related to the difference in the surface diffusivity of Al and Ga adatoms. Ga adatoms have a greater diffusivity compared to Al adatoms.

Figure 21.7f and h and their respective higher magnification versions (Figure 21.7e and g) are AFM images of the AlN with nominal thicknesses of ~9 and ~270 nm, respectively. The 9 and 270 nm thicknesses correspond to the growth stages of initial nucleation and complete islands coalescence, respectively. The AlN clearly exhibits island nucleation behavior with an average island size of ~10 nm (see Figure 21.7e); striations along Si$\langle\bar{1}10\rangle$ step edges are clearly seen in the lower magnification image (see Figure 21.7f). As the thickness increases to 270 nm (Figure 21.7g and h), columnar AlN islands increase in size and coalesce with adjacent neighbors to form a continuous film with a roughness of 0.75 nm RMS at 20 × 20 μm^2. Pits as large as 70 nm in diameter are observed on the surface of the 270 nm thick AlN layer due to an incomplete coalescence of islands (Figure 21.7g). These pits could be formed due to the limited diffusivity of Al adatoms, and this is further discussed in Section 21.3.5. The lack of observable surface striations in Figure 21.7h, unlike those seen in Figure 21.7f, indicates that the 270 nm thick AlN layer grows in a manner independent of the original surface atomic steps of the substrate.

21.3.2.4 Overview

In this section, a mixture of Si_2H_6 and H_2 was found to be an effective method to remove the native oxide layer and SiC islands on the Si substrate surface before the AlN growth. With Si_2H_6 treatment, the Si substrate before the AlN growth consisted of vicinal atomic planes with regularly aligned bilayer steps.

21.3.3 AlN/Si Interface

21.3.3.1 Introduction

The crystal quality of the entire GaN-on-Si structure critically depends on the initial AlN/Si interface. In this section, we address two of the most important questions about the AlN/Si interface as discussed in Section 21.2.2.1, that is, the existence of amorphous SiN_x and the AlN/Si interfacial structure.

21.3.3.2 Interfacial Amorphous SiN_x

Figure 21.9 depicts typical high-resolution TEM (HRTEM) cross-sectional images with zone axis (ZA) of AlN$\langle\bar{1}\bar{1}20\rangle$ and Si$\langle110\rangle$ recorded at the AlN/Si interface of the same sample that yielded the AFM images in Figure 21.7g and h. There is a highly distorted AlN region formed within approximately 3–4 bilayers (~1 nm) of the AlN/Si interface as seen in Figure 21.9b and c. While it might appear to be a very thin amorphous SiN_x interlayer, this region can be resolved as a crystallographically abrupt region at different locations of the AlN/Si interface as shown in Figure 21.9a. Thus, the amorphous SiN_x interlayer in the sample is actually discontinuous.

The scanning transmission electron microscopy (STEM) energy-dispersive x-ray spectroscopy (EDX) data in Figure 21.10 are used to study the interdiffusion at the AlN/Si interface. In Figure 21.10a, the STEM image of the AlN/Si interface and the location of the EDX scanning line are shown. The 50 nm EDX scanning line is perpendicular to the AlN/Si interface. In Figure 21.10b, the EDX spectra corresponding to the

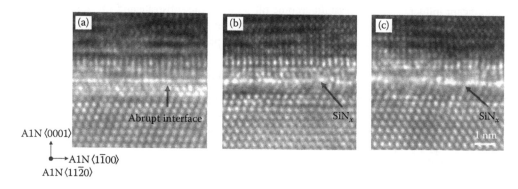

FIGURE 21.9 AlN $\langle11\bar{2}0\rangle$ zone axis HRTEM of AlN/Si interface at different locations (a)–(c) of the TEM sample, showing that formation of the amorphous SiN_x is discontinuous.

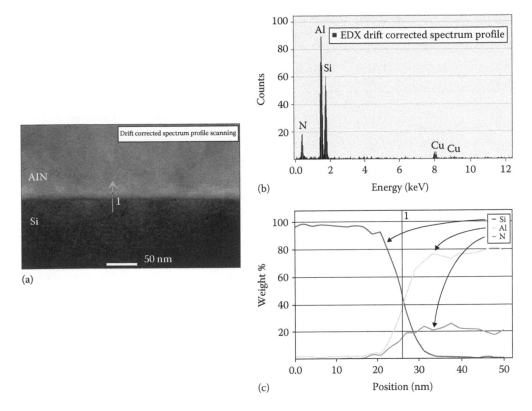

FIGURE 21.10 STEM EDX data of the AlN/Si interface. (a) STEM image of the EDX scan line. (b) EDX spectrum profiles of N, Al, and Si at the position marked as "1" in (a). (c) Normalized elemental concentrations along the scan line reveal that the interdiffusion region in the AlN/Si can be at least 8–9 nm thick after the growth of the AlN layer at 1080°C.

location marked as "1" in Figure 21.10a are shown. The elements presented in the AlN/Si interface are identified as Al, N, Si, and Cu. Cu is the contamination originating from the copper grid used to mount the TEM sample. Finally, normalized elemental concentrations along the scan line are plotted in Figure 21.10c. It can be seen clearly that the interdiffusion region at the AlN/Si interface extends at least 8–9 nm.

In Table 21.1, the growth conditions and observations of the interfacial amorphous SiN_x in the literature and this chapter are summarized. A general observation is that samples grown with NH_3 nitridation are less susceptible to the formation of amorphous SiN_x compared to samples grown with TMAl treatment. The epitaxial transfer from Si to AlN through an ultrathin crystalline β-Si_3N_4 is not favored from our TEM analysis in the crystallographically abrupt region of the AlN/Si interface, since crystalline β-Si_3N_4 is not observed. Our TEM observation favors the hypothesis that the formation of SiN_x at the AlN/Si interface is due to the interdiffusion of AlN and Si at high growth temperature (>1000°C), as described in Section 21.2.2.1. At the high growth temperature of 1070°C, the LT-AlN nucleation layer that was grown at 980°C readily interdiffused with the Si substrate to form amorphous SiN_x and highly Si-doped AlN. Thus, the amorphous SiN_x layer is not continuous and there must be regions where crystalline AlN was grown directly on Si.

21.3.3.3 Effect of Nitridation Duration on TDD in GaN-on-Si

From the previous section, we concluded that the amorphous SiN_x layer is not continuous and there are regions where crystalline AlN was grown directly on Si. In this section, the effect of nitridation duration on the TDD in the GaN layer is discussed. GaN heteroepitaxy can be described by the model of mosaic crystals. The large lattice mismatch can cause a twist and tilt in the mosaic blocks [38]. The average absolute values of tilt and twist are directly related to the full width at half maximum (FWHM) of GaN

TABLE 21.1 Summary of the Growth Conditions and Observations of the Interfacial Amorphous SiN$_x$ in the Literature and This Chapter

Reference	Surface Treatment	Layers of AlN	Treatment Temperature (°C)	Growth Temperature of AlN (°C)	Thickness of AlN (nm)	Existence of SiN$_x$
[37]	TMAl	1	720	720	20	No
[23]	TMAl	1	1040	1100	200	Continuous at 1.5–2 nm
[22]	TMAl	1	735	735	20	Discontinuous at 1.5–2 nm
	TMAl	2	735	735	20	Continuous at 2 nm
				1100	20	
[26]	TMAl	1	860	860	230	No
	TMAl	2	860	860	23	Continuous at 2 nm
				920	230	
	TMAl	2	860	860	23	Continuous at 25 nm
				1010	230	
[33]	NH$_3$	1	1020	1100	150	Continuous at 2 nm
This chapter	NH$_3$	2	980	980	40	Discontinuous at 1 nm
				1080	220	

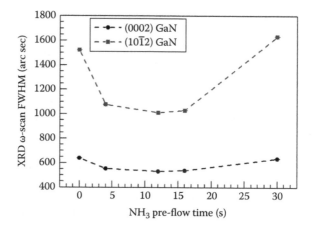

FIGURE 21.11 Effect of nitridation duration on FWHM of XRD rocking curve. A smaller FWHM indicates better crystal quality.

symmetric and skewed-symmetric peaks as measured by XRD. In Figure 21.11, the FWHM of GaN (0002) symmetric and $\left(10\bar{1}2\right)$ skewed-symmetric peaks of the grown GaN epilayer are plotted against the duration of Si substrate NH$_3$ nitridation.

It is seen that the optimal Si nitridation duration is in the range from 5 to 15 s. According to Arslan et al., nitridation duration affects the lateral and vertical coherence lengths of the mosaic blocks in GaN films, and they found that the optimal nitridation duration was 120 s for their nitridation process [33]. We attribute the effect of nitridation duration on TDD to the fact that nitridation could change the extent of coverage of the observed discontinuous amorphous SiN$_x$ layer. There is a possibility that the SiN$_x$ is acting as an antisurfactant to prohibit AlN nucleation so there is an ELOG effect arising from the nitridation step [50]. When the nitridation time is short, a large part of the Si substrate surface is exposed to AlN nucleation and the dislocation reduction effect of ELOG is not obvious. When the optimum coverage ratio is achieved before AlN growth, the AlN buffer shows a superior crystal quality due to the ELOG effect, resulting in correspondingly higher-quality GaN. However, when the SiN$_x$ coverage gets beyond the optimal point, polycrystalline AlN nucleation on the SiN_x layer becomes significant, which increases the TDD. However, in order to conclusively prove the preceding hypothesis, further TEM study of the interface is necessary.

21.3.3.4 AlN/Si Interfacial Structure

The selective area electron diffraction (SAED) pattern as shown in Figure 21.12 with a spot size of 60 nm reveals the epitaxial relationship of the AlN/Si heterostructure. In Figure 21.12a, the image is taken with a ZA of $\left[\bar{1}10\right]_{Si}$ and $\left[11\bar{2}0\right]_{AlN}$, while in Figure 21.12b, the ZA is $\left[11\bar{2}\right]_{Si}$ and $\left[1\bar{1}00\right]_{AlN}$, and the following epitaxial relations can be established: $(0001)_{AlN}||(111)_{Si}$, $\langle1\bar{1}00\rangle_{AlN}||\langle11\bar{2}_{Si}\rangle$, and $\langle11\bar{2}0\rangle_{AlN}||\langle\bar{1}10\rangle_{Si}$.

A typical HRTEM image of the AlN/Si interface is shown in Figure 21.13a. Epitaxy according to the domain matching condition is observed in Figure 21.13a. On average, the regular periodic array of MDs occurs at almost every five $\left\{1\bar{1}00\right\}_{AlN}$ as described in Section 21.2.2.2. The AlN/Si interface becomes highly distorted due to the large, dense array of MDs.

We have confirmed that our AlN layer is Al polar by using the Potassium hydroxide (KOH) etch test as shown in Figure 21.14. An Al-polar AlN layer on a Si substrate produces a surface that consists of hexagonal surface pits after the KOH etch test, whereas the surface of an N-polar AlN layer would consist of hexagonal pyramids after the same test. As a result, only six of the Al-polar interfacial structures are possible in the case of our MOCVD-grown AlN/Si, as illustrated in Figure 21.4.

The diamond structure of the Si substrate is centrosymmetric, having a two-atom basis with atoms at $(1/8)[111]$ and $\left(1/8\right)\left[\bar{1}\bar{1}\bar{1}\right]$. The 3C stacking sequence of Si (111) can be represented as $...A\alpha B\beta C\gamma A\alpha B\beta C\gamma$ ..., where capital letters correspond to the layers formed by Si atoms at the $(1/8)[111]$ position while the Greek letters correspond to the layers formed by Si atoms at the $\left(1/8\right)\left[\bar{1}\bar{1}\bar{1}\right]$ position. The surface morphological study in Section 21.3.2.3 showed well-aligned Si bilayer steps across the entire 200 mm Si substrate after annealing in $H_2 + Si_2H_6$. It is also well known that Si (111) forms hydrogen terminated 1×1 reconstructed surfaces [67]. As a result, the topmost layer of Si substrates should be formed by Si atoms at the $(1/8)[111]$ position of the basis. We rule out cnids 3–6 in Figure 21.4 as the interfacial structure of AlN/Si because in those structures the first layer of Al atoms has three of their tetragonal bonds attached to Si atoms at the $\left(1/8\right)\left[\bar{1}\bar{1}\bar{1}\right]$ positions of the basis.

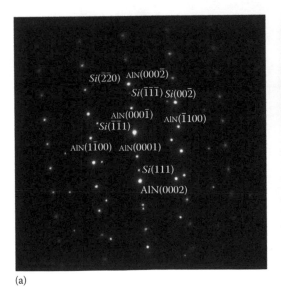

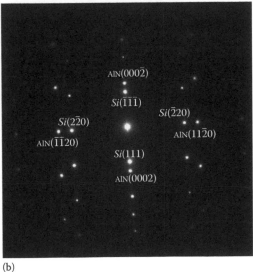

(a)　　　　　　　　　　　　　　　(b)

FIGURE 21.12　TEM SAED pattern of the AlN/Si interface (a) with ZA of $\left[\bar{1}10\right]_{Si}$ and $\left[11\bar{2}0\right]_{AlN}$ and (b) with ZA of $\left[11\bar{2}\right]_{Si}$ and $\left[1\bar{1}00\right]_{AlN}$.

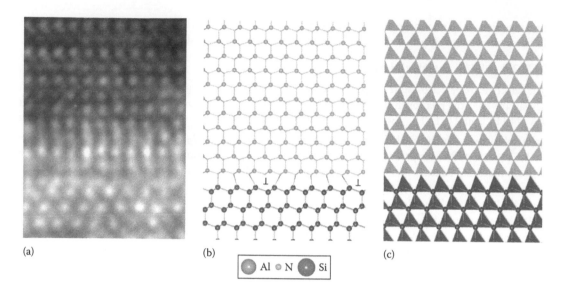

(a)   (b)   (c)

Al ○ N ● Si

FIGURE 21.13 HRTEM of the AlN/Si with ZA of $\left[\bar{1}10\right]_{Si}$ and $\left[11\bar{2}0\right]_{AlN}$ (a) regions without any amorphous SiN_x. (b) Schematic of the atomic arrangement in the cross section. (c) Schematic showing the stacking sequence of the bond tetrahedrals.

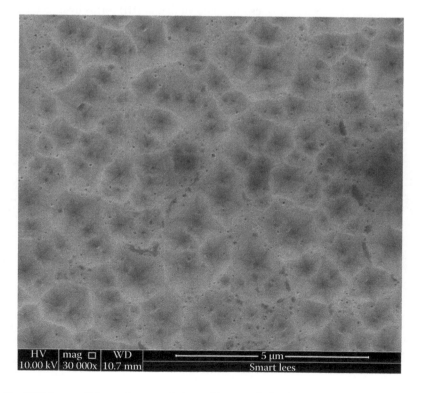

FIGURE 21.14 SEM image of the surface morphology of 350 nm thick AlN on Si (111) substrate after 1800 s KOH etch at 80°C. This AlN film on Si is Al polar.

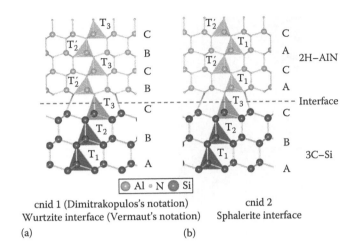

cnid 1 (Dimitrakopulos's notation) cnid 2
Wurtzite interface (Vermaut's notation) Sphalerite interface
(a) (b)

FIGURE 21.15 Interfacial structural models in the proposed cells of nonidentical displacements (cnids) in AlN/Si [36]. (a) cnid1 in Dimitrakopulos notation. The growth can start at the wurtzite position in Vermaut's notation or at (b) cnid2 in Dimitrakopulos notation. The sphalerite on the substrate surface termination in Vermaut's notation.

Thus, there are two remaining possible stacking structures of AlN on Si substrates, as depicted in Figure 21.15. Both of them have a first layer of Al-polar AlN consisting of one Si–N bond per Si atom on the Si substrate. The main difference is how the first Al-polar Al–N tetrahedron in 2H–AlN stacks with respect to the 3C–Si (111). The same problem has been addressed in the case of the 2H–AlN/6H–SiC interface by Vermaut et al. [68]. There are only two possible interfacial stacking sequences: one of them is a first layer of Al–N tetrahedrons starting at the wurtzite position of the substrate (see Figure 21.15a) and the other one is a first layer of tetrahedrons starting at the sphalerite position of the substrate (see Figure 21.15b). They have very different implications on the stacking sequence of the 2H–AlN epilayer. If cnids1 (wurtzite) is the interfacial structure, the stacking sequence of the AlN layer would repeat the stacking sequence of the topmost two layers of the Si substrate, that is, **ABC**:BCBC (in Pirouz and Yang's notation [69], i.e., $\mathbf{T_1T_2T_3}$:$T_2'T_3T_2'T_3$). The bold letters indicate the stacking sequence of the Si (111) substrate. If cnids2 (sphalerite) is the interfacial structure, the first layer of Al–N tetrahedrons would stack at the sphalerite position of the substrate and the subsequent layers would repeat the stacking sequence of the topmost layer of the Si and bottommost layer of the AlN, that is, **ABC**:ACAC (in Pirouz and Yang's notation [69], i.e., $\mathbf{T_1T_2T_3}$:$T_1T_2'T_1T_2'$).

The initial AlN/6H–SiC heterostructure stacking sequence is reported to follow cnids1 (wurtzite), that is, AlN repeats that of the topmost two bilayers of the 6H–SiC substrate [68,70]. This phenomenon was observed and verified in HRTEM analysis of the AlN/6H–SiC interfaces. However, this is not the case in our MOCVD-grown AlN on Si. The schematic of the atomic arrangement and stacking sequence of the HRTEM shown in Figure 21.13b is based on the HRTEM image of our MOCVD-grown AlN/Si interface without the amorphous SiN_x layer (Figure 21.13a). Figure 21.13b and c is derived based on the fact that the center of the bright spots corresponds to Si dumbbells in the lower part of the image, while in the upper part of the image, the center of the bright spots corresponds to Al in the Al–N bonds as Al has greater Z-contrast. We have scanned several locations in our MOCVD-grown AlN/Si interface and found that the interfacial model as described earlier seems to be global. Another example is shown in Figure 21.16, and thus we believe that the same interfacial structure exists across the entire wafer.

In conclusion, our HRTEM analysis of the AlN/Si interface suggests that Dimitrakopulos cnid2 is the most likely interfacial structure of AlN/Si for AlN grown on Si with MOCVD. The first layer of Al–N tetrahedrons would stack at the sphalerite position of the substrate and the subsequent layers would repeat the stacking sequence of the topmost layer of Si and bottommost layer of the AlN, that is, **ABC**:ACAC.

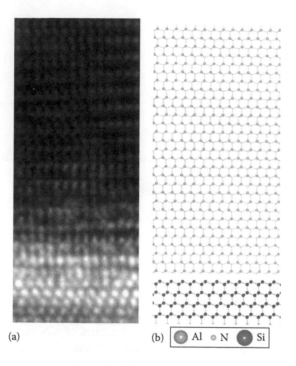

(a)   (b) ⬤ Al ○ N ⬤ Si

FIGURE 21.16   HRTEM of the AlN/Si with ZA of $\left[\bar{1}10\right]_{Si}$ and $\left[11\bar{2}0\right]_{AlN}$ (a) regions without any amorphous SiN$_x$. (b) Schematic of the atomic arrangement in the cross section.

21.3.3.5 Overview

In this section, we have observed that the amorphous SiN$_x$ layer is not continuous and there are regions where crystalline AlN was grown directly on Si in our samples. We found that samples grown with NH$_3$ nitridation are less susceptible to the formation of amorphous SiN$_x$ compared to samples grown with TMAl treatment. The effect of nitridation duration on the TDD in the GaN layer was also discussed, with the optimal nitridation time being about 10–15 s under our growth conditions. Finally, Dimitrakopulos cnid2 was determined to be the AlN/Si interfacial structure.

21.3.4 Strain Engineering

21.3.4.1 Introduction

In this work that we have described, step-graded AlGaN layers are employed as *in situ* strain compensation buffers. The compressive stress in AlGaN layers will cause a convex wafer bow during high-temperature growth, which will compensate the concave wafer bow that occurs during cooling at room temperature [3].

21.3.4.2 Experimental Details

The growths of GaN-on-Si samples covered in this section were performed under the same conditions as described earlier. A new susceptor design was used, in which the wafer was suspended by 12 protrusions (400 μm in height and radially separated by 30°) around the edge of the pocket. This is illustrated in Figure 21.17b. The SiN$_x$ masking layer is introduced by introducing the dopant precursor Si$_2$H$_6$ and NH$_3$ in the midst of growing the GaN layer. Real-time curvature, reflectance, and true surface temperature measurements were taken by a commercial Epicurve®TT sensor. Data from a standard 200 mm

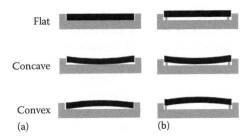

FIGURE 21.17 Illustration of the wafer bow change and its conduction with different susceptor designs: (a) flat pocket susceptor and (b) shaped susceptor.

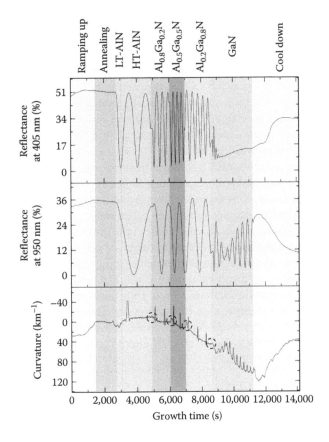

FIGURE 21.18 *In situ* true surface temperature, reflectance at 405 and 950 nm, and curvature of a standard 200 mm GaN-on-Si run.

MOCVD GaN run on 725 μm thick Si substrate with a three-step-graded AlGaN strain compensation buffer are shown in Figure 21.18.

21.3.4.3 Stress Evolution with Step-Graded AlGaN Strain Compensation Buffer

In situ curvature monitoring is an essential characterization tool in the growth of GaN on large-diameter Si substrates. From Figure 21.18, the effects of the strain compensating layers are clearly observed. The starting 200 mm Si substrate has a (convex) curvature of 26 km⁻¹ at 400°C. During the reactor ramping up to the H₂ and H₂ + Si₂H₆ annealing temperature of 1030°C, the center of the 200 mm Si

(111) substrate has a temperature of 10°C–20°C higher compared to that at the edge of the wafer and the bottom of the wafer is always higher in temperature than the top surface. This in-plane and out-of-plane temperature nonuniformity causes the substrate to have an increasing concaveness during the temperature ramp up. When the temperature reaches and stabilizes at the annealing temperature, the curvature is mostly constant. There are exceptions, however; if a large radial temperature nonuniformity (>30°C) is present, it will cause the curvature to change during the annealing period. This is due to the plastic deformation that occurs as a result of the thermal stress across the wafer, which manifests as slip lines on the Si substrate. We will discuss the consequences of this happening later. After the *in situ* Si single bilayer step recovery is completed, the reactor is cooled down to 980°C for 10 s NH_3 nitridation and LT-AlN growth. The temperature is then ramped up to 1080°C for the growth of the HT-AlN layer. Domain matching between the growing AlN film and the Si (111) substrate occurs via matching of four $(220)_{Si}$ planes with five $(2\bar{1}\bar{1}0)_{AlN}$ planes to reduce the 23.4% tensile strain to less than 1% residual strain in the first monolayer of the epitaxy. Due to this 1% residual tensile strain, the deposition of HT-AlN layer causes a small increase in curvature by 10 km^{-1} (i.e., more concave). Despite the long growth duration (2000 s), the final thickness of the HT-AlN is only 240 nm due to the slow growth rate of AlN.

The first layer of the three-step-graded AlGaN strain compensation buffer typically has an Al content around 80% (as determined by *ex situ* HRXRD), depending on the growth temperature and curvature change during the growth. $Al_{0.8}Ga_{0.2}N$, which is partially pseudomorphic (compressively strained) on the AlN, initiates the convex bowing of the entire 200 mm wafer. The subsequently grown $Al_{0.5}Ga_{0.5}N$, $Al_{0.2}Ga_{0.8}N$, and GaN layers are all under compressive strain and the wafer maintains its convex bowing until cooling from the GaN growth temperature at 1020°C to room temperature, during which the curvature changes from convex to concave. The changes in curvature of the different layers are summarized in Table 21.2. κ is the curvature and positive curvature indicates convex bow. $\Delta\kappa/h_f$ represents the change in curvature per unit thickness of the deposited layer. According to the Stoney equation (Equation 21.1), $\Delta\kappa/h_f$ is proportional to the change in strain in the entire wafer (i.e., substrate plus all epilayers). The $Al_{0.2}Ga_{0.8}N$ layer exhibits the largest $\Delta\kappa/h_f$ value of the entire growth process. There are two main factors that contribute to it possessing the largest change in curvature per unit thickness.

First, as discussed earlier in this section, the amount of compressive strain that a strain compensation layer can introduce is related to its crystal quality. If the TDD of the deposited layer is high, dislocation bending [48], which causes strain relaxation or introduces additional tensile strain due to island coalescence, could limit the incorporation of compressive strain significantly. In the $Al_{0.8}Ga_{0.2}N$ and $Al_{0.5}Ga_{0.5}N$ layers, as the total film thickness to that point is low, a high TDD density is expected. Thus, dislocation bending across the $Al_{0.8}Ga_{0.2}N$/HT-AlN and $Al_{0.5}Ga_{0.5}N$/$Al_{0.8}Ga_{0.2}N$ interfaces reduces the amount of compressive strain that can be built up. At the $Al_{0.2}Ga_{0.8}N$/$Al_{0.5}Ga_{0.5}N$ interface, the TDD is

TABLE 21.2 Summary of Layer Thickness and Curvature Change for a Standard GaN/3 Step-Graded AlGaN/AlN/200 mm Si Substrate

Material	h_f (nm)	$\Delta\kappa$ (km^{-1})	$\dfrac{\Delta\kappa}{h_f}$ ($km^{-1}\,nm^{-1}$)	Δa (%)
Ramping up	—	+26	—	—
Annealing	—	0	—	—
LT-AlN	30	+1.7	+0.057	+19
HT-AlN	240	+10	+0.042	0
$Al_{0.8}Ga_{0.2}N$	245	−11	−0.045	−0.51
$Al_{0.5}Ga_{0.5}N$	250	−12	−0.048	−0.76
$Al_{0.2}Ga_{0.8}N$	300	−34	−0.11	−0.75
GaN	1200	−65	−0.054	−0.50
Cooling down	—	+72	—	—

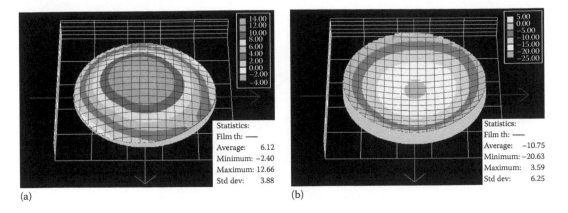

(a)

Statistics:
Film th: ——
Average: 6.12
Minimum: −2.40
Maximum: 12.66
Std dev: 3.88

(b)

Statistics:
Film th: ——
Average: −10.75
Minimum: −20.63
Maximum: 3.59
Std dev: 6.25

FIGURE 21.19 *Ex situ* bow measurement of (a) starting 200 mm Si (111) substrate (convex), (b) after the growth of a standard GaN/3 step-graded AlGaN/AlN/200 mm Si (concave). The unit of bow is μm.

lower compared to the previous two interfaces, and thus more compressive strain can be built up in the upper layers. Another reason why $Al_{0.2}Ga_{0.8}N$ introduces more compressive strain per unit thickness is the large (30%) Al compositional change, which results in a large compressive lattice mismatch of −0.75%. As shown in Figure 21.19, *ex situ* bow mapping of a starting Si (111) substrate shows a starting convex bow of +15.1 μm, whereas after the growth of the GaN/3 step-graded AlGaN/AlN/200 mm Si heterostructure, the bow becomes −24.2 μm concave, which is sufficiently small for GaN-on-Si integration purposes.

Another interesting observation is that there is always a small tensile strain component after the transition to the next layer epilayer. This is indicated by the upward curve within the dashed circles in Figure 21.18. We believe that this is caused by the strain relaxation due to dislocation bending at the interfaces at the onset of growing a new layer.

21.3.4.4 Shaped Susceptor

The change in thermal conduction from the heated susceptor to the wafer as the wafer bows during the growth induces in-plane temperature nonuniformity. As illustrated in Figure 21.17a, conventional growth of GaN-on-Si is performed with flat pocket SiC-coated graphite susceptors, where the wafer is in full contact with the susceptor surface during loading. However, during temperature ramping up, annealing, growth of LT- and HT-AlN, the wafer is going to experience concave bowing in which the edge of the wafer becomes suspended above the susceptor surface. This leads to difficulty in temperature uniformity control because thermal conduction from the heated susceptor to the wafer becomes significantly greater at the center of the wafer (which is in direct contact with the susceptor), as compared to that at the edges of the wafer. The situation is reversed during the growth of the step-graded AlGaN buffers and GaN layer when the wafer has a convex bow, and therefore the edge of the wafer experiences greater conduction relative to the center. Given that the wafer curvature changes continuously during growth, the thermal conduction profile between the susceptor and the wafer is constantly changing throughout the entire growth run.

The new susceptor design in which the wafer is suspended by 12 protrusions (400 μm in height and radially separated by 30°) around the edge of the pocket is illustrated in Figure 21.17. This modified design ensures that during the entire growth run, the wafer is not in contact with the susceptor except at the 12 protrusions. As a result, the wafer is isolated in terms of conductive thermal heating and at the same time the increased gap between the wafer and susceptor minimizes the change in thermal convection during the growth run as well, since the fractional change in gap size between the wafer and susceptor is smaller. Figure 21.20 gives the comparison of the curvature change during two GaN-on-Si growth runs with the

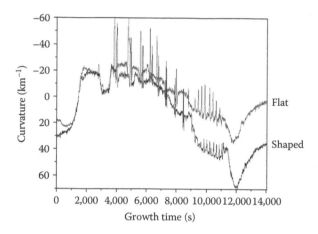

FIGURE 21.20 Curvature change of during two GaN-on-Si runs with different susceptor designs.

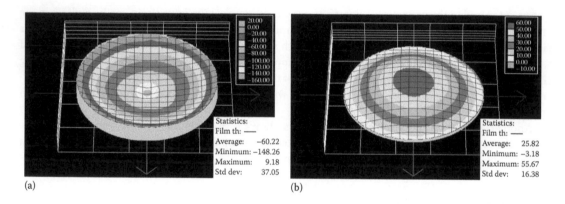

FIGURE 21.21 *Ex situ* bow measurement of a standard GaN/3 step-graded AlGaN/AlN/200 mm Si (111) substrate with 725 μm substrate thickness (a) with a flat susceptor and (b) with a shaped susceptor. The unit of bow is μm.

different susceptors. Clearly, with the use of the shaped susceptor, it is much easier to create a greater amount of compressive strain in the $Al_{0.2}Ga_{0.8}N$ buffer layer and GaN layer. The *ex situ* bow measurement is shown in Figure 21.21. The 200 mm GaN-on-Si wafer grown using the flat susceptor has a concave bow of −157.4 μm, while the wafer grown with the shaped susceptor has a convex bow of +58.9 μm. In conclusion, the shaped susceptor is able to reduce the change in thermal conduction experienced by the wafer as the curvature of the wafer changes during the growth run. As a result, maintaining a uniform temperature ($\Delta T_{radial} < 20°C$) across a 200 mm wafer is possible. At the same time, more compressive strain is observed in the $Al_{0.2}Ga_{0.8}N$ buffer with the use of a shaped susceptor.

21.3.4.5 Control of Wafer Bow with AlGaN Buffer Thickness

The automated processing facilities developed for 200 mm Si VLSI generally require wafers with SEMI-spec thickness of 725 μm and bow less than 50 μm. The two separate specifications mean that we are unable to reduce wafer bow by using thicker substrates. In this section, we introduce the bow control method that we have employed to meet this stringent bow requirement. Unlike traditional cubic semiconductors (e.g., Si, III-As/P), the wurtzite structure of the group III nitrides has unusual slip systems for in-plane strain relaxation [38]. As mentioned earlier, the MDs with mixed-type Burgers vector $b = a + c$ can glide on the $\{11\bar{2}2\}$ planes. In Sections 21.3.4.3 and 21.3.4.4, we have learnt that on a shaped susceptor,

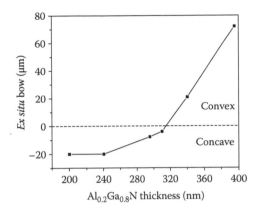

FIGURE 21.22 *Ex situ* bow of standard GaN/3 step-graded AlGaN/AlN/200 mm Si (111) substrate with 725 μm thick substrate with different $Al_{0.2}Ga_{0.8}N$ thicknesses.

three-step-graded AlGaN strain compensation buffers can introduce continuous compressive strain with the last layer—$Al_{0.2}Ga_{0.8}N$—being the most effective in terms of the amount of compressive strain built up per unit thickness. Thus, the first method we employed to adjust the final bow of the 200 mm Si wafer was to adjust the thickness of the $Al_{0.2}Ga_{0.8}N$ layer to control the amount of compressive strain introduced by this layer. Figure 21.22 is the plot of *ex situ* bow measurements of standard GaN/3 step-graded AlGaN/ AlN/200 mm Si (111) wafers with different $Al_{0.2}Ga_{0.8}N$ thicknesses. Clearly, the thicker the $Al_{0.2}Ga_{0.8}N$ layer, the more compressive strain was built up in that layer. With a thickness of 310 nm, the final bow is almost flat, with the wafer having a concave bow of about 7 μm. However, it is important to note that these bow-free wafers were grown by a strain compensation buffer technique, and thus bow free does not equate to strain-free within individual layers. The 200 mm wafer is in a metastable strain state that is very sensitive to mechanical and thermal shock.

While controlling the final wafer bow with the thickness of the $Al_{0.2}Ga_{0.8}N$ layer, we also noticed that the crystal quality of the final GaN layers changed as well. As illustrated in Figure 21.23, as the thickness of the $Al_{0.2}Ga_{0.8}N$ layer increases, the FWHM of the GaN(0002) peak increases and the FWHM of

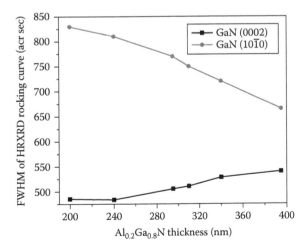

FIGURE 21.23 FWHM of HRXRD ω-scan of standard GaN/3 step-graded AlGaN/AlN/200 mm Si (111) substrate with 725 μm substrate with different $Al_{0.2}Ga_{0.8}N$ thicknesses.

the GaN$(10\bar{1}2)$ peak decreases. Here, we adopt the formula proposed by Kaganer et al. to estimate the dislocation density from HRXRD FWHM of the ω-scan [71]. The observed trend can be interpreted as a reduction of the total TDD as the layer thickness increases. The screw component of the TDD increases slightly while the edge component of the TDD decreases with increasing layer thickness.

In conclusion, thicker AlGaN buffer layers introduce more compressive stress. The optimized $Al_{0.2}Ga_{0.8}N$ layer thickness (for minimizing wafer bow) also depends on the final GaN layer thickness because the thickness of the GaN layer also contributes significantly to the tensile stress from cooling down.

21.3.4.6 Strain Decoupling with SiN$_x$ *In Situ* Masking

In this section, we emphasize an alternative application of SiN$_x$ *in situ* masking—decoupling the strain from the masked GaN layer. As discussed in Section 21.2.4.1, when an *in situ* SiN$_x$ masking layer is applied, the exact mask coverage (i.e., relative proportion of masked and unmasked regions) of the surface depends on the duration of "growth" of the SiN$_x$ mask (i.e., when silane and NH_3 are flowing, but no TMGa). In general, there is only partial coverage, and it is reasonable to assume that for GaN regrown (i.e., grown after SiN$_x$ mask treatment) in the unmasked regions, the GaN locally inherits the strain state from the exposed underlying GaN film, while the GaN that forms due to lateral growth over the SiN$_x$ mask is strain-free. It is apparent that the parameter $\Delta\kappa/h_f$ (curvature change per unit film thickness) of GaN from Table 21.2 can thus be adjusted by the growth duration of the SiN$_x$ mask. Four standard GaN/3 step-graded AlGaN/AlN/200 mm Si (111) substrates were grown with different SiN$_x$ growth duration. Since the introduction of the SiN$_x$ interlayer would make the GaN layer less compressive during growth, we fixed the thickness of $Al_{0.2}Ga_{0.8}N$ at 350 nm so that the final convex bow without SiN$_x$ interlayer would be around +40 μm. The SiN$_x$ layer was inserted after 80 nm of GaN growth on the $Al_{0.2}Ga_{0.8}N$ layer. The *in situ* curvature change is plotted in Figure 21.24, and the key parameters are summarized in Table 21.3.

Without the SiN$_x$ layer, the GaN layer is compressively grown on the $Al_{0.2}Ga_{0.8}N$ layer with a $\Delta\kappa/h_f$ value of −0.047 km^{-1} nm^{-1}. This $\Delta\kappa/h_f$ value is consistent with that from another sample (with different $Al_{0.2}Ga_{0.8}N$ layer thicknesses) that was grown without SiN$_x$ interlayer in Table 21.2. As the SiN$_x$ growth duration increases, the magnitude of $\Delta\kappa/h_f$ decreases since there is an increased fraction of regrown GaN that is decoupled from the compressive strain resulting from the $Al_{0.2}Ga_{0.8}N$ layer. When the coverage of SiN$_x$ is high (duration of SiN$_x$ = 60 s), the coalescence thickness of GaN is approximately 660 nm. The regrown GaN layer is almost fully decoupled from the compressive strain resulting from the $Al_{0.2}Ga_{0.8}N$ layer. The $\Delta\kappa$ of this sample (+100 km^{-1}) at cooling down is about 154% of the $\Delta\kappa$ (+65 km^{-1}) of the rest of the samples. This is because its GaN thickness is 150% of the rest of the samples. The $\Delta\kappa$ at cooling is

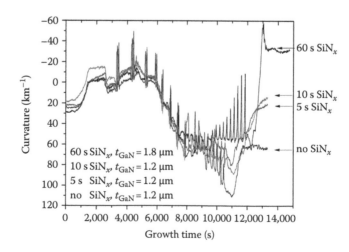

FIGURE 21.24   Curvature change during four GaN-on-Si runs with different SiN$_x$ *in situ* masking duration.

TABLE 21.3 Summary of Four GaN/3 Step-Graded AlGaN/AlN/200 mm Si Substrate Samples with Different SiN$_x$ Duration

SiNx Duration (s)	0	10	15	60
GaN coalescence thickness (nm)	0	150	190	660
$\Delta\kappa/hf$ of GaN (km^{-1} nm^{-1})	−0.047	−0.013	−0.010	0
Final bow (μm)	+49 Convex	−7 Concave	−19 Concave	−179 Concave
HRXRD ω-scan FWHM GaN(0002) (arc sec)	529	515	507	430
HRXRD ω-scan FWHM GaN$(10\bar{1}2)$ (samples)	711	622	613	512

approximately proportional to the GaN thickness (i.e., $\Delta\kappa/h_f$ is roughly constant). From the *ex situ* bow measurement, the final wafer bow becomes more concave when the duration of SiN$_x$ increases and the wafer is almost bow-free at 10 s SiN$_x$ duration.

In conclusion, a longer SiN$_x$ growth duration decouples the compressive strain of the Al$_{0.2}$Ga$_{0.8}$N layer from the regrown GaN layer. Additional tensile strain (post-cooldown) can be introduced by growing a thicker GaN layer. Thus, SiN$_x$ *in situ* masking can be used as a tool for strain engineering to adjust the amount of compressive or tensile stress in the wafer during cooling down, thereby allowing one to control the final wafer bow.

Another benefit of employing the SiN$_x$ interlayer as a strain engineering method is its dislocation reduction ability. As discussed earlier, the SiN$_x$ interlayer causes TD bending and promotes TD annihilation. A comparison of the two strain engineering methods (control of the Al$_{0.2}$Ga$_{0.8}$N layer thickness vs. adjustment of SiN$_x$ mask coverage) in terms of HRXRD GaN(0002) and GaN$(10\bar{1}2)$ ω-scan FWHM and final wafer bow is provided in Figure 21.25. FWHM of both GaN(0002) and GaN$(10\bar{1}2)$ peaks decreases monotonically with increasing SiN$_x$ duration as illustrated in Figure 21.25b. Thus, both screw and edge components of the TDD are reduced by SiN$_x$ masking. In practice, we combine these two methods to produce a bow-free wafer with minimum TDD, where the Al$_{0.2}$Ga$_{0.8}$N layer thickness and SiN$_x$ mask growth time are optimized for a given device heterostructure.

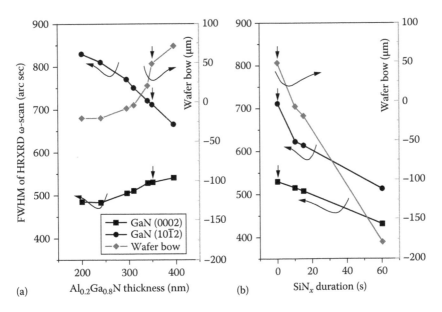

FIGURE 21.25 Comparison of HRXRD ω-scan FWHM of standard GaN/3 step-graded AlGaN/AlN/200 mm Si (111) substrate with 725 μm thick substrate with different strain engineering methods: (a) thickness of Al$_{0.2}$Ga$_{0.8}$N layer and (b) SiN$_x$ growth duration. (a) and (b) share the same y scale for both FWHM and bow and the points marked by the arrow illustrate identical data shared by both (a) and (b) (350 nm Al$_{0.2}$Ga$_{0.8}$N without SiN$_x$ interlayer).

21.3.4.7 Overview

In Section 21.3.4, we introduced the principles of strain compensation in GaN-on-Si heteroepitaxy. The basic idea is to introduce compressive strain during epitaxial growth by employing the in-plane lattice mismatch in the AlGaN material system to compensate for the large tensile stress generated during cooling from growth temperature to room temperature due to CTE mismatch. The stress evolution of step-graded AlGaN was discussed in detail based on the basic strain engineering principles. We have demonstrated that the difficulty in strain engineering on a 725 μm thick 200 mm diameter Si wafer can be solved by employing a shaped susceptor that decouples the change in thermal conduction from wafer curvature change. After changing the susceptor from a flat one to a shaped one, $Al_{0.2}Ga_{0.8}N$ layer thickness and SiN_x *in situ* masking can be applied to fine-tune the final bow of the GaN-on-Si wafers. Usually, a more convex wafer is produced from an increase in $Al_{0.2}Ga_{0.8}N$ layer thickness and TDD, in general, improves as well. However, the composition of screw and edge TD changes with $Al_{0.2}Ga_{0.8}N$ layer thickness. In the case of SiN_x *in situ* masking, it decouples the compressive strain in the regrown GaN from the $Al_{0.2}Ga_{0.8}N$ layer, so the GaN layer is less compressively strained. TDD monotonically improves with the coverage of SiN_x. In practice, these two methods are combined to produce wafers that are simultaneously optimized for bow and TDD.

21.3.5 Dislocation Reduction with MEE AlN Layer

21.3.5.1 Introduction

In this section, the effect of the AlN buffer layer surface morphology on the TDD and the stress evolution in GaN epilayers deposited by MOCVD on 200 mm Si (111) substrates are investigated. The MEE AlN buffer layers were grown using alternately pulsed flows of the group III and group V precursors. The enhanced lateral growth rate in migration-enhanced AlN (ME-AlN) improved the extent of coalescence of the films and eliminated the high density of surface pits in the AlN buffer layer. *In situ* curvature measurements and post-growth AFM measurements of TDD revealed that strain relaxation during GaN growth is correlated with the density of TD with edge components (i.e., pure edge + mixed threading dislocations). GaN epilayers grown on optimized ME-AlN buffer layers exhibited a significant reduction in TDD compared to GaN epilayers grown on conventional continuously grown AlN. This indicates that ME-AlN can be one of the many buffer engineering strategies to reduce dislocation densities and control strain, thereby improving GaN film crystal quality.

21.3.5.2 Experimental Details

The epitaxial growths of the GaN-on-Si wafers were performed in an AIXTRON CRIUS® close-coupled-showerhead (CCS) MOCVD reactor on 725 μm thick 200 mm diameter Si substrates. Prior to the LT-AlN growth, the Si (111) substrate was annealed in H_2 ambient for 600 s at 1050°C, followed by $H_2 + Si_2H_6$ annealing for another 600 s at the same temperature. Then, a 25 nm low-temperature AlN nucleation layer was grown at 995°C with both group III and group V precursors flowing simultaneously with a V/III ratio of 250. The temperature was increased to 1080°C for HT-AlN buffer growth with continuous AlN growth during the temperature ramping (15 nm of AlN deposition took place). For the conventional HT-AlN buffer, the group III and group V precursors were injected simultaneously whereas for the HT-AlN buffer by MEE, alternately pulsed group III and group V precursor flows were adopted. As illustrated in Figure 21.26, in MEE, when the NH_3 (group V precursor) is turned on, the TMAl (group III precursor) is turned off and vice versa. For simplicity, t_{NH3} is set to be equal to t_{TMAl}. Therefore, in one MEE cycle, NH_3 and TMAl are separately injected 50% of the time. The carrier gas is always H_2.

21.3.5.3 Surface Morphology of Conventional and ME-AlN

First, 135, 220, and 350 nm thick HT-AlN buffer layers were grown using the conventional continuous-flow epitaxy method. Each sample was grown on top of 25 nm of LT-AlN and 15 nm of the AlN grown during the temperature transition from LT to HT to give a total AlN thickness of 175, 260, and 390 nm.

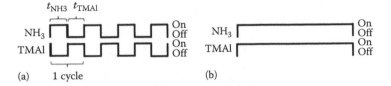

FIGURE 21.26 Comparison of NH$_3$ and TMAl flow in (a) MEE and (b) conventional epitaxy.

All exhibited characteristics of three-dimensional (3D) island growth mode as shown in Figure 21.27. Large densities (~10^9 cm^{-2}) of hexagonal pits (~50 nm diameter) were observed on the HT-AlN surfaces due to incomplete coalescence of the AlN mosaic islands. As illustrated in Figure 21.28, the density of the pits increased with HT-AlN thickness from 5 × 10^8 cm^{-2} and saturated at 2–3 × 10^9 cm^{-2}. The average pit size decreased gradually from 65 to 53 nm when the thickness of the conventional HT-AlN buffer was increased from 135 to 350 nm. As for the surface roughness over a 25 μm^2 area, AFM RMS of the 350 nm thick conventional HT-AlN layer was 2.4 nm, which was higher compared to the 135 and 220 nm samples (RMS ~1.0 nm). Here, we conclude that the pits in a conventional HT-AlN layer cannot be eliminated or reduced by increasing the thickness of the HT-AlN layers. The surface roughness in fact increases with the layer thickness.

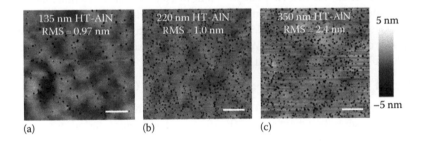

FIGURE 21.27 AFM scans of conventional HT-AlN, (a) 135 nm, (b) 220 nm, and (c) 350 nm. The scale bars are 1 μm.

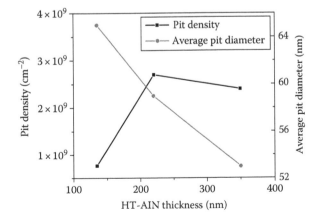

FIGURE 21.28 Comparison of pit density and average pit diameters with different conventional HT-AlN thicknesses.

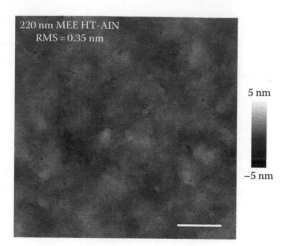

FIGURE 21.29   AFM scan of 220 nm MEE HT-AlN. The scale bar is 1 μm.

In contrast, the 220 nm thick ME-AlN layer consisted of 25 nm of LT-AlN grown conventionally, 15 nm of AlN grown conventionally during the temperature transition from LT to HT, and 180 nm of HT-AlN grown with MEE method with $t_{NH3} = t_{TMAl} = 6$ s. Unlike the conventionally grown LT-/HT-AlN layer, the ME-AlN layer underwent complete coalescence. Comparing the 220 nm thick AlN samples, the step-flow growth mode experienced by the ME-AlN sample results in a pit-free surface as illustrated in Figure 21.29, in comparison to the sample grown conventionally as shown in Figure 21.27b. The AFM RMS value of 220 nm ME-AlN over an area of 25 μm² is 0.35 nm, in comparison to an RMS value of 1.0 nm for the conventional AlN layer of the same thickness. We believe that the transition to a step-flow-like growth mode in ME-AlN is due to the enhancement in Al adatom mobility.

21.3.5.4 Effect of MEE AlN on Dislocations and Stress in GaN-on-Si

A GaN/3 step-graded AlGaN/ME-AlN/ Si (111) wafer with total III-N film thickness of 800 nm was grown on a 200 mm Si (111) substrate, whereby the ME-AlN layer had a pulse period of $t_{NH3} = t_{TMAl} = 6$ s. A similar wafer was grown, with the only difference being that the AlN layer was grown conventionally, to study the difference made by the ME-AlN layer versus the conventionally grown AlN layer. For simplicity, the GaN-on-Si with the conventional AlN buffer is designated as sample A and the GaN-on-Si with the ME-AlN buffer is designated as sample B in the following analysis. The growth rate of the ME-AlN layer in terms of nm/h was slightly more than half of the growth rate of the conventional AlN layers. The small discrepancy (based on the 50% gas flow duty cycle, the ME-AlN growth rate would be expected to be half that of the conventional AlN growth rate) may be due to the elimination of parasitic gas-phase reactions between TMAl and NH₃ in the MEE process.

In TEM, the effect of a dislocation on the image depends on its Burgers vector **b**. In the previous section, we briefly introduced the Burgers vectors of different types of dislocations in III-nitrides. If the Burgers vector **b** of a particular dislocation type is perpendicular to the TEM diffraction vector **g**, these dislocations will not exhibit any contrast under a two-beam condition with diffraction vector **g**. Thus, diffraction conditions for which $g \cdot b = 0$ will not lead to contrast in the image [72]. Figure 21.30 shows cross-sectional TEM images taken under a $g = (0002)$ two-beam diffraction condition. Screw dislocations and mixed dislocations are visible in the image, as they have Burgers vectors of $b_{screw} = c = \langle 0001 \rangle$ and $b_{mixed} = a + c = 1/3\langle 11\bar{2}3 \rangle$, respectively, and thus $g \cdot b \neq 0$. However, since pure screw dislocations are rare compared to mixed dislocations [38], the dislocations observed in Figure 21.30 can all be regarded as mixed dislocations. In contrast, the TEM images in Figure 21.31 were taken under a $g = (1\bar{2}10)$ two-beam diffraction condition. Thus, in these images, edge and mixed dislocations that have Burgers vectors of $b_{edge} = a = 1/3\langle 11\bar{2}0 \rangle$ and $b_{mixed} = a + c = 1/3\langle 11\bar{2}3 \rangle$, respectively, are visible. All TDs in GaN-on-Si

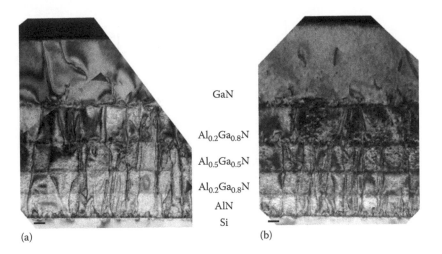

GaN

Al$_{0.2}$Ga$_{0.8}$N

Al$_{0.5}$Ga$_{0.5}$N

Al$_{0.2}$Ga$_{0.8}$N

AlN

Si

(a)   (b)

FIGURE 21.30 Bright-field cross-sectional composite images of GaN-on-Si near the $\langle 11\bar{2}0 \rangle$ ZA direction using $g = (0002)$ two-beam diffraction conditions. Mixed dislocations are in contrast. (a) Sample A (GaN thickness = 850 nm). (b) Sample B grown with MEE (GaN thickness = 1100 nm). The scale bar is 100 nm.

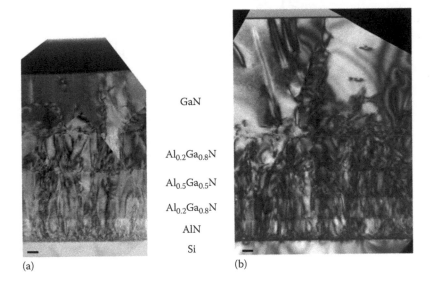

GaN

Al$_{0.2}$Ga$_{0.8}$N

Al$_{0.5}$Ga$_{0.5}$N

Al$_{0.2}$Ga$_{0.8}$N

AlN

Si

(a)   (b)

FIGURE 21.31 Bright-field cross-sectional composite images of GaN-on-Si near the $\langle 11\bar{2}0 \rangle$ ZA direction using $g = (1\bar{2}10)$ two-beam diffraction conditions. Mixed and edge dislocations are in contrast. (a) Sample A (GaN thickness = 850 nm). (b) Sample B grown with MEE (GaN thickness = 1100 nm). The scale bar is 100 nm.

are revealed in Figure 21.31, since, as mentioned earlier, pure screw dislocations with Burgers vector $b_{screw} = c = \langle 0001 \rangle$ are rare.

The TDD and average bending angles of mixed TDs are obtained from the analysis of Figure 21.30, while the TDD of pure edge and mixed TDs can be found by analyzing Figure 21.31. Thus, to obtain the pure edge TDD, we subtract the two TDD values. The results of the TEM analysis are summarized in Table 21.4. The TDDs of the different layers are calculated from the square of the line densities in each of the layers as illustrated in Figure 21.32, which was derived from Figure 21.30. The average bending angle is measured with respect to the TDs in the bottom layer (AlN) as shown in Figure 21.32, so there are no average bending angle values reported for the AlN layer.

TABLE 21.4 Summary of Dislocation Densities and Strain Values in GaN-on-Si with Conventional HT-AlN (A) and MEE HT-AlN (B) Buffers

Material	h_f (nm)	$\rho_{mixed\,TDD}$ ($\times10^9$ cm^{-2})		$\theta^{avg}_{mixed\,TDs}$ (°)		$\rho_{edge\,TDD}$ ($\times10^9$ cm^{-1})		$\theta^{avg}_{total\,TDs}$ (°)		$\dfrac{\Delta\kappa}{h_f}$ (km^{-1} nm^{-1})	
		A	B	A	B	A	B	A	B	A	B
LT-AlN	40	—	—	—	—	—	—	—	—	+0.057	+0.063
HT-AlN	240	7.9	8.1	—	—	11	11	—	—	+0.042	+0.037
Al$_{0.8}$Ga$_{0.2}$N	245	7.9	8.1	2.7	4.1	11	11	2.5	4.3	−0.052	−0.049
Al$_{0.5}$Ga$_{0.5}$N	250	7.9	6.1	3.9	7.3	11	8.9	4.6	7.1	−0.048	−0.074
Al$_{0.2}$Ga$_{0.8}$N	300	6.3	4.7	5.2	16	7.9	7.6	4.3	17	−0.11	−0.17
GaN	850 to 1100	1.4	0.7	8.9	21	2.7	1.9	5.6	19	−0.054	−0.048

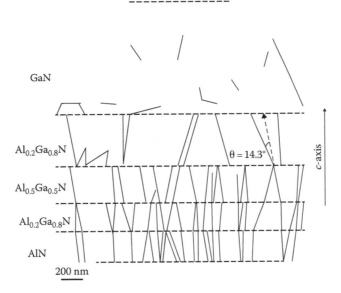

FIGURE 21.32 Schematic illustration of mixed dislocations from the TEM image of sample B (see Figure 21.30b). The TDDs of different layers are calculated from the square of line densities. θ^{avg} is the average angle of the dislocation lines with respect to the *c*-axis.

Both samples A and B are grown with the same 40 nm LT-AlN buffer. The TD bending angle is measured with respect to the *c*-axis of the GaN layer. The conventional AlN and the ME-AlN buffer have almost identical edge and mixed TDDs. This agrees well with our HRXRD ω-scan FWHMs of AlN(0002) and AlN$(10\bar{1}5)$ peaks of both conventional AlN and ME-AlN buffers. The HRXRD results for samples A and B were 2100–2300″ and 3200–3400″, respectively. The difference between the conventional AlN and the ME-AlN buffers starts when Al$_{0.8}$Ga$_{0.2}$N layers are grown on top of them. There are many surface pits in the conventional AlN buffer, whereas the ME-AlN buffer has a smooth surface. Although the TDDs of mixed and edge TDs are similar in both samples, the average bending angle of both mixed and total (mixed + edge) TDs in the Al$_{0.8}$Ga$_{0.2}$N layer in sample B is approximately double of that in sample A. In the Al$_{0.5}$Ga$_{0.5}$N layers, TDDs of both mixed and edge TDs are reduced in sample B, and the average bending angles of both mixed and total TDs are about three to four times of those in sample A. We believe that the Al$_{0.5}$Ga$_{0.5}$N layer in sample B has lower TDDs because there is more bending of TDs from the Al$_{0.5}$Ga$_{0.5}$N/ Al$_{0.8}$Ga$_{0.2}$N interface. In both the Al$_{0.2}$Ga$_{0.8}$N and GaN layers, both edge and mixed TDDs are significantly lower and the bending of TDs is much greater in sample B. Compared with sample A, the reduction in TDD in sample B is likely due to the greater bending of TDs in the buffer layer, which in turn originated from the pit-free AlN surface morphology at the Al$_{0.2}$Ga$_{0.8}$N/ME-AlN interface.

In order to study the effect of the ME-AlN buffer on the stress evolution of the GaN-on-Si heterostructure, *in situ* reflectance and curvature change data obtained during the growth of samples A and B are plotted in Figure 21.33. $\Delta\kappa/h_f$ of each layer is summarized in Table 21.4. From Equation 21.3, the average relaxation ε^{avg} is proportional to the product of $\rho_{mixed + pure\ edge}$ and $h\tan\theta^{avg}$. Due to the dislocation reduction ability of the MEE AlN buffer, magnitudes of $\Delta\kappa/h_f$ in the Al$_{0.5}$Ga$_{0.5}$N and the Al$_{0.2}$Ga$_{0.8}$N layers are significantly greater for sample B than for the same layers in sample A. As shown in Figure 21.34, *ex situ* bow measurements reveal that the final bow of sample A is +58.9 μm convex whereas the final bow of sample B is +77.2 μm convex. In conclusion, with MEE of the HT-AlN layer, the strain compensation AlGaN buffer layers are able to build up more compressive strain during the growth.

From the *in situ* reflectance measurement at 405 nm of samples A and B shown in Figure 21.33, it can be deduced that the initial growth of GaN on the Al$_{0.2}$Ga$_{0.8}$N layer in sample B is rougher than that of

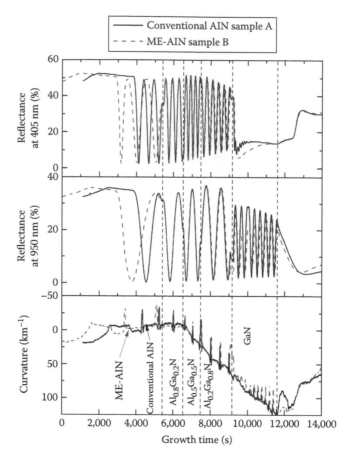

FIGURE 21.33 *In situ* reflectance at 405, 950 nm, and curvature of two 200 mm GaN-on-Si runs with conventional HT-AlN or MEE HT-AlN buffer.

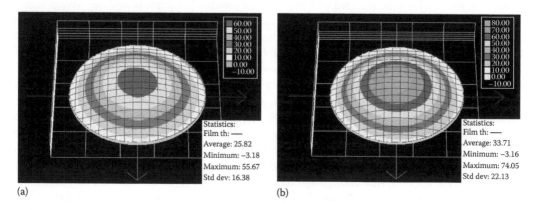

(a)   (b)

FIGURE 21.34 *Ex situ* bow measurement of a standard GaN/3 step-graded AlGaN/AlN/200 mm Si (111) substrate with 725 µm substrate thickness: (a) sample A with conventional HT-AlN buffer and (b) sample B with MEE HT-AlN buffer. The unit of bow is µm.

sample A. At 1000°C during the growth of GaN layers, 405 nm light is absorbed in the GaN due to the temperature-dependent bandgap narrowing effect. A sharp decrease in reflectance at the 405 nm wavelength was observed during the first 50 nm of GaN growth in sample B and it gradually reversed after 500 nm of GaN growth. The recovery of reflectance at 405 nm took a longer time in sample B. From the *in situ* reflectance measurement at 950 nm, a similar phenomenon is observed. The oscillation in the GaN layer in sample A has an almost constant amplitude, whereas the GaN layer in sample B oscillates with increasing amplitude. Cantu et al. showed that $Al_xGa_{1-x}N$ epilayers of the same compositions have greater dislocation bending with increased surface roughness [44]. Our observation of increased roughness of GaN is also correlated with increased dislocation bending in the GaN layer. Therefore, our findings strongly suggest a surface-mediated climb model as previously proposed by Follstaedt et al. [47] as the dislocation bending mechanism. However, this is not borne out at the AlN surface, where the increased roughness caused by surface pits in the conventional AlN surface appears to induce a smaller amount of dislocation bending. The exact mechanism for this dislocation behavior is not well understood at the moment.

21.3.5.5 Effect of ME-AlN Pulse Duration on GaN-on-Si

A series of GaN/3 step-graded AlGaN/ME-AlN/ Si (111) wafers with total III-N film thickness of 800 nm were grown on 200 mm Si (111) substrates. The various samples had ME-AlN layers that were grown with varying pulse periods of $t_{NH3} = t_{TMAl} = 2$–8 s and were compared with a GaN-on-Si wafer with conventionally grown AlN buffer and the same total III-N film thickness. The GaN TDD values, both measured by AFM and calculated from the FWHM of HRXRD ω-scans, are plotted in Figures 21.35 and 21.36, respectively. At the optimized 6 s single pulse duration, the GaN epilayer exhibited a significant reduction in edge TDD compared to the GaN epilayer grown with conventionally grown AlN. This was confirmed by both AFM and XRD FWHM measurements.

We have concluded that the ME-AlN buffer reduces the TDD in the GaN layer compared to a conventional HT-AlN buffer. The vertical breakdown voltages of conventionally grown AlN and ME-AlN buffers were measured by an Agilent B1500A semiconductor analyzer using Ti/Al/Ni/Au ohmic contacts. We found that breakdown (defined as when current exceeded 1 A cm^{-2}) occurred at 360 and 420 V for the conventional HT-AlN and ME-AlN buffers, respectively. We believe that by eliminating the surface pits on the AlN buffer, the vertical breakdown voltage was improved (see Figure 21.37).

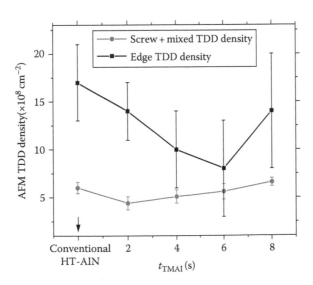

FIGURE 21.35 AFM TDD measurements of GaN layers with different pulse durations of ME-AlN buffer.

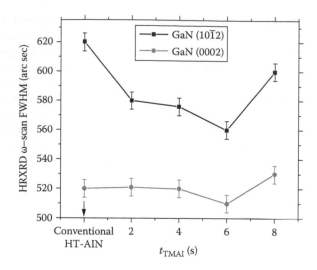

FIGURE 21.36   FWHM of HRXRD ω-scans of GaN layers with different pulse durations of ME-AlN buffer.

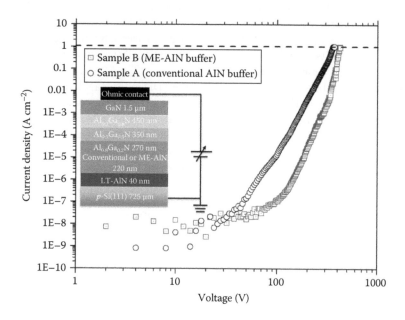

FIGURE 21.37   Measured forward biased log J–log V characteristics in GaN-on-Si structures with different AlN buffers at room temperature. The inset shows the measurement setup.

21.3.5.6 Overview

In conclusion, the reduction in edge TDD in GaN on ME-AlN buffer was attributed to dislocation annihilation caused by surface roughening during the initial GaN growth on the ME-AlN buffer. The smooth and pit-free surface of ME-AlN results in increased dislocation bending at the $Al_{0.8}Ga_{0.2}N$/AlN interface. This was further confirmed by cross-sectional TEM, where increased bending and annihilation of mixed- or edge-type dislocations were observed for GaN on the ME-AlN buffer. In addition, larger amounts of compressive strain were built up in the ME-AlN buffer as compared to conventional AlN buffers due to the reduction in strain relaxation as a result of the reduction in TDs.

21.3.6 Wafer Fragility

21.3.6.1 Introduction

For the growth of GaN on 200 mm diameter 725 μm thick Si (111) wafers, we often noticed that although the tensile stress due to CTE mismatch was carefully compensated by the strain-engineered buffer, the wafer was very fragile. There were several incidences of the GaN-on-Si wafers cracking into multiple large pieces during the wafer bonding processes described in Section 1.1. We noticed that the fragility of the 200 mm diameter 725 μm thick GaN-on-Si (111) wafers increased significantly when slip lines were formed in the Si substrate during the substrate annealing step before LT-AlN deposition. The slip line formation was found to be caused in part by large vertical and radial temperature variations across the 200 mm Si substrate during the annealing step.

Si crystal slip takes place if the local stress exceeds the yield strength at the annealing temperature (1050°C). In the growth of GaN on 200 mm diameter 725 μm thickness Si (111) wafers with a shaped susceptor (this was discussed in Section 21.3.4.4), the Si substrate is always suspended by 12 protrusions on the shaped susceptor. As a result, there are two major sources of stress on the Si substrate in our MOCVD growth. They are the contact stresses between the protrusions and the wafer and the thermal stress due to temperature nonuniformity in the vertical and radial directions [73].

21.3.6.2 Experimental Details

Samples were grown with the same conditions as described in Section 21.3.4.2. The observation of slip lines across the entire 200 mm GaN-on-Si wafer was achieved by the following process. Prior to the hydrophilic SiO$_2$–SiO$_2$ bonding step in the GaN/Si-CMOS integration process, plasma-enhanced chemical vapor deposition (PECVD) is used to deposit a layer of SiO$_2$ on the 200 mm diameter GaN-on-Si wafer. The wafer then goes through chemical–mechanical polishing (CMP) to reduce the surface roughness of the PECVD SiO$_2$ in preparation for the actual bonding step. As illustrated in Figure 21.38, the CMP process creates a thickness difference in the conformally deposited PECVD SiO$_2$ across Si slip lines. As a result, there is clear optical contrast in the slip line regions of the GaN-on-Si wafer.

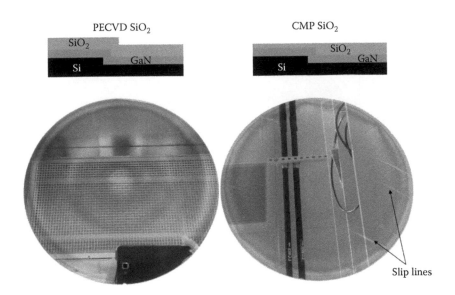

FIGURE 21.38 Illustration of slip lines being observed in the Si substrate after CMP planarization of the deposited SiO$_2$. The pattern on the wafers is due to reflection.

21.3.6.3 Slip Line Reduction

As illustrated in Figure 21.38, the slip lines originate from the edge of the wafer and propagate toward the center of the wafer. Similar observations have been made by Zhu et al. who concluded that radial temperature differences were the most likely cause of the formation of slip lines in 150 mm GaN-on-Si wafers [74]. Another important observation is that the origins of the slip lines on the edge of the wafer coincide with the positions of the protrusions on the shaped susceptor. The surface temperature profiles of two GaN-on-Si runs during the Si substrate annealing step are illustrated in Figure 21.39. The difference in ΔT_{radial} for the two separate runs is due to differences in heater zone settings.

It is clear that there are 12 regions with higher temperature corresponding to the 12 protrusions in the shaped susceptor. The thermal conduction between the protrusions on the shaped susceptor and the Si wafer produces extra radial and vertical thermal stress. From Figure 21.39, it has been estimated that regions on the protrusions are about 13.7°C higher in temperature compared to the suspended regions. There is additional contact stress exerted on the wafer by the protrusions as well. In Figure 21.39a, the average ΔT_{radial} is 34.9°C whereas it is much lower in Figure 21.39b due to better heater control and temperature

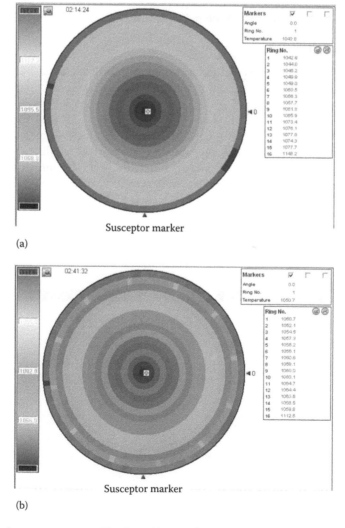

(a)

(b)

FIGURE 21.39 Surface temperature profile of two 200 mm diameter GaN-on-Si runs with (a) radial temperature difference, ΔT_{radial} = 34.9°C and (b) ΔT_{radial} = 9.1°C.

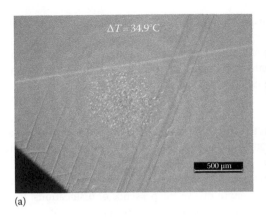
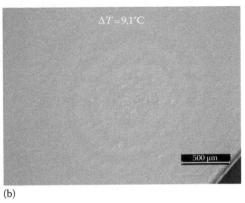

(a)

(b)

FIGURE 21.40 Nomarski microscopy images of regions of GaN-on-Si on top of the protrusions in the shaped susceptor with (a) ΔT_{radial} = 34.9°C and (b) ΔT_{radial} = 9.1°C.

balancing. The corresponding Nomarski microscopy images of the regions of GaN-on-Si on top of the protrusions in the shaped susceptor with ΔT_{radial} of 34.9°C and 9.1°C are shown in Figure 21.40. The regions with concentric circular patterns in Figure 21.40 correspond to where the protrusions in the shaped susceptor were located. Short (up to 1 mm) lines at the edge of the wafer are cracks formed in the GaN films, while the long lines are slip lines on the Si substrate (seen only in Figure 21.40a). The slip lines are usually a group of parallel lines that extend up to a few centimeters into the center of the wafer. In Figure 21.40b, slip lines are not observed.

We believe that the shear stress induced by the ΔT_{radial} across the 200 mm diameter Si wafer is enhanced by the local stress produced by the protrusions on the shaped susceptor. This is the reason why protrusions on the shaped susceptor are the origins of the slip lines. With smaller ΔT_{radial} (9.1°C), local stress produced by the protrusions on the shaped susceptor via thermal stress is not large enough to generate slip lines. The *in situ* curvature measurements of the aforementioned two GaN-on-Si runs are plotted in Figure 21.41. Large ΔT_{radial} (34.9°C) induces slip line formation, which is also a manifestation of plastic deformation in the Si wafer. The slip lines cause the Si wafer to plastically deform with an increasing concave bow. However, if ΔT_{radial} is carefully minimized (9.1°C) by optimizing heater zone

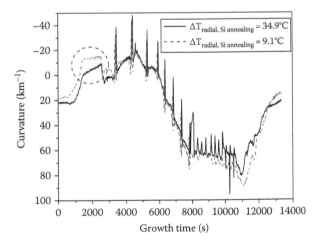

FIGURE 21.41 *In situ* curvature of two 200 mm GaN-on-Si runs with different radial temperature differences. The dashed circle highlights the curvature changes during Si annealing step.

settings, there is almost no change in the curvature during the substrate annealing step as indicated within the circle in Figure 21.41.

In conclusion, in a shaped susceptor, protrusions of the susceptor can result in extra thermal and contact stresses at the edge of the wafer. ΔT_{radial} can be enhanced by the presence of the protrusions. However, slip line–free wafers can be achieved by minimizing ΔT_{radial}.

21.3.6.4 Advanced Engineered Substrate

In order to improve the yield in the bonding process to form GaN/Si-CMOS engineered wafers, we have minimized the number of slip lines formed on the 200 mm diameter Si wafer during the annealing step by optimizing the heater zone settings to reduce ΔT_{radial}. Another approach to further minimize slip line formation is to passivate the edge of the Si wafer before the GaN growth. The edges of the wafer often experience the highest stress while having reduced strength due to the presence of dislocations. It was observed in Figure 21.40 that most of the slip lines in the Si wafer originate from the edge of the wafer. An engineered Si (111) substrate as illustrated in Figure 21.42 is prepared by the deposition of 140 nm of thermal oxide, and then the thermal oxide was patterned and dry etched to form a 190 mm diameter growth window on a standard 200 mm Si (111) wafer.

Since the entire edge of the Si wafer is passivated, nucleation of slip lines from the edge of the Si wafer is impossible. As a result, there are no slip lines found in the GaN growth window in Figure 21.42. Furthermore, the protrusions on the susceptor are now in contact with the thermal oxide masked region where there is no epitaxial growth of GaN. This again minimizes the stress on the edge of the wafer.

21.3.6.5 Overview

Wafer fragility in GaN-on-Si is associated with Si substrate slip lines that are formed due to large ΔT_{radial} and exacerbated by the protrusions on the shaped susceptor via thermal stress and contact stress. In order to improve the yield in the bonding process to form GaN/Si-CMOS engineered wafers, we have minimized slip line formation on the 200 mm diameter Si wafer during the annealing step by optimizing the

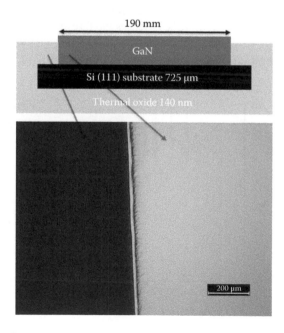

FIGURE 21.42 Schematic illustration of the structure of GaN-on-engineered Si substrate and Nomarski microscopy images of the edge region of the wafer.

heater zone settings to reduce ΔT_{radial}. Another approach to further minimize the slip line formation is to passivate the edge of the wafer in engineered substrates.

21.3.7 Chapter Overview

In summary, a detailed discussion about the growth of III-nitrides on 200 mm Si substrates is presented including studies on the AlN/Si interface, crystallographic tilting introduced by substrate vicinal steps, strain engineering with $Al_{0.2}Ga_{0.8}N$ buffer layers and SiN_x interlayers, dislocation reduction with migration-enhanced epitaxy of AlN buffer, and Si slip line reduction for III-nitride/Si integration.

21.4 Conclusion and Outlook

It has been widely accepted that GaN-on-Si technology will be very important in the coming decades. GaN-on-Si always starts with the growth on an AlN layer on Si. In this chapter, the initial focus is on understanding the interface between the AlN and the Si substrate because high-quality AlN is the necessary foundation for subsequent growth of high-quality GaN films. Prior to the growth of the LT-AlN nucleation layer, the Si (111) surface is annealed in H_2 at 1030°C to remove impurities. Subsequently, the Si substrate is annealed in $H_2 + Si_2H_6$ at the same temperature. After these pretreatment steps, we have shown that the Si substrate before AlN growth consists of vicinal atomic planes with regularly aligned bilayer steps according to the miscut of the substrate.

After the substrate pretreatment, the substrate then goes through an optimized 10 s NH_3 nitridation step followed by 30 nm of LT-AlN deposition at 980°C. From the HRTEM images of the AlN/Si interface, we draw the conclusion that the thin amorphous SiN_x interlayer exists with crystallographically abrupt zones within the thickness of the TEM lamella. Thus, the amorphous SiN_x layer is not continuous and there must be regions where crystalline AlN was grown directly on Si. At the same time, our HRTEM analysis of the AlN/Si interface suggests that Dimitrakopulos cnid2 is the most likely interfacial structure of AlN/Si for AlN growing on Si with MOCVD, where the first layer of Al–N tetrahedrons would stack at the sphalerite positions of the substrate and the subsequent layer would repeat the stacking sequence of the topmost layer of Si and bottommost layer of the AlN, that is, **ABC:ACAC**.

During the transition from LT-AlN (980°C) to HT-AlN (1080°C), 15 nm of continuous AlN growth takes place, after which 220 nm of HT-AlN is best grown with the MEE method. We find that there is a reduction in edge TDD in the subsequently grown GaN layers on ME-AlN buffers. This is attributed to dislocation annihilation caused by surface roughening during the initial GaN growth on the ME-AlN buffer. The smooth and pit-free surface of ME-AlN results in increased dislocation bending at the $Al_{0.8}Ga_{0.2}N$/AlN interface. This was further confirmed by cross-sectional TEM, where increased bending and annihilation of mixed- or edge-type dislocations were observed for GaN on the ME-AlN buffer. In addition, larger amounts of compressive strain were built up in the ME-AlN buffer as compared to conventional AlN buffers due to the reduction in strain relaxation caused by the reduction in TDs.

Next, three-step-graded AlGaN layers with Al content of 20%, 50%, and 80% are grown on top of the AlN/Si structure at a temperature of 1050°C. We have introduced the principles of strain compensation in GaN-on-Si heteroepitaxy. The basic idea is to introduce compressive strain during epitaxial growth by employing the in-plane lattice mismatch in the AlGaN material system to compensate for the large tensile stress that arises during cooldown from growth temperatures to room temperature, due to CTE mismatch between the III-N materials and Si. The stress evolution of step-graded AlGaN is discussed in detail based on the strain engineering principles. We have demonstrated that the difficulty in strain engineering on a 725 μm thick 200 mm diameter Si wafer can be solved by employing a shaped susceptor that decouples the change in thermal conduction from wafer curvature change. After changing the susceptor from a flat one to a shaped one, adjustment of the $Al_{0.2}Ga_{0.8}N$ layer thickness and SiN_x *in situ* masking can be applied to fine-tune the final bow of the GaN-on-Si wafers. Usually, a more convex wafer is produced from an increase in $Al_{0.2}Ga_{0.8}N$ layer thickness, and TDD, in general, improves as well. However, the relative proportions of screw and edge TDs change with $Al_{0.2}Ga_{0.8}N$ layer thickness.

Finally, 1.2 μm of GaN can be deposited with a SiN_x *in situ* masking layer inserted after 80 nm of GaN at 1000°C. In the case of SiN_x *in situ* masking, it decouples the compressive strain from the $Al_{0.2}Ga_{0.8}N$ layer so that the GaN layer is less compressively strained. The TDD monotonically improves with the coverage of SiN_x. In practice, ME-AlN buffer and SiN_x *in situ* masking are combined to produce a bow-free wafer with minimum TDD on 725 μm thick 200 mm diameter Si wafers. New problems associated with the scaling of Si substrate size used for GaN growth such as Si slip lines are dealt with based on the understanding of GaN-on-Si system.

References

1. M. Bockowski, High nitrogen pressure solution growth of GaN, *Jpn. J. Appl. Phys.*, 53, 100203, (2014).
2. H. F. Liu, S. B. Dolmanan, L. Zhang, S. J. Chua, D. Z. Chi, M. Heuken, and S. Tripathy, Influence of stress on structural properties of AlGaN/GaN high electron mobility transistor layers grown on 150 mm diameter Si (111) substrate, *J. Appl. Phys.*, 113, 23510, (2013).
3. D. Zhu, D. J. Wallis, and C. J. Humphreys, Prospects of III-nitride optoelectronics grown on Si, *Rep. Prog. Phys.*, 76, 106501, (2013).
4. J. Wan, R. Venugopal, M. R. Melloch, H. M. Liaw, and W. J. Rummel, Growth of crack-free hexagonal GaN films on Si(100), *Appl. Phys. Lett.*, 79, 1459, (2001).
5. S. Nakamura, GaN growth using GaN buffer layer, *Jpn. J. Appl. Phys.*, 30, L1705–L1707, (1991).
6. A. Dadgar, Sixteen years GaN on Si, *Phys. Status Solidi B*, 252, 1063–1068, (2015).
7. M. F. Schubert, S. Chhajed, J. K. Kim, E. F. Schubert, D. D. Koleske, M. H. Crawford, S. R. Lee, A. J. Fischer, G. Thaler, and M. A. Banas, Effect of dislocation density on efficiency droop in GaInN/GaN light-emitting diodes, *Appl. Phys. Lett.*, 91, 231114, (2007).
8. M. Ťapajna, U. K. Mishra, and M. Kuball, Importance of impurity diffusion for early stage degradation in AlGaN/GaN high electron mobility transistors upon electrical stress, *Appl. Phys. Lett.*, 97, 23503, (2010).
9. A. Dadgar, J. Bläsing, A. Diez, A. Alam, M. Heuken, and A. Krost, Metalorganic chemical vapor phase epitaxy of crack-free GaN on Si (111) exceeding 1 μm in thickness, *Jpn. J. Appl. Phys.*, 39, L1183–L1185, (2000).
10. T. N. Bhat, S. B. Dolmanan, Y. Dikme, H. R. Tan, L. K. Bera, and S. Tripathy, Structural and optical properties of $Al_xGa_{1-x}N$/GaN high electron mobility transistor structures grown on 200 mm diameter Si(111) substrates, *J. Vac. Sci. Technol. B Microelectron. Nanom. Struct.*, 32, 21206, (2014).
11. K. Cheng, H. Liang, M. Van Hove, K. Geens, B. De Jaeger, P. Srivastava, X. Kang et al., AlGaN/GaN/AlGaN double heterostructures grown on 200 mm silicon (111) substrates with high electron mobility, *Appl. Phys. Express*, 5, 11002, (2012).
12. D. Christy, T. Egawa, Y. Yano, H. Tokunaga, H. Shimamura, Y. Yamaoka, A. Ubukata, T. Tabuchi, and K. Matsumoto, Uniform growth of AlGaN/GaN high electron mobility transistors on 200 mm silicon (111) substrate, *Appl. Phys. Express*, 6, 26501, (2013).
13. T. Mukai and S. Nakamura, Ultraviolet InGaN and GaN single-quantum-well-structure light-emitting diodes grown on epitaxially laterally overgrown GaN substrates, *Jpn. J. Appl. Phys.*, 38, 5735–5739, (1999).
14. D. S. Wuu, W. K. Wang, W. C. Shih, R. H. Horng, C. E. Lee, W. Y. Lin, and J. S. Fang, Enhanced output power of near-ultraviolet InGaN-GaN LEDs grown on patterned sapphire substrates, *IEEE Photon. Technol. Lett.*, 17, 288–290, (2005).
15. J.-Y. Kim, Y. Tak, J. Kim, H.-G. Hong, S. Chae, J. W. Lee, H. Choi et al., Highly efficient InGaN/GaN blue LED on 8-inch Si (111) substrate, *Proc. SPIE OPTO*, 8262, 82621D-1–82621D-9, (2012).
16. K. Y. Zang, L. S. Wang, S. J. Chua, and C. V. Thompson, Structural analysis of metalorganic chemical vapor deposited AlN nucleation layers on Si (111), *J. Cryst. Growth*, 268, 515–520, (2004).

17. A. Dadgar, M. Poschenrieder, J. Bläsing, O. Contreras, F. Bertram, T. Riemann, A. Reiher et al., MOVPE growth of GaN on Si(111) substrates, *J. Cryst. Growth*, 248, 556–562, (2003).

18. A. Strittmatter, A. Krost, J. Bläsing, and D. Bimberg, High quality GaN layers grown by metalorganic chemical vapor deposition on Si(111) substrates, *Phys. Status Solidi*, 176, 611–614, (1999).

19. C. I. Park, J. H. Kang, K. C. Kim, E. K. Suh, K. Y. Lim, and K. S. Nahm, Characterization of GaN thin film growth on 3C–SiC/Si(111) substrate using various buffer layers, *J. Cryst. Growth*, 224, 190–194, (2001).

20. L. Wang, X. Liu, Y. Zan, J. Wang, D. Wang, D. Lu, and Z. Wang, Wurtzite GaN epitaxial growth on a Si(001) substrate using γ-Al$_2$O$_3$ as an intermediate layer, *Appl. Phys. Lett.*, 72, 109, (1998).

21. R. Dargis, R. Smith, F. E. Arkun, and A. Clark, Epitaxial rare earth oxide and nitride buffers for GaN growth on Si, *Phys. status solidi*, 11, 569–572, (2014).

22. G. Radtke, M. Couillard, G. A. Botton, D. Zhu, and C. J. Humphreys, Structure and chemistry of the Si(111)/AlN interface, *Appl. Phys. Lett.*, 100, 11910, (2012).

23. G. Radtke, M. Couillard, G. A. Botton, D. Zhu, and C. J. Humphreys, Scanning transmission electron microscopy investigation of the Si(111)/AlN interface grown by metalorganic vapor phase epitaxy, *Appl. Phys. Lett.*, 97, 251901, (2010).

24. M. K. Ozturk, E. Arslan, İ. Kars, S. Ozcelik, and E. Ozbay, Strain analysis of the GaN epitaxial layers grown on nitridated Si(111) substrate by metal organic chemical vapor deposition, *Mater. Sci. Semicond. Process.*, 16, 83–88, (2013).

25. V. Wagner, O. Parillaud, H. J. Bühlmann, M. Ilegems, S. Gradecak, P. Stadelmann, T. Riemann, and J. Christen, Influence of the carrier gas composition on morphology, dislocations, and microscopic luminescence properties of selectively grown GaN by hydride vapor phase epitaxy, *J. Appl. Phys.*, 92, 1307, (2002).

26. P. Y. Lin, J. Y. Chen, Y. C. Chen, and L. Chang, Effect of growth temperature on formation of amorphous nitride interlayer between AlN and Si(111), *Jpn. J. Appl. Phys.*, 52, 08JB20, (2013).

27. N. Yamabe, Y. Yamamoto, and T. Ohachi, Epitaxial growth of β-Si3N4 by the nitridation of Si with adsorbed N atoms for interface reaction epitaxy of double buffer AlN(0001)/β-Si3N4/Si(111), *Phys. status solidi*, 8, 1552–1555, (2011).

28. J. W. Kim and H. W. Yeom, Surface and interface structures of epitaxial silicon nitride on Si(111), *Phys. Rev. B*, 67, 35304, (2003).

29. Y. Morita, T. Ishida, and H. Tokumoto, Self-limiting formation of silicon-nitride monolayer on Si(111) surface using N$_2$/H$_2$ mixture gas, *Surf. Sci.*, 486, L524–L528, (2001).

30. C. L. Wu, W. S. Chen, and Y. H. Su, N$_2$-plasma nitridation on Si(111): Its effect on crystalline silicon nitride growth, *Surf. Sci.*, 606, L51–L54, (2012).

31. X. S. Wang, G. Zhai, J. Yang, L. Wang, Y. Hu, Z. Li, J. C. Tang, X. Wang, K. K. Fung, and N. Cue, Nitridation of Si(111), *Surf. Sci.*, 494, 83–94, (2001).

32. J. R. Chang, T. H. Yang, Y. C. Chen, J. T. Ku, and C. Y. Chang, Growth and characterization of MBE-GaN on Si (111) using AlN/α-Si3N4 buffer structure, *ECS Trans.*, 13, 29–37, (2008).

33. E. Arslan, M. K. Ozturk, Ö. Duygulu, A. A. Kaya, S. Ozcelik, and E. Ozbay, The influence of nitridation time on the structural properties of GaN grown on Si (111) substrate, *Appl. Phys. A*, 94, 73–82, (2008).

34. J. Narayan and B. C. Larson, Domain epitaxy: A unified paradigm for thin film growth, *J. Appl. Phys.*, 93, 278, (2003).

35. L. Wang, F. Huang, Z. Cui, Q. Wu, W. Liu, C. Zheng, Q. Mao, C. Xiong, and F. Jiang, Crystallographic tilting of AlN/GaN layers on miscut Si (111) substrates, *Mater. Lett.*, 115, 89–91, (2014).

36. G. P. Dimitrakopoulos, A. M. Sanchez, P. Komninou, T. Kehagias, T. Karakostas, G. Nouet, and P. Ruterana, Interfacial steps, dislocations, and inversion domain boundaries in the GaN/AlN/Si (0001)/(111) epitaxial system, *Phys. Status Solidi*, 242, 1617–1627, (2005).

37. R. Liu, F. A. Ponce, A. Dadgar, and A. Krost, Atomic arrangement at the AlN/Si (111) interface, *Appl. Phys. Lett.*, 83, 860, (2003).

38. X. J. Ning, F. R. Chien, P. Pirouz, J. W. Yang, and M. A. Khan, Growth defects in GaN films on sapphire: The probable origin of threading dislocations, *J. Mater. Res.*, 11, 580–592, (2011).

39. S. Zamir, B. Meyler, and J. Salzman, Thermal microcrack distribution control in GaN layers on Si substrates by lateral confined epitaxy, *Appl. Phys. Lett.*, 78, 288, (2001).

40. A. Dadgar, M. Poschenrieder, J. Bläsing, K. Fehse, A. Diez, and A. Krost, Thick, crack-free blue light-emitting diodes on Si(111) using low-temperature AlN interlayers and in situ SixNy masking, *Appl. Phys. Lett.*, 80, 3670, (2002).

41. T. Soga, S. Hattori, S. Sakai, and M. Umeno, Epitaxial growth and material properties of GaAs on Si grown by MOCVD, *J. Cryst. Growth*, 77, 498–502, (1986).

42. E. Feltin, B. Beaumont, M. Laügt, P. de Mierry, P. Vennéguès, H. Lahrèche, M. Leroux, and P. Gibart, Stress control in GaN grown on silicon (111) by metalorganic vapor phase epitaxy, *Appl. Phys. Lett.*, 79, 3230, (2001).

43. E. A. Fitzgerald, Dislocations in strained-layer epitaxy: Theory, experiment, and applications, *Mater. Sci. Rep.*, 7, 87–142, (1991).

44. P. Cantu, F. Wu, P. Waltereit, S. Keller, A. E. Romanov, U. K. Mishra, S. P. DenBaars, and J. S. Speck, Si doping effect on strain reduction in compressively strained Al 0.49Ga0.51N thin films, *Appl. Phys. Lett.*, 83, 674–676, (2003).

45. D. M. Follstaedt, S. R. Lee, P. P. Provencio, A. A. Allerman, J. A. Floro, and M. H. Crawford, Relaxation of compressively-strained AlGaN by inclined threading dislocations, *Appl. Phys. Lett.*, 87, 121112, (2005).

46. P. S. Hsu, E. C. Young, A. E. Romanov, K. Fujito, S. P. DenBaars, S. Nakamura, and J. S. Speck, Misfit dislocation formation via pre-existing threading dislocation glide in (1122) semipolar heteroepitaxy, *Appl. Phys. Lett.*, 99, 81912, (2011).

47. D. M. Follstaedt, S. R. Lee, A. A. Allerman, and J. A. Floro, Strain relaxation in AlGaN multilayer structures by inclined dislocations, *J. Appl. Phys.*, 105, 83507, (2009).

48. A. E. Romanov and J. S. Speck, Stress relaxation in mismatched layers due to threading dislocation inclination, *Appl. Phys. Lett.*, 83, 2569, (2003).

49. H. Lahrèche, P. Vennéguès, B. Beaumont, and P. Gibart, Growth of high-quality GaN by low-pressure metal-organic vapour phase epitaxy (LP-MOVPE) from 3D islands and lateral over-growth, *J. Cryst. Growth*, 205, 245–252, (1999).

50. T. Markurt, L. Lymperakis, J. Neugebauer, P. Drechsel, P. Stauss, T. Schulz, T. Remmele, V. Grillo, E. Rotunno, and M. Albrecht, Blocking growth by an electrically active subsurface layer: The effect of Si as an antisurfactant in the growth of GaN, *Phys. Rev. Lett.*, 110, 36103, (2013).

51. L. S. Wang, K. Y. Zang, S. Tripathy, and S. J. Chua, Effects of periodic delta-doping on the properties of GaN:Si films grown on Si (111) substrates, *Appl. Phys. Lett.*, 85, 5881, (2004).

52. P. R. Hageman, S. Haffouz, V. Kirilyuk, A. Grzegorczyk, and P. K. Larsen, High quality GaN layers on Si(111) substrates: AlN Buffer layer optimisation and insertion of a SiN intermediate layer, *Phys. Status Solidi*, 188, 523–526, (2001).

53. E. Arslan, M. K. Ozturk, A. Teke, S. Ozcelik, and E. Ozbay, Buffer optimization for crack-free GaN epitaxial layers grown on Si(1 1 1) substrate by MOCVD, *J. Phys. D. Appl. Phys.*, 41, 155317, (2008).

54. H. Hirayama, N. Maeda, S. Fujikawa, S. Toyoda, and N. Kamata, Recent progress and future prospects of AlGaN-based high-efficiency deep-ultraviolet light-emitting diodes, *Jpn. J. Appl. Phys.*, 53, 100209, (2014).

55. F. Yan, M. Tsukihara, A. Nakamura, T. Yadani, T. Fukumoto, Y. Naoi, and S. Sakai, Surface Smoothing Mechanism of AlN Film by Initially Alternating Supply of Ammonia, *Jpn. J. Appl. Phys.*, 43, L1057–L1059, (2004).

56. H. Hirayama, T. Yatabe, N. Noguchi, T. Ohashi, and N. Kamata, 231–261 nm AlGaN deep-ultraviolet light-emitting diodes fabricated on AlN multilayer buffers grown by ammonia pulse-flow method on sapphire, *Appl. Phys. Lett.*, 91, 71901, (2007).

57. D.-B. Li, M. Aoki, H. Miyake, and K. Hiramatsu, Improved surface morphology of flow-modulated MOVPE grown AlN on sapphire using thin medium-temperature AlN buffer layer, *Phys. Status Solidi*, 5, 1818–1821, (2008).

58. R. G. Banal, M. Funato, and Y. Kawakami, Initial nucleation of AlN grown directly on sapphire substrates by metal-organic vapor phase epitaxy, *Appl. Phys. Lett.*, 92, 241905, (2008).

59. J. C. Gagnon, J. M. Leathersich, F. (Shadi) Shahedipour-Sandvik, and J. M. Redwing, The influence of buffer layer coalescence on stress evolution in GaN grown on ion implanted AlN/Si(111) substrates, *J. Cryst. Growth*, 393, 98–102, (2014).

60. H. Hirayama and T. Tatsumi, Si (111) surface cleaning using atomic hydrogen and SiH_2 studied using reflection high-energy electron diffraction, *J. Appl. Phys.*, 66, 629, (1989).

61. J. P. Becker, Reflection high-energy electron diffraction patterns of carbide-contaminated silicon surfaces, *J. Vac. Sci. Technol. A Vac. Surf. Film.*, 12, 174, (1994).

62. K. Kitahara and O. Ueda, Observation of atomic steps on vicinal Si(111) annealed in hydrogen gas flow by scanning tunneling microscopy, *Jpn. J. Appl. Phys.*, 32, L1826–L1829, (1993).

63. R. C. Henderson, Silicon cleaning with hydrogen peroxide solutions: A high energy electron diffraction and auger electron spectroscopy study, *J. Electrochem. Soc.*, 119, 772, (1972).

64. G. Tarrach, Laser and thermal annealed Si(111) and Si(001) surfaces studied by scanning tunneling microscopy, *J. Vac. Sci. Technol. B Microelectron. Nanomed. Struct.*, 9, 677, (1991).

65. K. Balakrishnan, A. Bandoh, M. Iwaya, S. Kamiyama, H. Amano, and I. Akasaki, Influence of high temperature in the growth of low dislocation content AlN bridge layers on patterned 6H–SiC substrates by metalorganic vapor phase epitaxy, *Jpn. J. Appl. Phys.*, 46, L307–L310, (2007).

66. T. Detchprohm, S. Sano, S. Mochizuki, S. Kamiyama, H. Amano, and I. Akasaki, Growth mechanism and characterization of low-dislocation-density AlGaN single crystals grown on periodically grooved substrates, *Phys. Status Solidi*, 188, 799–802, (2001).

67. A. Dadgar, A. Strittmatter, J. Bläsing, M. Poschenrieder, O. Contreras, P. Veit, T. Riemann et al., Metalorganic chemical vapor phase epitaxy of gallium-nitride on silicon, *Phys. Status Solidi*, 6, 1583–1606, (2003).

68. P. Vermaut, P. Ruterana, G. Nouet, and H. Morkoç, Structural defects due to interface steps and polytypism in III-V semiconducting materials: A case study using high-resolution electron microscopy of the 2H-AlN/6H-SiC interface, *Philos. Mag. A*, 75, 239–259, (1997).

69. P. Pirouz and J. W. Yang, Polytypic transformations in SiC: The role of TEM, *Ultramicroscopy*, 51, 189–214, (1993).

70. B. N. Sverdlov, G. A. Martin, H. Morkoç, and D. J. Smith, Formation of threading defects in GaN wurtzite films grown on nonisomorphic substrates, *Appl. Phys. Lett.*, 67, 2063, (1995).

71. V. M. Kaganer, O. Brandt, A. Trampert, and K. H. Ploog, X-ray diffraction peak profiles from threading dislocations in GaN epitaxial films, *Phys. Rev. B*, 72, 45423, (2005).

72. D. Hull and D. J. Bacon, Observation of dislocations, in *Introduction to Dislocations*, Butterworth-Heinemann, Oxford, U.K., Elsevier (2011).

73. T. Y. Wang, Microscopic contact and slip in Si epitaxy, *J. Cryst. Growth*, 280, 16–25, (2005).

74. D. Zhu, C. McAleese, K. K. McLaughlin, M. Häberlen, C. O. Salcianu, E. J. Thrush, M. J. Kappers et al., GaN-based LEDs grown on 6-inch diameter Si (111) substrates by MOVPE, *Proceedings of the SPIE OPTO: Integrated Optoelectronic Devices*, pp. 723118-1–723118-11, San Jose, CA (2009).

22

Terahertz Spectroscopy Study of III–V Nitrides

Xinhai Zhang
*Department of Electrical
and Electronic Engineering*

Huafeng Shi
*Department of Electrical
and Electronic Engineering*

Abstract　Nowadays, most of the frequency intervals of electromagnetic spectrum have been intensively studied. However, there still is a gap between microwave and infrared light in the electromagnetic spectrum, named terahertz (THz) gap because of the lack of reliable radiation sources. The frequency of this gap covers the range from 0.1 to 10 THz, and the wavelength spans from 30 μm to 3 mm. The terahertz region of the electromagnetic spectrum is of critical importance in the spectroscopy of condensed matter systems. The electronic properties of semiconductors are greatly influenced by bound states whose energies are resonant with terahertz photons. In this chapter, we have reviewed terahertz spectroscopy study of III–V nitrides.

22.1 Introduction

Nowadays, most of the frequency intervals of electromagnetic (EM) spectrum have been intensively studied. However, there still is a gap between microwave and infrared light in the electromagnetic spectrum, named terahertz (THz) gap, because of the lack of reliable radiation sources [1,2]. The frequency of this gap covers the range from 0.1 to 10 THz, and the wavelength spans from 30 μm to 3 mm. The terahertz region of the electromagnetic spectrum is of critical importance in the spectroscopy of condensed matter systems. The electronics properties of semiconductors and metals are greatly influenced by bound states whose energies are resonant with terahertz photons. The terahertz region also coincides with the rates of inelastic processes in solids, such as confinement energies in artificially synthesized nanostructures like quantum dots, which lie in the terahertz region (Figure 22.1).

Many efforts have been made in developing THz sources. The optically pumped THz laser is one type of continuous wave THz source, with applications in astronomy, environmental monitoring, and plasma diagnostics [3]. Backward wave oscillators (BWO) are electron tubes that can be used to generate tunable output at the long-wavelength end of the terahertz spectrum, which require a highly homogeneous magnetic field [4]. Frequency multiplied sources directly multiply millimeter wave up to THz frequencies, which can produce substantially high output power at lower frequencies. Quantum cascade

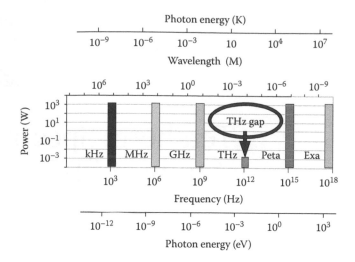

FIGURE 22.1 THz gap in the electromagnetic spectrum.

semiconductor lasers could become an important source of commercial THz systems in the future, but these lasers currently work best at a very low temperature [5]. THz time-domain spectroscopy (THz-TDS) is a useful spectroscopic technique, in which the properties of a material are probed with short pulses of THz radiation [6,7]. The generation and detection schemes are sensitive to the sample material's effect on both the amplitude and the phase of the THz radiation. The choice of the THz source will determine the type of detection scheme required [8,9].

This chapter will describe the mechanism of the terahertz time-domain spectroscopy (THz-TDS) and optical pump terahertz probe spectroscopy (OPTP) and present a review of THz spectroscopy study on the basic optical properties and carrier dynamics of III–V nitride materials.

22.2 Terahertz Spectroscopy

In conventional THz-TDS and OPTP setups, THz pulses are generated and detected using ultrafast Ti:sapphire lasers with a pulse width ranging from ~10 to 100 fs and a center wavelength of around 800 nm. THz pulses have time duration of ~1 ps. The coherent detection scheme in the time domain allows measurement of the transient electric field, not only of the amplitude but also of the phase. Fourier transform of the transient electric field enables direct determination of both the amplitude and the phase of each of the spectral components that make up the pulse. In addition to the advantages of coherent detection, THz spectrometers based on pulsed Ti:sapphire lasers enable time-resolved THz spectroscopy with sub-picosecond time resolution. Hence, dynamics induced by near-instantaneous photoexcitation or heating with a laser pump pulse can be investigated [10,11].

A new era of THz spectroscopy development followed with the emergence of ultrafast lasers and semiconductor technologies. The synergy between these two advances led to the first THz photoconductive antenna by Auston et al. [12,13]. Years later, subsequent developments led to an efficient approach to coherent THz wave generation and detection, now widely known as THz-TDS. Its broadband measurement capability and strong immunity to background noise facilitate observation of THz wave–material interactions, which are attractive to researchers in a wide range of fundamental disciplines.

THz-TDS is based on the availability of highly coherent radiation in pulse form of extremely short duration [14–16]. The frequency-dependent dielectric properties of the sample can be extracted from the waveform of THz pulses transmitted the sample measured by the THz-TDS or OPTP. The method has been used to study the properties of a wide range of solids, from semiconductors to superconductors.

The emerging field of THz-TDS typically relies on a broadband short-pulsed THz wave, usually generated with a photoconductive (PC) antenna. A PC antenna is usually fabricated on low-temperature-grown GaAs. A dc bias is applied across the antenna, and an ultra-short pump laser pulse (<100 fs) is focused onto the gap of the antenna. The bias-laser pulse combination allows electrons to rapidly cross the gap, and the resulting transient current in the antenna produces THz electromagnetic wave. This THz radiation is collected and collimated with the appropriate optical system.

22.2.1 Terahertz Time-Domain Spectroscopy (THz-TDS)

In a THz-TDS setup, the THz emitters are usually excited by Ti: sapphire ultrafast lasers. Specifically, for a fast scan THz-TDS system, a Ti: sapphire oscillator with 80 MHz repetition rate is used. For a low-speed system, the laser output has 1 kHz repetition rate, but with a much higher pulse energy. Illustrated here are three different types of THz-TDS, including THz-TDS based on GaAs photoconductive antenna (PCA), THz-TDS based on electro-optic (EO) crystal, and THz-TDS based on air plasma.

22.2.1.1 THz-TDS Based on PCA

The structure diagram of THz-TDS based on PCA is shown in Figure 22.2. The ultrafast laser beam is divided into two beams by a beam splitter to excite the emitter and detector [17,18]. After the THz pulse being collimated by the silicon lens on the emitter, it is focused onto the sample under test and re-collimated by a pair of off-axis parabolic reflectors and reaching the PCA-based detector [19–23]. Meanwhile, the detection beam travels through a delay line and incident onto the same detection PCA. The THz pulse-induced photocurrent is proportional to the THz electric field strength at the specific relative time referring to the probe beam [24,25].

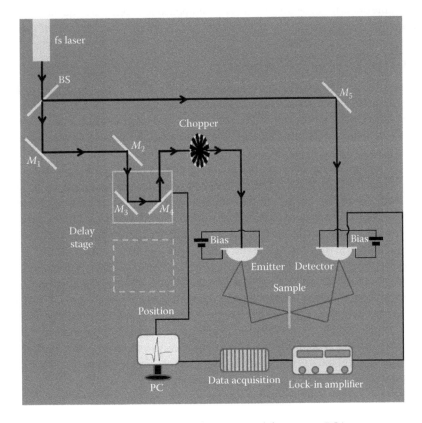

FIGURE 22.2 Schematic diagram of the THz-TDS. The emitter and detector are PCAs.

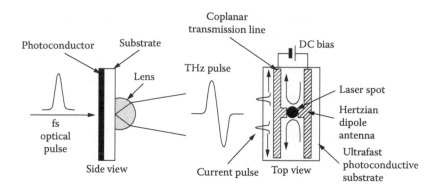

FIGURE 22.3 Structure diagram of a typical THz photoconductive antenna.

The diagram of a THz PCA emitter is shown in Figure 22.3. Two dipoles are fabricated on semiconductor wafers and separated by a small gap (5 μm typically). An AC or DC voltage source is applied to the dipoles. Due to the high resistance of semiconductor antennas—typically a few mega ohms and above—the current through the channel is very low. When the femtosecond laser is focused in between the channel, a large amount of free electrons are excited in a short time and accelerated by the bias voltage. Such a dynamic process at picosecond timescale radiates THz waves into the free space. The side view shows that a focusing lens is placed after the antenna chips, which collimates the THz wave to the designed direction. The PCA detector is very similar to the emitter, but the bias voltage is replaced by current-sensing devices to detect the THz wave–induced photocurrent.

By scanning the optical path difference between the detection and THz generation beams, the THz pulse profile in time domain can be constructed. FFT is applied to access the frequency-domain information, as shown in Figure 22.4. Generally, the THz beam path is purged with dry air or nitrogen gas to avoid absorption from atmospheric water vapors.

22.2.1.2 THz-TDS Based on EO Crystal

EO crystal is another popular candidate for building up THz-TDS, as shown in Figure 22.5. Here, the femtosecond laser oscillator is amplified by a regenerative amplifier. The pump laser shines onto the EO crystal, such as ZnTe (500 μm thick with <110> crystalline), and generates THz wave by optical rectification. After two parabolic mirrors, the THz wave is focused onto the other EO crystal with the detection beam for signal detection. Due to the Pockels effect, the THz wave induces birefringence in the EO crystal, and this birefringence changes the polarization of the probe beam when it passes through the crystal. After the detection crystal, the probe is manipulated by a quarter wave plate, and the beam polarization changes from linear to circular. Later, a Wollatson prism separates the P and S components of the beam and distributes the beams to a balanced photodetector module. Any polarization change induced by the THz wave results in an imbalance between the S- and P-polarized probe beams. By scanning the optical delay line, the time-domain spectroscopy of the THz wave is able to be read out [11,26].

THz generation from EO crystal is induced by optical rectifications, which is described by the nonlinear susceptibility, and the THz detection is based on the Pockels effect in the EO crystal [27,28]. When the THz pulse is present, the phase shift leads to a slightly elliptical polarization, as shown in Figure 22.6. In practice, a quarter wave plate is placed behind the EO crystal to make the initially linear polarization of the probe beam circular [29,30]. A Wollatson prism separates its y and z components and sends them to a differential detector that is connected to a preamplifier and a lock-in amplifier. With no THz present, the components have equal intensity and the differential signal is zero. When the THz field is applied, a nonzero phase difference of the two components appears [31–33].

A wide range of EO crystals are studied in the community such as ZnTe, GaAs, GaP, and GaSe. The systems, construction with different EO crystals is similar; therefore, only ZnTe-based system curve is shown in Figure 22.7.

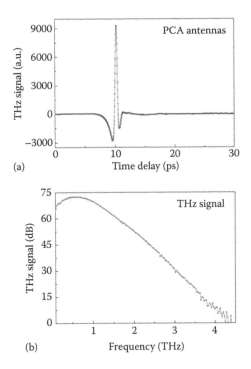

FIGURE 22.4 THz signal: (a) THz E-field strength in time domain and (b) THz-field strength in frequency domain.

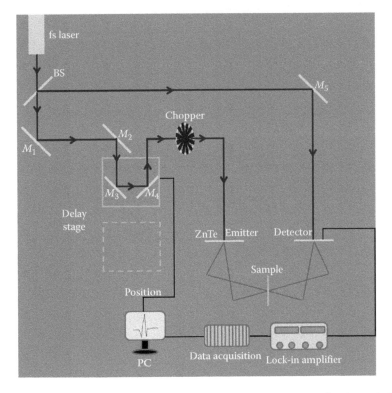

FIGURE 22.5 Schematic diagram of the THz-TDS based on EO crystal BS (beam splitter), PM (off-axes parabolic mirror), and WP (Wollaston prism).

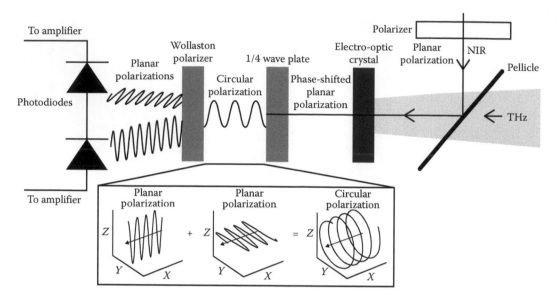

FIGURE 22.6   Schematic diagram of the free space EO crystal detection.

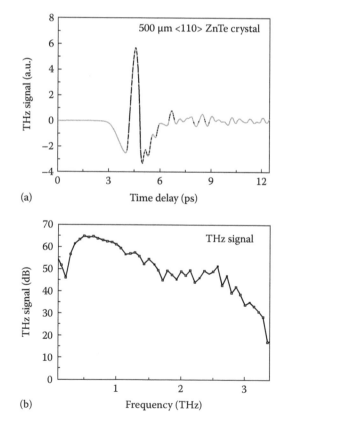

FIGURE 22.7   THz signal of ZnTe: (a) THz E-field strength in time domain and (b) THz-field strength in frequency domain.

22.2.1.3 THz-TDS Based on Air Plasma

The THz-TDS requires much higher laser specification for THz generation. Generally, an fs laser oscillator and a regenerative amplifier are used in the system. And the laser pulse is compressed carefully for THz generation and detection. In this system, air plasma is also used for THz detection. The schematic diagram of THz-TDS based on air plasma is shown in Figure 22.8. When the focused THz is collimated with the focused probe beam (800 nm), and if a bias electric field is applied through the focal point of the detection beam, the second harmonic of the probe beam will be greatly enhanced, which is proportional to the *E*-field strength of the THz signal at a programmed delay time. By referring to the frequency of the bias electric field, the THz signal is able to be acquired by a lock-in amplifier [34,35].

Dual-color-mixing air plasma has been considered as one of the most optimum solutions for THz emission with high-frequency bandwidth and high peak intensity. The schematic diagram of dual-color air plasma is shown in Figure 22.9. The basic idea is to focus the fs laser pulse at a high power density, which is able to generate ionized air plasma. A beta barium borate (BBO), installed between the focus lens and the plasma point, converts a part of the pump beam into its second harmonic and forms dual-color-mixed plasma [36,37]. The asymmetric electrical field induced by the superposition of the dual-color beams accelerates the ionized electrons, resulting in transient charge current enhancement and strong THz emission [38,39].

The signal of THz-TDS is shown in Figure 22.10. It is easy to see that the time-domain pulse is a very fast process, while the frequency-domain curve shows the signal from 0.3 up to more than 20 THz, which is even beyond the upper limit of the THz region range from 0.1 to 10 THz. However, the signal to noise

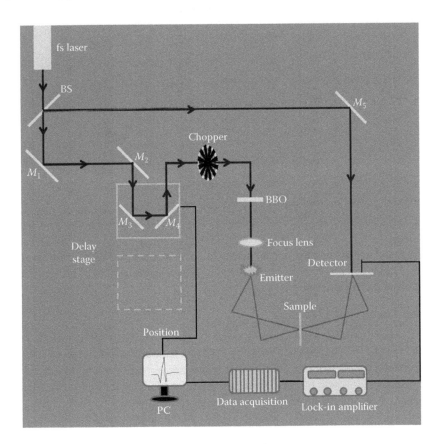

FIGURE 22.8 Schematic diagram of the THz-TDS based on air plasma.

FIGURE 22.9   Schematic of THz generation via dual-color air plasma.

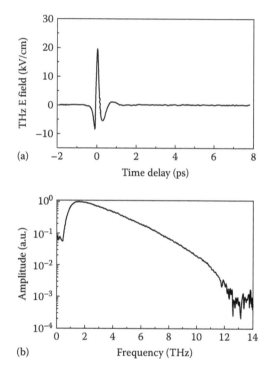

FIGURE 22.10   THz signal from the THz-TDS based on air plasma in (a) time domain and (b) frequency domain.

ratio (SNR) is limited. Other than the low sensitivity of the air plasma detection, the poor stability of the fs laser amplifier should be also responsible for the low SNR. When broadband THz spectrum is desired, the THz-TDS based on air plasma will be a good choice.

22.2.1.4 Data Analysis in THz-TDS

In a THz-TDS system, the THz pulse passes through the sample medium and it records a signal that is proportional to the electric field of the THz pulse $E(t)$. In a transmitted spectroscopic experiment, two signals are recorded. One, $E_{sam}(t)$, denotes the pulse propagates through the sample and the other, $E_{ref}(t)$, denotes the reference pulse. And it is easy to extract information about the dielectric function $\varepsilon(\omega)$, the index of refraction $n(\omega)$, and the index of absorption $\alpha(\omega)$ of the sample.

There are some proper assumptions to analyze the data. In the simplest case, it is only considered transmission and reflection at the first and second interface [40]. It is assumed that planar interfaces are without scattering and normal incidence, as shown in Figure 22.11. It is also assumed that the medium surrounding

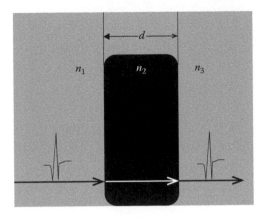

FIGURE 22.11 Transmission in THz-TDS.

the sample is dry air. This means the refractive index is 1, as shown in Figure 22.12. Employing the Fresnel equations, the electric field pulse transmitted to the sample can be expressed as

$$E_t(\omega) = E_0(\omega) \frac{2}{1+n(\omega)} \frac{2n(\omega)}{1+n(\omega)} e^{-\frac{\alpha d}{2}} e^{\frac{in\omega d}{c}} \tag{22.1}$$

where

$E_0(\omega)$ is the incident THz field
$E_t(\omega)$ is the transmitted THz field
The electric field of the THz pulse propagating through the sample, $E_{sam}(\omega)$, and dry air as reference, $E_{ref}(\omega)$, is expressed as

$$T(\omega)e^{i\varphi(\omega)} = \frac{E_{sam}(\omega)}{E_{ref}(\omega)} = \frac{2}{1+n(\omega)} \frac{2n(\omega)}{1+n(\omega)} e^{-\frac{\alpha d}{2}} e^{\frac{i(n-1)\omega d}{c}} \tag{22.2}$$

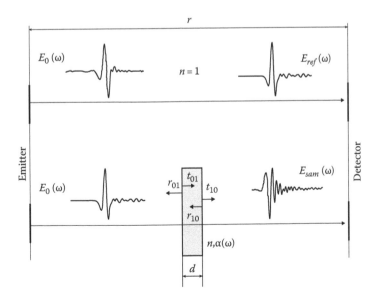

FIGURE 22.12 THz pulse propagating through dry air (reference) and the sample.

and it could be solved that

$$n(\omega) = 1 + \frac{\varphi(\omega)c}{\omega d} \tag{22.3}$$

And

$$\alpha(\omega) = -\frac{2}{d}\ln\left[\frac{(n+1)^2}{4n}T(\omega)\right] \tag{22.4}$$

And the complex frequency-dependent refractive index, $n(\omega)$, could be extracted by

$$\alpha(\omega) = \frac{\omega\kappa(\omega)}{2c} \tag{22.5}$$

However, when studying phonons in materials, it is more applicable to calculate the complex dielectric function, $\varepsilon(\omega)$, and the complex conductivity, $\sigma(\omega)$. Thus, the relationship between the refractive index, $n(\omega)$, dielectric function, $\varepsilon(\omega)$, and conductivity $\sigma(\omega)$ is summarized as

$$\omega(\omega) = n(\omega) + i\kappa(\omega) \tag{22.6}$$

$$\varepsilon(\omega) = n(\omega)^2 + \kappa(\omega)^2 + i2\,n(\omega)\kappa(\omega) \tag{22.7}$$

$$\sigma(\omega) = 2\varepsilon_0\omega n(\omega)\kappa(\omega) + i\varepsilon_0\omega\left(\varepsilon_\infty - n(\omega)^2 + \kappa(\omega)^2\right) \tag{22.8}$$

22.2.2 Optical Pump Terahertz Probe Spectroscopy (OPTP)

Since the emergence of modern THz-TDS system, THz radiation has been employed in pump–probe spectroscopy [41–45]. The femtosecond to nanosecond dynamics of materials can be studied using THz pulses that are delayed with respect to a photoexcitation beam from the same femtosecond laser. The pump and probe could be any combination of white light, infrared pulse, visible, THz pulses, or even microwave [46–49]. And in this chapter, we will focus on optical pump terahertz probe spectroscopy. It is the most commonly cited pump–probe THz spectroscopy configuration at present, as shown in Figure 22.13. Similar to other pump–probe spectroscopy techniques, the pump pulse induces a change in the sample and the

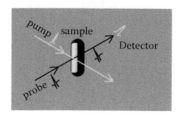

FIGURE 22.13 Schematic diagram of basic OPTP.

probe pulse measures the induced change in the sample. The temporal delay between the pump and probe pulse is achieved via a delay stage; varying the delay between the pump pulse and the probe pulse, one can monitor the temporal development of the induced change. This development is measured in transmission $\Delta T/T$, which contains information about the relaxation dynamics of the sample. Measurements of the dynamics of hot-electron relaxation in semiconductors give important information about the physics of nonequilibrium phenomena in these materials as well as information about the carrier–carrier and carrier–phonon interaction.

The transiently induced change in transmission $\Delta T/T$, in the case of small perturbations, is connected to the induced changes in the dielectric function, $\varepsilon = \varepsilon_1 + i\varepsilon_2$, of the sample through

$$\Delta T = \frac{\partial lnT}{\partial \varepsilon_1}\Delta\varepsilon_1 + \frac{\partial lnT}{\partial \varepsilon_2}\Delta\varepsilon_2 \tag{22.9}$$

which contains the electronic properties of the material. Here, $\Delta\varepsilon_1$ and $\Delta\varepsilon_2$ represent the induced change in the real and imaginary part of ε.

22.2.2.1 Mechanism of OPTP

OPTP is also similar to THz-TDS, in which the THz pulse that probes the sample can be coherently detected in time domain. This requires another delay stage to control the overlap between the optical sampling pulse and the THz probe pulse. Different types of TDS have been introduced previously [50,51]. The OPTP system composed of TDS based on air plasma is shown in Figure 22.14. An fs laser is separated into three beams by two splitters. Most of the power of the fs laser is allocated to the optical pump beam, and the other two beams are used for THz emission and detection. For a typical regenerative amplifier system, the fundamental wavelength is usually at 800 nm, and the fundamental frequency can be doubled via β-BBO for excitation at 400 nm. One delay stage in the optical pump beam controls the optical path length traveled by the pump beam, and the other delay stage is used in either the emitter or detector arm. Thus, essentially, there are two modes of operation for OPTP measurements: pump scans and probe scans [52,53].

In a probe scan, the optical pump delay stage is fixed, and the delay line in emitter arm is moved to decrease the optical path length traveled by the emission pulse. And the THz pulse is recorded, just as in

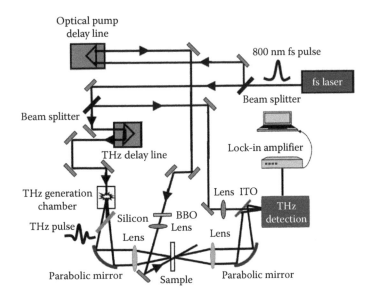

FIGURE 22.14 Schematic diagram of OPTP.

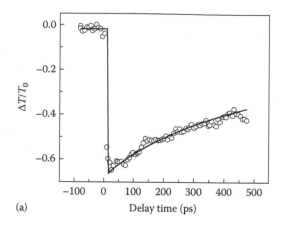

(a)

FIGURE 22.15   Representative example of the data resulting from a pump scan. Pump scans record the change in THz transmission at the two delays as compared to the transmission through the unexcited sample.

THz-TDS. The spot size of the optical pump beam should be large enough thus a uniform excited area could be probed. Anomalies can be found in the complex conductivity at high frequencies when the pump beam is not large enough. So, in order to avoid erroneous results, it is best to make spot size of the pump beam at least twice as large as the probe beam on the sample.

In a pump scan, the emitter delay line is fixed; thus, the THz waveform is being sampled. The optical pump delay stage is scanned so that the optical path traveled by the pump pulse is decreased and the pump pulse arrives earlier. When the pump and probe pulses arrive at the sample simultaneously, a sharp decrease in THz transmission occurs due to THz absorption by the excited state. As the pump pulse arrives earlier, the THz pulse samples the excited sample at longer delay time after excitation. Changes of the THz peak transmission are monitored as a function of this delay time, and the dynamics of the excited state is obtained, as shown in Figure 22.15.

22.2.2.2 Analysis Model of OPTP

For a thin-film sample ($\lambda \gg d$, thickness), there is an analytical approximation, Tinkham formula, to obtain the dynamic optical functions. The assumption is established that $n_{sample} \gg n_{substrate}$. In this case, the dielectric function is given by

$$\varepsilon(\omega) = \varepsilon_\infty + \frac{i\sigma(\omega)}{\varepsilon_0 \omega} \tag{22.10}$$

and the ratio of the transmission through the excited and unexcited samples is given by

$$\frac{E_{excited(\omega)}}{E_{unexcited(\omega)}} = \frac{n+1}{n+1+Z_0 d\sigma(\omega)} \tag{22.11}$$

where

    n is the refractive index of the substrate
    Z_0 is the impedance of free space
    d is the penetration depth at the pump wavelength
    $E_{excited(\omega)}$ and $E_{unexcited(\omega)}$ are the ratios of the transmission through the excited and unexcited samples, respectively

So the complex conductivity can be solved by the following equation:

$$\sigma(\omega) = \frac{n+1}{z_0 d} \frac{1}{1 - \left(-\Delta E(\omega)/E_{unexcited}(\omega)\right)} \tag{22.12}$$

where the change in transmission through the excited and unexcited samples $\Delta E(\omega) = E_{excited}(\omega) - E_{unexcited}(\omega)$. In order to calculate the complex frequency-dependent photoconductivity, the transmission and the change in transmission due to pumping through the unexcited sample are measured via OPTP. Under some experimental conditions, the photo-induced change in THz transmission is relatively small $\left(\Delta E(\omega)/E_{unexcited}(\omega) < 20\%\right)$. And the complex conductivity can be rearranged by

$$\sigma(\omega) = \frac{n+1}{z_0 d} \frac{-\Delta E(\omega)}{E_{unexcited}(\omega)} \tag{22.13}$$

yielding a linear relationship between the frequency-dependent complex photoconductivity and the negative, normalized, photo-induced change in THz transmission. By fitting the curve of the frequency-dependent complex photoconductivity of a material in THz region, some physical relevant quantities, such as mobility and carrier density, can be extracted using appropriate model.

22.2.2.3 Drude Model, Drude-Smith Model, and Lorentz Oscillator Model

The most common and simplest conductivity model is called Drude model, proposed in 1900 by Paul Drude [54]. In this model, there are no carrier–ion or carrier–carrier interactions, and the carriers are treated as free and independent. This motion is damped by carrier scattering at the rate of $1/\tau_s$. If an electric field is applied, the ensemble of carriers moves in one preferred direction leading to a drift velocity v. And the carrier propagation can be described by

$$m\dot{v}(t) = -\frac{mv(t)}{\tau_s} - eE(t) \tag{22.14}$$

where

  m is the carrier effective mass
  e is the electron charge

In a DC field, the drift velocity is constant, $v(t) = 0$, and the conductivity can be given by

$$\sigma = \frac{J}{E} = \frac{ne^2\tau_s}{m} = ne\mu = \omega_p^2 \varepsilon_0 \tau_s \tag{22.15}$$

where

  n is the carrier density
  μ is the carrier mobility
  ω_p is the plasma frequency
  the current density is given by $J(t) = -nev$

In an AC field with frequency ω, the form of the electric field is $E(\omega)e^{-i\omega t}$, and the Drude conductivity formula can be solved

$$\sigma(\omega) = \frac{ne^2\tau_s}{m(1-i\omega\tau_s)} = \frac{ne\mu}{1-i\omega\tau_s} = \frac{\omega_p^2\varepsilon_0\tau_s}{1-i\omega\tau_s} \tag{22.16}$$

For data fitting, it is necessary to separate $\sigma(\omega)$ into real part $\sigma_r(\omega)$ and imaginary part $\sigma_i(\omega)$:

$$\sigma_r(\omega) = \frac{ne^2\tau_s}{m}\frac{1}{1+(\omega\tau_s)^2} \tag{22.17}$$

$$\sigma_i(\omega) = \frac{ne^2\tau_s}{m}\frac{\omega\tau_s}{1+(\omega\tau_s)^2} \tag{22.18}$$

A modification of the simple Drude model was proposed by Smith, which was called the Drude-Smith model [55]. The possibility of n scattering events within time t is

$$P(0,t) = \frac{\left(\dfrac{t}{\tau_s}\right)e^{-\frac{t}{\tau_s}}}{n!} \tag{22.19}$$

where τ_s is the scattering time. And the current will decay as

$$\frac{j(t)}{j(0)} = e^{-\frac{t}{\tau_s}}\left[1+\sum_{n=1}^{\infty}c_n\frac{\left(\frac{t}{\tau_s}\right)^n}{n!}\right] \tag{22.20}$$

where constant c_n is the persistence of velocity, which represents the fraction of the original velocity retained after each collision. Applying Fourier transform, the conductivity can be rewritten as

$$\sigma(\omega) = \frac{ne^2\tau_s}{m(1-i\omega\tau_s)}\left[1+\sum_{n=1}^{\infty}\frac{c_n}{(1-\omega\tau_s)^n}\right] \tag{22.21}$$

In practice, it is usually just considered the first scattering event. Physically, this treats the first event as ballistic and all subsequent collisions as dispersive. For data fitting convenience, it is necessary to separate $\sigma(\omega)$ into real part $\sigma_r(\omega)$ and imaginary part $\sigma_i(\omega)$:

$$\sigma_r(\omega) = \frac{ne^2\tau_s}{m\left[1+(\omega\tau_s)^2\right]^2}\left\{1+(\omega\tau_s)^2+c\left[1-(\omega\tau_s)^2\right]^2\right\} \tag{22.22}$$

$$\sigma_i(\omega) = \frac{ne^2\tau_s}{m\left[1+(\omega\tau_s)^2\right]^2}\left\{\omega\tau_s\left[1+(\omega\tau_s)^2+2c\right]\right\} \tag{22.23}$$

where c is the persistence of velocity, which is allowed to be valued from 0 to −1, accounting for varying degrees of backscattering or carrier localization.

The Drude-Smith model is usually employed for describing materials with significant carrier localization, where $c \rightarrow 1$. In this case, the real part of conductivity, $\sigma_r(\omega)$, will have a peak at frequency $\omega = 1/2\pi\tau$, and the imaginary part of conductivity, $\sigma_i(\omega)$, equals to zero. For $c = 0$, the Drude-Smith model reduces to Drude model.

The Lorentz oscillator model can be employed to describe resonant features at a finite frequency with different origins [56]. The THz region covers a variety of fundamental excitations including rotations, vibrations in molecules, and collective modes in condensed matter such as phonon, plasma, and energy gaps associated with superconductivity. Lorentz oscillator presents the simplest expression of response at a finite frequency. The dielectric function can be described by

$$\varepsilon = \varepsilon_0 + \frac{A}{\omega_0^2 - \omega^2 - i\omega\gamma} \tag{22.24}$$

where

A is the oscillator amplitude
ω_0 is the resonance frequency
γ is the width of the resonance

The Lorentz oscillator model can be employed to characterize the charge oscillation accompanied with the polarization of excitons.

22.3 THz Spectroscopy Study of III–V Nitride Materials

In recent years, wide bandgap semiconductors, especially III–V nitride materials, have emerged as promising materials for developing electronic devices due to their unique physical and electronic properties, in particular, direct bandgap structure, high electric breakdown field, and high thermal conductivity [57,58]. The advantages associated with large bandgap include high breakdown voltages, the ability to sustain large electric field, radiation hardness, and high-temperature operation. Moreover, GaN is characterized by large thermal conductivity, an essential criterion for high-power operation. III–V compounds have nonpolar structure, resulting in no absorption in the THz region unless they possess specific phonon resonances. THz spectroscopy could probe physical phenomena such as low-energy excitations and carrier dynamics in electronic materials and collective vibrational or torsional modes in condensed-phase media. This chapter introduces some investigation in III–V nitride material using terahertz spectroscopy, including THz-TDS and OPTP.

22.3.1 Boron Nitride (BN)

Boron nitride is a III–V compound that is expected to have high transparency in the THz region, like many others in the series. It is manufactured as a ceramic and can be produced in large quantities and machined to the desired shape. As such, it is an attractive candidate for THz optics. Group of Mira Naftaly has done some work focusing on boron nitride using THz spectroscopy [59,60].

Boron nitride has some different possible modifications analogous to the polymorphs of carbon. The hexagonal boron nitride (h-BN), also called rhombohedra boron nitride (r-BN), has layered structures with lower density (2.18 g/cm³). It comprises $(BN)_3$ rings with very strong intralayer bonding and weak van der Waals bonding between layers. The interlayer boron and nitrogen atoms of BN are staggered in their arrangement, preventing π-orbital overlap. This structure explains both the high thermal conductivity and high electrical resistivity of boron nitride because it permits phonon conduction within layers, while electron localization inhibits electrical conduction. The hexagonal crystal structure gives rise to optical anisotropicity, the resulting birefringence being such that the extraordinary ray lies in the

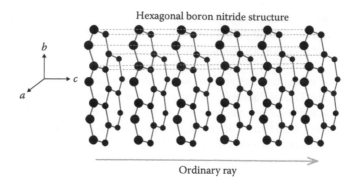

FIGURE 22.16 Schematic structure of hexagonal boron nitride (h-BN). (From Naftaly, M., Leist, J., and Fletcher, J.R., Optical properties and structure of pyrolytic boron nitride for THz applications, *Opt. Mater. Express*, 3, 260–269, 2013. With permission of Optical Society of America.)

ab-plane and the ordinary ray is aligned with the *c*-axis, as shown in Figure 22.16. The unique intralayer and interlayer bonding structure of born nitride is the source of its versatility for a number of diverse applications, including high-temperature electrical insulators for industrial and semiconductor processes, crucibles and furniture for molten metals and glass processing, inert setter plates for ceramic sintering, nozzles for metal powder generation, and furniture for epitaxial device growth. With such a diverse array of applications for BN powders and shapes, there exists a need to quantify many different properties. It is important to understand the effect of microstructure and chemistry on the end use of the material, as well as to control these properties in the manufacturing process. The van der Waals bonding between layers and the strong covalent bonding within layers allow adjacent layers to slide over one another with low interaction. Wurtzite structure (w-BN), with a high density (2.18 g/cm^3), and cubic zinc-blende structure (c-BN) are other different possible modifications of boron nitride.

Hot pressing methods have been developed to achieve nearly full-density shapes comprising boron nitride. They are produced by hot pressing submicron, turbostratic boron nitride, in the presence of a binder phase under temperature approaching 2000°C and pressures of up to 2000 psi (14 MPa). The grades described later—designated as HBN, HBR, HBC, and HBT—are manufactured by Momentive Performance Materials and are representative of commercially available forms of hot-pressed boron nitride.

22.3.1.1 Grades of Boron Nitride

Grade HBN's binder phase is constituted by boric oxide (B_2O_3). The molten phase is also known to mediate growth and refinement of turbostratic BN [61,62]. A peculiarity of the hot pressing of BN is that individual grains or platelets tend to "stand up proud" and align themselves along the direction of the pressing axis. It means that there is a strong tendency for the normal to the c-plane in each platelet to assume a preferential orientation perpendicular to the pressing axis, as shown in Figure 22.17. And this produces a relatively narrow distribution of orientations around perfect alignment. This preferential orientation is also reflected in the physical properties of grade HBN: data must be reported with respect to the pressing axis, and orientation must be specified when designing parts based on this grade. Upon cooling, the boric oxide phase in the resulting cylindrical "plug" forms a glassy phase. Due to the low surface activity of the basal plane in BN, boric oxide is preferentially segregated at the edges of platelets in the pressed plug. As boric oxide is known to react readily with moisture to form boric acid, the HBN grade is used when moisture resistance and thermal shock are not a concern.

Grade HBC is manufactured by removing the boric oxide binder phase from the hot-pressed BN shape. The purity gain is offset by a loss of mechanical strength and an increase in porosity. However, the resulting material is also more resistant to moisture and thermal shock. Orientation effects of the original hot-pressed BN material are maintained.

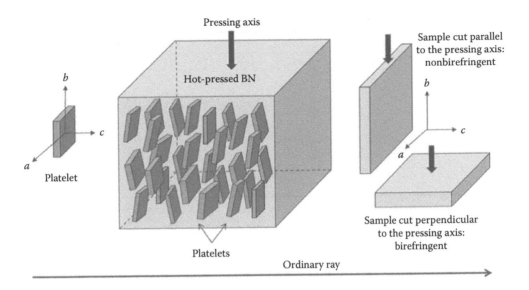

FIGURE 22.17 Relationship between boron nitride sample planes and crystal growth. (From Naftaly, M., Leist, J., and Fletcher, J.R., Optical properties and structure of pyrolytic boron nitride for THz applications, *Opt. Mater. Express*, 3, 260–269, 2013.With permission of Optical Society of America.)

Grade HBT is an economical, high-purity material similar to grade HBC. The binder phase in this material is removed by a different method from that used to manufacture HBC. Mechanical strength suffers and porosity increases to achieve the cost benefit. Again, the absence of boric oxide phase makes the material more resistant to moisture and thermal shock. Orientation effects of the starting hot-pressed BN material are maintained.

Grade HBR was developed for applications where higher mechanical strength, higher temperatures, and moisture resistance are important. The binder phase in this material is calcium borate ($Ca_3(BO_3)_2$), an inorganic material with a nominal melting point of 1150°C. Calcium borate is not water soluble, so moisture sensitivity is not an issue. The hot pressing process for this grade is similar to that for grade HBN, and it displays the same orientation effects.

Grade PBN was made by introducing a boron source gas (boron trichloride) and a nitrogen source gas (ammonia) into a reduced pressure chamber at elevated temperatures (~1970°C), with boron nitride forming as a deposit on a high-temperature substrate (graphite). Depending on the flow rates, temperatures, and pressures, three different type structures of PBN can be achieved: a purely turbostratic structure analogous to pyrolytic carbon (Type I), a more crystalline structure with higher density (Type II), and a crystalline structure with a columnar component (Type III). The form investigated in this study was Type III material. Two of the samples were cut along the c-axis, while the third was cut orthogonally to it.

Grade BIN77 is nonbirefringent because it is isotropic on the macroscopic scale, being formed as a mixture of randomly oriented crystals of BN and AlN. The refractive index of AlN is 2.92, while the mean index of fully dense BN is 2.11. The refractive index of BIN77 is 2.53 and is therefore consistent with its composition of 50BN : 50AlN.

Table 22.1 summarizes the composition and properties of the four boron nitride grades. The porosity was calculated from the measured weight of a given volume of the material and the known density of crystalline BN (2.18 g/cm³), including binders, if applicable. Any weight deficit was assumed to be attributable to porosity. Sample plates for each grade were fabricated with dimensions of 30 mm diameter and two thicknesses: 2.5 and 5 mm. One set of samples was cut with their thickness parallel to the pressing axis, resulting in the face being composed of edges of c-planes. Another set of samples was cut with their thickness perpendicular to the pressing axis, resulting in the face being composed of c-planes.

TABLE 22.1 Properties of Different Grades of Boron Nitride

	pBN	HBN	HBR	HBC	HBT	BIN77
BN mol. %	100	>95	>94	>99	>99	47
AIN mol. %	—	—	—	—	—	47
Binder	None	B_2O_3	$Ca_3(BO_3)_2$	None	None	$Ca_3(BO_3)_2$
Density (g/cm³)	2.18	2.10	2.00	1.95	1.75	2.43
Porosity (%)	0	7	11	13	22	10
$n_e \pm 0.002$ at 2 THz	1.912	2.087	1.995	2.058	2.015	2.490[a]
$n_o \pm 0.002$ at 2 THz	2.292	2.197	2.128	2.2	2.202	2.490[a]
$n_e - n_o$ at 2 THz	−0.392	−0.11	−0.133	−0.142	−0.187	0
Loss ± 5% (cm⁻¹) at 2 THz	0.62(e)4.1(o)	2.2	11	3	13(e)5.2(o)	7.7[a]

Source: Reproduced from Naftalya, M., Leistb, J., and Dudleya, R., Investigation of ceramic boron nitride by terahertz time-domain spectroscopy, *J. Eur. Ceram. Soc.*, 30, 2691–2697, 2010. With permission of Optical Society of America.
  [a] At 1 THz.

22.3.1.2 Experiment and THz Spectroscopy of Boron Nitride

The THz time-domain spectrometer (TDS) used a standard configuration incorporating a femtosecond laser, four off-axis parabolic mirrors, a biased GaAs emitter, and electro-optic detection with a ZnTe crystal, and balanced photodiodes. The maximum dynamic range of the system was 5000 in amplitude, and the frequency resolution in the experiments was 7.5 GHz. The samples were placed in the collimated part of the THz beam, which had a diameter of 25 mm and was vertically polarized. Measurements were carried out in dry air in order to eliminate water absorption lines from the recorded spectra [59]. The amplitude and phase of the THz signal as a function of frequency are obtained from the measured time-domain data of THz electric field by applying the Fourier transform using a standard FFT application, as shown in Figure 22.18.

The axis directions of the perpendicular-cut samples were determined by observing the time-domain trace of the transmitted THz beam. The sample was positioned in the beam and was rotated so as to obtain a single-peak trace at one of two delay positions. Particular attention was given to minimizing the residual features of the other trace. This is because the absence of such features signals good alignment of the THz beam polarization to the c-axis, either parallel (papa) or orthogonal (pape). However, it was not possible to eliminate these residues completely, indicating an imperfect mutual alignment of the platelets in the sample, as shown in Figure 22.19.

Refractive indices (*n*) and loss coefficients (*α*) of the samples were measured by comparing THz transmission through two different thicknesses of the sample material. In comparing the values for different grades, it may be expected that the refractive index will decrease linearly with increasing porosity, since less material lies in the optical path, as shown in Figure 22.20. It could be seen that the extraordinary refractive index does, but the ordinary index does not. This supports the view that porosity in these materials is located predominantly at the edges of the c-planes, rather than at the c-plane faces, which have low-surface energy. The extraordinary ray is polarized in the plane of the platelets, therefore experiencing an effective refractive index that is modified (reduced) by the interstices between the platelet edges. The ordinary ray, which is polarized transversely to the platelets, would be similarly affected by interstices between platelet layers. The absence of such effect indicates the absence of porosity between platelet faces.

The values for pe and papa orientations are similar for all samples except HBN, where the pe value is slightly lower. It is because the beam cross section is much larger than the sample thickness and more platelet edges are incorporated in the interaction volume of the beam in the pe orientation than in papa. So the transmitted THz beam could interact with a greater volume of binder in the pe than in the papa orientation. And since these have lower refractive indices than BN, this would explain the observed low values in the HBR samples.

The losses in the pe and papa orientations are similar. Notably, the loss in the pape orientation rises linearly with porosity, and the value extrapolated to fully dense (zero porosity) is close to zero, which indicates that THz absorption in BN is very low. In the HBT grade, losses in the pe and papa orientations are much

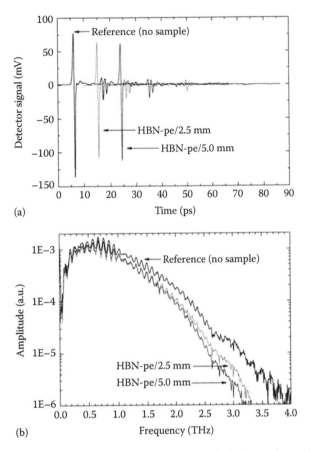

(a)

(b)

FIGURE 22.18 THz signal of reference, transmitted through HBN with thickness of 2.5 and 5 mm cut perpendicular in time domain (a) and frequency domain (b). (Reprinted from *J. Eur. Ceram. Soc.*, 30, Naftalya, M., Leistb, J., and Dudleya, R., Investigation of ceramic boron nitride by terahertz time-domain spectroscopy, 2691–2697. Copyright 2010, with permission from Elsevier.)

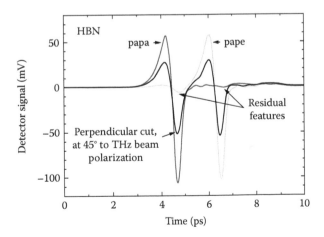

FIGURE 22.19 Time-domain traces of the THz pulses transmitted through the HBN samples of 5 mm thickness cut parallel to the pressing axis; papa = aligned parallel to the THz beam polarization; pape = aligned perpendicular to the THz beam polarization. (Reprinted from *J. Eur. Ceram. Soc.*, 30, Naftalya, M., Leistb, J., and Dudleya, R., Investigation of ceramic boron nitride by terahertz time-domain spectroscopy, 2691–2697. Copyright 2010, with permission from Elsevier.)

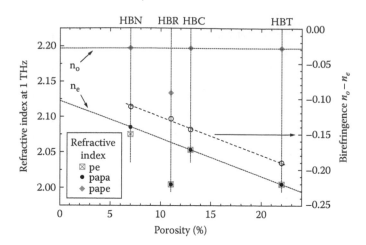

FIGURE 22.20 Refractive index of different grades BN at 1 THz. (Reprinted from *J. Eur. Ceram. Soc.*, 30, Naftalya, M., Leistb, J., and Dudleya, R., Investigation of ceramic boron nitride by terahertz time-domain spectroscopy, 2691–2697. Copyright 2010, with permission from Elsevier.)

higher than in pape, as shown in Figure 22.21. This is consistent with the view that porosity is segregated at the edges of the platelets, thus causing greater scattering of the beam polarized in the ab-plane. In ceramic materials, loss is caused by combined contributions of absorption and scattering. If absorption is the dominant mechanism, then samples with higher porosity will have reduced losses because less material lies in the beam path. Conversely, if scattering predominates, then loss will rise with porosity due to greater presence of scattering centers. Among the four grades of BN studied, loss increases with porosity, indicating that scattering is the dominant cause. The HBR grade exhibits higher than expected loss. All such compounds are polar to a much greater degree than BN, and therefore highly absorbing at THz frequencies, giving rise to absorption losses. Higher than calculated values of porosity would also contribute to increased scattering. Scattering loss is frequency dependent; the nature of the dependence indicates the size of the scatters.

THz time-domain spectroscopy was demonstrated as a technique capable of providing insight into the structure of optically opaque ceramic materials. Significant differences of different grades of BN were

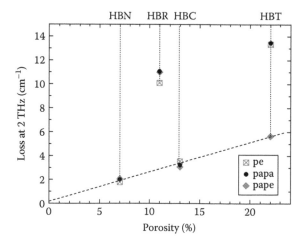

FIGURE 22.21 Loss coefficient of different grades BN at 2 THz. (Reprinted from *J. Eur. Ceram. Soc.*, 30, Naftalya, M., Leistb, J., and Dudleya, R., Investigation of ceramic boron nitride by terahertz time-domain spectroscopy, 2691–2697. Copyright 2010, with permission from Elsevier.)

observed in THz-TDS, in terms of their refractive indices and transmission loss. The low-porosity HBN grade was shown to have good transparency at THz region. And the results indicate that by reducing porosity further, transparency can be greatly improved. However, when selecting the appropriate material for an application, one must consider the slightly lower loss in HBN versus the lack of moisture sensitivity in HBC. Both HBN and HBC grades have potential uses as optical materials at THz region, especially in applications requiring high thermal resistance and/or hardness.

22.3.2 Aluminum Nitride (AlN)

Aluminum nitride (AlN) is a wide bandgap III–V semiconductor with promising optical and optoelectronic properties. Its potential optical applications range from the THz to ultraviolet spectral region. Also, due to a high thermal conductivity exceeding 200 W/(m·K), it is commonly used in polycrystalline form for thermal management applications. Recent improvements in the AlN bulk and thin-film crystal quality have sparked interest in nonlinear optical devices. In order to explore the possibility of using the material for terahertz applications, the ordinary and extraordinary refractive indices and the corresponding optical power loss of undoped AlN single crystals are measured in the 1–8 THz region. The availability of good quality bulk single crystals allowed us to determine experimentally the dielectric properties in the THz frequency range for both eigen polarizations. Group of Aleksej Majkić has done some work with the focus on aluminum nitride using THz spectroscopy [63,64].

The optical characterization was performed on m-plane <1100> single crystalline wafers, cut out from a single-crystal AlN boule grown in the -c-direction <0001>. The crystal was grown by physical vapor transport from a solid AlN source and a nitrogen atmosphere. Two wafers were used for these measurements, with thicknesses of 610 ± 10 μm (sample I) and 1110 ± 10 μm (sample II), respectively. The crystals were mounted on an aluminum holder with a 3 mm diameter hole, which defined the optical aperture for the terahertz beam. The optical characterization was performed using THz-TDS at room temperature in nitrogen atmosphere. Each waveform consisted of 800 points, recorded over a delay of 41 ps. An identical holder without the sample was used to measure the reference signals. By rotating the crystal around the probe beam axis, the optical axis of the crystal was aligned parallel or perpendicular to the polarization of the incoming THz wave, thereby probing the extraordinary and ordinary refractive index, respectively.

The evaluation procedure is illustrated in an example of determining the refractive index and optical power loss coefficient for the ordinary beam in sample I, as shown in Figure 22.22. The peak

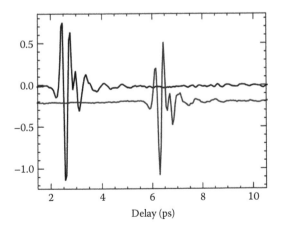

FIGURE 22.22 THz pulse of reference (up) and sample (down) in time domain. (From Majkić, A., Puc, U., Franke, A., Kirste, R., Collazo, R., Sitar, Z., and Zgonik, M., Optical properties of aluminum nitride single crystals in the THz region, *Opt. Mater. Express*, 5, 2106–2111, 2015. With permission of Optical Society of America.)

amplitude of the pulse transmitted through this 610 µm thick sample was delayed for ~3.7 ps, giving a rough estimate of an average ordinary refractive index of ~2.8. The pulse was attenuated by only 20%, suggesting low absorption.

A Fourier transform of the measured reference and sample pulse was calculated, including the echoes repeating at multiples of 13 ps after the main pulse, as shown in Figure 22.23. The oscillations in the reference and sample spectra are due to the mentioned multiple reflections inside the filter and, in addition, due to the internal reflections inside the sample, respectively. The observed noise floor was at ~5 × 10⁻³ of the amplitude maximum. The apparent higher-than-1 transmission above 7 THz was due to signal-level drift between both measurements. Variations in the measured transmission amplitude were always present, therefore limiting the useful range for fitting the amplitude to around 1.5–6 THz, while a range of 1–8 THz could still be used for fitting the phase. Its indices exhibit no pronounced absorption resonances.

The refractive indices, including ordinary and extraordinary, along with shaded uncertainty intervals, are shown in Figure 22.24. For comparison, the indices estimated in W. J. Moore's work are added. Its indices exhibit normal dispersion. By extrapolation, it could be estimated that the AlN static dielectric constants $\varepsilon_{DC,o} = 7.84 \pm 0.05$ and $\varepsilon_{DC,e} = 9.22 \pm 0.15$ for the ordinary and extraordinary polarization, respectively.

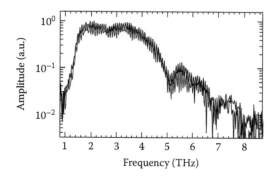

FIGURE 22.23 THz signal of reference (up) and sample (down) in frequency domain. (From Majkić, A., Puc, U., Franke, A., Kirste, R., Collazo, R., Sitar, Z., and Zgonik, M., Optical properties of aluminum nitride single crystals in the THz region, *Opt. Mater. Express*, 5, 2106–2111, 2015. With permission of Optical Society of America.)

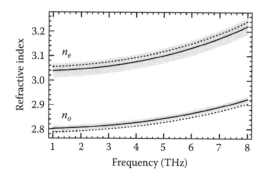

FIGURE 22.24 Ordinary and extraordinary refractive indices determined in AlN in frequency domain (solid lines). The shaded regions represent the estimated uncertainties. The indices estimated in W. J. Moore's work for comparison (dotted lines). (From Majkić, A., Puc, U., Franke, A., Kirste, R., Collazo, R., Sitar, Z., and Zgonik, M., Optical properties of aluminum nitride single crystals in the THz region, *Opt. Mater. Express*, 5, 2106–2111, 2015. With permission of Optical Society of America.)

22.3.3 Gallium Nitride (GaN)

As a direct wide bandgap (~3.4 eV) semiconductor, gallium nitride (GaN) has attracted much attention and has been intensively studied in the past decades. Most important are the applications in blue and ultraviolet light emitting diodes and laser diodes. On the other hand, GaN has many excellent electron transport properties, such as large breakdown voltage, good mobility, and high saturation drift velocity, which make GaN one of the promising materials for high-power and high-speed transistors operated at frequencies in the GHz and THz frequency regions. Moreover, THz emission was also reported in InGaN/GaN structures excited by blue femtosecond laser [65,66].

The THz dielectric properties of unintentionally doped *n*-type bulk GaN crystal have been studied using THz-TDS by group Xinhai Zhang. This group focuses on the characterization of the THz free carrier dynamics and photon–phonon interaction of epitaxial grown *n*-type GaN thin films with different carrier concentrations. The measured complex conductivity is well fitted by simple Drude model. The free carrier density N_0 obtained by fitting the THz-TDS measurement data is in good agreement with that from four-probe Hall measurement. On the other hand, when considering the dielectric contribution from the lattice vibration for the case of lower free carrier concentrations, a better analysis of the frequency-dependent complex dielectric response was obtained.

The GaN samples used in this study are two *n*-type GaN epilayers grown on (0001)-oriented sapphire substrate by metal–organic chemical vapor deposition (MOCVD) with different doping concentrations [67,68]. The layer thickness, free carrier concentration, and damping constant, from both Hall measurement and fitting of THz-TDS date, are listed in Table 22.1. The transmission THz spectra were measured using a conventional THz-TDS system (TeraView Spectra 3000). The 100 fs optical pulses centered at 800 nm at repetition rate of 76 MHz from a mode-locked Ti:sapphire laser was focused onto a photoconductive antenna of low-temperature-grown GaAs to generate THz wave. The emitted THz wave was collimated and focused onto the sample by a pair of off-axis paraboloidal mirrors and then directed to the photoconductive detector. All the optical components and sample were enclosed in a chamber purged with dry nitrogen to reduce water vapor absorption.

The transmitted THz pulse through the sample and the reference (sapphire substrate) were recorded in the time domain by varying the delay time between the THz pulse and gating optical pulse, as shown in Figure 22.25a. For the two GaN films, the peak intensity of the main transmitted pulse shows continuous decrease with increasing free carrier concentration, as compared to the reference signal transmitted through sapphire substrate. It is easy to obtain the corresponding amplitude spectra of two *n*-type GaN samples and the sapphire substrate after applying Fourier transform on the time-domain data, as shown in Figure 22.25b.

To analyze the experimental data, it is appropriate to use Drude model and damped oscillator model. And the incident THz filed not only drives the free carriers and loses energy by free carriers damping, but also couples to the lattice vibration and loses energy due to creation of phonons. The measured refractive index and power absorption are fitted using both the simple Drude model and the combined Drude and damped oscillator model. It can be seen that free carriers' transportation is the dominant process for the dielectric response and Drude model provides the overall fitting profile. However, comparing sample 1 with sample 2, it could be seen that with the decrease of free carrier concentration, the effects of phonon contribution become more evident and the combined model provides a better fitting curve as compared to the simple Drude model, as shown in Figure 22.26.

In semiconductors, the free carrier contribution to the optical and electrical properties of material at long wavelength below the fundamental band gap is determined by the intraband transitions of the partially filled band. With the classical dispersion theory and omitting the restoring force for free carriers, this intraband transition finally leads to the Drude model. The measured complex conductivities of two *n*-type GaN epilayers with different free carrier concentrations are analyzed using the simple Drude model, as shown in Figure 22.27. It can be seen that both the real and imaginary parts of the complex conductivity of the two GaN films can be relatively well fitted using simple Drude model. It is known that when an

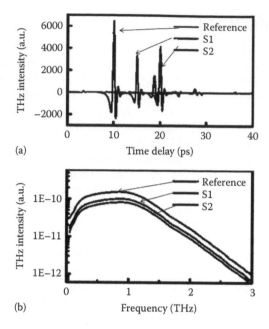

(a)

(b)

FIGURE 22.25 THz signal transmitted through reference and n-type GaN samples in time domain (a) and in frequency domain (b). (Reprinted with permission from Guo, H.C., Zhang, X.H., Liu, W., Yong, A.M., and Tang, S.H., Terahertz carrier dynamics and dielectric properties of GaN epilayers with different carrier concentrations, *J. Appl. Phys.*, 106, 063104. Copyright 2009, American Institute of Physics.)

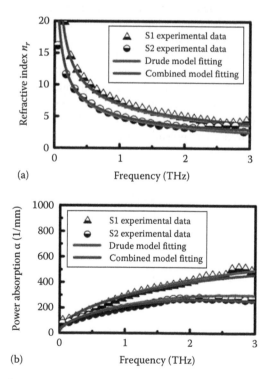

(a)

(b)

FIGURE 22.26 Refractive index (a) and power absorption (b) of GaN samples. (Reprinted with permission from Guo, H.C., Zhang, X.H., Liu, W., Yong, A.M., and Tang, S.H., Terahertz carrier dynamics and dielectric properties of GaN epilayers with different carrier concentrations, *J. Appl. Phys.*, 106, 063104. Copyright 2009, American Institute of Physics.)

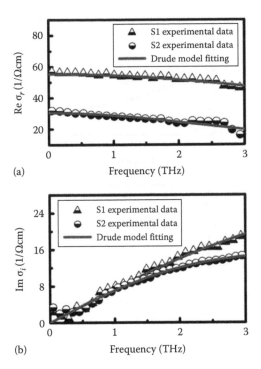

FIGURE 22.27 Real part (a) and imaginary part (b) complex conductivity of GaN samples. (Reprinted with permission from Guo, H.C., Zhang, X.H., Liu, W., Yong, A.M., and Tang, S.H., Terahertz carrier dynamics and dielectric properties of GaN epilayers with different carrier concentrations, *J. Appl. Phys.*, 106, 063104. Copyright 2009, American Institute of Physics.)

electromagnetic wave passes through a sample, the electric field not only drives the free carriers to move but also couples with the materials transverse optical phonon modes and then propagates as phonon polaritons in the crystal and gives rise to the dielectric response. Hence, it is necessary to take into consideration the lattice vibration contribution together with the Drude model to investigate THz complex dielectric function of the sample.

22.3.4 Indium Nitride (InN)

Indium nitride (InN) has attracted tremendous research attention recently because of its narrow bandgap (0.7 eV) and superior electronic transport properties. The good thermal and chemical stability, small bandgap, and high electron mobility and high saturated drift velocity make this material very attractive for applications in high-speed electronic devices, infrared light emitters, and solar cells [69–73]. The knowledge of ultrafast carrier dynamics is essential for optimizing performance of these applications. The ultrafast carrier dynamics in semiconductors is typically investigated by using OPTP measurements, which are sensitive primarily to the carrier occupation of specific energy levels in the energy bands. In contrast, OPTP spectroscopy is sensitive to photo-induced changes in conductivity, which contains the information of carrier density and mobility. The latter reflects the energy distribution of free carriers. OPTP is therefore able to study both intraband relaxation and interband recombination dynamics of photogenerated electrons and holes with a temporal resolution of sub-picosecond.

Group H. Ahn has done some work with the focus on InN [69–72]. The ultrafast optical pump was provided by a Ti:sapphire regenerative amplifier laser system, which delivers about 50 fs optical pulses

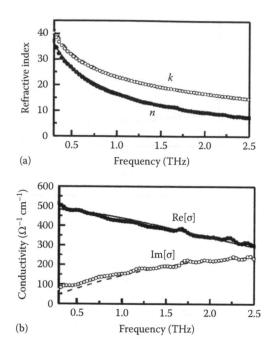

(a)

(b)

FIGURE 22.28 Refractive index (a) and conductivity (b) of InN film. Solid and dashed lines correspond to the calculated results based on Drude model (Reprinted with permission from Ahn, H., Ku, Y.-P., Wang, Y.-C., Chuang, C.-H., Gwo, S., and Pan, C.L., Terahertz spectroscopic study of vertically aligned InN nanorods, *Appl. Phys. Lett.*, 91, 163105. Copyright 2007, American Institute of Physics.)

at a center wavelength of 800 nm with a repetition rate of 1 kHz. The terahertz probe beam was generated from a photoexcited (100) InAs surface and detected by free-space electro-optic sampling in a 2 mm-thick ZnTe crystal. In the optical pump–terahertz probe experiment, the transient behavior of the photoexcited carriers was monitored by measuring the transmitted peak amplitude of terahertz waveforms at normal incident angle as a function of delay time between the terahertz probe and the optical pump pulses. The static electrical properties of the samples were separately measured by a terahertz-TDS system based on low-temperature-grown GaAs photoconductive dipole antennas, which were excited and probed by a Ti:sapphire laser at a repetition rate of 82 MHz. All the measurements were done under dry nitrogen purge.

A wurtzite InN epitaxial film and vertically aligned InN nanorod arrays were grown on Si (111) substrates by plasma-assisted molecular-beam epistaxis. The InN epilayers were grown on Si (111) using the epitaxial AlN/ double-buffer layer technique. The InN nanorods were grown at a sample temperature of 330°C on β-Si$_3$N$_4$/ Si (111) without the AlN buffer layer. The scanning electron microscopy (SEM) image of the hexagonal-shaped nanorods exhibits nanorods with a uniform diameter of about 130 nm and an average aspect ratio (height/diameter) of about 6. The nanorod arrays have an aerial density of about 5×10^9 cm^{-2}. The thicknesses of the InN epilayer and nanorods are about 1.0 and 0.75 μm, respectively. The morphology and size distribution of InN nanorods were analyzed using SEM.

In THz-TDS, the frequency-dependent dielectric constants of the InN samples are calculated from the spectral amplitude and phase difference between the samples and the substrate. Figure 22.28 shows the real (solid circles) and imaginary (open circles) parts of the frequency-dependent refractive index of the InN film. The real part of conductivity Re[σ] decreases as the frequency increases, while the imaginary part

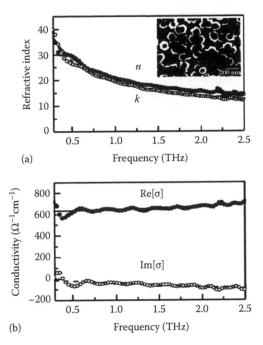

FIGURE 22.29 Refractive index (a) and conductivity (b) of InN film. Solid and dashed lines correspond to the calculated results based on Drude-Smith model. (Reprinted with permission from Ahn, H., Ku, Y.-P., Wang, Y.-C., Chuang, C.-H., Gwo, S., and Pan, C.L., Terahertz spectroscopic study of vertically aligned InN nanorods, *Appl. Phys. Lett.*, 91, 163105. Copyright 2007, American Institute of Physics.)

of conductivity Im[σ] increases slowly. This frequency dependence is typically observed for Drude-like materials below the plasma frequency.

Since the InN nanorod film consists of loosely packed InN nanorods and air, the physical parameters directly measured from the data of THz-TDS contain the contributions from both the air and the pure InN nanorods. To obtain the physical parameters of the pure InN nanorods, a simple effective medium approximation (EMA) was employed. It illustrates frequency-dependent refractive index and conductivity of the pure InN nanorods extracted under EMA, respectively, as shown in Figure 22.29. The trend of monotonic decrease of the complex refractive index of the nanorods is similar to the case of InN film. The complex conductivity response of the nanorods, however, is different from that of InN film. The real part Re[σ] gradually increases with increasing frequency, while the imaginary part Im[σ], with a negative value, decreases with increasing frequency. This frequency dependence cannot be explained by Drude model, in which the frequency-dependent conductivity has a maximum at zero frequency and monotonically decreases with frequency. Using the Drude-Smith model, an excellent fit of complex conductivity of the InN nanorods could be obtained.

Figure 22.30 shows the time-dependent differential transmission signals of InN nanorods (up) and the epilayers (down). Each sample is excited at the laser fluence of 1.1 mJ/cm². As soon as the pump pulse arrives, transmission responses of both samples instantaneously drop and the sample independent sharp fall time is measured to be 0.6–0.7 ps. Due to our relatively broad pulse width of terahertz probe (0.6 ps) and the slow detector response time, sample-dependent fall time cannot be monitored. The transmission response of InN film gradually recovers from 70% to about 22% within 200 ps, while that of nanorods quickly recovers to its steady value within 2 ps and persists at this value over 200 ps. The peak values of the carrier density near zero time delay indicate that during the pump pulse, photoexcited carriers of about 1.0×10^{19} cm^{-3} are generated for InN film, which is higher than

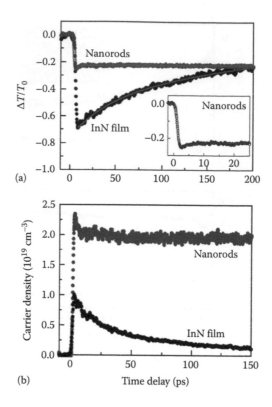

FIGURE 22.30 OPTP signal of InN film and InN nanorods (a) 800 nm excitation for nanorods (up) and the InN epilayer (down). The solid lines are the results of biexponential fitting. (b) The photoexcited carrier density of InN film and nanorods calculated. (Reprinted with permission from Ahn, H., Chuang, C.-H., Ku, Y.-P., and Pan, C.-L., Free carrier dynamics of InN nanorods investigated by time-resolved terahertz spectroscopy, *J. Appl. Phys.*, 105, 023707. Copyright 2009, American Institute of Physics.)

its unintentionally doped carrier concentration$(2.5 \pm 0.2) \times 10^{18}$ cm^{-3}. On the contrary, pump pulse generates 2.3×10^{19} cm^{-3} of photoexcited carrier density for nanorods, which is of the same order to or lower than its free electron concentration.

Observed biexponential relaxation of InN film agrees with other experimental results, in which the carrier dynamics is found to be due to the hot carrier cooling through phonon emission followed by the defect-related nonradiative recombination process. The fast relaxation time of 30 ps measured for our InN film is consistent with that of 48 ps for InN film ($N_e = 1.2 \times 10^{19}$ cm^{-3}) measured by the optical pump–probe technique. The carrier scattering (or damping) time related to the carrier conduction mobility through the Drude model is indeed shorter (13 fs) for nanorods than that of film (52 fs). The observed shorter scattering time of nanorods suggests a faster capture rate to the defect states, which further supports the existence of the high defect concentration of nanorods. Therefore, observed shorter initial relaxation time constant of nanorods compared to film can be explained by the inverse relation between the carrier lifetime and the free electron density due to the increased electron trapping by the defects.

It is well known that nonradiative defect–related recombination has the carrier density–independent lifetime. To identify the nature of the recombination in InN nanorods, differential transmission is measured for the pump fluence range of 0.32–0.96 mJ/cm^2. The general behavior of transmission trace for each sample is the same and the average fast relaxation time constant of nanorods is 2.1 ± 0.3 ps. The slow relaxation components for both samples also do not show any observable pump fluence

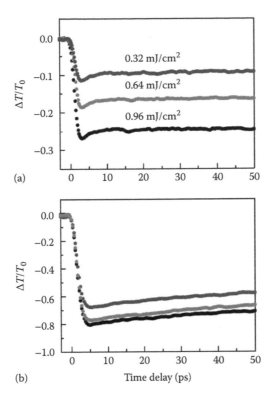

(a)

(b)

FIGURE 22.31 Pump fluence dependence of differential terahertz transmission for (a) InN nanorods and (b) InN film excited at the pump fluences of 0.32 (up), 0.64 (middle), and 0.96 mJ/cm² (down). (Reprinted with permission from Ahn, H., Chuang, C.-H., Ku, Y.-P., and Pan, C.-L., Free carrier dynamics of InN nanorods investigated by time-resolved terahertz spectroscopy, *J. Appl. Phys.*, 105, 023707. Copyright 2009, American Institute of Physics.)

dependence. The pump fluence–independent carrier lifetime suggests that the defect-related nonradiative recombination rather than Auger recombination is the common recombination process for nanorods and InN film. At high pump fluence (>0.6 mJ/cm²), the maximum negative change in transmission of InN film shows saturation, as shown in Figure 22.31, while that of nanorods scales linearly with the pump fluence.

References

1. B. Ferguson and X.-C. Zhang, Materials for terahertz science and technology, *Nat. Mater.* 1, 26–33 (2002).
2. S. H. Siegel, Terahertz technology, *IEEE T. Microw. Theory* 50, 910–928 (2002).
3. R. Huang, Q. Meng, X. Guo, and B. Zhang, Analysis of impact factors of output characteristics for optically pumped THz lasers, *Opt. Laser Technol.* 82, 63–68 (2016).
4. C. Paoloni, D. Gamzina, L. Himes, B. Popovic, R. Barchfeld, L. Yue, Y. Zheng et al., THz backward-wave oscillators for plasma diagnostic in nuclear fusion, *IEEE Trans. Plasma Sci.* 44, 369–376 (2016).
5. M. J. Suess, P. Jouy, C. Bonzon, J. M. Wolf, E. Gini, M. Beck, and J. Faist, Single-mode quantum cascade laser array emitting from a single facet, *IEEE Photon. Technol. Lett.* 28, 1197–1200 (2016).
6. S. L. Dexheimer, *Terahertz Spectroscopy: Principles and Applications*, CRC Press, Boca Raton, FL, 360pp, 2007.
7. L. Ho, M. Pepper, and P. Taday, Terahertz spectroscopy: Signatures and fingerprints, *Nat. Photon.* 2, 541–543 (2008).

8. X. Yin, B. W.-H. Ng, and D. Abbott, *Terahertz Imaging for Biomedical Applications*, Springer Press, New York, 316pp, 2012.

9. B. Sensale-Rodriguez, R. Yan, M. M. Kelly, T. Fang, K. Tahy, W. S. Hwang, D. Jena, L. Liu, and H. G. Xing, Broadband graphene terahertz modulators enabled by intraband transitions, *Nat. Commun.* 3, 780 (2012).

10. X.-C. Zhang, Y. Jin, and X. F. Ma, Coherent measurement of THz optical rectification from electro-optic crystals, *Appl. Phys. Lett.* 61, 2764–2766 (1992).

11. A. Nahata, A. S. Weling, and T. F. Heinz, A wideband coherent terahertz spectroscopy system using optical rectification and electro-optic sampling, *Appl. Phys. Lett.* 69, 2321–2323 (1996).

12. G. Mourou, C. V. Stancampiano, A. Antonetti, and A. Orszag, Picosecond microwave pulses generated with a subpicosecond laser-driven semiconductor switch, *Appl. Phys. Lett.* 39, 295–296 (1981).

13. M. van Exter, Ch. Fattinger, and D. Grischkowsky, High-brightness terahertz beams characterized with an ultrafast detector, *Appl. Phys. Lett.* 55, 337–339 (1989).

14. D. Grischkowsky, S. Keiding, M. van Exter, and Ch. Fattinger, Far-infrared time-domain spectroscopy with terahertz beams of dielectrics and semiconductors, *J. Opt. Soc. Am. B* 7, 2006–2015 (1990).

15. M. R. Leahy-Hoppa, M. J. Fitch, X. Zheng, L. M. Hayden, and R. Osiander, Wideband terahertz spectroscopy of explosives, *Chem. Phys. Lett.* 434, 227–230 (2007).

16. S. D. Ganichev and W. Prettl, *Intense Terahertz Excitation of Semiconductors*, Oxford University Press, London, U.K., 432pp, 2006.

17. R. M. Smith and M. A. Arnold, Terahertz time-domain spectroscopy of solid samples: Principles, applications, and challenges, *Appl. Spectrosc. Rev.* 46, 636–679 (2011).

18. D. Dragoman and M. Dragoman, Terahertz fields and applications, *Progr. Quantum Electron.* 28, 1–66 (2004).

19. I. S. Gregory, C. Baker, W. R. Tribe, I. V. Bradley, M. J. Evans, E. H. Linfield, A. Giles Davies, and M. Missous, Optimization of photomixers and antennas for continuous-wave terahertz emission, *IEEE J. Quantum Electron.* 41, 717–728 (2005).

20. C. W. Berry, N. Wang, M. R. Hashemi, M. Unlu, and M. Jarrahi, Significant performance enhancement in photoconductive terahertz opto-electronics by incorporating plasmonic contact electrodes, *Nat. Commun.* 4, 1622 (2013).

21. Y. C. Shen, P. C. Upadhya, H. E. Beere, E. H. Linfield, A. G. Davies, I. S. Gregory, C. Baker, W. R. Tribe, and M. J. Evans, Generation and detection of ultrabroadband terahertz radiation using photoconductive emitters and receivers, *Appl. Phys. Lett.* 85, 164–166 (2004).

22. S. Kono, M. Tani, Ping Gu, and K. Sakai, Detection of up to 20 THz with a low temperature grown GaAs photoconductive antenna gated with 15 fs light pulses, *Appl. Phys. Lett.* 77, 4104–4106 (2000).

23. M. Beck, H. Schäfer, G. Klatt, J. Demsar, S. Winnerl, M. Helm, and T. Dekorsy, Impulsive terahertz radiation with high electric fields from an amplifier-driven large-area photoconductive antenna, *Opt. Express* 18, 9251–9257 (2010).

24. Z. Piao, M. Tani, and K. Sakai, Carrier dynamics and terahertz radiation in photoconductive antennas, *Jpn. J. Appl. Phys.* 39, 96–100 (2000).

25. Y. Zhang, X. Zhang, S. Li, J. Gu, Y. Li, Z. Tian, C. Ouyang, M. He, J. Han, and W. Zhang, A broadband THz-TDS system based on DSTMS emitter and LTG InGaAs/InAlAs photoconductive antenna detector, *Sci. Rep.* 6, 26949 (2016).

26. G. Gallot and D. Grischkowsky, Electro-optic detection of terahertz radiation, *J. Opt. Soc. Am. B Opt. Phys.* 16, 1204–1212 (1999).

27. N. C. J. van der Valk and P. C. M. Planken, Electro-optic detection of subwavelength terahertz spot sizes in the near field of a metal tip, *Appl. Phys. Lett.* 81, 1558–1560 (2002).

28. Q. Chen, M. Tani, Z. Jiang, and X.-C. Zhang, Electro-optic transceivers for terahertz-wave applications, *J. Opt. Soc. Am. B Opt. Phys.* 18, 823–831 (2001).

29. K. L. Vodopyanov, Optical generation of narrow-band terahertz packets in periodically-inverted electro-optic crystals: Conversion efficiency and optimal laser pulse format, *Opt. Express* 14, 2263–2276 (2006).

30. T. Löffler, T. Hahn, M. Thomson, F. Jacob, and H. Roskos, Large-area electro-optic ZnTe terahertz emitters, *Opt. Express* 13, 5353–5362 (2005).

31. A. M. Sinyukov and L. M. Hayden, Generation and detection of terahertz radiation with multilayered electro-optic polymer films, *Opt. Express* 27, 55–57 (2002).

32. J. Kröll, J. Darmo, and K. Unterrainer, High-performance terahertz electro-optic detector, *Electron. Lett.* 40, 763–764 (2004).

33. J. Kim, O.-P. Kwon, F. D. J. Brunner, M. Jazbinsek, S.-H. Lee, and P. Günter, Phonon modes of organic electro-optic molecular crystals for terahertz photonics, *J. Phys. Chem. C* 119, 10031–10039 (2015).

34. M. D. Thomson, V. Blank, and H. G. Roskos, Terahertz white-light pulses from an air plasma photo-induced by incommensurate two-color optical fields, *Opt. Express* 18, 23173–23182 (2010).

35. H. Zhong, N. Karpowicz, and X.-C. Zhang, Terahertz emission profile from laser-induced air plasma, *Appl. Phys. Lett.* 88, 261103 (2006).

36. P. Klarskov, A. C. Strikwerda, K. Iwaszczuk, and P. U. Jepsen, Experimental three-dimensional beam profiling and modeling of a terahertz beam generated from a two-color air plasma, *New J. Phys.* 15, 075012 (2013).

37. B. Clough, J. Liu, and X.-C. Zhang, All air-plasma terahertz spectroscopy, *Opt. Letters* 36, 2399–2401 (2011).

38. V. A. Andreeva, O. G. Kosareva, N. A. Panov, D. E. Shipilo, P. M. Solyankin, M. N. Esaulkov, P. González de Alaiza Martínez et al., Ultrabroad terahertz spectrum generation from an air-based filament plasma, *Phys. Rev. Lett.* 116, 063902 (2016).

39. J. Das and M. Yamaguchi, Terahertz wave excitation from preexisting air plasma, *J. Opt. Soc. Am. B Opt. Phys.* 30, 1595–1600 (2013).

40. P. U. Jepsen, D. G. Cooke, and M. Koch, Terahertz spectroscopy and imaging—Modern techniques and applications, *Laser Photon. Rev.* 5, 124–166 (2011).

41. B. N. Flanders, D. C. Arnett, and N. F. Scherer, Optical pump-terahertz probe spectroscopy utilizing a cavity-dumped oscillator-driven terahertz spectrometer, *IEEE J. Sel. Top. Quantum Electron.* 4, 353–359 (1998).

42. K. P. H. Lui, and F. A. Hegmann, Ultrafast carrier relaxation in radiation-damaged silicon on sapphire studied by optical-pump-terahertz-probe experiments, *Appl. Phys. Lett.* 78, 3478–3480 (2001).

43. F. Kadlec, C. Kadlec, P. Kužel, P. Slavíček, and P. Jungwirth, Optical pump-terahertz probe spectroscopy of dyes in solutions: Probing the dynamics of liquid solvent or solid precipitate, *J. Chem. Phys.* 120, 912–917 (2004).

44. F. Kadlec, H. Němec, and P. Kužel, Optical two-photon absorption in GaAs measured by optical-pump terahertz-probe spectroscopy, *Phys. Rev. B* 70, 125205 (2004).

45. R. P. Prasankumar, A. Scopatz, D. J. Hilton, A. J. Taylor, R. D. Averitt, J. M. Zide, and A. C. Gossard, Carrier dynamics in self-assembled ErAs nanoislands embedded in GaAs measured by optical-pump terahertz-probe spectroscopy, *Appl. Phys. Lett.* 86, 201107 (2005).

46. G. L. Dakovski, K. Brian, L. Song, and S. Jie, Finite pump-beam-size effects in optical pump-terahertz probe spectroscopy, *J. Opt. Soc. Am. B Opt. Phys.* 23, 139–141 (2006).

47. P. D. Cunningham and L. Michael Hayden, Carrier dynamics resulting from above and below gap excitation of P3HT and P3HT/PCBM investigated by optical-pump terahertz-probe spectroscopy, *J. Phys. Chem. C* 112, 7928–7935 (2008).

48. M. C. Hoffmann, J. Hebling, H. Y. Hwang, K.-L. Yeh, and K. A. Nelson, Impact ionization in InSb probed by terahertz pump-terahertz probe spectroscopy, *Phys. Rev. B* 79, 161201 (2009).

49. J. H. Strait, H. Wang, S. Shivaraman, V. Shields, M. Spencer, and F. Rana, Very slow cooling dynamics of photoexcited carriers in graphene observed by optical-pump terahertz-probe spectroscopy, *Nano Lett.* 11, 4902–4906 (2011).

50. H. Liu, J. Lu, H. F. Teoh, D. Li, Y. Ping Feng, S. H. Tang, C. H. Sow, and X. Zhang, Defect engineering in CdSxSe1-x nanobelts: An insight into carrier relaxation dynamics via optical pump-terahertz probe spectroscopy, *J. Phys. Chem. C* 116, 26036–26042 (2012).

51. G. Li, D. Li, Z. Jin, and G. Ma, Photocarriers dynamics in silicon wafer studied with optical-pump terahertz-probe spectroscopy, *Opt. Commun.* 285, 4102–4106 (2012).

52. H. W. Liu, L. M. Wong, S. J. Wang, S. H. Tang, and X. H. Zhang, Ultrafast insulator-metal phase transition in vanadium dioxide studied using optical pump-terahertz probe spectroscopy, *J. Phys. Condens. Matter* 41, 415604 (2012).

53. H. W. Liu, L. M. Wong, S. J. Wang, S. H. Tang, and X. H. Zhang, Effect of oxygen stoichiometry on the insulator-metal phase transition in vanadium oxide thin films studied using optical pump-terahertz probe spectroscopy, *Appl. Phys. Lett.* 103, 151908 (2013).

54. D. Paul, Zur Elektronentheorie der metalle, *Annalen der Physik* 306, 566–613 (1900).

55. N. V. Smith, Classical generalization of the Drude formula for the optical conductivity, *Phys. Rev. B* 64, 155106 (2001).

56. R.D. Averitt and A.J. Taylor, Ultrafast optical and far-infrared quasiparticle dynamics in correlated electron materials, *J. Phys. Condens. Matter* 14, R1357–R1390 (2002).

57. M. Mukherjee and S. K. Roy, Optically modulated III-V nitride-based top-mounted and flip-chip IMPATT oscillators at terahertz regime: Studies on the shift of avalanche transit time phase delay due to photogenerated carriers, *IEEE Trans. Electron Dev.* 56, 1411–1417 (2009).

58. M. Mukherjee and S. K. Roy, Wide-bandgap III-V nitride based avalanche transit-time diode in Terahertz regime: Studies on the effects of punch through on high frequency characteristics and series resistance of the device, *Curr. Appl. Phys.* 10, 646–651 (2010).

59. M. Naftalya, J. Leistb, and R. Dudleya, Investigation of ceramic boron nitride by terahertz time-domain spectroscopy, *J. Eur. Ceram. Soc.* 30, 2691–2697 (2010).

60. M. Naftaly and J. Leist, Investigation of optical and structural properties of ceramic boron nitride by terahertz time-domain spectroscopy, *Appl. Opt.* 52, B20–B25 (2013).

61. M. Janek, A. Vincze, J. Darmo, V. Szöcs, M. Matejdes, T. Zacher, Š. Kavecký et al., Dielectric properties of boron nitride in THz region synthesized with nonenergetic CVD, *Int. J. Appl. Ceram. Technol.* 10, E167–E176 (2013).

62. M. Naftaly, J. Leist, and J. R. Fletcher, Optical properties and structure of pyrolytic boron nitride for THz applications, *Opt. Mater. Express* 3, 260–269 (2013).

63. R. B. Jaculbia, M. H. M. Balgos, N. S. Mangila IV, M. A. C. Tumanguil, E. S. Estacio, A. A. Salvador, and A. S. Somintac, Enhanced terahertz emission from GaAs substrates deposited with aluminum nitride films caused by high interface electric fields, *Appl. Surf. Sci.* 303, 241–244 (2014).

64. A. Majkić, U. Puc, A. Franke, R. Kirste, R. Collazo, Z. Sitar, and M. Zgonik, Optical properties of aluminum nitride single crystals in the THz region, *Opt. Mater. Express* 5, 2106–2111 (2015).

65. C. E. Martinez, N. M. Stanton, P. M. Walker, A. J. Kent, S. V. Novikov, and C. T. Foxon, Generation of terahertz monochromatic acoustic phonon pulses by femtosecond optical excitation of a gallium nitride/aluminium nitride superlattice, *Appl. Phys. Lett.* 86, 221915 (2005).

66. D. M. Moss, A. V. Akimov, A. J. Kent, B. A. Glavin, M. J. Kappers, J. L. Hollander, M. A. Moram, and C. J. Humphreys, Coherent terahertz acoustic vibrations in polar and semipolar gallium nitride-based superlattices, *Appl. Phys. Lett.* 94, 011909 (2009).

67. H. C. Guo, X. H. Zhang, W. Liu, A. M. Yong, and S. H. Tang, Terahertz carrier dynamics and dielectric properties of GaN epilayers with different carrier concentrations, *J. Appl. Phys.* 106, 063104 (2009).

68. A. Hamano, S. Ohno, H. Minamide, H. Ito, and Y. Usuki, High resolution imaging of electrical properties of a 2-inch-diameter gallium nitride wafer using frequency-agile terahertz waves, *Jpn. J. Appl. Phys.* 49, 022402 (2010).

69. H. Ahn, Y.-P. Ku, Y.-C. Wang, C.-H. Chuang, S. Gwo, and Ci-Ling Pan, Terahertz spectroscopic study of vertically aligned InN nanorods, *Appl. Phys. Lett.* 91, 163105 (2007).

70. H. Ahn, C.-H. Chuang, Y.-P. Ku, and C.-L. Pan, Free carrier dynamics of InN nanorods investigated by time-resolved terahertz spectroscopy, *J. Appl. Phys.* 105, 023707 (2009).
71. H. Ahn, Y.-J. Yeh, Y.-L. Hong, and S. Gwo, Background and photoexcited carrier dependence of terahertz radiation from Mg-doped nonpolar indium nitride films, *Appl. Phys. Express* 3, 122105 (2010).
72. H. Ahn, J.-W. Chia, H.-M. Lee, Y.-L. Hong, and S. Gwo, Mg-induced terahertz transparency of indium nitride films, *Appl. Phys. Lett.* 99, 232117 (2011).
73. I. Wilke, Y. J. Ding, and T. V. Shubina, Optically- and electrically-stimulated terahertz radiation emission from indium nitride, *J. Infrared Millim. Terahertz Waves* 33, 559–592 (2012).

23

Internal Luminescence Mechanisms of III-Nitride LEDs

Shijie Xu
University of Hong Kong

Abstract III-nitride multi-quantum-well (MQW) LEDs play a key role in emerging solid-state lighting technologies. However, the current luminescence efficiency of III-nitride LEDs is insufficient to support this solid-state lighting technology revolution. Moreover, the drop in efficiency under a large injection current is also a severe technical hindrance. To solve these challenging problems, gaining a deeper and better understanding of the internal luminescence mechanisms of III-nitride MQW LEDs will prove crucial. In this chapter, we attempt to summarize our understanding of the internal luminescence mechanisms of III-nitride MQW LEDs with regard to the reverse quantum confinement Stark effect, localized excitons and exciton-phonon coupling, super lateral diffusion and recombination of carriers, and the non-exponential decaying dynamics of localized carriers. To conclude, suggestions for further studies are briefly discussed.

23.1 Introduction

As predicted by Shuji Nakamura et al. in their book *The Blue Laser Diode*,[1] III-nitride LED-based solid-state lighting (SSL) sources with which to replace conventional incandescent light bulbs and luminescent tubes are developing rapidly. According to a recent report from the U.S. Department of Energy, for example, compared with conventional lighting technologies, by 2025 advanced SSL technologies will be much more energy efficient, have longer lifetimes, and be cost competitive, which will be achieved by targeting a product system efficiency of 50% with lighting that accurately reproduces the sunlight spectrum.[2] To achieve this goal, the steady development of cost-effective high-efficiency III-nitride LEDs must be implemented. Insufficient luminescence quantum efficiency, the high cost, and the drop in efficiency under a large injection current are the current major challenges to III-nitride LEDs. Growing III-nitride LEDs on low-cost Si wafer substrates[3] may solve the problem of high cost since the majority of current III-nitride LEDs are grown on sapphire and SiC substrates. Recent significant progress was seen in the demonstration of a room-temperature continuous-wave electrically injected InGaN-based laser directly grown on Si.[4] Regarding the problems of insufficient quantum efficiency and the drop in quantum efficiency under a large injection current, the investigation and elucidation of the mechanisms and origins of this drop will be a key step toward solving the problems associated with quantum efficiency. From the viewpoint of physics, III-nitride material systems and LEDs exhibit certain mysterious behaviors compared with traditional III–V compounds and LEDs. They are truly extraordinary in terms of output and reliability, in spite of the existence of a high density of crystal dislocations (i.e., up to 1×10^{12} cm^{-2}). For instance, the lifetime of laser diodes based on InGaN/GaN multi-quantum-wells (MQW) already exceeds 10,000 h under room-temperature continuous wave operation.[1,5] However, the large number of dislocations still affect the device performance, that is, they lead to a significant increase in the threshold current density of the III-nitride laser diodes.[5] Several interconnected factors, including indium content fluctuation and nanoscale assembling, stress and a strong piezoelectric field, carrier localization and electron–phonon coupling, and so on, make the luminescence mechanism of III-nitride LEDs very complicated. These factors have proven to be a challenging issue, even though various groups have invested considerable effort in trying to resolve these issues. For almost 20 years, the author and his coworkers have also put extensive effort into investigating the optical properties of III-nitride materials and LED structures, with the particular aim of gaining an increased understanding of the complex light emission mechanisms in InGaN/GaN MQW LEDs.[6–16] We have made some important progress on several key points, that is, the piezoelectric field and the quantum confinement Stark effect (QCSE),[8,14] the abnormal hydrostatic pressure behaviors of luminescence,[17] exciton localization and exciton-phonon coupling,[13] super anisotropic ambipolar diffusion and recombination of carriers,[10,18] non-exponential decaying dynamics of carriers, and so on. It appears to be the right time to address and review the major findings of these studies.

The following sections of this chapter present our experimentation and discuss our results with regard to the role of the piezoelectric field and the reverse quantum confinement Stark effect, the effect of the piezoelectric field on the hydrostatic pressure PL of $In_{0.13}Ga_{0.87}N/In_{0.03}Ga_{0.97}N$ MQW, the super lateral anisotropic ambipolar diffusion of carriers in InGaN/GaN MQW, state-filling and carrier localization effects in InGaN/GaN MQW LEDs with Si-doped barriers, and the role of electron–phonon coupling in the internal luminescence mechanisms of InGaN/GaN MQW LEDs. The chapter closes with a brief outlook.

23.2 Experimental Results and Discussions

23.2.1 The Role of the Piezoelectric Field and the Reverse Quantum Confinement Stark Effect

Presently, the majority of III-nitride MQW LEDs are epitaxially grown on sapphire substrates by means of metal–organic chemical vapor deposition (MOCVD). Typically, trimethylindium, trimethylgallium, and ammonia (NH$_3$) have been used, respectively, as indium, gallium, and nitrogen precursors for the growth of III-nitride LEDs. Bicyclopentadienyl magnesium (Cp$_2$Mg) and silane (SiH$_4$) have been employed as the sources of p-type and n-type dopants, respectively. In such hetero-epitaxial MQW layered structures, the InGaN well layers are pseudomorphically strained, and a very strong piezoelectric field usually exists in the well layers.[19] This strong built-in electric field may markedly affect the electronic states of QW structures, for example, resulting in the QCSE effect, as well as Franz–Keldysh oscillations. On the other hand, free-carrier screening of the piezoelectric field in InGaN/GaN MQWs was theoretically investigated by Sala et al. using a self-consistent tight-binding approach.[20] Such a screening effect may lead to a reverse QCSE effect in which a blueshift of optical transition energy in MQWs can be observed and has important device applications, such as for some optical switching devices.[21] Actually, the screening effect of a high density of electrically injected carriers on the piezoelectric field in III-nitride QW structures has been observed by several groups.[22–24] For example, Peng et al. observed an energy blueshift of 80 meV in the electroluminescence from a GaN/InGaN/AlGaN single QW LED when the injection current was increased from 1 mA to 1 A.[24]

In order to investigate the screening effect of photogenerated carriers on the piezoelectric field or reverse QCSE effect, we employed a sample of $In_{0.13}Ga_{0.87}N/In_{0.03}Ga_{0.97}N$ MQW grown on a sapphire substrate by MOCVD. The structure of the sample was achieved by, first, growing a 30 nm low-temperature GaN buffer layer, followed by a 1 µm GaN epilayer. Then 10 periods of 3 nm $In_{0.13}Ga_{0.87}N$/5 nm $In_{0.03}Ga_{0.97}N$ MQW were grown, which were finally capped by a 20 nm GaN top layer.[8] During photoluminescence (PL), measurements were taken at various temperatures, and the sample was mounted on the cold finger of a Janis closed-cycle cryostat with a varying temperature range of 4–300 K. The excitation light source was the 325 nm line (photon energy ~3.82 eV) of a Kimmon He–Cd continuous-wave laser with a maximum output of 40 mW. The emission signal from the sample was dispersed by a Spex 750M monochromator and was detected with a Peltier-cooled Hamamatsu R928 photomultiplier. The standard lock-in amplification technique was employed to enhance the signal-to-noise ratio.

The PL spectra (semi-logarithmical scale) dependent on the excitation power of the sample at 4 K is shown in Figure 23.1. The dominant peak in all the spectra is the intrinsic transition between the confined levels in the MQWs. A large blueshift of ~83 meV was observed for the dominant peak, when the excitation power was increased from 0.13 to 16.4 mW. As shown in the inset figure, no observable shift was found for the PL peak of the barrier layers.

The excitation power dependence of the MQW PL peak at 4 K is plotted in Figure 23.2, where the horizontal axis is in logarithmical scale. It should be noted that the blueshift of the PL peak does not seem to saturate even when the excitation power reaches to 16.4 mW. This means that the photogenerated carriers screen only a portion of the piezoelectric field even under excitation of the laser using maximum available power. Self-consistent calculations, which will be described later, estimate a maximum blueshift of 160 meV.[8]

As the temperature increases, the PL from the barrier layers decreases quickly and almost completely quenches at 40 K. However, the MQW PL decreases in intensity at a relatively slower rate. Its position redshifts more rapidly than that predicted by Varshni's well-known empirical formula. For example, the extent of redshift for the MQW PL peak is 42 meV when the temperature is increased from 4 to 80 K.[8] This phenomenon is attributed to thermal redistribution of the carriers in the localized states with a broad energy distribution.[7] The localized states may be induced by indium-rich cluster spontaneously formed in the well layers.[5,25] Figure 23.3 shows the extracted peak positions of the MQW PL at 80 K for different excitation powers. From the figure, it can be seen that the blueshift of the MQW PL peak is as

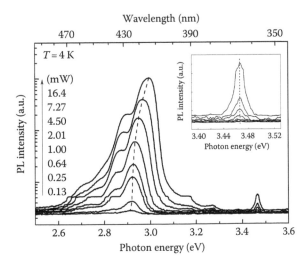

FIGURE 23.1 PL spectra of the sample dependent on excitation power recorded at 4 K. A large blueshift of the dominant PL peak can be seen. The inset shows the PL peak from the barrier layers. No observable shift of this peak is found. (APEX/JJAP Copyright Division, reproduced with permission of the publisher.)

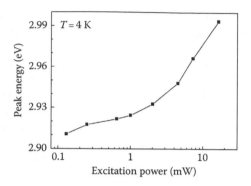

FIGURE 23.2 Energetic positions (filled squares) of the MQW PL peak of the sample at 4 K versus the excitation power. The solid line is drawn to guide the eye. (APEX/JJAP Copyright Division, reproduced with permission of the publisher.)

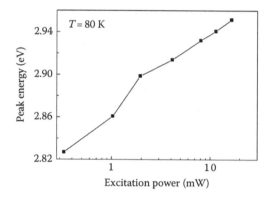

FIGURE 23.3 Energetic positions (filled squares) of the MQW PL peak of the sample at 80 K versus the excitation power. The solid line is drawn to guide the eye. (APEX/JJAP Copyright Division, reproduced with permission of the publisher.)

large as 120 meV when the excitation power is increased from 0.33 to 16.4 mW.[8] The PL spectra of the sample dependent on excitation power were also measured at other temperatures. It was found that the extent of blueshift for the MQW PL peak depends on temperature. Possible origins for such temperature dependence will be discussed later.

In order to form a quantitative interpretation of the observed blueshift of the QW transition peak, a self-consistent theoretical calculation taking into account the QCSE and the screening effect of photogenerated carriers was carried out. Under the effective mass approximation, the confined states in QW can be described by the Schrödinger equation[26]

$$\left[-\frac{\hbar^2}{2m^*}\frac{\partial^2}{\partial z^2} + V(z) \right]\psi(z) = E\psi(z), \tag{23.1}$$

where
   $\hbar$ is the Planck constant over 2pai
   m^* is the effective mass of the carrier
   E is the energy eigenvalue
   $\psi(z)$ is the eigen wavefunction
   $V(z)$ is the potential energy

The potential energy profile of a QW under an applied electrical field is given by

$$V(z) = \begin{cases} V_0 & (z < 0) \\ -qFz & (0 < z < L_w), \\ V_0 - qFz & (z > L_w) \end{cases} \tag{23.2}$$

where

V_0 is the band offset

q is the carrier charge

$q = \mp e$ for the electron and hole, respectively

F is the piezoelectric field

L_w is the width of the well layers

Parameter values used in the calculations are tabulated in Table 23.1. A valence band offset of 0.098 eV was adopted here.[27,28]

The effective-mass Schrödinger equation (23.1) can be numerically solved using the Numerov method.[26,29] It should be noted that Equation 23.1 is usually transformed into a reduced form before it is numerically solved.[26] Figure 23.4a shows the energy of transition from the first electron state (E_1) and the first hole state (H_1) as a function of the electrical field. It is clearly seen that a redshift of the $E_1 - H_1$ transition occurs with an increasing electrical field regardless of whether the field is the built-in piezoelectric field or the externally applied field. This is known as the QCSE effect. Compared with the experimental data, the unscreened piezoelectric field in the 10 periods of 3 nm $In_{0.13}Ga_{0.87}N$/5 nm $In_{0.03}Ga_{0.97}N$ MQW sample at 4 K is theoretically estimated to be 1.36 MV cm^{-1}, which is consistent with the values reported in the literature.[19,29,30] It is known that the built-in piezoelectric field may spatially separate the electrons and holes photogenerated within the MQW. In turn, the spatial separation of electrons and holes will produce an electric field that screens out the original piezoelectric field. In other words, it results in a reduction of the net field. As a result, a reverse QCSE effect may be observed due to the efficient screening of photogenerated carriers on the original piezoelectric field. Obviously, the screening extent depends on the concentration of the photogenerated carriers, which is proportional to the excitation power. The screening field due to the spatial separation of electrons and holes can be derived from the Poisson equation

$$-\varepsilon \frac{d^2}{dz^2} V_{scr}(z) = e(p - n), \tag{23.3}$$

where

V_{scr} is the screening potential

ε is the dielectric constant

p and n are the electron and hole charge distribution, respectively

TABLE 23.1 Material Parameters of $In_{0.13}Ga_{0.87}N$/$In_{0.03}Ga_{0.97}N$

	$In_{0.13}Ga_{0.87}N$	$In_{0.03}Ga_{0.97}N$
Bandgap energy E_g (eV)	2.971[a]	3.466[b]
Effective mass (in m_0)		
m_e^*	0.19	0.19[c]
m_h^*	1.734	1.734[c]
Dielectric constant ε (in ε_0)	10.4	10.4[c]

If not noted, the parameters' values are obtained from the linear interpolation of those of the binary GaN and InN from Reference 28;

[a] For In0.13Ga0.87N: Eg = 3.507 eV(GaN) × 0.87 + 1.994 eV(InN) × 0.13 − 3.0 × 0.13 × 0.87 = 2.971 eV, where a 3.0 eV of the bandgap bowing parameter for InGaN is used.

[b] Obtained directly low-temperature PL spectra (Figure 23.1).

[c] Treated as the same as those of a $In_{0.13}Ga_{0.87}N$ well layer.

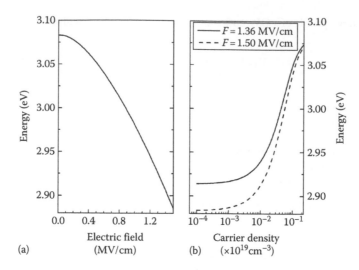

FIGURE 23.4 (a) Calculated transition energies as a function of the electrical field and (b) self-consistently calculated transition energies as a function of carrier density for the built-in piezoelectric field of 1.36 and 1.50 MV cm^{-1}. (APEX/JJAP Copyright Division, reproduced with permission of the publisher.)

Self-consistent calculations taking into account the QCSE and the screening effect of the photogenerated carriers were performed by solving simultaneously Equations 23.1 and 23.3. Two cases for the original piezoelectric field of $F = 1.36$ MV cm^{-1} and $F = 1.50$ MV cm^{-1} were considered. The calculated energies of the $E_1 - H_1$ transition are plotted in Figure 23.4b as a function of the carrier concentration. The calculated maximum energy blueshift of the $E_1 - H_1$ transition can be as large as 160 meV for $F = 1.36$ MV/cm and 190 meV for $F = 1.50$ MV/cm, respectively.[8] This means that, if the original piezoelectric field is fully screened by the photogenerated carriers, an energy blueshift of 160 ~ 190 meV for the $E_1 - H_1$ transition will be observed. The calculated results also show that the stronger the original built-in field, the larger the blueshift. The observed blueshift amounts for the MQW PL peak are 83 meV and 120 meV at 4 and 80 K, respectively, as shown in Figures 23.2 and 23.3. It is not surprising that the observed amount of the blueshift is smaller than the calculated maximum value because a full screening of the piezoelectric field in the sample never occurs. This is also consistent with the lack of saturation in Figures 23.2 and 23.3. Regarding the observed temperature dependence of the peak energy blueshift, it may be attributed to a change of the original piezoelectric field with temperature. It is known that lattice mismatch strain in InGaN-based MQW induces a piezoelectric field. A theoretically estimated value for the piezoelectric field in the sample investigated here due to such lattice mismatch strain is about 1.71 MV cm^{-1}. Depending on growth conditions, such as substrate temperature, the lattice mismatch strain may be different. Moreover, the residual strain after growth is also a function of temperature due to thermal mismatch in the thermal expansion coefficient between the sapphire and III-nitrides. Therefore, the built-in piezoelectric field in the sample depends on temperature. Our calculations show that the amount of the energy blueshift for the MQW intrinsic transition depends directly on the strength of the original piezoelectric field. This is why the observed blueshift is temperature dependent.

In summary, a blueshift as large as 120 meV is observed for the dominant PL peak in the In$_{0.13}$Ga$_{0.87}$N/ In$_{0.03}$Ga$_{0.97}$N MQW heterostructure when the excitation power is increased. The reverse quantum confinement Stark effect due to the efficient screening of the photogenerated carriers on the piezoelectric field was identified as being responsible for the phenomenon. By numerically solving the coupled effective-mass Schrödinger equation and the Poisson equation, we conducted self-consistent calculations on the QCSE effect of a built-in piezoelectric field and the screening effect of photogenerated carriers on the piezo-electric field in the MQW. Good agreement between experiment and theory was achieved, enabling us to have a deep insight into the significant role of the piezoelectric field in the luminescence mechanisms of III-nitride MQW structures.

23.2.2 Effect of the Piezoelectric Field on the Hydrostatic Pressure PL of $In_{0.13}Ga_{0.87}N/In_{0.03}Ga_{0.97}N$ MQW

In addition to the QCSE effect, a large piezoelectric field may also lead to anomalous PL of III-nitride MQW structures, which is dependent on hydrostatic pressure. Surprisingly small hydrostatic pressure coefficients have been observed in InGaN MQW structures.[31-33] The reported hydrostatic pressure coefficients of the luminescence peak of InGaN/GaN MQW range from a few meV/GPa to less than 30 meV/GPa. In contrast to the case of MQW, the measured pressure coefficients for bulk GaN are as high as 39–42 meV/GPa.[34] It was also found that the pressure coefficients of bulk-like $In_xGa_{1-x}N$ ($0 < x < 0.15$) epitaxial layers do not substantially deviate from those of bulk GaN. The small pressure coefficients of InGaN MQW were originally attributed to the low dimensionality of recombination processes involving deep-level localized states.[31] However, more and more evidence has been obtained to show the role of the built-in piezoelectric field in the hydrostatic pressure PL of InGaN-based MQW. Due to the difference in compressibility between barrier material and well material, extra tensile strain is usually induced by the externally applied hydrostatic pressure, which may result in an increase of the piezoelectric field. For instance, an increase in the piezoelectric field from 1.4 to 2.6 MV cm^{-1} in InGaN/GaN MQW was reported by Vaschenko et al. when they applied 8.7 GPa hydrostatic pressure to the sample.[31] The increment in the piezoelectric field will induce an additional reduction in the transition energy due to the QCSE effect, which compensates to some extent for the increase in the transition energy due to external hydrostatic pressure. Small pressure coefficients of the PL peak in InGaN-based MQW are thus expected. Further evidence for the role of the piezoelectric field in the pressure-dependent PL of GaN/AlGaN MQW is the recently reported well-width dependence of the pressure coefficients.[35] Since the QCSE effect depends on quantum well width, the PL peak position of strained GaN MQW under hydrostatic pressure will thus depend on the well width.

As argued in Section 23.2.1, on the other hand, a reverse QCSE effect may occur in InGaN-based MQW structures due to the screening effect of the piezoelectric field by photogenerated or electrically injected carriers. This reverse QCSE effect will be a function of carrier concentration. Therefore, the hydrostatic pressure coefficient of the PL peak of InGaN MQW may show dependence on the carrier concentration and, hence, the excitation power. As will be shown and argued later, indeed, the pressure coefficient of the PL peak of InGaN MQW is a strong function of the excitation power. Fox example, the low-temperature pressure coefficient is found to change from 26.9 to 32.1 meV GPa^{-1} when the excitation power is increased from 1 to 33 mW.[17] Also, an emission related to an excited state is also observed with increasing hydrostatic pressure when the excitation power is greater. Again, self-consistent calculations on the screening effect of free carriers and QCSE are performed by solving the Schrödinger and Poisson equations together. These results indicate the important role of the piezoelectric field in the explanation of small pressure coefficients.[17]

The sample used for the low-temperature hydrostatic pressure PL experiment has been described in Section 23.2.1. PL measurements under pressure were performed with a diamond anvil cell mounted on a cold finger of a closed-cycle cryostat with a varying temperature range from 11 K to room temperature. Condensed argon gas was employed as the pressure-transmitting medium at low temperatures. The 325 nm line from an He–Cd laser was used for excitation. The emission lines from the sample and a ruby were dispersed by a JY-HRD1 double-grating monochromator and detected by a charge-coupled device detector. The ruby fluorescence line was used to calibrate the pressure applied to the sample.

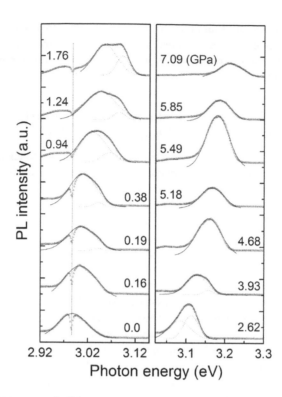

FIGURE 23.5 Measured PL spectra (hollow triangles) of the $In_{0.13}Ga_{0.87}N/In_{0.03}Ga_{0.97}N$ MQW at 11 K under different hydrostatic pressures. The excitation power was kept at 33 mW. The dip at 415.2 nm indicated by a vertical dash line is due to impurity-related absorption in diamond in the pressure cell. The spectral curves can be fitted with either single or double Gaussian line shape functions. (Wiley-VCH Verlag GmbH & Co. KGaA., reproduced with permission of the publisher.)

Figure 23.5 shows the PL spectra of the sample measured at 11 K under different hydrostatic pressures while the excitation power is fixed at 33 mW. Under atmospheric pressure, the dominant emission (~2.99 eV) is from the MQW.[8] As seen in Figure 23.5, the PL band shifts toward higher energy as the applied external pressure is increased. The dip at 415.2 nm (2.987 eV) indicated by a vertical dashed line is due to an impurity absorption line of the diamond in the anvil cell. As shown in Figure 23.5, the whole PL peak can be fitted very well with single or double Gaussian line shape functions (dotted lines). The energetic positions of various luminescence peaks are then determined.

When the applied pressure is increased to 0.19 GPa, a shoulder appears at the higher-energy side. This shoulder grows with pressure and even develops into a dominant structure when the applied pressure reaches 2.62 GPa. However, when the applied pressure is further increased, its intensity decreases and gradually diminishes. Its energy separation (~63 meV) from the original dominant peak does not change appreciably with increasing pressure. It should be further noted that this new peak cannot be resolved in the PL spectra under low excitation power. It is thus attributed to the optical transitions related to the excited electronic states of the InGaN MQW.

Figure 23.6a shows the peak positions of the optical transition between the ground electron and hole states in the MQW as a function of the applied pressure when the excitation power was 33 meV. The pressure coefficient of 32.1 ± 0.6 meV/GPa is obtained by a linear fitting to the data. Previously, the efficient screening effect of the photogenerated carriers on the built-in piezoelectric field in the sample has already been demonstrated.[8] If the QCSE effect induced by the piezoelectric field plays an important role in the behavior of hydrostatic pressure PL, the pressure coefficients should be dependent on carrier density and, hence, excitation power. To test this idea, the low-temperature hydrostatic pressure PL spectra of the same

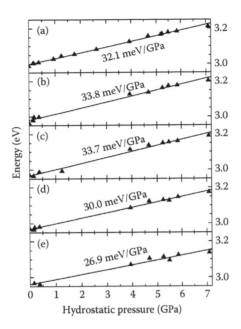

FIGURE 23.6 Energetic positions (solid triangles) of the ground-state transition of the sample versus applied pressures. The excitation power was (a) 33 mW, (b) 20 mW, (c) 11 mW, (d) 5 mW, and (e) 1 mW. The linear fitting (solid lines) to the data gives rise to the pressure coefficients. (Wiley-VCH Verlag GmbH & Co. KGaA., reproduced with permission of the publisher.)

sample under different excitation powers were measured. Peak positions of the main luminescence band are summarized in Figure 23.6. Similarly, linear fitting to the experimental data was done to obtain the pressure coefficients of the main luminescence peak at various excitation powers. The obtained pressure coefficients and errors are illustrated in Figure 23.7. It can be seen that the pressure coefficients are obviously dependent on the excitation power. For example, when the excitation power is increased from 1 to 11 mW, the pressure coefficient increases from 26.9 meV GPa^{-1} to 33.7 meV/GPa. However, eventually, the pressure coefficient exhibits typical saturation behavior.

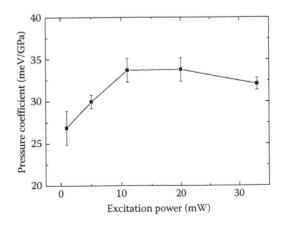

FIGURE 23.7 Obtained pressure coefficients versus excitation power. The pressure coefficient, at first, increases with the increase in the excitation power and then shows a tendency for saturation. (Wiley-VCH Verlag GmbH & Co. KGaA., reproduced with permission of the publisher.)

Clearly, the observed increase of the pressure coefficient with excitation power and its saturation behavior both support the assertion that the built-in piezoelectric field plays a central role in determining the small pressure coefficients of InGaN-based MQW. In the self-assembled InAs/GaAs quantum dot system, we also demonstrated the same phenomenon both experimentally and theoretically.[36]

23.2.3 Super Lateral Anisotropic Ambipolar Diffusion of Carriers in InGaN/GaN MQW

In the preceding sections, we have demonstrated the important role played by a built-in piezoelectric field in the internal luminescence mechanisms of III-nitride MQW structures via examining the PL spectra of InGaN-based MQW that are dependent on the excitation power and on hydrostatic pressure-at different temperatures. In this section, we present an in-depth investigation of the role of the piezoelectric field in the luminescence mechanisms of InGaN-based MQW structures. It is found that the vertical piezoelectric field in strained InGaN/GaN MQW is the original "driving force" for significant enhancement of the lateral diffusion of photogenerated carriers, which is directly observed by luminescence images.[10] By solving the two-dimensional drift-diffusion equation and using an ambipolar diffusion coefficient, derived by Gulden et al.,[37] we are able to give a quantitative interpretation to the phenomenon. The simulation results clearly demonstrate the original and central role of the piezoelectric field in the enhanced lateral diffusion of photogenerated carriers in the structures. Moreover, our results also reveal that the density and average mobility of photogenerated carriers have important influence on enhanced lateral diffusion.[10,18]

Two InGaN/GaN MQW samples, which were grown on sapphire with MOCVD, were investigated in this study. The typical growth sequence of both samples is a 30 nm GaN nucleation layer grown at 500°C, a 1500 nm GaN buffer at 1050°C, 4 periods of InGaN/GaN MQW at 740°C, and a 20 nm GaN cap layer at 740°C. Sample B has 3.4 nm wells and 3 nm barriers, while sample B has 2 nm thick wells and 12 nm barriers. The variable-temperature PL spectroscopy system used in the study was described in Section 23.2.1.[6]

Figure 23.8 shows luminescence photos of samples A and B at 5 K. We also photographed luminescence images of the samples at room temperature. Very similar images were obtained (not shown here). The central white circular regions in the two images are the laser illuminating spots on the surface of the samples, while the gray-white areas surrounding them are the emission areas of the samples. Within the illuminating spots, the intensities of the luminescence signals are basically constant along their radial directions due to the uniform generation of the photogenerated carriers in these regions. Obviously, the light emission areas are much larger than the excitation spots, unambiguously indicating that strong lateral

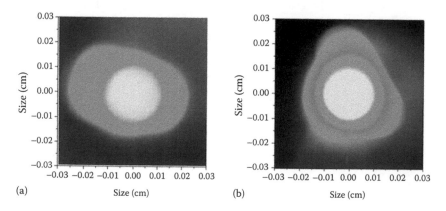

FIGURE 23.8 Photos of the luminescence of two InGaN/GaN MQW samples at 5 K under the excitation of a 325 nm UV laser (a) for sample A and (b) for sample B. (AIP Publishing, reproduced with permission of the publisher.)

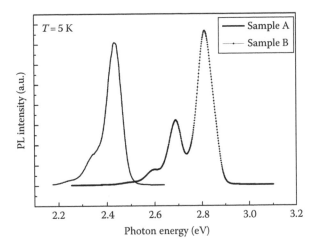

FIGURE 23.9 PL spectra of the two InGaN/GaN MQW samples recorded at 5 K. (AIP Publishing, reproduced with permission of the publisher.)

diffusion of photogenerated carriers takes place. Furthermore, the enlarged luminescent areas exhibit irregular shapes, implying the nature of anisotropic ambipolar diffusion of the photogenerated carriers. As pointed out by Yairi and Miller,[38] a lateral electric field created by the vertical separation of the photo-generated carriers due to the vertical built-in or externally applied electric field in semiconductor *p–i–n* or *n–i–p–i* structures may result in a giant lateral diffusion. It is obvious that in the strained InGaN/GaN MQW structures the giant lateral ambipolar diffusion may be observed due to the existence of a built-in strong piezoelectric field[29,30] and efficient screening effect of photogenerated carriers on the piezoelectric field.[8] As seen in Figure 23.8, the strong enhancement of the lateral diffusion of photogenerated carriers is, indeed, observed in the InGaN/GaN MQW samples. The corresponding PL spectra of the two samples are depicted in Figure 23.9.

When the InGaN/GaN MQW sample is optically excited with a beam of He–Cd 325 nm from the laser, the incident photons will be efficiently absorbed by the sample, and excess electrons and holes will be created. These photogenerated carriers will be vertically separated by the vertical piezoelectric field E_P induced by the residual strain. The separated carriers will produce an opposite electric field E_S to screen the piezoelectric field. In the region where there is no incident light, the difference between the electron and hole quasi-Fermi levels, ϕ_{np}, will be determined by the vertical piezoelectric field E_P. On the other hand, within the region of illumination where the incident light is significantly absorbed, the vertically separated carriers then screen the piezoelectric field E_P to some extent.[8] As a result, ϕ_{np} will change significantly because it is an approximately linear function of the separated photogenerated carrier density.[38] Moreover, it has a lateral dependence due to the lateral intensity variation of the incident light beam. The resulting lateral gradients of both the electron and hole quasi-Fermi level, ϕ_n and ϕ_p, will produce electric fields in the top and bottom layers, respectively. These fields push both electrons and holes away laterally and make enhanced diffusion possible. In the following equations employed for the quantitative simulation of the lateral diffusion, the subscripts *n* and *p* refer to the electron and the hole, respectively. The symbols *j*, *e*, *n*, *p*, D_{am}, μ_n, μ_p, L_D, and L_Q represent current density, the electronic charge, electron density, hole density, the ambipolar diffusion coefficient, electron mobility, hole mobility, the thickness of the electrically charged layer due to the existence of the strong piezoelectric field, and the total thickness of the quantum wells including the barrier layers, respectively. Following Gulden et al.'s work,[37] the ambipolar diffusion coefficient D_{am} is defined by

$$j_n(\vec{r}) = eD_{am}\nabla n(\vec{r}) = -j_p(\vec{r}) = eD_{am}\nabla p(\vec{r}) \tag{23.4}$$

Equation (23.4) also reflects the fact that no net electrical current is generated in the structures. An analytical expression of the ambipolar diffusion coefficient D_{am} was derived as[37]

$$D_{am} = \frac{\sigma_n \sigma_p}{\sigma_n + \sigma_p} \frac{\partial \phi_{np}}{\partial n} \frac{1}{e^2} \tag{23.5}$$

where

$\sigma_n = n\mu_n e$ is the electron conductivity
$\sigma_p = p\mu_p e$ is the hole conductivity

From Equation (23.5), it can be seen that the ambipolar diffusion coefficient depends directly on how rapidly the energy difference between the electron and hole quasi-Fermi levels changes with the density of the separated photogenerated carriers. Once the quantity $\partial \phi_{np}/\partial n$ is known, the ambipolar diffusion coefficient D_{am} can be obtained. The energy difference between the quasi-Fermi levels of electrons and holes has an approximated linear relationship with the screening electric field of photogenerated carriers.[37] That is, $\phi_{np} \approx E_S L_Q e$.

In optically illuminated InGaN/GaN MQW structures, the net vertical electric field E in the MQW region will be given by $E_P - E_S$ and can be calculated by numerically solving Poisson's equation.[8] Here, we adopt an analytical expression of E_S derived by Poole et al.[39] to calculate ϕ_{np} and then $\partial \phi_{np}/\partial n$. Equations of these parameters are as follows:

$$E_s = \frac{n}{2\varepsilon_r \varepsilon_0} L_D e, \tag{23.6}$$

$$\phi_{np} \approx \frac{n}{2\varepsilon_r \varepsilon_0} L_D L_Q e^2, \tag{23.7}$$

and

$$\frac{\partial \phi_{np}}{\partial n} = \frac{L_D L_Q e^2}{2\varepsilon_r \varepsilon_0}, \tag{23.8}$$

where ε_r and ε_0 are the relative dielectric constant of the InGaN and the permittivity of vacuum, respectively.

The diffusion equation describing the steady-state distribution of photogenerated carriers in InGaN/GaN MQW can be written as

$$\frac{dn(\vec{r})}{dt} = -D_{am} \nabla n(\vec{r}) + \frac{n(\vec{r})}{\tau} = 0, \tag{23.9}$$

where τ is the life of photogenerated carriers. Because the lateral distribution of photogenerated carriers is only of concern in the present work, Equation (23.9) can be simplified. Furthermore, in the derivation of the ambipolar diffusion coefficient,[37,38] it is assumed that the photogenerated electrons and holes have already separated vertically across the intrinsic region of the p-i-n structures. Therefore, n and p are functions of the lateral dimensions only, (ρ, θ). Here, ρ and θ represent the in-plane radial distance and polar angle, respectively. Equation (23.9) can thus be reduced to a second-order ordinary differential equation[10]

$$n''(\rho) + \frac{1}{\rho} n'(\rho) - \frac{n(\rho)}{D_{am}\tau} = 0. \tag{23.10}$$

The finite difference method was employed for numerically solving Equation (23.10). The boundary conditions were defined by

$$n(\rho)=1.0\times10^{17}\,\text{cm}^{-3} \quad \text{at} \quad 0\leq\rho\leq0.01 \text{ cm}; \quad n(\rho)=0 \quad \text{at} \quad \rho=\infty.$$

Figure 23.10 depicts the calculated in-plane radial distribution of photogenerated carriers in a InGaN/GaN MQW structure. The parameters adopted in the calculation are listed in Table 23.2. Clearly, the distribution area of photogenerated carriers is much larger than that of the excitation spot (i.e., the central white region). Here, we define the diffusion radius as the distance from the center to the edge at which the density of the carrier declines to zero after a steady state builds up. The calculated results are in good agreement with the experimental data shown in Figure 23.8. If anisotropic mobility of carriers is under consideration, of course, one can obtain anisotropic lateral diffusion results.[18]

The influence of photogenerated carrier density and average mobility—defined as $\mu_n\mu_p/(\mu_n+\mu_p)$—on the enhanced diffusion is also examined. Figure 23.11 shows the calculated diffusion radius of photogenerated carriers in InGaN/GaN MQW versus the average mobility of carriers. It can be seen that the diffusion radius super-linearly increases with an increase in the average mobility of carriers. As expected, the larger the photogenerated carrier density, the longer the in-plane diffusion radius. Substituting Equation (23.8)

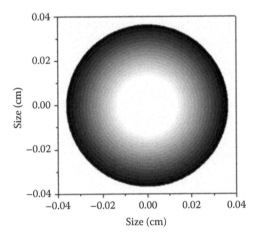

FIGURE 23.10 Calculated in-plane diffusion region of photogenerated carriers in comparison with the excitation spot (white circular region). (AIP Publishing, reproduced with permission of the publisher.)

TABLE 23.2 Material Parameters Used in Calculations

Bandgap energy of GaN E_g (eV)	3.507[a]
Dielectric constant of GaN ε (in ε_0)	8.9[b]
Mobility of electrons in GaN μ_n (cm^2 V^{-1} s^{-1})	1000[c]
Mobility of holes in GaN μ_p (cm^2 V^{-1} s^{-1})	200[c]
Lifetime of carriers in InGaN/GaN MQW τ (ns)	12.5[d]
Thickness of electrically charged layer L_D (nm)	20
Thickness of quantum wells' region L_Q (nm)	56

[a] Vurgaftman et al. [28]

[b] Bougrov et al. [40]

[c] Chow and Ghezzo [41]

[d] Zhang [42]

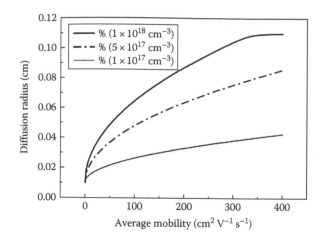

FIGURE 23.11 Calculated diffusion radius of photogenerated carriers as a function of the average mobility of electrons and holes for different photogenerated carrier densities. (AIP Publishing, reproduced with permission of the publisher.)

into Equation (23.5) yields $D_{am} \propto (nL_Q)$, which may theoretically interpret the phenomenon observed by Huang et al.[43] For example, the ambipolar diffusion coefficient depends strongly on the thickness of the quantum well.

According to the experimental results and theoretical calculations presented, we conclude that the piezoelectric field in InGaN-based quantum well structures can have a decisive impact on the lateral diffusion of carriers and, hence, on the luminescence of carriers in these nanoscale heterostructures.

23.2.4 State-Filling and Carrier Localization Effects in InGaN/GaN MQW LED with Si-Doped Barriers

As addressed and argued in previous sections, the significant role of the piezoelectric field in the luminescence mechanisms of InGaN-based MQW structures has been demonstrated both experimentally and theoretically. Another idiosyncratic feature in present InGaN-based MQW structures could be the carrier localization due to the apparent local compositional fluctuations in the In and Ga compositions.[5] In fact, the piezoelectric field can also lead to carrier localization to some extent. Furthermore, the two factors of compositional fluctuation and piezoelectric field may interplay and jointly obscure the luminescence mechanisms of III-nitride MQW LED structures.[44] Si doping in the GaN barrier layers could be a good way to reduce the influence of the piezoelectric field due to a partial screening of the strain-induced piezoelectric field by free carriers induced by Si dopants.[45] In this section, we show that heavy Si doping in barrier layers can result in bandgap renormalization (BGR) and phase space filling (PSF) effects and may also significantly affect the recombination dynamics of photogenerated carriers in InGaN/GaN MQW LED structures.[11,12]

The conclusions are obtained from a comparative study on the optical properties of two kinds of InGaN/GaN MQW LED of an identical geometric structure and composition, the barriers of one of which were doped with Si, while the barriers in the other LED were not. The two samples, labeled C and D, were grown on a *c*-plane sapphire substrate by MOCVD. Each sample consists of a 3 μm n^+-GaN layer, an active layer of 5-period InGaN/GaN MQW, and a 200 nm p^+-GaN layer. The thickness of the well (barrier) layer is 3 nm (12 nm). The only difference between the two samples is that the barrier layers of MQW in sample C were intentionally undoped, while the barrier layers of sample D were doped with Si to about 10^{18} cm^{-3}. Steady-state PL spectroscopic measurements of the two samples were carried out on a home-made high-resolution PL system already described earlier and elsewhere.[6] The excitation light source was the 325 nm laser line of a He–Cd laser (Kimmon) for the steady-state PL measurements. For time-resolved

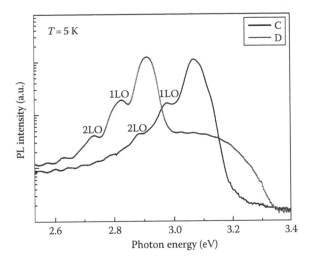

FIGURE 23.12 Measured 5 K PL spectra of samples C and D under the excitation of a 33 mW UV laser. (AIP Publishing, reproduced with permission of the publisher.)

PL (TRPL) measurements, we employed the frequency doubled femtosecond pulses (λ_{exc} = 355 nm) from a Ti:sapphire laser (Spectra-Physics Tsunami system) with a repetition rate of 82 MHz to excite the samples, which were loaded into a 77 K home-made dewar. The transient luminescence signal was dispersed in a 0.25 m monochromator and was captured by a Hamamatsu streak camera (C4334).[46] The overall temporal resolution of the entire system was about 20 ps.

Figure 23.12 shows the semi-logarithmic PL spectra of the two samples measured at 5 K under identical experimental conditions (i.e., the same excitation power of 33 mW). Several spectral features can be identified. First, the dominant PL peak intensities of the two samples are almost equal. Furthermore, the PL spectra of both samples have very similar structures on the lower-energy side of their principal peak. At least two longitudinal optical (LO) phonon sidebands (referred to as 1LO and 2LO) can be resolved. The energy separations between the principal peaks and 1LO as well as between 1LO and 2LO are identical and can be determined to be 90.5 ± 0.5 meV, which is very close to the characteristic LO phonon energy of GaN (~91 meV).[6] Second, the principal peak (located at ~2.910 eV) and its LO phonon sidebands in sample D exhibit a redshift of about 158 meV with respect to the corresponding peaks of sample C. This redshift should be the signature of the BGR effect due to the many-body interactions.[47] The BGR effect was observed in Si-doped GaN epilayers.[48] In the InGaN/GaN MQW sample with Si-doped barriers under study here, a large number of electrons donated by the Si donors in the barrier layers may form a high density of Fermi gas in the wells. Exchange and correlation of electrons result in renormalization of the bandgap, which is manifested by a redshift of the luminescent bandgap. An empirical expression, $\Delta E_{PL} = -Kn^{1/3}$, where K is a proportionality constant depending on materials and n is the carrier concentration, is usually used to evaluate the BGR shift. If ΔE_{PL} is simply assumed to be proportional to the Si-doping concentration, K of InGaN/GaN MQW is estimated to be 1.58×10^{-4} meV cm, which is several times higher than that of doped GaN epilayers.[48] The larger BGR effect is attributed to the enhancement of the carrier localization degree in the well layers due to the Si doping in the barrier layers, as will be proved later.

Another spectral feature in Figure 23.12 is that a broad shoulder appears at the higher-energy side of the PL spectrum of sample D. This shoulder is attributed to the filling effect of the higher-energy states by electrons releasing from the Si donors in the sample. It is well known that, as the carrier concentration increases, electronic states at higher energies will be occupied according to the Pauli Exclusion Principle, and thus the Fermi level goes up. That is the so-called "phase space filling" (PSF) effect. A possible spectral feature of the PSF effect is that the high-energy edge of the PL peak expands toward the high-energy direction, or

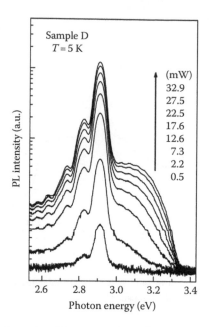

FIGURE 23.13 Excitation power-dependent PL spectra of sample D at 5 K. (AIP Publishing, reproduced with permission of the publisher.)

some new structures may appear on the higher-energy side of the PL peak.[49] In order to verify that the PSF effect is actually observed in the Si-doped sample, excitation power-dependent PL measurements were performed on both samples. The results of sample D are shown in Figure 23.13, since no spectral variation was observed for sample C except the increased intensity.[12] Indeed, the broad shoulder gradually appears at the higher-energy side of the PL dominant peak of sample D with the increase in the excitation power. The broad shoulder is thus attributed to the radiative recombination of electrons occupying higher-energy states and the optically generated holes. The electrons involved may be mainly from the Si donors. We also note that the PL dominant peak energies of both samples remain almost unchanged, even when the excitation power was increased from 0.5 to higher than 30 mW. This means that the piezoelectric field frequently observed in strained InGaN/GaN MQW is successfully suppressed in the samples under study.

To obtain further information, the TRPL spectra of both samples were measured at 77 K, as indicated in Figure 23.14. It is clear that the decay time of the luminescence signal increases with increasing wavelength.

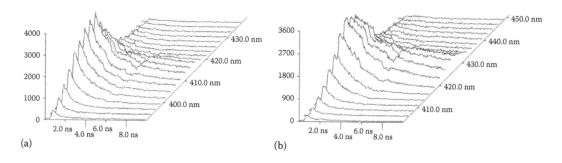

FIGURE 23.14 3D time-resolved photoluminescence spectra of the samples at 77 K (a) for undoped sample C and (b) for doped sample D. (Optical Society of America (OSA), reproduced with permission of the publisher.)

In fact, this phenomenon is frequently observed in localized-state ensemble luminescent systems. Very recently, we developed an analytical model to interpret this phenomenon quantitatively.[50] The key physical mechanism causing the shorter luminescence life on the higher-energy side is the much higher likelihood of the thermal escape of localized carriers occupying higher-energy states. Here, we are more interested in the phenomenon of the peculiarly different decaying dynamics of photogenerated carriers in the two samples. The main PL peak of sample C displays a non-exponential decay behavior, which is a typical characteristic of a disorder system. In contrast to sample C, the time evolution of the dominant PL signal from sample D puzzlingly obeys the signal exponential decay law. To interpret this interesting observation, we employed a theoretical model developed by Rubel et al.[51] to simulate the experimental curves.

In this model, for a semiconductor system with a sufficiently strong disorder, charge carriers are assumed to be instantaneously captured into localized states of the system after their photogeneration. From these states, the carriers can either radiatively recombine or perform a phonon-assisted hopping transition to other localized states. Since both these processes take place in the localized states, their likelihood depends exponentially on the distances involved.[51] The behaviors of recombination dynamics in carriers are thus essentially determined by the localization lengths of the carriers. Such dynamic behavior could be described by the rate equations originally proposed by Marshall.[52] The energy range where localized states are distributed is divided into a set of m energy slices of a given width. For the system, an exponential density of states (DOS) $g(\varepsilon) = N_0/\varepsilon_0 \exp(-\varepsilon/\varepsilon_0)$ with a total concentration of localized states N_0 and energy scale ε_0 was adopted. For simplicity, we treat the densities of localization states and localization lengths for electrons and holes as being equal. The rate equation for carrier density n_k in those energy slices k is formulated as follows:[51]

$$\frac{dnk}{dt} = \sum_{\substack{j=1 \\ j \neq k}}^{m} \left(n_j \Gamma_{j \to k} - n_k \Gamma_{k \to j} \right) - n_k \Gamma_r, \tag{23.11}$$

where Γ_r denotes the recombination rate for a localized electron to recombination with a localized hole and $\Gamma_{j \to k}$ ($\Gamma_{k \to j}$) is the rate for a charge carrier to perform a nonradiative hopping transition from an occupied state $j(k)$ to an empty localized state $k(j)$ over a distance r_{jk}. In general, the rate for hopping transitions depends exponentially on the distance involved[51]

$$\Gamma_{j \to k} = v_0 \exp\left[-\frac{2r_{jk}}{\alpha} - \frac{\varepsilon_k - \varepsilon_j + |\varepsilon_k - \varepsilon_j|}{2K_B T} \right], \tag{23.12}$$

where

ε_j and ε_k are the energies of states j and k, respectively
α is the localization length
v_0 is the attempt to escape frequency

For the transition from the slice k downward in energy, $\varepsilon_k > \varepsilon_j$, only the tunneling term remains and the downward transition rate can be considered as[51]

$$\Gamma_{k \to j} = v_0 \exp\left(-\frac{2R_k}{\alpha} \right) \frac{d_j - n_j(t)}{\sum_{i=k}^{m} d_i - n_i(t)} \tag{23.13}$$

where R_k is the typical hopping distance, determined by the concentration of unoccupied states with energy below ε_k

$$R_k = \left\{ \pi \sum_{j=k}^{m} \left[d_i - n_i(t) \right] \right\}^{-1/2} \tag{23.14}$$

where d_j is the concentration of localized states in the energy slice. On the other hand, the upward transition from slice j to k can be derived from the downward transition rate

$$\Gamma_{j\to k} = \Gamma_{k\to j} \frac{d_k - n_k(t)}{d_j - n_j(t)} \exp\left(-\frac{\varepsilon_k - \varepsilon_j}{K_B T}\right) \tag{23.15}$$

The rate for the recombination of a localized electron with a localized hole also depends exponentially on the distance R between them[51]

$$\Gamma_r(R) = \tau_0^{-1} \exp(-2R/\alpha), \tag{23.16}$$

where τ_0 is a time constant that depends on the particular recombination mechanism and is of the order of the excitonic radiative life. The most efficient recombination is found in the pairs of localized states in which the state for the electrons is as close to the state for the holes as the localization length. Correspondingly, the recombination time does not contain the exponential factor and is close to τ_0. The concentration of such pairs is the product of the density of filled electron states n and the probability α^2. The recombination rate can be considered as $\Gamma_r(R) = \tau_0^{-1} n(t)\alpha^2$. The time-resolved luminescence spectrum is calculated as a convolution of carrier densities obtained by solving the equations given earlier.

As indicated by Equations (23.12) and (23.16), the likelihood of transport and recombination of localized carriers depends exponentially on the distances involved. For a system with a broad localized state distribution, therefore, the recombination times of carriers will take a wide range of values. Both competitive processes jointly determine the decay behavior of the PL intensity in the disordered system. It is therefore not difficult to understand that the decay time of the PL intensity in a disordered system will have a somewhat broad distribution and will probably show a non-exponential or multiexponential character.

From Equations (23.12) and (23.16), it can be seen that the localization length α is a key parameter characterizing the disorder degree. It can be viewed as the average spreading length of the wave functions of the localized carriers in the localized states. This length essentially determines the dynamic behaviors of localized carriers. Using the model briefly described earlier, the localization length can be obtained by fitting to the experimental decay curve of the emission intensity. The solid curves in Figure 23.15 represent

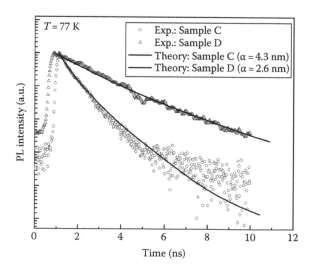

FIGURE 23.15 Measured PL decay curves (empty symbols) at 77 K for samples C and D. The solid lines represent the fitting curves using the model developed by Rubel et al.[51] (Optical Society of America (OSA), reproduced with permission of the publisher.)

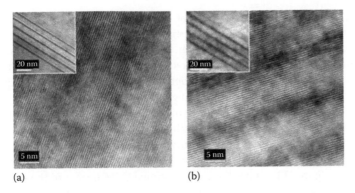

FIGURE 23.16 Cross-sectional TEM images of InGaN/GaN MQW (a) for undoped sample C and (b) for doped sample D. The inset shows a low-magnification image showing five periods of InGaN/GaN MQW. (Optical Society of America (OSA), reproduced with permission of the publisher.)

the fitting results when parameters $\tau_0 = 1$ ns, $\nu_0 = 10^{10}$ s^{-1}, $kT = 6.64$ meV, and $\varepsilon_0 = 8$ meV were adopted. Note that the obtained localization length is 4.3 nm for sample C and 2.6 nm for sample D. The initial concentration of electron–hole pairs just after the excitation pulse was estimated as 4×10^{12} cm^{-2}. Clearly, sample D (Si-doped) has a shorter localization length, indicating that the Si doping in the barrier layers of InGaN/GaN MQW may lead to further localization of carriers and thus reduce the hopping rate of carriers between different localized states. In fact, the localization length depends on the density of localized states. The lower the density of the localized states, the further the distance the wave function can decay, hence the longer the localization length. This result is supported by transmission electron microscopy (TEM) observation, as shown in Figure 23.16 which shows that sample D has the higher density of In clusters than sample C. According to Equation (23.16), the recombination time is exponentially dependent on the inverse of the localization length. The longer life of the Si-doped sample, as observed in Figure 23.15, is consistent with the theoretical prediction. The decay curve will tend to exhibit an exponential variation if the life is sufficiently long.

In short, Si doping in barrier layers may have significant influence on carrier localization and luminescence mechanisms in InGaN/GaN MQW LED structures. It can lead to an apparent band-gap renormalization effect (i.e., large redshift in the dominant PL peak of the Si-doped sample), and unusual transient luminescence behaviors. Localization length could be a key parameter with which to interpret the peculiar luminescence dynamics of uncorrelated electrons and holes involved in the luminescence process. In Section 23.2.5, electron–phonon coupling, especially coupling with an LO phonon, will be shown to have an important impact on the luminescence of InGaN/GaN MQW LEDs.

23.2.5 The Role of Electron–Phonon Coupling in the Internal Luminescence Mechanisms of InGaN/GaN MQW LEDs

As seen in Figure 23.12, LO phonon sidebands due to electron–LO phonon coupling appear at the lower-energy side of the dominant PL peaks of InGaN/GaN MQW LED structures, unambiguously indicating the role of lattice vibrations (i.e., LO phonons) in the luminescence process of carriers in these functional quantum structures.[53,54] Moreover, compared with the case of GaN bulk,[6] the electron–LO phonon coupling strength, that is, the Huang–Rhys factor S, is stronger by at least one order of magnitude.[13,55] Obviously, electron–LO phonon coupling strength is a function of dimensionality, and even the characteristic scale of the Bohr radius localization length, of localized carriers. In addition to electron–LO phonon coupling, the role of acoustic phonons in various interband optical processes in low-dimensional semiconductor systems has also been discussed at length.[56–58] It is currently recognized that the interactions of strongly confined electrons with acoustic phonons result in a broadened zero-phonon line (ZPL) in luminescence spectra at low temperatures. It is also concluded that acoustic phonons should be taken into consideration for a quantitative interpretation of the luminescence spectra of confined semiconductor systems. However, a microscopic description remains elusive to account quantitatively for the contribution of acoustic phonons to the broadening of the ZPL of the fundamental optical transitions.[57] Therefore, achieving a deeper insight into the role of electron–phonon interactions in luminescence processes in III-nitride MQW LED structures is of both fundamental and technological interest.

Adopted for this study is five periods of narrow InGaN MQW with Si-doped GaN barrier layers grown by MOCVD on a sapphire substrate. The barrier layers are 12 nm thick and the Si-doping density is 10^{18} cm^{-3}. Figure 23.17 shows the cross-sectional TEM images of the sample: on the left, the low-magnification image; on the right, the high-resolution image. From these images, one can justify that In-rich nanoscale dots spontaneously form within the well layers and have two groups of average size distribution: ~5 and ~2 nm. The bimodal size distribution of In-rich nanoscale dots gives a natural and reasonable explanation for the double sets of ZPL lines and their LO phonon sidebands, as shown in Figure 23.18.

As widely recognized, Huang and Rhys published the first detailed quantum–mechanical calculation of the absorption of light in *F* centers in solids by addressing the central role of electron–phonon coupling.[59] In their seminal work, Huang and Rhys assumed that one phonon model (i.e., the optical phonon mode) is involved in the optical transition of electrons. Due to the fundamental importance of electron–phonon coupling, the work by Huang and Rhys was followed by Lax,[60] O'Rourke,[61] Kubo,[62] and other scholars for the theoretical study of more general electron–phonon coupling cases (i.e., all vibrational modes including optical and acoustic modes and a general frequency distribution). On the basis of these outstanding studies, Mukamel and his coworkers developed the multimode Brownian Oscillator (MBO) model to describe the electronic relaxation of a two-level system attached to several primary oscillators which were, in turn, linearly coupled with a bath of secondary phonons.[63] We successfully applied the model to achieve the quantitative reproduction of the structured green emission band in ZnO for the first time.[64] We now discuss a further application of the MBO model in the quantitative simulation of the luminescence spectra of a narrow InGaN MQW sample.[13]

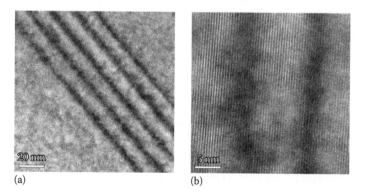

(a)   (b)

FIGURE 23.17 Cross-sectional TEM images of the narrow InGaN MQW sample. Left: low-magnification image clearly showing the five periods of well layers; right: high-resolution image. Note that indium-rich nanoscale dots spontaneously form, possessing two representative average sizes: (a) 5 nm and (b) 2 nm. (AIP Publishing, reproduced with permission of the publisher.)

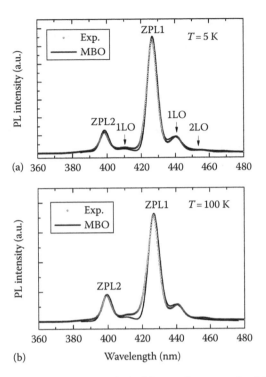

FIGURE 23.18 Representative PL spectra (open circles) of the sample measured at (a) 5 K and (b) 100 K. The solid lines are the theoretical curves with the MBO model taking into account inhomogeneous broadening due to dot size distribution. (AIP Publishing, reproduced with permission of the publisher.)

The main formula of the MBO model should be recalled here. Considering the inhomogeneous broadening effect of the natural dot size distribution to a PL spectrum of many dots,[13] the PL lineshape formula of the MBO model may be written as

$$I_{PL}(\omega) = \frac{1}{\pi} \int_0^\infty f(\omega_{eg}) d\omega_{eg} \operatorname{Re} \int_0^\infty \exp\left[i(\omega - \omega_{eg} + \lambda)t - g*(t)\right] dt, \tag{23.17}$$

where $\lambda = S\hbar\omega_{LO}$, with S being the Huang–Rhys factor characterizing the electron–LO phonon coupling strength and $\hbar\omega_{LO}$ being the energy of the LO phonon. In general, the treatment of inhomogeneous broadening in optical transitions of an ensemble of two-level systems is to assume a Gaussian-type density-of-states distribution with a width W

$$f(\omega_{eg}) = \exp\left[-2W^{-2}\left(\omega_{eg} - \omega_{eg}^0\right)^2\right] \tag{23.18}$$

$g*(t)$ is the complex conjugate of the line broadening function $g(t)$[63]

$$g(t) = -\frac{1}{2\pi} \int_{-\infty}^\infty d\omega \frac{C''(\omega)}{\omega^2}\left[1 + \coth(\beta\hbar\omega/2)\right]\left(e^{-i\omega t} + i\omega t - 1\right) \tag{23.19}$$

where $\beta = 1/K_B T$ and K_B is the Boltzmann constant. The spectral response function for a Markovian bath coupled linearly to the primary oscillator of frequency ω_{LO} with a damping strength γ has the form[63]

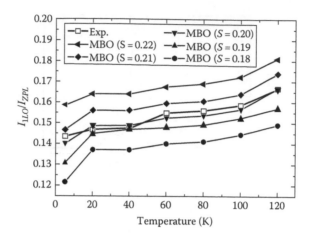

FIGURE 23.19 Measured and calculated intensity ratios of the first-order LO phonon sideband (denoted as 1LO) and the zero-phonon line (ZPL1) versus temperature. Direct comparison of experiment with theory determines that the Huang–Rhys factor is S = 0.20. Solid lines are drawn to guide the eye. (AIP Publishing, reproduced with permission of the publisher.)

$$C''(\omega) = \frac{2\lambda\omega_{LO}^2\omega\gamma}{\omega^2\gamma^2 + \left(\omega_{LO}^2 - \omega^2\right)^2} \tag{23.20}$$

The damping strength γ, which accounts for the coupling strength of the primary oscillator and the bath modes, controls the broadening of the zero phonon peak and its phonon sidebands.

For the InGaN narrow MQW and quantum dot system under study, it is obviously a case of relatively weaker coupling due to the Huang–Rhys factor $S < 1$. The solid lines in Figure 23.18 represent the theoretical curves of the MBO model. The upper panel and bottom panel show the representative PL spectra (open circles) of the sample measured at 5 and 100 K, respectively. The peaks denoted by ZPL_1 and ZPL_2 are assigned to the zero-phonon lines associated with the InGaN dots with a bimodal size distribution,[13] which is supported by time-resolved PL measurements. Their LO phonon sidebands are marked by downward arrows on the upper panel figure. The energy separations between the two ZPL lines and their corresponding LO phonon sidebands are ~90 meV, matching the characteristic energy of the A1-LO mode of GaN.[6] The parameters used in the calculations are $\omega_{LO} = 740$ cm^{-1}, $\gamma = 200$ cm^{-1}, and $S = 0.20$. Calculated PL spectra (solid lines) at 5 K (upper panel) and 100 K (bottom panel) are found to be in good agreement with those measured (open circles).

As argued earlier, the Huang–Rhys factor S is a key parameter characterizing electron–LO phonon coupling strength, which has been proven to govern the relative intensities of the ZPL line and its successive sidebands. Accurate determination of the Huang–Rhys factor from the experimental spectra is thus crucial to understanding the electron–phonon coupling system. Here, we test a number of Huang–Rhys factors for the calculation of theoretical PL spectra, while other parameters are kept unchanged.[13] The measured and calculated integrated intensity ratios of the first-order LO phonon sideband and the ZPL_1 are depicted in Figure 23.19 as a function of temperature. It is found that, for $S = 0.2$, a satisfactory agreement between theory and experiment is achieved in the temperature range of interest. It is worth noting that both theoretical and experimental curves tend to increase with increasing temperature, although the Huang–Rhys factor is fixed at 0.20. Increase of the relative intensity of the phonon sideband with temperature means an increase in the likelihood of phonon generation at higher temperatures, which may cause a severe heating effect and degrade the performance of LED devices.[14]

23.3 Conclusions

In conclusion, the internal luminescence mechanisms of InGaN-based MQW heterostructures have been investigated, placing emphasis on the roles of several interconnected factors, such as screening of the piezoelectric field by photogenerated carriers and the reverse QCSE effect, carrier localization, and unusual luminescence decay, as well as electron–phonon coupling, induced broadening, phonon side-bands, and so on. It has been demonstrated, theoretically and experimentally, that the piezoelectric field, carrier localization, and electron–phonon coupling play crucial roles in the spontaneous light emission mechanisms in InGaN-based MQW LED structures. These physical factors have to be taken into account in order to interpret luminescence mechanisms and for the design and development of high-efficiency blue and green LED sources.

It should be pointed out that, despite the great effort that has been made by many groups around the world over time, there is still a considerable way to go to achieve full understanding of the complex luminescence mechanisms in III-nitride LED heterostructures. In particular, nonradiative recombination of kinetic carriers via multi-phonon emission (heating) around defects and dislocations in the nitride functional nano-structures needs to be studied comprehensively, both theoretically and experimentally. With such knowledge and understanding at our disposal, we will be able to achieve the successful design and production of massive III-nitride LED-based lighting sources to support the emerging solid-state lighting revolution.

Acknowledgments

The work was financially supported by HK-RGC-GRF Grants (No. HKU 705812P), National Natural Science Foundation of China (NSFC) Grants (No. 11374247), and HKU SRT on New Materials, as well as, in part, by HK-UGC AoE Grants (No. AoE/P-03/08).

I would like to express my sincere gratitude to all my research collaborators, students, and friends for their important contributions and stimulating discussions over the years. My special thanks are given to Dr. Q. Li, Dr. Y. J. Wang, Dr. G. Q. Li, Dr. S. L. Shi, Prof. J. Q. Ning, Dr. C. C. Zheng, Mr. Z. C. Su, and Mr. M. Z. Wang of my group at the University of Hong Kong for their hard work and outstanding contributions to the study. I am indebted to Prof. D. G. Zhao, Prof. R. X. Wang, Prof. X. H. Zhang, Dr. W. Liu, Prof. H. Yang, Prof. D. P. Yu, Prof. J. N. Wang, Prof. S. J. Chua, Prof. Y. Zhao, and Prof. G. H. Chen for their extensive collaboration and friendship over many years. Last, but not least, I want to thank Prof. Z. C. Feng, the editor of this book, for his kind invitation and long-term friendship.

References

1. S. Nakamura, S. Pearton, and G. Fasol, *The Blue Laser Diode*, 2nd edn., Springer-Verlag, Berlin/Heidelberg, Germany, 2000.
2. J. R. Brodrick, 2015, DOE Solid-State Lighting Program, Aug. 2016. http://energy.gov/eere/buildings/downloads/solid-state-lighting-overview-2015-bto-peer-review.
3. D. Zhu, D. J. Wallis, and C. J. Humphreys, Prospects of III-nitride optoelectronics grown on Si, *Rep. Prog. Phys.* 76, 106501 (2013).
4. Y. Sun, K. Zhou, Q. Sun, J. P. Liu, M. X. Feng, Z. C. Li, Y. Zhou et al., Room-temperature continuous-wave electrically injected InGaN-based laser directly grown on Si, *Nat. Photon.* 10, 595 (2016).
5. S. Nakamura, The roles of structural imperfections in InGaN-based blue light-emitting diodes and laser diodes, *Science* 281, 956 (1998).
6. S. J. Xu, W. Liu, and M. F. Li, Effect of temperature on longitudinal optical phonon-assisted exciton luminescence in heteroepitaxial GaN layer, *Appl. Phys. Lett.* 77, 3376 (2000).
7. Q. Li, S. J. Xu, W. C. Cheng, M. H. Xie, S. Y. Tong, C. M. Che, and H. Yang, Thermal redistribution of localized excitons and its effect on the luminescence band in InGaN ternary alloys, *Appl. Phys. Lett.* 79, 1810 (2001).

8. Q. Li, S. J. Xu, M. H. Xie, S. Y. Tong, X. H. Zhang, W. Liu, and S. J. Chua, Strong screening effect of photo-generated carriers on piezoelectric field in $In_{0.13}Ga_{0.87}N/In_{0.03}Ga_{0.97}N$ quantum wells, *Jpn. J. Appl. Phys.* 41, L1093 (2002).

9. D. G. Zhao, S. J. Xu, M. H. Xie, S. Y. Tong, and H. Yang, Stress and its effect on optical properties of GaN epilayers grown on Si(111), 6H-SiC(0001), and c-plane sapphire, *Appl. Phys. Lett.* 83, 677 (2003).

10. S. J. Xu, Y. J. Wang, Q. Li, X. H. Zhang, W. Liu, and S. J. Chua, Direct observation and theoretical interpretation of strongly enhanced lateral diffusion of photogenerated carriers in InGaN/GaN quantum well structures, *Appl. Phys. Lett.* 86, 071905 (2005).

11. Y. J. Wang, S. J. Xu, D. G. Zhao, J. J. Zhu, H. Yang, X. D. Shan, and D. P. Yu, Non-exponential photo-luminescence decay dynamics of localized carriers in disordered InGaN/GaN quantum wells: The role of localization length, *Opt. Express* 14, 13151 (2006).

12. Y. J. Wang, S. J. Xu, Q. Li, D. G. Zhao, and H. Yang, Band gap renormalization and carrier localiza-tion effects in InGaN/GaN quantum-wells light emitting diodes with Si doped barriers, *Appl. Phys. Lett.* 88, 041903 (2006).

13. S. J. Xu, G. Q. Li, Y. J. Wang, Y. Zhao, G. H. Chen, D. G. Zhao, J. J. Zhu, H. Yang, D. P. Yu, and J. N. Wang, Quantum dissipation and broadening mechanisms due to electron-phonon interactions in self-formed InGaN quantum dots, *Appl. Phys. Lett.* 88, 083123 (2006).

14. J. Li, S. L. Shi, Y. J. Wang, S. J. Xu, D. G. Zhao, J. J. Zhu, H. Yang, and F. Lu, Violet electroluminescence of AlInGaN–InGaN multiquantum-well light-emitting diodes: Quantum-confined stark effect and heating effect, *IEEE Photon Technol. Lett.* 19, 789 (2007).

15. J. H. Zhu, J. Q. Ning, C. C. Zheng, S. J. Xu, S. M. Zhang, and H. Yang, Localized surface optical pho-non mode in the InGaN/GaN multiple-quantum-wells nanopillars: Raman spectrum and imaging, *Appl. Phys. Lett.* 99, 113115 (2011).

16. W. Bao, Z. C. Su, C. C. Zheng, J. Q. Ning, and S. J. Xu, Carrier localization effects in InGaN/GaN multiple-quantum-wells LED nanowires: Luminescence quantum efficiency improvement and "Negative" thermal activation energy, *Sci. Rep.* 6, 34545 (2016).

17. Q. Li, Z. L. Fang, S. J. Xu, G. H. Li, M. H. Xie, S.Y. Tong, X. H. Zhang, W. Liu, and S. J. Chua, Large excitation-power dependence of pressure coefficients of $In_xGa_{1-x}N/In_yGa_{1-y}N$ quantum wells, *Phys. Stat. Sol. (b)* 235, 427 (2003).

18. Y. J. Wang, S. J. Xu, and Q. Li, Anisotropic ambipolar diffusion of carriers in InGaN/GaN quantum wells, *Phys. Stat. Sol. (c)* 3, 1988 (2006).

19. P. Lefebvre, A. Morel, M. Gallart, T. Taliercio, J. Allegre, B. Gil, H. Mathieu, B. Damilano, N. Grandjean, and J. Massies, High internal electric field in a graded-width InGaN/GaN quantum well: Accurate determination by time-resolved photoluminescence spectroscopy, *Appl. Phys. Lett.* 78, 1252 (2001).

20. F. D. Sala, A. Carlo, P. Lugli, F. Bernardini, V. Fiorentini, R. Schoz, and J.-M. Jancu, Free-carrier screening of polarization fields in wurtzite GaN/InGaN laser structures, *Appl. Phys. Lett.* 74, 2002 (1999).

21. D. A. B. Miller, D. S. Chemla, and S. Schmitt-Rink, Electric field dependence of optical properties of semiconductor quantum wells: Physics and applications, in *Optical Nonlinearities and Instabilities in Semiconductors*, H. Haug (ed.), Academic Press, Boston, MA, Ch. 13, pp. 325–360, 1988.

22. S. Chichibu, T. Azuhata, T. Sota, and S. Nakamura, Spontaneous emission of localized excitons in InGaN single and multiquantum well structures, *Appl. Phys. Lett.* 69, 4188 (1996).

23. M. Osinski, P. Perlin, P. G. Elliseev, J. Lee, and V. A. Smagley, Comprehensive studies of light emis-sion from GaN/InGaN/AlGaN single-quantum-well structures, *J. Cryst. Growth* 189/190, 803 (1998).

24. L.-H. Peng, C.-W. Chuang, and L.-H. Lou, Piezoelectric effects in the optical properties of strained InGaN quantum wells, *Appl. Phys. Lett.* 74, 795 (1999).

25. K. O'Donnell, R. Martin, and P. Middleton, Origin of luminescence from InGaN diodes, *Phys. Rev. Lett.* 82, 237 (1999).

26. S. J. Xu, J. S. Luo, and M. Q. Chen, Analysis of energy levels in GaAs quantum wells, *J. Electron.* 15, 445 (1993) (in Chinese).

27. C. G. Van de Walle and J. Neugebauer, Small valence-band offsets at GaN/InGaN heterojunctions, *Appl. Phys. Lett.* 70, 2577 (1997).

28. I. Vurgaftman, J. R. Meyer, and L. R. Ram-Mohan, Band parameters for III–V compound semiconductors and their alloys, *J. Appl. Phys.* 89, 5815 (2001).

29. T. Takeuchi, C. Wetzel, S. Yamaguchi, H. Sakai, H. Amano, I. Akasaki, Y. Kaneko, S. Nakagawa, Y. Yamaoka, and N. Yamada, Determination of piezoelectric fields in strained GaInN quantum wells using the quantum-confined Stark effect, *Appl. Phys. Lett.* 73, 1691 (1998).

30. C. Wetzel, T. Takeuchi, H. Amano, and I. Akasaki, Electric-field strength, polarization dipole, and multi-interface band offset in piezoelectric $Ga_{1-x}In_xN$/GaN quantum-well structures, *Phys. Rev. B* 61, 2159 (2000).

31. P. Perlin, V. Iota, B. A. Weinstein, P. Wiśniewski, T. Suski, P. G. Eliseev, and M. Osiński, Influence of pressure on photoluminescence and electroluminescence in GaN/InGaN/AlGaN quantum wells, *Appl. Phys. Lett.* 70, 2993 (1997).

32. W. Shan, J. W. Ager III, W. Walukiewicz, E. E. Haller, M. D. McCluskey, N. M. Johnson, and D. P. Bour, Pressure dependence of optical transitions in $In_{0.15}Ga_{0.85}N$/GaN multiple quantum wells, *Phys. Rev. B* 58, R10191 (1998).

33. G. Vaschenko, D. Patel, C. S. Menoni, N. F. Gardner, J. Sun, W. Götz, C. N. Tomé, and B. Clausen, Significant strain dependence of piezoelectric constants in $In_xGa_{1-x}N$/GaN quantum wells, *Phys. Rev. B* 64, 241308(R) (2001).

34. T. Suski, P. Perlin, H. Teisseyre, M. Leszczyński, S. Porowski, and T. D. Moustakas, Mechanism of yellow luminescence in GaN, *Appl. Phys. Lett.* 67, 2188 (1995).

35. S. P. Lepkowski, H. Teisseyre, T. Suski, P. Perlin, N. Grandjean, and J. Massies, Piezoelectric field and its influence on the pressure behavior of the light emission from GaN/AlGaN strained quantum wells, *Appl. Phys. Lett.* 79, 1483 (2001).

36. Y. Wen, M. Yang, S. J. Xu, L. Qin, and Z. X. Shen, Effects of internal strain and external pressure on electronic structures and optical transitions of self-assembled $In_xGa_{1-x}As$/GaAs quantum dots: An experimental and theoretical study, *J. Appl. Phys.* 112, 014301 (2012).

37. K. H. Gulden, H. Lin, P. Kiesel, P. Riel, G. H. Döhler, and K. J. Ebeling, Giant ambipolar diffusion constant of n-i-p-i doping superlattices, *Phys. Rev. Lett.* 66, 373 (1991).

38. M. B. Yairi and D. A. B. Miller, Equivalence of diffusive conduction and giant ambipolar diffusion, *J. Appl. Phys.* 91, 4374 (2002).

39. P. J. Poole, C. C. Phillips, M. Henini, and O. H. Hughes, All-optical measurement of the giant ambipolar diffusion constant in a hetero-nipi reflection modulator, *Semicond. Sci. Technol.* 8, 1750 (1993).

40. V. Bougrov, M. E. Levinshtein, S. L. Rumyantsev, and A. Zubrilov, Gallium Nitride (GaN), in *Properties of Advanced Semiconductor Materials GaN, AlN, InN, BN, SiC, SiGe*, M. E. Levinshtein, S. L. Rumyantsev, M. S. Shur (eds.), John Wiley & Sons, Inc., New York, pp. 1–30, 2001.

41. T. P. Chow and M. Ghezzo, SiC power devices, in *III-Nitride, SiC, and Diamond Materials for Electronic Devices*, D. K. Gaskill, C. D. Brandt, and R. J. Nemanich (eds.), *Material Research Society Symposium Proceedings*, Vol. 423, Pittsburgh, PA, pp. 69–73, 1996.

42. X. H. Zhang, Private communication.

43. Y.-C. Huang, J.-C. Liang, C.-K. Sun, A. Abare, and S. P. DenBaars, Piezoelectric-field-enhanced lateral ambipolar diffusion coefficient in InGaN/GaN multiple quantum wells, *Appl. Phys. Lett.* 78, 928 (2001).

44. N. M. Johnson, A. V. Nurmikko, and S. P. DenBaars, Blue diode lasers, *Phys. Today* 53, 31 (2000).

45. C. K. Choi, Y. H. Kwon, B. D. Little, G. H. Gainer, J. J. Song, Y. C. Chang, S. Keller, U. K. Mishra, and S. P. DenBaars, Time-resolved photoluminescence of $In_xGa_{1-x}N$/GaN multiple quantum well structures: Effect of Si doping in the barriers, *Phys. Rev. B* 64, 245339 (2001).

46. S. J. Xu, M. B. Yu, Rusli, S. F. Yoon, and C. M. Che, Time-resolved photoluminescence spectra of strong visible light-emitting SiC nanocrystalline films on Si deposited by electron-cyclotron-resonance chemical-vapor deposition, *Appl. Phys. Lett.* 76, 2550 (2000).

47. H. C. Casey and F. Stern, Concentration-dependent absorption and spontaneous emission of heavily doped GaAs, *J. Appl. Phys.* 47, 631 (1976).

48. E. F. Schubert, I. D. Goepfert, W. Grieshaber, and J. M. Redwing, Optical properties of Si-doped GaN, *Appl. Phys. Lett.* 71, 921 (1997).

49. S. J. Xu, S. J. Chua, X. H. Zhang, and X. H. Tang, Strong interaction of Fermi-edge singularity and exciton related to $N = 2$ subband in a modulation-doped $Al_xGa_{1-x}As/In_yGa_{1-y}As/GaAs$ quantum well, *Phys. Rev. B* 54, 17701 (1996).

50. Z. C. Su and S. J. Xu, A generalized model for time-resolved luminescence of localized carriers and applications: Dispersive thermodynamics of localized carriers, *Sci. Rep.*, 7, 41598 (2017).

51. O. Rubel, S. D. Baranovskii, K. Hantke, J. D. Heber, J. Koch, P. Thomas, J. M. Marshall, W. Stolz, and W. W. Rühle, On the theoretical description of photoluminescence in disordered quantum structures, *J. Optoelectron. Adv. Mat.* 7, 115 (2005).

52. J. M. Marshall, Computer-assisted study of carrier thermalization by hopping in disordered semiconductors, *Philos. Mag. Lett.* 80, 691 (2000).

53. M. Smith, J. Y. Lin, H. X. Jiang, A. Khan, Q. Chen, A. Salvador, A. Botchkarev, W. Kim, and H. Morkoc, Exciton-phonon interaction in InGaN/GaN and GaN/AlGaN multiple quantum wells, *Appl. Phys. Lett.* 70, 2882 (1997).

54. S. Kalliakos, X. B. Zhang, T. Taliercio, P. Lefebvre, B. Gil, N. Grandjean, B. Damilano, and J. Massies, Large size dependence of exciton-longitudinal-optical-phonon coupling in nitride-based quantum wells and quantum boxes, *Appl. Phys. Lett.* 80, 428 (2002).

55. M. Z. Wang and S. J. Xu, Band-edge optical transitions in a nonpolar-plane GaN substrate: exciton-phonon coupling and temperature effects, *Semicond. Sci. Technol.* 31, 095004 (2016).

56. P. Borri, W. Langbein, S. Schneider, U. Woggon, R. L. Sellin, D. Ouyang, and D. Bimberg, Ultralong dephasing time in InGaAs quantum dots, *Phys. Rev. Lett.* 87, 157401 (2001).

57. I. Favero, G. Cassabois, R. Ferreira, D. Darson, C. Voisin, J. Tignon, C. Delalande, G. Bastard, Ph. Roussignol, and J. M. Gérard, Acoustic phonon sidebands in the emission line of single InAs/GaAs quantum dots, *Phys. Rev. B* 68, 233301 (2003).

58. K. J. Ahn, J. Förstner, and A. Knorr, Resonance fluorescence of semiconductor quantum dots: Signatures of the electron-phonon interaction, *Phys. Rev. B* 71, 153309 (2005).

59. K. Huang and A. Rhys, Theory of light absorption and non-radiative transitions in F-centers, *Proc. R. Soc. Lond., Ser. A* 204, 406 (1950).

60. M. Lax, The Franck-Condon principle and its application to crystals, *J. Chem. Phys.* 20, 1752 (1952).

61. R. C. O'Rourke, Absorption of light by trapped electrons, *Phys. Rev.* 91, 265 (1953).

62. R. Kubo, Note on the stochastic theory of resonance absorption, *J. Phys. Soc. Jpn.* 20, 935 (1954).

63. S. Mukamel, *Principles of Nonlinear Optical Spectroscopy*, Oxford University Press, Oxford, U.K., p. 226, 1995.

64. S. L. Shi, G. Q. Li, S. J. Xu, Y. Zhao, and G. H. Chen, Green luminescence band in ZnO: Fine structures, electron-phonon coupling, and temperature effect, *J. Phys. Chem. B* 110, 10475 (2006).

24

Fabrication of Thin-Film Nitride-Based Light-Emitting Diodes

Ray-Hua Horng
Institute of Electronics,
National Chiao Tung
University, Taiwan

Dong-Sing Wuu
Department of Materials
Science and Engineering,
National Chung Hsing
University, Taiwan

Chia-Feng Lin
Department of Materials
Science and Engineering,
National Chung Hsing
University, Taiwan

Abstract   In this chapter, the thin-film LED structures are introduced in three parts. The first part is the transferring of GaN LED epistructure to Cu substrate. The second describes how to fabricate thin-film GaN LEDs with embedded electrodes to solve the electrode shading problems. The last part describes the thin-film GaN LEDs with photoelectrochemically treated structures to enhance light extraction.

24.1 Introduction

Gallium nitride (GaN)-based materials are important direct bandgap semiconductors for high-performance light-emitting diodes (LEDs), especially applied to the wavelength of the ultraviolet to the green light wavelength range [1–3]. Generally, these materials are grown on sapphire substrates due to a lack of low-price substrates that match their lattice structure. It is well known that the sapphire substrate is an insulator, so it is necessary to put both p- and n-type electrodes on top of the chip. There exist current crowding problems and electrodes covering the light output in the lateral electrodes structure. The electrode shading problem can be overcome through flip-chip processing [4]. Nevertheless, contact pads on the Si substrate should be precisely aligned. Moreover, flip-chip processing uses the Au bump to establish contact between the chip and the Si substrate. It cannot provide a well thermal dissipation due to thermal dissipation only of discrete Au bumps. Normally, there are only 16 Au bumps in 1 mm LED chip [4]. Moreover, the mirror was directly deposited on the p-doped GaN layer. The mirror (after annealed metal) plays two roles: as ohmic contact and reflector. The reflectivity of annealed metal is lower than that of an omnidirectional reflector made of insulator and metal [5]. On the other hand, thin-film GaN LEDs fabricated by laser liftoff (LLO) are a good solution for current crowding problems [5–8]. In this chapter, thin-film LEDs will be described in three parts. The first part describes the transfer of the GaN epilayer structure to Cu substrate to provide high thermal dissipation. The second part describes how a novel thin-film structure is used to solve the electrode shading problem. The last part describes the use of a photoelectrochemical process on the thin-film GaN LEDs for light extraction.

24.2 Thin-Film GaN LEDs

24.2.1 Thin GaN LEDs with Cu Substrates

Thin-film LEDs can be obtained by epitaxial layer-transferring technology. In this part, fabrication of large-area (1×1 mm²) vertical conductive GaN-mirror-Cu LED using laser liftoff and electroplating techniques are demonstrated [9]. Selective p-GaN top areas are first electroplated by thick copper films, and then an excimer laser is employed to separate the GaN from sapphire substrates. Each LED structure (dominant wavelength at 470 nm) consists of a low-temperature 200 nm-thick GaN buffer layer, a 1 µm-thick undoped GaN layer, a 2 µm-thick highly conductive n-type GaN layer, an InGaN–GaN multiple-quantum-well active layer, and a 0.5 µm-thick p-type GaN layer. The samples are cleaned using a standard solvent, immersed in HCl solution for 1 min, in boiling aqua regia (3:1, HCl/HNO₃) for 10 min, and rinsed in running deionized water. After the cleaning process, a square mesa structure with a dimension of 1×1 mm² is fabricated using an inductively coupled plasma etcher for electrical current isolation shown in Figure 24.1a.

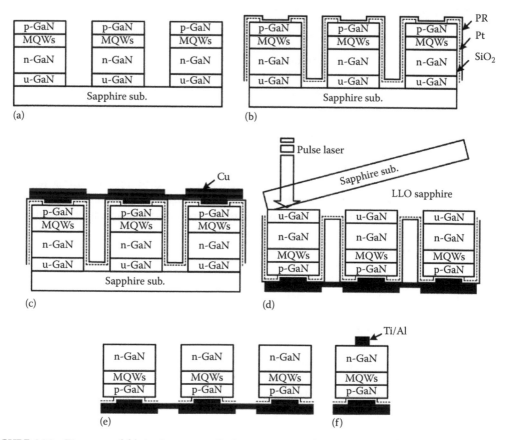

FIGURE 24.1 Diagrams of fabrication process for large-area vertical n-GaN-mirror-Cu LED. (a) LED structure after ICP etching, (b) SiO₂ deposition, Pt mirror coating, photolithography, (c) Cu electroplating, (d) LLO, (e) U-GaN removal, and (f) N-pad metallization and free-standing n-GaN-mirror-Cu LED chip.

(a)   (b)

FIGURE 24.2 SEM micrographs of fabrication process for large-area vertical GaN-mirror-Cu LED. (a) Pattern of electroplated copper substrate, (b) GaN epitaxial layer on copper substrate. (After Lin, W.Y., Wuu, D.S., Pan, K.F., Huang, S.H., Lee, C.E., Wang, W.K., Hsu, S.C., Su, Y.Y., Huang, S.Y., and Horng, R.H., High-power GaN-mirror-Cu light-emitting diodes for vertical current injection using laser liftoff and electroplating techniques, *IEEE Photon. Technol. Lett.*, 17, 1809–1811. Copyright 2005 IEEE.)

Plasma-enhanced chemical-vapor-deposited SiO_2 films are used to preserve the sidewall of LEDs. Then 200 nm-thick Pt films are deposited onto the p-GaN top layers as ohmic contacts, reflective mirrors, and conductive layers for the subsequent electroplating process. Before the electroplating process, the device sidewall and separation channel are defined by thick photoresist via conventional photolithography to prevent copper from being electroplated into this area, as shown in Figure 24.1b. 100 μm-thick copper films are electroplated as metallic substrates for LEDs, shown in Figure 24.1c. Figure 24.2a shows the corresponding micrograph examined by an SEM. Samples are then subjected to the LLO process, where an excimer pulse laser is irradiated from the back surface of the sapphire substrates. GaN epitaxial layers are locally heated and dissociated close to the sapphire/GaN interfaces. After entire LED wafers are illuminated by the laser beam, the sapphire substrates are separated from the LED structures, as shown in Figure 24.1d. The remaining GaN surfaces are smoothed using chemical–mechanical polishing. At the same time, undoped GaN epitaxial layers are also removed in the last process, as shown in Figure 24.1e. The corresponding SEM micrograph is shown in Figure 24.2. Finally, the electron-beam evaporated Ti–Al films on the n-GaN epitaxial layers are used as n-contacts, as shown in Figure 24.1f. High-power GaN-mirror-Cu LED structures for vertical current injection are now completed. For comparison, conventional GaN/sapphire LEDs with the same chip size (1×1 mm²) are fabricated using Ni–Au as transparent p-electrodes. Ti–Al are deposited on the exposed n-GaN surfaces as n-electrodes. No additional transparent conductive layer is employed in both types of GaN-sapphire and GaN-mirror-Cu samples.

Figure 24.3 shows the output power versus injection current characteristics for the GaN-mirror-Cu and original GaN-sapphire LEDs, where the chips are encapsulated in lamp form. It is well known that the device performance in the lamp form condition degrades much faster due to excessive heating. At 500 mA, the light output power for the GaN-mirror-Cu LED is about twofold stronger. It is evident that the output power of GaN-mirror-Cu LEDs increases lineally compared to original GaN-sapphire LEDs, increasing the current. This result suggests that the GaN-mirror-Cu LEDs have higher current capability than original GaN-sapphire LEDs due to their higher thermal conductivity of copper substrates.

24.2.2 Thin GaN LEDs with Embedded Electrodes

From the previous part, it is found that the electrodes always exist on the top layer of the traditional sapphire based and vertical thin-film GaN LEDs. Thin-film GaN LEDs prepared on Si substrates, pretreated by surface roughening and embedded electrodes, are studied to overcome the electrode shielding problem. LEDs are fabricated with roughened p-GaN layers, followed by epitaxial layers, transfer. There are four types of LEDs: the first type is n-side-up thin GaN LEDs with embedded electrodes and

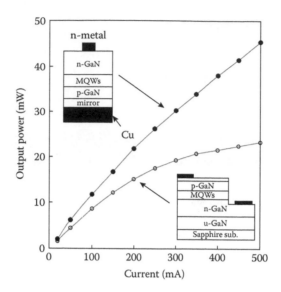

FIGURE 24.3 Output power versus injection current characteristics of a large-area vertical n-GaN-mirror-Cu and GaN-sapphire LED samples. (After Lin, W.Y., Wuu, D.S., Pan, K.F., Huang, S.H., Lee, C.E., Wang, W.K., Hsu, S.C., Su, Y.Y., Huang, S.Y., and Horng, R.H., High-power GaN-mirror-Cu light-emitting diodes for vertical current injection using laser liftoff and electroplating techniques, *IEEE Photon. Technol. Lett.*, 17, 1809–1811. Copyright 2005 IEEE.)

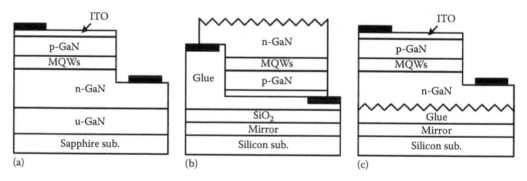

FIGURE 24.4 Diagrams of (a) C-LED structure; the fabrication process of (b) first-type LEDs, and (c) second-type LEDs.

KOH-roughened n-GaN. The second is p-side-up thin GaN LEDs with KOH-roughened n-GaN. The third is n-side-up thin GaN LEDs with embedded electrodes and without KOH-roughened n-GaN. The fourth type is conventional LEDs (C-LED) [10]. The electrical properties of LEDs are almost the same. Standard chips are fabricated with both p- and n-type electrodes on the top. Each chip has an area of 1×1 mm². The schematic illustration of the cross section of the conventional structure is shown in Figure 24.4a. To avoid run-by-run epitaxial layer quality difference, structures are fabricated using the same two epitaxial layers. After chip processing on the two wafers, the wafers are divided into four pieces. One piece is fabricated into thin-film n-side up LEDs by wafer bonding once. Contact electrodes of n-side up thin GaN LEDs are embedded under the epitaxial layers. Epitaxial layers on both n-side up and p-side up LED structures are roughened on both sides and bonded to the Si structure with mirror, called first-type LED. The second piece is fabricated into thin-film p-side up LEDs by wafer bonding twice, called second-type LED. One roughened surface is the p-layer obtained by low-temperature p-GaN roughening. The other roughened surface is the n-GaN layer obtained by KOH surface treatment after the sapphire substrate is removed by LLO. To compare the n-GaN roughening effect on light extraction, the third of the four pieces is fabricated into an n-side up GaN thin-film LED without n-GaN roughening, called third-type LED. The last piece is fabricated into a conventional sapphire-based LED with mirror coating on the sapphire substrate (C-LED). The fabrication flowchart of these LEDs is shown in Figure 24.4. Thin-film GaN LEDs are all transferred to the same mirror substrate (Al/Ti/Si substrate) by low-temperature glue bonding. Thus, degradation of the mirror's reflectivity, a problem with vertical-type LEDs made by metal bonding, did not occur in our fabrication process. It is worth mentioning that the glue/metal structure can form a high reflective omnidirectional mirror [11]. Electrodes are fabricated before the epitaxial layer transfer for the n-side and p-side up LEDs. Electrodes are processed to expose the contact pads to the n-side up LEDs and to avoid electrode shading. Note that it is not necessary to precisely align the n- and p-electrodes with the circuit substrate like the flip-chip LED.

Figure 24.5 shows SEM images of three types of LEDs. Emitter areas for the C-LED and second-type LEDs are the same at about 1.04 mm^2, and the p-pad electrode is about 73,990 μm^2. The electrode shading ratio is 7%. The emitter area is increased to 1.25 mm^2 for the first- and third-type LEDs using the same chip structure to serve as C-LEDs and second-type LEDs. The electrode shading ratio generated is 0%. The designed first- and third-type LEDs are with an increased emitter area and are also provided with a surface without electrode covering.

Figure 24.6 presents the luminance intensity of packaged LEDs (without epoxy) as a function of injection current (L–I). It is found that the intensity of all thin GaN LEDs is higher than that of C-LEDs. The luminance intensities (at 350 mA injecting DC current) for C-LED, first-, second-, and third-type LEDs are 3468, 9105, 7610, and 7047 mcd, respectively. The brightest LED is the first-type LED, followed by the second-, third-, and C-LED. The intensity of the first-type LED is 162% higher than that of C-LEDs at an injection current of 350 mA. It is worth mentioning that the second-type LEDs are brighter than the third type.

The main difference among the four types of LEDs is the surface roughness, in particular, the degree of roughening of the light emissive surface. The degree of p-GaN roughness is smaller compared to that of the n-GaN treated by KOH etching, as shown in Figure 24.7a and b. The root-mean-square (rms) roughness of p-GaN layer (light escapes surface) of the second-type LED is 100–200 nm, which is less than the emission wavelength. The surface of n-GaN etched by KOH showed a pyramidal-textured surface with 300–1000 nm rms roughness. It is very close to its optical wavelength so that it could deteriorate optical confinement in n-GaN/MQW/p-GaN waveguide structures and emit lights out of the surface for first-type LED. Light could easily escape from the semiconductor on the top layer of LEDs with pyramidal-textured surface as compared with that of the top layer of an LED with a little rough p-layer. On the other hand, the n-GaN surface without KOH etching (third-type LED) is very smooth after LLO, as shown in Figure 24.7c. This resulted in light extraction being less than that of the low-temperature p-GaN. Therefore, optimized light extraction efficiency depends not only on electrode shading but also on a pyramidal-textured surface.

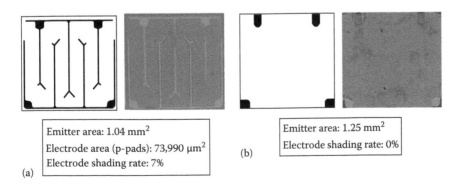

(a)

Emitter area: 1.04 mm^2
Electrode area (p-pads): 73,990 μm^2
Electrode shading rate: 7%

(b)

Emitter area: 1.25 mm^2
Electrode shading rate: 0%

FIGURE 24.5 SEM surface images for (a) C- and second-type LEDs and (b) first- and third-type LEDs.

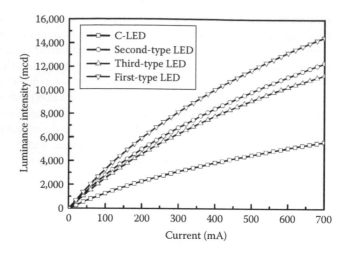

FIGURE 24.6 Luminance intensity as a function of injection current for C-LED, first-type, second-type, and third-type LEDs. (After Horng, R.H., Lu, Y.A., and Wuu, D.S., Light extraction Investigation for thin-film GaN light-emitting diodes with imbedded electrodes, *IEEE Photon. Technol. Lett.*, 23, 54–56. Copyright 2011 IEEE.)

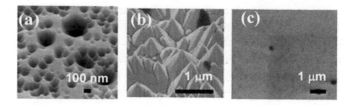

FIGURE 24.7 SEM surface morphologies of (a) low-temperature p-GaN, (b) n-GaN after roughening, and (c) n-GaN before roughening. (After Horng, R.H., Lu, Y.A., and Wuu, D.S., Light extraction Investigation for thin-film GaN light-emitting diodes with imbedded electrodes, *IEEE Photon. Technol. Lett.*, 23, 54–56. Copyright 2011 IEEE.)

The chips process has been designed for four LED structures using the same epiwafer and LED fabrication processes. Under no ohmic contact and wafer bonding issues, first-type LED with omnidirectional reflector/Si substrates and no electrode shading can produce optimized light extraction performance. This confirms that pyramidal-textured surface regions (n-GaN roughening) can deteriorate optical confinement and results in more light extraction. Thus, optimized light extraction efficiency is not only affected by electrode shading but also by a pyramidal-textured surface.

24.2.3 Thin-Film GaN LEDs with Photoelectrochemically Treated Structures

Porous semiconductors [12] have been widely studied due to their unique optical properties compared to conventional bulk materials. Porous structure can reduce the defeat density and relieve strain caused by lattice mismatch between GaN and sapphire. Fabricating a nanoporous structure on a p-GaN layer is more difficult than on an n-GaN layer, because the holes are harder to accumulate on a p-GaN surface to form a Ga_2O_3 layer. Here, PEC wet oxidation and oxide-removing techniques are used to form nanoporous GaN:Mg and GaN:Si structures, which are different from the previous reports [13]. PEC oxidation and oxide-removing processes with pure water solution are used to form nanoporous p-type GaN:Mg surface on InGaN-based LED structures [14]. This technology created a rough p-type GaN:Mg surface above the MQW active layer, which can increase both η_{int} and η_{out} remarkably. Surface morphology and optical properties for InGaN-based LED with nanoporous structures are also discussed here.

For PEC wet oxidation process, DI water solution was used as the oxidation solution without stirring it. Accompanied by an illumination exposure of a 400 W Hg lamp for 30 min, an external dc bias fixed at 20 V is applied to the n-type GaN:Si layer surface as the anode contact and platinum as the cathode. A Ga_2O_3 layer was formed on the p-type GaN:Mg mesa surface, mesa sidewall, and n-type GaN:Si ICP etching surface. After removing the Ga_2O_3 layer from the GaN surface, the nanoporous p-type GaN:Mg and n-type GaN:Si structures were observed on the top mesa and the bottom ICP etching surfaces.

The uniformly distributed nanoporous p-type GaN:Mg and n-type GaN:Si surfaces were observed on the top mesa and the bottom ICP etching surfaces after PEC oxidation and oxide-removing processes shown in Figure 24.8. The Ga_2O_3 layer was formed through this PEC oxidation process, which was removed with a diluted HCl solution ($HCl/H_2O = 1:1$). SEM micrographs of nanoporous structures of the p-type GaN:Mg and n-type GaN:Si layers are shown in Figure 24.8. The roughened GaN surfaces on the top GaN:Mg mesa surface, the mesa sidewall, and the bottom GaN:Si surface can lead to a higher light extraction from a GaN to air interface. Magnified SEM micrographs of the nanoporous p-type GaN:Mg surface are observed in Figure 24.8. The average pore size and GaN:Mg grain boundaries are found to be 75–85 and 180–200 nm, respectively. The pore size of a GaN:Mg grain boundary is also measured from the section line scan of the atomic force microscopy showing the similar value of about 75–85 nm.

PL emission spectra for standard and nanoporous InGaN/GaN MQW LED samples are measured under different temperatures varying from 10 to 300 K, shown in Figure 24.9. PL peak intensities of the InGaN/GaN MQW active layer and the GaN layer in LEDs in this study are both enhanced as compared to standard LEDs. The light enhancement ratio of an MQW active layer and a GaN layer are 1.42 times and 7.4 times that of standard ones at 300 K. PL emission peaks of the InGaN/GaN active layer are observed at 465.5 nm (2.664 eV) for standard LED and 456.0 nm (2.720 eV) for nanoporous LED samples measured at 300 K. The emission peak of a nanoporous LED structure shows a blueshift phenomenon of 9.5 nm (56 meV), which is caused by a partial reduction of the piezoelectric field. The strain of the InGaN/GaN active layer caused by the lattice mismatch between GaN and InGaN layers is partially released from the nanoporous GaN:Mg layer above the MQW active region. Dividing to 3 nm width of the InGaN well, the 56 meV blueshift corresponded to the reduction of the piezoelectric field at around 0.187 MV/cm. The band structure of the InGaN/GaN MQW active layer became flatter and increased the wavefunction

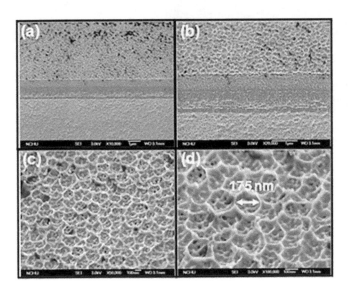

FIGURE 24.8 Surface morphology of nanoporous GaN:Mg on a typical InGaN-based LED structure: (a and b) Step profile between n-type and p-type and (c and d) nanoporous GaN:Mg through PEC wet oxidation and Ga_2O_3 removal process.

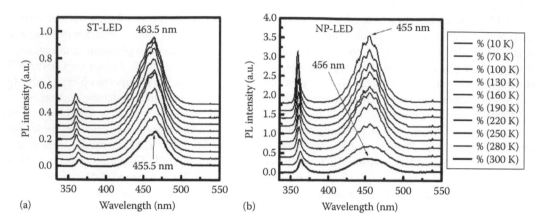

FIGURE 24.9 PL emission spectra of (a) standard and (b) nanoporous InGaN-based LED samples measured under varying temperatures ranging from 10 to 300 K.

overlap of electrons and holes. By partially reducing the piezoelectric field in an InGaN/GaN active layer, the internal quantum efficiency increases leading to a higher PL emission intensity in nanoporous LEDs.

InGaN-based LEDs with a top pattern-nanoporous p-type GaN:Mg surface are fabricated using a PEC process [15]. A selective area underwent PEC oxidized reaction for 10 min on the p-type GaN:Mg surface without Ti metal protection. The ultraviolet light was blocked by Ti layer during PEC oxidation process. Unprotected p-type GaN:Mg surface was oxidized as the GaO_x layer. After removing the GaO_x layer in a diluted HCl solution, the flat p-type GaN:Mg surface became a nanoporous structure. The LED structure with pattern-nanoporous p-type GaN:Mg surface is defined as pattern-nanoporous LED, shown in Figure 24.10a and c. The nanoporous structure is uniformly distributed on the p-type GaN:Mg mesa region, which is clear in Figure 24.10b. The surface coated by a TCL metal layer is studied by an

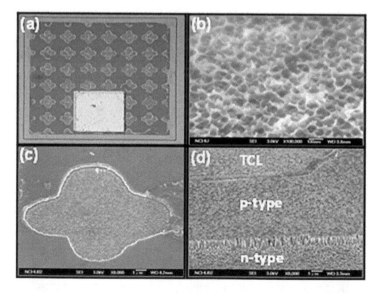

FIGURE 24.10 (a) Microscopy image of a typical patterned nanoporous LED. SEM micrographs of a patterned nanoporous structure are shown in (b) and (c) with 85 and 95 nm porous sizes, respectively, (d) the cross-sectional image of the nanoporous LED is observed at 45° tilted angle. The ITO TCL layer, mesa edge region, mesa sidewall, and ICP-etched n-type GaN:Si surface are observed in this SEM micrograph.

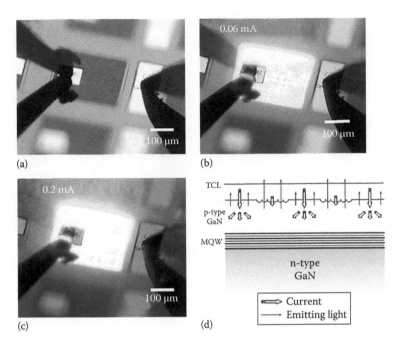

FIGURE 24.11 (a) The OM image of the pattern-porous LED. EL microscopy images of a pattern-nanoporous LED operating at (b) 0.06 mA and in (c) 0.2 mA. (d) Schematic diagram of current spreading process and light extraction process of pattern-nanoporous LEDs.

SEM system. The PEC etching depth is 129 nm without damaging the InGaN/GaN MQW active layers. The porous size of the nanoporous structure is 85–95 nm, as shown in Figure 24.10d. A lower contact resistance occurred between nonnanoporous regions of the p-type GaN:Mg and the TCL layers.

The OM image of the pattern-nanoporous LED structure is shown in Figure 24.11a. A higher EL emission intensity is observed in Figure 24.11b at the pattern-nanoporous region at 0.06 mA. When the injection current is increased to 0.2 mA in Figure 24.11c, the light emission crisscrosses the mesa region, which is observed in the EL microscopy image. A higher light scattering process in pattern-nanoporous LED occurred in the pattern-nanoporous and pattern-boundary regions, thereby increasing light extraction efficiency. A higher light scattering process and a lower operating voltage are seen in pattern-nanoporous LEDs as compared to the nanoporous LEDs. The internal quantum efficiency of InGaN active layers is increased due to the partial compression strain release that occurred during the formation of the top nanoporous p-type GaN:Mg surface. In Figure 24.11d, the current spreading process and light extraction process in pattern-nanoporous LED are shown. When the injection current flowed into the pattern-nanoporous structure, a large part of the injection current flowed into the lower resistance region, which is not a nanoporous structure. The injection current spread to the p-type GaN:Mg surface through the ITO TCL layer, and a higher light scattering process occurred in the pattern-nanoporous region.

Small self-assembled inverted hexagonal pyramids of GaN:Mg and InGaN/GaN (MQW) structures are formed using photoelectrochemical wet etching [16]. The formation mechanism of the pyramids during the etching process sequentially as lateral etching, bottom-up etching, and anisotropic etching is studied. The (0001) Ga-face and {10-1-1} faces are exposed to achieve the lowest possible surface energy.

Figure 24.12 shows SEM images of a GaN-based LED layer structure undergoing the PEC wet etching process. The 20 min etched mesa sidewall profile and the lateral etching rate of the InGaN/GaN MQW layer were found obviously higher than those of the top GaN:Mg and bottom GaN:Si layers. Rapid lateral etching of the MQW layer is similar to that attributed to the bandgap-selective etching, attributed to the removal of thick InGaN layers for optical devices reported by Youtsey et al. [17]. Proposed etching steps

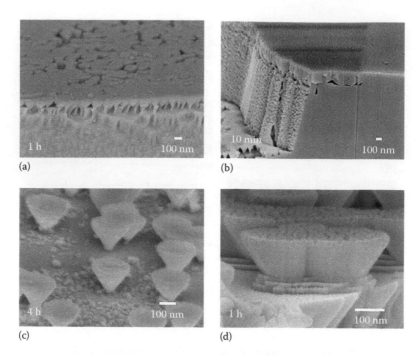

FIGURE 24.12 SEM images of a GaN-based LED structure after PEC wet etching process, (a) 1 h etched surface and side view of the etched GaN:Mg layer, (b) 20 min etched mesa sidewall profile, and (c) 4 h etched and (d) 3 h etched inverted hexagonal pyramid structures.

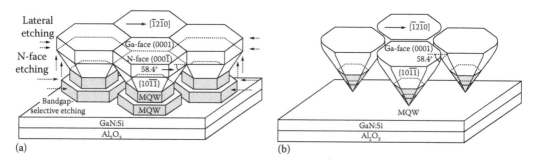

FIGURE 24.13 (a) Schematic diagram of the formation mechanism of GaN:Mg inverted hexagonal pyramids under a PEC wet etching process including InGaN bandgap-selective etching effect, bottom-up N-face GaN:Mg etching, and anisotropic etching of the GaN:Mg layer. (b) The stable crystallographic plans were formed on the inverted hexagonal pyramids.

are schematically illustrated in Figure 24.13 of the formation mechanism of GaN:Mg inverted hexagonal pyramids under a PEC wet etching process. A large number of photogenerated holes under UV light illumination accumulated at the n-type InGaN/GaN MQW layer under the applied bias of positive 1 V on the GaN-based LED sample at the beginning stage. Due to enhancement of photogenerated holes on the decomposition of GaN [17], rapid lateral etching of the InGaN/GaN MQW layer gets placed. Immediately after the lateral removal of the MQW layer, channels were left between the top (GaN:Mg) and bottom layers (GaN:Si). These channels provide paths for the KOH solution to further flow and etch the exposed bottom N-face surface of the top GaN:Mg layer.

Etching process increases the light extraction efficiency in InGaN-based LEDs. After laser scribing and selective lateral wet etching processes on LEDs' chip edge region, stable crystallographic $\{10\bar{1}2\}$ planes are formed, which have 40.3° including angle with top GaN (0001) plane [18]. A 2 in. LED wafer

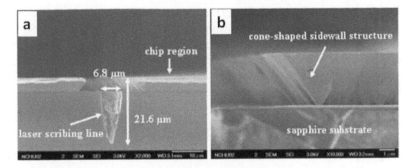

FIGURE 24.14 CSS-LED structures are observed in SEM images. (a) Laser scribing lines between each CSS-LED chip after wet etching process. The laser cutting width and depth are about 6.8 and 21.6 μm, respectively, (b) CSS structure shows 4.57 μm in height and 5.35 μm in width. The stable crystallographic etching planes formed at the inclined GaN $\{10\bar{1}2\}$ planes and the top GaN (0001) plane are both included at the 40.3° angle. The continuous CSS structure close to the laser scribing lines is observed in the CSS-LED structure.

is cleaved into two half-wafers with a mesa region of 380 × 380 μm². The LED chips were isolated using a triple frequency ultraviolet Nd:yttrium aluminum garnet (355 nm) laser in scribing process. The laser scribing depth is 21.6 μm to expose the AlN/sapphire interface for all LED samples shown in Figure 24.14. The dimension of LED chips is 420 × 420 μm², as defined by laser scribing process. Half of the LED wafer is immersed in hot phosphoric acid solution (H_3PO_4, 170°C) for 20 min; wet etching process is a selective lateral process on an AlN buffer layer and a bottom-up N-face to form CSS structure around LED chips. Laser scribing depth in CSS-LED, measured from its cross-sectional SEM images, is 21.6 μm that isolated InGaN-based LED chips. Then, the AlN buffer layer is exposed to a selective lateral wet etching process. After etching process in a hot H_3PO_4 solution for 20 min, CSS structures are formed at LED chip's edge regions. The lateral wet etching width from the laser scribing line is 9.17 μm measured. Lateral etching rate of AlN buffer layer is calculated as 27.5 μm/h. The lateral etching process on AlN buffer layer and bottom-up N-face is used to form continuous CSS structures. From the SEM image shown in Figure 24.14, the dimensions of the CSS structure were found to be 4.57 μm in height and 5.35 μm in width.

Stocker et al. [19,20] reported that the GaN epitaxial layers have been crystallographically etched as $\{10\bar{1}2\}$ planes in H_3PO_4 at 132°C. The stable and controllable CSS structure is formed when GaN $\{10\bar{1}2\}$ plane and top GaN (0001) plane met at the continuous cone-shaped edge. After the laser scribing and crystallographic wet etching process, the continuous CSS structure is observed in the CSS-LED structure shown in Figure 24.14.

The light intensity profiles of both LED samples at a 20 mA operation current measured by a beam profiler are shown in Figure 24.15. The top view and 45° bird's-eye view of light intensity patterns of both LED

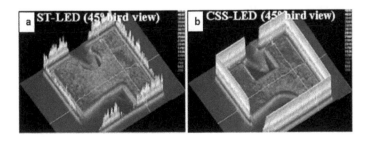

FIGURE 24.15 Light intensity profiles of (a) the ST-LED and (b) the CSS-LED samples at 20 mA operation current were measured by beam profiler for 45° bird's-eye view.

samples are shown in Figure 24.15. In the ST-LED structure, a higher light intensity is observed around the chip edge caused by light scattering process from the laser scribing lines. In the CSS-LED structure, a much higher light intensity is observed around the LED chip that has a larger light scattering process occurring at the CSS structure close to the laser scribing lines. In both LED structures, the light intensity in the mesa regions on the ITO layer is almost the same. A lower light-intensity region on the ST-LED is observed in the ICP-etched n-type GaN:Si layer.

References

1. R. H. Horng, Y. L. Tsai, T. M. Wu, D. S. Wuu, and C. H. Chao, Investigation of light extraction of InGaN LEDs with surface-textured indium tin oxide by holographic and natural lithography, *IEEE J. Sel. Top. Quantum Electron.* 15, 1327–1331 (2009).
2. S. Y. Huang, R. H. Horng, J. W. Shi, H. C. Kuo, and D. S. Wuu, High performance InGaN-based green resonant-cavity light-emitting diodes for plastic optical fiber applications, *J. Lightw. Technol.* 27, 4084–4090 (2009).
3. C. T. Chang, S. K. Hsiao, E. Y. Chang, Y. L. Hsiao, J. C. Huang, C. Y. Lu, H. C. Chang, K. W. Cheng, and C. T. Lee, 460-nm InGaN-based LEDs grown on fully inclined hemisphere-shape-patterned sapphire substrate with submicrometer spacing, *IEEE Photon. Technol. Lett.* 21, 1366–1368 (2009).
4. O. B. Shchekin, J. E. Epler, T. A. Trottier, T. Margalith, D. A. Steigerwald, M. O. Holcomb, P. S. Martin, and M. R. Krames, High performance thin-film flip-chip InGaN–GaN light-emitting diodes, *Appl. Phys. Lett.* 89, 071109 (2006).
5. R. H. Horng, S. H. Huang, C. Y. Hsieh, X. Zheng, and D. S. Wuu, Enhanced luminance efficiency of wafer-bonded InGaN-GaN LEDs with double-side textured surfaces and omnidirectional reflectors, *IEEE J. Quantum Electron.* 44, 1116–1123 (2008).
6. W. S. Wong, T. Sands, N. W. Cheung, M. Kneissl, D. P. Bour, P. Mei, L. T. Romano, and N. M. Johnson, Fabrication of thin-film InGaN light-emitting diode membranes by laser lift-off, *Appl. Phys. Lett.* 75, 1360–1362 (1999).
7. Z. S. Luo, Y. Cho, V. Loryuenyong, T. Sands, N. W. Cheung, and M. C. Yoo, Enhancement of (In, Ga)N light-emitting diode performance by laser liftoff and transfer from sapphire to silicon, *IEEE Photon. Technol. Lett.* 14, 1400–1402 (2002).
8. M. K. Kelly, O. Ambacher, B. Dahlheimer, G. Groos, R. Dimitrov, H. Angerer, and M. Stutzmann, Optical patterning of GaN films, *Appl. Phys. Lett.* 69, 1749 (1996).
9. W. Y. Lin, D. S. Wuu, K. F. Pan, S. H. Huang, C. E. Lee, W. K. Wang, S. C. Hsu, Y. Y. Su, S. Y. Huang, and R. H. Horng, High-power GaN-mirror-Cu light-emitting diodes for vertical current injection using laser liftoff and electroplating techniques, *IEEE Photon Technol. Lett.* 17, 1809–1811 (2005).
10. R. H. Horng, Y. A. Lu, and D. S. Wuu, Light extraction Investigation for thin-film GaN light-emitting diodes with imbedded electrodes, *IEEE Photon. Technol. Lett.* 23, 54–56 (2011).
11. R. H. Horngg, S. H. Huang, C. Y. Hsieh, X. Zheng, and D. S. Wuu, Enhanced luminance efficiency of wafer-bonded InGaN-GaN LEDs with double-side textured surfaces and omnidirectional reflectors, *IEEE J. Quantum Electron.* 44, 1116–1123 (2008).
12. M. Mynbaeva, A. Titkov, A. Kryganovskii, V. Ratnikov, K. Mynbaev, H. Huhtinen, R. Laiho, and V. Dmitriev, Structural characterization and strain relaxation in porous GaN layers, *Appl. Phys. Lett.* 76, 1113 (2000).
13. Y. D. Wang, S. J. Chua, M. S. Sander, P. Chen, S. Tripathy, and C. G. Fonstad, Fabrication and properties of nanoporous GaNGaN films, *Appl. Phys. Lett.* 85, 816 (2004).
14. C. F. Lin, J. H. Zheng, Z. J. Yang, J. J. Dai, D. Y. Lin, C. Y. Chang, Z. X. Lai, and C. S. Hong, High-efficiency InGaN-based light-emitting diodes with nanoporous GaN:Mg structure, *Appl. Phys. Lett.* 88, 083121 (2006).

15. C. C. Yang, C. F. Lin, C. M. Lin, C. C. Chang, K. T. Chen, J. F. Chien, and C. Y. Chang, Improving light output power of InGaN-based light emitting diodes with pattern-nanoporous p-type GaN:Mg surfaces, *Appl. Phys. Lett.* 93, 203103 (2008).

16. C. F. Lin, J. J. Dai, Z. J. Yang, J. H. Zheng, and S. Y. Chang, Self-assembled GaN:Mg inverted hexagonal pyramids formed through a photoelectrochemical wet-etching process, *Electrochem. Solid State Lett.* 8, C185–C188 (2005).

17. C. Youtsey, I. Adesida, and G. Bulman, Highly anisotropic photoenhanced wet etching of n-type GaN, *Appl. Phys. Lett.* 71, 2151 (1997).

18. C. F. Lin, C. M. Lin, C. C. Yang, W. K. Wang, Y. C. Huang, J. A. Chen, and R. H. Horng, InGaN-based light-emitting diodes with a cone-shaped sidewall structure fabricated through a crystallographic wet etching process, *Electrochem. Solid State Lett.* 12, H233–H237 (2009).

19. D. A. Stocker, I. D. Goepfer, K. S. Boutros, and J. M. Redwing, Crystallographic wet chemical etching of p-type GaN, *J. Electrochem. Soc.* 147, 763 (2000).

20. D. A. Stocker, E. F. Schubert, and J. M. Redwing, Crystallographic wet chemical etching of GaN, *Appl. Phys. Lett.* 73, 2654 (1998).

Index

P